Flow Assurance

Oil and Gas Chemistry Management Series

Flow Assurance

Volume II

Edited by

Qiwei Wang

Saudi Aramco, Dhahran, Saudi Arabia

ELSEVIER

G P P

Gulf Professional Publishing

An imprint of Elsevier

Gulf Professional Publishing is an imprint of Elsevier
50 Hampshire Street, 5th Floor, Cambridge, MA 02139, United States
The Boulevard, Langford Lane, Kidlington, Oxford, OX5 1GB, United Kingdom

ISBN: 978-0-12-822010-8

For Information on all Gulf Professional Publishing publications
visit our website at https://www.elsevier.com/books-and-journals

Publisher: Charlotte Cockle
Senior Acquisitions Editor: Katie Hammon
Senior Editorial Project Manager: Sara Valentino
Production Project Manager: Sojan P. Pazhayattil
Cover Designer: Miles Hitchen

Typeset by MPS Limited, Chennai, India

Contents

List of contributors

Gaurav Bhatnagar
Shell International Exploration and Production, Houston, TX, United States

Mashhad Fahes
The University of Oklahoma, Norman, OK, United States

Dario Marcello Frigo
Scaled Solutions Ltd., Livingston, Scotland, United Kingdom

Shawn Gao
Shell International Exploration and Production, Houston, TX, United States

Gordon Michael Graham
Scaled Solutions Ltd., Livingston, Scotland, United Kingdom

Fenfen Huang
CandV Consulting and Services, Sugar Land, TX, United States

David W. Jennings
Baker Hughes, Sugar Land, TX, United States

Priyanka Juyal
ChampionX LLC, Sugar Land, TX, United States

Dharmendr Kumar
TCS Research, Tata Consultancy Services Ltd., Pune, Maharashtra, India

Jeremy Moloney
ChampionX, Sugar Land, TX, United States

Venkata Muralidhar K
TCS Research, Tata Consultancy Services Ltd., Pune, Maharashtra, India

Mike Newberry
Baker Hughes, Sugar Land, TX, United States

Philip Nguyen
Halliburton, Houston, TX, United States

Thunyaluk Pojtanabuntoeng
Curtin University, Perth, WA, Australia

Mike Sanders
Halliburton, Houston, TX, United States

Kees C.A.M. Veeken
Mature Gas Well Consultancy, Gieten, Drenthe, The Netherlands

Jonathan J. Wylde
Heriot-Watt University, Edinburgh, Scotland, United Kingdom; Clariant Oil Services, Clariant Corporation, Houston, TX, United States

Andrew T Yen
ENNOVA LLC, Stafford, TX, United States

Xiangyang Zhu
Saudi Aramco, Dhahran, Saudi Arabia

CHAPTER

Gas hydrate management

1

Gaurav Bhatnagar and Shawn Gao

Shell International Exploration and Production, Houston, TX, United States

Chapter outline

Flow Assurance. DOI: https://doi.org/10.1016/B978-0-12-822010-8.00009-X

1.1 Introduction

First discovered in 1810, gas hydrates are nonstoichiometric crystalline ice-like compounds composed of cooperative hydrogen-bonded water molecules forming nano-scale clathrate cage structures that trap smaller guest molecules. They can form naturally under certain conditions of pressure and temperature within a gas/water mixture. Gas hydrates attract interest from many different fields of study since naturally occurring gas hydrates have been found in diverse conditions and locations, such as arctic permafrost, under the ocean floor, and even suspected on some other bodies in the solar system. Gas hydrates can also form during extraction/transport/process of oil/gas under certain pressure/temperature conditions in presence of water, which can cause production interruption, asset integrity, and safety issues. This type of hydrates will be the focus of this chapter, which covers the fundamentals of gas hydrates and hydrate management strategies in oil/gas industry. Hydrates in nature play a big role in the global carbon and climate cycles and are also considered an alternative source of energy.

Once a hydrate plug is formed in the oil/gas production system, it can be very challenging to remediate, especially in deepwater, and can bear significant implications in terms of deferred production, asset integrity, and safety. The formation of hydrate can result in significant production loss. Depending on the options available, remediating hydrate can take a long time, during which the production can be either compromised or completely stopped. For example, a hydrate plug in a deepwater flowline took a year to dissociate and the production was lost during the entire time. In case the hydrate plug is formed in gas export line, the entire production of a platform will be shut down in that scenario.

Formation of hydrate is also integrity and safety concern. The risks are mainly related to the remediation process. Due to the large amount of gas release when dissociated, adding heat to dissociate hydrate in confined volume can create extremely high pressure that can burst the containment barrier, which have resulted in fatalities. A recent study even indicated that gas hydrate may be one of the root causes for the Deepwater Horizon explosion [1]. When depressurization is applied to dissociate a hydrate plug, the hydrate plug can be suddenly dislodged and pushed toward to the low-pressure end at high speed. The resulting hydrate projectile has caused asset damage in the past.

1.2 Fundamentals of hydrate

1.2.1 Definition

In the oil/gas industry, gas hydrates, clathrate hydrates, and hydrates are often used interchangeably. They are ice-like solid compounds that typically form under high pressure and low-temperature conditions in presence of both water and light hydrocarbon molecules [2]. Natural gas hydrates are composed of approximately 85 mol.% water, therefore they have many physical properties similar to those of ice. For instance, the appearance and mechanical properties of hydrates are comparable to those of ice. The densities of hydrates vary somewhat due to the nature of the guest molecule(s) and the formation conditions but are generally comparable to that of ice.

Although hydrates were first discovered over 200 years ago, no practical implication was realized until 1934 when it was discovered that it was gas hydrates, not ice, that plugged natural gas pipelines. This brought a renewed burst of research interests on gas hydrates, especially focusing on determining thermodynamic and structural properties and preventing hydrate plugs.

Naturally occurring natural gas hydrates were predicted and found by Russian researchers in 1960s. This brought another surge of research interest that considered gas hydrates as a potential energy resource and as an important factor affecting global climate changes. The cumulative efforts, beginning with Humphrey Davy in 1810, provided tremendous amounts of knowledge about the thermodynamic, physical, and structure properties of gas hydrates and a rich collection of hydrate formers, including nitrogen, carbon dioxide, hydrogen sulfide, methane, ethane, propane, *iso*-butane, n-butane, and some branched or cyclic C5–C8 hydrocarbons.

1.2.2 Structures

In hydrate structures, water molecules are hydrogen bonded, while gaseous molecules are bonded to those only via van der Waals forces. Though the energy required to dissociate one hydrogen bond is about 5 kcal/mol, only 0.3 kcal/mol is needed to break one van der Walls bond, suggesting that gaseous molecules are only considered physically but not chemically entrapped into crystalline water cages [2]. Depending on the sizes of the guest molecules included in the gas hydrates, three hydrate structures are traditionally found, which are structure I, II, and H (Table 1.1, Figs. 1.1 and 1.2). The basic repeating unit in structure I is a primitive cubic lattice consisting of two pentagonal dodecahedra (5^{12}) (5 is the number of edges in a face and 12 is the total number of this type of faces in a cage) and six tetra-decahedra ($5^{12}6^{2}$) clathrate cages with a total number of 48 water molecules and a dimension of 1.2 nm. The average cavity radius of each type of cage is 3.95 and 4.33 Å, respectively. Methane, ethane, CO_2 and Xenon are typical structure I hydrate formers. While methane and Xenon occupy both small (5^{12}) and large ($5^{12}6^{2}$) cages, CO_2 and ethane are only small enough to dwell in large cages.

Table 1.1 Characteristics of hydrate cages of different structures.

	Structure I (sI)		Structure II (sII)		Structure H (sH)		
	Small	**Large**	**Small**	**Large**	**Small**	**Medium**	**Large**
Cages	5^{12}	$5^{12}6^2$	5^{12}	$5^{12}6^4$	5^{12}	$4^35^66^3$	$5^{12}6^8$
No. of cages per unit cell	2	6	16	8	3	2	1
No. of waters per unit cell	48		136		34		
Average cage radius (Å)	3.95	4.33	3.91	4.73	3.91	4.05	5.71
Coordination number	20	24	20	28	20	20	36

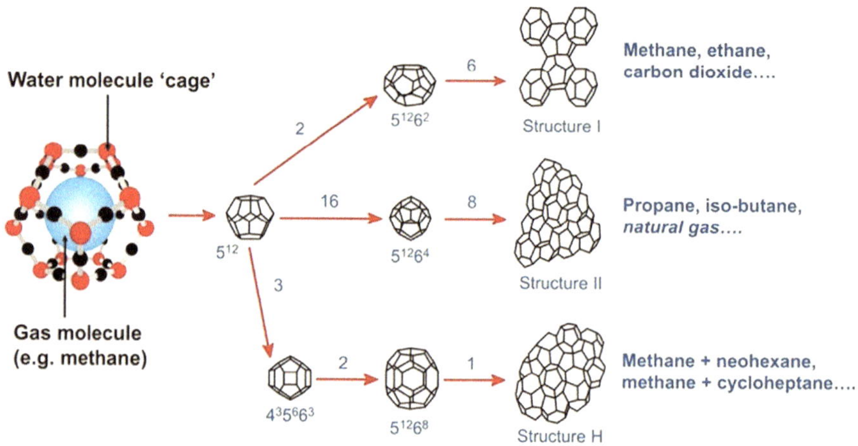

FIGURE 1.1

Different types of cavities found in hydrates and respective unit structures of sI, sII and sH structures [3].

Reproduced with permission from E.D. Sloan, C.A. Koh, Clathrate Hydrates of Natural Gases, third ed., CRC Press, New York, 2008.

The repeating unit of structure II hydrate also contains two types of cavities 16 pentagonal dodecahedra 5^{12} (3.91 Å) and 8 hexadecahedra $5^{12}6^2$ (4.37 Å) composed of 136 water molecules. Its lattice type is face-centered cubic and its unit dimension is 1.7 nm. Most hydrates in the oil/gas industry are expected to be structure II hydrates. In structure H, a layer of 5^{12} (3.91 Å) cavities connects a layer of $5^{12}6^8$ (4.06 Å) and $4^35^66^3$ (5.71 Å) cavities. In its hexagonal unit cell (a = 1.21 nm, c = 1.01 nm), 34 water molecules form three 5^{12}, two $5^{12}6^8$, and one $4^35^66^3$. One unique feature of structure H is that both small and large sizes of molecules are required to stabilize the structure. For example, neohexene and cycloheptane, which cannot form hydrates alone, form structure H hydrates with the help of methane. These three hydrate structures are important for the oil/gas industry because the types of hydrocarbons encountered in the field can form all these three types of hydrates.

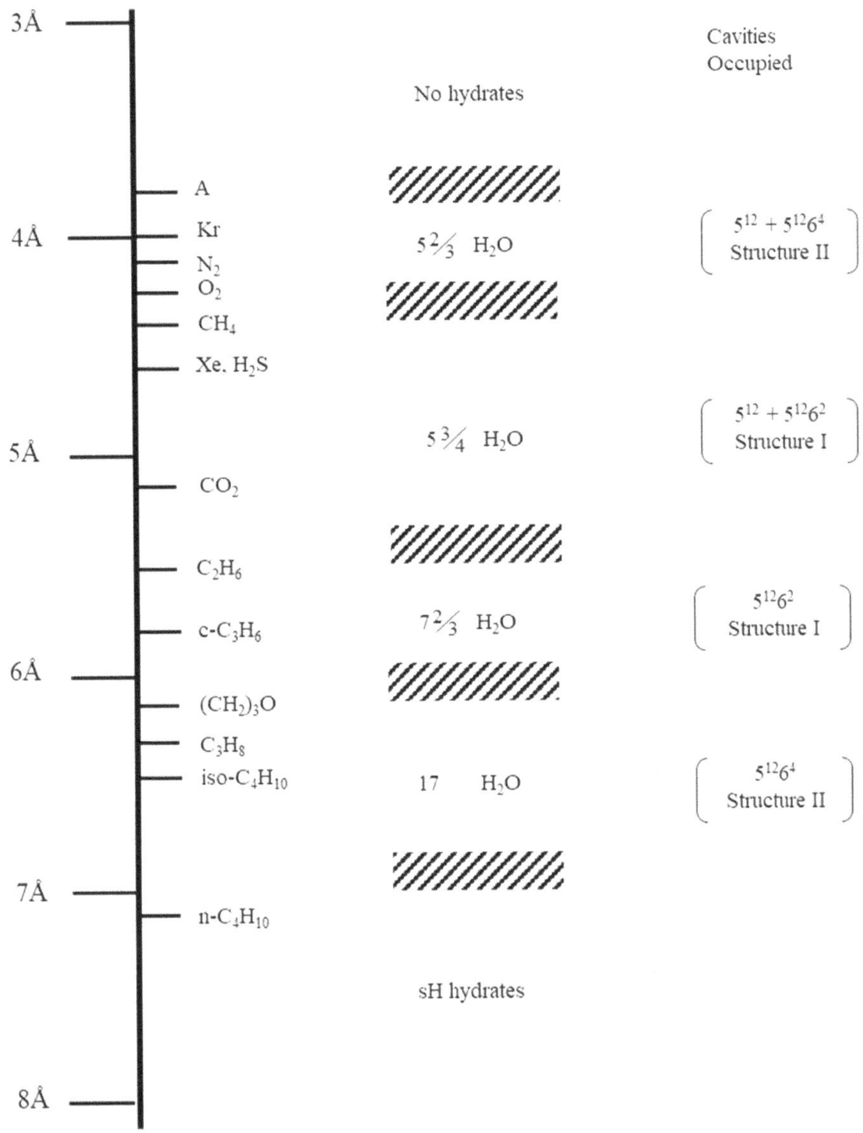

FIGURE 1.2

Comparison of guest molecule sizes and cavities occupied as simple hydrates.

With the advancement of experimental technologies and continuous research efforts on clathrate hydrates, some new types of hydrate structures at high pressures have been identified, mostly not relevant to oil/gas industry. It was discovered that under a pressure of 0.8 GPa, tetrahydrofuran (THF) and deuterium oxide (D_2O)

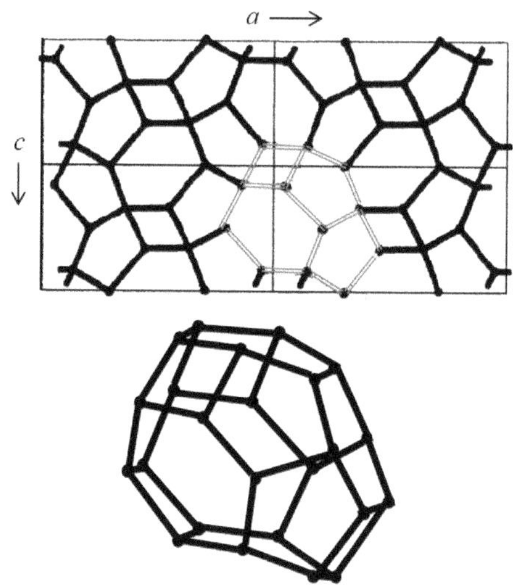

FIGURE 1.3

Packing and schematic view of the space-filling polyhedron.

Reproduced with permission from A. Kurnosov, V. Komarov, V. Voronin, et al., New clathrate hydrate structure: high-pressure tetrahydrofuran hydrate with one type of cavity, Angewandte Chemie International Edition, 43 (2004): 2922–2924.

form an orthorhombic structure, in which water molecules form 14-hedra cages with four tetra-, four penta, and six hexagonal faces ($4^4 5^4 6^6$) that are able to pack three-dimensionally without the need for other types of polyhedrons [4]. The stoichiometric composition of a unit cell can be presented as $4T_3{\cdot}24D_2O$, where T_3 is a $4^4 5^4 6^6$ cage. Projection of the structure along the b axis is presented in Fig. 1.3.

While investigating dimethyl ether (DME) hydrate using X-ray diffraction, Udachin et al. [5] identified another new hydrate structure T that is dense and highly complex (Fig. 1.4). It does not have 5^{12} polyhedra and can contain $5^{12}6^3$(P), $5^{12}6^2$(T), $4^1 5^{10} 6^3$(T') and $4^2 5^8 6^1$(U) cages. This hydrate structure is trigonal, space group P321, a = 34.995 Å, c = 12.368 Å, and stoichiometry can be described as $12P{\cdot}12T{\cdot}24T{\cdot}12U{\cdot}348H_2O$. The DME molecules are accommodated in all three types of large cages (P, T, T') (Fig. 1.5), giving an overall composition of $DME{\cdot}7.25H_2O$.

1.2.3 Phase behavior

The conditions required for hydrate formation and the resulting hydrate phase diagram vary based on the types of hydrate formers. The typical four pillars of hydrate formation are presence of free water, hydrate former(s), high pressure, and low temperature with only a few exceptions, e.g., THF and ethylene oxide can form hydrate with water under ambient pressure at ~4.5°C and ~11°C, respectively.

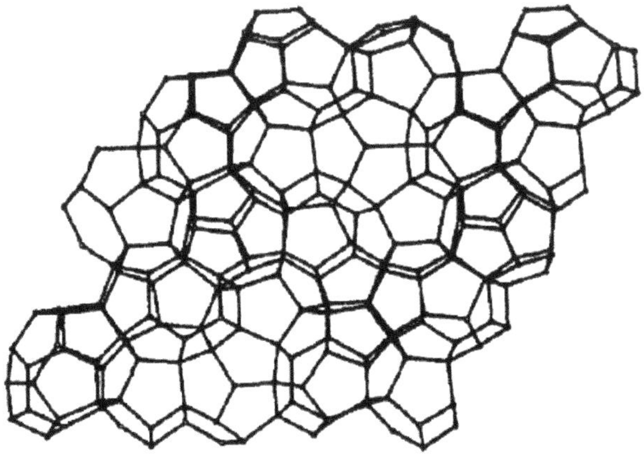

FIGURE 1.4

General view of the structure T hydrate as determined by single crystal.

Reproduced with permission from K. Udachin, C. Ratcliffe, J. Ripmeester, A dense and efficient clathrate hydrate, Angewandte Chemie International Edition, 40(2001): 1303–1305.

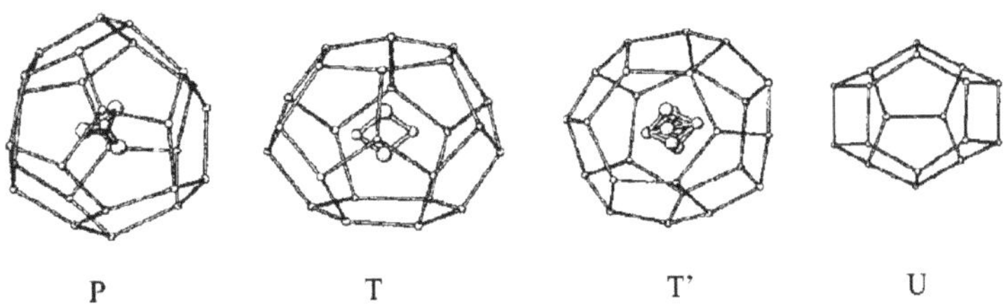

FIGURE 1.5

View of the cages in the str. T hydrate.

Reproduced with permission from K. Udachin, C. Ratcliffe, J. Ripmeester, A dense and efficient clathrate hydrate, Angewandte Chemie International Edition, 40(2001): 1303–1305.

Most hydrocarbons found in the oil/gas industry require high pressure to form hydrate. At the same temperature, lighter hydrocarbon hydrate formers typically require higher pressure to form hydrate than heavier hydrocarbon hydrate formers. For example, at 10°C (50°F) with fresh water, methane hydrate's equilibrium pressure is about 1000 psia while it would only require about 250 psia to form ethane hydrate. In addition, the salinity of water plays a key factor in the hydrate phase diagram. From a molecular level, the presence of ions in the water can cause disruption in the formation process of hydrogen bonding structure of clathrate cages. The higher the salinity, the higher the pressure required to form hydrate. Since the hydrate structure itself is salt free, the hydrate formation process in saltwater would extract water in the solution and convert them to hydrate while causing the salinity of the remaining water to increase. For this reason, the hydrate formation in salty

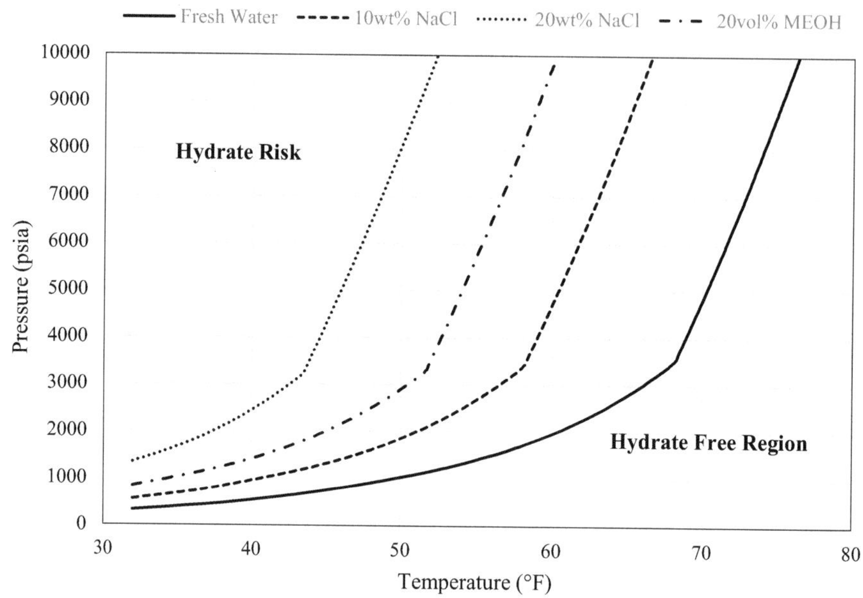

FIGURE 1.6

Typical hydrate phase diagram of a black oil system.

water will become self-inhibiting at a certain point of the conversion process. The addition of water-soluble compounds such as methanol (MeOH) or glycol into the water will require higher pressure to form a hydrate, a similar effect is observed with higher salinity. This is the foundation to manage hydrate risk with thermodynamic inhibitors (mostly MeOH, glycol, and to a lesser extent ethanol). Fig. 1.6 is a typical hydrate phase diagram of a black oil system with different salinities. The hydrate phase diagrams for natural gas are similar but without the inflection points along the curve where the black oil system typically goes through the bubble point. Obtaining accurate hydrate phase diagrams is a key step in proper hydrate risk assessment and risk management and will be discussed in more details later.

1.2.4 Properties

All the hydrates implications and applications are rooted in their unique properties under various conditions of temperature and pressure. The following are a few properties that are particularly important to energy industry.

1.2.4.1 Mechanical properties

Experimental determination of the mechanical properties of clathrate hydrates is difficult due to the challenge of making pure nonporous hydrate samples. In addition, the presence of residual water/ice and free gas in the system due to incomplete hydrate formation process can also contribute to the measurement reliability. Therefore there are considerable uncertainties associated with all the hydrate mechanical property measurements. The following are a few general conclusions about the mechanical properties of gas hydrates based on experimental/theoretical studies and field observations.

- Generally speaking, the hydrate mechanical properties are similar to that of ice. Once formed, they can be hard to remove mechanically and cannot be scrapped off by sending down a pig like removing wax deposit. Doing so will only make matters worse. Therefore the main hydrate management strategy is focused on hydrate prevention in the first place and hydrate dissociation once formed.
- The elastic properties of gas hydrate depend on temperature, pressure, and hydrate composition, including structure, guest molecule, and cage occupancy. Lower temperature and higher pressure both contribute to an increased bulk modulus and consequently make the hydrate harder.
- When well within the stability zone, the compressive strength of hydrates is higher than that of ice however, the strengths become closer in value when hydrate is less supercooled. This difference is attributed to the special hydrate lattice structure and the host, guest and host–guest interactions.

1.2.4.2 Self-preservation during dissociation

Hydrate dissociation is an endothermic process similar to ice melting, but its heat of dissociation (~54 kJ/mol for methane hydrate) is much higher than that of ice (6 kJ/mole). During dissociation process, hydrates separate into water and guest molecules by breaking up hydrogen bonding networks of water molecules and the van der Waals interaction forces between guest and host water molecules. This process takes up a significant amount of heat from the environment and causes the temperature to drop, which contributes to the stability of the remaining hydrate, that is, a self-preserving/limiting phenomenon. Without active supply of heat from the environment, a hydrate plug can take up to a year to dissociate by itself. Therefore continuous supply of heat is key to modulate the rate of hydrate dissociation.

1.2.4.3 Large gas-to-hydrate volume ratio

A large volume of hydrate former gas can be released upon hydrate dissociation. For example, one cubic foot of methane hydrate releases about 160 cubic feet of gas under ambient condition. This has two key implications for oil/gas industry applications:

- Low gas-to-oil ratio (GOR) fluid can become gas-starved during the hydrate formation process and become self-limiting. Consequently, the hydrate plugging risk during shut-in in such system can be much lower than a high GOR fluid.
- Gas release during hydrate dissociation can quickly build pressure up in confined volume, which can either cause pressure containment rupture or rapid acceleration of dislodged hydrate plugs. This can pose asset integrity risks and personnel safety risks.

1.3 Hydrate formation

1.3.1 Hydrate formation scenarios

As long as the requisites for hydrate formation are present, that is, high pressure, low temperature, presence of hydrate formers and water, there are risks of hydrate formation. The following are a few common scenarios to consider for the potential risk of hydrate formation in upstream production. Hydrate management strategies will be covered in later half of this chapter.

1.3.1.1 Shut-in

Fluid temperature typically drops after production is shut-in and the flow has stopped in cases where the environmental temperature is lower than the production fluid temperature, e.g., deepwater, winter time. The pressure can also start building up after shut-in. Both these changes can eventually push the system into the hydrate formation conditions in either wells or flowlines.

1.3.1.2 Cold restart

If the shut-in period is long enough to push the system into hydrate conditions, restarting the production from this state can pose a high hydrate formation risk since the flow turbulence can greatly accelerate the mass transfer and cause rapid hydrate formation. This risk is especially significant for long single flowline tiebacks.

1.3.1.3 Steady state

Some production system can enter hydrate formation condition during steady state. This is less common for black oil system, but more common for gas system due to the much smaller heat capacity and lack of insulation.

1.3.1.4 Gas injection, gas lift, or gas export

These gas systems rely on dehydration to control the risk of hydrate. However, the dehydration can sometime underperform, or the water content measurement/reading can be faulty. Hydrate can and have formed and plugged the line in these scenarios.

1.3.2 Hydrate formation mechanism

The molecular mechanism of hydrate formation is till yet to be fully understood [6]. One theory is that hydrate formation starts with some labile clusters with hydrate former molecules at the center (Fig. 1.7). These clusters then agglomerate and start to from initial hydrate nucleation nuclei, which grows into bulk hydrate crystalline phase.

From a macroscopic perspective, hydrate is believed to form at water/oil or gas interface. This has been the basis for the hydrate plugging model in oil/gas industry. Jeong et al. [8] experimentally studied and imaged the hydrate formation process of acoustically levitated water droplets in high pressure natural gas environment. The experimental results match the existing conceptual model (Fig. 1.8).

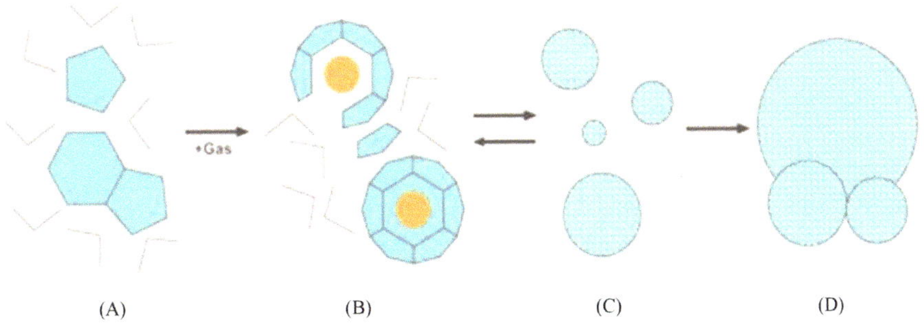

FIGURE 1.7

Labile-cluster model of hydrate nucleation: (A) labile clusters, (B) agglomeration of clusters, (C) primary nucleolus, and (D) hydrate crystal [7].

Reproduced with permission from A. Hassanpouryouzband, E. Joonaki, M. Farahani, et al., Gas hydrates in sustainable chemistry, Chemical Society Reviews, 49(2020): 5225–5309.

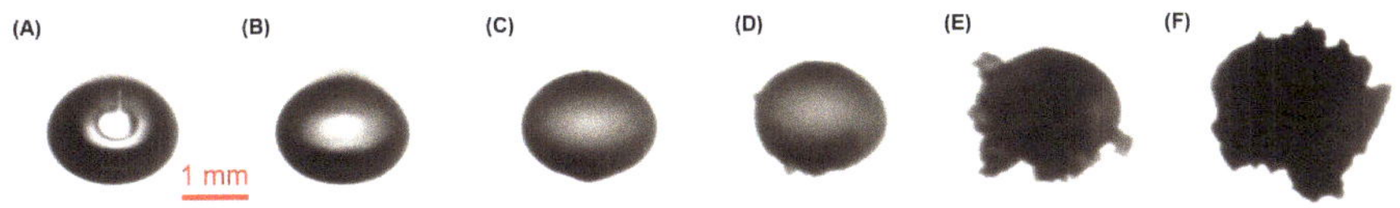

FIGURE 1.8

A series of back-illuminated optical images showing the evolution from a levitated water droplet to a hydrate particle. (A) Initial water droplet. (B) Hydrate formation onset detected via change in optical transmission relative to previous panel. (C) Shell thickening [2050 s after panel (B)]. (D) Initial outward growth (14650 seconds after panel (B)). (E) and (F) Hydrate propagation with a major roughness increase (respectively 25320 and 49410 s after panel B).

Reproduced with permission from K. Jeong, P. Metaxas, A. Helberg, et al., Gas hydrate nucleation in acoustically levitated water droplets, Chemical Engineering Journal, 433 (2021) 133494.

1.4 Hydrate management in production systems

Gas hydrate management in hydrocarbon production systems has conventionally relied on complete prevention of hydrate formation, which leads to a predominantly thermodynamics (phase behavior)-based operating strategy. As a result, system design guidelines and operating strategies ensured that the production system is never exposed to hydrate forming conditions during steady-state production as well as during transient operations, such as production shut-ins and re-starts. Over the last decade, fundamental improvements in understanding of hydrate formation, including nucleation, kinetics of agglomeration and deposition have led the industry to increasingly take a more risk-based approach in managing hydrates. The general direction has been to identify the key risk factors that control hydrate formation and then exploit them to optimize system design and operating strategies where the risk is demonstrated to be low and manageable. The following sections discuss some of these key risk factors involved. The strategies for prevention, mitigation, and remediation will be presented in Sections 1.5–1.8.

1.4.1 Risk assessment

In hydrocarbon production systems, the following four conditions must typically be met for gas hydrates to start forming:

- high pressure
- low temperature
- hydrate forming guest (gas) components
- water

These conditions provide a basic framework to assess the hydrate risk envelope for the overall production system. Eliminating any of the above listed conditions/variables effectively means that gas hydrates cannot thermodynamically form in the system. Conventional hydrate management strategies often remove hydrate risk through either eliminating one of the above conditions or through injection of chemical inhibitors.

The pressure–temperature-based hydrate phase envelope defines the thermodynamic risk region for a given fluid mixture. The region to the left of the hydrate curve represents conditions where hydrates are thermodynamically stable, whereas conditions to the right of the curve defines the region where hydrates cannot form. This region also defines the operating envelope where hydrocarbon production systems are typically designed to operate during steady-state production as well as during transient events (shut-ins and re-starts).

1.4.2 Hydrate modeling

1.4.2.1 Thermodynamic modeling

As part of overall hydrate risk assessment, thermodynamic modeling often forms the first step in evaluating whether any part of the production system is at risk of entering or residing within the hydrate stable region. The theoretical framework underlying thermodynamic modeling of gas hydrates was first postulated by van der Waals and Platteeuw [9]. Their approach modeled the interaction between the gas molecules and the clathrate lattice structure using a Lennard-Jones type potential. This model was further refined by Parrish and Prausnitz [10] using Kihara parameters to characterize gas-water interaction. Importantly, they also developed a simple computational algorithm to calculate the hydrate dissociation pressure at a given temperature that formed the basis for several subsequent modeling approaches. Since those early days of hydrate thermodynamic modeling, huge improvements have been made in the modeling approaches and the data regression parameters. This has allowed for these models to provide an accurate calculation of hydrate equilibrium curves for complex hydrocarbon-water-salt/inhibitor mixtures. Most of the current thermodynamic models still rely on this basic formalism along with a Gibbs free energy minimization-based technique to perform hydrate-related phase equilibrium for flash calculations [11]. Recent advances in thermodynamic modeling of gas hydrates are summarized in [12].

In current practice, hydrate equilibrium curves for oilfield applications, such as those shown in Fig. 1.6, are almost always generated by commercial pressure-volume-temperature (PVT) software. Some of the commonly used thermodynamic software used include Multiflash (from KBC), PVTSim (from Calsep) and CSMGem (from the Colorado School of Mines (CSM)). Over the years, these models have been continuously updated and refined with additional experimental data to the point that experimental generation of hydrate equilibrium curves is rarely utilized, unless dealing with some special components or inhibitors/solvents. Benchmarking across a variety of these commercial software [13] indicates that the average absolute errors for single and binary gas mixtures are typically < 1 deg K (Fig. 1.9). For black oil and gas condensate systems, the errors are generally higher, but these often stem from errors in characterization of heavier components in these hydrocarbon systems. If the phase envelope of the black oil

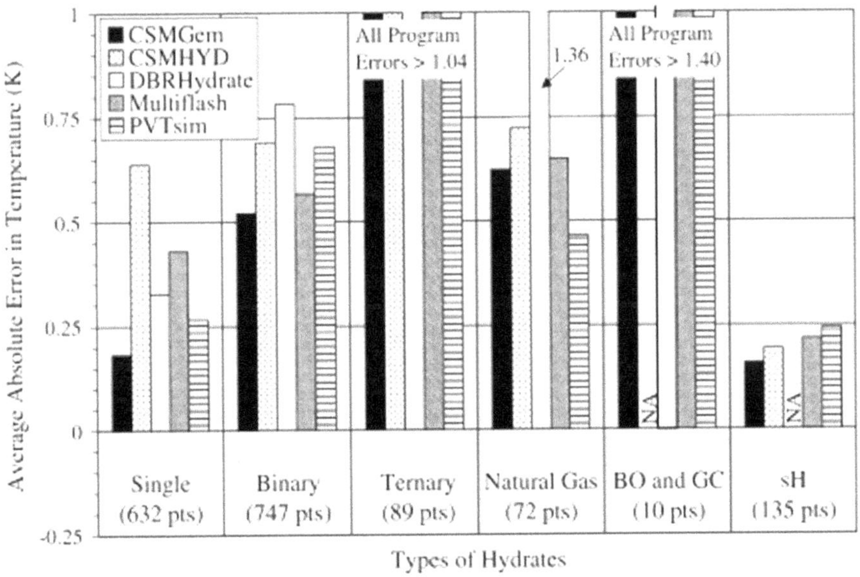

FIGURE 1.9

Average absolute errors in predicting hydrate equilibrium temperatures for a range of hydrate types.

Reproduced with permission from A.L. Ballard, E.D. Sloan, The next generation of hydrate prediction IV a comparison of available hydrate prediction programs, Fluid Phase Equilibria, 216(2004): 257–270.

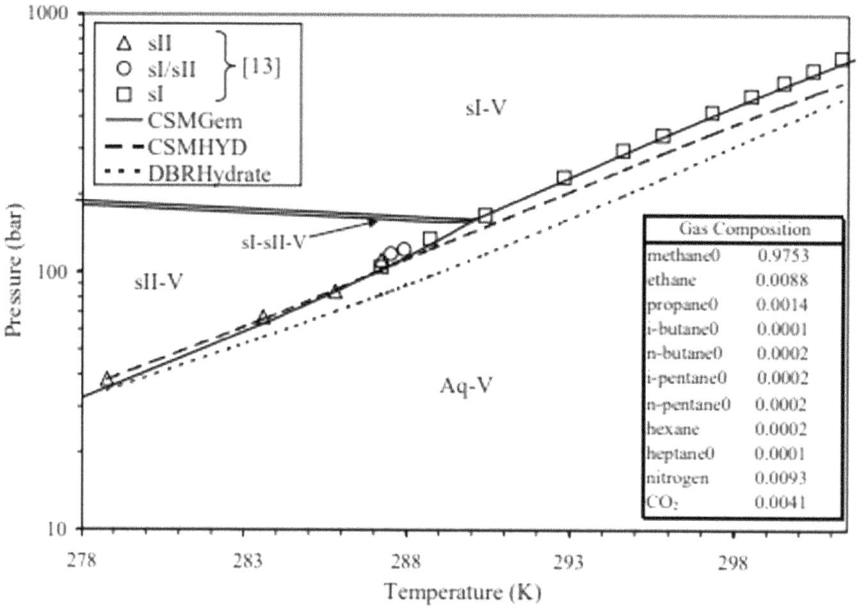

FIGURE 1.10

Hydrate phase equilibrium curve for a natural gas mixture benchmarked against various commercial software [13]. Authors report predictions from Multiflash and PVTsim to be similar to CSMGem.

or gas condensate system is well tuned with respect to the gas-oil ratio and/or bubble points, errors in thermodynamic hydrate curve prediction can be further minimized. Fig. 1.10 compares the hydrate equilibrium data for a natural gas mixture with predictions from various commercial software and shows very good agreement between experimental data and modeling results [13].

Generation of hydrate equilibrium curves for multcomponent black oil and gas condensate systems requires the following set of inputs:

- Hydrocarbon composition - Obtained from standard PVT reports.
- Formation water composition - total salinity, or if available, detailed ionic composition of the aqueous phase.
- Fluid properties - GOR and bubble point information can help match the overall phase envelope by tuning of pseudo-component properties, which leads to more accurate hydrate equilibrium curves.

In addition, most modeling software also require the specification of thermodynamic models (equation of states (EoS)) for various fluid phases. Commonly used EoS to model phase equilibria of various gas hydrate formers, other hydrocarbon components, inhibitors and water include Peng-Robinson, Soave-Redlich-Kwong, and the Cubic Plus Association (CPA). In particular, the CPA EoS has proven to be quite successful in predicting hydrate phase equilibrium for such multiphase multi-component mixtures containing hydrocarbons, MeOH/glycol and water, because of its ability to account for hydrogen bonding between water and inhibitors [14–16]. Phase equilibria predictions modeled using the CPA EoS show very good agreement with experimentally measured data [16] for pure methane (Fig. 1.11, top) and pure

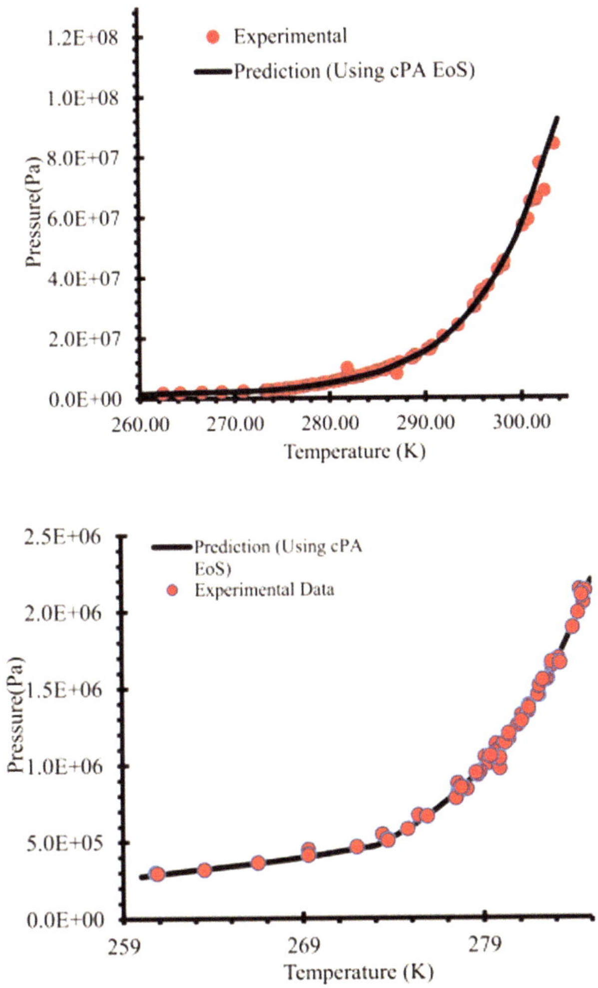

FIGURE 1.11

Hydrate phase equilibria predictions for pure methane (top) and pure ethane (bottom) using CPA EoS.

Reproduced with permission from M.N. Khan, P. Warrier, J.L. Creek, et al., Vapour-liquid equilibria (VLE) and gas hydrate phase equilibria predictions using the cubic-plus association equation of state: CSMGem extension to association EoS model, Journal of Natural Gas Science and Engineering, 94(2021): 104083.

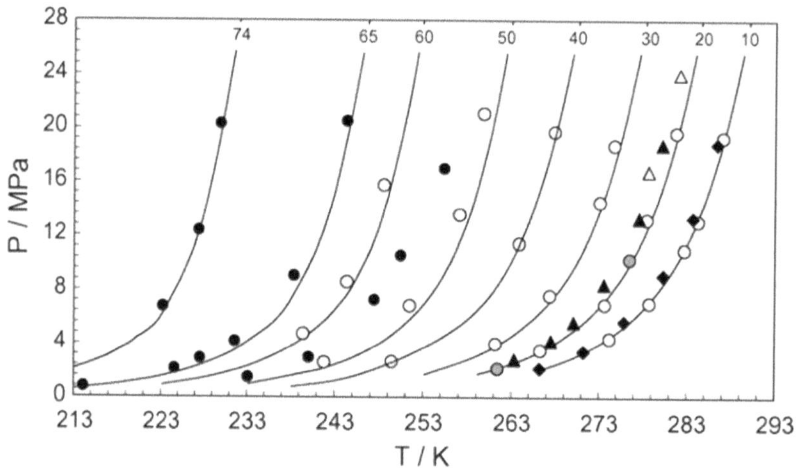

FIGURE 1.12

Experimental and predicted methane hydrate dissociation conditions using CPA EoS for varying methanol concentrations.

Reproduced with permission from H. Haghighi, A. Chapoy, R. Burgess, et al., Phase equilibria for petroleum reservoir fluids containing water and aqueous methanol solutions: Experimental measurements and modelling using the CPA equation of state, Fluid Phase Equilibria, 278(2009): 109–116.

ethane (Fig. 1.11, bottom). Good agreement between CPA-based models and experimental data is also observed for methane hydrate equilibrium across a range of MeOH concentrations (Fig. 1.12) [14].

Thermodynamic risk assessment utilizes these phase equilibrium curves to check whether the production system enters the hydrate stable region or not. If the system is at risk of entering the hydrate thermodynamic region, suitable mitigation methods, discussed below, are designed to eliminate or manage the hydrate risk.

1.4.2.2 Kinetic modeling

Hydrate formation is a phase change process that, similar to all phase transformations, is characterized by a kinetic behavior. This means that hydrate formation will not happen instantaneously upon crossing the thermodynamic phase boundary. Rather, a time lag is usually observed between the time the system enters the thermodynamically stable hydrate region and by the time hydrate formation is observed at the macroscopic or detectable scale. This time lag has been found to be dependent on a number of system parameters, such as degree of subcooling, fluid phases present, chemistry of the oil phase, flow behavior (regime) and mixing between different fluid phases, and even the type and scale of apparatus where kinetic behavior is being evaluated. These large number of variables involved make modeling kinetic behavior of hydrates much more difficult compared to thermodynamic phase behavior, which mostly depends on the composition of the fluid phases.

From a kinetic perspective, hydrate formation has been divided into two main steps—nucleation of stable hydrate nuclei followed by growth of these nuclei into bigger hydrate crystals [2,17–19], as illustrated in Fig. 1.13. Nucleation of gas hydrates, as measured through induction times, has been observed to be a stochastic process based on the variability in measured data. This variability tends to be higher at lower subcooling or supersaturation conditions. Classical nucleation theory has been applied by Kaschiev and Firoozabadi [20] to relate the rate of nucleation to the driving force in terms of supersaturation and to the work required to form a stable cluster of hydrate building units. Their work suggests that once hydrate nuclei reach a certain critical size, they are energetically favored to sustainably grow into bigger crystals, whereas nuclei smaller than the critical size will not have a chance at crossing the energy barrier. The energy barrier, and consequently the critical radius, required for nuclei to grow has been shown to be inversely dependent on the subcooling. This implies that higher subcooling in any given system will lead to a smaller critical nuclei radius and smaller free energy barrier required to be exceeded for hydrates to nucleate and grow [17,19,20]. This theoretical framework by Kaschiev and Firoozabadi [20] was applied to both homogeneous nucleation that occurs in bulk fluid phases without any impurities as well as to heterogeneous nucleation that is favored in systems containing solid substrates, impurities or interfaces between fluid phases. Consistent with experimental observations, their theory indicates free energy barrier for heterogeneous nucleation to be lower than that for homogeneous nucleation, generally leading to significantly higher nucleation rates for heterogeneous nucleation for the same pressure and temperature conditions (subcooling).

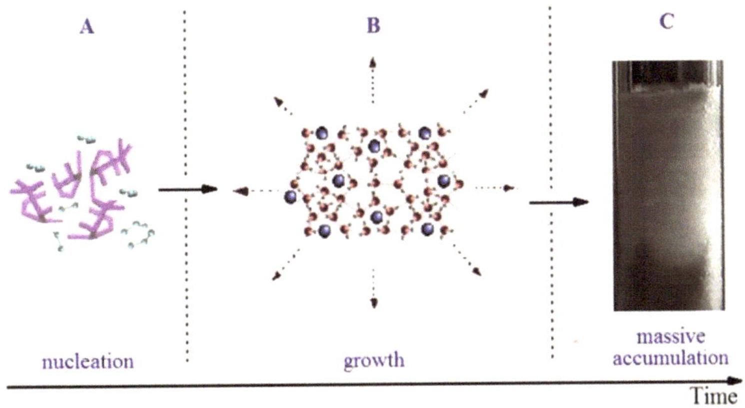

FIGURE 1.13

Illustration of the sequential nature of nucelation (A) and growth (B) steps that lead to bigger hydrate accumulations (C).

Reproduced with permission from W. Ke, T.M. Svartaas, D. Chen, A review of gas hydrate nucleation theories and growth models, Journal of Natural Gas Science and Engineering, 61(2019): 169–196.

This classical nucleation-based theory can be used to calculate the cumulative probability of nucleation, P, as [20]:

$$P(t) = 1 - \exp(-Jt)) \tag{1.1}$$

where J is the total nucleation rate and t is time elapsed prior to nucleation.

Recently, high-resolution nucleation data from experiments, using a structure II gas mixture, has been used to fit the predicted probability distribution function (PDF) to calculate the nucleation rate J [19,21]. Fig. 1.14 shows cumulative hydrate nucleation probability distributions as a function of time compared with model predictions with J, the nucleation rate, being the only fitting parameter. As expected, experimental data indicate higher subcooling to result in shorter nucleation times leading to the probability distribution progressively shifted to the left. At any given instant of time, this also implies higher overall probability of nucleation being detected in the system. Nucleation rates calculated from fitting this dataset also indicate the parameter J to increase with the subcooling. Although the nucleation rate-based model correctly captures the functional form of the distribution, fitting measured data across a range of experiments shows significant variability in the value of the nucleation rate J. The original theory [20] proposed the nucleation rate J to be further decomposed into a kinetic parameter A and a thermodynamic parameter B. When regressed to experimental data, large variations have been observed

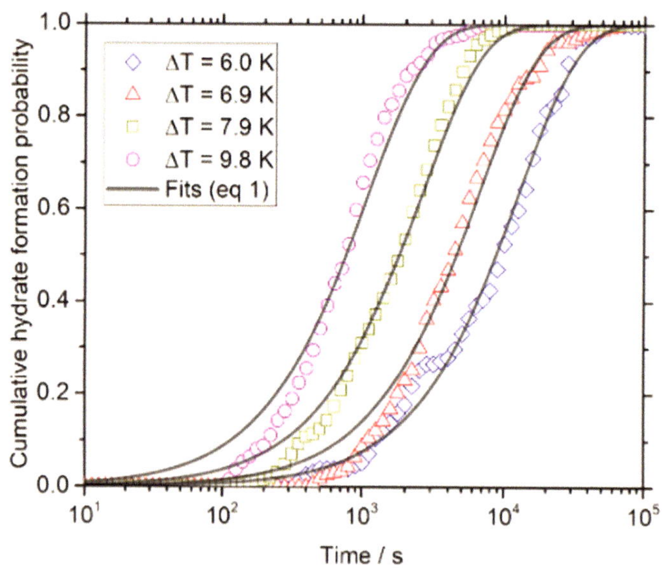

FIGURE 1.14

Experimentally measured cumulative hydrate nucleation probability distribution as a function of time compared with predictions from the nucleation theory.

Reproduced with permission from P.J. Metaxas, V.W.S. Lim, C. Booth, et al., Gas hydrate formation probability distributions: Induction times, rates of nucleation and growth, Fuel, 252(2019): 448–457.

between predictions from the nucleation theory and values of these fitted parameters deduced from measurements [21,22].

Based on these observations, it can be argued that nucleation models like the one proposed by Kaschiev and Firoozabadi can be fitted well to a particular dataset. However, one single set of regressed parameters will struggle to predict nucleation kinetics behavior across a wide range of experimental apparatus, test conditions and different fluid types. For industrial applications in oilfield systems, operators have typically chosen much simpler approaches and guidelines to quantify and analyze nucleation risk. For example, Matthews et al. [23] used field test data conducted at Werner Bolley well in Wyoming along with Texaco's high pressure flow loop data to find that a subcooling of ~6–6.5°F was required to see hydrate formation and plugging in these systems. Similar flow loop studies performed using Exxon Mobil's Friendswood flowloop and University of Tulsa's flowloop have supported a subcooling requirement of 6–6.5°F to observe hydrate nucleation in these larger scale systems [24]. More recently, Statoil (now Equinor) [25] and Total [26] described how operators design risk-based operability envelopes by studying the induction time as a function of subcooling. Analogous to a traffic-light system, their approach essentially divides the hydrate stable phase region into a low-risk green zone that extends from the equilibrium curve to a finite subcooling inside the stability region, an intermediate-risk yellow region followed by a high-risk red region that is characterized by very fast hydrate nucleation and formation. Kinnari et al. [25] report that the green low-risk region can provide operations greater than 12 hours of induction time under certain conditions, whereas the yellow intermediate-risk region can cover induction times from 1–2 hours up to 12 hours. Such operating envelopes can be generated for specific oils through testing in autoclave setups and provide a simpler alternative to apply induction time behavior to real production systems.

Post nucleation, hydrate nuclei have been hypothesized to grow in size and quantity through a more conventional reaction or kinetic growth model. From a modeling perspective, it is simpler to treat nucleation and the subsequent growth as separate phases. However, experimentally it is very difficult to distinguish these two steps, given they both still occur at the microscopic level. Vysniauskas and Bishnoi [27] were one of the first to quantify the kinetics of hydrate formation. They also reported experimental data for methane and ethane gases using a stirred semi-batch reactor. Compared to induction time data to measure nucleation statistics, measurement of hydrate growth rate often relies on tracking gas phase consumption in an isobaric and isothermal setup [27]. Using this consumption data to benchmark against, Vysniauskas and Bishnoi proposed a growth rate model that took the following form:

$$r = Aa_{\mathrm{s}}\exp\left(-\frac{\Delta E_{\mathrm{a}}}{RT}\right)\exp\left(-\frac{a}{\Delta T^{b}}\right)P^{\gamma}P^{\gamma} \tag{1.2}$$

where A is a pre-exponential constant, a_{s} is the interfacial gas–liquid area, ΔE is the activation energy, R is the gas constant, T is the absolute temperature, ΔT is the subcooling, P is the system pressure, a and b are empirical parameters, and γ is the order of the reaction.

The above rate equation models hydrate growth as a three-step process—an initial clustering process, formation of a critical size cluster and the growth of the hydrate crystal around the stable nucleus [27]. This semi-empirical model was regressed against growth rates for methane and ethane hydrates to obtain the relevant parameters in the rate equation stated above. Following this work, Englezos et al. [18] published their kinetic growth model based on crystal growth from a solution. A two-step process was hypothesized that included the first step of diffusion of gas from the bulk phase to the crystal-liquid interface followed by the second step of reaction at the interface. The growth rate per particle was described as [18]:

$$\left(\frac{dn}{dt}\right)_{\mathrm{p}} = K * A_{\mathrm{p}}\left(f - f_{\mathrm{eq}}\right) \tag{1.3}$$

where A_{p} is the surface area of hydrate particles, $\left(f - f_{\mathrm{eq}}\right)$ represents the gas fugacity difference between the bulk phase and its equilibrium value, and $K * \infty$ is a combined rate parameter describing the two combined resistances as

$$\frac{1}{K*} = \frac{1}{k_{\mathrm{r}}} + \frac{1}{k_{\mathrm{d}}} \tag{1.4}$$

where k_{r} is the reaction rate constant and k_{d} is the mass transfer coefficient around the particle. The term $\left(f - f_{\mathrm{eq}}\right)$ in effect represents the driving force for hydrate formation.

Fig. 1.15 illustrates various aspects of the Englezos model, namely the diffusion-controlled gas concentration profile in the interface and bulk phases (Fig. 1.15A) as well as the fugacity difference across the phases as the main driving force (Fig. 1.15B).

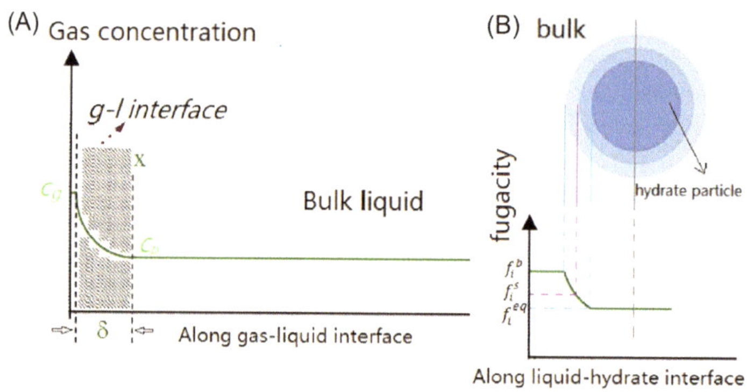

FIGURE 1.15

Schematic illustration of the Englezos model. (A) concentration profiles of the gas former along the gas–liquid interafcea d*n* in the bulk phase. (B) Gas fugacity profile along the liquid-hydrate interafce.

Reproduced with permission from W. Ke, T.M. Svartaas, D. Chen, A review of gas hydrate nucleation theories and growth models, Journal of Natural Gas Science and Engineering, 61 (2019) 169–196.

Since the publication of the Englezos model, various other improvements have been proposed and are discussed in several review papers [17,19]. However, most of the subsequent models are still based on the same underlying form that relates the kinetic growth rate to a rate constant times total interfacial area times the driving force [19]. These models were able to predict hydrate growth rates in smaller scale setups, such as autoclaves or stirred reactors, but were seldom applied to larger scale flowing systems.

For hydrocarbon production systems, understanding and utilizing kinetic information about hydrate formation has numerous important applications. In producing assets, this knowledge can lead to minimizing hydrate mitigation measures during oil and gas production, resulting in significant chemical savings (operating expenditures (OPEX)) as well as the reduction in production deferment related to faster restarts after a shut-in [25]. For greenfield or brownfield projects, an improved understanding of hydrate kinetics can lead to the optimized design of production facilities leading to lower cost (capital expenditures (CAPEX)) systems.

Developing a comprehensive transient kinetic model that can be applied to production systems has been an area of active research in the gas hydrate community. In particular, the Colorado School of Mines Hydrate Kinetics Model (CSMHyK) [28] comes closest to fulfilling that goal of developing a simple integrated tool that can be used by flow assurance engineers to assess the kinetic aspects of hydrate formation and plugging in real production systems. The CSMHyK model was originally developed for oil-dominated systems based on the conceptual model depicted in Fig. 1.16. The overall hydrate formation process is divided into four main sequential steps [24,28]:

1. Water entrainment—the model assumes that all water from the aqueous phase gets completely dispersed into the oil phase to create a water-in-oil emulsion. The resulting droplet size distribution depends not only on water and oil properties, such as interfacial tension, densities, viscosities, but also depend on the flow behavior (shear stresses experienced by the droplets). Mean water droplet sizes are then predicted using correlations with dimensionless Weber numbers.

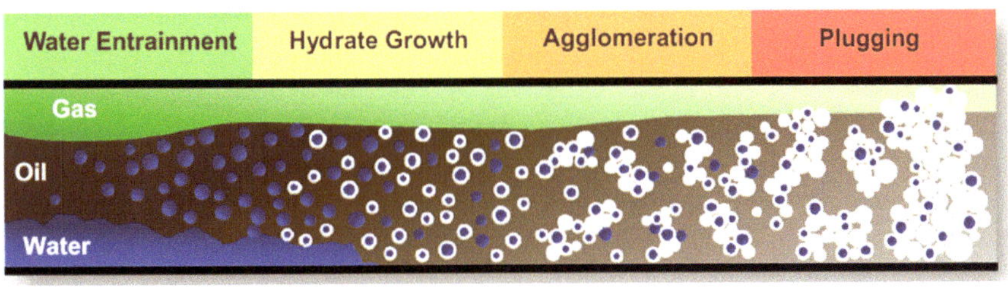

FIGURE 1.16

Conceptual model for hydrate formation in multiphase oil-dominated systems.

Reproduced with permission from L.E. Zerpa, E.D. Sloan, A.K. Sum, C.A. Koh, Overview of CSMHyK: a transient hydrate formation model, Journal of Petroleum Science and Engineering, 98–99 (2012) 122–129.

2. Hydrate growth—if pressure and temperature conditions are inside the hydrate region, a hydrate shell starts to form at the oil-water interface and grows radially inwards. The kinetic growth model used within CSMHyK is similar in essence to the rate equations described above, where hydrate growth is related to the subcooling and interfacial area through a first-order kinetic reaction constant.
3. Hydrate Agglomeration—this phase involves agglomeration of hydrate-encrusted particles into bigger masses. Adhesion of hydrate particles has been attributed to strong capillary attractive forces due to the presence of unconverted water on the surface.
4. Plugging—finally, if enough hydrate mass has agglomerated, it can either lead to jamming/deposition of hydrate on surfaces or cause significant increase in the slurry viscosity, resulting in a hydrate plug.

The model includes effects of mass transfer and heat transfer limitations across the boundary layer next to the hydrate particle as well as across the hydrate shell. To include effects of agglomeration, the original CSMHyK model utilized Camargo and Palermo's [29] approach that balances the interparticle cohesion force with the flow-based shear forces. The hydrate agglomerate size is used to calculate the effective hydrate volume fraction and the relative viscosity of the hydrate slurry. Details of the individual models are published elsewhere [28]. For oil-dominated systems, a hydrate plug was identified by a large increase in viscosity but not as wall deposition.

Of all the kinetic growth models developed and published over the years, CSMhyK appears to be the most frequently used by flow assurance engineers in the industry, especially to assess hydrate risk during transient events. A major reason behind its acceptance as an industry standardized tool is its integration into a well-known transient thermal-hydraulic simulator OLGA. Flow assurance engineers often use dynamic simulation tools like OLGA to design operability envelopes and associated hydrate mitigation measures, such as depressurization, fluid displacement or inhibitor injection. Having a comprehensive hydrate kinetic model fully integrated with the transient simulator allows them to quickly evaluate hydrate risk for various operation scenarios. This integration is illustrated schematically in Fig. 1.17, which shows that basic fluid and thermodynamic information within each discretized pipeline section in OLGA is input to the CSMHyK model. In return, the model outputs relevant hydrate properties such as amount of hydrate formed or dissociated, viscosity or temperature change within that pipeline section in a given time step back to OLGA.

The model can be applied to full-scale production systems, but it should be noted that there is not enough data available on hydrate growth or hydrate volume fraction changes from operating flowlines to benchmark model predictions. Instead, the CSMHyK model has often been benchmarked against industrial-scale flow loop data. Fig. 1.18 shows the CSMHyK model, that was tuned to ExxonMobil's flow loop data, successfully predicting hydrate volume fractions measured in a separate flow loop facility (University of Tulsa) but with the same fitted rate constant [24].

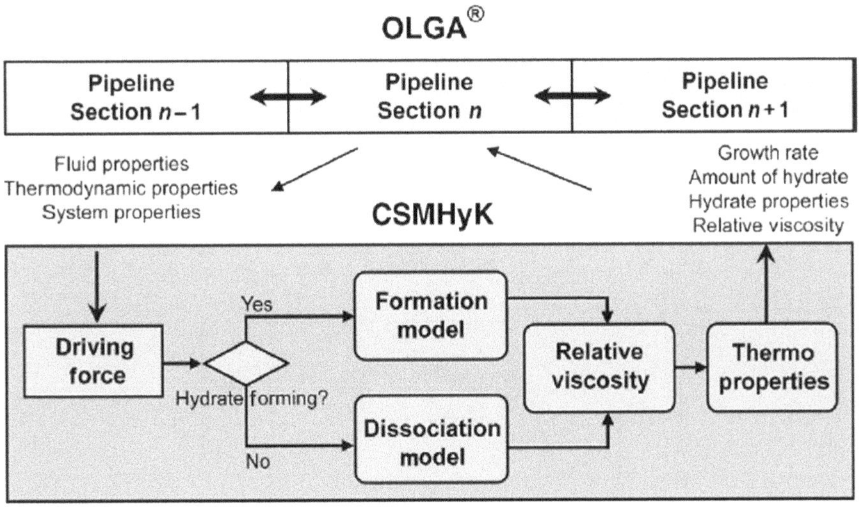

FIGURE 1.17

Schematic depicting the integration of CSMHyK into the transient thermal-hydraulic simulator OLGA.

Reproduced with permission from E.D. Sloan, J. Creek, A.K. Sum, Where and how are hydrate plugs formed?, in: E. D. Sloan, C.A. Koh, A. Sum (Eds.), Natural Gas Hydrates in Flow Assurance, Elsevier, 2010, pp. 13–36.

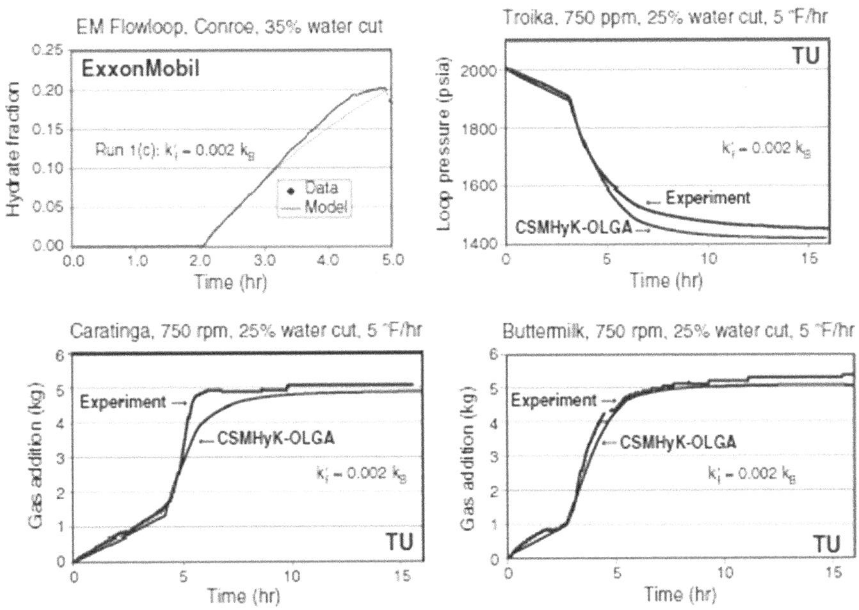

FIGURE 1.18

CSMHyK predictions tuned to ExxonMobil flow loop data were successfully able to predict University of Tulsa flow loop experiments.

Reproduced with permission from E.D. Sloan, J. Creek, A.K. Sum, Where and how are hydrate plugs formed?, in: E. D. Sloan, C.A. Koh, A. Sum (Eds.), Natural Gas Hydrates in Flow Assurance, Elsevier, 2010, pp. 13–36.

Since the first generation of the CSMHyK model that was developed for oil-dominated systems, improvements and extensions to other types of production systems have been proposed. These include model frameworks for gas condensate [28] and water-dominated systems [30]. Fig. 1.19 shows good agreement between model predictions and experimentally measured hydrate volume fractions for oil-dominated (top) and water-dominated system (bottom) using the same high pressure multiphase flow loop facility [30].

As the models and underlying mechanisms for different fluid systems continue to improve, it is expected that such transient and integrated kinetics models will be able

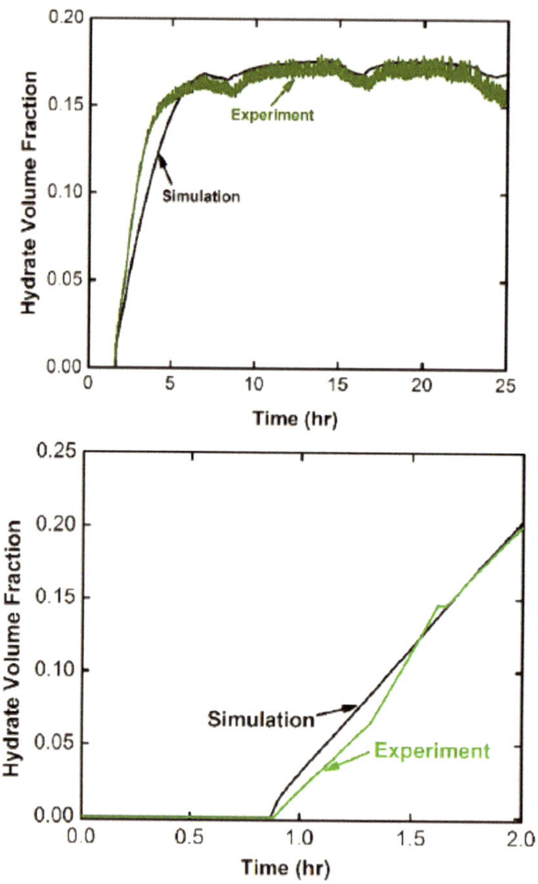

FIGURE 1.19

Model predictions compared with experimentally measured hydrate volume fraction for an oil-dominated system (top) and a water-dominated system (bottom).

Reproduced with permission from Y. Wang, C.A. Koh, J.A. Dapena, et al., A transient simulation model to predict hydrate formation rate in both oil- and water-dominated systems in pipelines, Journal of Natural Gas Science and Engineering, 58 (2018) 126–134.

to predict hydrate volume fraction more reliably for real production systems. Having that level of detailed hydrate fraction profile is useful, but the more important and logical next step is to understand what hydrate volume fraction means in terms of deposition or plugging risk. In other words, at what hydrate volume fraction does the risk of hydrate plugging in a given system transition from low to high?

Recent work has started to address that linkage between hydrate volume fraction and plugging risk through modeling of high-pressure flow loop experiments. Joshi et al. [31] performed experiments in a gas and 100% water-cut system to evaluate transportability of hydrates as a function of increasing hydrate volume fraction. Their conceptual model Fig. 1.20 shows good transportability of hydrates in the flow loop up to a certain concentration, denoted as $\varphi_{transition}$, that is characterized by hydrates being well dispersed in the water phase. Beyond this transition point (Fig. 1.20), rapid increase in pump pressure drop is observed as the hydrate volume fraction further increases in what is characterized as the heterogeneous distribution phase. In the final third stage, flow is described to be dominated by gas due to significant bedding of hydrate particles at the bottom of the pipe or due to wall deposition. For maintaining low hydrate plugging risk in such a system, it would be recommended to stay below this $\varphi_{transition}$, which serves an indicator for eventual hydrate plugging. Based on this work on transition hydrate volume fraction [31], Chaudhari et al. [32] have proposed an index called Hydrate Risk Evaluator (HRE) that attempts to quantify plugging risk in oil-dominated systems. This HRE index is a function of hydrate volume fraction and

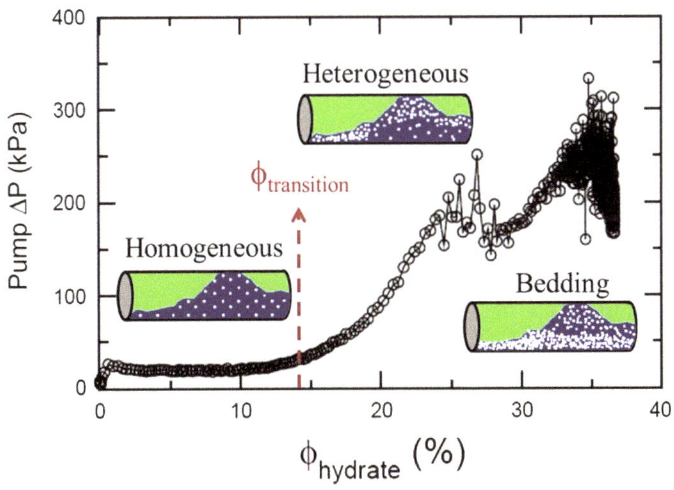

FIGURE 1.20

Conceptual model of hydrate plug formation in 100% water-cut systems.

Reproduced with permission from S.V. Joshi, G.A. Grasso, P.G. Lafond I, et al., Experimental flowloop investigations of gas hydrate formation in high water cut systems, Chemical Engineering Science, 97 (2013) 198–209.

other flow properties in oil-dominated systems, such as liquid loading and mixture velocity. Experimental flow loop data for oil-dominated systems was used to generate HRE index (Fig. 1.21), which suggests that for a given system, hydrates remain in homogeneous dispersion when HRE < 2.25 [32]. Beyond this value, they transition to an intermediate to high-risk region (Fig. 1.21).

The work described above represents active areas of research that are yet to be applied in real operating environments. Hence, more work is needed to fully understand how one could relate localized calculation of hydrate volume fraction in a pipeline to plugging risk. It should also be noted that most of these recent measurements and resulting plugging risk models were based on experiments performed at a couple of flow loop facilities. It remains to be seen how well they represent and cover a broader range of operating conditions and flow characteristics to allow applications of these hydrate growth and plugging models to a wider set of hydrocarbon production systems.

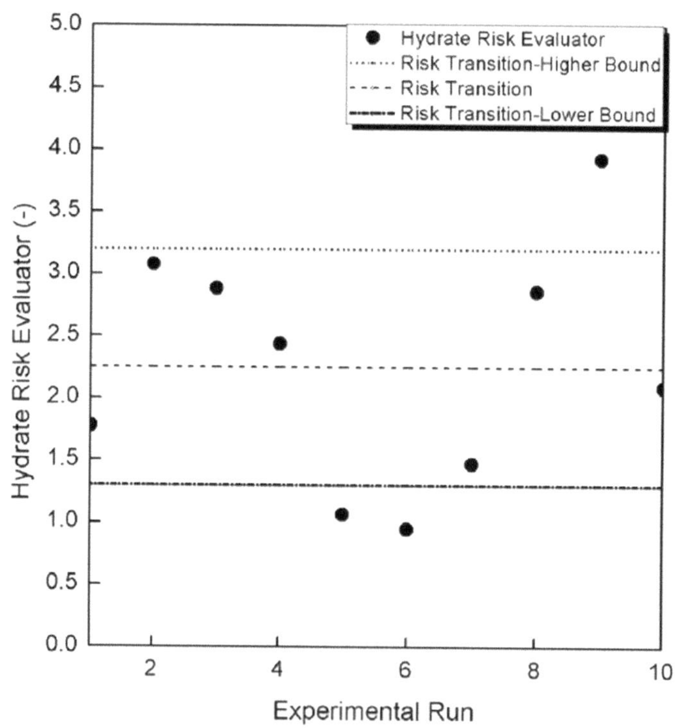

FIGURE 1.21

HRE index calculated for various flow loop experiments along with bounds for risk transition.

Reproduced with permission from P. Chaudhuri, L.E. Zerpa, A.K. Sum, A correlation to quantify hydrate plugging risk in oil and gas production pipelines based on hydrate transportability parameters, Journal of Natural Gas Science and Engineering, 58 (2018) 152–161.

1.5 Temperature control

This mitigation strategy involves keeping the temperature of the entire production system above the hydrate equilibrium temperature (HET) and can be achieved either through heat retention (passive insulation, flowline burial) or through active heating of the system (e.g., direct electrical heating (DEH) or heat tracing). For most subsea/deepwater production systems that operate at an ambient subsea temperature of about 40°F, passive insulation can prevent hydrate conditions during steady-state production depending on fluid properties, operating conditions, and flowline lengths. However, during a shut-in, such systems will eventually cool down to the ambient conditions, thereby entering the hydrate region. Hence, passive insulation systems typically need additional mitigation measures to prevent hydrate formation during such transient events.

1.5.1 Thermal insulation

Thermal insulation is often used to retain heat in production systems that are at risk of entering the hydrate region during steady-state flow. This strategy is predominantly used for oil production systems that tend to retain heat over longer distances due to their higher relative heat capacity. Gas-dominated production systems, on the other hand, lose thermal energy over relatively shorter distances, making heat retention not an attractive mitigation option (except for short tieback distances). For cases where heat retention is the primary hydrate mitigation strategy, thermal insulation is usually applied to all components of the production system where water and hydrocarbon phases can interact. In subsea production systems, this design choice implies that thermal insulation will be applied to all components downstream of the well, including the subsea trees, jumpers, manifolds, flowlines, and risers. This ensures that temperatures all along the production stream are always maintained higher than the HET.

As mentioned previously, a key drawback with thermal insulation-based hydrate mitigation is its inability to eliminate hydrate risk during production shut-ins. For production environments where ambient temperatures are relatively low (e.g., deepwater, cold onshore), the flowline and associated production system will eventually cool down to the ambient temperature. At this point, other mitigation measures are required to prevent the system from seeing hydrate forming conditions. The time it takes for a system that normally operates/produces outside the hydrate region to cool to the hydrate temperature, at the given shut-in pressure, is termed the cooldown time. The cooldown time is dependent not only on the thermal insulation parameters (type, thickness) but also on the system dependent variables, such as flowline length, flow rates, flowing wellhead temperatures, shut-in pressure, fluid GOR and water-cut. As a result, cooldown time is not typically defined as a static number, but dynamically evolves with the life cycle of production conditions.

With a thermal insulation-based strategy, the cooldown time is the maximum time duration available to the operator to secure the system and bring it outside the hydrate region. Two main mitigation options that are commonly used during such shut-in events include:

- Dead oil displacement—this is typically used if either a dual flowline system exists, or a service line is available. For offshore platforms, dead-oiling is performed from the host, which requires adequate storage for performing this operation for all subsea flowline systems. Sweep rates and displacement times are designed to ensure that most of the live production fluids are swept out of the flowlines before they cool down to the hydrate region.
- Depressurization - This strategy can be used for dual flowline systems but is more often used for single flowline tiebacks. The objective is to depressurize or blowdown the production system to a pressure below the hydrate equilibrium pressure corresponding to the ambient temperature. For subsea systems, and especially for deepwater environments, depressurization can become challenging due to the additional hydrostatic pressure head in the riser. Comparatively, blowdowns are much easier to execute in onshore environments. Several system and fluid dependent variables affect the feasibility of successful depressurization. Some key variables, more relevant to offshore production systems, are discussed below:
- Water-depth - as water-depths become deeper, liquid loading in the riser tends to ne higher. This will result in increased hydrostatic head on the flowline and makes depressurization challenging.
- GOR - higher GOR implies greater gas content in the system and leads to higher probability of achieving the target blowdown pressure. Conversely, lower GOR systems will have a higher liquid holdup, which makes them more challenging to successfully depressurize.
- Water-cut - higher water-cuts lead to greater water holdup in the riser and higher hydrostatic head at the riser base, thereby making blowdowns difficult.
- Salinity - produced water salinity has a mixed effect on feasibility of a successful blowdown. Higher salinity shifts the hydrate curves to increase the target blowdown pressures but the increased salinity also increases the density and hydrostatic head of the water present in the riser, leading to a higher riser base pressure.
- Flowline topography - the flowline topography dictates whether blowdown has been achieved across the entire length of the flowline. Even if the pressure at the riser base is successfully reduced to a value below the hydrate equilibrium pressure, some sections of a flowline might have trapped pressure that can put them at hydrate risk. For a mostly uphill flowline, even if riser base pressure is below hydrate equilibrium pressure post depressurization, liquid holdup along the length of the flowline could mean that the manifold end of the system might still be within hydrate forming conditions. Transient modeling along with measurements at the subsea manifold can be used to ensure all sections of the subsea system are below the hydrate equilibrium pressure.

Other factors include ability of the offshore platform to be able to handle all the liquids during depressurization and flowline length.

Dead oil displacement and depressurization are the main mitigation strategies to prevent hydrate formation in passive insulated systems during planned or unplanned shut-ins. Chemical inhibition can be an option to manage planned shut-ins, where the system is pre-treated with either a thermodynamic inhibitor or a low-dosage hydrate inhibitor prior to the shut-in. Unfortunately, such a strategy cannot be used during unplanned shut-ins. These mitigation options do not happen instantaneously and require finite time durations to safely execute. The time it takes to execute these mitigation strategies (dead oil displacement or depressurization) is subtracted from the cooldown time to define the no-touch time (NTT) of the system (Fig. 1.22). As the name suggests, NTT represents the time after a shut-in event during which no action is taken. If the shut-in duration (planned or unplanned) lasts smaller than the NTT, the system generally can be restarted without implementing any hydrate mitigation strategies. Post expiration of the NTT, if the shut-in situation is still not resolved, operations begin implementing the mitigation options discussed above. This ensures that by the time cooldown time is reached, all mitigation activities have been completed and the system is safely secured against hydrate forming conditions. As defined, NTT is always shorter than the cooldown time of the system (Fig. 1.22).

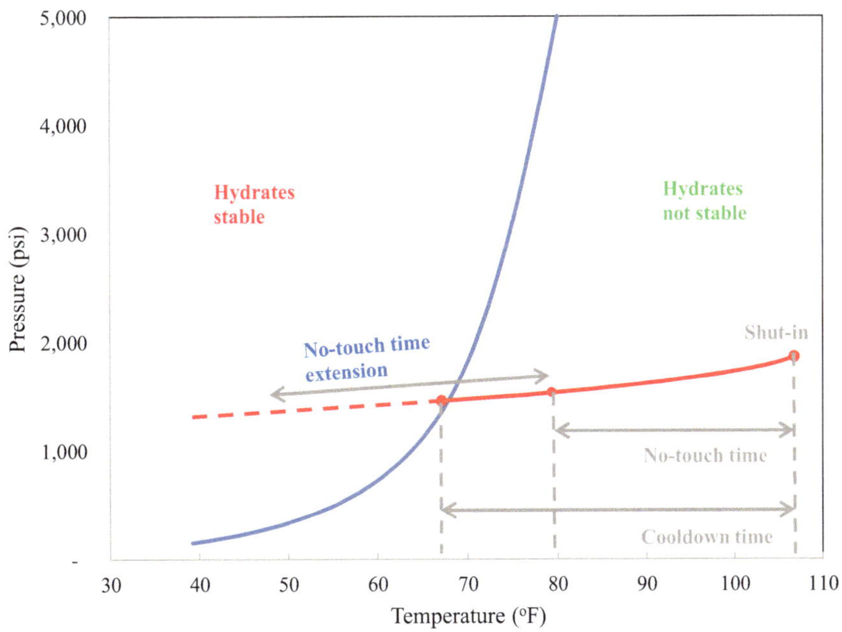

FIGURE 1.22

Hydrate equilibrium curve depicting no-touch time and cooldown time for a typical subsea system.

Recent advances in the understanding of hydrate risk have allowed operators to let the system enter the hydrate region without resulting in significant hydrate deposition or plugging. Factors that enable such excursions into the hydrate region depend on fluid properties and are an area of active research [25,33–36]. Some of these key risk factors include:

- Fluid GOR - generally, higher GOR has been related to increased rate of hydrate formation and severity. On the other hand, fluids with low GOR or systems with no or limited free gas phase have been shown to have mass transfer constraints that reduce the risk of significant hydrate formation [34,35].
- Water-cut - given other parameters stay fixed, higher water-cuts result in higher absolute amount of hydrates in the system and higher overall risk of deposition or plugging. Various operators have defined low water-cut regions at which hydrate risk is deemed to be low during a shut-in event [25,33,35].
- Salinity/Subcooling - Higher produced water salinity can limit the thermodynamic subcooling experienced by the system, which can limit the rate and total amount of hydrate formation. Higher salinity also results in lower overall conversion of produced water into hydrates because hydrate formation results in increased salinity of the residual unconverted brine.

Together, these factors allow for the expansion of hydrate operating envelope during planned or unplanned shut-ins that can lead to extension of the NTT past the conventional thermodynamic-based boundary (Fig. 1.22). Based on the risk analysis for a given system, this strategy allows transient operations inside the hydrate region, for example during shut-ins and re-starts [36], without requiring further mitigation procedures.

The aforementioned strategies and associated operating envelopes form the basis of designing a passive/thermal insulation strategy for hydrate mitigation. The three main thermal insulation solutions used in the industry are discussed below.

1.5.1.1 Wet insulation

As the name implies, this type of insulation is applied on the external surface of the pipe or subsea component and is in direct contact with the surrounding water. Wet insulation materials that are commonly used in the industry include epoxy, polyurethane, polypropylene and glass syntactic (reinforced) versions of these materials [37,38]. U-values for these materials average about 2 W/(m^2-K) (0.35 Btu/(hour-ft^2-F)). Often, this is not sufficient to prevent hydrate free operating conditions throughout the length of the flowline, especially for longer distances. Even for shorter flowline lengths, late-life production conditions (low flow rate, low flowing wellhead temperatures) might not be completely mitigated against hydrate forming conditions.

1.5.1.2 Dry insulation

Dry insulation materials offer better thermal performance (lower U-value) compared to wet insulation options, typically ranging around 1 W/(m^2-K) (0.18 Btu/hour-ft^2-F) [37]. This is enabled by the insulation material being encased in a protective layer that prevents it from coming in contact with water. A pipe-in-pipe (PIP) system can also be used to encase the insulation material with a second concentric (outer) pipe to achieve such low U-values. Fig. 1.23 shows schematics of various types of insulation that are commonly used in the subsea systems.

1.5.1.3 Burial

Flowline burial can also be used as an insulation strategy, although the U-values tend to be generally lower than the insulated options. Burial is achieved by trenching and backfilling the soil to create an insulated system. As expected, the thermal performance of a buried system will depend on parameters such as the thermal conductivity, density and heat capacity of the soil, burial depth and other soil related parameters like particle sizes, porosity, permeability [39]. However, flowline burial is generally cheaper in costs compared to a dry or wet insulated system, which might make it a suitable solution for some applications (e.g., shorter distances). Heat from fluid production warms up the soil surrounding the buried pipe and acts as a heat reservoir, which is most effective if there is no insulation surrounding the pipe. Hence, buried un-insulated pipelines typically have the longest cooldown times. On the other hand, in case of a long shut-in, the soil will also lose the stored heat and can take a long time to accumulate the thermal energy from production to provide meaningful insulation. This can require

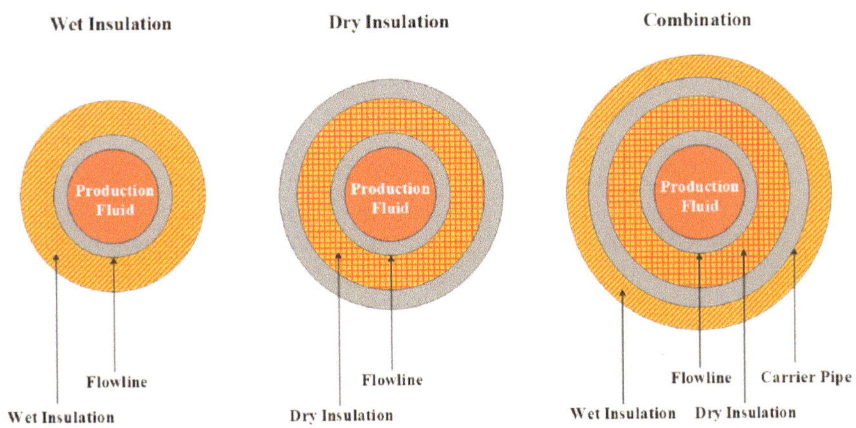

FIGURE 1.23

Various insulation configurations commonly used for heat retention.

Reproduced with permission from E. Bell, Y. Lu, N. Daraboina, Thermal methods in flow assurance: a review, Journal of Natural Gas Science and Engineering, 88 (2021) 103798.

additional chemical-based mitigation strategies during long transient periods of operations. In addition, lateral and upheaval buckling issues [40,41] often need to be addressed for a particular flowline before burial is chosen as the main heat retention option.

1.5.2 Active heating

Thermal insulation strategies discussed previously can be applied to flowlines as well as other subsea components like trees, manifolds, and jumpers. In contrast, active heating is mostly used for flowline applications, where temperatures are maintained above the HET either through electrical heating or by circulating hot fluids around a pipe bundle that encases the flowline.

1.5.2.1 Electrical heating

Electrical heating can be accomplished either through DEH, where electrical current flows directly through the pipe wall to generate heat [38,42,43], or through indirect heating, which heats up a heating element that is strapped to the outside of the pipe [44,45]. In this case, heat generated by this heating element is transferred to the pipe through conduction. Electrical-heating-based thermal management systems are generally higher in cost compared to thermal insulation, thereby making them more competitive for long distance flowlines [44]. DEH-based electrical heating systems can be further divided into the following types:

- Single pipe open loop - In open-loop DEH systems, a single piggy-back cable is used along the length of the flowline to carry the current, while the flowline acts as the return conductor (Fig. 1.24). The flowline is heated through resistance heating, but this design suffers from relatively large current losses to the seawater [38]. Such open-loop DEH systems have been installed at several fields (Asgard, Huldra, Kristin, Urd) operated by Statoil on the Norwegian Continental Shelf [42,43].
- End-fed PIP - In end-fed PIP DEH, the inner and outer pipes of the PIP system are used as the electrical conductors. A conducting bulkhead at the subsea end is used to electrically connect the two pipes [42,45]. Hence, current flows down outer carrier pipe and returns through the inner pipe (Fig. 1.25, middle). The key advantage compared to the open-loop DEH architecture is significantly lower electrical and heat losses to the environment compared to the open-loop DEH system. but the overall CAPEX of the PIP DEH system might be higher. Field installations using this configuration include Serrano and Oregano fields in the Gulf of Mexico operated by Shell [42].
- Center-fed PIP - this is another configuration of the DEH PIP system similar to the end-fed configuration. The main difference is that in center-fed systems, there are bulkheads on either end of the PIP system, while the power source is connected at the mid-point of the flowline (Fig. 1.25, bottom) [42,45,46].

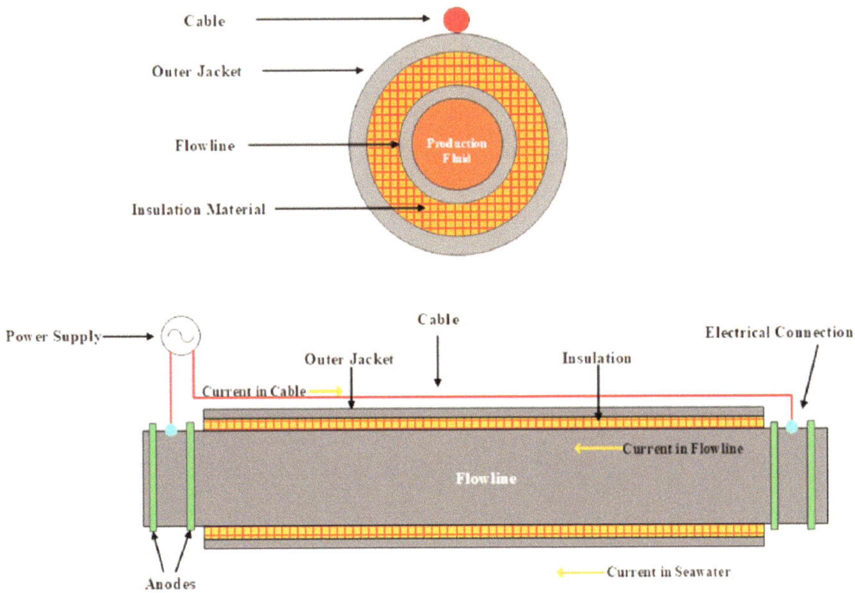

FIGURE 1.24

Schematic showing cross sectional view (top) and side view (bottom) of the open-loop DEH system.

Reproduced with permission from E. Bell, Y. Lu, N. Daraboina, Thermal methods in flow assurance: a review, Journal of Natural Gas Science and Engineering, 88 (2021) 103798.

As an alternative to DEH-based systems, electrically trace heated PIP has been more recently developed to provide high thermal performance with relatively low power requirements [46,47]. The system utilizes multiple trace heating cables that are directly laid on the outside wall of the inner pipe. This system has been designed both for continuous operations as well as transient/preservation operations. It has also been demonstrated to be effective in safely remediating hydrate blockages through active heating [47] and successfully deployed at the Islay field operated by Total in the Northern North Sea [48].

1.5.2.2 Pipeline bundles

Active heating through this strategy can involve several pipelines (including chemical injection lines) being bundled together inside a larger pipeline and kept warm by circulating hot fluids through the bundle. Similar to electrical heating [46], pipeline bundles can also work through direct heating, where the heating medium flows in the annular space outside the main production line, or through indirect heating, where the heating medium flows in its own dedicated line (Fig. 1.26). In this latter approach, the heating fluid does not come in direct contact with the production line, thereby eliminating corrosion and biofouling risks

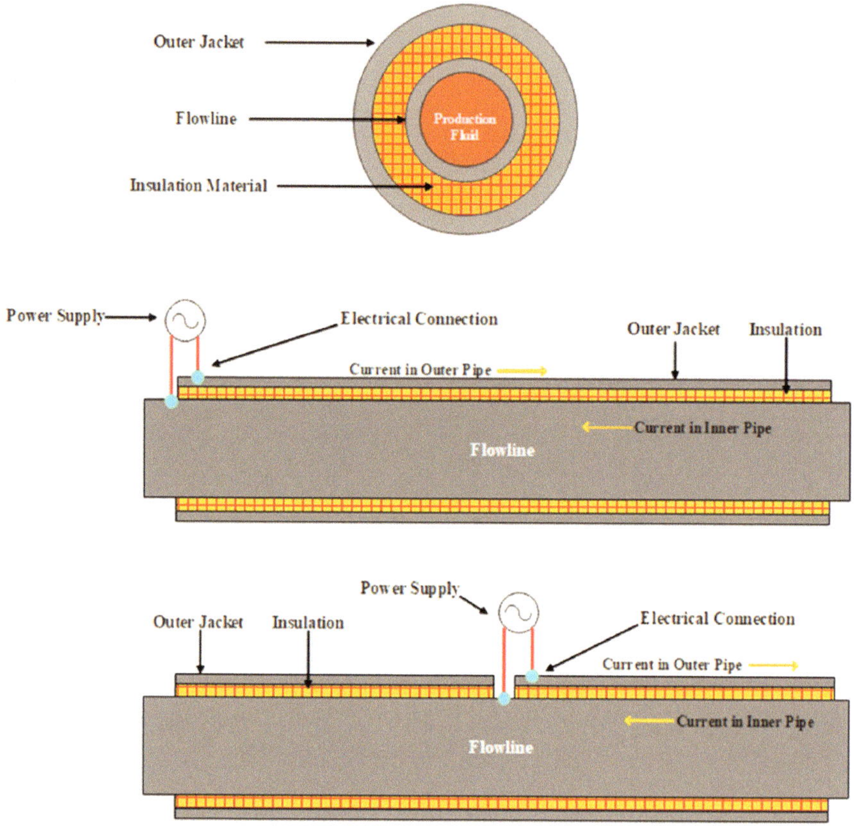

FIGURE 1.25

(Top) Cross sectional view of the PIP DEH system with side views of the end-fed (middle) and center-fed (bottom) configurations.

Reproduced with permission from E. Bell, Y. Lu, N. Daraboina, Thermal methods in flow assurance: a review, Journal of Natural Gas Science and Engineering, 88 (2021) 103798.

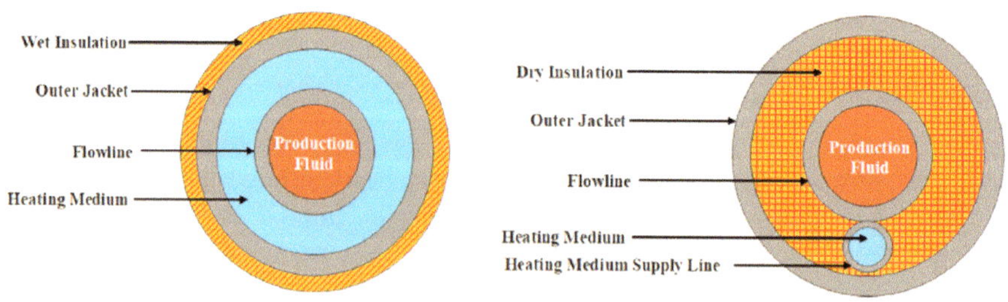

FIGURE 1.26

Schematic of active heating through pipeline bundles: Direct (left) and Indirect (right).

Reproduced with permission from E. Bell, Y. Lu, N. Daraboina, Thermal methods in flow assurance: a review, Journal of Natural Gas Science and Engineering, 88 (2021) 103798.

on the outside diameter of the lines in the bundle. The heating fluid can be a simple water-glycol mixture or a chemically treated water mixture that minimizes corrosion and fouling issues [38].

Table 1.2 summarizes various passive and active thermal management methods and compares their relative advantages and disadvantages.

Table 1.2 Comparison of various thermal management methods.

Thermal Methods		Pros	Cons
Insulation	Wet insulation	• Easy to mold insulation material directly around pipe • Avoids need for outer jacket	• Can be insufficient at preventing solids deposition on longer pipelines or during a shutdown • In event of a leak the product goes into the environment
	Dry insulation	• In the event of a leak the product is contained within the outer pipe, lowering environmental risks • Higher thermal efficiency than wet insulation	• More complex and expensive construction than wet insulation • Higher repair costs than wet insulation
	Pipeline burial	• Simple construction and low costs • High thermal efficiency	• In event of a leak the product goes into the environment • Limited to fairly shallow water
Electrical Heating	DEH open loop	• Remains effective across long pipelines • More effective than indirect heating at putting heat into the fluid • Flexible Installation Options • Capable of continuous heating	• Approx. 50% of electricity lost to the ambient • Low thermal performance from wet insulation • High operating costs • Continuous heating limited by water depth
	DEH pipe-in-pipe	• Remains effective across long pipelines • Dry insulation, high thermal efficiency	• No continuous operation • Approx. 50% of electricity lost to ambient
	Indirect electrical heating	• More heating energy goes into the fluid • Remains effective across long pipelines • Low power consumption and high electrical efficiency	• Limited inner pipeline diameter, may not be suitable for larger production pipes
	Integrated production bundle	• Remains effective across long pipelines • High electrical efficiency	• Low thermal performance

(Continued)

Table 1.2 Comparison of various thermal management methods. *Continued*

Thermal Methods		Pros	Cons
Hot Fluid Circulation	Pipe-in-pipe	• More efficient at putting heat into the fluid than bundle configurations • Capable of continuous heating	• Wet insulation, higher U-value • More heat is lost to the ambient
	Bundle	• Dry insulation, low U-value • Less heat is lost to the ambient	• Loses less heat to the ambient

Reproduced with permission from E. Bell, Y. Lu, N. Daraboina, Thermal methods in flow assurance: a review, Journal of Natural Gas Science and Engineering, 88(2021): 103798.

1.6 Chemical inhibition

Chemical inhibitors are commonly used to mitigate hydrate formation in production systems where the temperature cannot be maintained above the HET. This can occur during steady-state production or during transient events, like shut-ins and restarts. For subsea applications, continuous chemical inhibition is usually considered only for gas-dominated production because of the inability of such systems to maintain heat over significant tieback distances (even with insulation). Consequently, gas-dominated systems often rely on continuous inhibition through chemical injection, without being normally insulated. Oil-dominated production, however, does not normally use chemical inhibition as the main strategy because the OPEX associated with continuous injection often exceeds the CAPEX associated with insulating the oil flowline. Some exceptions to this general strategy can exist for oil systems where either the tieback length is too long for insulation to be feasible or where water-cut throughout field life is expected to be low enough for continuous chemical injection to be economically viable. There are two major types of chemical inhibitors that are used to manage hydrate risk—thermodynamic hydrate inhibitors (THIs) and low-dosage hydrate inhibitors (LDHIs) and are discussed separately below.

1.6.1 Thermodynamic hydrate inhibitors

THIs are the more commonly used chemical hydrate inhibitors and, as the name suggests, mitigate hydrate risk by shifting the thermodynamic hydrate equilibrium curve towards lower temperatures and higher pressures. Most polar solvents, such as alcohols, glycols, and ionic salts, work effectively as THIs by changing the activity of water molecules in the fluid mixture. In hydrocarbon production systems, MeOH and mono-ethylene glycol (MEG) are the two most commonly employed THIs due to their relatively low cost and wide availability. The molecular structure of MeOH and MEG (Fig. 1.27) depicts the hydrogen atoms that hydrogen-bond with water molecules and restrict them from forming the clathrate structure around the guest molecule [49]. This is the fundamental mechanism behind inhibition of hydrate formation through THIs.

Hydrate equilibrium curves are shown below (Fig. 1.28) for a simple gas mixture (C1:C2:C3−85:10:5 mol.%) with fresh water (no salinity) and various MeOH and mono-ethylene glycol (MEG) concentrations in the aqueous phase. Compared to the rightmost curve (fresh water), the addition of MeOH or MEG progressively shifts the hydrate curves to the left. As the dosage of the inhibitor (MeOH or MEG) increases, the corresponding hydrate curve is shifted towards lower temperatures and higher pressures, thereby expanding the available envelope for a

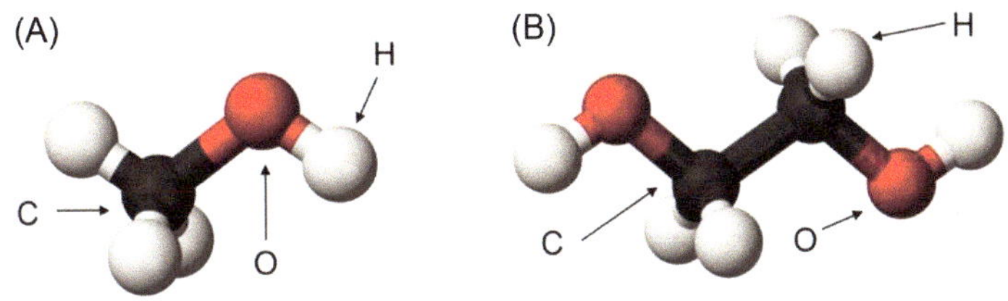

FIGURE 1.27

Molecular models of (A) methanol and (B) mono-ethylene glycol.

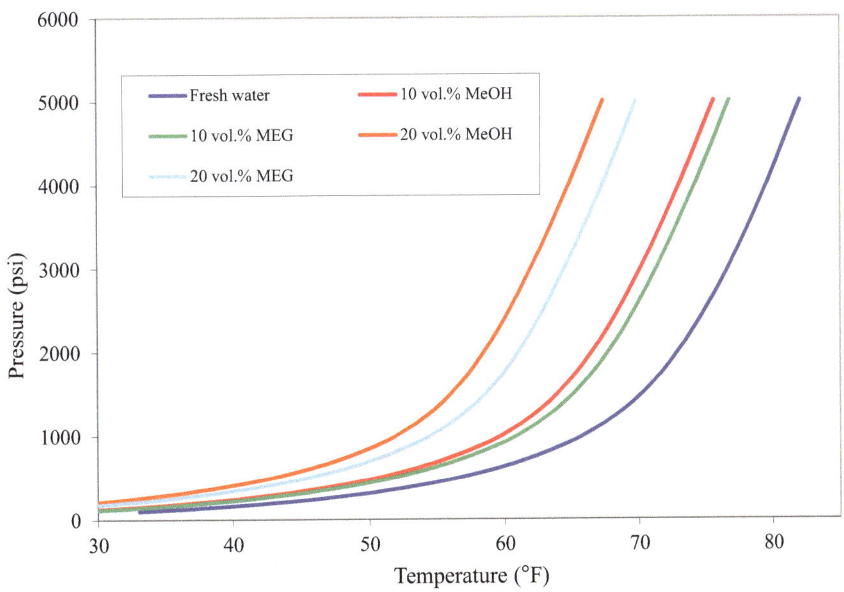

FIGURE 1.28

Effect of methanol (MeOH) and MEG on hydrate equilibrium curves compared to a freshwater system (no salinity).

production system to operate without any hydrate risk. It can also be seen that for the same volumetric inhibitor dosage, MeOH shifts the hydrate curve more than MEG (Fig. 1.27). This happens because lowering of the HET, similar to freezing point depression of water, is more dependent on the mole fraction of the inhibitor than the molecular structure of the inhibitor. Methanol has a lower molecular weight compared to all glycols, which results in a higher MeOH mole fraction in water compared to glycols for the same volumetric dosage.

In the past, simple analytical expressions were used to predict the effect of MeOH or MEG on lowering the HET for a given system. The Hammerschmidt equation [2,50] was commonly used to predict this depression, but to use such equations, one had to first estimate the hydrate curves without any inhibitor or salts in the system. The shift in the hydrate curve based on this equation can be calculated as:

$$\Delta T = \frac{KW\%}{M(100 - W\%)} \tag{1.5}$$

where M is the molar mass of the inhibitor in g/mol, $W\%$ is the weight % of inhibitor in the aqueous phase and K is an inhibitor dependent constant [2,50].

Once the hydrate curve is generated without the effect of inhibitor, the ΔT from the above equation can be used to include the effect of MeOH or MEG. It should be noted that this is an approximate method and that nowadays specialized thermodynamic software (Multiflash, PVTsim, CSMGem) are more commonly used to predict the inhibition effect of THIs and salts on the hydrate curves.

Methanol and MEG have distinct physical and chemical properties that should be carefully evaluated while designing an effective continuous injection-based inhibition strategy. Key properties to consider for each chemical are discussed below (Table 1.3).

With the above considerations, once MeOH or MEG has been identified as the THI of choice, the appropriate chemical dosage must be calculated. As mentioned above, specialized thermodynamic software is commonly used to calculate the required inhibitor dosage based on the specific hydrocarbon/brine composition and expected pressure-temperature conditions. Thermal-hydraulic simulations are often used to predict pressure and temperature profiles along the length of the subsea production system. THI dosages are then calculated to prevent hydrate formation representing the worst-case hydrate risk in terms of hydrate subcooling. Fig. 1.29 depicts this approach where the pressure and temperature trajectory are shown for a gas production system that is about 50 miles long [24]. It is seen that this subsea system enters the hydrate forming region at about 9 miles from the wellhead if no THI is injected to mitigate the hydrate risk. Between the 25–30-mile mark, the system is expected to experience the highest subcooling/hydrate risk. Hence, to adequately protect this system from forming hydrates, continuous MeOH dosage of about 23 wt.% is required.

The above example explains how dosage rates for a continuous injection case are calculated. As mentioned previously, this is typically the strategy of steady-state

Table 1.3 Properties and operational considerations of using methanol and mono-ethylene glycol (MEG).

Property	Methanol	MEG
Molar mass	Low: 32 g/mol	High: 62 g/mol
Safety issues [50]	High toxicity and flammable. NFPA rating: 3	Relatively lower toxicity and flammability. NFPA rating: 1
Flash point [50]	12°C	111°C
Viscosity (at 20°C) [50]	Low (0.59 cP)	High (21 cP)
Partitioning in hydrocarbon phase	Significant losses into hydrocarbon phases due to low flash point—makes regeneration very challenging	Low partitioning in nonaqueous phases makes regeneration and recycling of MEG an attractive option where high injection rates are needed
Corrosivity	Highly corrosive	Noncorrosive
Cost	Lower in cost per unit volume	Higher in cost per unit volume
Effect of salts	Methanol can cause salt precipitation if mixed with high salinity brines	Salt precipitation risk is not as high with MEG, but salt drop out can be a risk in regeneration systems during water removal
Operating strategy	For gas systems: Typically used as a single pass inhibitor without recycling when only saturation water is being produced or when water-cuts remain low. For oil systems: Typically used only during shut-ins or restarts to manage transient risk.	For gas systems: continuous injection during steady-state production with regeneration and recycle due to higher unit cost and ease of regeneration.

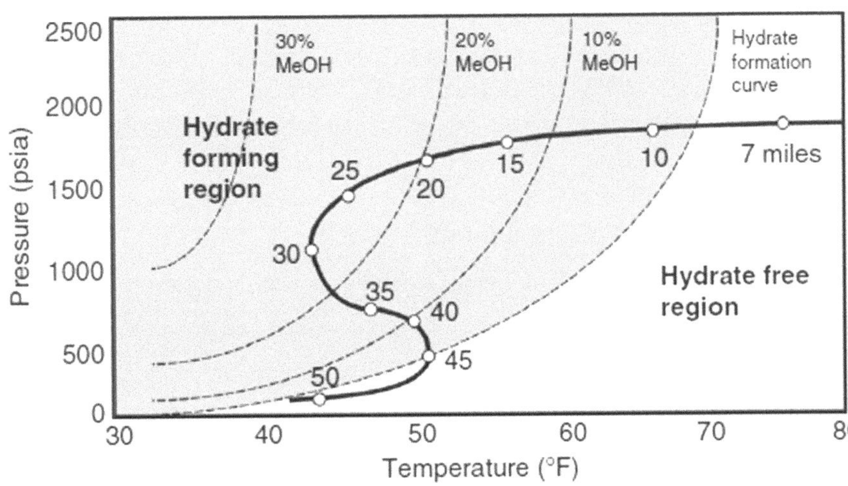

FIGURE 1.29

Pressure-temperature profile along a flow path superimposed on top of hydrate equilibrum curves with varying methanol dosages.

Reproduced with permission from E.D. Sloan, J. Creek, A.K. Sum, Where and how are hydrate plugs formed?, in: E. D. Sloan, C.A. Koh, A. Sum (Eds.), Natural Gas Hydrates in Flow Assurance, Elsevier, 2010, pp. 13–36.

production for gas-dominated systems, which then also protects the entire subsea system during a shut-in or restart event. For oil-dominated systems, THIs are commonly used to manage hydrate risk only during transient events. THI dosage for such operations is often calculated in terms of total volumes of inhibitor injected rather than a rate-based dosage that is commonly used during continuous injection. These transient injection operations can include the following depending on the specific fluid/system risk:

- Wellbore - Bullheading or displacement of the top part of the wellbore with MeOH to physically displace hydrocarbon fluids and water deeper into the wellbore where temperatures are warmer and outside the hydrate region. Due to its lower density, MeOH is usually the THI of choice while attempting a bullheading operation compared to the heavier MEG. THI or MeOH dosage is specified as bullheading volumes (in bbls) and calculated from the depth displacement required to secure the wellbore.
- Subsea tree/jumper/manifold - past expiration of NTT, subsea trees and jumpers/manifolds are often displaced with MeOH to mitigate hydrate risk during shut-ins and restarts. THI volumes are typically calculated to displace certain equivalent volumes of the flow path (e.g., 1–2 times the jumper volume for a subsea jumper).
- Subsea choke (Joule-Thomson cooling) for certain cases, hydrate mitigation during a shut-in might involve depressurization of the production system downstream of the subsea choke. If adequate backpressure is not applied, the system can experience significant drop in temperature downstream of the choke during restart due to Joule-Thomson cooling associated with the large pressure drop across the choke. For such cases, additional MeOH injection might be required to mitigate hydrate risk due to the increased subcooling, with the dosage rate depending on the worst temperature drop expected during the restart process.

1.6.2 Low-dosage hydrate inhibitors

LDHIs form the second class of chemical inhibitors commonly used for hydrate management in hydrocarbon production systems. As their name suggests, LDHIs are typically injected at much lower dosages (<1–2 vol.%, basis water rate) compared to THIs, that can require dosages as high as 50–60 vol.% of the produced water depending on local pressure and temperature conditions. Requiring such drastically lower dosage rates for LDHIs can result in significant advantages over THIs [51], some of which are discussed below:

- OPEX savings - although LDHIs are dosed at a much lower rate than THIs, they generally cost much more per unit volume compared to MeOH or MEG. Depending on the hydrate risk severity and specific dosages, this can lead to chemical cost savings for LDHIs over THIs, but often times the net OPEX is comparable, at least of the same order, for either chemical solution. Since dosages are typically based on water volumes, higher water production rates usually amplify the expected OPEX savings for the chemical solution. For offshore platforms, supplying large volumes of THIs can become a logistical issue that also impacts the overall OPEX of the production strategy. For MeOH-based continuous inhibition, penalties associated with crude contamination should also be assessed while comparing it with a LDHI-based injection strategy.

- CAPEX savings - compared to OPEX impact, LDHIs can offer much bigger CAPEX savings compared to THIs due to the significant reduction in dosage rates and volumes. Requiring lower absolute inhibitor dosage rates impacts several design elements, such as lower host storage volumes, smaller pumps, smaller/fewer umbilical lines for subsea delivery [51]. Combined, these design optimizations lead to large CAPEX savings that can make LDHIs a more attractive option compared to THIs. Life-extension of mature producing assets is another benefit attributed to LDHIs in cases where limited existing topsides storage or subsea umbilical capacity can make it difficult to continue production with THIs [51,52]. This is classified here as a CAPEX savings because, in theory, this constraint can be addressed by making major topsides or subsea (umbilical) modifications. However, in practice, the CAPEX costs associated with such changes in late-life production is often not economically justifiable.
- Production deferment - injecting relatively large volumes of THIs can lead to higher liquid loading in the subsea system that can start impacting hydrocarbon production, leading to production deferment. This is often the case in mature assets where higher water production and lower flowing wellhead pressures can cause higher deferment rates.

Compared to THIs, LDHIs do not significantly shift or affect the thermodynamic hydrate curve. Rather, they allow the system to enter the thermodynamic hydrate region but prevent hydrate particles from posing a plugging risk by either delaying their growth or preventing them from agglomerating into bigger particles or deposits. LDHIs can be subdivided into two main categories—Kinetic hydrate inhibitors (KHI) or antiagglomerants (AA), and are discussed in detail below.

1.6.2.1 Kinetic hydrate inhibitors

KHIs are chemical inhibitors that work by delaying/impeding the nucleation and/or the growth phase of hydrate formation [53]. The first generation of KHI molecules were mostly water-soluble polymers that were discovered by the CSM and utilized pyrrolidone and caprolactam-based structures [54]. The resulting KHI polymers were poly(vinylpyrrolidone) (PVP) and poly(vinylcaprolactam) (PVCap) [54,55] and were shown to be effective in delaying hydrate formation at field operating conditions. Chemical structures of some of the early tested KHI molecules are shown in Fig. 1.30. Although variations exist in the exact mechanism through which KHIs work, most hypotheses suggest KHI molecules to attach to the hydrate crystal through hydrogen bonding, thereby blocking further growth of that crystal [53,55].

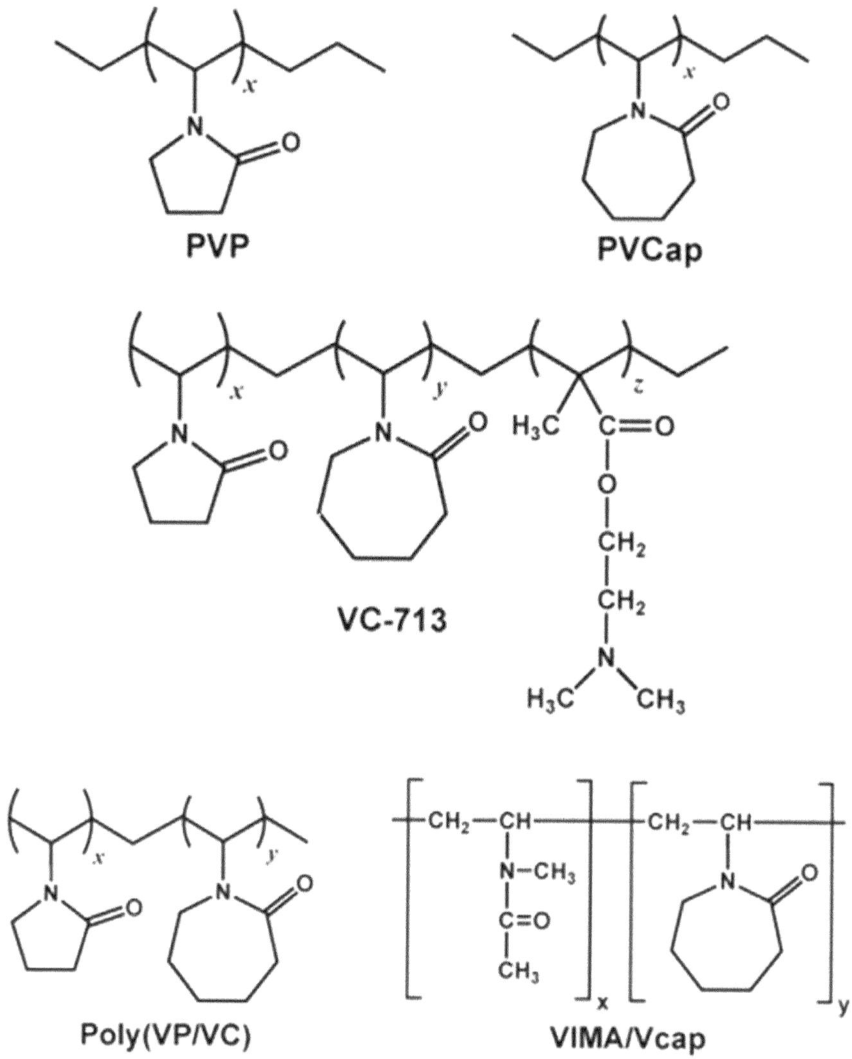

FIGURE 1.30

Chemical structure of early KHI molecules.

Reproduced with permission from T. Palermo, E.D. Sloan, Artificial and natural inhibition of hydrates, in: E.D. Sloan, C.A. Koh, A. Sum (Eds.), Natural Gas Hydrates in Flow Assurance, Elsevier, 2010, pp. 87–104.

Since this first generation of KHI molecules were evaluated, several other chemistries have been tried and tested by various academic and industrial R&D laboratories [53]. These include co-polymers of vinylpyrrolidone and vinylcaprolactam (Fig. 1.30), polyalkylacrylamides, *N*-methyl-*N*-vinylacetamide:VCap copolymer, polyacryloylpyrrolidine, mixtures of tetrabutylammonium bromide (TBAB) and PVCap, and more recently polyesteramide-based KHIs. Kelland [53] provides an excellent review of the history and evolution of various KHI chemistries for further details.

The kinetic descriptor in KHI refers to the fact that these molecules only prevent hydrate formation for a finite time duration and not indefinitely. This time duration is commonly referred to as the hold time or induction time of the KHI. Once this time duration is exceeded, KHI does not prevent further hydrate formation and will often lead to relatively rapid conversion of the available water to hydrates. Hence, while designing an effective hydrate mitigation strategy with KHIs, it is imperative to ensure that the fluid residence time within the production system is lesser than the qualified KHI hold time at the representative field conditions. The hold time for a particular KHI can depend on several parameters, some of which are listed below:

- chemical structure of the KHI molecule
- dosage rate of the KHI
- chemical composition of the produced hydrocarbon and brine
- hydrate subcooling expected
- structure of hydrate expected—structure I or structure II
- other chemical inhibitors being injected in the field, for example, corrosion inhibitor, scale inhibitor
- flow conditions-flow regime, mixing (shear)

The first step of hydrate formation is well understood to be the nucleation phase. It is also well recognized that nucleation of hydrates is probabilistic, with the stochasticity dependent on parameters such as subcooling, interfacial area between phases and homogeneous vs heterogeneous nucleation [2,20]. Following formation of stable hydrate nuclei, growth of these nuclei into crystals forms the subsequent phase [20]. Given that KHIs are believed to affect the nucleation and growth phases of hydrate crystals, it follows that the performance of KHIs, in terms of hold time, is also probabilistic in nature. This feature makes testing commercial KHIs for field applications relatively challenging due to the large number of tests needed to get statistically significant datasets and accurately quantify the performance envelope.

In terms of laboratory-based qualification, a variety of equipment has historically been utilized to evaluate the performance of a KHI, ranging from smaller scale equipment such as high-pressure differential scanning calorimetry, Peltier cells, autoclaves, rocking cells to larger scale apparatus like high-pressure multiphase flow loops. A common strategy has been to test different KHI chemistries with the smaller scale setup, which also allows for enough tests to be conducted to generate repeatable and statistically significant data in terms of induction time distributions. Following this phase of testing, multiphase flow loops are often employed as a final validation and to assure scale-up in terms of flow regimes expected in the field application.

Hydrate formation is detected in these setups through continuous monitoring of pressure and temperature. Fig. 1.31 shows hydrate onset in a typical autoclave setup using either a constant cooling rate (left plot) or using an isothermal test (right plot) [56]. Hydrate onset is detected through distinct changes in temperature trace, due to the exotherm associated with hydrate formation, along with a simultaneous decrease in pressure due to gas consumption. The time these systems spend from entering the hydrate region till the point of onset is identified as the induction time, which is followed by the hydrate growth phase (Fig. 1.31). These type of test methods are used with different KHIs to compare their relative performance as a function of cooling rate or hydrate subcooling (Fig. 1.32). Induction times for different products can then be used to select the ideal chemistry for the intended application, with the exact test protocol and conditions usually specified by the field operator.

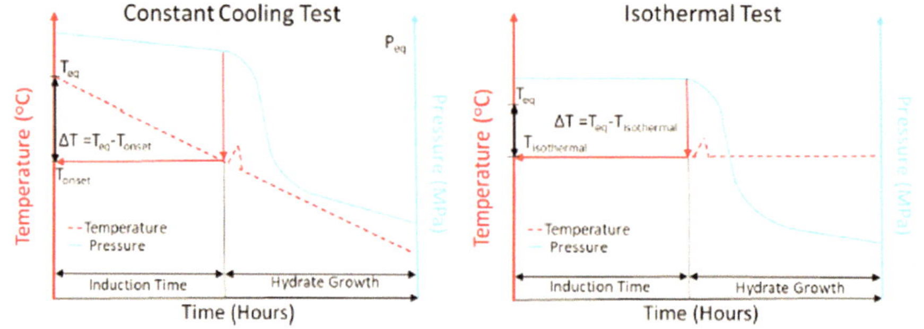

FIGURE 1.31

Constant cooling (left) and isothermal tests used for KHI evaluation.

Reproduced with permission from A. Singh, A. Suri, A review on gas hydrates and kinetic hydrate inhibitors based on acrylamides, Journal of Natural Gas Science and Engineering, 83 (2020) 103539.

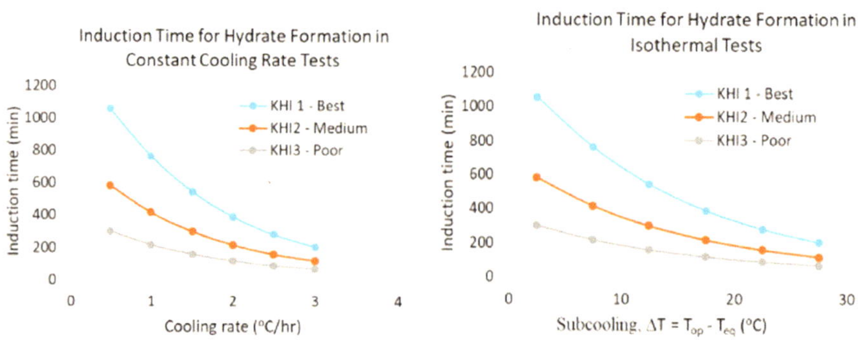

FIGURE 1.32

Dependence of induction times measured with constant cooling and isothermal tests on relevant test parameters.

Reproduced with permission from A. Singh, A. Suri, A review on gas hydrates and kinetic hydrate inhibitors based on acrylamides, Journal of Natural Gas Science and Engineering, 83 (2020) 103539.

Although rocking cells and autoclaves are the most common experimental apparatus used to evaluate LDHI performance, flow loops have also been utilized to further de-risk field applications [57,58]. Fig. 1.33 compares induction time data obtained from autoclaves with flow loop-based tests for CO_2 and H_2S containing systems with KHIs [57]. Results show that KHIs can offer significant induction times at lower subcooling ($< \sim 10-15°F$). However, induction times drop rapidly as the subcooling approaches a threshold, $\sim 15-16°F$ in this particular example.

For field application, KHI induction times are often also presented as a function of KHI dosage. Fig. 1.34 shows this dependence for PVP and PVCap for KHI dosages ranging from 0 to 2 wt.% for a natural gas mixture [59] using multiple high-pressure cells. Induction times offered by PVP and PVCap generally increase as KHI dosage increases, but the resulting curves are strongly dependent on subcooling. If the subcooling is too severe, increasing KHI dosage might not help in providing any significant induction time (Fig. 1.34).

Recent advances in KHI performance evaluation include high-pressure stirred automated lag time apparatus [60,61], which has been utilized to generate high-resolution PDFs. This setup allows for statistical evaluation of thousands of hydrate nucleation events in small-scale Peltier cells under controlled temperature conditions. Fig. 1.35 shows pressure-temperature data as well as pressure-time series data obtained from this setup that is used to identify the onset of hydrate formation

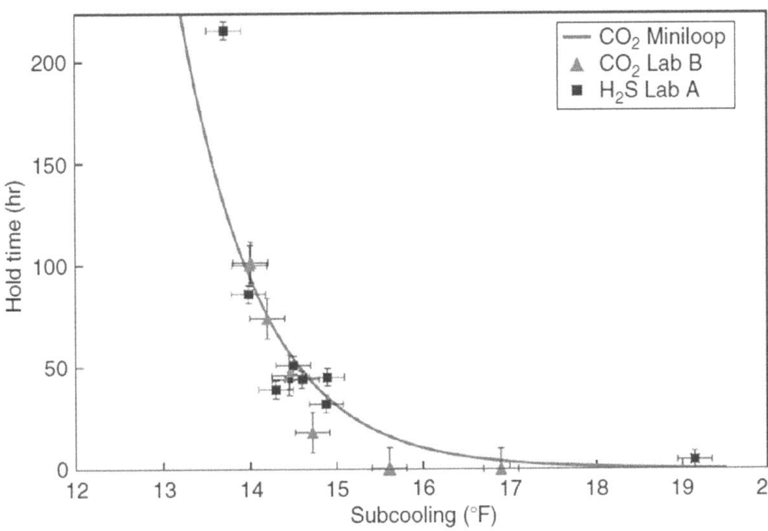

FIGURE 1.33

Comparison of KHI induction times between CO_2 and H_2S-containing autoclave tests with CO_2-containing mini-loop tests.

Reproduced with permission from M. Eaton, J. Lachance, E.D. Sloan, Kinetic hydrate inhibitors performance, in: E.D. Sloan, C.A. Koh, A. Sum (Eds.), Natural Gas Hydrates in Flow Assurance, Elsevier, 2010, pp. 105–143.

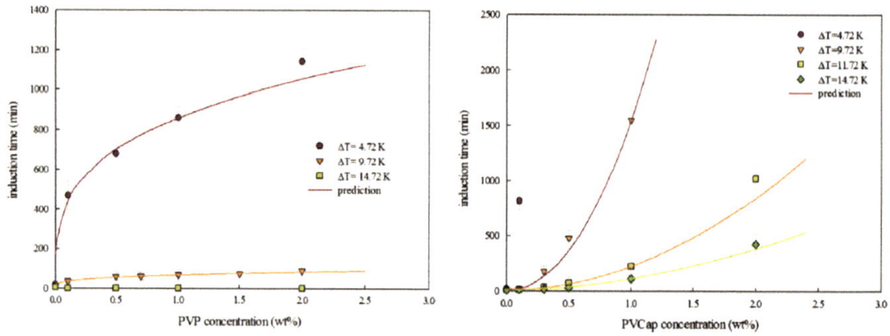

FIGURE 1.34

Average induction times measured with PVP and PVCap for a water-natural gas -mixture.

Reproduced with permission from S.P. Kang, J.Y. Shin, J.S. Lim, et al., Experimental measurement of the induction time of natural gas hydrate and its prediction with polymeric kinetic inhibitor, Chemical Engineering Science, 116 (2014) 817–823.

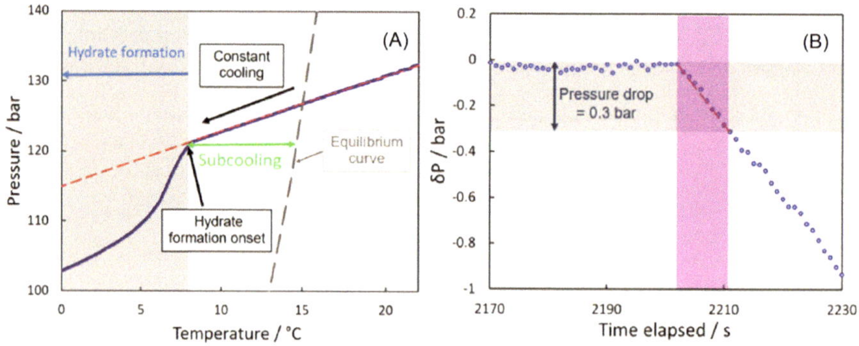

FIGURE 1.35

(A) Pressure-temperature trace showing method used to detect the onset of hydrate formation during constant cooling. (B) Measured pressure-time data, corrected for temperature changes, showing hydrate formation and the associated pressure drop.

Reproduced with permission from V.W.S. Lim, P.J. Metaxas, P.L. Stanwix, et al., Gas hydrate formation probability and growth rate as a function of kinetic hydrate inhbiitor (KHI) concentration, Chemical Engineering Journal, 388 (2020) 124177.

in the system [60]. Fig. 1.36 shows cumulative PDFs from this setup for methane hydrate (top) and for systems with varying dosages of a commercially available KHI (bottom). The distribution profiles confirm the natural variability in induction times of these systems, supporting the need to have high-quality repeatable data to validate KHI performance. These cumulative PDFs can also be plotted as a function of time for various KHI dosage rates to depict induction time distributions more clearly [61]. For a given KHI dosage, Fig. 1.37 shows that induction time increases

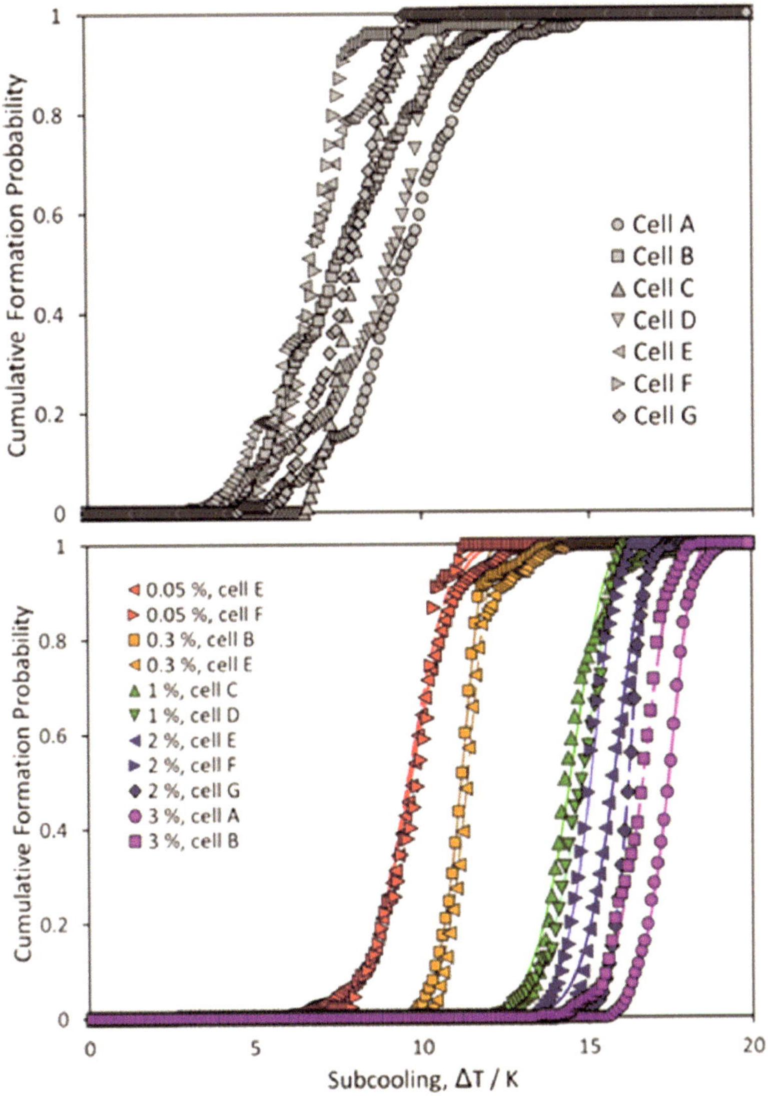

FIGURE 1.36

Cumulative probability distribution functions of methane hydrate formation from seven different cells without KHI (top) and with varying concentrations of commercial KHI (bottom).

Reproduced with permission from V.W.S. Lim, P.J. Metaxas, P.L. Stanwix, et al., Gas hydrate formation probability and growth rate as a function of kinetic hydrate inhbiitor (KHI) concentration, Chemical Engineering Journal, 388 (2020) 124177.

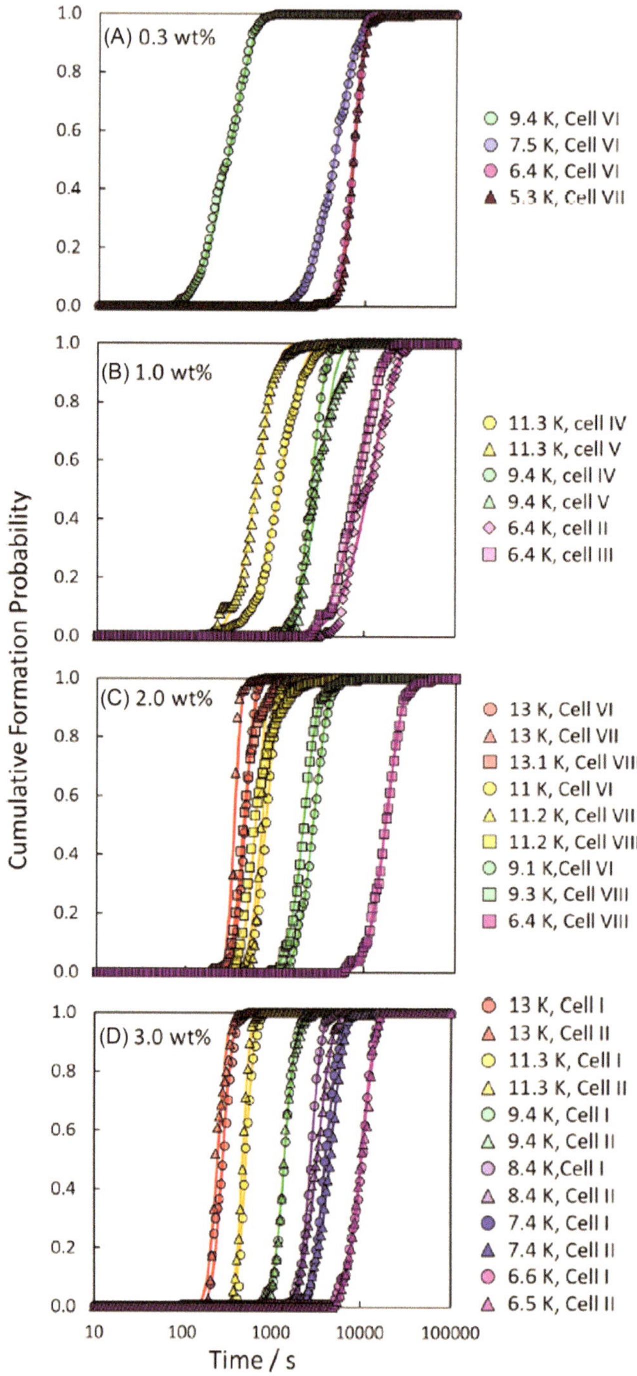

FIGURE 1.37

Cumulative PDFs depicting induction time distributions grouped by KHI dosages for different subcooling.

Reproduced with permission from V.W.S. Lim, P.J. Metaxas, M.L. Johns, et al., The delay of gas hydrate formation by kinetic inhibitors, Chemical Engineering Journal, 411 (2021) 128478.

significantly as the subcooling is lowered. Together with the field application information, these induction time distributions can be used to select appropriate KHI chemistries and dosage rates with a high degree of confidence.

As evident from the KHI performance data presented above, KHIs do have strong subcooling dependence. In general, it has been observed that KHIs struggle at subcooling >10°C [51,53,58], beyond which the induction times decrease quite rapidly to be of limited use in field applications. This characteristic limits KHI application for field deployment in cases where subcooling is relatively low (<10°C), but is should be noted that this is a generic rule of thumb. Each field application should be appropriately tested and evaluated by a hydrate specialist to understand the relevant KHI envelope. Sometimes, to get around the subcooling limit for applications, KHIs are used in conjunction with a THI, such as MeOH or MEG [53]. In principle, the THI reduces the subcooling experienced by the fluids to below the 10°C threshold, which can then enable a suitable KHI to be selected. However, it is critical to experimentally evaluate the compatibility of the KHI product with the THI before embarking on such an operating strategy. Two key KHI characteristics should also be noted while evaluating field application:

- Solubility/cloud point - KHIs can pose problems with respect to their solubility in the carrier phase. Early generations of KHIs had low solubility (or low cloud points) that resulted in the precipitation of the KHI polymer at higher temperatures, causing deposition in the subsea production system [53,58,62]. Any new product qualification should carefully evaluate the stability of the KHI chemistry at the warmest temperature condition expected along the production stream.
- Compatibility with corrosion inhibitors (CI) - CIs have been reported to have a significant negative effect on the performance of a KHI [58,63]. Hence, prior to field deployment, KHI performance should always be evaluated in the presence of a CI to ensure now adverse impact on KHI performance is observed in the field.

1.6.2.2 Antiagglomerants

AA type chemical inhibitors, as the name indicates, mitigate hydrate risk by preventing agglomeration of small hydrate particles into bigger agglomerates or plugs [2,53]. Compared to KHIs that delay nucleation/growth of crystals, AAs work through a fundamentally different mechanism by allowing relatively small sized hydrate particles to form. AA molecules then attach themselves to the hydrate crystals thereby making them nonagglomerating (Fig. 1.38). This feature helps in creating a dispersed hydrate slurry that becomes easier to transport with the oil phase.

Building on this conceptual model of adsorbed AA molecules on the hydrate particle surface preventing agglomeration, it can be argued that a minimum effective dosage is often required to attain full inhibition. This is depicted in Fig. 1.39 where underdosing with AA (left schematic) cannot achieve adequate coverage of

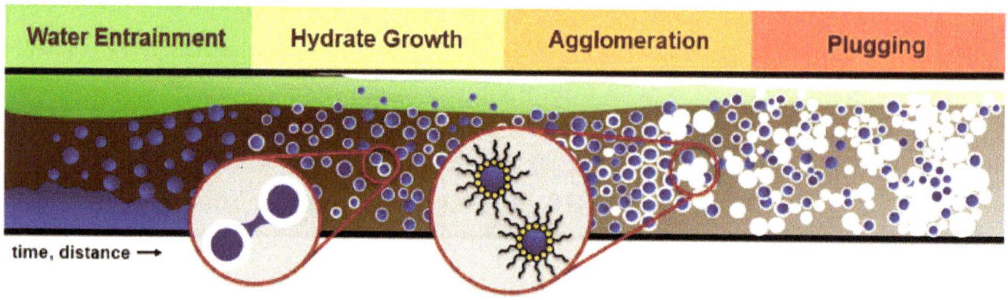

FIGURE 1.38

Conceptual model for hydrate formation in oil-dominated systems—without AA, hydrate coated water droplets can agglomerate due to capillary cohesion from unconverted water (left inset). With AA molecules binding to the hydrate surface, agglomeration of hydrates is minimized (right inset) [64].

Reproduced with permission from S. Hu, C.A. Koh, CH4/C2H6 gas hydrate interparticle interactions in the presence of antiagglomerants and salinity, Fuel, 269 (2020) 117208.

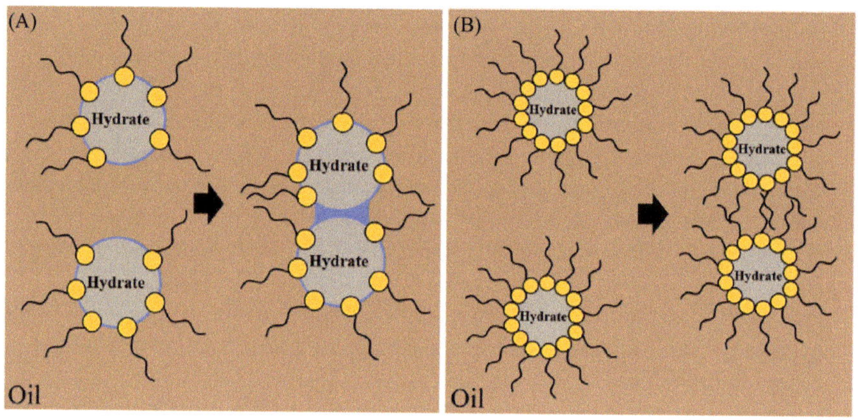

FIGURE 1.39

Conceptual model of hydrate particle interaction in the presence of AA in an under-dosed system (A), and in an adequately dosed system (B).

Reproduced with permission from S. Hu, C.A. Koh, CH4/C2H6 gas hydrate interparticle interactions in the presence of antiagglomerants and salinity, Fuel, 269 (2020) 117208.

the hydrate particle, thereby resulting in agglomeration due to strong cohesive forces from water layers bridging across the hydrate particles. If full surface coverage of the hydrate particle is achieved through adequate dosing (Fig. 1.39, right), the long hydrocarbon tail of the AA molecule prevents hydrate particles from coming too close to each other for capillary cohesion forces to be effective.

Recent studies have validated this conceptual framework by demonstrating that addition of AA to an uninhibited system leads to significant reduction in both cohesion forces between two hydrate particles as well as adhesion forces between a hydrate particle and a metal substrate, such as a pipe wall [65]. Such studies have used a micromechanical force (MMF) apparatus to measure the cohesive and adhesive forces between hydrate particles and surfaces under a range of conditions [65,66]. Experiments performed with cyclopentane hydrates [65] show significant reduction in cohesion forces between hydrate particles with commercial AAs compared to the case where no AA was used (Fig. 1.40). Adhesion forces between hydrate particle and stainless-steel surface also show reduction with addition of AA, but not as large as the decrease in cohesive forces (Fig. 1.40, right).

Similar MMF experiments performed with a structure II natural gas mixture for a range of AA dosage rates (Fig. 1.41) shows significant reduction in cohesive force between particles as the AA dosage increases [64]. Fig. 1.41 also shows that the reduction in cohesive force is relatively large when AA dosages increase from 0 to 0.5 vol.% at which point the cohesive forces reach a plateau.

Subsequent increases in AA dosage above 1 vol.% do not result in any significant decrease in the measured cohesive forces [64]. These observations support the conceptual model above (Fig. 1.39) where increasing AA coverage on hydrate particles initially leads to lower cohesion and prevents hydrates from agglomeration. Once the hydrate particle is sufficiently covered by AA molecules, likely corresponding to the minimum effective dosage, further increases in AA dosage rate does not lead to any additional adsorption on the surface and no further benefit in terms of preventing agglomeration.

From a molecular structure perspective, AAs are in general composed of two distinct chemical groups. The headgroup is the part of the AA molecule that gets

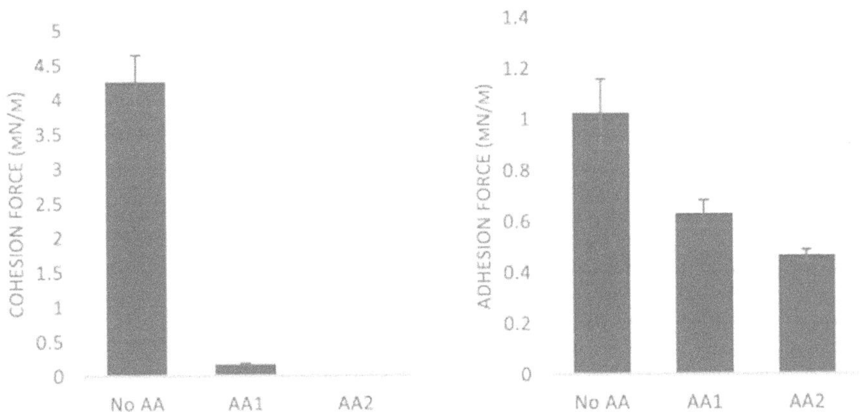

FIGURE 1.40

Cohesion force (left) and adhesion force (right) measured for cyclopentane hydrate particles without AA and with two commercially available AAs.

Reproduced with permission from E.P. Brown, D. Turner, G. Grasso, et al., Effect of wax/anti-agglomerant interactions on hydrate depositing systems, Fuel, 264 (2020) 116573.

incorporated into the hydrate crystal structure and, for most effective AAs, this group tends to be a quaternary ammonium or phosphonium salt [2,53,63]. The hydrophobic tail of the AA molecule is typically a long hydrocarbon chain, which helps repel other water molecules or hydrate particles. A generic quaternary ammonium salt-based AA structure is depicted in Fig. 1.42.

Research and development of different AA molecules has been going on since the early 1990s when Shell started looking at antifreeze fish proteins (AFPs) [53]. The basic idea was to find molecules that prevent agglomeration of gas hydrates similar to these AFPs that help prevent ice crystals in fish. After experimenting with various chemical structures, Shell decided to focus on quaternary ammonium surfactants [67], in part due to them being good particle dispersants and

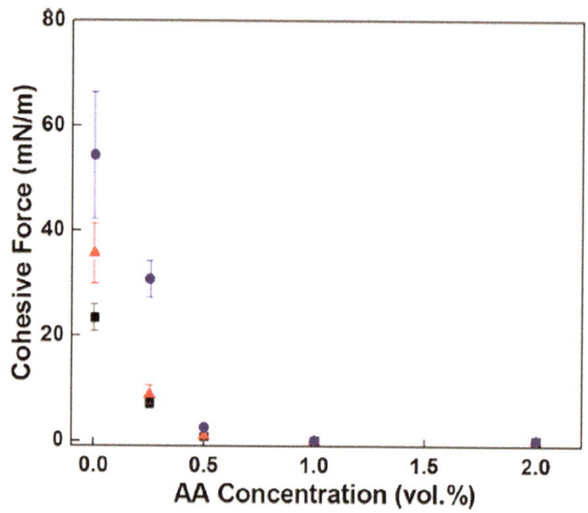

FIGURE 1.41

Cohesive force measurements as a function of AA dosage for various contact times—squares: 10 s; triangles: 1 min; circles: 30 min.

Reproduced with permission from S. Hu, C.A. Koh, CH4/C2H6 gas hydrate interparticle interactions in the presence of antiagglomerants and salinity, Fuel, 269 (2020) 117208.

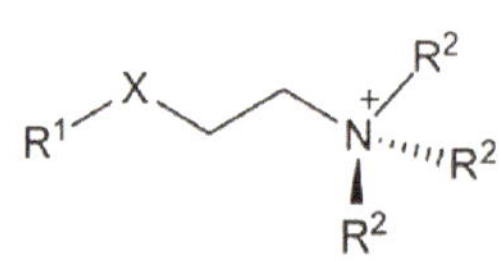

FIGURE 1.42

Generic quaternary ammonium salt-based AA molecule. R^1 denotes a long carbon chain, R^2 can be a butyl or pentyl group, and X is a spacer group.

Reproduced with permission from S. Hu, C.A. Koh, CH4/C2H6 gas hydrate interparticle interactions in the presence of antiagglomerants and salinity, Fuel, 269 (2020) 117208.

also because some of these ammonium salts like TBAB and tetrapentylammonium bromide (TPAB) formed the same $5^{12}6^4$ cage as a structure II hydrate (see Fig. 1.1) [53]. A key breakthrough was to replace one or two of the short-chained alkyl groups with a longer chain hydrophobic tail (Fig. 1.42). This resulted in the now common single-tailed or twin-tailed generic AA structure (Fig. 1.43).

The combination of the hydrophilic headgroup and the long chain hydrophobic tail(s) allowed these chemical structures to bind effectively to the hydrate cages as well as keep them well dispersed. Since Shell's early work on these AA chemistries, other research groups have tried different molecular structures (surfactant-based) as potential AAs. CSM evaluated different commercial Span surfactants along with synthesized structures, such as ethanoates, ethanamides and alkylamides (Fig. 1.44), and reported the chemical dodecyl-2-(2-caprolactamyl)ethanamide (Fig. 1.44B) to have

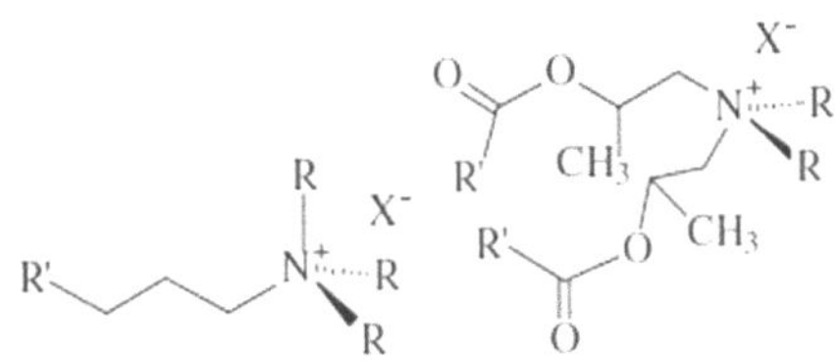

FIGURE 1.43

Generic structures of single-tail (left) and twin-tail (right) quaternary ammonium-based AA surfactants [68].

Reproduced with permission from M.A. Kelland, T.M. Svartaas, J. Ovsthus, et al., Studies on some zwitterionic surfactant gas hydrate anti-agglomerants, Chemical Engineering Science, 61 (2006) 4048–4059.

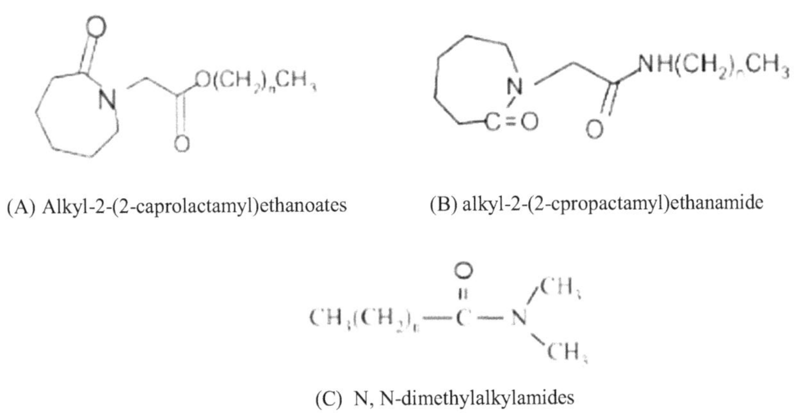

FIGURE 1.44

Synthesized chemicals evaluated as AAs by CSM. (A) Alkyl-2-(2-caprolactamyl) ethanoates. (B) alkyl-2-(2-cpropactamyl)ethanamide. (C) *N*,*N*-dimethylalkylamides.

Reproduced with permission from Z. Huo, E. Freer, M. Lamar, et al., Hydrate plug prevention by anti-agglomeration, Chemical Engineering Science, 56 (2001) 4979–4991.

the best AA performance. At a dosage of 0.75 wt.%, this chemical prevented hydrate agglomeration during a 5-day test at a subcooling of 11.5°C [69].

Still newer AA structures evaluated include zwitterionic surfactants [68], twin-tail diester quarternary AA [53], polypropoxylates [70]. Kelland [53] provides an excellent overview of the history and development of some of these AA chemistries for the reader interested in further details. A major drawback with the original quarternary ammounium-based AAs was its high toxicity, which prevented wide-scale implementation in the field. A lot of the subsequent research in developing novel AA chemistries was devoted to lower the toxicity of these chemicals. However, most of the currently available commercial AA chemistries are still based on the quaternary ammonium salts because of their field-proven effectiveness under high subcooling conditions.

Similar to KHIs, selection and qualification of AAs for field application is based on experimental testing with the hydrocarbon fluids. Key parameters that can affect the performance of an AA include:

- hydrocarbon fluid composition
- produced brine composition
- water-cut
- subcooling
- AA chemical structure and dose rate
- flow properties like flow regime, viscosities, flowing pressure at wellhead
- type of field application - steady-state production or shut-in/re-start
- potential interaction with other solids (wax, asphaltenes) or injected chemicals

Evaluating AA performance requires specialized test equipment along with field-specific testing protocol. This test protocol is typically provided by the operator based on the production conditions and field parameters listed above. Although high-pressure autoclaves, wheel loops and flow loops have been used to characterize AA performance [52,69,71,72], transparent sapphire rocking cells (Fig. 1.45) tend to be the most used equipment for evaluating different AA chemistries and dosage rates. These rocking cell setups can contain multiple sapphire cells submerged in a temperature-controlled bath that allows for full visual observations of the fluid contents inside the cell. They also contain a steel ball that rocks back and forth to provide some level of fluid mixing. The steel ball also serves as a good visual indicator of the severity of hydrate formation inside the cell. If the ball is stuck and does not move upon rocking (Fig. 1.46A), it is generally an indicator of hydrate deposition that was not adequately mitigated by the particular type and/or dosage of the AA being tested.

Visual observations from a fully transparent rocking cell test are very helpful in assessing the performance of a particular AA, especially understanding the morphology and distribution of the hydrate deposits or particles. Apart from the visual output, which is mostly qualitative in nature, pressure, temperature, and ball travel time within the rocking cell (Fig. 1.47) can also be recorded to quantify important hydrate-related parameters [73]. These can include the time at which hydrates start to form in the cell, hydrate volume fraction from the pressure trace and the overall flowability of the slurry based on changes in the ball travel time (Fig. 1.47).

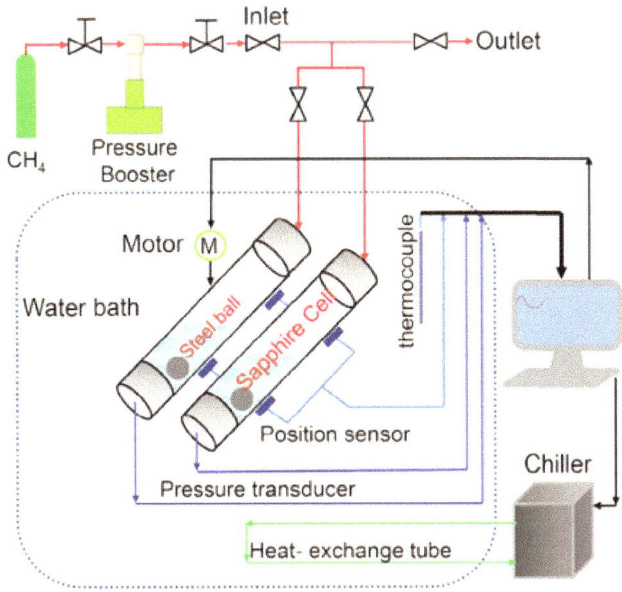

FIGURE 1.45

Schematic of a sapphire rocking cell setup used for AA qualification.

Reproduced with permission from S. Dong, A. Firoozabadi, Hydrate anti-agglomeration and synergy effect in normal octane at varying water cuts and salt concentrations, Journal of Chemical Thermodynamics, 117 (2018) 214–222.

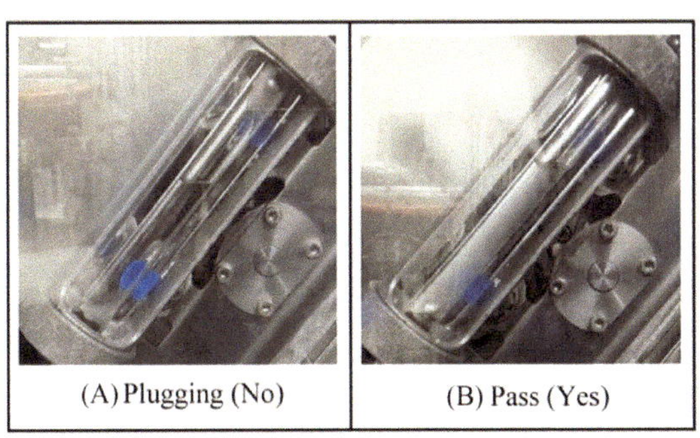

FIGURE 1.46

Visual observations from a rocking cell test showing (A) case where plugging was observed—ball stuck at top of cell and (B) where no pugging was observed.

Reproduced with permission from S. Dong, A. Firoozabadi, Hydrate anti-agglomeration and synergy effect in normal octane at varying water cuts and salt concentrations, Journal of Chemical Thermodynamics, 117 (2018) 214–222.

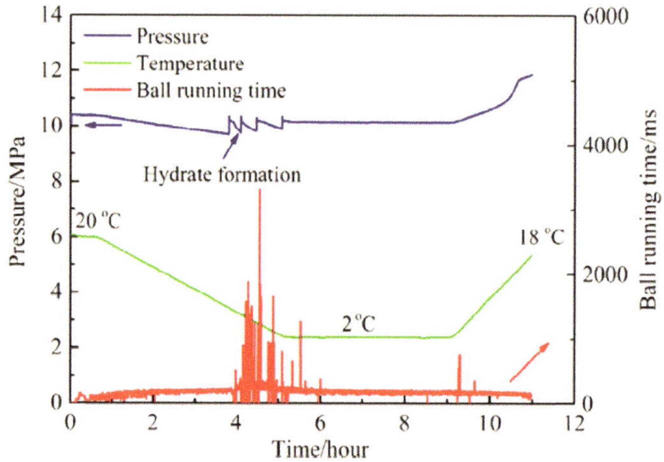

FIGURE 1.47

Pressure, temperature, and ball travel time recorded during a typical rocking cell test with AA.

Reproduced with permission from S. Dong, A. Firoozabadi, Hydrate anti-agglomeration and synergy effect in normal octane at varying water cuts and salt concentrations, Journal of Chemical Thermodynamics, 117 (2018) 214–222.

The relatively small oil volume requirement makes them quite efficient to run a full qualification program where multiple AA chemistries might be evaluated at different dosage rates and operating conditions. In that respect, they are well suited for initial screening of potential AA candidates for a field application. However, the rocking cell setup is not well suited to evaluate flow behavior of the resulting hydrate slurries. Rheological properties, such as slurry viscosity, are typically evaluated using larger scale flow loops and are dependent on system variables like shear, flow regime, liquid loading, water-cut, AA dosage. Understanding AA performance in terms of keeping the slurry viscosity manageable becomes more important in cases where water-cut is expected to be high and/or where the well flowing pressure is relatively low. These flow loops often require much larger sample volumes compared to rocking cell tests, which means that they are not commonly used for basic AA screening. In addition, only few operators and research facilities own and maintain such larger scale flow loop setups, which makes them more of a specialized application rather than a standardized technical service provided by external labs or vendors.

Camargo and Palermo [29] were one of the first to study the rheological properties of hydrate slurries in asphaltenic crude oils. The asphaltene component in the Brazilian deepwater crude oils they tested was hypothesized to act as natural AA that resulted in nonagglomerating hydrate slurries. A laboratory scale rheometer as well as a larger scale flow loop were used in the experiments to obtain viscosity data. Results show the hydrate suspensions to exhibit non-Newtonian shear-thinning behavior at higher hydrate volume fractions (>0.27) and Newtonian behavior at lower volume fractions [29]. Based on the experimental data, they proposed a phenomenological model that related the effective viscosity to the hydrate volume fraction using a hard sphere dispersion model. The hydrate aggregate structure was assumed to be fractal, which allowed them to relate effective hydrate volume fraction to the fractal dimension and to the diameters of the particles and aggregates [29]. Most of the subsequent rheological work on hydrate slurries [74–76], including the viscosity model used in CSMHyK [28], are based on this framework.

Fidel-Dufour et al. [75] supplemented the viscosity model with transient population balance models to account for distinct time-dependent populations of non-hydrated water droplets, crystallized water droplets and agglomerates. They also performed experiments in a large-scale flow loop setup with a water-dodecane emulsion containing an AA additive (from Institut Français du Pétrole). As expected, relative viscosity increases with time as more water gets converted to hydrate and reaches a plateau (Fig. 1.48). Moreover, this plateau in relative viscosity is a function of the water content—in general, higher water content leads

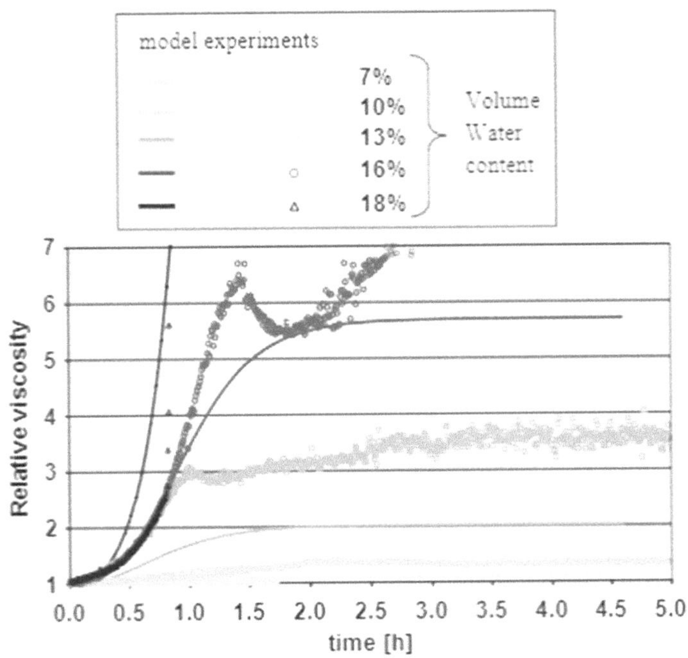

FIGURE 1.48

Model predictions compared with experimental data for relative viscosity of an AA-dosed hydrate slurry as function of time with different water-cuts.

Reproduced with permission from A. Fidel-Dufour, F. Gruy, J.M. Herri, Rheology of methane hydrate slurries during their crystallization in awater in dodecane emulsion under flowing, Chemical Engineering Science, 61 (2006) 505–515.

to higher relative slurry viscosity (Fig. 1.48). Their transient model was able to capture the overall trends in the data [75]. Recent flow loop experiments and rheological modeling have further improved understanding of viscosity dependence on hydrate volume fraction and helped constrain some of the relevant modeling parameters [76,77]. Majid et al. [76] compared predictions from various rheological models with published data available from several studies spanning a range of test equipment and gas hydrate formers (Fig. 1.49). Results show that relative viscosity of hydrate slurries increases with hydrate volume fraction gradually at lower hydrate fractions and rapidly at higher hydrate volume fractions. Similar results were also observed by Sakurai et al. [77] for flow loop experiments conducted with methane and water-dominated system (Fig. 1.50). A key learning from these viscosity data plots (Figs 1.48–1.50) covering a wide range of fluids and flow conditions is that relative viscosity of hydrate slurries can increase by an order of magnitude when hydrate volume fractions approach 20–30 vol.%. Extrapolating these models to higher hydrate volume fractions (>50%–60%) results in sharp increase in relative viscosity that can increase by almost two orders of magnitude. This rheological behavior aligns well with the often-quoted limit for AA application that constrains the maximum water-cuts for AAs to work to be <50%–60% [53].

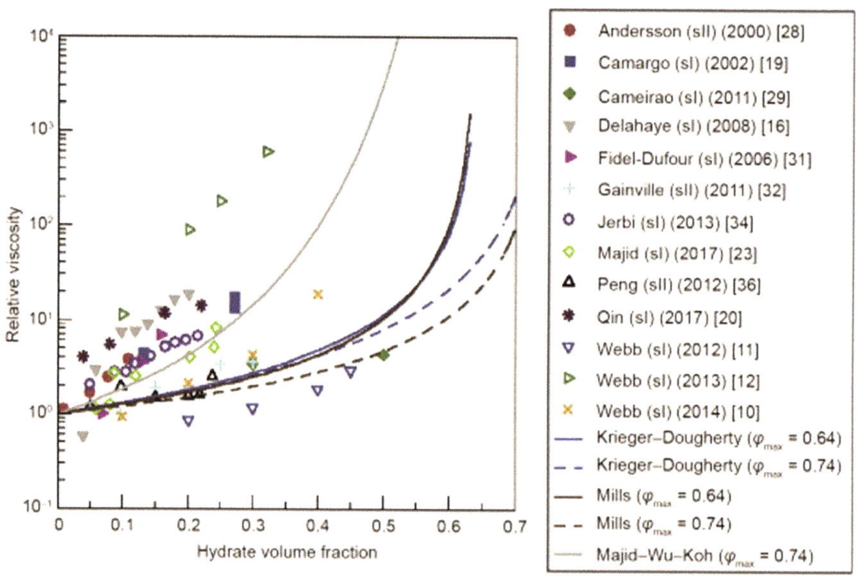

FIGURE 1.49

Comparison of relative viscosities of hydrate slurries as a function of hydrate volume fraction with various model predictions. For details on data and references, see.

Reproduced with permission from A.A. Majid, D.T. Wu, C.A. Koh, A perspective on rheological studies of gas hydrate slurry properties, Engineering, 4 (2018) 321–329.

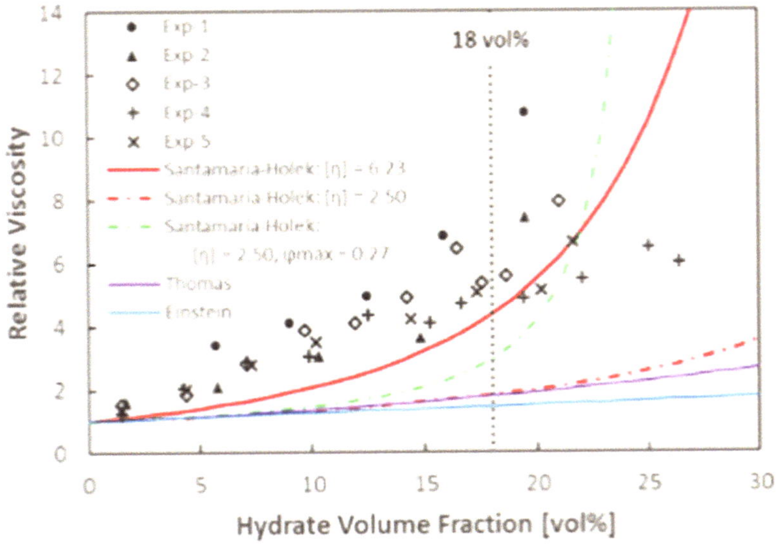

FIGURE 1.50

Comparison of relative viscosities of hydrate slurries as a function of hydrate volume fraction with various model predictions for a 100% water system.

Reproduced with permission from S. Sakurai, B. Hoskin, J. Choi, et al., Investigating hydrate formation rate and the viscosity of hydrate slurries in water-dominant flow: flowloop experiments and modelling, Fuel, 292 (2021) 120193.

Recent flow loop experiments have also improved understanding of morphology of hydrate slurries that form with the addition of AA in a flowing mixture [72,78]. Visual windows installed in these flow loop setups allow for visualization of hydrate particles as they flow through the system. Fig. 1.51 shows the morphology of emulsion (prior to hydrate formation) and hydrate slurries for a water + diesel + natural gas + AA mixture for water-cuts ranging from 5 to 30 vol.% during various stages of the flow loop test. At lower water-cuts, the dispersed hydrate particles can be seen settling at the bottom of the window due to density differences (Fig. 1.51), while the slurry appears more homogeneous as the water-cut increases. However, after restarting, the hydrate slurry becomes well dispersed again for all water-cuts [78], demonstrating that AAs can transport hydrate slurries successfully during steady flow as well as after shut-ins/restarts. Another type of test apparatus recently developed is a rock-flow cell [71] that allows for a more compact set up compared to a full-scale flow loop and better visualization along the length of the pipe. Experiments conducted with kerosene–gas–water mixtures show different performance and deposition behavior when evaluated with two different AAs (AA-A and AA-B) [71]. For ethane gas, AA-A results in formation of a nice flowable slurry (Fig. 1.52), whereas AA-B shows significant deposition at the bottom and side of the cell (Fig. 1.53).

FIGURE 1.51

Morphologies of emulsion/hydrate slurries observed at various water-cuts at different stages of flow loop tests.

Reproduced with permission from K.L. Yan, C.Y. Sun, J. Chen, et al., Flow characteristics and rheological properties of natural gas hydrate slurry in the presence of anti-agglomerant in a flow loop apparatus, Chemical Engineering Science, 106 (2014) 99–108.

Development and application of such new test equipment continue to provide new insights into the qualitative and quantitative aspects of AA performance in oilfield applications. As explained through the examples above, flow loop-based setups can be used to provide detailed rheological information about a particular oil/condensate + AA combination. However, it should be noted that most of these advanced rheological characterization techniques and equipment still reside within the academic and research domain—most commercial AA qualification still relies on simple rocking cell tests to pick the best-in-class product for a specific application.

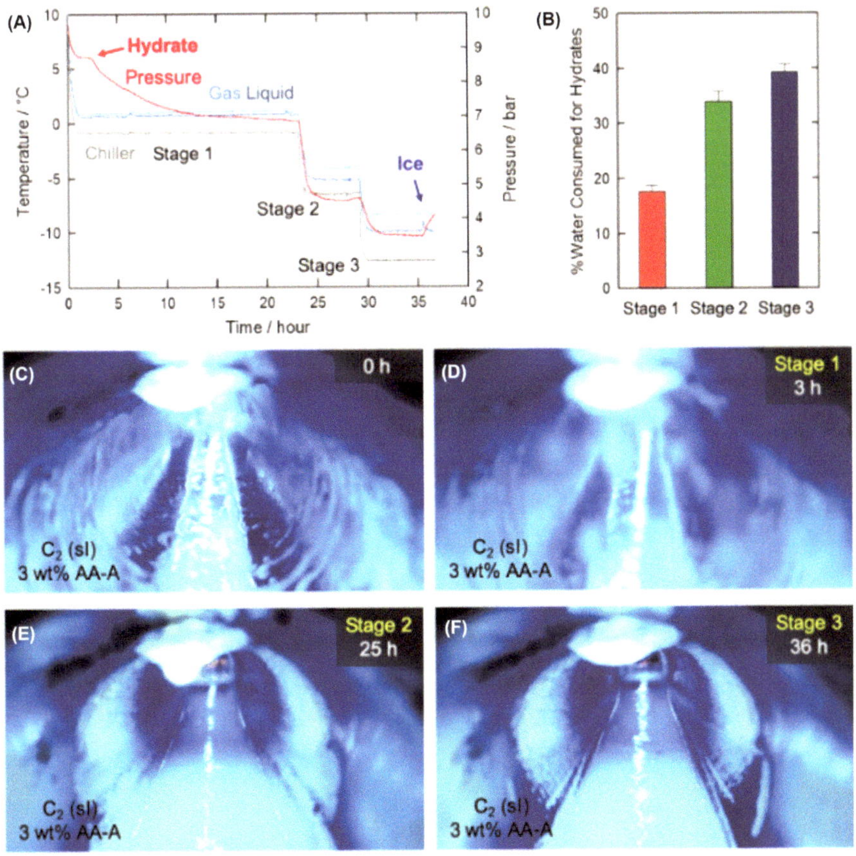

FIGURE 1.52

Temperature/pressure traces (A), water conversion (B) and images depicting hydrate formation behavior with AA-A at different stages of the experiment (C–F).

Reproduced with permission from J.-H. Sa, A. Melchuna, X. Zhang, et al., Investigating the effectiveness of anti-agglomerants in gas hydrates and iceformation, Fuel, 255 (2019) 115841.

AA operating envelope usually depends on a different set of test conditions and parameters compared to KHIs. Unlike KHIs, that commonly have a subcooling limit associated with their performance, AAs can work well up to relatively high subcooling (>20°C) [53]. Some other relevant aspects related to AA application are noted below:

- Requirement of a liquid carrier phase - since AAs are required to create nonagglomerating hydrate slurries, presence of a liquid carrier phase becomes an important requirement to keep the slurry flowing. Gas condensate systems with very low CGRs can struggle if there is not enough condensate phase to carry the hydrate particles. For oil-dominated systems, this transportability

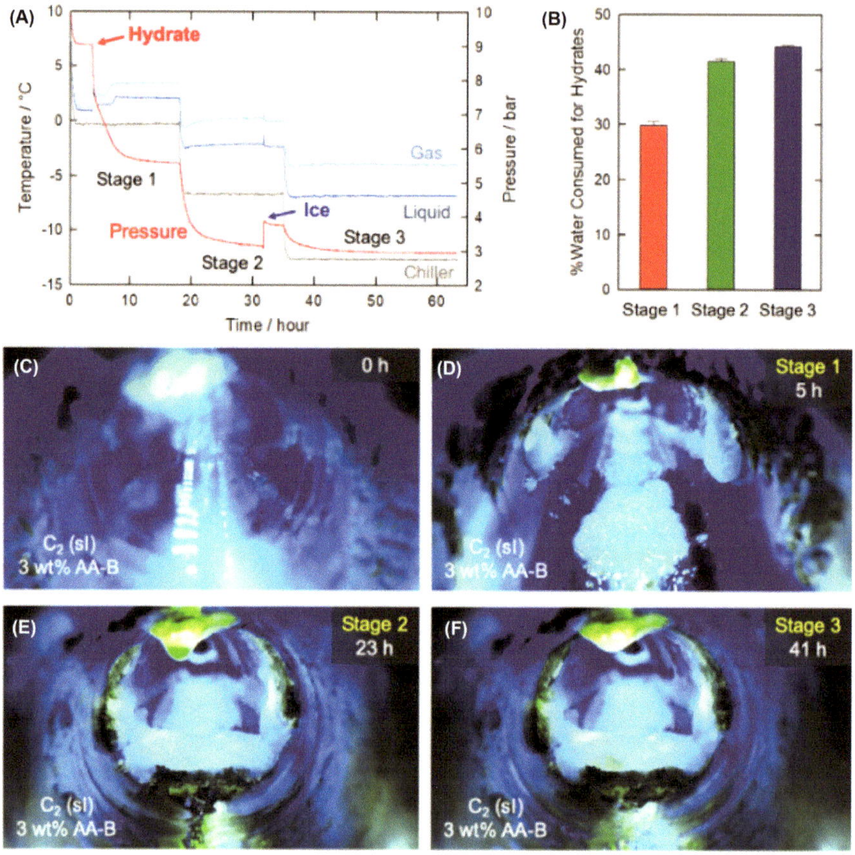

FIGURE 1.53

Temperature/pressure traces (A), water conversion (B) and images depicting hydrate formation behavior with AA-B at different stages of the experiment (C–F).

Reproduced with permission from J.-H. Sa, A. Melchuna, X. Zhang, et al., Investigating the effectiveness of anti-agglomerants in gas hydrates and iceformation, Fuel, 255 (2019) 115841.

requirement puts an indirect constraint on the total water-cut that a given AA can handle. As explained above, water-cuts >50%–60% can result in too high a viscosity increase for the hydrate slurry to keep flowing. Therefore application of AA can require some minimum amount of liquid hydrocarbon along with a maximum amount of produced water.

- Time-limit - AAs are not constrained by a time-limit under hydrate forming conditions that is typically associated with KHI application. This makes them suitable for cases where long planned or unplanned shut-ins can occur. During AA qualification testing, test protocols can involve simulating shut-ins and restart periods of up to several days in duration.

- Topsides water quality/separation - given many commercial AAs are surface-active in nature, continuous use of AAs can stabilize emulsions leading to separation issues topsides and poor overboard water quality [79,80]. This aspect of AA performance is commonly validated as part of the chemical qualification using simple bottle tests and/or through measurements of interfacial tension [79] upon adding the AA to the oil and water mixture.
- Chemical/material compatibility and stability - prior to being injected into a production system, most chemicals follow a standard compatibility testing program including AAs. Some of these tests can include compatibility with various metallic and nonmetallic materials, umbilicals, acceptable viscosity at high pressures and low temperatures, and stability over a range of temperatures.
- Toxicity - as mentioned previously, quaternary ammonium salts-based AAs are generally toxic in nature, which can limit their application in certain geographic locations. Regulations for toxicity and biodegradability vary substantially by region, for example Gulf of Mexico requirements are generally less stringent that North Sea requirements [53]. As such, local regulatory requirements should govern the type of AA chemistry being qualified. This toxicity constraint has spurred research into new AA products that aim to improve the biodegradability of these molecules while still preserving their key function of creating nonagglomerating hydrate slurries [81,82].
- Low water-cut/freshwater systems - based on lab qualification testing, AA dosages are usually prescribed based on the volume of water that is being produced in the field. Typical dosages can range from 0.5 to 2 vol.% AA (basis water volume) and are obtained from rocking cell tests that are run with a certain minimum water-cut. This minimum water-cut (often ~10 vol.%) is more of an equipment constraint as it becomes difficult to detect the hydrate formation and behavior of slurries accurately and reliably for water-cuts lower than this threshold. If field measured water-cuts stay above this minimum, AA dosages can be easily applied based on lab data. However, AA dosages for water-cuts less than this threshold can become tricky to extrapolate based on just lab tested data. Webber et al. [83] described an AA application where a field dose rate data were used to supplement lab data to account for low water-cut production scenarios. They found that at low water-cuts, AA dosage rate followed a power law type dependence on the water-cut. At an average basic sediment & water of 0.5%, field observed AA dosage rate was as high as 17 vol.% (basis water). Similar dosage trends were recently confirmed by Bodnar et al. [84]. It should be noted that AA dosage can appear to become significantly high at such low water-cuts, but that these high dose rates would still be based on water volumes, which implies that the absolute AA amount to be injected would still be manageable. Apart from this relatively high dosage requirement at low water-cuts, most commercial AAs also work best when some salts are present in the produced water. During early-life

production, a field might only produce saturation water, which tends to be fresh or very low in salinity. Hence, AAs that might have been qualified to prevent hydrate agglomeration with the formation brine salinity might not work best with the initial freshwater production. To manage such transitions in water salinity, one can either consider different AA chemistries that are optimized to the water salinity or increase the dosage rate of the previously qualified AA chemistry.

1.7 Dehydration

As mentioned in the Risk Assessment section, presence of water is one of the requirements to form gas hydrates. Consequently, one of the ways to mitigate hydrate risk in production systems could be to remove all water from the system. In subsea production systems, dehydrating the production stream could mean installing a separation system to remove bulk water, followed by a glycol-based contactor and/or scrubber to lower the water content enough for the production fluids to be exported without any hydrate risk. The removed water can be exported in a separate line, cleaned/discharged to ocean or re-injected into a formation. All these options can carry significant additional CAPEX costs, which has made subsea water removal/dehydration prior to long distance transport not a commonly used mitigation solution. Instead, hydrate mitigation through dehydration is mostly confined to gas export applications, where topsides facilities can manage to remove and discharge water from wet gas more easily compared to subsea applications.

For wet gas applications, hydrate equilibrium curves do not strongly depend on the amount of water in the system if it is in excess or as a free liquid phase. Once free water phase is removed from the stream, further reduction in water content leads to lower dissolved water in the gas phase that, at some point, starts shifting the hydrate curves to lower temperatures and higher pressures. Fig. 1.54 shows this effect of progressive dehydration on the hydrate equilibrium curves for a simple gas mixture. Compared to the hydrate curve for excess water (furthest to right hand side), decreasing water amounts in the gas phase from 15 lbs/ million standard cubic feet (MMscf) gas to 3 lbs/MMscf gas results in shifting the hydrate curve significantly. For offshore applications, typical gas export line pressures are in the range of a few thousand psi and the worst-case ambient temperature is often close to 40°F. Dew point requirements for gas export will obviously depend on the specific gas composition and local P-T conditions, but it can be seen from Fig. 1.54 that, in general, dehydrating the gas to $<3-5$ lbs/MMscf gas is a reasonable guideline in order for the export line to stay outside the hydrate equilibrium region.

The most common way to achieve the required dehydration levels in offshore facilities is using a glycol-based dehydration column, with tri-ethylene glycol (TEG) often being the most frequently used glycol Fig. 1.55. A key advantage of

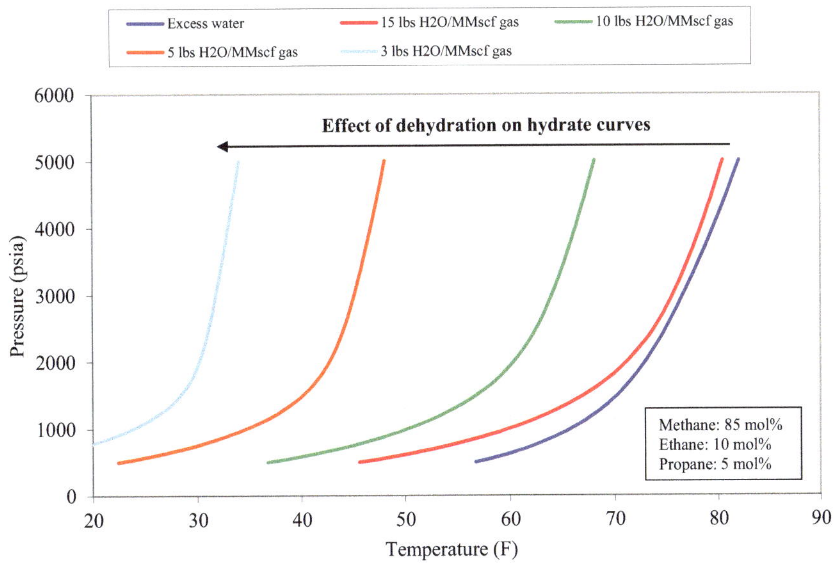

FIGURE 1.54

Effect of dehydration on hydrate equilibrium curves for a simple gas mixture.

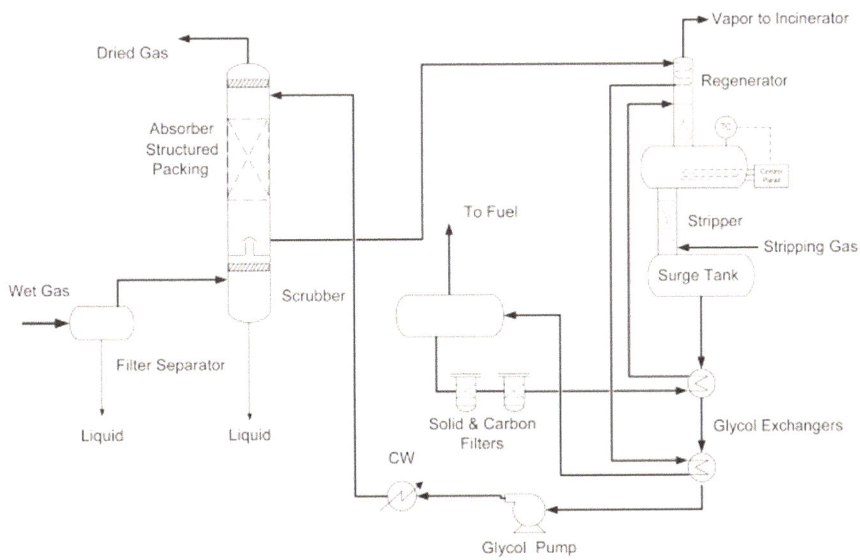

FIGURE 1.55

Process flow diagram for a conventional TEG dehydration system [85].

Reproduced with permission from S. Mokhatab, W.A. Poe, J.Y. Mak, Chapter 9: natural gas dehydration and mercaptans removal, in: Handbook of Natural Gas Transmission and Processing, fourth ed., Gulf Professional Publishing, Massachusetts, 2019, pp. 307–348.

this process is minimal pressure drops caused by the dehydration process compared to other separation techniques. Glycol-based dehydration is also able to economically handle relatively large volumetric throughput of gas compared to adsorption-based processes such as molecular sieves or membranes.

1.8 Hydrate remediation

Hydrate flow assurance strategies are commonly designed to eliminate the risk of plugging in production systems. Despite all the mitigation measures that are put in place during the design phase, hydrate plugs still occur frequently in hydrocarbon production systems. These can either be due to mitigation measures not being adequately followed—for example, not depressurizing or dead-oiling subsea systems properly or within the specified time limit, not injecting chemicals at the required dosage rates, etc., or due to change in production parameters or fluid properties that were not monitored properly. Examples of this latter category include changes in fluid GOR or water-cut, water breakthrough, changes in water salinity or P-T conditions along the production system. If a hydrate restriction or plug is suspected anywhere in the production system, it is recommended to carefully evaluate available remediation options with respect to potential safety issues and seek expert input if needed.

Given that hydrate formation requires the simultaneous presence of gas formers, water, high pressure, and low temperature, it follows that removing any of these four elements should result in the hydrate deposit or plug being dissociated. Unfortunately, once a hydrate plug or restriction has formed, it becomes very difficult to get the gas or the water phase out of the hydrate plug. The only available options are to use pressure, temperature, or chemical means to get the system out of the hydrate stability region. Some of these strategies are discussed below:

1.8.1 Depressurization

Operationally, the most often used strategy to remediate hydrates is to try and depressurize the hydrate blockage to a pressure below the hydrate equilibrium curve. The target pressure for depressurization depends on the hydrocarbon composition and ambient temperatures. For deepwater/subsea environments, an ambient temperature of close to 40°F could imply a target blowdown pressure as low as a few hundred psi. Feasibility of achieving such low target pressures is not always a given. In addition, hydrate dissociation through depressurization can be a slow process, with heat transfer through the system often being the limiting or controlling factor. Dissociation times will depend on several factors such as length and physical properties of a blockage, use of single or dual-sided depressurization, and type of insulation surrounding the blockage (which controls the heat transfer).

System constraints and some other important things to consider while evaluating this remediation strategy include:

- Type of production system - depressurization to a low target pressure where hydrates can dissociate can be strongly dependent on whether the production system is oil or gas-dominated. Oil-dominated production systems, especially in deeper water depths, can have significant liquid loading in the riser that prevents them from being adequately depressurized. Gas-dominated systems with low liquid loading, in general, do not suffer from this constraint. Hence, fluid GOR is an important variable that should be considered while designing a facility that can be successfully depressurized in the event of a hydrate blockage. This feasibility is often evaluated during the project design stage using transient thermal-hydraulic simulators. If results show that the system cannot be blown down due to low fluid GOR, high water-cut or deep water-depth, then depressurization cannot be used as a remediation strategy.
- Dual vs single flowline systems - it is well known and documented that hydrate plugs can release at fast speeds if depressurized only from one end [2,86]. Trapped pressure on the other side of the plug can lead to high pressure drop across the blockage that can result in the plug being launched as a projectile causing significant safety and integrity issues. Subsea production systems can either be tied back to host facilities using a single flowline tieback or through a dual flowline production loop. In terms of a hydrate blockage remediation, the dual flowline system allows for depressurization to be attempted from both ends of the flowline loop, thereby minimizing the risk of a high-speed plug after dissociation. A single flowline system, on the other hand, can be blown down only from the host end, without subsea intervention, which makes one-sided depressurization in these cases riskier. Hence, for a single flowline system, one-sided depressurization should only be considered if the blockage is confirmed to be at a safe distance from other subsea components, such as valves, sleds, riser. An additional point to consider is the possibility of multiple hydrate plugs in the flow path. In such a case, even if dual-sided depressurization is possible at the ends of the flow path, the presence of multiple plugs might not eliminate the risk of launching a hydrate plug due to trapped pressure between plugs. Unfortunately, once production is shut-in, there is no easy and accurate way of detecting either the length or the number of plugs in a flowline. This feature always results in more uncertainty while designing a safe remediation strategy for hydrate plugs.

1.8.2 Heating

Thermal or temperature-based remediation methods use some form of active heating to bring local conditions outside the hydrate equilibrium region. For cases where a restriction exists rather than a complete blockage, heat can be supplied internally by flowing warm fluids (e.g., dead oil) through the flow path. Feasibility of this technique will also depend on whether a flow loop exists that is able to allow circulation of fluids. For a complete blockage, heating will likely be achieved through externally applied thermal energy, for example through a heated bundle [87], electrical heating [47], or an electrically heated blanket [88]. Unfortunately, if a system is not designed and installed with such capability, many external heating options will not be available. A key risk to consider while externally heating the blockage is to ensure that heat is applied uniformly across the length of the blockage and not focused on a short section of the blockage that is trapped on either end by more hydrates. This risk arises because dissociation of hydrates leads to release of gas, which can result in significant trapped pressure increase if not allowed to vent to a lower pressure region. An oft-cited example in hydrate operations relates to a field example in Siberia, where a torch was used to heat a hydrate blockage from the middle that caused enough overpressure (due to localized hydrate dissociation) to cause a pipeline rupture [86]. With this constraint in mind, temperature-based remediation strategies that involve heating the entire flowline, such as a flowline bundle, DEH or electrically trace heated PIP will result in more uniform heating of the system and minimize the risk of locally over-pressurizing the flowline.

1.8.3 Chemical dissociation

Apart from being used to prevent hydrate formation, chemical thermodynamic inhibitors such as MeOH or glycol can also be used to dissociate hydrate plugs. The major problem associated with using chemicals for remediation is to get the inhibitor to reach the blockage and ensure good contact between the two. In a typical subsea system, these inhibitors are usually injected either from the subsea end or through the host. Hence, chemical injection-based remediation is generally more suited for plugs that occur close to these injection points. Hydrate blockages in subsea trees, jumpers, manifolds, risers or close to the riser base will be more successful in being remediated through chemical injection compared to a hydrate blockage in the middle of a flowline. To overcome distance constraints, delivery of the chemical inhibitor through coiled tubing can be considered, although that option also has a finite reach. Some other factors to keep in mind while using inhibitors to dissociate hydrates include:

- Density differences - both MeOH and MEG have been shown to be successful in dissociating hydrates, but MeOH, in general, can be lighter than the hydrocarbon liquid phase, whereas MEG is usually heavier. As such, these density differences should be noted while choosing the right inhibitor. For example, if a hydrate blockage in a wellbore or near the riser base is being remediated with a chemical inhibitor, MEG will have a much better chance at reaching the blockage due to its higher density. On the other hand, MeOH might just float on top of any oil phase present and not contact the blockage at all. This problem will obviously depend on the local geometry surrounding the blockage, but for vertical systems, MEG is usually the inhibitor of choice over MeOH when it comes to remediation.

- Dilution - given both MeOH and MEG are miscible in water, any water phase present around the blockage will result in dilution of the inhibitor at the contact area. Moreover, dissociation of hydrate releases water, which will further dilute any injected inhibitor. In practice, multiple batches of inhibitor are often injected to counter the effect of continuous dilution because of dissociation.

1.8.4 Model predictions for remediation

Numerical models have been published over the years that can assist in reasonably predicting time required to dissociate hydrate plugs. These models were developed for single-sided and dual-sided depressurization techniques [89–91] as well as for active heating-based methods [91,92]. Many of them rely on finite-difference-based discretization of the hydrate blockage to track the dissociation front as a function of time. The models predict observed dissociation times well with minimal fitting parameters [90,91]. A key learning from the experiments and modeling results is that for both depressurization and heating-based remediation techniques, plug dissociation is mostly radial in nature rather than axial, and that radial heat transfer through the wall controls the overall rate of dissociation [89]. CSM have developed a plug dissociation model called CSMPlug [2] that is commonly used by operators to calculate high-level plug dissociation times.

1.9 Case studies

Case studies and field experiences are presented here that depict scenarios where hydrate formation was observed or mitigated in hydrocarbon production systems.

1.9.1 Hydrate management in dry tree facility facilities

Hydrate management for dry tree wells typically involves keeping the production fluids out of hydrate forming conditions during continuous production as well as during shut-in and restarts. Mitigation barriers that are often used include:

- Tubing/riser insulation - production fluids are kept outside the hydrate region through passive heat retention/insulation. The portion of the wellbore beneath the seafloor does not see significant heat loss compared to the riser portion that is often exposed to cold seawater currents. To mitigate this risk, the riser portion of the well can be insulated to prevent hydrate conditions during steady-state production. However, the well/riser still needs to be secured during shut-ins due to eventual cooldown of the system to ambient temperatures. For such cases, the riser portion above the mudline tends to have a much shorter cooldown time compared to the part of the wellbore below the mudline. Post cooldown, some mitigation is required to secure the wellbore fluids against hydrate formation.

- Bullheading - one of the most common methods to mitigate against hydrate formation during a wellbore shut-in is to displace or bullhead production fluids deeper into the wellbore to a region where temperatures are warm enough to be outside the hydrate region. For subsea wells, MeOH is commonly used for bullheading, while for dry trees diesel or dead oil is often used for displacement [93]. Since the displacement volumes can become significant, optimizations have been proposed where fluids are displaced below the mudline during short shut-ins, whereas longer term shut-ins require bullheading of fluids all the way to the SCSSV depth. Further, operators have also proposed taking credit for gas caps that form in the wellbore upon shut-in and only bullheading the liquid volume in the riser to deeper depths [93]. This relies on the hypothesis that risk of a complete hydrate blockage is relatively low in a gas column with only saturated water. It should be noted that for unplanned shut-ins, bullheading of fluids is typically the only mitigation strategy available to operators.
- Chemical injection - THIs or LDHIs can also be continuously injected to mitigate hydrate formation but are not commonly used during steady-state production. Due to relatively high temperatures during production along with short transit times through the well, continuous chemical injection is often not economically justifiable. Therefore chemical injection is most commonly used during planned shut-ins, so that the production fluids are pre-treated and secured prior to the well being shut-in.

Kurup [94] discussed BP's hydrate management philosophy at a dry tree facility in the Gulf of Mexico, operating in water depths of ~4500 ft. The asset's original hydrate management strategy relied on complete avoidance of hydrate conditions through a combination of insulation for steady-state mitigation and bullheading + LDHI pre-treatment for managing shut-in hydrate risk. Along the usual mitigation options, such as insulation, chemical treatment and bullheading, this asset also benefited from the production fluids being characterized by low GOR. Lower GOR can limit the availability of gas, which can slow down the kinetics of hydrate formation and consequently reduce the risk of a complete hydrate blockage. Many wells at this facility were gas-lifted, which annuls the benefit of lower GOR on the overall hydrate risk.

In practice, cooldown times for treatment of these wells were relatively short due to rapid cooling of the well near the mudline. This problem was exacerbated for wells that were gas lifted due to the shift in the corresponding hydrate curve to higher temperatures, which led to several instances in the field where the wells could not be properly treated within the prescribed cooldown time. A review of such instances revealed that in many such cases, pressure and temperature conditions were well within the hydrate region, but no hydrate blockages were observed upon restart.

The authors quantified hydrate risk for these field events by calculating the maximum amount of hydrates that can thermodynamically form in the system as a function of water-cut for various GORs. This simple and relatively novel way of quantification leads to a "hydrate volume fraction" that is equivalent to a solids phase holdup in the system. It is still a conservative estimate because it is a thermodynamic estimation, which takes away any constraints the system might see in terms of heat and mass transfer limitations or related to kinetics of hydrate formation. For a defined GOR, the hydrate volume fraction initially tends to increase with water-cut, which implies that water phase is the limiting component at lower water-cuts. The hydrate volume fraction reaches a maximum at a certain water-cut, beyond which further increases in water-cuts result in lower hydrate volume fractions. This occurs due to insufficient hydrocarbon components in the system becoming limited, thereby restricting the amount of water that can be converted into the hydrate phase. As expected, as the fluid GOR increases, this peak hydrate volume fraction is observed at progressively higher water-cuts.

Past field events at this BP facility that had excursions into the hydrate zone were mapped to the hydrate volume fraction plot. It was noted that when the fluid GOR and/or water-cut were low, the resulting hydrate volume fraction was also predicted to be lower (<20%). This led to the scenario where some hydrates can form in the tubing but are not enough to completely block the tubing upon restart.

To better understand the critical hydrate volume fraction and the resulting NTTs, BP also executed some deliberate tests with well C for varying gas lift rates that resulted in GORs ranging from ~1000–1700 scf/bbl. These tests were also mapped on to the hydrate volume fraction and corresponded to hydrate volume fractions in the 30%–50% range. Some of these tests showed resistance during dead oil displacement, while others indicated no resistance. Based on these field tests, it was decided to expand the operating envelope for these wells as long as water-cut and GOR were within the hydrate test conditions. This allowed operations to dead oil displace within 5 hours, instead of the original thermodynamic-based <1 hour cooldown time.

In June 2015, well C that was gas-lifted had an unplanned shut-in that resulted in production being shut-in for more than a day. The operating conditions prior to this trip were also outside the hydrate test conditions—higher water-cut (60%), higher GOR (1800 scf/bbl) and insufficient dead oil displacement. As a result, the well could not be restarted after recovering from the trip and a hydrate blockage formed in the tubing. The hydrate volume fraction corresponding to this event was calculated to be >60%, which was significantly higher than what past field events and planned tests were predicted to experience. This event represents a critical threshold for the hydrate volume fraction, beyond which the amount of hydrate that forms in the system is likely to result in a complete blockage.

This case study is a good example of hydrate risk management in practice. It not only shows how hydrate risk and exposure were quantified across a range of operating conditions and parameters, but also illustrates how past field events and data analysis can provide an expanded operating envelope for various production scenarios. More importantly, this study was able to identify an upper critical limit, above which hydrate risk can become too high if the system is left unmitigated. Although, this study was restricted to hydrate formation in dry tree wells, similar approaches can be adapted and extended for application to other subsea components, such as jumpers, manifolds, flowlines, and risers. It is expected that such hydrate risk management approaches will become more commonplace in the industry, as other operators share their hydrate experiences in the field leading to further refinement and optimization of the hydrate operating envelope.

1.9.2 Low-dosage hydrate inhibitor field application

This case study discusses one of the earliest applications of AA for a continuous field application at a Shell-operated deepwater asset [52]. The field is in 2000 ft of water in the Gulf of Mexico and consists of a 6-in. dual flowline system tied back 24 miles to an existing platform. It was designed to produce up to 120 MMscf of gas per day along with associated gas condensate. During initial years of production, three wells (A1, A2 and A3) were producing at this asset without any production water. Operating conditions at that time were deep inside the hydrate region, with hydrate subcooling in the 33°F −35°F range. Consequently, saturation water still had to be treated to prevent hydrate formation given the system was continuously flowing in the hydrate region. The operator chose MeOH injection into each flowline for hydrate mitigation at a dosage rate of ~1.5 bbl MeOH/bbl of condensed water.

In the early 2001, well A2 started seeing formation water break through, which started requiring significantly higher volumes of MeOH to achieve full thermodynamic inhibition. Even at this highest possible continuous daily injection of 120 bbl of MeOH, the subsea system still operated inside the hydrate region at ~10°F subcooling.

The flowline pressures were monitored for evidence of hydrate deposition through the well choke discharge pressures. As soon as the onset of hydrate deposition was observed through these pressure traces, additional MeOH was injected into the flowline through the annulus vent line. This resulted in dissociation of the hydrates in the flowline causing the pressure traces to revert to baseline levels, thereby restoring gas production to normal rates. Although MeOH had been widely used before to prevent hydrate formation or to dissociate blockages, this field example represents one of the first instances of using intermittent batch treatment with MeOH to dissociate hydrates during steady-state production.

This production scenario also provided impetus to replace continuous MeOH injection with an AA-LDHI-based mitigation strategy, due to the MeOH capacity eventually becoming limited by the increasing water production rates. Consequently, an implementation plan was developed to select and test an appropriate AA chemistry based on laboratory testing, followed by field preparation that included topsides storage and injection facilities for the LDHI and chemical compatibility tests.

For laboratory testing, condensate samples from the field were tested in high-pressure, rolling-ball hydrate cells maintained in a temperature-controlled bath. Test protocols for LDHI testing often involve starting with blank solutions (i.e., without addition of any active LDHI chemistry) to get baseline results with the given oil or condensate sample. Initial tests for this application were also performed with a blank solution and as expected, resulted in large hydrate deposits within the cell.

Laboratory tests run with the AA-LDHI created a fine dispersion of hydrate particles in the condensate phase that did not agglomerate. The test protocol involved rocking the cell for 12 hours to simulate stead-state production, which resulted in most of the "hydrateable" water being converted to hydrates. After this rocking stage, the cells were shut-in for 24-hours, followed by continuous rocking to simulate a restart. No hydrate deposits were observed during the entire test and, upon restart, the hydrate particles were able to redisperse into the condensate phase. Based on these test results, the AA-LDHI was qualified for field deployment.

During field deployment, it was decided to initially co-inject the AA-LDHI chemistry with MeOH. This phase-in approach is common while switching from one type of inhibitor to another. The AA-LDHI was initially injected at the vendor-recommended dosage rate along with MeOH, while monitoring the system for any pressure drop increases. When no changes were observed in the pressure traces, the LDHI and the MeOH dosage rates were gradually decreased over a period of 24 hours. At this point, MeOH injection was completely stopped, and the flowline was solely operating on the AA-LDHI injection. This transition showed no increases in flowline pressures with the LDHI chemistry. Topsides sampling during this field trial showed milky dispersions of hydrate particles, which provided evidence of hydrate slurry formation in the flowline. To demonstrate efficacy of LDHI during extended shut-in periods, production was shut-in for 24 hours. Upon restart, no hydrate deposits or blockages were encountered.

This successful conclusion of the field trial led the operator to eliminate continuous MeOH injection and switch the field to LDHI-based inhibition on a continuous basis. Since this field implementation in early 2000s, AA-LDHIs have been increasingly used as a key hydrate mitigation strategy, especially in deepwater systems where high pressures and low ambient temperatures result in significant hydrate subcooling. For such conditions, KHIs usually do not meet the subcooling threshold or the long hold time requirements related to extended shut-ins (e.g., hurricane-related), making AAs an attractive solution.

1.9.3 Tommeliten-gamma field

Statoil performed field tests at the Tommeliten-Gamma field in 1994 where hydrates were intentionally allowed to from in a 6-in service line. The subsea layout of the production lines and the service lines are shown in Fig. 1.56, with the associated condensate and water holdup in the service line predicted using OLGA depicted in Fig. 1.57. These holdup profiles show that even with the relatively small amount of liquids present in the line, localized spikes in holdup were seen near the elevation changes. These local low spots allowed for collection of sufficient amounts of water to form hydrates. Eight tests were conducted and four of

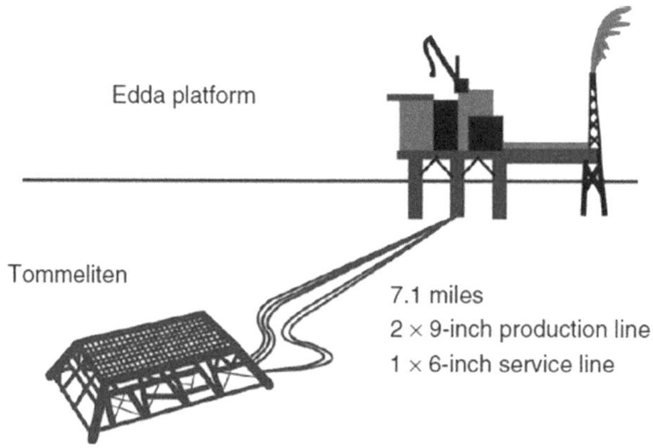

FIGURE 1.56

Subsea layout of the Tommeliten-Gamma lines tied back to the Edda platform.

Reproduced with permission from E.D. Sloan, J. Creek, A.K. Sum, Where and how are hydrate plugs formed?, in: E.D. Sloan, C.A. Koh, A. Sum (Eds.), Natural Gas Hydrates in Flow Assurance, Elsevier, 2010, pp. 13–36.

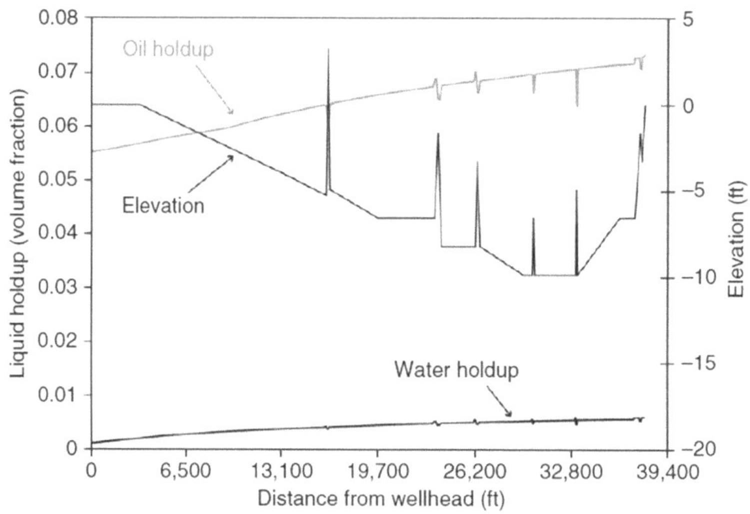

FIGURE 1.57

Elevation and holdup profiles of the condensate and water phases for the Tommeliten service line.

Reproduced with permission from E.D. Sloan, J. Creek, A.K. Sum, Where and how are hydrate plugs formed?, in: E.D. Sloan, C.A. Koh, A. Sum (Eds.), Natural Gas Hydrates in Flow Assurance, Elsevier, 2010, pp. 13–36.

them resulted in subsea hydrate plugs close to these low spots. Some of the tests where hydrates formed were either uninhibited or under inhibited with respect to MeOH injection. Test conditions include steady-state flow as well as during restarts following pressurized shut-in of the service line [24].

Due to the controlled nature of these tests, the Tommeliten field trial has become one of the most well studied hydrate formation cases. Several modeling studies have attempted to describe the field observed hydrate events. Early studies used the oil-dominated conceptual model of CSMHyk coupled with OLGA transient models to successfully predict the approximate time scales of the hydrate events but through viscosity increases rather than deposition-based plugs [95]. More recently, transient hydrate kinetic models developed for gas-dominated systems have overcome this issue by incorporating hydrate deposition based on flow shear and wall adhesion [96]. These models were able to predict the approximate location as well as time of observing hydrate events in the field. Their ability to include deposition represents a key improvement in the modeling of hydrate risk in oil and gas production systems.

1.9.4 Remediation of hydrate plug in west Africa deepwater floating production storage and offloading

A hydrate plug was formed and remediated in a 12-inch gas export line of a deepwater Floating production storage and offloading operated by total in West Africa [97]. Hydrate management for gas export lines typically relies on dehydrating the gas phase to a dew point below which hydrates cannot thermodynamically form. This is commonly achieved through a TEG dehydration unit. However, process upsets in terms of water content of the export gas going off-spec frequently occurs due to operational issues with the dehydration units. In such cases, a common strategy is to turn on continuous THI injection (mostly MeOH) if the export gas–water content is not being met.

During summer 2018, Total stated in their paper [97] that the TEG unit was not operational for about 10 days, during which MeOH was continuously being injected to mitigate hydrate risk in the gas export line. When the TEG unit was back in service, export of dry gas resumed but after 5 days of dry gas export, the compressor was stopped for maintenance. During restart, a hydrate blockage was suspected to form rapidly at the base of the riser.

As soon as a hydrate plug was suspected, as often is the case in such operations, a multi-disciplinary team was assembled to develop and execute a remediation plan. The authors report the following three objectives to start the remediation program [97]:

- Plug nature and location identification - multiphase flow simulations were used to identify the nature and location of the plug. Results indicated gas rates prior to the event were low and below the turndown rate, which would have resulted in significant liquid accumulation in the system. Moreover, due to the uphill export line bathymetry, most of this accumulated liquid (water + MeOH) would be prone to collect near the riser base. For the location of the plug, a commonly used method for gas-dominated systems is to pressure cycle the system and measure the subsequent pressure increase or decrease, depending on whether the cycle involved pressurization or depressurization. Along with gas rate measurements over time, the total amount of gas injected or released can be used to get a reasonable estimation of the total volume behind the plug. Knowing the volume and the line size then leads to calculation of the likely plug location. Total reported performing two depressurization cycles from the riser top that constrained the plug location to about 1000–1500 m from the host, which coincides with the location of the riser base.

- Plug removal - once the plug location was identified, the actual remediation strategy was defined and executed. The authors report using a combination of several depressurization cycles along with MeOH/MEG injection from topsides being utilized as part of the remediation strategy. In addition to single-sided depressurization from topsides, a dual-sided depressurization was attempted from the downstream end of the plug through a subsea hot stab. Although the riser end of the plug was presumably depressurized to below the hydrate equilibrium pressure, the depressurization through the subsea hot stab was limited to about 150 bar, which is significantly above the equilibrium pressure needed for dissociation. It is suspected that this was done to minimize pressure drop across the blockage that could lead to fast moving projectile upon dissociation. Methanol and MEG were also injected from topsides and as expected, the authors report MEG to being much more effective in clearing the plug due to its higher density compared to MeOH. As mentioned in the remediation section, MEG is usually the preferred chemical inhibitor for plug remediation especially in vertical systems where it has a much higher chance of contacting the blockage face. Following these multiple remediation strategies, the plug was successfully dissociated, and communication was restored in about 5 days from the plugging event.
- Return to normal export - after clearing a hydrate plug, restarting the system can be a complicated process. This is because the hydrate forming components are still present in the line and any resumption of flow, causing an increase in pressure without additional mitigation, can lead to the re-formation of a hydrate plug. Strategies for the safe restart of a line post dissociation often depend on local system design and resources available to the operator and can involve either flushing the system or having a sufficient volume of inhibitors to prevent hydrates from forming again. In this case, the operator first used dry gas to flush the liquid content in the export line to a gas injection well further downstream. This step was followed by backflushing the remaining liquid up the riser by importing gas from the downstream gas network. Following these flushing steps, gas export was resumed without any issues.

This case study is a good example of hydrate remediation in practice where multiple strategies are often evaluated in parallel to remove the blockage quickly and safely. One of the key learnings from this experience was the reason why the hydrate plug formed in the gas export line when all mitigation measures were being properly followed. Post-plug investigation revealed that even if the wet gas was adequately inhibited with MeOH, the low gas export rate meant that there was significant accumulation of liquid near the riser base. When dry gas was exported without additional inhibition, the MeOH in the accumulated water was continually being stripped by the gas phase, which eventually led to under-inhibited water and a hydrate plug.

1.10 Summary

Gas hydrates can pose great production, asset integrity and safety risks in oil/gas production. This chapter covers a wide range of areas on this topic with a focus on hydrate risk management. The fundamental rationales for hydrate risk management are either to keep the condition outside of hydrate formation region or keep the hydrates dispersed if they do form. The two main approaches are either mechanical (insulation, electrical heating, pressure control, water removal etc.) or chemical (THI, LDHI, etc.). Many examples and case histories were included in the chapter to illustrate typical field problems and performance of different chemical treatments.

It is the hope of the authors that this chapter will provide a fundamental understanding and practical knowledge as a foundation to manage your own field-specific hydrate challenges, at least to ask the right questions. The optimal and/or most cost-effective hydrate management program for an operation depends on the specifics of the production system and often require a trial-and-error process.

Nomenclature

AA	anti-agglomerant
bbl	barrels
AFP	antifreeze fish proteins
CAPEX	capital expenditures
CI	corrosion inhibitors
CPA	Cubic Plus Association
CSMHyK	Colorado School of Mines hydrate kinetics model
D_2O	deuterium oxide
DEH	direct electrical heating
DME	dimethyl ether
EoS	equation of states
ft	Feet
GOR	gas-to-oil ratio
GPa	gigapascals
h	hours

HET	hydrate equilibrium temperature
HRE	hydrate risk evaluator
K	temperature in kelvin
KHI	kinetic hydrate inhibitor
lbs	pounds
LDHI	low-dosage hydrate inhibitor
MEG	mono-ethylene glycol
MeOH	methanol
min	minutes
MMscf	million standard cubic feet
NTT	no-touch time
OPEX	operating expenditures
PDF	probability distribution functions
PIP	pipe-in-pipe
PolyAP	polyacryloylpyrrolidine
PVCap	polyvinylcaprolactam
PVP	polyvinylpyrrolidone
PVT	pressure-volume-temperature
TBAB	tetrabutylammonium bromide
TEG	tri-ethylene glycol
THF	tetrahydrofuran
THI	thermodynamic hydrate inhibitor
TPAB	tetrapentylammonium bromide

References

[1] D. Vavik, Implications of Gas Hydrates in Drilling and Completion: A Scientific Root Cause Analysis of the Deepwater Horizon Disaster (Ph.D.), Norwegian University of Science and Technology, Trondheim, 2020.

[2] E.D. Sloan, C.A. Koh, Clathrate Hydrates of Natural Gases, CRC Press, New York, 2008.

[3] S. Das, K.M. Tadepalli, S. Roy, et al., A review of clathrate hydrate nucleation, growth and decomposition studied using molecular dynamics simulation, Journal of Molecular Liquids 348 (2021) 118025.

[4] A. Kurnosov, V. Komarov, V. Voronin, et al., New clathrate hydrate structure: high-pressure tetrahydrofuran hydrate with one type of cavity, Angewandte Chemie International Edition 43 (2004) 2922–2924.

[5] K. Udachin, C. Ratcliffe, J. Ripmeester, A dense and efficient clathrate hydrate, Angewandte Chemie International Edition 40 (2001) 1303–1305.

[6] S. Gao, Formation and Dissociation Mechanisms of Clathrate Hydrates (Ph.D. dissertation), Rice University, Houston, 2006.

[7] A. Hassanpouryouzband, E. Joonaki, M. Farahani, et al., Gas hydrates in sustainable chemistry, Chemical Society Reviews 49 (2020) 5225–5309.

[8] K. Jeong, P. Metaxas, A. Helberg, et al., Gas hydrate nucleation in acoustically levitated water droplets, Chemical Engineering Journal 433 (2021) 133494.

[9] J.H. van der Waals, J.C. Platteeuw, Clathrate solutions, Advances in Chemical Physics, vol. 2, Interscience Publishers Inc, New York, 1959, pp. 1–57.

[10] W.R. Parrish, J.M. Prausnitz, Dissociation pressures of gas hydrates formed by gas mixtures, Industrial & Engineering Chemistry Process Design and Development 11 (1972) 26–35.

[11] A.L. Ballard, E.D. Sloan, The next generation of hydrate prediction: part III. Gibbs energy minimization formalism, Fluid Phase Equilibria 218 (2004) 15–31.

[12] A. Medeiros, I.S.V. Segtovich, F.W. Tavares, et al., Sixty years of the van der Waals and Platteeuw model for clathrate hydrates—a critical review from its statistical thermodynamic basis to its extensions and applications, Chemical Reviews 120 (2020) 13349–13381.

[13] A.L. Ballard, E.D. Sloan, The next generation of hydrate prediction IV a comparison of available hydrate prediction programs, Fluid Phase Equilibria 216 (2004) 257–270.

[14] H. Haghighi, A. Chapoy, R. Burgess, et al., Phase equilibria for petroleum reservoir fluids containing water and aqueous methanol solutions: Experimental measurements and modelling using the CPA equation of state, Fluid Phase Equilibria 278 (2009) 109–116.

[15] A.M. Palma, A.J. Queimada, J.A.P. Coutinho, Modeling hydrate dissociation curves in the presence of hydrate inhibitors with a modified CPA EoS, Industrial & Engineering Chemistry Research 58 (2019) 19239–19250.

[16] M.N. Khan, P. Warrier, J.L. Creek, et al., Vapour-liquid equilibria (VLE) and gas hydrate phase equilibria predictions using the cubic-plus association equation of state: CSMGem extension to association EoS model, Journal of Natural Gas Science and Engineering 94 (2021) 104083.

[17] C.P. Ribeiro Jr., P.L.C. Lage, Modelling of hydrate formation kinetics: State-of-the-art and future directions, Chemical Engineering Science 63 (2008) 2007–2034.

[18] P. Englezos, N. Kalogerakis, P.D. Dholabhai, et al., Kinetics of formation of methane and ethane gas hydrates, Chemical Engineering Science 42 (1987) 2647–2658.

[19] W. Ke, T.M. Svartaas, D. Chen, A review of gas hydrate nucleation theories and growth models, Journal of Natural Gas Science and Engineering 61 (2019) 169–196.

[20] D. Kashchiev, A. Firoozabadi, Nucleation of gas hydrates, Journal of Crystal Growth 243 (2002) 476–489.

[21] P.J. Metaxas, V.W.S. Lim, C. Booth, et al., Gas hydrate formation probability distributions: Induction times, rates of nucleation and growth, Fuel 252 (2019) 448–457.

[22] E.F. May, V.W.L. Lim, P.J. Metaxas, et al., Gas hydrate formation probability distributions: The effect of shear and comparisons with nucleation theory, Langmuir 34 (2018) 3186–3196.

[23] P.N. Matthews, P.K. Notz, M.W. Widener, et al., Flow loop experiments determine hydrate plugging tendencies in the field, Annals of the New York Academy of Sciences 912 (2006) 330–338.

[24] E.D. Sloan, J. Creek, A.K. Sum, Where and how are hydrate plugs formed? in: E.D. Sloan, C.A. Koh, A. Sum (Eds.), Natural Gas Hydrates in Flow Assurance, Elsevier, 2010, pp. 13–36.

[25] K. Kinnari, J. Hundseid, X. Li, et al., Hydrate management in practice, Journal of Chemical & Engineering Data 60 (2015) 437–446.

[26] P. Glenat, R. Devoisselle, L. Pagezy, et al., Natural kinetic inhibitors of crude oils, in: Proceedings of the 9th International Conference on Gas Hydrates, Denver, Colorado, 25–30 June, 2017.

[27] A. Vysniauskas, P.R. Bishnoi, A kinetic study of methane hydrate formation, Chemical Engineering Science 38 (1983) 1061–1072.

[28] L.E. Zerpa, E.D. Sloan, A.K. Sum, et al., Overview of CSMHyK: a transient hydrate formation model, Journal of Petroleum Science and Engineering 98–99 (2012) 122–129.

[29] R. Camargo, T. Palermo, Rheological properties of hydrate suspensions in an asphaltenic crude oil, in: Proceedings of the 4th International Conference of Gas Hydrates, Yokohama, Japan, 19–23 May, 2002.

[30] Y. Wang, C.A. Koh, J.A. Dapena, et al., A transient simulation model to predict hydrate formation rate in both oil- and water-dominated systems in pipelines, Journal of Natural Gas Science and Engineering 58 (2018) 126–134.

[31] S.V. Joshi, G.A. Grasso, P.G. Lafond I, et al., Experimental flowloop investigations of gas hydrate formation in high water cut systems, Chemical Engineering Science 97 (2013) 198–209.

[32] P. Chaudhuri, L.E. Zerpa, A.K. Sum, A correlation to quantify hydrate plugging risk in oil and gas production pipelines based on hydrate transportability parameters, Journal of Natural Gas Science and Engineering 58 (2018) 152–161.

[33] D.J. Turner, G. Mahadevan, J. Lachance, Hydrate stable start-up and restart of oil and gas production systems, in: Proceedings of the 25th International Ocean and Polar Engineering Conference, Kona, Hawaii, 21–26 June, 2015.

[34] Y. Kapoor, I.N. Alves, K. Ramachandran, et al., Hydrate formation tendency of lower tertiray fluid (LTF) production, OTC-25683-MS, in: Offshore Technology Conference, Houston, Texas, 4–7 May, 2015.

[35] J. Creek, S. Subramanian, D. Estanga, New method for managing hydrates in deepwater tiebacks OTC-22017-MS, in: Offshore Technology Conference, Houston, Texas, 2–5 May, 2011.

[36] G. Bhatnagar, D.L. Crosby, J.C. Guillory, et al., Advanced hydrate management strategies in deepwater production systems, in: Proceedings of the 9th International Conference on Gas Hydrates, Denver, Colorado, 25–30 June, 2017.

[37] J. Lee, Design and installation of deepwater petroleum pipelines, 2003. <https://cupdf.com/document/52394013-design-and-installation-of-deepwater-petroleum-pipelines.html>.

[38] E. Bell, Y. Lu, N. Daraboina, Thermal methods in flow assurance: a review, Journal of Natural Gas Science and Engineering 88 (2021) 103798.

[39] Y. Bai, Q. Bai, Chapter 19: heat transfer and thermal insulation, Subsea Pipelines and Risers, Elsevier, Amsterdam, 2005, pp. 317–356.

[40] G.E. Harrison, M.S. Harrison, D.A.S. Bruton, King flowlines - thermal Expansion design and implementation, OTC-15310-MS, in: Offshore Technology Conference, Houston, Texas, 5–8 May, 2003.

[41] T. Golczynski, Deepstar case study: flowline burial for insulation, Journal of the Australian Petroleum Production & Exploration Association 50 (2010) 684. -684.

[42] R.F. Roth, Direct electrical heating of flowlines - Guide to uses and benefits, OTC-22631-MS, in: Offshore Technology Conference Brasil, Rio de Janeiro, Brazil, 4–6 October, 2011.

[43] A. Nysveen, H. Kulbotten, J.K. Lervik, et al., Direct electrical heating of subsea pipelines—Technology development and operating experience, IEEE Transactions on Industry Applications 43 (2007) 118–129.

[44] J.K. Lervik, H. Kulbotten, A. Nysveen, Flow assurance by electrical heating of long pipelines, in: Proceedings of The 23rd International Offshore and Polar Engineering Conference, Anchorage, Alaska, June 30–July 5, 2013.

[45] P. McDermott, R. Sathananthan, Active heating for life of field flow assurance, OTC-25107-MS, in: Offshore Technology Conference, Houston, Texas, 5–8 May, 2014.

[46] B. Ansart, A. Marret, T. Parenteau, et al., Technical and economical comparison of subsea active heating technologies, OTC-24711-MS, in: Offshore Technology Conference-Asia, Kuala Lumpur, Malaysia, 25–28 March, 2014.

[47] C. Tzotzi, T. Parenteau, D. Kaye, et al., Safe hydrate plug dissociation in active heating flowlines and risers – full scale test, OTC-27051-MS, in: Offshore Technology Conference, Houston, Texas, 2–5 May, 2016.

[48] F. Nebell, H. de Naurois, T. Parenteau, et al., Flow assurance modeling using an electrical trace heated pipe-in-pipe: From qualification to offshore testing, OTC-24060-MS, in: Offshore Technology Conference, Houston, Texas, 6–9 May, 2013.

[49] T. Palermo, E.D. Sloan, Artificial and natural inhibition of hydrates, in: E.D. Sloan, C.A. Koh, A. Sum (Eds.), Natural Gas Hydrates in Flow Assurance, Elsevier, 2010, pp. 87–104.

[50] J. Carroll, Natural Gas Hydrates - A Guide for Engineers, Elsevier, 2020.

[51] L.M. Frostman, V. Thieu, D.L. Crosby, et al., Low-dosage hydrate inhibitors (LDHIs): reducing costs in existing systems and designing for the future, SPE-80269-MS, in: International Symposium on Oilfield Chemistry, Houston, Texas, 5–7 February, 2003.

[52] A.P. Mehta, P.B. Hebert, E.R. Cadena, et al., Fulfilling the promise of low-dosage hydrate inhibitors: journey from academic curiosity to successful field implementation, SPE Production & Facilities 18 (2003) 73–79.

[53] M.A. Kelland, History of the development of low dosage hydrate inhibitors, Energy & Fuels 20 (2006) 825–847.

[54] E.D. Sloan, Method for controlling clathrate hydrates in fluid systems. Patent WO 94/12761, 1994.

[55] J.P. Lederhos, J.P. Long, A. Sum, et al., Effective kinetic inhibitors for natural gas hydrates, Chemical Engineering Science 51 (1996) 1221–1229.

[56] A. Singh, A. Suri, A review on gas hydrates and kinetic hydrate inhibitors based on acrylamides, Journal of Natural Gas Science and Engineering 83 (2020) 103539.

[57] M. Eaton, J. Lachance, E.D. Sloan, Kinetic hydrate inhibitors performance, in: E.D. Sloan, C.A. Koh, A. Sum (Eds.), Natural Gas Hydrates in Flow Assurance, Elsevier, 2010, pp. 105–143.

[58] J. Jardine, S. Jackson, J. Longfield, et al., History and evolution of the product selection criteria for selecting kinetic hydrate inhibitors, IPTC-18164-MS, in: International Petroleum Technology Conference, Kuala Lumpur, Malaysia, 10–12 December, 2014.

[59] S.P. Kang, J.Y. Shin, J.S. Lim, et al., Experimental measurement of the induction time of natural gas hydrate and its prediction with polymeric kinetic inhibitor, Chemical Engineering Science 116 (2014) 817–823.

[60] V.W.S. Lim, P.J. Metaxas, P.L. Stanwix, et al., Gas hydrate formation probability and growth rate as a function of kinetic hydrate inhbiitor (KHI) concentration, Chemical Engineering Journal 388 (2020) 124177.

[61] V.W.S. Lim, P.J. Metaxas, M.L. Johns, et al., The delay of gas hydrate formation by kinetic inhibitors, Chemical Engineering Journal 411 (2021) 128478.

[62] P. Webber, N. Morales, A. Madden, Development and implementation of a high temperature and salinity tolerant kinetic hydrate inhibitor to replace an existing, ten-year continuous application, OTC-25336-MS, Offshore Technology Conference, May 5–8, 2014, Houston, Texas, USA.

[63] L.W. Clark, J. Anderson, Development of effective combined hydrate inhibitor/corrosion inhibitor (KHI/CI) products, in: Proceedings of the 17th International Oil Field Chemistry Symposium, Geilo, Norway, 19–22 March, 2006.

[64] S. Hu, C.A. Koh, CH4/C2H6 gas hydrate interparticle interactions in the presence of antiagglomerants and salinity, Fuel 269 (2020) 117208.

[65] E.P. Brown, D. Turner, G. Grasso, et al., Effect of wax/anti-agglomerant interactions on hydrate depositing systems, Fuel 264 (2020) 116573.

[66] Z.M. Aman, L.E. Dieker, G. Aspenes, et al., Influence of model oil with surfactants and amphiphilic polymers on cyclopentane hydrate adhesion forces, Energy & Fuels 24 (2010) 5441–5445.

[67] U.C. Klomp, V.C. Kruka, R. Reijnhart, A method for inhibiting the plugging of conduits by gas hydrates, Patent WO 95/17579, 1995.

[68] M.A. Kelland, T.M. Svartaas, J. Ovsthus, et al., Studies on some zwitterionic surfactant gas hydrate anti-agglomerants, Chemical Engineering Science 61 (2006) 4048–4059.

[69] Z. Huo, E. Freer, M. Lamar, et al., Hydrate plug prevention by anti-agglomeration, Chemical Engineering Science 56 (2001) 4979–4991.

[70] M.A. Kelland, T.M. Svartaas, L.D. Andersen, Gas hydrate anti-agglomerant properties of polypropoxylates and some other demulsifiers, Journal of Petroleum Science and Engineering 64 (2009) 1–10.

[71] J.-H. Sa, A. Melchuna, X. Zhang, et al., Investigating the effectiveness of anti-agglomerants in gas hydrates and iceformation, Fuel 255 (2019) 115841.

[72] B.Z. Peng, J. Chen, C.Y. Sun, et al., Flow characteristics and morphology of hydrate slurry formed from (natural gas + diesel oil/condensate oil + water) system containing anti-agglomerant, Chemical Engineering Science 84 (2012) 333–344.

[73] S. Dong, A. Firoozabadi, Hydrate anti-agglomeration and synergy effect in normal octane at varying water cuts and salt concentrations, Journal of Chemical Thermodynamics 117 (2018) 214–222.

[74] A. Sinquin, T. Palermo, Y. Peysson, Rheological and flow properties of gas hydrates suspensions, Oil & Gas Science and Technology 59 (2004) 41–57.

[75] A. Fidel-Dufour, F. Gruy, J.M. Herri, Rheology of methane hydrate slurries during their crystallization in awater in dodecane emulsion under flowing, Chemical Engineering Science 61 (2006) 505–515.

[76] A.A. Majid, D.T. Wu, C.A. Koh, A perspective on rheological studies of gas hydrate slurry properties, Engineering 4 (2018) 321–329.

[77] S. Sakurai, B. Hoskin, J. Choi, et al., Investigating hydrate formation rate and the viscosity of hydrate slurries in water-dominant flow: flowloop experiments and modelling, Fuel 292 (2021) 120193.

[78] K.L. Yan, C.Y. Sun, J. Chen, et al., Flow characteristics and rheological properties of natural gas hydrate slurry in the presence of anti-agglomerant in a flow loop apparatus, Chemical Engineering Science 106 (2014) 99–108.

[79] P.A. Webber, Fundamental understanding on the effects of anti-agglomerates towards overboard water quality, OTC-20841-MS, in: OTC Offshore Technology Conference, Houston, Texas, 3–6 May, 2010.

[80] J. Tian, C. Walker, Non-emulsifying, new anti-agglomerant developments, OTC-23144-MS, in: Offshore Technology Conference, Houston, Texas, April 30–May 3, 2012.

[81] H. Delroisse, G. Barreto, J.-P. Torre et al., Evaluation of the performance of a new biodegradable AA-LDHI in cyclopentane hydrate and CH4/C3H8 gas hydrate

systems, SPE-195054-MS, in: SPE Middle East Oil and Gas Show and Conference, Manama, Bahrain, 18–21 March, 2019.

[82] D.S. Monteiro, L.K. Vo, P. Prince, et al., Biodegradable anti-agglomerant chemistry for hydrate plug prevention in various production conditions, OTC-29660-MS, in: Offshore Technology Conference, Houston, Texas, 6–9 May, 2019.

[83] P. Webber, P. Conrad, D. Lewis, et al., Continuous anti-agglomerant LDHI application for deepwater subsea tieback: A study on water quality and low water cut scenarios, OTC-23397-MS, in: Offshore Technology Conference, Houston, Texas, April 30–May 3, 2012.

[84] S. Bodnar, Z.T. Ward, A. Steinocher, et al., Challenges of hydrate risk management at low water cuts using anti-agglomerants, SPE-207668-MS, in: Abu Dhabi International Petroleum Exhibition & Conference, Abu Dhabi, UAE, November 15–18, 2021.

[85] S. Mokhatab, W.A. Poe, J.Y. Mak, Chapter 9: natural gas dehydration and mercaptans removal, Handbook of Natural Gas Transmission and Processing, fourth ed., Gulf Professional Publishing, Massachusetts, 2019, pp. 307–348.

[86] C.A. Koh, J. Creek, Safety in hydrate plug removal, in: E.D. Sloan, C.A. Koh, A. Sum (Eds.), Natural Gas Hydrates in Flow Assurance, Elsevier, 2010, pp. 37–48.

[87] K.A. Easklul, G. Fung, G. Harrison, et al., Active heating for flow assurance control in deepwater flowlines, OTC-15188-MS, in: Offshore Technology Conference, Houston, Texas, 5–8 May, 2003.

[88] A. Marret, P. Espinasse, T. Parenteau, et al., The electrically trace heated blanket: a new concept for efficient subsea flowline unplugging, OTC-27043-MS, in: Offshore Technology Conference, Houston, Texas, 2–5 May, 2016.

[89] D. Peters, M.S. Selim, E.D. Sloan, Hydrate dissociation in pipelines by two-sided depressurization: experiment and model, Annals of the New York Academy of Sciences 912 (2000) 304–313.

[90] P. Bollavaram, E.D. Sloan, Hydrate plug dissociation by method of depressurization, OTC-15257-MS, in: Offshore Technology Conference, Houston, Texas, 5–8 May, 2003.

[91] S.R. Davies, M.S. Selim, E.D. Sloan, et al., Hydrate plug dissociation, AIChE Journal 52 (2006) 4016–4027.

[92] J.A. Boxall, S.H. Ng, Z.M. Aman, et al., Hydrate plug dissociation via active heating: uniform heating and a simple predictive model, Energy & Fuels 30 (2016) 9275–9284.

[93] L.A. Dykhno, G.J. Hatton, Bullheading optimization for hydrate management in deepwater wells, OTC-17587-MS, in: Offshore Technology Conference, Houston, Texas, 2–5 May, 2005.

[94] A.S. Kurup, O. Hernandez, T. Idstein, et al., Pushing conventional boundaries of hydrate management in a dry tree facility, OTC-27780-MS, in: Offshore Technology Conference, Houston, Texas, 1–4 May, 2017.

[95] S.R. Davies, J.A. Boxall, C.A. Koh, et al., Predicting hydrate-plug formation in a subsea tieback, SPE Production & Operations 24 (2009) 573–578.

[96] T.B. Charlton, L.E. Zerpa, C.A. Koh, et al., Predicting hydrate blockage formation in gas-dominant systems, OTC-28311-MS, in: Offshore Technology Conference Asia, Kuala Lumpur, Malaysia, 20–23 March, 2018.

[97] F. Papot, C. Alexandre, M. Abdulssalam, et al., Remediation of hydrate plug in gas export line – a West Africa deep water case, OTC-29946-MS, in: Offshore Technology Conference Brasil, Rio de Janeiro, Brazil, 29–31 October, 2019.

CHAPTER

Paraffin management

2

Mike Newberry and David W. Jennings
Baker Hughes, Sugar Land, TX, United States

Chapter outline

2.1 History of paraffin management developments

Webster's dictionary defines paraffin as "a waxy crystalline flammable substance obtained especially from distillates of wood, coal, petroleum or shale oil that is a complex mixture of hydrocarbons and is used chiefly in coating and sealing,

Flow Assurance. DOI: https://doi.org/10.1016/B978-0-12-822010-8.00003-9

in candles, in rubber compounding, and in pharmaceuticals and cosmetics." Petroleum is the main supply source for paraffin wax. Hence, paraffin wax is a valuable by-product from petroleum refining that is used in many consumer goods. Paraffin can cause problems in the production of petroleum though. Some production operations have no issues while others have problems that range the gamut from minor operational hindrances to severely challenging the ability to produce. The severity depends on the specific, production system and its respective temperature profile, crude oil chemistry, and the management programs implemented.

In the early days of crude oil production, paraffin deposition was little more than a nuisance. Accordingly, not much attention was initially given to wax deposition and little was understood about it; beyond being aware of the problems it caused and physical characteristics of the deposits. In 1932, Reistle [1] stated that "the deposits of crude wax or paraffin that accumulate consist of very small wax crystals that tend to agglomerate and form granular particles of wax about the size of ordinary table salt. Deposited paraffin also contains gums, resins, asphaltic material, crude oil, sand silt, and in many instances water, in addition to wax crystals. The deposit varies in consistency from a mushy liquid to a firm hard wax, depending primarily upon the amount of oil present" [1]. It was not until the 1960s and the 1970s that more in-depth paraffin deposition studies were undertaken, and technical articles began to appear in the literature. These were driven by new challenges in the petroleum industry and better realization of the economic incentives that existed for improved understanding and management of paraffin problems. For example, the installation of the Trans Alaska Pipeline System contributed to the undertaking of some of these studies. During this period, a variety of test methods were developed to study the deposition and characterize wax. And in the next ensuing decade, it is fair to say that a good understanding of the mechanism of wax deposition was developed [2,3]. It is also fair to say that further advancements in expanding and applying these learnings toward the development of deposition modeling programs were driven by the petroleum industry's move into deepwater production in the 1990s. The cold seafloor temperatures in deepwater (~4°C/40°F) require that deepwater operators with long subsea flowlines need to assess paraffin deposition concerns and likely implement management programs. During this time the Deepstar consortium was formed to support operator needs in addressing deepwater development issues, including many related to flow assurance. The magnitude of concern over paraffin deposition in deepwater is illustrated in a Deepstar survey during this period that reported 51 severe plugs occurred in the Gulf of Mexico (GOM) due to paraffin deposition between 1992 and 2002. Since the industry continued the move into deepwater, certainly improvements in insulation (more related to hydrates) and pigging followed. The concept of using dual flowlines for round-trip pigging for paraffin deposition control is one mainstay.

What was known early on was that paraffin deposits would melt at elevated temperatures. Industry exploited this characteristic and developed several thermal techniques to address deposition problems. These included hot oiling, hot watering, and heat tracing. Physical intervention was another obvious approach and a number of mechanical means were also developed including cutting and pigging to remove deposits. Chemical applications were primarily limited to solvent treatments generally consisting of readily available diesel. As such, management of deposition at the time was mostly limited to remediating or removing deposition after the fact. Remediation as a paraffin management strategy can be effective and, in some instances, may be the most practical approach. However, in many instances, preventing (or reducing) deposition has significant economic incentives and may be more optimal. Similarly, incentives exist to improve handling properties of waxy high pour point crude oils.

In the 1960s and 1970s when more attention was starting to be given to paraffin problems, it was the time when the first nonsolvent chemical paraffin treatments were being developed which introduced new methods to reduce or prevent deposition as management strategies. In fact, in 1975, one of this chapter's authors joined the new Paraffin and Asphaltene Group of Petrolite (the petroleum industry's first oilfield specialty chemical company) with the goal to develop new chemical treatment products and expand paraffin technical service applications. At that time, only one nonsolvent paraffin chemical treatment product (a surfactant-based product) existed in the company's product line. In the next decade, some of the industry's first nonsolvent paraffin chemical treatments were developed. In the ensuing decades, hundreds of products have been developed and applied globally by Petrolite and its successor. Similarly, other chemical providers have since developed many products as well. The result of these efforts are nonsolvent chemical treatment options now exist to augment thermal and mechanical control methods.

The first nonsolvent chemical treatments were primarily dispersant/surfactant-type products for remediation of wax deposition. However, some of these products could be applied continuously or with regular periodic backside batch treatment allowing some control or mitigation of deposition without halting production to perform remedial operations. Later, the development of paraffin inhibitor chemistries closely followed allowing the ability to better inhibit or reduce wax deposition. These significantly improved the ability to control paraffin deposition.

Paraffin inhibitors are chemistries that incorporate into wax crystal growth to either: (1) reduce paraffin deposition or (2) improve flow characteristics of waxy crude oils. When the paraffin inhibitor is used for reducing pour points and improving the flow characteristics of high-wax crude oils, it is called a pour point depressant (PPD). In this chapter the term paraffin inhibitor will be primarily reserved to designate products used to control deposition. The application of paraffin inhibitors and PPDs has resulted in countless increased production of crude oil through the decades. Through the ensuing decades since the first products were developed, incremental advancements and/or better customization of products for specific applications have been and continue to be made.

2.2 Crude oil and paraffin chemistry

Crude oils are some of the most chemically complex fluid mixtures known to exist. Literally, hundreds of thousands of different chemical species can be present in a crude oil. Further each crude oil is unique as its chemical makeup is determined by the initial hydrocarbon source and the maturation process that it endures over millions of years within the respective reservoirs. Crude oils can vary significantly ranging from extra heavy bitumen to near water-white condensate fluids. In general, chemical complexity increases as crude oils become heavier, more asphaltic. The ability to fully analyze and characterize crude oils similarly becomes more challenging as complexity increases. Fig. 2.1 illustrates the unique nature of crude oils showing variation even within the same field producing "less-chemically complex", condensate and low-gravity conventional crude oil. Each bottle shown is a sample from a different well in the field.

Fortunately for understanding and managing paraffin issues, crude oil complexity is not a major issue. Paraffin chemistries are some of the simpler structures within crude oil and analysis methods exist to provide adequate characterization for most paraffin management purposes—like performing wax deposition modeling. That is not to say that improvements in the methods and management systems are not still desired.

Before discussing paraffin chemistry, it is useful to take a simplified look at overall crude oil chemistry. The most widely used simple method to characterize crude oils is with saturates, aromatics, resins, and asphaltenes (SARA) analyses [4]. In SARA analyses, physical and chromatographic separations are used to characterize crude oils as containing four major fractions: **S**aturates, **A**romatics, **R**esins, and **A**sphaltenes. The most chemically complex ones are the asphaltene and resin fractions. These fractions are not defined chemically but rather by the separation methods used to isolate and quantify these species. Chemical similarities exist within the distribution of species in these fractions though. Discussion of which will be left to other sources. However, note these species are highly aromatic but contain both saturate and aromatic chemical functionality and a large percentage of the N, S, and O polar species in a crude oil. In addition, these species are solid when isolated at ambient temperature. Hence, these components are largely dissolved in the remaining saturate and aromatic fractions. The saturate

FIGURE 2.1

Example of potential variation in crude oils from a single field.

and aromatic fractions are also the "solvent source" for dissolving the paraffin species present in the crude oil.

In addition to containing solvent chemistries, the saturate fraction also contains the paraffin species. Saturates, of course, meaning molecular species saturated with hydrogen. The aromatics are the molecular species containing aromatic benzene rings that are not fractionated and included in the resin and asphaltene fractions.

The paraffin wax species present in crude oils include *n*-alkanes, *iso*-alkanes, alkyl cyclic compounds, and alkyl aromatics. In most crude oils, *n*-alkanes are the predominant wax species. The amount of non *n*-alkanes present depends on the specific crude oil. Generally, this amount is small, but for some crude oils, it can be significant. *Iso*-alkanes are the second most prevalent species followed by alkyl cyclic structures. The *iso*-alkane structures present usually have only few branches that are short (methyl, ethyl). In the alkyl cyclic structures, mono-ring structures are dominant. The alkyl aromatics generally are present only in very small amounts.

N-alkanes have a chemical formula C_nH_{2n+2}. This series displays different ambient physical states depending upon the length of the carbon chain:

- C_1-C_4: Gas
- C_5-C_{17}: Liquid
- C_{18+}: Solid

Besides being the most prevalent paraffin in most crude oils, the *n*-alkanes for a given carbon number have higher melting points and corresponding lower solubility versus the other waxes, which makes them more prone to deposit. The length of the carbon chains and the melting point can reach elevated levels. The carbon chain length greatly affects the melting point and solubility; which in turn, affects propensity to precipitate and deposit. Melting point data for various *n*-alkane chain lengths are listed in Table 2.1.

As the carbon number increases, the solubility decreases dramatically. As a rough rule of thumb, the solubility decreases by a factor of 10 for each increase in 10 carbons in the chain length. This is shown in Fig. 2.2. The differences in

Table 2.1 *N*-Alkane melting points.

Compound	Formula	Melting point (°C)
Eicosane	$C_{20}H_{42}$	36.5
Triacontane	$C_{30}H_{62}$	65.9
Tetracontane	$C_{40}H_{82}$	82
Pentacontane	$C_{50}H_{102}$	94
Hexacontane	$C_{60}H_{122}$	98
Heptacontane	$C_{70}H_{142}$	105
Hectane	$C_{100}H_{202}$	115

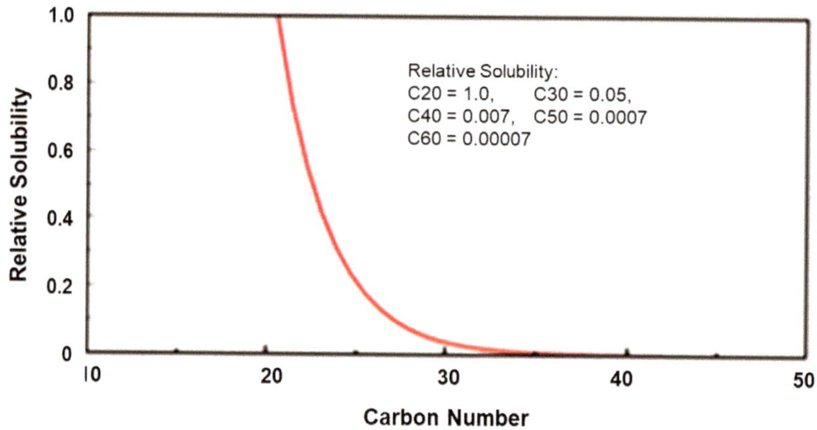

FIGURE 2.2

Relative solubility of *n*-alkanes.

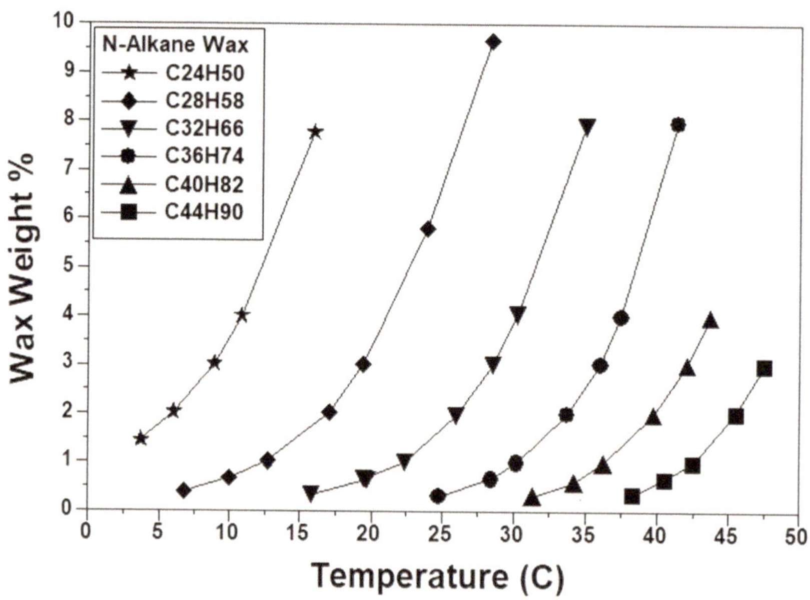

FIGURE 2.3

Solubility of *n*-alkanes in toluene.

solubility are further illustrated in Fig. 2.3, which shows solubility data of *n*-alkanes ranging from chain lengths C_{24} to C_{44} in toluene [5]. Note the higher temperatures required to "hold in-solution" an equal weight of the higher carbon number *n*-alkanes. For example in comparing C_{44} versus C_{24} data, the solvent needs to be approximately 40°C hotter to fully dissolve 2 wt.% wax. Fig. 2.3 also shows the large influence in temperature on solubility. Particularly, note the very low solubility of waxes at low temperatures even in a good solvent like toluene.

2.3 Paraffin analysis and crude oil characterization

2.3.1 Paraffin analysis methods

As mentioned previously, paraffin wax is a mixture of normal, branched and cyclo-alkanes. Methods have been developed to measure the wax content of crude oils and paraffin deposits. The first methods developed were "wet" extractions of paraffin wax from crude oil. One such method is the Universal Oil Products Method 46 [6]. This method uses an alkane solvent such as heptane to dilute the crude oil, precipitate asphaltenes and dissolve all the other components. An activated clay is added to the heptane solution to remove the asphaltenes and asphaltic resins. These components adsorb to the clay and the color is removed. Enough clay needs to be added to render the resulting solution colorless. The clay is filtered out of the solution and discarded. The heptane solution is then heated to evaporate the solution down to about 50 mL in volume. This remaining solution is then loaded with an excess of methyl ethyl ketone to precipitate the wax from solution. The solution is cooled to crystalize as much wax as possible and then filtered to remove and collect the wax. The filtered wax is dried and weighed to determine the percent wax in the crude oil. Note that wet extraction values can be higher than other methods that only measure *n*-alkane content as branched and cyclo-paraffin species are captured along with the *n*-alkanes.

A second main method used to determine paraffin content is high-temperature gas chromatography (HTGC). One such method is described in American Society for Testing and Materials (ASTM) D5442 [7]. This is a quantitative measurement of the higher boiling *n*-alkanes with the *n*-alkane weight percentage determined by calibrating the HTGC method using internal standards. An example HTGC for a crude oil is shown in Fig. 2.4.

A challenge for this method is obtaining complete separation and resolution of species through the HTGC column allowing all the wax to be quantified. Analyzing samples with hundreds of thousands of species that are present in crude oils is not an easy feat. HTGC analyses often do not quantify all the wax due to resolution issues. The shifting baseline shown in the chromatogram in Fig. 2.4 is indicative of some resolution problems. The benefit of HTGC analyses though is the distribution of individual *n*-alkanes is obtained, not just a total wax-content number. If deposition modeling is to be performed using a molecular approach for the thermodynamic calculations, the wax distribution is required. Additionally, knowledge of the highest molecular weight waxes present in the crude oil is valuable as these waxes strongly influence the wax deposition tendencies of crude oils.

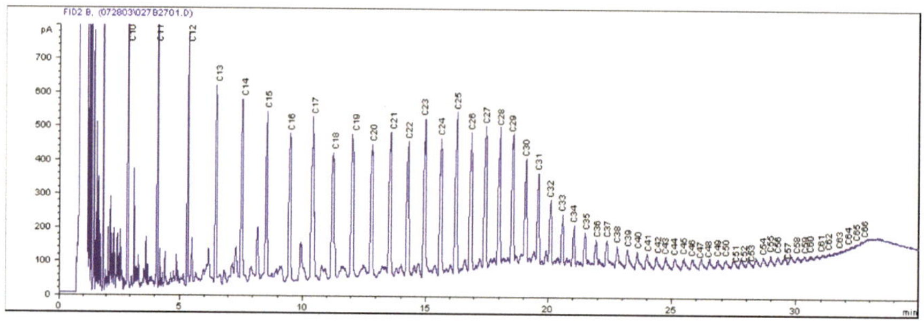

FIGURE 2.4

Example HTGC of crude oil.

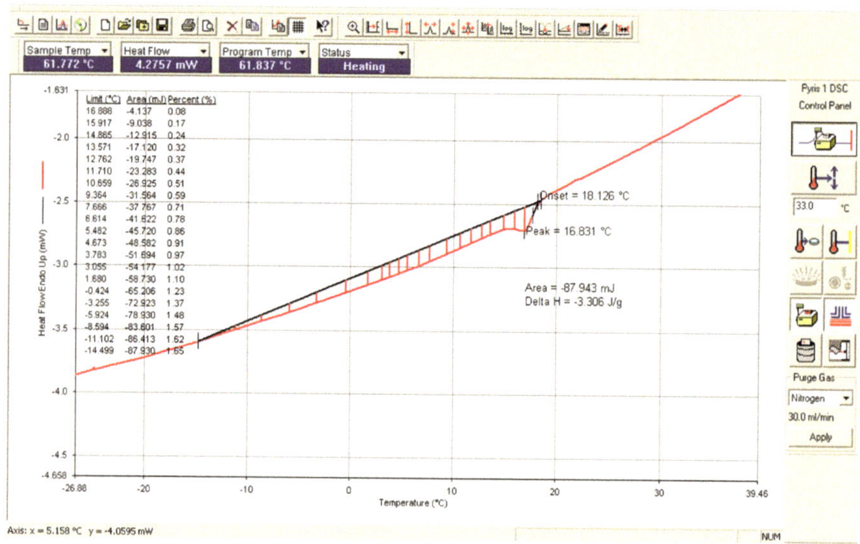

FIGURE 2.5

Example DSC wax content measurement. *DSC*, Differential scanning calorimetry.

The third method to determine wax content was developed using differential scanning calorimetry (DSC). Wax crystallization is an exothermic process that releases heat as wax crystals are formed. In DSC measurements, crude oil samples are cooled to low temperatures to precipitate the wax in the crude oil. The heat flow generated from the enthalpy of crystallization is measured and used to determine the mass of wax present in the sample. And with knowledge of the total sample weight, quantification of the wax content is obtained. Fig. 2.5 shows a DSC crude oil analysis. The cross-hashed area is the total heat measured for the measurement. In addition to measuring the total wax content, DSC can also be used to measure crude oil wax appearance temperature (WAT) - discussed below.

2.3.2 Crude oil characterization

The WAT, pour point, and viscosity profile are three measurements that are used to characterize crude oils to help assess paraffin flow assurance concerns. The WAT and pour point can be considered as physical properties of the crude oil.

2.3.2.1 Wax appearance temperature

The WAT is the temperature at which the first wax crystal precipitates from a crude oil. The WAT is also called "cloud point" since with cooling translucent fluids like diesel, the fluids become hazy or cloudy shortly after reaching the WAT. The WAT is a thermodynamic phase equilibrium property of the crude oil. It is a vitally important property of crude oils for deposition. Knowing the WAT is the first step in determining the wax deposition tendencies in a production system. Wax deposition will not occur at temperatures above the WAT. Therefore the WAT (along with the system temperature profile) dictates where in a production system wax deposition will start to occur. Similarly, if a paraffin management strategy for avoiding deposition through thermal management is planned, it will fix production rates, insulation, and/or heating required in flowlines and well tubulars. As such, considerable discussion on WAT measurement is given next.

Note that a distinction exists between the true thermodynamic WAT and an experimentally measured WAT value. Experimental measurements have a bias to yield values slightly below the true thermodynamic WAT, due to the kinetics involved in the crystallization process and in the detection limits of the individual measurement methods. Each measurement method requires a certain mass of wax to precipitate to be detected beyond the point where the absolute first crystallization occurs. Also note that the discussion here is on dead crude oil measurements. In production systems under pressure, live crude oils still contain additional light end gas species, which provide additional wax solvency. Therefore their WATs are typically slightly lower than the dead crude oil.

DeepStar conducted a rather rigorous investigation of WAT measurements [8]. References to their work will be cited in the following section. Seven different WAT measurement methods are discussed.

1. Visual method ASTM D2500 [9] or Institute Of Petroleum (IP) 219 [10]
2. Viscosity/temperature profile
3. DSC
4. Cross-polarized microscopy (CPM)
5. Filter plugging (FP)
6. Fourier transform infrared spectroscopy (FTIR)
7. Cold finger WAT measurement

"Method 1" was initially applied only to clear petroleum products but has been used for crude oils that were not opaque, like gas condensates and light crude oils. Basically, the crude oil/condensate is cooled at a controlled rate until wax crystals appear causing the fluids to become cloudy. This was an easy test

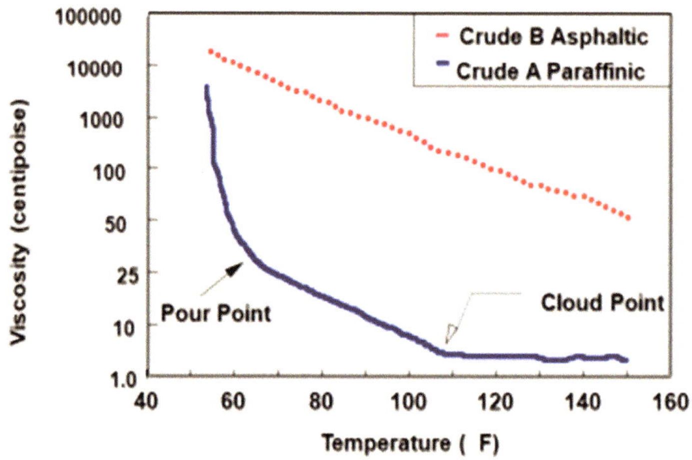

FIGURE 2.6

Viscosity-temperature profiles for crude oils.

for petroleum products like distillate fuels since these have a boiling point cut-off that limits the smaller amounts of high carbon number paraffin. Products such as diesel fuels display a sharp cloud point where large volumes of wax rapidly appear within a relatively narrow temperature range. For petroleum products like gas condensates, the WAT is more difficult to determine. With a wider and higher carbon number range for the wax species, paraffin precipitates more slowly over a wider temperature range. For light crude oil samples, it is even more difficult to get a reliable value. And with dark crude oils it is even not possible to use the visual method.

"Method 2" uses the viscosity-temperature profile of the crude oil to find an inflection point of viscosity increase at the WAT. Prior to the WAT, the crude oil behaves as a Newtonian fluid. At and below the WAT the crude oil can behave as a thixotropic fluid. The viscosity increases at an increasing rate as the temperature decreases [11]. An example of this behavior is shown in Fig. 2.6.

Fig. 2.6 shows viscosity profiles for an asphaltic and paraffinic crude oil. The blue profile for the paraffinic crude oil is distinct because the viscosity is low at elevated temperatures. As the crude oil cools, the viscosity increases only slightly (or is relatively flat on semilog plot like Fig. 2.6) until the WAT is reached. At the WAT, wax crystals begin to form and a change in the rate of viscosity increase occurs due to the solids in the crude oil. With further cooling in moderate and high wax crude oils, like the paraffinic crude oil shown in Fig. 2.6, wax crystal networks can form and become so intensive that the crude oil gels. This is the point where the viscosity curves make a sudden upward right-hand turn with increase in viscosity. Section 2.5 discusses this in more detail. The red trace shows an asphaltic oil. For this crude oil the viscosity is much higher at elevated temperatures due to the higher asphaltene content and lower API gravity. As the crude oil cools, the viscosity increases, but there is no inflection point from the formation of wax crystals because low API gravity oils generally lack appreciable paraffin content. The main drawback to this WAT detection method is the sensitivity. A sufficient amount of wax needs to precipitate (not just a few crystals) to cause a notable change in the rate of viscosity increase. It is also subject to different experimental factors: equipment sensitivity, cooling rate, shear rate, and operator interpretation. For low wax content crude oils, it is particularly challenging as the viscosity rate change will be very gradual and hard to detect even with viscometers and rheometers with high sensitivity.

"Method 3" uses DSC [12,13]. The first onset of heat flow from wax crystallization can be related to the WAT. In Fig. 2.5 the first onset (and hence WAT) occurs at 18.126°C. DSC is a considerably more sophisticated method with better accuracy than Methods 1 and 2. Variations in measurements can still occur depending on the equipment used, cooling rates, sample size, wax content of the crude oil, and data processing. DeepStar recommended using a 90-mg sealed sample preheated to 82°C and cooling at a rate below 0.5°C/minute, but initial screening attempts can be run at 2°C. Differential calorimeters with high-pressure cells can also be used to measure WATs of live crude oils [14].

"Method 4" uses a cross-polarized microscope to view the formation of wax crystals as they form when crude oil is cooled. The wax crystals rotate the polarized light and appear white. The crude oil does not rotate the polarized light and thus appears as a dark background. The operator can view the crude oil as it is cooled and see the first crystals appear. The presence of any inorganic solids needs to be discounted. If present, these should be observable even when the crude oil reaches its maximum temperature. Extensive operational experience indicates that all the wax should be melted at 82°C.

An example of a CPM WAT measurement is shown in Fig. 2.7. This is a CPM WAT measurement run at a cooling rate of 0.5°C/minute. At 32.2°C the image is completely dark and no wax crystals are visible. At 31.7°C the formation of a few small wax crystals is observable (at least in the original full-scale photo). At 31.2°C, some increased crystallization has occurred, and wax crystals are more visible. In the bottom row pictures at temperatures of 30.7°C and 30.2°C, further increased wax crystallization is apparent with the growth more prevalently driven from the right-hand side of the photo. Hence, the WAT reported for this measurement would be ~31.7°C. But note with a slower cooling rate and more sensitive detection of smaller crystal sizes, a slightly higher WAT closer to the true thermodynamic WAT would be reported.

DeepStar recommended heating the crude oil to at least 8°C above the anticipated WAT and cooling at 0.2°C/minute for CPM measurements, but initial screening could be run at 1°C/minute.

"Method 5" measures the increase in differential pressure across a filter. Preheated and hot-filtered crude oil is passed through a capillary and across a filter. As the crude oil is cooled, the differential pressure is monitored until an

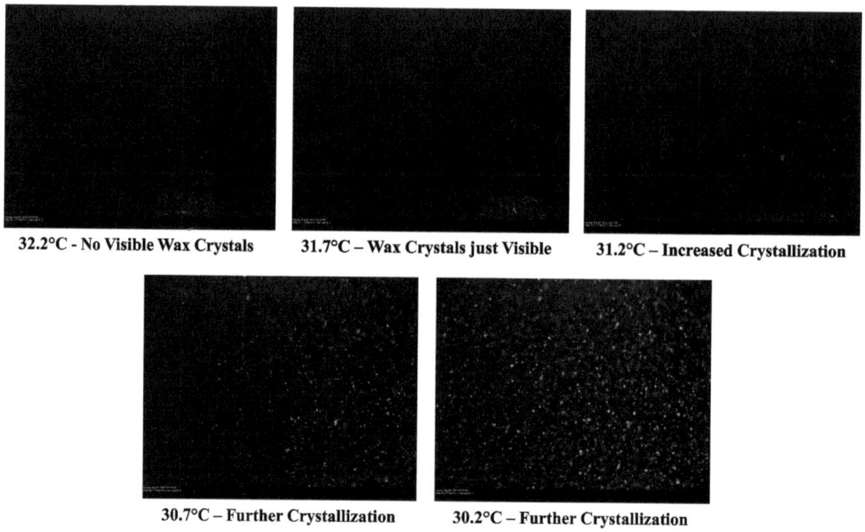

FIGURE 2.7

Cross-polarized microscopy for WAT measurement. *WAT*, Wax appearance temperature.

Table 2.2 WAT measurements of synthetic wax solutions.

Technique	Standard A (°C)	Standard B (°C)
DSC	9.4	15.0, 27.8, 43.3
CPM	6.7	52.2
FP	15.6	17.8, 30.0
FTIR	12.2	18.9

increase is observed. DeepStar recommended a 0.5-μm filter, 0.5 mL/minute flowrate, crude oil preheating to 60°C, and a 0.25°C cooling rate.

"Method 6" uses FTIR to measure wax crystallization by measuring the increase in energy scattering. DeepStar used midinfrared wavenumbers between 4000 and 650 cm^{-1} (wavelengths of 2.5–15 μm) and a cell pathlength of 0.025 mm. The crude oil sample was cooled at a cooling rate of 0.17°C/minute after preheated to 82°C, and the pressure was maintained during loading and testing [8].

In the round-robin testing performed, DeepStar used two synthetic wax solutions to tests the different methods [8]. The first solution Standard A was 5 wt.% (n-C_{24}) in dodecane. The second solution Standard B was a mixture of 2.0 wt.% n-C_{28}, 0.055 wt.% n-C_{30}, 0.038 wt.% n-C_{40}, 0.019 wt.% n-C_{50} in dodecane. The results of the testing are shown in Table 2.2.

From the abovementioned table, it can be observed that the agreement between the various WAT methods was not good for Standard A. The spread was

6.2°C between the methods. Surprisingly, CPM unexpectedly gave the lowest value. However, the kinetics with low molecular weight C_{24} wax and WAT could be in involved. The FP method gave the highest value and this was also suspect because this should be less sensitive than some of the other methods. It should take more wax crystals to see a differential pressure increase across the filter versus the other methods. This phenomenon was associated to supercooling. The values were considerably closer for Standard B for selected wax components. Here, the DSC was able to determine individual WATs for three individual wax components in the mixture. The FP was able to determine two. The FTIR could only pick up one. The CPM had the greatest sensitivity of the methods and showed a WAT of 52.2°C. The DSC missed picking-up signal from the n-C_{50} wax due to its low concentration. As a result, DSC would show a WAT at 43.3°C where the combined wax content between the n-C_{40} and n-C_{50} was high enough to produce a signal. The FP method missed both the n-C_{50} and n-C_{40} waxes and did not show a WAT until the n-C_{30} component began to crystalize around 30°C. The FTIR appeared to be the least sensitive method and did not pick up a signal until the final n-C_{28} wax contributed to the bulk wax coming out of solution. This WAT would then be 18.9°C.

Based on the abovementioned information, it appears that sensitivity of the method becomes very important for WAT measurements for low wax content solutions. The CPM can register a WAT at 0.019 wt.% wax, the DSC at 0.051 wt.% wax, the FP at 0.112 wt.% wax, and finally the FTIR at 2.112 wt.% wax. These numbers would be maximums because not all the wax of a certain carbon number is going to come out of solution at a given temperature.

DeepStar then went on to measure the WATs of 13 crude oil samples with 9 samples run in duplicates. There seemed to be inconsistencies in the data and a conclusion was to assure accuracy, multiple measurement methods may be desired.

"Method 7" is procedure not considered in the DeepStar assessment. It is a method using a series of cold finger tests to measure WAT. In this procedure, cold finger temperatures are progressively set at lower temperatures to find temperatures which will bracket having no deposition and having the start of thin wax deposition. The method is relatively simple, but some care does need to be taken with the procedure to rinse surface oil from the cold finger to ensure material on the surface is deposited wax and not dried oil. To help confirm, material can be analyzed with HTGC analyses. Also, as the temperature is progressively lowered, the amount of deposition should increase from the point where deposition was first observed. This method can be very sensitive; particularly if the tests are performed over a long test duration (period of hours). This helps alleviate kinetic issues that can affect many WAT measurements, especially in measurements with lower wax-content crude oils. Fig. 2.8 provides an example of using cold fingers to measure WATs. These tests were run over $\sim$16 hours and determined that the WAT for the respective crude oil sample was between $\sim$46°C and 44°C.

As discussed earlier, n-alkane solubility decreases dramatically with increasing chain length. As such, the WAT is highly depended on the higher molecular

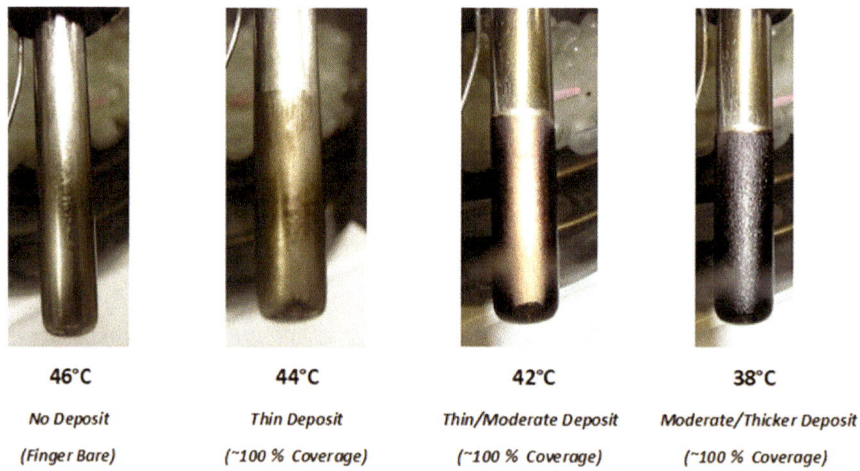

FIGURE 2.8

Example of using cold fingers to measure WAT. *WAT*, Wax appearance temperature.

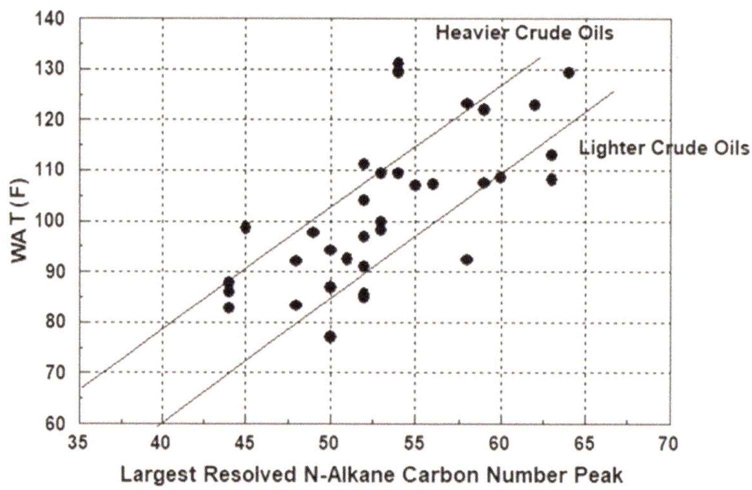

FIGURE 2.9

WAT correlation with highest carbon length *n*-alkanes detected from HTGC analyses from a wide range of different West Texas crude oils. *WAT*, Wax appearance temperature.

weight waxes present in a crude oil. Fig. 2.9 illustrates how the WAT of crude oils can be correlated with the highest molecular weight waxes present [15] or, correspondingly, how WAT measurements of crude oils can be checked for consistency. Shown is a data set from a project involving West Texas crude oils. The crude oils' measured WATs are plotted against their highest carbon number

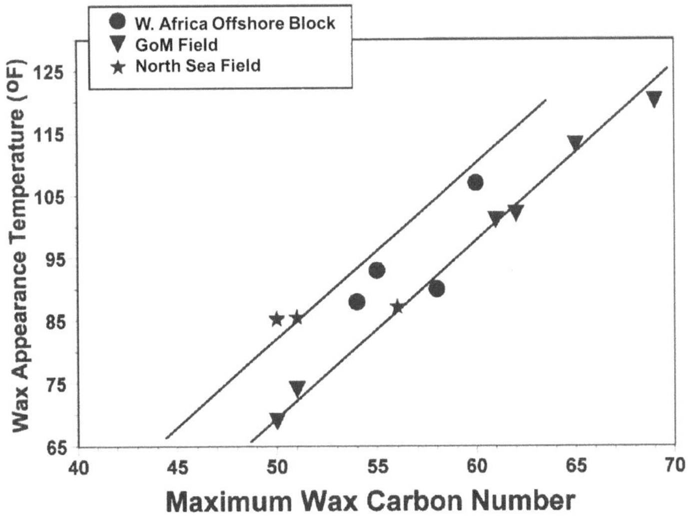

FIGURE 2.10

WAT correlation restricted to crude oils from same field/region. *WAT*, Wax appearance temperature.

n-alkanes detected in HTGC analyses. Note the correlation is dependent on the resolution of the specific HTGC instrument used. In general, the WATs here correlate within a band with the highest carbon number *n*-alkanes detected. If the plot below were restricted to include crude oils from the same field and/or with similar API gravity or asphaltene contents, a narrower band would be expected. This is because the inherent "solvent properties" of the crude oil also play a role. Generally, the heavier crude oil data points lie in the upper portion of the band and lighter crude oil data points in the lower portion. Lighter crude oils provide better solvency due to containing more chemical species that are better wax solvents; thereby keeping the wax solvated to lower temperatures. In addition, their aromatic solvent species will be less "tied-up" solvating resin and asphaltene species. Fig. 2.10 shows an example restricting the crude oils to the same field or geographical region. In this figure the data from the GOM field crude oils lie in a straight line. For the North Sea field and West Africa offshore block crude oils, the crude oils have similar correlation with the exception of one outlier point each. However, the outlier points can be explained though as these crude oils had notable higher saturate and aromatic fractions which lead to the lower WATs versus the trend from the other crude oils.

2.3.2.2 Pour point

The pour point of crude oil is defined as the temperature at which a crude oil gels plus 3°C as determined via a specific test procedure. This procedure was initially designated under ASTM D97 covering petroleum products [16] and later modified to ASTM D5853 for crude oils [17]. The ASTM D5853 method has two different procedures varying in the thermal conditioning procedure used on the crude oil. There is a maximum and minimum method. For the maximum pour measurement, the crude oil is first preconditioned by heating to 60°C, then allowed to cool to room temperature and held for 24 hours. This is to allow equilibrium of wax crystallization in the sample. Afterwards, the maximum pour point is determined by heating the crude oil to 45°C and immediately thereafter cooling to measure the pour point. This provides the highest or "maximum" pour point value for crude oils because all the wax crystals may not dissolve back into solution with the heating to 45°C. Any undissolved wax crystals provide nucleation sites for promoting and accelerating crystal growth. By contrast, the minimum pour point procedure calls for heating the crude oil to 105°C for at least 30 minutes in a pressure vessel. Then cool the pressure vessel at room temperature for exactly 20 minutes and transfer the sample into a pour point tube and begin the pour point evaluation. A modified version of the minimum procedure will often be performed avoiding needing a pressure vessel with simply heating the crude oil to ~70°C instead of 105°C. For most crude oils, this dissolves all the wax crystals into solution and erases the thermal history of the crude oil. If the crude oil has an extremely high pour point (>37°C), then heating to a higher temperature and/or extending the time of heating may be appropriate. Again, note that the exact procedure used can affect measurement results.

The pour point measurement procedures are relatively simple but do require special equipment. The pour point tubes, thermometers, gaskets, and cooling bath temperatures have detailed specifications. Most pour point cooling bathes can accommodate multiple pour point tests to allow multiple measurements to be performed concurrently. The procedure is designed to generate a cooling rate of about 1°C per minute during testing. During the cooling process, the samples are reviewed every 3°C starting at 45°C. The samples start in a 27°C cooling vessel and are moved to a 0°C cooling vessel, if the sample is still liquid at 30°C. When the sample reaches 9°C, it is moved to a −18°C cooling vessel. If the sample is still liquid at −9°C, it moves to a −33°C vessel. Finally, if still liquid at −24°C, it is moved to a −54°C vessel. The sample is determined to be solid if it has no movement when held horizontally for 5 seconds. Care should be taken when evaluating samples at each observation temperature. The tube should only be tilted just until movement is observed and then immediately returned to the cooling vessel. The last temperature at which the crude oil moved prior to solidification is defined as the pour point, that is, 3°C above the gel point. Fig. 2.11 summarizes the test process.

The ASTM pour point procedure is a manual test method with sample handling and assessment of fluidity performed by a person. There are also commercial automatic pour point apparatuses that mechanically handle tilting and assessing fluidity. However, note that these apparatuses are generally single units

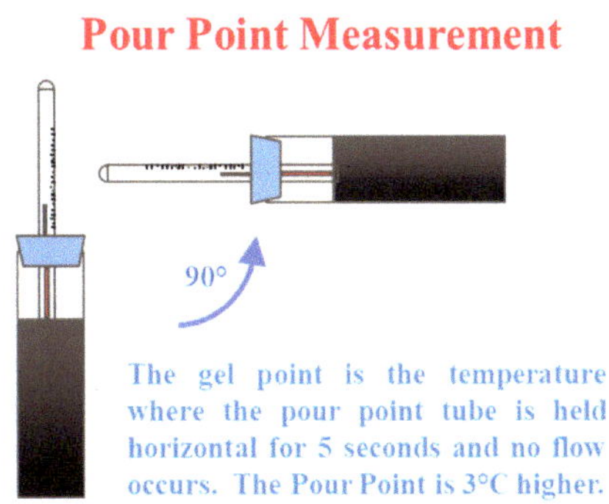

FIGURE 2.11

Pour point measurement procedure.

and are more aggressive when turning samples to mechanically evaluate fluidity. Therefore these measurements are prone to yield slightly lower pour points than manual testing.

Although pour point measurements are highly useful for crude oil characterization, pour point is not a perfect measurement by any means. There are issues in using pour point specifications for crude oil transportation that can be problematic and need to be highlighted [18]. The pour point was originally designed for refined petroleum fluids. Petroleum fluids like diesel gel quickly, usually within 3°C of the WAT and the pour point is easy to determine. For crude oils the measurement sometimes can be more difficult. The difference between the WAT and pour point temperature can be dramatic. During sample observations, samples can become considerably more viscous at each temperature reading until the sample gels. This can extend over a 6°C to 30°C range in many instances. As crude oils become increasingly viscous, more and more tilting is required to evaluate whether samples have gelled. This can make measurements between different operators subject to variations just due to differences in handling the samples while tilting. Another challenge that can occur is de-waxing of the top layer of crude oil during cooling. In this instance, wax precipitates and falls from the top layer. Or similarly the wax network in the top portion is broken during tilting samples for observation. It is possible that the majority of crude oil beneath the top level can solidify at a significantly higher temperature than the top fluid portion that is being observed. Sometimes this results in pour points which are misleading with respect to reflecting the true gelling nature of the crude oil.

Another factor that must be understood is the formal pour point measurement for a crude oil is the result of a test performed under specific cooling and shear test conditions. These conditions do not represent real-world conditions in production or transportation systems. The pour point value does not mean that a crude oil will stay fluid indefinitely at all temperatures above the pour point measurement value. Wax crystal growth and subsequent gel formation is time and shear dependent. Solidification of a crude oil at or above the pour point temperature can occur at temperatures slightly or even moderately above the pour point, particularly if the crude oil is left sitting static for long time periods. In this instance, the pour point measurement provides an indication of a safe nongelling temperature which is erroneously too low. In other instances when accounting for shear in a production system, using pour point as an indication of a safe nongelling temperature may be too conservative as a baseline for an operator. Paraffinic crude oil below the WAT is thixotropic in nature. If the crude oil is sheared during the wax crystallization process (as occurs in varying extents during flow through flowlines) the actual gelling point of the crude oil in the process can be reduced considerably due to breaking crystal network linkages. This is particularly true for crude oils that have been chemically treated with PPDs [19]. This will be discussed in detail in Section 2.5.

2.3.2.3 Viscosity profile

Fig. 2.6 was used previously to show differences between paraffinic and asphaltenic crude oils and to show how the viscosity-temperature profile could potentially determine WATs. It also illustrated how the viscosity profile could provide another means to determine gelling/pour points for crude oils. As with pour point measurements, viscosity measurements also vary depending on the specifics of the exact measurement procedure used: crude oil preconditioning, instrument geometry, cooling rate, shear profile, etc. Many different types of viscometers and rheometers exist for making measurement. Rotational viscometers or rheometers are more commonly used. Various instrument geometries (cup and bob, Taylor–Couette, parallel plate, cone and plate, etc.) can be used. Rheometers also provide the capability to make yield strength and other rheological measurements. These extra capabilities make rheometers preferred instruments for many users over the less-expensive viscometers.

In Section 2.5, examples will be shown for using viscosity profile measurements to evaluate PPD performance. Viscosity measurements have some advantages over pour point measurements as viscosity measurements tend to avoid the "top-fluid de-waxing" problem. Also, measurements can be made with different shear rates.

In this section the intent is to illustrate the use of viscosity-temperature profile to help characterize waxy crude oils. Fig. 2.12 provides a comparison of different crude oils taken from around the globe that have used PPD treatment to address pour point issues. Shown are the untreated crude oil viscosity profiles. These crude oils range from high pour-point onshore crude oils to crude oils from deepwater offshore operations with much lower pour-points, yet higher than subsea shut-in temperatures. Note how sharp the upward turn in viscosity is for some of these crude oils.

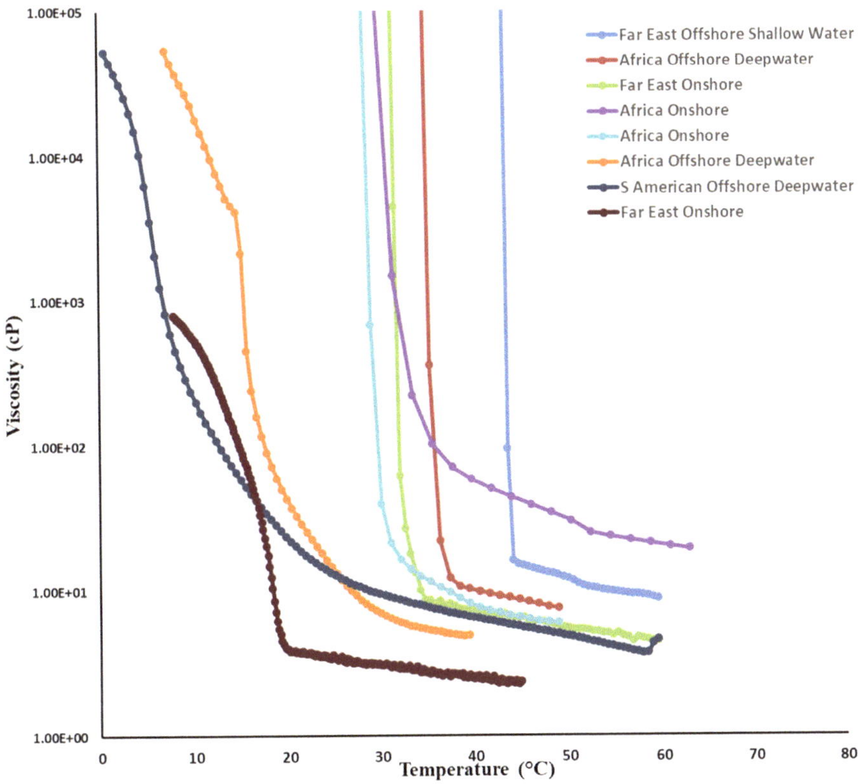

FIGURE 2.12

Example untreated viscosity profiles of different crude oils that have used PPDs. *PPD*, Pour point depressants.

2.4 Paraffin deposition

2.4.1 Paraffin deposition mechanisms

Paraffin deposition takes place by three mechanisms which transport both dissolved and precipitated wax crystals laterally. When the crude oil is cooled, a concentration gradient leads to the transport, precipitation, and deposition of wax at the wall by molecular diffusion. Additionally, small particles of previously precipitated wax can be transported laterally by Brownian diffusion and shear dispersion [2,3]. The molecular diffusion mechanism can result in very hard, tenacious deposits, whereas the shear dispersion mechanism tends to give deposits which are soft and shear sensitive. Increased shear at the pipe wall typically decreases deposition, thus limiting the contributions to deposition by the shear dispersion mechanism [20]. Even before the works from Bern and colleagues, it was determined that deposition will only take place if the deposition surface is below the WAT temperature and the crude oil solution temperature [21,22]. This is the driving mechanism for molecular diffusion.

Note there are some exceptions to the generalized statements above on the deposition mechanism based on field observations. The various flow regimes play a major role in paraffin deposition. Single, two-phase, and three-phase flows play a roll. The variables between laminar, turbulent, slug, stratified, annular, and dispersed bubble/mist flow all contribute to variations in paraffin deposition. Single-phase oil flow is the simplest system to explain. This system is common in crude oil export pipelines. In this system, paraffin deposition is controlled primarily by molecular diffusion. As long as a temperature gradient exists between the colder pipeline wall surface and the crude oil, paraffin deposition will occur if the system is below the WAT. Once steady-state temperature is achieved where the bulk crude oil temperature is equal to pipeline wall, the deposition process ceases. Therefore the temperature profile in the pipeline will dictate where in the flowline deposition occurs. In long lines, if the system cools relatively fast, deposition will be localized toward the beginning of the line and cover a smaller distance. If the system cools slower, the deposition will be further downstream in the line and extend over a larger distance. Deposition in single-phase flow will generally be evenly distributed across the pipe wall circumference.

Two-phase flow systems show more variation in their deposition profile. Turbulent flow conditions for lower gas-to-oil ratio (GOR) wells will experience even paraffin deposition across the flowline wall circumference. Under laminar stratified flow, the deposition will primarily be on the bottom portion of the line depending on the GOR in the line. The larger the volume of the line occupied by gas, the smaller the deposition area on the bottom of the line. The deposition mechanism will still be driven by molecular diffusion. Invariably, laminar flow in two-phase production systems leads to slugging occurring with terrain changes. If the slugging occurs before the steady-state temperature is reached, then deposition will take place across the entire circumference of the line. This occurs because the top of the line is constantly wetted and drained with oil during the churning effects of slugging on an uphill section of the line. This thin film of crude oil draining down the top of the line results in rapid temperature loss and accelerated paraffin deposition from molecular diffusion. There is also a potential contribution from the shear dispersion mechanism. In this instance previously precipitated wax crystals that did not deposit in the line or had been sheared off the wall have the opportunity to deposit. These wax particles can exist in the thin film of crude oil draining down the top of the flowline wall and can be trapped in the matrix of newly depositing wax. The gas flow also strips away the liquid crude oil and can leave solid wax particles behind.

This combined deposition mechanism becomes particularly problematic under mist flow conditions. In this instance, the GOR ratio is so high that the liquid travels down the line in the form of a mist. The liquid mist impacts the pipe wall where it slows down and forms small droplets of crude oil. There is a rapid cooling of these droplets and wax deposits by molecular diffusion. The droplets may also contain previously precipitated wax crystals that can be trapped in the developing wax matrix. The gas flow pushes the remaining liquid out of the deposit and further dries the wax deposit. Even in regions where steady-state temperature has been reached and no molecular diffusion takes place, paraffin deposition can occur from the shear dispersion mechanism.

An excellent example of this shear dispersion mechanism was a gas/condensate well in Northern Michigan in the United States. This well was experiencing a severe temperature drop across the choke. The temperature drop was so severe that the choke was encased in a block of ice in the middle of the summer. The flowline after the choke was at a higher temperature than the produced condensate coming through the choke. Therefore no molecular diffusion deposition could take place in the flowline. However, the line experienced severed paraffin deposition from the mist flow and shear dispersion mechanism.

Three-phase flow under turbulent conditions will have even wax deposition across the pipewall circumference. Under laminar flow, the deposition will tend to be along the center in the line with the stratified crude oil layer. Again, the surface area of deposition will depend upon the volume of crude oil in the line. Slugging will also be possible with the same deposition considerations listed earlier.

As molecular diffusion being the predominant mechanism, let's discuss this mechanism in more detail with respect to flowlines. And more specifically within the boundary layer at the flowline wall as this is the region where diffusion controls the deposition process. Diffusion occurs due to a concentration gradient being created by the precipitation and potential deposition of wax at the flowline surface when: (1) the surface is below the WAT and (2) a radial temperature gradient exists between the colder wall temperature and crude oil temperature inside the flowline. As wax precipitates at the colder wall surface, the remaining wax concentration dissolved in the crude oil adjacent to the surface is lower than the wax concentration of the crude oil radially outward from the wall surface. This creates a concentration gradient causing diffusion of wax radially inward toward the wall. The rate of deposition has been determined to be related to the diffusion rate. Mathematically if the rate of deposition is set equal to the diffusion rate, it can be numerically expressed as a product of—crude oil density, wax molecular diffusion coefficient, deposition surface area, and the radial wax concentration gradient:

$$\mathrm{d}m_{\mathrm{wax}}/\mathrm{dt} = -\rho \times D_{\mathrm{wax}} \times A_{\mathrm{dep}} \times \left(\mathrm{d}C_{\mathrm{wax}}/\mathrm{dr}\right)$$

where $\mathrm{d}m_{\mathrm{wax}}/\mathrm{dt}$ is the wax deposition rate, ρ is the density of crude oil, D_{wax} is the wax molecular diffusion coefficient, A_{dep} is the deposition surface area, and $\mathrm{d}C_{\mathrm{wax}}/\mathrm{Dr}$ is the wax radial concentration gradient at the wall.

Using the chain rule of differentiation, the radial wax concentration gradient $\mathrm{dC_{wax}/Dr}$ can be expressed as a product of (1) the partial derivative of the wax concentration with temperature and (2) the partial derivative of temperature with radial position at the wall. Therefore the deposition rate can be expressed as:

$$dm_{wax}/dt = -\rho \times D_{wax} \times A_{dep} \times \left(\partial C_{wax}/\partial T\right) \times \left(\partial T/\partial r\right)$$

where $\partial c_{wax}/\partial T$ is the partial derivative of wax concentration gradient with temperature, and $\partial T/\partial r$ is the partial derivative of temperature gradient with radial length.

Hence, the deposition rate can be related to a function of crude oil and wax physical properties, wax thermodynamic equilibrium, and thermal conditions. This provides a convenient means to be able to model deposition from the diffusion mechanism. The physical properties can be measured or estimated. Wax equilibrium can be calculated using various thermodynamic models. Likewise, flowline thermal properties are readily available with different hydrodynamic fluid flow models.

The other pieces of the puzzle needed for modeling are means to account for determining how much of the wax that diffuses to the wall and precipitates actually adheres or deposits to the surface and remains adhered versus being removed by shear forces of flow. Surface roughness plays a role here. Kinetics are also involved in the precipitation and deposition process. In addition, the diffusion mechanism only accounts for the wax that is physically precipitating and depositing. It does not account for entrained crude oil that is captured in the deposit wax matrix. This entrained crude oil is a part of the wax deposit and a means to appropriately estimate the entrained crude oil content is required. It is a function of the flow regime, shear stress, and thermal/concentration gradients. In addition, when flow moves from single-phase flow to multiphase flow, other complications on wax deposition from the different flow regimes will need to be accounted for as discussed earlier.

Another phenomenon not accounted for in modeling is deposit aging. The aging phenomenon is called Ostwald ripening. Basically, this process squeezes the crude oil out of the deposit and causes the wax crystals to grow larger over time. The deposits are then harder to remove and less prone to shear effects from production flow. For more in depth discussion on wax deposition mechanisms and modeling, complete reviews are available in the literature [23].

2.4.2 Paraffin deposition modeling

2.4.2.1 Flow loop testing

Many flow loop systems have been constructed and used for studying wax deposition. These have ranged from large multiphase flow loops to small benchtop laboratory units. Scale-up difficulties with smaller units in matching the flow and heat transfer conditions of actual production systems have limited their use though. Flow loops have been constructed by many major oil companies and research institutes in past years [24–30]. These activities were primarily to support development of deepwater assets particularly in the late 1990 and 2000s. Studies with the flow loops have been made to generate deposition data to make improvements in and validate wax deposition modeling programs. In addition, operators may use asset specific flow loop testing to tune modeling programs to assist in the design of deepwater developments and their paraffin management programs.

Flow loops consist of two main sections: a deposition section and fluids handling/pumping section. For the deposition section a portion of the flow loop is jacketed and connected to a cooling system to allow temperature control at desired deposition temperatures. As the majority of studies have been related to deepwater, most flow loops have been designed to be chilled to at least 4°C/40°F. The deposition test sections will have differential pressure transducers connected across the deposition section(s) to allow monitoring the wax deposition thickness build-up overtime. Often removeable sections or spool pieces will be included to facilitate examination and collection of wax deposit samples after completion of the testing. Different means are used to measure and quantify the final amount of wax deposited. Calipers, probes, and borescope cameras have been used to measure thickness at different locations in the pipe. The total mass can be determined simply by removing and weighing the deposit. In addition, photo documentation and other analyses of the deposition are often made. For example, HTGC analyses for measuring wax and entrained oil content.

The fluids handling and pumping section will recondition the returning fluids and pump them back through the deposition test section of the flow loop. If testing is with single-phase flow, the reconditioning may be as simple as reheating the crude oil to redissolve suspended precipitated wax and return the crude oil temperature to starting inlet temperatures. The fluids handling and pumping section requires a sufficiently sized tank and hold-up volume to ensure enough crude oil is used in the testing to avoid significant wax depletion effects during the testing. If the testing is with multiphase flow, then the fluid handling is exceedingly more complicated. Fluids may need to be separated and recombined for return pumping to get the same flow profile. Similarly, for multiphase flow attention is required in the flow loop design and operation to get the specific flow regimes desired.

The operating procedure and other physical aspects of the apparatus are important. For example, whether fully developed flow profiles are achieved and entrance effects avoided in the deposition section can affect test results. Some apparatuses have included other features and capabilities. For example, different external sensors have been trialed as other nondestructive means to monitor deposition along with differential pressure measurements. Pigging capabilities/studies have been incorporated in some flow loops. For specific details on different flow loops, the reader is referred to individual references from the different organizations.

One of the more active research programs has been through the Tulsa University Paraffin Deposition Projects (TUPDP) joint industry consortium which has been active for approximately 25 years. Work has focused on better

FIGURE 2.13

Examples of paraffin deposition flow loops (Courtesy TUPDP).

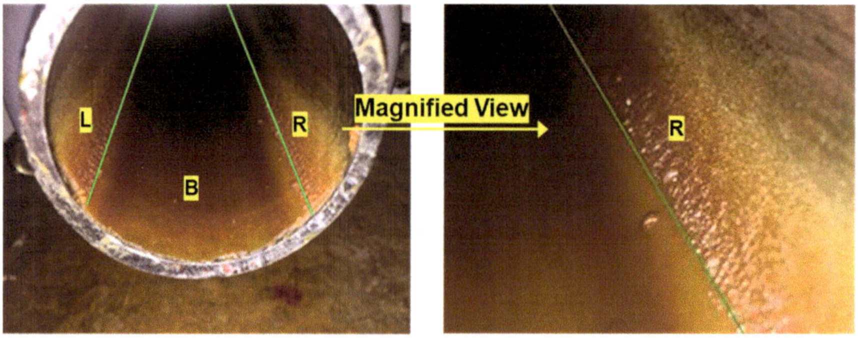

FIGURE 2.14

Paraffin deposition from flow loop test under stratified flow (Courtesy TUPDP).

understanding paraffin deposition in different flow regimes and translating learnings from flow loop testing into the development and improvement of paraffin deposition modeling. Studies of deposition have been performed in flow regimes ranging from laminar and turbulent single-phase flow, two-phase gas/oil stratified flow, two-phase gas/oil slug flow, and even some three-phase flow testing [29,31–35].

Fig. 2.13 shows pictures of single- and multiphase TUPDP flow loops. Fig. 2.14 is a picture of deposition from a TUPDP test under gas-oil stratified flow. Shown is deposition in a removal spool section. Note deposition is present only on the wetted portions of the flowline. Deposition is most heavily on the bottom, but some graded transition on the left and right sides just above the nominal liquid–gas interface is present too.

2.4.2.2 Modeling programs

Different software packages have been developed to model paraffin deposition. These include commercial simulation packages (such as Olga Wax, FloWax,

Leda Flow), proprietary in-house operator programs (ExxonMobil, Shell, etc.), and programs from research institutes (Tulsa University and Michigan University). A full deposition simulation package will require coupling models for the different processes contributing to the deposition: (1) hydrodynamic model, (2) wax solubility thermodynamics, (3) fluid physical properties, and (4) the wax deposition mechanism. The hydrodynamic modeling establishes thermal profiles, pressure drops, flow regimes, and fluid holdups in the flowline. The thermodynamic modeling provides calculations on wax solubility which therefore sets the maximum amount of wax available for deposition. Different thermodynamic models are available [36,37]. Fluid properties such as the wax diffusion coefficient, crude oil viscosity, and crude oil density are required in the calculations as a function of changing temperatures and pressures. The wax deposition model provides the actual deposition mechanism used in the simulation along with the various assumptions, correlations, and/or empirical adjustment factors included. Most simulation models assume wax deposition occurs from molecular diffusion and shear dispersion. However various means are required to handle entrained oil, kinetics, shear stripping, and even deposit aging. As it is recognized that many uncertainties are still not fully understood, the ability to tune models to match field or flow loop data exist. More details on modeling programs and modeling studies are available in the literature [23,26,31,32,38,39].

As indicated earlier, most modeling efforts are focused on deepwater subsea flowlines. The modeling is used to help design the paraffin management programs for production systems. If paraffin deposition is a concern, pigging is the most common control strategy employed. In this case, deposition modeling is used to determine pigging frequencies anticipated prior to start-up. However, if simulations indicate heavy wax deposition, other paraffin control methods often are employed to assist pigging efforts. If the volume of wax deposition becomes too large, the risk of sticking a pig during pigging operations is high. Therefore correctly identifying an appropriate pigging frequency is extremely important. In turn, knowing the pigging frequency will help determine the amount of deferred production that will be encountered during operation. This helps determine the value of using other deposition control means such as paraffin inhibitor application in conjunction with pigging. Similarly, if dealing with large amounts of deposition, paraffin inhibitor applications can help reduce risk for sticking a pig. More discussion on pigging and paraffin inhibitors is given later in the deposition control methods section.

Figs. 2.15 and 2.16 show typical wax deposition simulation output data. Shown are example simulations for a deepwater subsea flowline comparing development options using uninsulated and insulated flowlines. Fig. 2.15 shows wax deposition build-up overtime for the uninsulated flowline option. Deposition simulation information such as shown in Fig. 2.15 is used to determine approximate pigging frequency required for the asset. In addition to the maximum thickness, the deposition simulation shows where in the flowline the deposition would be expected. And with integrating the curves, the total amount of deposition

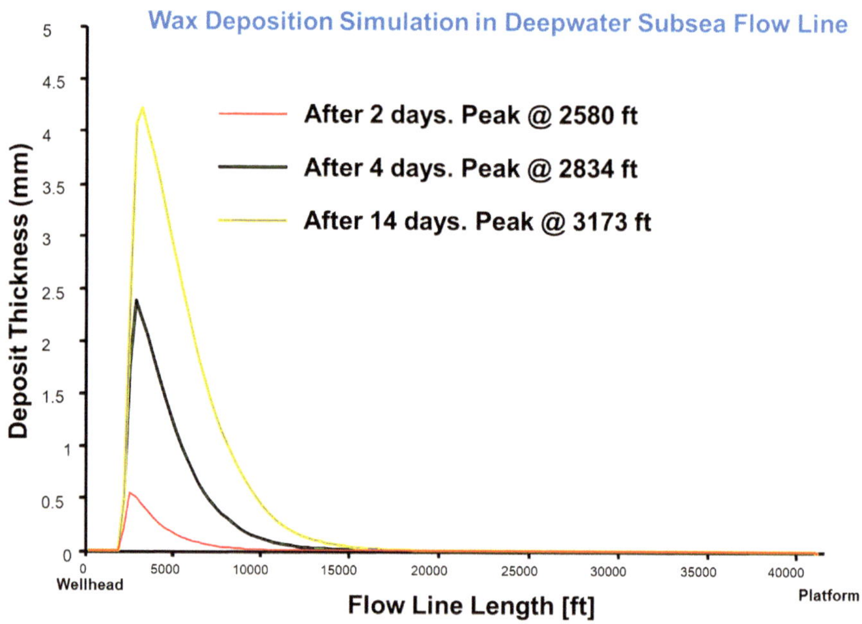

FIGURE 2.15

Paraffin deposition modeling for deepwater subsea flowline.

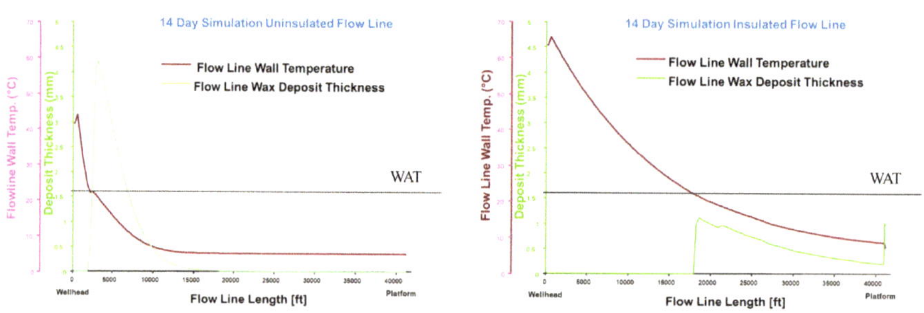

FIGURE 2.16

Paraffin deposition modeling for a deepwater subsea flowline examining the effect of insulation.

expected is obtained. Typically, operators have guidelines to pig after the deposition thickness reaches a specific height, usually on the order of few millimeters thickness. The amount of total deposition will also need to be factored into the pigging frequency decision though. Again, if the total deposition amount is excessive, there is high danger of sticking the pig. Also, there are constraints in the volume of wax the pig receiver can retain. Fig. 2.16 compares the two options for using uninsulated and insulated flowlines showing deposition simulation results after 14 days. Note, in this example, the insulation was not sufficient to completely eliminate deposition in the flowline. It just delayed deposition further downstream. Although the insulation did reduce the total amount of deposition, using insulation solely for paraffin deposition control in an application such as this one is not economical as insulation cost is significant and pigging or paraffin inhibitor application would still be required. However, if insulation were installed for another reason (such as hydrate control), a side benefit would be the reduction in total amount of wax deposition and a corresponding longer interval between pigging.

No matter the ultimate system design for a specific asset, having good capabilities to model deposition are necessary to make optimal decisions for the design of the system and paraffin management program. Equally important is the analysis of the crude oil wax distribution as it is critical input data to the thermodynamic and deposition modeling.

2.4.3 Paraffin deposit characteristics

Paraffin deposits can vary in color from white to black. Actual paraffin is white, but the deposit will take on the color of the crude oil under production. The dark color of crude oils is due to the resins and asphaltenes in the crude oil. Crude oil is incorporated into deposits in varying amounts as entrained oil. Figs. 2.17–2.20 show some different paraffin deposits from onshore and offshore operations. As seen in the figures, paraffin deposition is fully capable to completely plug

FIGURE 2.17

Severe paraffin deposition in various onshore flowlines.

FIGURE 2.18

Severe paraffin deposition in two offshore subsea flowlines.

FIGURE 2.19

Paraffin deposition on sucker rod in rod-pump well.

flowlines and stop production. Certainly, paraffin deposition continues to be responsible for restricting production in many fields globally.

Wax deposits contain both the solid wax species that are physical depositing and entrained crude oil contained within the deposit matrix. The amount of entrained crude oil can vary widely. It is a function of the thermal differential driving force for deposition, the shear forces on the deposit, and aging of the deposit. The hardness of the deposit is highly related to the entrained crude oil content. Higher shear forces (higher flowrate) and aging promote harder deposits. Fig. 2.21 shows an HTGC of a paraffin deposit illustrating the two main constituents of deposits: (1) wax species physically depositing and (2) entrained crude oil. The hump at the longer retention time is the wax species physically depositing. The hump at earlier retention time along with a trailing portion is the entrained

FIGURE 2.20

Paraffin deposition removed from onshore flowline.

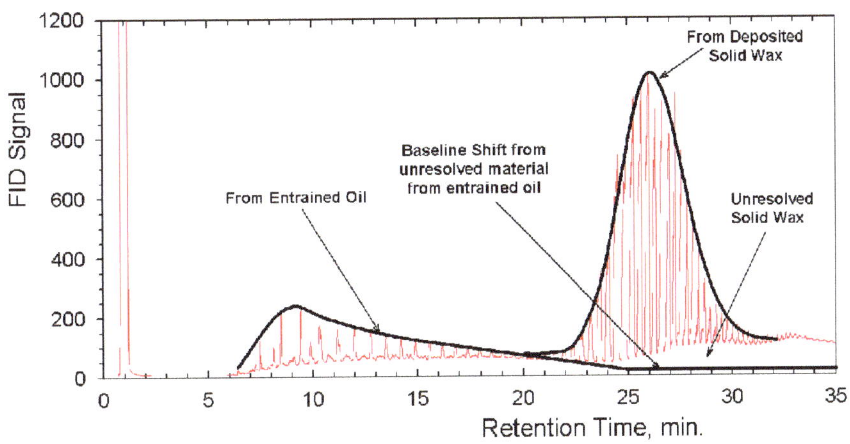

FIGURE 2.21

HTGC of paraffin deposit illustrating portions of deposited wax and entrained crude oil.

crude oil within the deposit. Within this deposit, the amount of entrained crude oil is relatively small as the first hump is significant smaller than the second hump. In deposits with high amounts of entrained crude oil, this entrained oil hump would be even larger than the depositing wax species portion.

Fig. 2.22 shows another example of an HTGC analysis of a wax deposit (right side) from a deepwater subsea GOM flowline along with the HTGC of the crude oil (left side) from one of two wells produced into the flowline. With this deposit sample, the depositing wax species are distributed around a carbon number of C_{36}

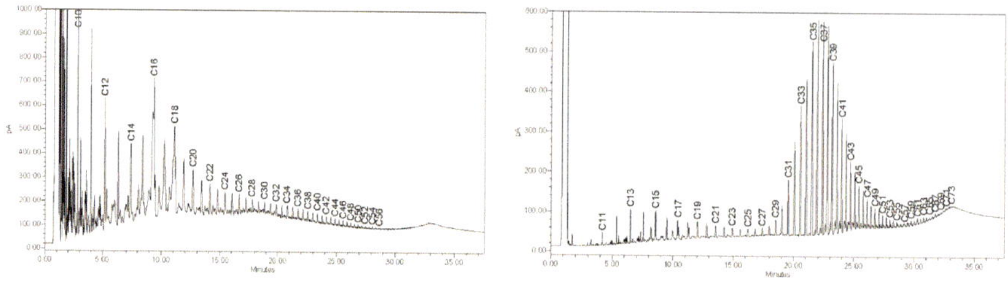

FIGURE 2.22

HTGCs of crude oil and paraffin deposit from deepwater GOM field.

with a range from approximately C_{28} to C_{57+}. Note as this deposit was a relatively hard, aged deposit, it had a low entrained crude oil content, even lower than the example in Fig. 2.21. A remediation application on cleaning paraffin deposition in this flowline with dispersants is discussed in Section 2.6.

The distribution of wax species that deposit depend on the location in the production system and the particular thermal conditions at each specific location. Starting at the point where the WAT reached, only the higher molecular weight waxes will precipitate and deposit. At cooler locations downstream, lower molecular weight waxes precipitate and will more preferentially appear in deposits. Fig. 2.23 illustrates changes in the deposit characteristics at different locations in one production system. Shown are two HTGC analyses of deposits taken from different locations in the production system. Also shown is an HTGC analysis of the crude oil from the well. The yellow trace is for the crude oil. The red trace is for a deposit taken downhole at a depth of 3000 ft. (914 m). Hence, for this well the WAT was reached below the 3000-ft. level. Being the closest deposit sample to the WAT, this deposit had higher chain length waxes with a peak carbon number about C_{51}. These higher molecular weight waxes are also higher melting and harder to remove. The blue trace is for a deposit taken further downstream in the flowline. For the flowline deposit, crude oil temperature decreased, and a lower range of carbon number waxes were deposited. The peak carbon number of this deposit was approximately C_{41}.

2.4.4 Paraffin deposition control

The following sections discuss various thermal, mechanical, and chemical methods used for managing paraffin deposition. These include methods for prevention and remediation.

2.4.4.1 Thermal methods

Thermal methods can be used for prevention or remediation of paraffin deposition. For prevention, if tubular and flowline conditions can be maintained above

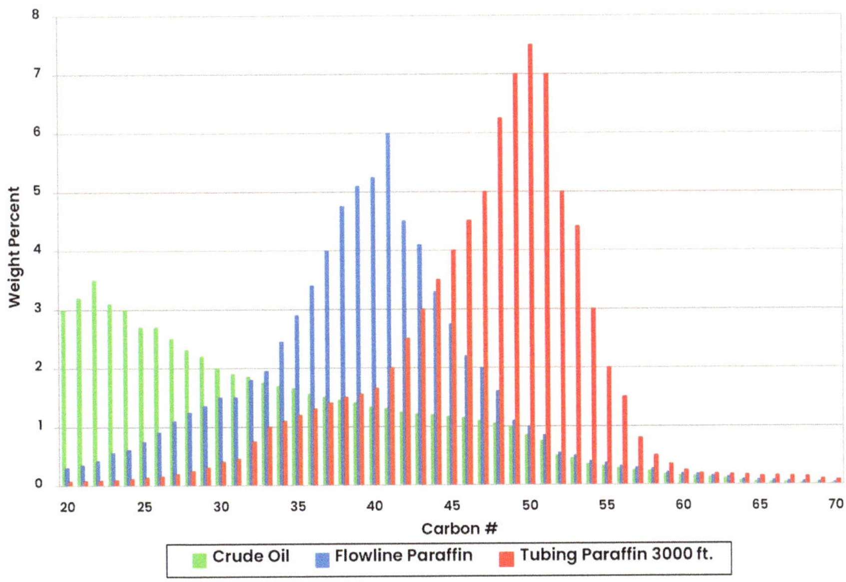

FIGURE 2.23

Comparison of *n*-alkane compositions in paraffin deposits taken from different locations in a production system.

the WAT, no deposition will occur. In some production systems, flowing temperatures can be maintained such that flowing wellhead temperature (FWHT) and receiving or boarding temperatures are above deposition conditions. Whether appropriate temperatures can be maintained depends on the production system specifics: reservoir temperature, well design, production rates (through well/field lifetime), flowline diameter, flowline length, insulation (if used), heat transfer characteristics of flowlines and associated production system components, and the ambient conditions cooling the system. Of course, if active heating is installed into the production system, it plays a role too.

Whether insulation can be effective or is economically worthwhile also depends on individual operations specifics. With respect to active heating, active heating systems have been used but are rare. Generally, it is expensive to continually heat entire production lines to prevent deposition.

Insulation

Insulation has been used to protect production in well tubing and flowlines. In deepwater subsea wells, the common practice for paraffin management design is to have FWHTs maintained above the WAT to avoid deposition within subsea wells until production significantly declines late in the well lifetime. If the natural reservoir temperature and production drive (along with any planned artificial lift) will not sufficiently maintain heat in early and midlife, then insulation is desired. Insulation sometimes may also be considered in cases where the temperature decline is not as severe, to decrease the abandonment flowrate for the well thereby extending well lifetime. Insulation options include vacuum insulated tubing (VIT), thin-film insulation coatings, insulating packer fluids, or combinations thereof. VIT has been used within wells for wax deposition control and other flow assurance concerns [40–43]. Although VIT is effective, application in deepwater wells is not extremely common. Higher cost, longer supply lead times, and increased difficulty and expense for installation have limited consideration. The use of insulating packer fluids is more common. Gas in the annulus, gelled oil insulating fluids, and aqueous insulating fluids have all been used. Or a combination of multilayer thin film insulation with packer fluids has been suggested as intermediate option to VIT if the low thermal conductivity provided by VIT is not required [44,45]. In 2006 Wang et al. referenced 10 well applications in the GOM using insulating aqueous-based insulating fluids applied specifically for paraffin control [46].

For deepwater subsea flowlines, consideration of insulation requires a complete flow assurance assessment examining the effectiveness and cost in comparison to competing paraffin deposition management solutions (pigging and paraffin inhibitors). Note that costs for an insulated flowline are generally more than twice the costs for an uninsulated flowline. The assessment is not only limited to paraffin management concerns though, as use of insulation is a potential strategy for hydrate control. Hence, the assessment must be made as part of the overall flow assurance management strategy.

For long deepwater subsea flowlines, insulation is not sufficient to protect the entire flowline. It would only be effective in pushing the deposition further downstream in the flowline. However, with the different thermal and deposition profile downstream with insulation, insulation still might reduce the total amount of deposition for long flowlines. But again, a complete flow assurance assessment is needed to judge economic merits for each potential flow assurance strategy.

The different types of flowline insulation are pipe-in pipe, wet insulation, and flexible insulated pipe [47]. Flowlines can also be buried to provide some insulation from seafloor currents. Pipe-in-pipe insulation can offer the highest insulation performance. However, the construction and installation costs are the highest. The heavier weigh may limit suitability of installation equipment too. Wet insulation is the most common insulation used in subsea flowlines and is available from different manufacturers. Syntactic foam insulation using polyurethane, polypropylene, and epoxy are common insulation materials. Insulation design with hollow microspheres in the syntactic structure increases the insulation efficiency and strength. Flexible insulated pipes are wrapped with syntactic tape and are more common for riser use.

When evaluating using insulation, consideration is also needed for the other components of the production system: trees, jumpers, manifolds, and risers.

Another consideration with insulation within wells is a potential decrease in the "no-touch" cool-down time for hydrate management during shut-ins.

Heating

Many different methods are available and have been used to supply heat, both for preventing and remediating wax deposition or pour point concerns. For preventing deposition, these include applications for heating well tubulars and flowlines, both onshore and offshore. Applications onshore are easier and less expensive. However, deposition problems offshore are more difficult and costly to remediate. And since deepwater offshore production rates are generally higher than onshore operations, the economic incentives for preventative solutions are often higher in offshore operations. Options for heating flowlines are electrical heating or with circulated hot fluids through jacketed or bundled flowlines. Insulation is commonly included with the heating. These require major capital expenditures (CAPEX) and are generally designed during the development stage of the project. In addition, significant operating costs are required to supply the energy for heat. Just as with evaluating insulation (without heating), consideration of installing heating systems requires a complete flow assurance assessment examining the effectiveness and cost in comparison to competing flow assurance management strategies for paraffin and hydrate control.

For remediating deposition, different means can be used for applying heat in production systems that do not have electrical heating or jacketed/bundled flowlines.

Heating for deposition prevention. Electrical heating methods include direct electrical heating (DEH) and electrically trace heated pipe-in-pipe (EHT-PIP). The DEH means are pipe-in-pipe and earthed current (used with wet insulation). Several subsea DEH systems have been installed in Norway and at least three Shell installations in the GOM [48]. The driver for most subsea applications has been hydrate control though. The high CAPEX [and operating expenditures (OPEX)] expense is usually not favorable for paraffin deposition control compared to pigging. Most early subsea flowline heating applications were with DEH; however, EHT-PIP heating appears to be more favored now due to improved efficiency with reported potential large OPEX savings over DEH and active heating with hot fluid circulation [48,49].

Electrical heating within production tubing is also not common. Fortunately, natural production temperatures or use of insulating packer fluids can push deposition into flowlines for a majority of operations. For wells that experience deposition in well tubing, various mechanical, thermal, and chemical remedial operations can be applied fairly easily for onshore and dry-tree offshore wells. For deepwater susbsea wells, it is a greater concern with higher incentive for preventative measures though. Successful use of skin-effect heat tracers applied internally within the production tubing string has been reported for an offshore New Zealand application [50].

Active heating of subsea flowlines with circulating hot fluids through jacketed or bundled flowlines is not common but has been done. Three applications have been reported in the literature for applications at Conoco Britannia, BP King, and Statoil Gullfaks fields [51–53]. The heating at Britannia and King fields was installed reportedly for both hydrate and paraffin deposition control. Britannia and Gullfaks bundled systems were with inhibited hot water and BP King was a jacketed system using water–glycol as the heat transfer fluid. As with electrical heating, the main limiting factor in using active heating with hot fluid circulation is the cost. Particularly with just paraffin deposition control, pigging is often more economical as the cost of jacketed system is easily more than doubled the cost of an uninsulated flowline.

Heating for deposition remediation. To remediate paraffin deposition, heating has been used since the early days of the petroleum industry. Different means exist to provide heating. In this section, two means will be discussed: hot oiling or hot water injection and thermochemical.

The active heating methods just discussed for preventive measures can also be used for remediation. Active heating systems do not have to be used full time for prevention. Systems can be operated to melt deposition periodically for remedial control or to heat lines to avoid pour point issues during shut-in. This approach would save on operating expenses; however, the capital expense would likely not be significantly different than installing a system for prevention. These systems generally require design before development of a field and have high installation costs. And, as previously mentioned, applications are limited due to expense.

Hot oiling/hot water injection. As previously mentioned, thermal methods were developed early in the history of the petroleum industry. Hot oil treatments were one of these methods and continue to be a major method still used today. Rod-pump wells are a major category for this control method. The operation uses a truck with a thermal exchange unit to heat the hot oil treatment to the desired temperature. The hot oil is then pumped down the annulus at a specified rate. Many papers have been written covering the advantages and disadvantages for hot oil applications. Strab and co-workers found that pumping 142 barrels of hot oil at 135°C and 1.7 barrels/minute actually decreased the bottom-hole temperature from 52°C to 46°C [54]. No production increase was observed. The only advantage for the treatment was the thermal exchange of heat from the oil to the upper area of the tubing. A second attempt was made where the pump was unseated and the treatment was pumped directly down the tubing. This time 400 barrels of oil at 135°C and a rate of 1.7 barrels/minute were utilized. This treatment resulted in an increase in bottom-hole temperature from 53°C to 62°C. Although this gives some advantage of thermal exchange in the tubing to melt paraffin deposits to a greater depth than the annular application treatment, this generally is considered a minimal increase in temperature given the additional expense for unseating the pump.

The loss of heat to the metal casing, tubing, and rock behind the casing is substantial. A hot oil temperature calculation program was developed to calculate the

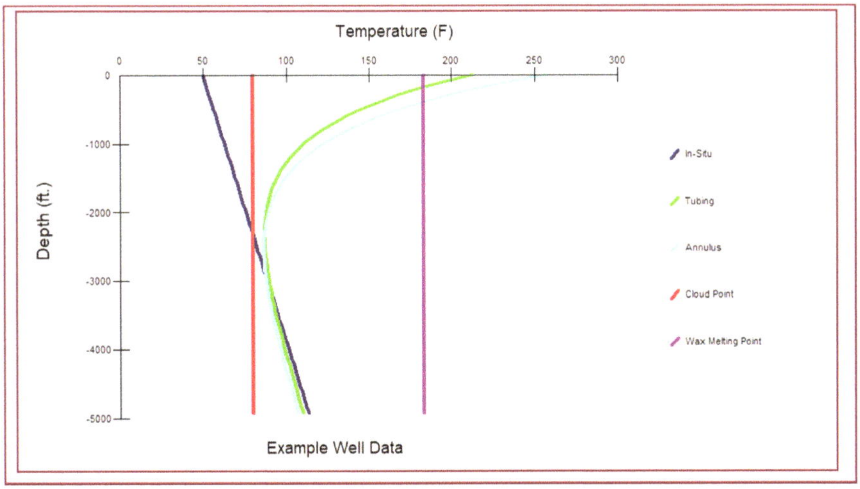

FIGURE 2.24

Example from hot oil temperature calculation program.

heat transfer during hot oiling jobs [55,56]. An example of these calculations is shown in Fig. 2.24.

The example in Fig. 2.24 indicates that the melting point of the deposited wax in the tubing is 180°F (82°C) and the crude oil WAT is 80°F (27°C). For hot oil pumped at 250°F (121°C) down the annulus, wax in the upper area of the tubing can be melted only down to about 300 ft. (91 m). Any wax below this level in the tubing will not be removed.

Additional publications have discussed the disadvantages of hot oil treatments [56–60]. A number of interesting observations have been made. First many wells have been later found to not have needed hot oil treatments and found the applications of these treatments were actually promoting formation damage and production decline. The reason for this is the source of hot oil and conditions it encounters in treatments. The source of oil for hot oil applications is stock tank oil. The highest WAT crude oil in any system exists in the stock tank. The crude oil at this location has lost the majority of light ends which previously acted as paraffin solvents to reduce the WAT. The stock tank oil WAT can be further increased if a paraffin tank bottom sludge exists. Many systems gravity settle precipitated wax particles to the bottom of stock tanks and these waxes can be picked up when the hot oil treatment is sourced. The paraffin content in stock tank oil is further concentrated during the hot oil process by heating the oil to an elevated temperature which drives additional light ends out of the crude oil when it enters the casing. Further gas from a rod-pump well is vented up the casing. Depending on the gas production level, crude oil may be carried with the gas up the casing. The authors have observed locations where paraffin wax has actually bridged-off the casing at the surface in such systems. Hot oil treatments down the backside in such systems can pick-up additional wax and further increase hot oil WAT. For the various reason discussed earlier, hot oil can be often become saturated with wax and/or carry wax into the well and/or formation. In many cases hot oiling has promoted formation damage and production decline.

As wax deposition in wells are typically higher molecular weight waxes applying a treatment with limited heat transfer ability at lower depths and limited or no spare solubility is not viable. For example, paraffin deposits with very higher molecular weight waxes have been observed in many fields. In one field, deposits have been reported with melting points up to 108°C and *n*-alkanes up to $C_{82}H_{166}$ identified in the deposits [58]. These fields with very high-molecular-weight waxes can experience formation damage and the production of skin in the immediate near-wellbore area from deposition. These wells require different treatments. Formation damage and remediation of paraffin will be addressed further in later sections. It should be noted that chemical additives have been developed to reduce formation damage from hot oil treatments.

To combat the formation damage effects of hot oil treatments, many producers have switched to hot water treatments for remediating deposition in the well. Here the hot oil is replaced typically with produced water. The water carries more heat and can achieve elevated temperatures to greater depths in the well compared to hot oil. The use of water also limits formation damage as high WAT stock tank crude oil is avoided.

Thermochemical Methods. The primary heat generation process that has been used in the petroleum industry has been based on the reaction of sodium nitrite with ammonium nitrate (or ammonium chloride) in an aqueous solution. This reaction produces nitrogen, water, sodium nitrate (or sodium chloride), and a large heat exotherm. The nitrogen heat generation process has been used over three decades within the petroleum industry. It was originally patented by Shell for unloading gas wells but has been used effectively for paraffin deposition. Baker Hughes applied some of the first applications in a discontinued process called N-Situ [61,62], but Petrobras has been the primary user of this process with their nitrogen generation system process [63,64] with numerous applications performed in the Campos Basin. It should be mentioned that extreme care needs to be undertaken with such processes due to the large amount of energy generated. Think of the recent horrific fertilizer incidents in West, TX and Beirut, Lebanon. Additionally, the generation of nitrogen gas in process lines can cause pressure build-up and create significant safety concerns.

Potentially safer options using organic fluids exist but do not generate as much heat. One reacts an organic acid with an organic or inorganic base [65]. Another is a process with reaction of terpene solvents such as limonene [66]. The heat generation in this process is coupled with the limonene reactant also being a good wax solvent. Hence, the process is a hybrid solvent and heating application. It also does not appear to create gas pressure concerns.

FIGURE 2.25

Cutting tools for paraffin wax.

From https://production-technology.org/slickline-service-tools/.

2.4.4.2 Mechanical methods

Wireline cutting

Open tubing production systems often employ wireline cutting tools (gauge ring, gauge cutter) to remove wax from the tubing. These units require less cost and equipment compared to a pulling unit. Generally, the cutter is a heavy metal object that is only slightly smaller in diameter than the tubing. These units are employed on either a schedule or "as needed" basis. Examples of cutting tools are shown in Fig. 2.25.

Coiled tubing operations

Mechanical interventions can also be run through coiled tubing operations. These possibilities include milling/drilling and high-pressure jetting. With coiled tubing the ability for extended reach exists for access into flowlines and wells. However, reach is not infinite due to limitation from friction and helical and sinusoidal buckling. Limitation may also exist with the coiled tubing bend radii that can make it difficult to negotiate bends in lines. Efforts in the 1990 were undertaken to extend capabilities of coiled tubing reach to around 5 miles with the rational that most flowline/pipeline blockages would likely occur within 5 miles of the well or host facility [67,68].

Coiled tubing is also utilized for supply of chemicals so coiled tubing operations may not always solely be mechanical-based approaches. Combination jetting/dispersant applications are recommended for use in wax deposition jetting

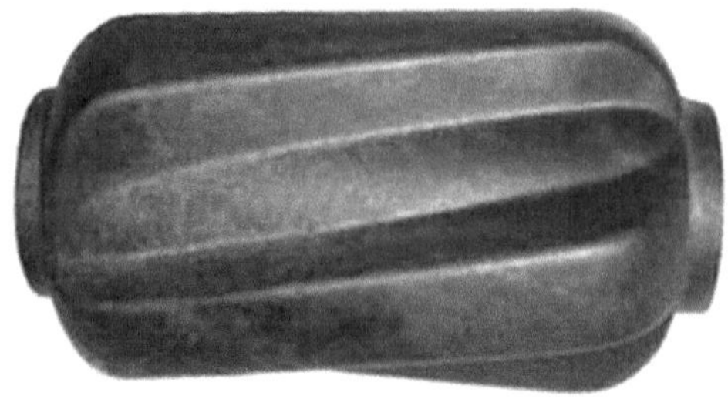

FIGURE 2.26

Rod scraper for wax control and rod centering.

remediation applications versus water alone. The effect of dispersant helps the deposit removed with high-pressure jetting from readhering to pipe walls. This was demonstrated in large-scale coiled-tubing jetting test study [69]. In another report, a case history on successfully cleaning-out a deepwater flowline was presented [70,71]. This operation required lifting the flowline to a workboat with significant expense though.

Rod scrapers

Rod-pump wells often use scrapers designed to attach to the rod string at intervals set by the pump stroke to keep the tubing clean of wax. These are also called rod centralizers to keep the rods from rubbing against the tubing. The scrapers come in a number of designs and materials of construction. An example is shown in Fig. 2.26.

Plunger-lift well design

This method of producing crude oil up the tubing employs a free-floating plunger that is lifted up the well by gas. The gas from the formation is directed into the tubing and not the casing. The plunger is held in a top assembly and dropped when the valve is opened. The plunger is tripped at the bottom by the bumper which closes the valve in the plunger. With each pass up and down the tubing the plunger scrapes deposited wax off the tubing. This effectively controls the build-up of paraffin deposition. There are problems, however, with the top assembly plugging with the removed paraffin and sticking the plunger. An example plunger-lift design is shown in Fig. 2.27.

Pigging

Flowlines for production systems and pipelines employ pigging operations to handle paraffin deposition. The pigs utilized come in many different designs. Foam pigs are composed of various weights of open-cell polyurethane from 16 to 160 kg/m^3. The lower the weight the softer the pig and more compressibility. Higher weight pigs are harder and less compressible. Cleaning pigs are generally

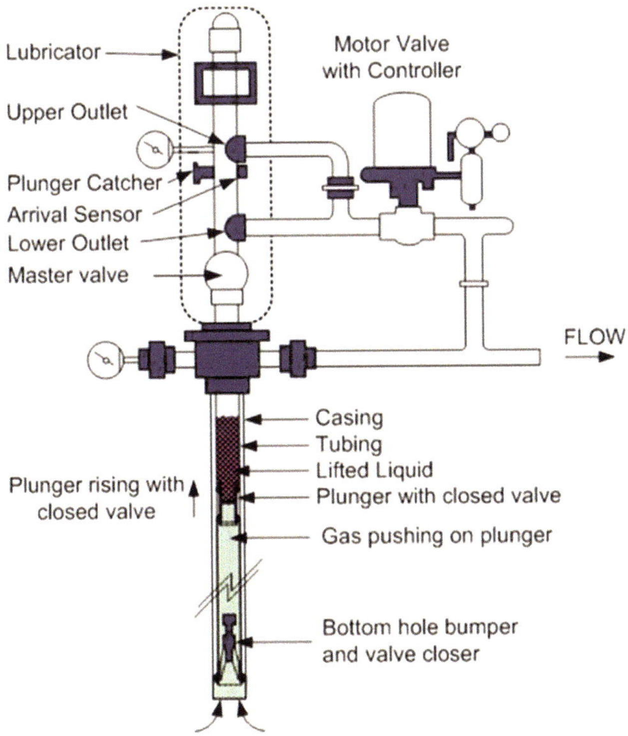

FIGURE 2.27

An example of plunger-lift system and the key components.

From https://www.croftsystems.net/oil-gas-blog/what-is-a-plunger-lift/.

cast polyurethane or neoprene which have no compressibility. Pigs can also include different types of disks, swabs, and scrappers to improve cleaning. Bypass pigs exists which are designed to allow some bypass flow to help allow dispersal of some of the deposition removed in the bypass flow. In pigging operations, the volume of wax removed from each pig run needs to be closely managed to not exceed the volume capacity of the pig receiver or create too much accumulation ahead of the pig causing the pig to stick. Fig. 2.28 shows examples of some typical pigs.

An arena in which pigging is commonplace is in offshore production systems. As mentioned, deepwater subsea flowlines often employ pigging as their paraffin deposition management strategy. Pigging is also common in export pipelines as well. Many deepwater production systems are often designed with production from various wells manifolded and produced through dual flowlines. Dual flowlines allow for round-trip pigging where the pig can be launched and received from the host facility. Pigging single deepwater flowlines is more challenging. Subsea pig launchers exist but are not common. Reloading pigs to a subsea pig

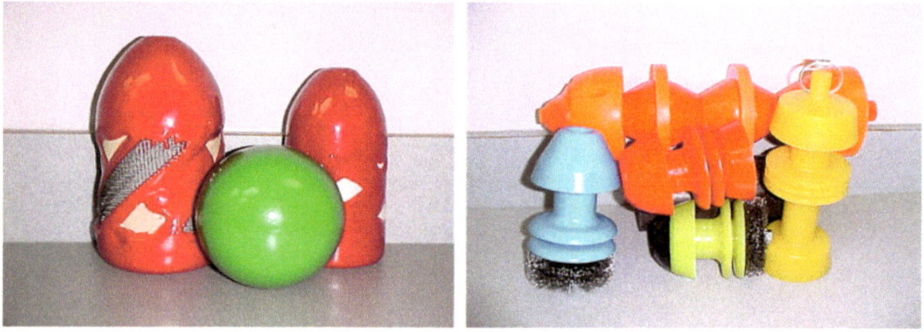

FIGURE 2.28

Examples of different pipeline pigs.

launcher is the main challenge to this technology. Note in pigging dual flowlines production must be stopped, at least for a portion of the pigging interval, to perform the pigging operation. Hence, each pig run incurs some deferred production. In some instances, pigging will be combined with paraffin inhibitor application to reduce the amount of deposition which reduces the required pigging frequency. In these instances, the reduction of deferred production is primary economic driver for paraffin inhibitor use. Some value may also be derived from the paraffin inhibitor helping to reduce the risk of sticking the pig.

As discussed, paraffin deposition modeling is relied on to help predict pigging frequencies required to maintain safe operations prior to start-up. After production start-up, pigging is closely monitored and pigging studies should be made to optimize pigging frequencies. This allows the operator to select an optimal pigging interval that minimizes downtime to reduce deferred production, but also appropriately cleans the flowline without creating a high risk of sticking pigs in future runs.

Fig. 2.29 shows pig returns for a GOM operation during trials performed to optimize paraffin inhibitor dosage and pigging frequency. Paraffin inhibitor dosage was reduced to determine whether dosage could be lowered to reduced chemical expenditures. Pigging frequencies were varied to determine whether a longer pigging interval could be used to reduce deferred production. Prior to the paraffin inhibitor/pigging optimization trials, the inhibitor was being applied at 300 ppm and pigging was performed every one to two days. During the optimization trials, it was found the pigging frequency could be increased. With the paraffin inhibitor applied at 300 ppm, pig returns were largely absent of paraffin deposition as indicated in left side picture of Fig. 2.29. When the paraffin inhibitor dosage was lower to 100–150 ppm over a 4-day trial, the pig returned with deposition ahead of the pig and the pig covered in deposition as shown in the right-side picture. Hence, indications were the paraffin inhibitor was highly effective at 300 ppm but had reduced performance at lower dosage. This was consistent with cold finger paraffin inhibitor testing results. However, the amount of deposition was

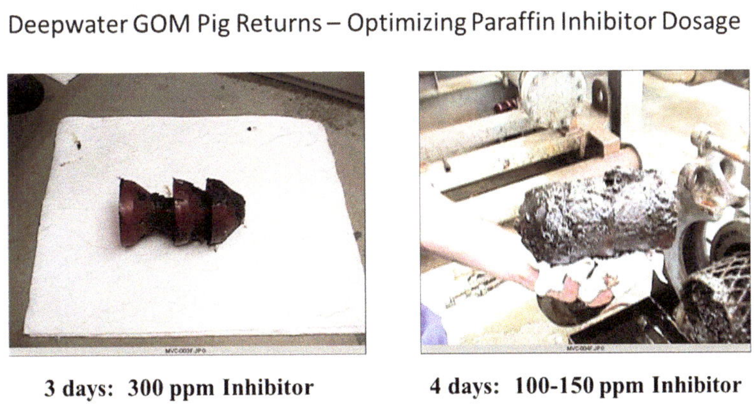

FIGURE 2.29

Pig returns from GOM field during trials for optimizing paraffin inhibitor dosage and pigging frequency.

manageable and within operator's comfort level for not posing a serious risk of sticking the pig. The outcome of the paraffin inhibitor/pigging trials was the operator elected to extend the pigging interval to 3–4 days and reduce paraffin inhibitor dosage to 150 ppm.

More discussion on pigging is continued in Section 2.4.4.3.

2.4.4.3 Chemical methods

The three major classes of paraffin control chemicals are: solvents, dispersants/detergents, and paraffin inhibitors.

Solvents

Solvents were the first chemicals used for paraffin control and are still heavily used. Paraffin is soluble in both aliphatic and aromatic solvents. Diesel was commonly used in the early days of oil production due to its wide availability. Better solvents exist though. Some more exotic solvents have been previously used over the years without understanding critical side effects. Chlorinated hydrocarbons such as carbon tetrachloride achieved widespread use in many areas before effects on refinery operations were well understood. Currently chlorinated solvents are banned for use in crude oil production due to the severe corrosion related issues in the refinery. This, however, has not totally eliminated their use because issues will occasionally arise with contaminated stock appearing in the system. Carbon disulfide was another highly used paraffin solvent due to its high paraffin solvation. Carbon disulfide is one of the best paraffin solvents. The use of carbon disulfide has been discouraged though due to a very low flash point (−30°C) and low auto-ignition temperature (125°C). In addition to the flammability concerns, there are major health hazards associated with carbon disulfide. As other

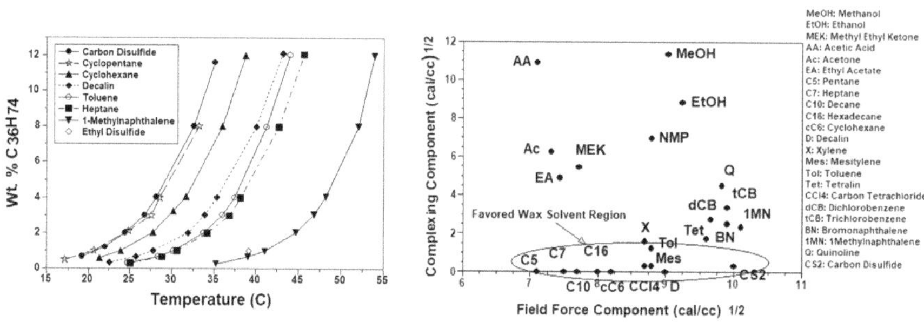

FIGURE 2.30

Comparison of wax solvent chemistries with two-dimensional solubility parameter plot.

alternative solvents or combination solvent products exist at lower costs with solvent power approaching carbon disulfide, there is little use of carbon disulfide today.

Fig. 2.30 shows the solubility of pure n-hexatriacontane (C_{36} *n*-alkane) in various solvents along with a representation of solvents on two-dimensional solubility parameter plot [5]. Highlighted on the two-dimensional solubility parameter plot is a region of solvents that are good wax solvents. Basically, good wax solvents have little complexing associations, else the solvent will self-associate rather than dissolve the nonpolar wax. Further note that smaller molecules tend to have enhanced solvent power over near-equivalent larger molecules due to ability to surround solutes more compactly.

Paraffin solvency can be measured in a number of ways [5,54,72–74]. Measurements can be made with actual deposit samples or with pure wax species. Since most wax deposit samples are dark in color from the entrained crude oil present in the deposit, it is easier (and sometimes appropriate) to use pure waxes or refined wax with known carbon length distributions for solubility measurements. The exception is if the deposit contains a significant amount of other nonwax material like asphaltene deposition. Then using actual deposit samples is desired. For solubility measurements usually a wax appearance or dissolution temperature is measured as a function of percent wax. From these measurements various solvency curves can be established like shown in Figs. 2.3 and 2.30.

A second type of measurement accounts for a solvents ability to not only dissolve the wax deposit, but to also cause it to fall apart into smaller pieces [54]. Once broken up the deposit can be produced through the system. Actual field deposit samples are desired in this type of test. The test usually relies on some method to measure both the speed and amount of break-up of the deposit in a given solvent. One method takes a paraffin deposit and attaches it to a mesh screen which is suspended in the solvent. This can be performed either static or with some level of agitation. The break-up of the deposit over time can be measured by the amount of residual wax remaining on the screen. Many other measurement procedures have also been used.

Solvent studies have been performed to quantity how common solvents and specialty oilfield paraffin solvents compare. Being able to quantify solvent power and compare against solvent prices allow operators to design the most cost-effective solvent remediation. A past solvent study with field paraffin deposit samples found that aromatics such as xylene and toluene were better than alkane solvents such as condensate, kerosene, and diesel [54]. This finding was likely due in part to the entrained crude oil in the deposits since condensate would be expected to have similar or better solvent power on pure waxes than xylene and toluene. However, solvents having some portion of aromatic species is often preferable as contacting whole crude oil with aliphatic-based solvents can destabilize asphaltenes from the crude oil.

Fig. 2.31 shows solubility measurements for another study performed with pure refined wax comparing three commercial paraffin solvents against kerosene, diesel, toluene, and condensate. Shellwax 300 was the refined wax used as it provided a good representation of alkanes in typical pipeline deposits. The carbon number range for the Shellwax 300 was $C_{21}-C_{54}$ with a peak carbon number of C_{37}. The solvent studies were conducted by measuring the WAT of solutions of different concentrations of wax in the solvents to develop a solvent index number. The solvent index was defined as the grams of Shellwax 300 dissolved at 37.8°C (100°F) per 100 grams of solvent [74]. Note in contrast to the aforementioned study with field deposit samples, condensate had a slightly higher solvent power

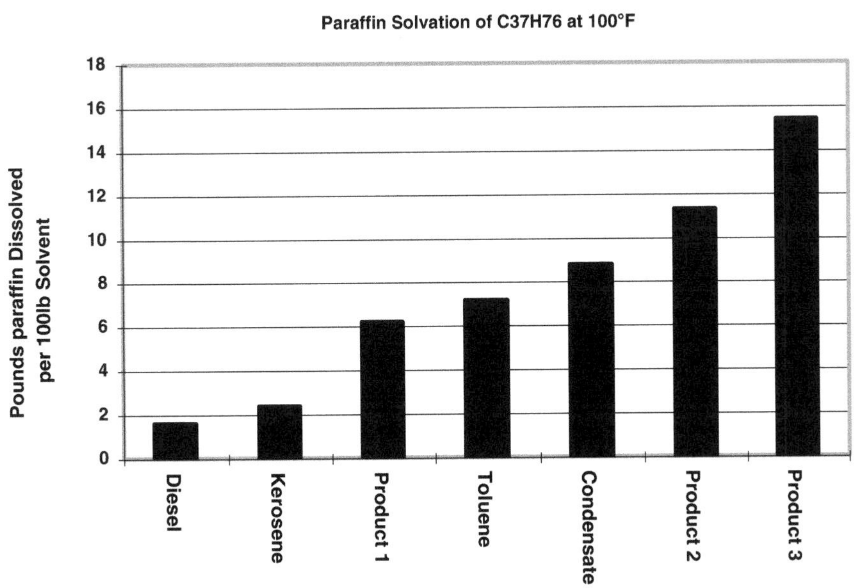

FIGURE 2.31

Solvent Index values for various solvents.

than toluene which is a good wax solvent. Further note that two of the specialty paraffin solvent products had significantly greater solvent power than toluene and the condensate. Depending on the application and potential increase in production from the solvent remediation, it can be worthwhile to use slightly higher-cost but more effective solvents.

With regard to wax solvent strength, for pure waxes the relative solvent performance will not change regardless of temperature, agitation or carbon number range. The amount of wax that may be dissolved may vary, but the relative order of solvent performance will remain the same. When dealing with actual field deposits this rule may not hold. The impurities in a paraffin deposit can lead to a change in performance. The aforementioned difference between condensate and toluene results is an example. This may have been from interaction with entrained crude oil in the deposit, however there are other impurities which can be present in deposits as well. Solvents have been formulated to address the potential impurities that may be in a given wax deposit such as water, asphaltenes, asphaltic resins, inorganic scales, formation fines, and other material. Due to these potential variations in deposit compositions, a variety of alcohols, acids, surfactants, dispersants, penetrants, and mutual solvents may be combined in small amounts with the selected base-solvent to improve performance. Again, note that it is not necessary to actually dissolve all the wax in a paraffin deposit to cause it to break up and move along in the system. Normally it is only necessary to break up the wax matrix to dislodge it from a problematic location in the system. The age and purity of the deposit will play a role. The more the deposit has aged, less entrained crude oil will be trapped in the matrix and the more difficult the deposit will be to dissolve.

Dispersants

Dispersants are surfactant/detergent type product formulations that are used to remove paraffin deposits in conjunction with water and mixing energy. Dispersant products are also used for removing asphaltene deposition and to aid separation issues such separator interface pads and tank bottoms. Different dispersant packages are used for the different applications though. For paraffin deposition control, dispersants are only a viable treatment option though with the presence of water and sufficient mixing energy (flow). These formulations are designed to attack the crude oil wetting and binding (sticky) components in the paraffin deposit. The chemistries work with water to create macro, micro, or nano-emulsions. They de-oil the paraffin matrix and internally emulsify the oil and binding components in an oil internal/water external emulsion. With continual attack, this mechanism essentially causes the deposit to break or shatter into small deposit fragments which can be swept-away in the aqueous or mixed water/ crude oil flow stream as shown schematically in Fig. 2.32.

In some applications, dispersants are applied into production systems (via batch or continuous injection) which use the natural produced water and production flow. Water cuts greater than 10% are generally required. Similarly, the higher the

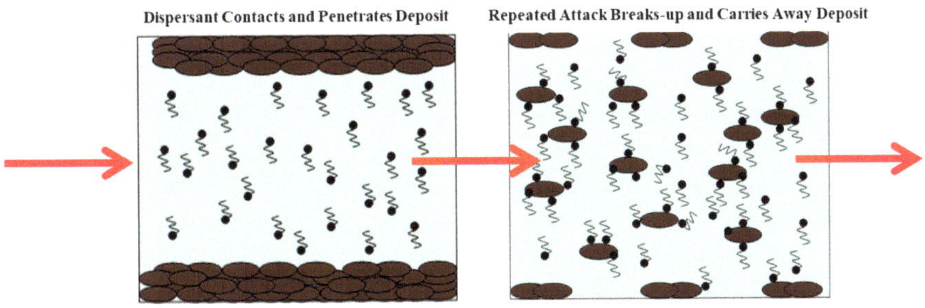

FIGURE 2.32

Schematic of dispersant mechanism.

production flow or shear generated at the deposit surface, the more effective will be the application. In these applications, the dispersants are used to manage or control paraffin deposition by providing continual or regular periodic remediation with little or no disruption to production. It should be mentioned that these types of dispersant treatments generally are less effective for managing deposition than paraffin inhibitor applications. Usually, whenever continuous paraffin inhibitor application is possible, it would be recommended. However, not all operations can economically accommodate installing continuous paraffin inhibitor applications and dispersant treatment may be the more economically viable option.

Other dispersant applications are more straightforward remediation operations where the dispersant is applied to remediate significant deposition to restore or improve production levels. In these applications the dispersant is pumped with water. Produced water from the field or nearby fields is the most common water source for land operations. For offshore operations, seawater is the most common water source. It is important that the dispersant formulation be matched to the total dissolved solids (TDS) content of the water source. Performance of some dispersant products can drop significantly with higher TDS brines. Typically, the higher the TDS, the greater the surfactant concentration required in the formulation.

Dispersant concentrations are normally applied at a concentration range of 2%–4% with the water. Sometimes higher concentrations may be required though. An attractive feature of dispersant applications is cost. As treatments are typically 90 + % water, chemical costs are generally low. Therefore if the dispersant is providing acceptable performance, dispersant applications can be very economical. Note, that other costs besides chemical costs can be involved in the total application cost though. In Section 2.6 an application is presented on a deepwater dual flowline dispersant remediation. In this application, the majority of the operation cost was associated with pumping the application and capturing & disposing the returns. Two marine vessels and large storage tanks for holding the removed deposition were required. In addition, extra cost was incurred for necessary hydrate prevention steps. Despite the extra cost, the application was highly economically successful.

Another attractive feature for dispersant applications versus solvent treatments is at cold temperatures dispersants can actually "carry" significantly more paraffin than a solvent can dissolve. Solvents work well at moderate and hot temperatures, but at low temperatures (as in deepwater flowlines at 4°C or below-freezing wintertime environments) even the best solvents have limited solvent power. To illustrate Fig. 2.33 provides a comparison. Shown are *n*-alkane solubilities in toluene (presented earlier in Fig. 2.3) compared against the loading capacity measured in testing at 0°C for the dispersant used in the abovementioned deepwater case history application. Testing indicated the dispersant/seawater paraffin deposit capacity was approximately 1–1.5 wt.%. As Fig. 2.33 shows, this is significantly greater than the solubility of even the lowest molecular weight alkanes in toluene. The paraffin deposit for this flowline (HTGC shown in Fig. 2.22) had a median *n*-alkane carbon number of C_{36}.

Selection of dispersants is generally performed with flask testing (also called bottle testing). In flask testing, weighed amounts of deposits are placed in a flask or bottle along with the source water and dispersant(s). The deposit may be coated on the glass or placed as larger sample pieces (sphere-shaped, cylinder-shaped, etc.) in the dispersant-water mixture. The samples are then agitated to obtain paraffin break-up and dispersion. The dispersant package that results in the best break-up and dispersion of the deposit, while keeping the dispersed paraffin water-wet, is considered to be the best dispersant for application. Other factors such as ease of separation of the paraffin dispersion, water quality, and propensity to foam may also be considered. Assessment of the paraffin break-up is usually

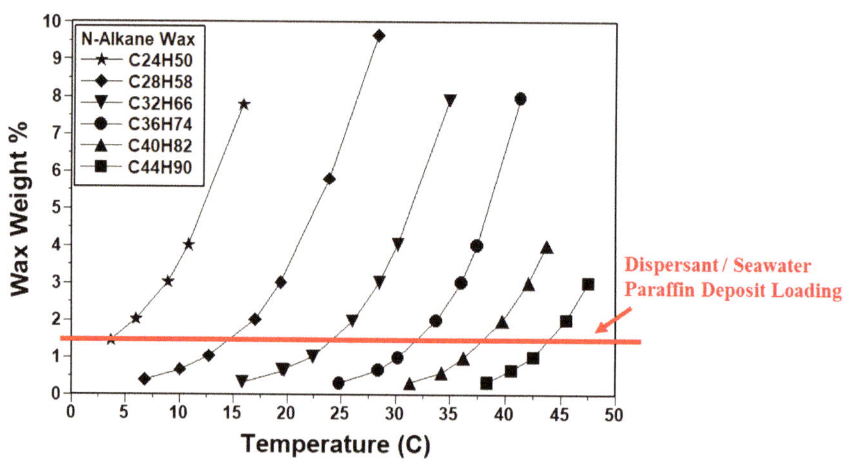

FIGURE 2.33

Comparison of paraffin loading capacity of dispersant versus solvent at cold temperatures.

made with a visual qualitative assessment. However, it can also be quantified with measuring the amount of paraffin broken-down below a certain particle size range.

Flask testing is used for hot and cold-water applications. Testing is best performed at the field temperature of the application. However, often tests are run only at room temperature. As implied earlier, use of the actual source water that is to be used in the treatment is highly recommended since the TDS of the brine affects dispersant performance. Usually tests are run for a period between 2 and 4 hours with agitation.

A simplified schematic the flask test procedure is shown in Fig. 2.34. Figs. 2.35 and 2.36 show examples of varying dispersant performance with three different dispersants used in the flask testing. Fig. 2.35 is from testing for an application with a West Africa paraffin deposit. The dispersant used in the left-side bottle was ineffective as the ball-shaped paraffin deposit remained almost completely intact in the bottle after agitation. It was not penetrated and broken into smaller pieces. The dispersant in the right-side bottle broke-up the deposit but did not have enough surfactant power to keep the paraffin water-wet and from sticking all over the side of the jar. The dispersant in the middle bottle had a good match of surfactant and dispersant properties. It resulted in complete break-up of the deposit into smaller pieces and kept the paraffin water-wet and well-dispersed not sticking to the bottle. Fig. 2.36 shows examples providing similar type performance on a North America paraffin deposit. However, note in the left-side bottle with good dispersant performance, the dispersant not only broke the deposit into finely dispersed water-wet particles, but provided clean water with

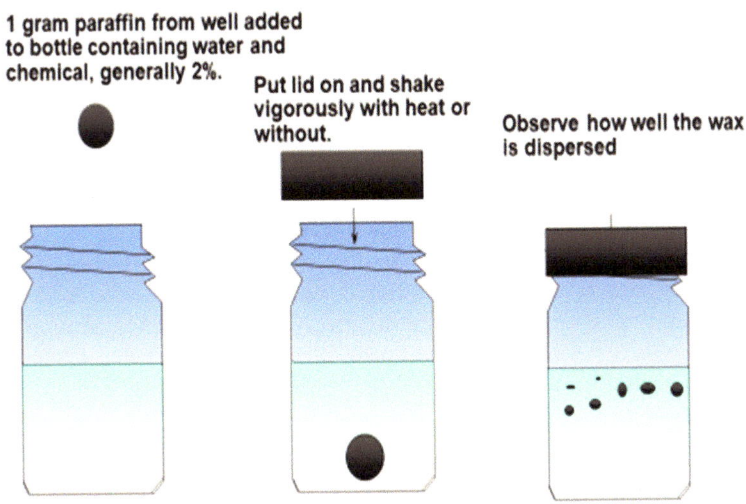

FIGURE 2.34

Typical flask test procedure.

FIGURE 2.35

Example dispersant performance in flask test with West Africa paraffin deposit.

FIGURE 2.36

Example dispersant performance in flask test with North America paraffin deposit.

inorganic solids (formation fines and other inorganics) dropped to the bottom of the bottle.

More discussion is included the following paraffin control section on dispersant applications and in the deepwater dual flowline application in Section 2.6.

Paraffin inhibitors / crystal modifiers

Paraffin inhibitor chemistries are polymers in which the polymer structure contains wax-like portions that allow the molecules to incorporate into wax crystal growth, yet also has other structural features which alter and disrupt the crystallization/crystal network growth. When effectively incorporated into a paraffin deposit network, the net result is a reduction in the ability of paraffin to deposit. In addition, the altered deposit network is weakened which makes it more prone for removal from flow shear forces which further contributes to reduced deposition.

The performance of paraffin inhibitors can vary widely depending on the specific application. Although some applications can provide near complete inhibition of deposition, generally paraffin inhibitor applications do not totally eliminate deposition and are instead relied upon to only reduce deposition.

Due to the mechanism on how paraffin inhibitors function, paraffin inhibitors are also called wax crystal modifiers. As mentioned previously, when paraffin inhibitors are used for reducing the pour point and improving the flow characteristics of high wax content crude oils, they are called PPDs. In this instance, interest is not just with interacting with the deposit but functioning to alter the entire bulk crystal growth network. More discussion on PPDs is given in Section 2.5.

For reducing deposition, paraffin inhibitors alter the crystal network structure development in the deposit as it is forming. The inhibitor's goal is to reduce the contact angle of the wax crystal to reduce the ability of the wax to adhere to a surface. This crystal modification results in a more compact size arrangement for a given mass of wax. The preferred crystal growth structure for untreated paraffin wax is a plate-type structure. This structure has large *X* and *Y* dimensions with a very small *Z* direction. An analogy would be a sheet of paper. The inhibitor increases crystal growth in the *Z* direction and thus reduces the surface area of the wax crystal. This effectively reduces the adhesive strength of the wax. This simplified explanation is shown schematically in Fig. 2.37.

For a paraffin inhibitor to function effectively, it needs to have an appropriate chemical structure that can incorporate into the deposit crystal network and

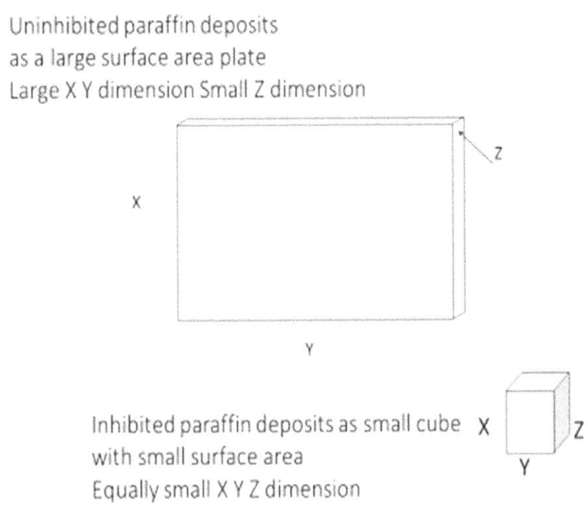

FIGURE 2.37

Paraffin inhibitor crystal modification mechanism.

effectively alter the network. To incorporate into the deposit the paraffin inhibitor needs to co-crystalize with the depositing wax. In the initial deposition region of a well or flowline, this means co-crystalizing with the higher molecular weight waxes present in the crude oil (as shown in Figs 2.21–2.23). Generally, it is optimal to select paraffin inhibitors which provide the best performance on this region of deposition. This is the region where the greatest amount of deposition occurs. Additionally, deposition here will often be harder to remove since it contains the highest melting point waxes. More discussion on paraffin inhibitor selection is given in the cold finger testing section below.

In regards to paraffin inhibitor chemical structures, hundreds of different chemistries have been developed over the years. The most common paraffin inhibitor structures are comb-like polymers with alkyl side chains. The major chemical polymer classes are: ethylene vinyl acetates, alkyl acrylates, alkyl methacrylates, styrene maleic esters, olefin maleic esters, olefin maleic imides, alkyl fumarate vinyl acetates, and alkyl phenol formaldehyde resins. The same chemistry classes are used as PPDs.

Product formulations and applications. Paraffin inhibitors are injected in various locations in different production systems [75]. Paraffin inhibitors are injected into wells through capillary injection lines, backside down the annulus, and into gas-lift systems. Paraffin inhibitors are also injected further downstream at wellheads and into flowlines. Some of these applications are applied subsea through umbilical systems. Other applications are applied topsides for treating export lines or applied into transportation tankers or storage tanks. Paraffin inhibitors are even squeezed into formations and absorbed in frac proppant. The paraffin inhibitor formulations for each of these specific applications must be qualified for its respective conditions.

As with most oilfield chemical products, the most common paraffin inhibitor product formulations are formulations in which the active constituents are dissolved in a solvent to allow easy delivery with chemical injection pumps. For solvent formulations, because the active paraffin inhibitor constituents are wax-like polymers, the polymers have similar solubility issues as higher molecular weight waxes with respect to remaining soluble at low temperatures. This creates challenges to applying solvent-based product formulations in cold environments that exist in various geographical regions and in subsea deepwater applications, or at least in applying with an economically favorable cost. Many global locations experience cold temperatures. These can range from near-freezing/subfreezing wintertime temperatures in temperate climates to extreme freezing temperatures below −40°C in artic regions.

Deepwater temperatures are not as extreme but are still cold at ~4°C/40°F and have an added challenge of being at high pressure. High pressure adversely decreases paraffin inhibitor solubility. As a result, paraffin inhibitor products for deepwater umbilical application need to be evaluated for stability at the temperature and maximum pressure the products will experience in the umbilical injection line. Several instances of plugging of deepwater umbilical lines have occurred

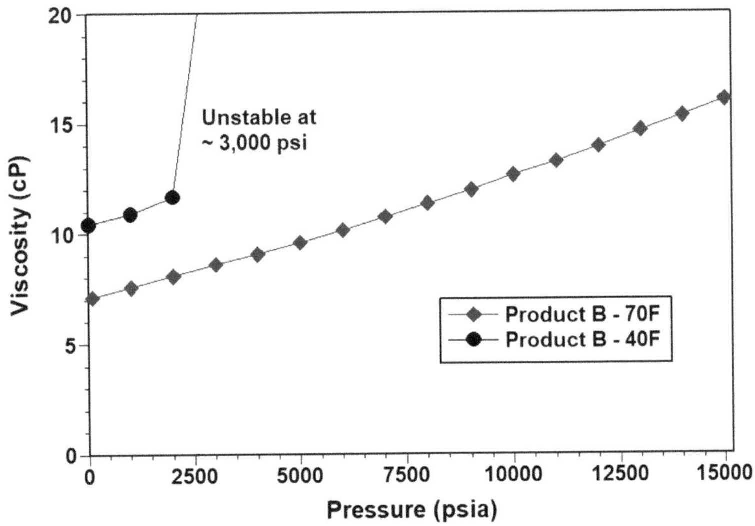

FIGURE 2.38

Example of effect of pressure on paraffin inhibitor stability at deepwater umbilical temperature.

due to application of paraffin inhibitors with inappropriate stability. Fig. 2.38 shows the effect of pressure on paraffin inhibitor stability at cold temperatures. Shown are high-pressure viscosity measurements of a paraffin inhibitor product at 70 and 40°F. At 70°F increasing pressure up to 15,000 psi causes an increase in the product viscosity but does not destabilize the product. In contrast, at deepwater temperature of 40°F, pressures above 2500 psi causes precipitation of the paraffin inhibitor polymer causing a rapid increase in the viscosity measurement. The destabilization creates a concern of the product gelling and/or causing plugging within the umbilical system. As a result, the product shown would not be recommended for applications in deepwater umbilicals at pressures above 2000 psi.

The consequence of the solubility limitations is that solvent-based paraffin inhibitor formulations are often not suitable for extreme cold temperatures (below −25°C). It is not possible to load enough of the active polymer into the product to be economically viable and still be liquid and pumpable. Solvent-based products are used for moderate freezing temperatures (−25 to 0°C). However, even with using optimal formulations to maximize polymer activity, the lower the application temperature the more dilute is the product activity becomes. Similarly, less effective inhibitor polymer chemistries are often required to be substituted in place of what could be applied in warmer locations.

The more dilute the product activity, the higher the required product dosage and the worse the overall treatment cost will be. To help illustrate the impact of

Table 2.3 Comparison of the effect of product activity on dosage.

Product activity %	Typical minimum temperature (°C)	Active dosage (ppm)	Product dosage (ppm)
100	N/A—heated injection	50	50
50	25	50	100
25	4	50	200
10	−10	50	500
2	−25	50	2500
0.5	< −25	50	10,000

product activity on treatment economics, Table 2.3 provides an example of product dosage requirement for applying the same active inhibitor dose, but with different finished product activities. The product activities range from using heated injection with a 100% active product to using 0.5% active product used in subfreezing environment. Also listed is ballpark estimate for the minimum temperature that the products could be applied. Of course, minimum application temperatures will depend on the specific product and chemistry, but estimates given in Table 2.3 provide a typical range.

Although lower active products are less expensive on a cost per gallon basis, the lower unit cost does not make up for significant increase in product dosage. Hence, as Table 2.3 attempts to emphasize, the ability to use higher active products can greatly improve application economics. Two means exist to circumvent problems with temperature limiting the activity achievable in solvent-based paraffin inhibitor products. These are using: (1) heated injection or (2) winterized dispersed paraffin inhibitors.

Heated injection allows very concentrated products (potentially up to 100% active) to be used. In addition, the full range of paraffin inhibitor polymer chemistries can generally be considered since chemistries are not rejected due to their inability to be sufficiently formulated in traditional solvent-based formulations. As such, heated injection will give the lowest treating rates and yearly chemical costs. Note that evaluation of the overall chemical cost will, however, need to factor in the cost of the injection system. For heated injection, heating is often required for chemical storage, injection tanks, injection pump, and injection lines. Moderate temperatures (~50 to 70°C) are usually sufficient to melt the inhibitor chemistries. As needed, small amounts of solvent (usually already present in the polymer reaction process) can be added to the product to help lower product melting points and viscosity to help handling properties.

Note there are some applications treating low production wells where solid high-active paraffin inhibitors in pelletized or stick-form are applied that do not use any heating. These applications batch treat by simply dropping the inhibitors into the well or backside annulus without requiring heating injection systems. The number of these-type applications is very small though.

The other method not requiring any heating is use of winterized dispersed paraffin inhibitors. Winterized dispersed paraffin inhibitors are paraffin inhibitors in which the active polymer is suspended as micron-sized solid particles in a nonfreezing (usually alcohol or aqueous alcohol-based) carrier fluid. The term winterized is used here since in warm climates solvent-based products can be used with relatively high activities. Hence, the primary market for dispersed paraffin inhibitor products is for colder regions. Over the past ~15 years different dispersed paraffin inhibitor products have been introduced to the marketplace. The use of these products has not been as great as would be expected yet from their potential advantages for applications in cold environments. This has been due to products not meeting performance expectations or providing problem-free pumping. Applications with these-type products are anticipated to increase though.

Cold finger testing for paraffin inhibitor evaluation. Cold finger testing was developed in the 1960s as a simple means to simulate deposition inside flowlines. Refinements have been made to testing and apparatus over the years, but the basic premise of testing remains the same. In cold finger testing, a cylindrical probe is inserted into stirred crude oil maintained at a fixed temperature. The probe is controlled at a temperature below the crude oil temperature (hence the name cold finger) to maintain a negative temperature differential against the crude oil temperature. This provides the driving force for deposition. Of course, the cold finger temperature must be below the WAT. And if the crude oil temperature is greater than the WAT, the driving force for deposition is WAT − cold finger temperature, not the entire temperature difference between the crude oil and cold finger.

In the cold finger apparatuses, the finger surface simulates the inside flowline wall and the stirred crude oil simulates crude oil flow inside tubulars, flowlines, or pipelines. Although the swirling, helical annular flow profile in most cold finger testing is different than the flow in production systems, useful simulations for deposition can be obtained—provided testing conditions are appropriately chosen. If there is a limitation in cold finger testing, it is in matching the shear rates in high production flowlines. In most cold finger apparatus, the maximum shear at the cold finger surface can be more than an order of magnitude lower than that present in high-rate flowlines. For example, it is not uncommon for deepwater subsea flowlines to produce 10,000–40,000 BOPD or more as the fields need to be prolific to support the cost of the deepwater developments.

Different means can be used to control temperature in cold finger apparatuses. Typically for the crude oil, it is contained in a bottle/beaker with a magnetic stirrer on the bottom and heated in a water bath or electric block heater. For the cold fingers, temperature is often controlled with coolant from a refrigerated circulating bath continually flowing through the finger. Other means of temperature control can be used though. Similarly, other variations of stirring or agitation have been used.

Different cold finger apparatus versions have varying levels of precision over temperature measurement and control; ranging from units with manifolded systems for control of multiple cold fingers to units with individual measurement and temperature control over each cold finger. Initially cold finger apparatuses were custom-built by users. Commercial units are available now from different vendors.

Fig. 2.39 shows an example schematic of one apparatus. It is dual unit with individual temperature control and temperature logging. The crude oil temperatures are individually controlled at the wall of each cold finger bottle and the two cold finger temperatures are maintained with uniform regulated split flow from a refrigerated circulating bath. The inlet and outlet temperature of coolant into the cold fingers can be measured which provides the ability to determine heat transfer from the fingers. The stirrers have tachometer controllers to provide a constant stirring/shear rate. However, the shear rate does vary from the bottom to the top of the finger with decaying annular flow field create from the magnetic stirring bar.

Examples of cold finger test results are shown in Figs. 2.40 and 2.41. The figures are presented not only to show typical cold finger results, but also to illustrate how changing test conditions can affect the deposition [76]. The effect of varying temperature differentials and shear/stirring rate is shown. As shown, the temperature differential and shear selected affect both the amount of deposition and characteristics of the deposit. Correspondingly, this illustrates why it is important to appropriately select cold finger test conditions to represent (as best

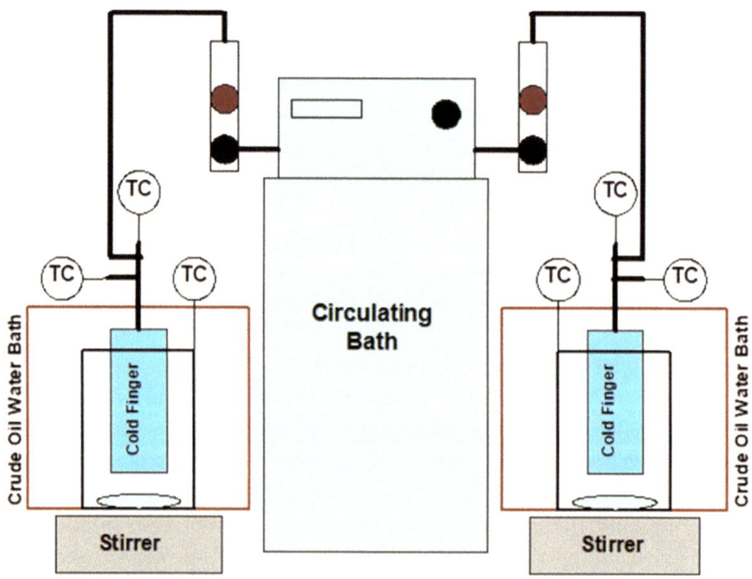

FIGURE 2.39

Example cold finger apparatus.

FIGURE 2.40

Effect of temperature differential on paraffin deposit composition ($\omega = 750$ rpm).

FIGURE 2.41

Effect of shear on paraffin deposit composition (ΔT = 15°F).

possible with a simple laboratory test procedure) the field conditions the cold finger testing is trying to simulate.

Fig. 2.40 shows results for testing with a GOM crude oil where the temperature differential (ΔT) between the crude oil and cold finger was varied between 8°F and 25°F. The crude oil temperature was held at 105°F slightly below the WAT for all tests. HTGC analysis on the cold finger deposits provided a measure of the minimum C20 + n-alkane wax content. Note the actual wax content was higher as the HTGC analysis does not fully resolve and account for all the wax. Regardless, a notable difference in the appearance and characteristics of the deposit exist. ~30% more deposition occurred in the 25°F ΔT-test. With the larger ΔT, more entrained crude oil is captured in the deposit matrix due to a faster forming and greater wax deposit network. This can be observed visually with the darker color and softer nature of the deposit and was further confirmed with HTGC analysis indicating the wax content was less than half the 8°F ΔT-test.

Similarly, Fig. 2.41 shows increasing shear rate leads to harder deposits with less entrained crude oil. In the tests shown, doubling the cold finger stirring rate (ω) from 500 to 1000 rpm resulted in reducing entrained crude oil and approximately doubling the wax content of the deposit.

Cold finger testing is most often used to evaluate paraffin inhibitors. Testing can be performed for other purposes though such as for WAT measurement (discussed earlier) and for predicting single-phase flow paraffin deposition using heat mass transfer analogies [77,78]. Since this chapter has special focus on chemical treatments for paraffin management, discussion on testing for paraffin inhibitor evaluation is highlighted here. Cold finger testing is favored over flow loop testing due to the minimization of sample requirements, screening rapidity, and ease of results evaluation. For paraffin inhibitor evaluation, cold finger test conditions should be selected to attempt to simulate (as best possible) the flowline conditions to be treated.

Note with cold finger testing, the goal is to simulate a section of the flowline, not just a single point in the flowline. Therefore the temperature differential used in cold finger testing is set to a larger value than the temperature differential present at any point in the flowline, which is typically small (less than 5°C). However, it is desirable to not exaggerate the temperature differential too much. For this reason, temperature differentials around 7.5°C–15°C are preferable. In some instances, using larger differentials may be necessary if the amount of cold finger deposition obtained from the untreated crude oil sample is small. As indicated earlier, it is recommended to simulate the initial section of deposition in the well or flowline. Again, this portion of the flowline experiences the greatest amount of deposition. As such, optimizing the paraffin inhibitor selection and performance on this section of the flowline will have the biggest impact in reducing deposition. For this instance the crude oil temperature should be set at or slightly above the WAT and the cold finger at ~7.5°C–15°C lower than the WAT.

Fig. 2.42 illustrates how these cold finger test conditions relate to simulating deposition in a hypothetical deepwater flowline for a long subsea well tieback. Shown are the flowline temperature and wax deposition profiles along with the cold finger test conditions superimposed. In this flowline the bulk crude oil temperature steadily decreases from the wellhead until it reaches the flowline wall and seafloor temperature. Similarly, the temperate differential between the bulk crude oil and flowline wall steadily decreases until everything is at seafloor temperature and no differential exists. In this example within the first mile, the flowline wall temperature reaches the WAT and deposition commences. It commences until the bulk oil temperature and wall temperature are equal at ~5 miles. As shown with the deposition profile, the largest amount of deposition occurs in the initial region where deposition begins as this region has the largest flowline temperature differential driving force for deposition. As deposition build-ups over time, a shift in deposition slightly downstream will occur as deposition starts to insulate the flowline. With setting the cold finger oil temperature at the WAT and the finger ΔT at ~15°C, it can be seen that the cold finger in this example is

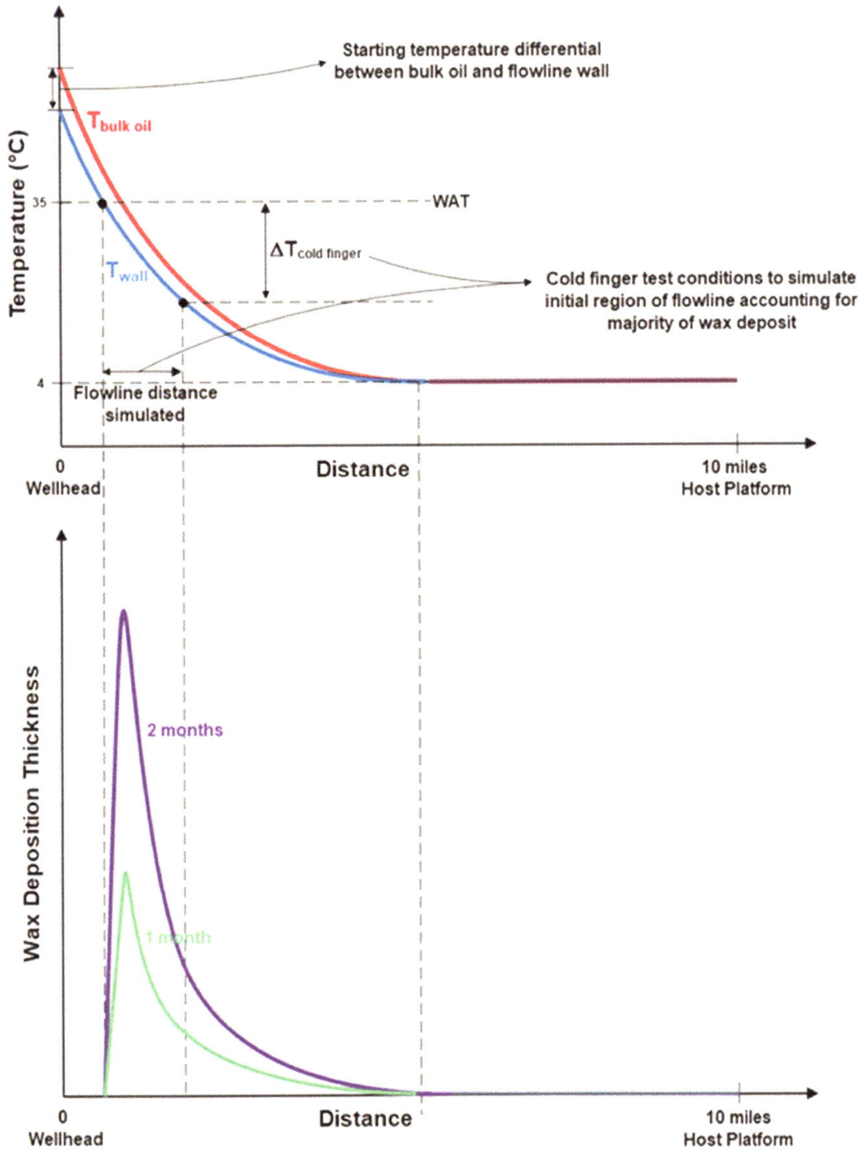

FIGURE 2.42

Example of selecting cold finger test conditions to simulate a flowline.

simulating approximately 25% of the deposition region, but this region contributes to approximately two-thirds of the deposition.

Numerous papers have been published detailing cold finger test results on specific crude oils [19,79–86]. Each of these papers demonstrates the effectiveness of paraffin inhibitor programs for controlling paraffin deposition under a variety of circumstances. The environments ranged from cold climate land-based wells to deepwater subsea applications. Cold finger data generally provide a gravimetric comparison of deposition weight between untreated and treated crude oil to gauge inhibitor performance. The cold fingers provide easy access to the deposit samples which can be removed and weighed. A percent inhibition can be determined based on the deposit weights with the following equation:

$$\%\textbf{Inhibition} = \frac{\text{Untreated deposit Wt.} - \text{Treated deposit Wt.}}{\text{Untreated deposit Wt.}} \times 100\%$$

In addition to deposit weight measurements, HTGC analysis can be performed on the deposit samples to quantify the waxes and wax carbon distribution in the deposit. For best accuracy (if time and sample size permits), the "blank" untreated cold finger deposits should be obtained for each of the individual cold fingers used in testing.

Below is an example of cold finger testing performed for selecting a paraffin inhibitor to reduce deposition in the production tubing of deepwater dry-tree wells in a GOM field. In this field, temperatures in the well production tubing strings decreased as production rates declined. This allowed wax deposition to occur in the tubing leading to the wellheads on the platform. Wireline cutting was used periodically to remove the deposition. Paraffin inhibitor application was pursued as a means to reduce the deposition to decrease the wireline cutting thereby reducing the downtime from wireline operations and increasing total annual production. The cost for the paraffin inhibitor application was significantly less than the deferred production. When it was decided to initiate a paraffin inhibitor program to reduce the deposition in the problem wells, cold finger testing was performed to select a paraffin inhibitor for the application. At that time for the problem wells, the temperature differential between the crude oil WATs and the FWHTs ranged from approximately 20°F to 34°F. To simulate the deposition in the tubing, cold finger testing was performed with the crude oil temperature held at approximately the WAT and the cold finger temperature 20°F lower. The exception was testing for one well with the higher 34°F temperature differential. Testing for this well was actually performed with two test series (a higher and lower temperature test) each with 15°F ΔT between the crude oil and cold finger temperatures. As such, the cold finger test conditions were selected that provided a good simulation of the deposition occurring in each production tubing string.

Fig. 2.43 shows some of the cold finger test results with the paraffin inhibitor chosen and applied in the field. Shown is the effect of product dose rate on inhibition performance on crude oil from one of the wells—a 20°F ΔT-test. Note, at 300 and 400 ppm, the inhibitor is excellent at inhibiting deposition. At 200 ppm the performance drops off more than 50% though. Results with other wells were similar.

It should also be mentioned that the inhibitor chosen was not deepwater umbilical-approved. Although the product is stable at 40°F and ambient pressure (such that it does not freeze in injection tanks in GOM during winter), high-pressure viscometer testing indicated potential concerns at elevated pressures

FIGURE 2.43

Example cold finger testing for product selection for a deepwater GOM field.

above 2000 psi at 40°F. With the dry-tree capillary injection application, the product was not exposed to seafloor temperatures under pressure during injection though. The capillary injection string was strapped to the outside of the production tubing and therefore was warmed by production flow. The only possible scenario that the product would experience deepwater seafloor temperature would be during an extended shut-in. And even in this instance, much of the product within the capillary line (downhole beneath the mudline and above toward the surface) would be warmer. In particular, the region of greatest concern toward plugging is with the check-valves. These were downhole at the injection point in this application (as with most capillary lines) and at elevated temperatures.

Another paraffin inhibitor application with example cold finger testing for a deepwater subsea flowine is included in Section 2.6.

Note cold finger–type deposition testing has been performed with other apparatuses with different geometries, principally cold disk [19,87] and concentric cylinders or Taylor–Couette-type [88] geometries. Testing with standard cold finger apparatuses is much more prevalent though. The Taylor–Couette systems are notable in that the shear rates in the apparatus flow fields can better match the field conditions with high-rate production systems. Some systems have also been designed to handle pressurized live fluids.

2.4.4.4 Treatment considerations for different problem locations

The complexity and variability of production systems throughout the world require multiple application techniques to address paraffin deposition. A total systems analysis is required to identify paraffin deposition locations, management options, and economics. In many locations a combination of mechanical, thermal, and chemical approaches will be employed to arrive at an optimum economical approach. Typically, the earlier in the production system that paraffin problems occur, the more costly the problem to the operator. This trend holds for formation damage, tubing deposition, flowline deposition, separator interface issues, tank bottoms, and pipeline transportation. Treatments for each of these paraffin issues are discussed next.

Formation damage

There are a variety of problems leading to formation damage and the development of "skin" i.e. plugging of the crude oil producing pores with a variety of solids or viscous materials. This damage can consist of formation fines, scale, paraffin, asphaltenes, and emulsion blocks. There is often confusion between what is considered the field's natural production decline and that of the development of skin damage. For any given well the skin damage can be a combination of various inorganic and organic damages. Damage is often addressed with acidizing, fracturing, or reperforating the well. Many of these options are expensive. If the majority of the damage is inorganic, then acidizing generally helps. If the damage is primarily organic, then this approach generally fails. Many published articles have covered this topic [89–93]. Formation damage can be assessed by running a pressure build-up study determined by measuring fluid levels during the shut-in period. The longer it takes to reach a steady-state fluid level, the more formation damage exists.

For paraffin formation damage the common approach is to use solvents for deposition removal. These are normally batch treated at a volume sufficient to provide an 18-in. penetration across the perforated interval. Note that formation damage only at the near-wellbore has a significant impact at reducing production, so only treating the immediate near-wellbore area is required. Further, if paraffin deposition is the cause of the formation damage it generally is not present within the reservoir, only at near-wellbore. Hence, an 18-in. treatment provides near optimal economic impact. An appropriate amount of crude oil overflush is used to hold the treatment in place against the bottom-hole pressure. The well is generally shut-in overnight before production is resumed. These applications can be applied regardless of the production type. Rod-pump and electric submersible pump (ESP) wells are treated down the annulus. Flowing and gas-lift wells are treated down the tubing. Sometimes, it is necessary to pretreat the tubing if substantial wax exists to avoid pushing this wax back into the formation for the open tubing applications. A formation damage profile is shown in Fig. 2.44.

One problem for high GOR condensate wells is wax deposition in the perforations. Joule–Thomson cooling can occur across the perforations resulting in localized cooling above 20°C [55,90]. An example of this for a gas/gas condensate well is shown in Fig. 2.45.

The gas/gas condensate wells in this field had a tendency to plug the perforations for a period of time. The multiple perforations would alternate between a

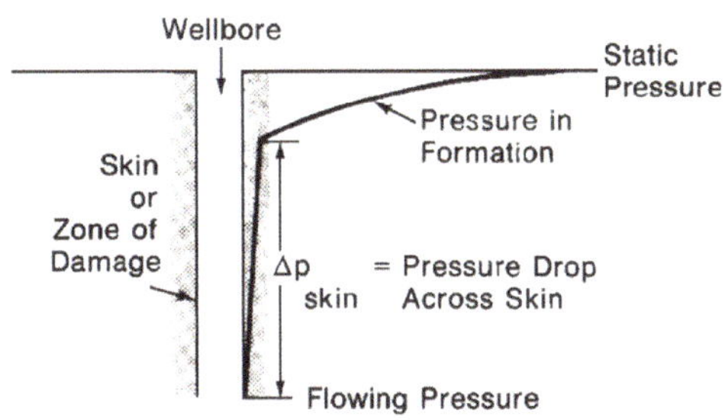

FIGURE 2.44

Formation damage pressure profile.

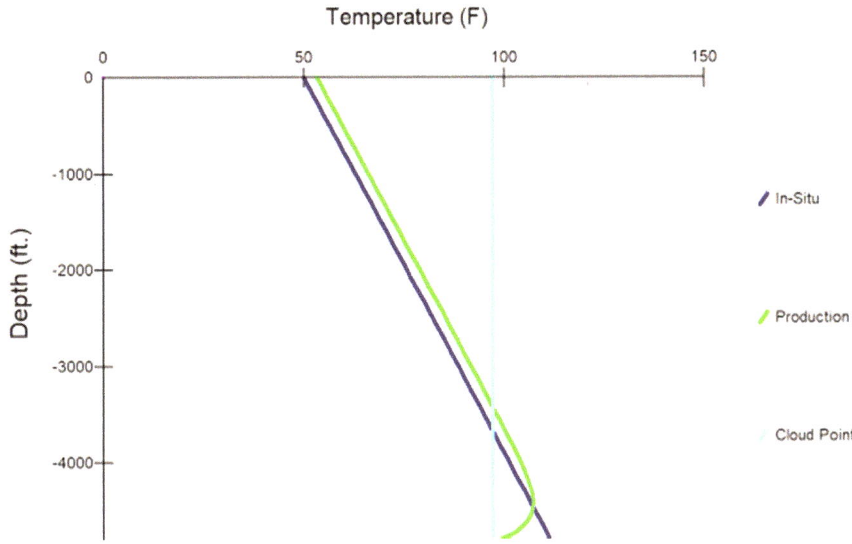

FIGURE 2.45

Gas production cooling effects in a well.

flowing and plugged status. When the perforations were plugged with paraffin deposition from the cooling, the production flow would stop. With no flow the wax deposits would warm to bottom-hole temperature and in some instances start producing again. During other times the perforation would remain plugged and not melt. Eventually, the production would drop such that remediation was required. Similar effects are observed in CO_2 flooding operations. Paraffin deposition caused by Joule–Thomson cooling in wells with CO_2 flooding operations is common.

Squeeze treatments of paraffin inhibitors into formations that suffer rapid damage from paraffin deposition have been tried with variable success rates. The major problem with a paraffin inhibitor squeeze treatment is the high minimum effective concentration (MEC) required to treat the problem. A MEC of 100 ppm or higher is often required to effectively treat a paraffin deposition problem. This is 10–20 times higher than that required for a scale inhibitor squeeze treatment. A 7-ft. formation penetration is generally used. This requires a substantial volume of overflush which is generally crude oil. This results in a higher placement cost than a scale inhibitor squeeze and further reduces the chance of economic success. In addition, most paraffin inhibitors are not designed to absorb to the formation, so a significant portion of the squeeze may be returned rapidly rather than extended overtime like scale squeezes. This further decreases squeeze performance and increases cost for paraffin inhibitor squeezes. Variable success has also been seen with incorporating paraffin inhibitors into fracking operations. There are applications that have demonstrated success though, especially in shale wells.

The majority of squeeze treatments performed have been in vertical wells with a limited number of production zones. For an increased chance of success, multizone wells require isolation and treatment of each producing zone. This adds to the cost of the treatment. Extended laterals in horizontal wells with multiple fractures are a particular problem for treatment if damage occurs. The use of isolation tools or diverters is required for proper placement. The volume of chemical required is also substantial. Great care in the design of these treatments is required to improve the chances for success.

Well tubing deposition

Paraffin deposition in the tubing string represents a major cost to producers. Typically, subsea wells will be designed to produce at a level where the WAT is not reached until the flowline. This eliminates downhole paraffin deposition concerns until the pressure has been depleted and the well is potentially abandoned. If the well is not abandoned and is producing at diminished rates (with or without artificial lifting techniques), then paraffin deposition is a concern. Most deepwater wells with paraffin deposition concerns are treated with a paraffin inhibitor on a continuous basis. Cold finger studies are usually conducted to determine the best inhibitor. The dry-tree GOM application discussed earlier is an example. For open tubing production the inhibitor is applied via capillary or into the gas lift gas. For subsea wells, the inhibitor will need to be pumped through an umbilical line before reaching the capillary line. Specific product qualification tests are necessary to qualify a chemical for application via these methods. These tests focus on evaluating product stability/cleanliness, physical properties (viscosity, density), and compatibility (chemical and material) across a wide range of temperature and pressures.

For deepwater subsea wells, if deposition does not occur until the flowlines, then injection at the wellhead is preferred over injection into a capillary line down the well. The paraffin inhibitor would experience lower pressure in the umbilical lines thereby reducing pressure effects on the paraffin inhibitor stability.

This potentially could allow a different, better performing paraffin inhibitor to be used. For deepwater dry-tree wells, depending on capillary line installation, injection downhole may not have any downside.

Paraffin inhibitors are often the most cost-effective approach for treatment of paraffin deposition. However, in cases where paraffin inhibitor application is not feasible, an alternative to paraffin inhibitors is the use of dispersants. This is especially applicable for high population-density onshore well locations like the Permian Basin in the United States where many wells are treated with batch treating trucks. Rod-pump and ESP wells are the targets for this application method. Treater trucks are equipped with water tanks, chemical compartments, and pumps to apply the chemicals. The water capacity is in the range of 20–40 barrels. The tank is filled with produced water from the field and driven to the well. A high-pressure hose or string of pipe is connected to the well and the treatment is pumped down the annulus. The paraffin dispersant is slip-streamed into the water for even chemical distribution. Multiple chemical applications including scale inhibitor, corrosion inhibitor, and paraffin dispersant treatments can be applied during a single stop. This has the advantage of eliminating all chemical injection pumps and in-field chemical placement at each well site. These treatments are conducted on a scheduled basis every week or two. Sometimes, the problem severity allows treatment on a less frequent basis.

Paraffin dispersants are selected using the flask test with field produced water source. The field treatment volume required is generally set at ~2 hours of contact time based on the total fluid production rate of the well. If the well produces 24 barrels per day, then the treatment would be a 2% chemical solution in 2 barrels of water. The treatment going down the annulus will fall through any fluid level in the annulus and stack-up at the pump. There it will be picked up by the pump and moved up the tubing as a treating solution. This application method has proven to be highly effective [55,79,81].

Monitoring the success of paraffin treating applications for tubing deposition can be challenging. Batch treating applications can be monitored by catching fluid samples as the treatment reaches the wellhead. The volume of wax removed with the treatment can be viewed. Amperage loads on rod-pump wells can be tracked to potentially spot an increase in power required to lift the rods due to wax deposition. Production rates or well productivity indices can be monitored to identify a potential drop in production. Power requirements for ESP operations versus production levels can be monitored for increases. Cutting operations for open tubing applications can be reviewed for frequency, run time, and ease of cleaning.

The major goals for paraffin control in the tubing are the maintenance of production levels and the avoidance of high-cost tubing pulling jobs. Both rod-pump and ESP production wells pose particular difficulties and have the potential for high-cost operations. Rod-pump wells require the removal of the rod string for a variety of reasons. Rod failures due to corrosion and pump replacements are common. Pulling the pump out of the well poses difficulty because the pump acts as a pig coming up the tubing. This will cause removed paraffin to stack on top of the pump and increase resistance on the pulling unit. The resistance could increase to the point that the unit capabilities are exceeded and the pump is stuck. If the pump is further down in the well below about 300 m, then hot oiling down the annulus will not unstick the pump. The tubing will then need to be pulled. The cost of one tubing pull can exceed the cost of an effective paraffin treatment program for a year. The same case exists for change-out of ESPs, but the pull frequency is much lower.

A wide variety of costs for tubing-related paraffin issues are often not considered. Downtime, production loss, increased power usage, retreatment costs, and environmental impacts are a few examples. The actual costs can be 2–5 times higher than that allocated to hot oiling or cutting costs alone.

Flowline deposition

Flowline deposition problems are less costly to address than problems further back in the production system. Subsea deposition is primarily handled with pigging and/or paraffin inhibitor applications. If paraffin deposition is substantial, deepwater developments will often be designed with dual flowlines to allow pigging to be the primary deposition control method. If deposition is less severe and the crude oil is amenable to paraffin inhibitor treatment, reliance solely on paraffin inhibitors may be considered. As expenditures for a single flowline (albeit a slightly larger diameter flowline) will save significant CAPEX over dual flowline installation, operators have significant incentive to consider the single flowline option. This is especially the case with developments for single subsea wells tied-back to an existing host facility. For this production scenario a single flowline is the primary design. With single flowlines, paraffin inhibitor applications are the primary deposition control means. Note paraffin inhibitor applications generally do not provide 100% inhibition. Therefore the inhibitor performance needs to be evaluated and considered in the paraffin management program employed. For single flowline without pigging, it can often become a race between the production decline of the well(s) versus the build-up of paraffin deposition over time.

Subsea paraffin inhibitor applications require installation of umbilical injection lines. Ideally, injection is preferred at the wellhead. However, if deposition protection downhole is also needed, then a capillary line into the well would be required too. Inhibitor treatments utilized for tubing deposition will generally carry over to handle flowline deposition. Paraffin inhibitor treatments are often combined with pigging when significant deposition occurs. The paraffin inhibitor's function here is to reduce the deposition allowing the pigging frequency to be extended thereby reducing the amount of deferred production. If acceptable paraffin inhibitor performance is obtained, the increased production revenue from the reduced downtime can more than pay for the paraffin inhibitor application cost.

For deepwater pigging operations, concerns over sticking a pig are justifiably high. If a pig is stuck within a dual flowline system, then capacity is essentially cut in half. Since production rates in most deepwater operation are relatively high, this means significant production and revenue is lost. In addition, unsticking a pig in a deepwater operation is not an easy task. Most deepwater flowlines are extended long lines which can pose a challenge to reaching the stuck pig with coiled tubing depending where in the system a pig becomes stuck. Relatedly, if a stuck pig plugs the flowline, there is major cost associated with remediation by any method. Perhaps the most important practice to avoid sticking a pig is to maintain pigging on a regular interval. Combining with paraffin inhibitor applications can help as well. Other operation practices such as adding a slug of solvent loaded ahead of the pig are sometime employed too.

If a pig becomes stuck during a pigging operation, the time of the run and flowrate should be used to calculate the approximate location of the pig in the line. It is common practice to blow down the pressure on the pushing side and pressure up from the receiving side in an attempt to push the pig back to the launcher. This may take multiple pressure up and bleed-down cycles to get the pig moving. If this fails, other more complex methods would be required. Different controlled pressurization methods or means to sonically transmit energy and resonant or vibrate plugs exist. The use of extended reach coiled tubing may be a possibility too.

Many subsea systems are designed for pigging operations, but once installed there can be a general reluctance to utilize the capability due to plugging concerns. Once scheduled, pigging operations are skipped (for whatever reason) and there is also reluctance to restart pigging later. Sometimes, a progressive pigging program starting with smaller diameter and softer pigs then progressing gradually stepwise with larger and harder pigs will be attempted to reset to normal baseline. In many instances though, pigging capability is effectively lost and chemicals are used solely to mitigate the problem. Two detailed case histories from BP on managing progressive pigging programs in their Azerbaijan operations in the Caspian Sea illustrate the complexity and engineering that can be required to successfully carry out progressive pigging programs [94,95]. The case histories also illustrate the usefulness of wax deposition modeling to help understand what is occurring in the entire production system through the field lifetime. The value of deposition modeling is not only limited to estimating pigging frequencies prior to start-up. In contrast, another BP case history illustrates the concerns involved in progressive pigging programs. The case history discusses the sticking of a pig (and causes behind the incident) during a progressive pigging program carried out in the BP GOM Marlin export line [96]. Fortunately, the stuck pig was able to be released by back pumping to the Marlin platform. It did, however, take 10 days to plan and execute the operation. Hence, in addition to the cost of the operations, significant lost production was incurred.

For subsea flowlines that have not been pigged, paraffin deposition often leads to loss in throughput capacity. The pressure drop (normalized with production volume) can be monitored to calculate the extent of paraffin deposition over time. Cleaning a subsea flowline to remove deposited wax is a challenging exercise. Many attempts in using solvent have failed in the past. The low temperature of deepwater subsea lines greatly limits the solubility of the wax in the solvent. In addition, the deposits in the line are often aged over time and contain little entrapped oil and are harder to dissolve. Total line displacements with xylene and extended soak periods of 2 weeks have resulted in no observable increase in throughput for a number of locations. Some successful applications have been achieved using dispersants in water [70,97,98]. The low deepwater temperatures do not affect this method to the degree observed with solvents. Details of this approach are discussed more in Section 2.6.

Flowlines located on land or in shallow waters are more easily addressed for paraffin control. As previously discussed, hot oil treatments are a major remediation method used despite drawbacks. Paraffin inhibitor and dispersant chemical applications are also used to a great extent. An appreciable amount of water production is necessary with dispersant applications. While paraffin inhibitors offer the most benefit if they are added to the crude oil above the WAT, many flowline applications are below the WAT of the crude oil. Many flowing wells for these systems do not have capillaries available that would allow injection of the paraffin inhibitor into the crude oil before the WAT. Paraffin inhibitors are still relied on for control of flowline deposition as it is acceptable to add an inhibitor after the WAT for a flowline application because the wax to be inhibited is a lower molecular weight than that depositing in the well tubing. As long as the paraffin inhibitor is selected properly to inhibit the wax depositing in the flowline temperature range, the product can be effective. The dispersant in water application technique can also be highly effective for the removal of wax in flowlines and an attractive option for locations that have treater trucks readily available.

Separation equipment

Wax particles sheared off in the system or present as suspended precipitated particles from the crude oil have a tendency to create interfaces in separation equipment. These interfaces build-up over time and interfere with the separation of the crude oil and water. This is not a "deposition" problem but is a common paraffin problem encountered by many operators and therefore included here. The lower the temperature of the separation equipment, the more severe the problem. Interface wax comes from bulk wax precipitated from the crude oil, wax sheared from flowline deposition, or wax from cutting and hot oiling operations. Wax precipitation from shale oils can be particularly problematic. The volume of interface in the separation equipment can be determined through thermal imaging or sampling vertically down the separator. The build-up of paraffin interface is a maintenance issue. It is not possible to appreciably reduce the WAT of crude oil. The crude oil chemistry and thermodynamics determine the amount of wax that can precipitate from solution. There are different methods to address interface problems. The first method is thermal where the contents of the separator are hot oiled to melt and dissolve all the wax into the separator oil stream. This will solve the problem in the separator but move it downstream to the storage tanks. This will add to the volume of tank bottoms. A second method would be to dump the crude oil portion of the separator to a slop oil tank where it can be worked off either with using thermal or chemical treatment. The third method would be to treat the separator chemically to reduce or eliminate the interface. Paraffin detergent/dispersant combinations are typically used to treat interfaces. These dispersant formulations are different than those used for paraffin removal with water previously discussed. The interface control agents will typically contain less surfactant, more dispersant, and a demulsifier component. For interface clean-up, it is more economical to batch treat the separator rather than try to treat at a lower chemical rate continuously. The volume of crude oil in the separator is calculated and the system is treated with about 500- to 1000-ppm batch treatment of interface dispersant based on the crude oil volume. The dispersant can often resolve the complex emulsion in the interface (crude oil, paraffin, water, inorganic solids, etc.) and allow the wax to disperse into the crude oil and move to the storage tank.

A secondary issue in separation equipment that occurs in gas separation systems is actual paraffin deposition. This typically occurs when there is liquid condensate carry-over from the high-pressure separator. The gas carries the liquid over in the system from foaming that occurs with high rates of gas production. Any liquid condensate carried over in the system will be subject to paraffin deposition from molecular diffusion when a temperature gradient exists. There can also be shear dispersion deposition from precipitated wax particles. Typically, these problems are addressed with a paraffin inhibitor/dispersant application. Chemical dosage rates are generally 500–1000 ppm based on the volume of condensate carry-over. Foaming is also addressed with defoamer application.

Tank bottoms

Tank bottoms are somewhat similar to separator interfaces and are handled in a similar manner. It is also strictly speaking not a deposition problem, but rather a solids-settling issue. Paraffin in a tank bottom is a valuable saleable product. It will pass through a local account custody transfer unit if residual water has been removed. For tanks that are sold for truck transportation considerable problems often occur. Generally, the tank is sampled at various levels for basic sediment and water (BS&W) content and the average BS&W calculated. If below a certain value, the crude oil is accepted. If the thief (sampling vessel) used to capture the sample fails to trip on the bottom of the tank sample, the load is generally rejected. The thief has a stinger on the bottom of the unit which trips the sample vessel closed when the stinger hits the bottom of the tank. If a paraffin bottom exists, the stinger cannot hit the bottom of the tank and trip the sample vessel closed. Other problems exist with the proper determination of the BS&W values for all the samples. Care must be taken to ensure that paraffin wax is not considered basic sediment. The proper procedure for BS&W determination is ASTM D4007 [99] or IP 359 [100]. The critical aspects of this test which are often overlooked can result in tank rejection due to improper testing. The solvent used in the test is toluene, not Varsol, diesel, gasoline, or other solvent. The ratio is 50%

toluene and 50% sample volume. A demulsifier is used to separate the crude oil and water. The sample is heated to 60°C and allowed to equilibrate for 15 minutes. The sample is then centrifuged for 10 minutes at a minimum centrifugal force of 600. The values are then read and recorded. The procedure is then repeated until the values remain constant for two consecutive samples. If any of these steps are not followed, the potential exists to record paraffin as BS&W. The worst-case scenario is that the tank is rejected. A secondary effect is that the producer gets short-changed on volume because the wax is subtracted from the volume load.

Tank bottom control is also a maintenance issue. To avoid a tank rejection the bottoms needs to be managed on a regular basis. Precipitated wax will gravity settle in the tank over time. This settling will accelerate when the wax is combined with water, residual emulsion, and inorganic solids. The bottoms can be thermally treated with hot oil just prior to sale. There is also the option to chemically treat the bottom with the same type of products used for interface treatment. This combination chemical is designed to resolve the residual emulsion, water-wet inorganic solids, and disperse the wax up into the crude oil. Following a successful treatment, the free water and inorganic solids need to be drained from the tank. Proper agitation is a major factor in allowing the treatment chemicals to resolve the bottoms. If tank stirrers have been installed this is a plus for handling these problems. If stirrers are not available, then use of a vacuum truck to provide mixing would be a secondary option. Crude oil is drawn from the bottom of the tank and returned through the top. Often a length of pipe is used to jet the crude oil directly into the tank bottom. Another option would be to use a hot oiler to both circulate and heat the bottom to melt the paraffin. Again, the use of a pipe to direct flow to the tank bottom would help break up the bottoms and retain the heat. The worst-case scenario is no available agitation. For this situation the level of bottom needs to be managed before a tank is rejected. If monitoring indicates the bottom is approaching the point of rejection the following procedure would be used. Immediately following the sale of a tank of crude oil, batch treat 500–1000 ppm of chemical to the remaining tank bottom. Produce crude oil into the tank under normal operations. The addition of the crude oil to the tank will provide a minimum level of agitation to the bottom during the initial loading stage. This is by far the least efficient application choice, but unfortunately the most often practiced method.

For very large tanks requiring inspection, the cleaning operation can be quite involved. Mechanical cleaning is always an option but often proves to be quite expensive. An alternative method involves a multistep process using water as a heat transfer medium. The crude oil is removed from the tank down to the tank bottom deposit. A volume of water containing 2000 ppm of a selected chemical is pumped into the tank. The volume of water should equal the volume of bottom present. The water is then circulated through a heat exchanger and back into the tank until the temperature of the water in the tank reaches 65°C. This water temperature is maintained until sampling of the bottom indicates all the bottom has floated to the top of the water and the top layer temperature is 60°C. Only the inorganic solids would remain below the water surface. Measure the volume of crude oil and melted paraffin floating on top of the water. Add a volume of clean oil or diluent equal to the volume of the crude oil/paraffin layer slowly while maintaining the water temperature and continue to circulate. After 4 hours of run the BS&W and pour point on the crude oil layer. Continue to test until the values are acceptable. If the pour point is high, add more diluent. If the values are acceptable for sale, drain the water from the bottom of the tank until the crude oil layer reaches the drain valve. Then empty the crude oil to the sales line. Empty all water from the tank. Then spray the inside of the tank with a solution of water and 1000 ppm of a surfactant selected to emulsify any remaining crude oil and degas the tank. Empty this wash water from the tank. Finally, then the tank can be entered to clean-out the remaining inorganic solids and inspect the tank.

Export pipelines

Pipelines are the easiest paraffin problems to handle as long as the lines are properly constructed with pigging capabilities. As mentioned previously, various modeling programs are available to ascertain pigging frequencies. Many pipelines can manage paraffin deposition problems with pigging alone. In some instances the travel time of the pig may be so long that the deposited wax may exceed the volume available in the pig catcher. In this instance, paraffin inhibitors may be used to reduce the deposition volume and increase the period between pig runs. In all instances the pig used in the run needs to be a scraper pig to remove the wax down to the pipewall. It is imperative to maintain a pigging program at the optimum run rate and not skip pig runs. Loss of the ability to pig a pipeline can have a major economic impact. Even a small amount of deposition per barrel of crude oil in a pipeline can overtime result in a large volume of wax deposition especially when production rates are high. For pipelines not pigged, this could impact throughput and result in multiple tons of wax in the system. Any attempt to clean these lines to gain throughput would be very costly.

2.5 Pour point/crude oil gelling problems

2.5.1 Crude oil gelling mechanism

As indicated earlier, waxy crude oils can become viscous and, in some cases, completely gel solid due to the formation of wax crystal network lattices within the bulk fluid. The network formation first starts to impede crude oil fluidity and later, if the network growth becomes extensive enough, completely solidifies the crude oil.

Crude oils with pour points above ambient conditions are not uncommon. In certain global regions, high pour points crude oils are prevalent due to the maturation process with the hydrocarbon source material present within the region millions of years ago and the reservoir conditions creating the crude oil. Fig. 2.46 shows an example of three waxy crude oils with pour points significantly greater than room temperature. Pour point may also become a concern in crude oils which do not necessarily have "high" pour points but become subjected to cold temperatures. Operations during wintertime and deepwater subsea production are two cold temperature environments which can cause pour point concerns. Fig. 2.12 includes a couple crude oils produced from subsea deepwater wells which fall into this category. Fig. 2.47 shows an onshore operation experiencing issues in wintertime. Shown is gelled crude oil from a separator. Notice the snow in the picture.

FIGURE 2.46

Example high pour-point waxy crude oils.

FIGURE 2.47

Congealed crude oil from separator in wintertime.

FIGURE 2.48

Cross-polarized microscopy picture showing wax crystal lattice in waxy crude oil.

Fig. 2.48 shows a cross-polarized micrograph view of the wax crystal lattice network in a waxy crude oil. If the picture is observed closely, the three-dimensional structure of the network can be distinguished due to variations in image intensity and clarity at different focal planes of the microscope view. Notice that the network is throughout the fluid. How various wax crystal networks form is dependent not only on the crude oil and paraffin chemistry, but also on the thermal and shear history the crude oil experiences. The thermal and shear history influence the formation and breakage of network linkages, thereby affecting the structure and strength.

The result of increasing crystal network growth on the impact of increasing viscosity and gelling of crude oils was shown earlier in Figs. 2.6 and 2.12. With regards to production, both the increase in viscosity and gelling are a concern. A significant increase in viscosity translates into reduced production and revenue. And of course, a gelled flowline means no production. Therefore there are significant economic incentives for treating high pour point crude oils.

2.5.2 Gelled flowline characteristics

The concern for high pour point crude oil gelling exists all the way through the production and delivery chain (wells, flowlines, processing equipment, export pipelines, tankers, rail tank cars, and storage tanks). If issues exist, it is important to fully understand the nature of the problem and have a means to rectify it. Considerable attention has been given to understand problems within flowlines. Specifically, on the criteria for restarting production in flowlines with gelled crude oil. Various factors affecting restarts and estimating restart pressures are discussed in various articles in the literature [101–104].

The gel strength and length of the gel plug play prominent roles in dictating the force or restart pressure required to restart production. As discussed earlier, the specific thermal and shear history during cooling affect the gel network formation and, along with aging, determine gel strength. The length of the gel plugging flowline may not be continuous. If dealing with a single-phase export lines, the gel plug will generally be continuous. If the flowline has multiphase production, then sections of the flowline might be filled (or partially filled) with water or gas depending on production ratios and topography of the flowline. In these cases, various sections or lengths of gelled crude may alternate with water/gas filled sections. Temperature differences in long flowlines can be another scenario where different sections of the flowline are gelled or not. The simplest restart pressure ($\Delta P_{\text{restart}}$) estimate is:

$$\Delta P_{\text{restart}} = 4\,\tau\,D_{\text{f}}\,L_{\text{plug}}$$

where τ is the gel yield stress, D_{f} is the flowline diameter, and L_{plug} is the length of the gel plug. If there are discontinuous sections of different gel plugs (e.g., as in multiphase flow situation discussed previously) with each having equivalent yield stress, then L_{plug} would be the length of the longest gel plug section.

In reality, the situation is more complex [101–104]. It has been noted that estimated restart pressures using the abovementioned equation with laboratory measurements are often conservative and significantly overestimate restart requirements. Variations in yield behavior measurements in different rheometer geometries and other laboratory restart/yield stress apparatuses are common. Similarly, variation in gel characteristic and strength can exist within different sections of the flowline. There are also complexities how the applied restart forces propagate through the flowline as well. The propagation of pressure through the gel(s) is not instantaneous. Even if a sufficiently high restart pressures are applied, it may take time for the force to ultimately cause the gels to yield.

2.5.3 Pour point treatment

Chemical treatment with PPDs is the most common means used to assist production of crude oils with pour point issues. Thermal control with insulation and heating are also used. In some cases, some management of production and operational processes can be used to help control problems.

2.5.3.1 Production management

If nearby low pour point crude oils or condensates are present, diluting high pour point crude oils with commingling production to reduce pour point concerns might be a simple and least costly means for dealing with pour point problems. Unfortunately, having suitable crude oil/condensate sources that can be used for this purpose is often not an option for many production systems. And using other dilution fluids such as diesel generally is cost-prohibitive as well as poses logistic challenges. In cases where diluent fluids are available, the effect of dilution will need to be evaluated at different levels of dilution. If the available amount of the lower pour point fluid is not sufficient to totally eliminate the pour point issue, it may still be useful as part of a treatment solution. Dilution when used in conjunction with other treatment methods can still provide a significant impact. One example with combining dilution with PPD treatment is given in Section 2.5.3.

In a much different embodiment of dilution, there has been a reported application of pumping a high pour point Indonesian crude oil as a slurry with water through a long pipeline [105]. This is not a common practice.

If a field's pour point concern is only to protect flowlines or pipelines from the crude oil solidifying during shut-in, displacement of the lines with a nongelling fluid/crude oil during shut-ins may be a possibility. This requires that a sufficient volume of the nongelling fluid be stored/available for unexpected shut-ins and can be pumped within the necessary operational window though.

2.5.3.2 Thermal methods

Insulation and/or heating are thermal methods that can potentially be used to assist handling high pour point crude oils. These have been discussed previously so only minimal discussion is included here. A concern with use of insulation is the potential for unexpected shut-ins if the potential exists for the crude oil to gel and plug the production system. If an unexpected shut-in occurs for extended period, insulation will not be sufficient to prevent high pour point crude oils from gelling.

The additional concern for insulation and heating is often cost. Insulation and heating can require significant added capital expense. With heating (if planned for continuous heating), significant operation cost is also added and generally is not economical. If only required to periodically heat lines such as during shut-in, OPEX economics will be greatly different. Pipelines transporting high pour crude oil represent a particular challenge. Some systems will employ periodic line heaters to keep the crude oil above the pour point temperature. This is a high expense unless there is also stranded gas present that has no viable market. Even in these situations, there is still the issue of equipment maintenance and transportation of the gas to multiple heating stations.

Whether thermal management is practical or optimal needs to be determined on individual basis. For very high pour point oils beyond the range of chemical treatment, thermal methods although expensive might be only technical possibility.

2.5.3.3 Chemical methods (Pour Point Depressants)

PPD treatments are generally applied for the following reasons or combination thereof.

1. To reduce viscosity to increase production, which in turn increases revenue.

2. To eliminate or minimize the possibility of crude oil solidifying in the production system (wells, flowlines, export pipelines, storage tanks, tanker cars, etc.) thereby reducing risk of complete production shutdown. If the field production rate is high and/or the means to remediate plugging is difficult, lengthy, and/or costly; then the incentive for treatment is high.
3. In some treatment schemes the goal is to not completely eliminate gelling during shut-in, but to apply only enough PPD treatment to reduce the gelled crude oil yield strength to a range low enough that would allow restart should extended shut-in occur. In this approach, PPD treatment cost is minimized. In production systems that are at sufficiently high temperatures under normal flowing conditions and no benefit exists from a PPD to increase production by reducing viscosity, this treatment strategy can be attractive. The challenge is being certain the PPD treatment will achieve its goal.
4. To meet crude oil specifications to transport crude oil into another entity's pipeline or tank farm. In this instance, the pipeline or tank owner/operator desires treatment for the above reasons. But for the individual producing companies supplying crude oil, their treatment goal may simply be to meet specs to allow transport and sale of their production with less concern of shut-in consequences. Particularly if the producer believes treatment is a cautious conservative measure for preventing risk. Consequently, these producers often do not have incentive to improve treatment beyond the bare minimum to meet specs.

PPD applications are common for treating crude oils with pour point issues. The success of applications is dependent on the crude oil, level of performance required, and PPD selected. PPD performance can be highly selective for specific crude oils. Performance is dependent on matching the correct PPD chemistry/chemistries to the crude oil wax distribution profile being treated. Unlike paraffin inhibitor treatments for reducing deposition, this requires the PPDs to interact with waxes besides the high-molecular-weight species.

In PPD treatments the same wax crystal modification occurs as with paraffin inhibitors for deposition control except modification of the networks in the entire bulk fluid must occur, not just in the deposit on the flowline surface. The untreated wax crystallization can be thought as a stacked plate structure interconnected in a honeycomb-type structure which traps the remaining crude oil in the matrix. This causes the crude oil to gel once the network is extensive enough. Crude oils treated with PPDs have smaller crystal network growth, which have difficulty forming a honeycomb-type structure. Less crude oil is trapped in the wax network leading to increased fluidity at reduced temperatures. If further cooled to even lower temperatures, eventually enough new wax crystallization may occur to overcome the effect of the PPD and the crude oil might yet gel. A schematic representation of PPD effect is shown in Fig. 2.49.

The chemistries of PPDs are the same as those used for inhibition of deposition. However, certain chemistries may function better as PPDs than as paraffin

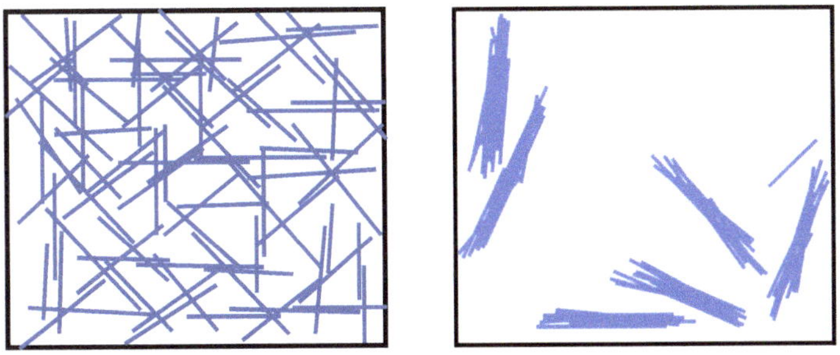

Gelled Uninhibited Crude Oil **PPD Inhibited Crude Oil**

FIGURE 2.49

PPD alteration of wax crystal network lattices. *PPD*, Pour point depressant.

inhibitors for preventing deposition. As with deposition inhibition, obtaining effective performance depends on matching correct PPD chemistries with the specific crude oil and its production conditions. Thermal and process history can also significantly affect performance. PPD selection is most often made with laboratory testing. Selection is often based on rheology and/or ASTM pour-point (or similar type) testing. The rheology testing can include measurement of crude oil viscosity and yield strength of gelled crude oil. For certain applications the difference between treated and untreated crude oils can be very dramatic. As PPD performance can be highly specific, testing is required on the exact crude oil to be treated to select the most optimal product.

Even with the most optimal product selection, performance can vary. For example, with most crude oils having extremely high wax contents and high pour points greater than 100°F/37.7°C, PPDs usually provide only minimal viscosity or pour point reduction even at high dosage. These crude oils simply have too much wax for the PPD to be effective. However, when these crude oils are diluted 15%–30% with a low pour point diluent, PPD treatment can be significantly more effective. Fig. 2.50 shows an example with an untreated crude oil that gelled at ~42°C in viscosity-temperature ramp measurements. With ~20% dilution and PPD treatment, an almost 30°C reduction in gel point was obtained. Again, often suitable dilution fluids are not available for this option though.

As the crude oil pour point and wax content become lower, the chance for PPDs to provide effective and economical treatment increases. Some applications may still require high PPD dosages in 1000–3000 ppm range though. Generally, as the wax content and pour point decrease, so too does required PPD dosage. The mechanism by which PPDs function is by incorporating into the forming wax network causing a reduction in network branching and growth. A weakened network structure also occurs from the incorporation of the PPD which renders the network more easily broken by shear forces. Both these effects lead to improved flow properties with reduced viscosity and reduced pour point (gelation temperature). And if gelled, the yield strength of the gel is weakened.

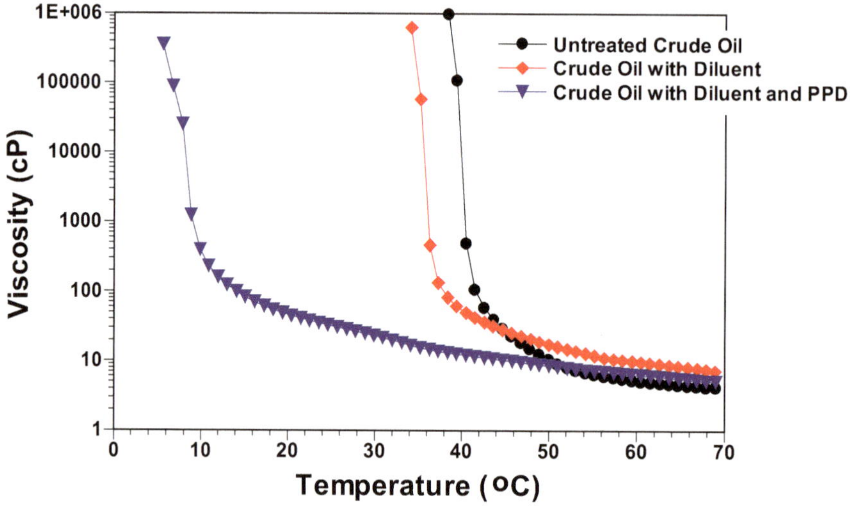

FIGURE 2.50

Diluent + PPD performances on high pour point crude oil. *PPD*, Pour point depressant.

PPD testing: pour point, viscosity, and yield stress

PPD evaluation generally involves performing testing with pour point measurements (ASTM method or variations thereof) and/or rheological measurements. Many producers that operate fields with high pour point crudes will often specify minimum requirements for a combination of pour point, viscosity, and yield value to be achieved with PPD treatments. Pour point measurements have been discussed in detail previously. The only comment here on pour point measurements is to remember that these are defined by specific test procedures and results can vary with the exact procedure used. The rheology measurements performed are either viscosity-temperature profile measurements or yield stress measurements. Like pour point measurements, these measurements are also dependent on the test procedure and preconditioning of the crude oil prior to performing measurements. Viscosity-temperature profiles have been discussed previously too. The temperature profiles are normally performed at cooling rates of 0.5°C–1°C per minute at a very low shear/stress rate [19,80,82].

Yield stress measurements are important to determine if production shuts down and crude oil gels in a flowline whether it will be possible to restart the flow. Restart requires that pumping pressure can deliver enough force to break the strength of the gel in the flowline. Hence, the strength of the gelled crude oil or yield stress is the main integral property determining whether restart can be achievable. Two methods are used to measure yield stress: rheometer measurements and model pipeline restart apparatus. Rheometer measurements are vastly easier and more commonly used. Again, results from both are dependent on the exact measurement procedures and equipment used. Different results can be obtained with different instrument geometries and/or instrument gaps used. Generally, the crude oil is loaded into the test equipment chosen for the evaluation and then cooled at a controlled rate and shear profile over a specified amount of time. The crude oil is then held or aged at the final minimum temperature for a period of time. After this shut-in time a force is applied stepwise until the gelled crude oil "yields" and the applied force value drops. Examples of model pipeline restart apparatuses can be found in the literature [101,102,106]. Similarly, with restart apparatuses different results with different diameter tubing can be obtained.

Fig. 2.51 shows an example of rheometer yield stress measurements made on a parallel-plate geometry with treated and untreated waxy crude oils. Crude oil was gelled between the rheometer plates with a specific cooling procedure and aging of the sample. Plotted is applied shear stress versus the rheometer rotational rate. If the sample does not yield, the rotation rate is zero. The yield stress is the stress at which the rheometer starts turning. In this testing, the untreated crude oil yield stress was ~20,000 dynes/cm^2. The yield stress of the PPD treated samples was considerably less, greater than 97% reduction. The gelled crude oil yielded at ~500 dynes/cm^2 with 500 ppm PPD. For 2000 ppm, it yielded at the first commanded stress ramp step (~100 dynes/cm^2 or less).

Fig. 2.52 shows the combined effect PPDs can have at reducing the viscosity, gel point, and yield strength of high pour point crude oils. Shown are measurements with a high pour point Brazilian crude. On the rheometer test, it can be

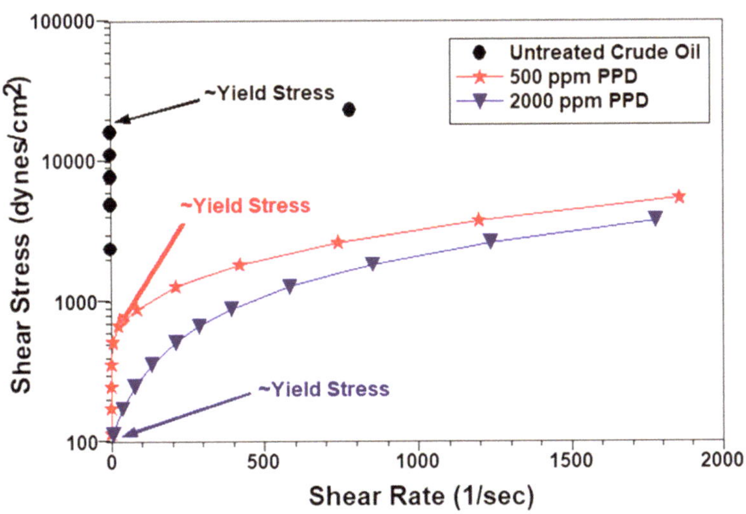

FIGURE 2.51

Rheometer yield stress measurements for evaluation of PPD performance in reducing gel strength. *PPD*, Pour point depressant.

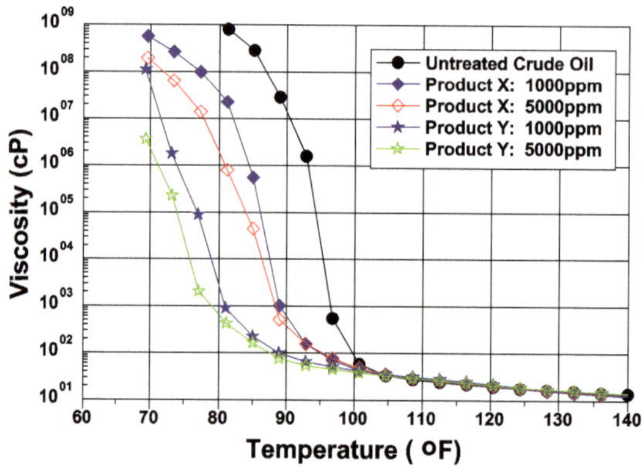

Gelled Samples at 77°F

Sample	Yield Stress (Dynes/cm2)	% Reduction
Untreated	10000	----
Product X:		
1000ppm	3000	70
5000ppm	500	95
Product Y:		
1000ppm	1800	82
5000ppm	250	97.5

FIGURE 2.52

Example PPD effect on crude oil viscosity, gelling, and yield strength. *PPD*, Pour point depressant.

seen that the untreated crude oil gels at ~95°F. Hence, it is a high wax content crude oil with a pour point in the range of 100°F. As mentioned, crude oils with these characteristics are harder to treat than crude oils with slightly lower pour points and wax contents. However, with this crude oil, PPDs provided improvement with viscosity reduction and gelling at 1000 ppm. Further increase in dosage did not yield much additional improvement. It is typical with PPD treatment for performance to plateau at some point with increasing dosage. Product Y was significant better yielding an ~17°F reduction in the gel point compared to ~7°F for Product X. Note that both PPDs caused a significant reduction in the yield strength. Even the lesser effective Product X gave ~70% reduction in yield strength at 1000 ppm over the untreated crude oil.

Another example of a PPD treatment is shown in Fig. 2.53. It is from testing performed to select a PPD for application in a deepwater well test in West Africa. The pour point of the crude oil was ~35°C, significantly greater than seafloor temperature. As a result, concerns existed with regards to flowing the crude oil during the well test; especially if the well test were to experience a shut-in. Rheology and pour-point testing were performed to determine whether a PPD/paraffin inhibitor could aid the well test and, if so, to select the best product for the application. There were some questions initially as to the actual constraints on the PPD for injection in the well test. Would the PPD be subjected to deepwater seafloor temperature like a subsea umbilical application? Or would it be more like the dry-tree capillary injection? Ultimately, it was decided that a deepwater umbilical-approved paraffin inhibitor was needed which effectively reduced the activity of products available meeting injection conditions.

Testing showed that PPDs could have a significant effect on lowering the crude oil pour point and changing the crude oil rheology. Relatively high-dose rates were required with the selected umbilical-approved PPD though, especially

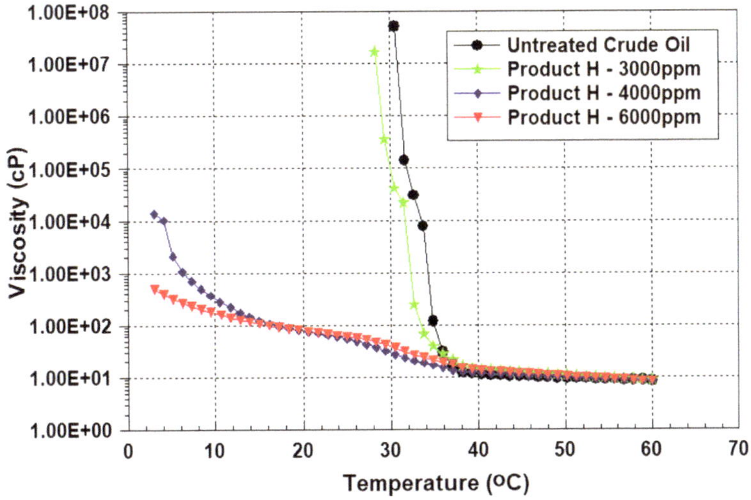

FIGURE 2.53

Viscosity-temperature profile for PPD applied in offshore well test. *PPD*, Pour point depressant.

if compared to an equivalent-chemistry product applied onshore or even product suitable for deepwater dry-tree capillary application like discussed in Section 2.5.3.3. If the dry-tree acceptable product could have been used at least 30%, lower dosage would be possible.

Fig 2.53 shows the performance of the selected product as a function of dose rate. Shown is the viscosity of the crude oil as a function of temperature as measured with a rheometer with a parallel-plate geometry cooling the crude oil from 60°C at a rate of 0.5°C/minute. Note, the crude oil was conditioned at hotter temperature. As shown, the untreated crude oil gels solid at ~32°C. The crude oil treated with 3000 ppm shows only slight improvement depressing the gelling point. At 4000 ppm though the effect of the PPD is tremendous. The complete gelling-point had not been reached by the end of the rheometer procedure at ~4°C. An increase in dose rate to 6000 ppm provided only slight further improvement. Increasing the dose rate further to 8000 ppm (not shown) did not provide additional benefit in changing the viscosity profile.

2.5.3.4 Pour point depressant treatment considerations

Pour point problems are not isolated to any specific location in a production system. Problems can start to occur after the WAT is reached and crude oil is cooled further beyond this point. Practically this means that issues are, however, generally downstream from the well. It is more an issue in flowlines, tanks, and pipelines. Geographically pour point issues exist globally affecting onshore and offshore applications. However, issues are more prevalent in certain geographical locations. Different regions in the Far East and Africa have a greater share of high-wax content, high pour point crude oils. This is reflective of the original hydrocarbon source present millions of years ago in these regions.

Many of these crude oils are treated with PPDs to assist production. Since many applications require relatively high dosages (thousands of ppm), there is significant incentive to apply the most cost-effective product. In addition to having effective chemistries providing optimal treatment performance, the most cost-effective PPDs must be formulated to have maximum activity. This requires the product is formulated to load as much of the PPD chemistries as possible in the product formulation while still meeting customer product physical specifications. If the customer requires a traditional liquid product for pumping, often the main product physical spec is having the PPD product pour point equivalent to the minimum ambient temperature.

As discussed earlier, a means to apply more active products is with heated injection. In fact, most heated injection applications are with PPD applications. Fig. 2.54 shows a heated injection skid used in Africa. The skid is located in close proximity to the injection point to minimize heating tracing needs with long injection lines. The skid is approximately 10 m behind and to the right of the wellhead. Fig. 2.55 illustrates the dosage reduction potential using heated injection systems comparing the performance of two PPD products on a high pour point African crude oil. The two products shown are formulations for heated injection and liquid surface injection. Note the heated injection product requires less than half the dosage to achieve almost equivalent performance.

FIGURE 2.54

Example of heated injection application.

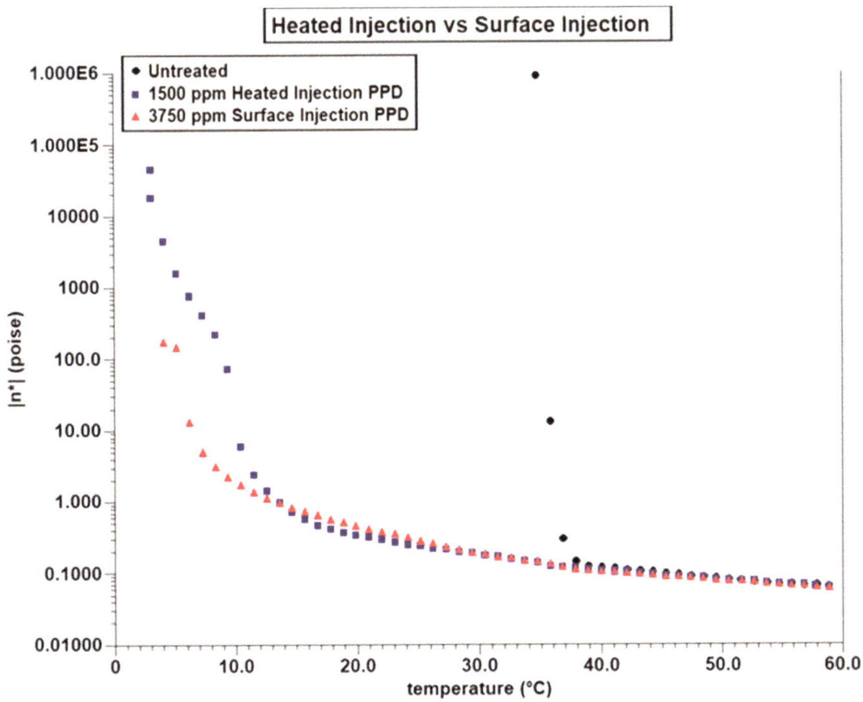

FIGURE 2.55

Example of advantage of heated injection system.

2.6 Case histories

A series of case histories have been selected to cover many of the issues previously discussed. In the interest of brevity, each will consist of a summary with limited details. These case histories were selected to provide additional examples of field chemical applications for treating the different types of paraffin problems that occur in the upstream petroleum industry. The case histories help complete the paraffin management overview presented in the chapter.

2.6.1 Formation damage

2.6.1.1 Case history no. 1

A well in the Ardmore Basin of Oklahoma was initially drilled in 1955, recompleted in 1974, and reperforated in 1994. The well experienced paraffin deposition in the tubing and had to be cut down to 1000 m once every 3 months. The WAT of the crude oil was 35°C and the reservoir temperature was 44°C.

Production had decreased from 48 to 17 m^3/day. Suspecting paraffin damage in the formation, the well was fractured with 102 m^3 of gelled oil and 17.24 metric tons of 20–40 sand. Production increased to 70 m^3/day but dropped to 46 m^3/day in 7 months. Following the steep production decline, it was decided to investigate the problem. Based on the analysis of the crude oil and paraffin samples, it was determined that formation damage could be occurring from paraffin deposition. A series of six solvent treatments were conducted over the next 21 months using various techniques [107]. The treatments and results consisted of the following:

1. February 1997: 22.22 m^3 of crude oil with 1-m^3 solvent and 1-m^3 paraffin inhibitor placed with tubing packers at 2- to 5-m intervals with each interval receiving 1.75 m^3 treatment and 5.4 m^3 overflush. These treatments were then followed by 10.5 m^3 treatment and 5.4 m^3 overflush placed across the entire zone. The well was immediately put back on production after the treatment and averaged 55.56 m^3/day over the next 6 months.
2. August 1997: Switched to a two-stage treatment to try to further increase the benefits. The first stage clean-up program consisted of 19.8 m^3 lease crude oil followed by 7.5-m^3 crude oil with 1.7-m^3 solvent and 0.4-m^3 paraffin inhibitor split into five intervals with 1–2 m^3 gelled oil pads between intervals. Overflush with 15.5-m^3 lease crude oil and shut-in for 4 hours, then returned to production for five days. The second stage was paraffin inhibitor squeeze consisting of 21.4-m^3 lease crude oil with 2500-ppm paraffin inhibitor followed by 14-m^3 oil with 2.5-m^3 paraffin inhibitor and a final overflush of 117-m^3 lease crude oil with 0.35-m^3 paraffin inhibitor. It was shut-in overnight and then resumed production. The well was produced for 2 months and reached a peak production of 71.4 m^3/day. This level of production was above that first observed after the fracture treatment.
3. November 1997: Two-stage treatment was based on successful August treatment design. The first stage was 21.4 m^3 crude oil preflush containing 2800-ppm paraffin dispersant, followed by 11.1 m^3 treating mixture split into ten intervals with 0.5-m^3 gelled oil pads followed by 19-m^3 overflush containing 4000-ppm paraffin dispersant. It was shut-in for 4 hours and then produced for 5 days. The second stage was 21.4-m^3 preflush with 3300-ppm paraffin inhibitor followed by 15.9-m^3 lease crude oil with 2.5-m^3 paraffin inhibitor followed by 108.6-m^3 overflush. Production reached a peak following this treatment of 77.8-m^3/day but slipped back to 63.5 m^3/day in February.
4. February 1998: Two-stage treatment design with the first stage cut back to 21.4-m^3 oil preflush containing 2800-ppm paraffin dispersant, followed by 9.5-m^3 treating mixture split into ten intervals but with no gelled pads followed by 19 m^3 overflush containing 4000-ppm paraffin dispersant. It was shut-in for 6 hours and then produced for 3 days. The second stage consisted of 21.4 m^3 preflush with 3300-ppm paraffin inhibitor followed by 15.9-m^3 lease oil with 2.5-m^3 paraffin inhibitor followed by 56 m^3 overflush. This

treatment seemed to give a longer squeeze life of 5 months with production greater than 63.5 m^3/day.

5. August 1998: Two-stage treatment design, same as February 1998 design except: in the first stage 7.6-m^3 condensate was used instead of crude oil for the treatment volume and in the second stage the overflush volume was increased back to 108.6 m^3. The results were slightly below that of the previous treatment with an increase to a peak of only 76.2 m^3/day and a period of 4 months above 63.5 m^3/day crude oil production.
6. November 1998: Duplicate of August 1998 treatment with no appreciable increase in crude oil production observed.

The results of these treatments are shown in Fig. 2.56.

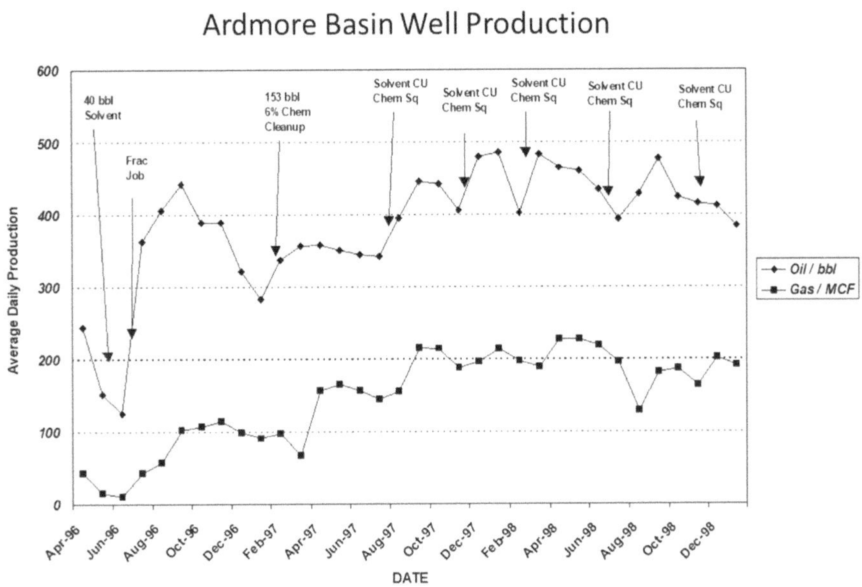

FIGURE 2.56

Paraffin remediation results in Ardmore Basin well.

2.6.1.2 Case history no. 2

A pumping well in the Permian Basin of the United States was producing 9.5 m^3/day crude oil (36° API) and 1.3 m^3/day water. The completion on the well was at 1280-m depth. Paraffin formation damage was suspected because the well had no tendency to develop scale. The well was treated with 0.4 m^3 paraffin solvent with an overflush of 3.2 m^3 lease crude oil to hold the treatment in place in the immediate wellbore area. The well was shut-in for 8 hours and then resumed production. The well averaged 15.9 m^3/day crude oil and 2.4 m^3/day water for a 3-month period before production began to drop back to 9.5 m^3/day crude oil.

A decision was made to retreat the well to increase production. The treatment was repeated with a 0.4 m^3 of a new solvent/dispersant combination and 3.2 m^3 lease crude oil overflush. Following the 8-hour shut-in soak period, the well was put back on production. The well average 15.9 m^3/day crude oil production is for 5 months. Each of the treatments was paid for with just three days of increased production. The well made an additional 1524 m^3 crude oil over the 8-month period over the baseline production level of 9.5 m^3/day crude oil production [89].

2.6.1.3 Case history no. 3

Five fields in the Permian Basin area of the United States were investigated for potential paraffin formation damage [90]. All the fields had a history of hot oil treatments downhole. Of the five fields, three were identified as potential candidates for remediation of paraffin formation damage. The physical properties of the crude oils produced from these fields are shown in Table 2.4.

A total of 22 wells were selected from the three fields to treat for paraffin formation damage. The average crude oil production from these wells had decreased from 7.94 m^3/day down to 3.17 m^3/day over a 2.5-year period. Each of the wells was treated with 4.16 m^3 solvent/dispersant product and enough overflush lease crude oil to hold the treatment in place with a 0.3- to 0.45-m radial penetration. The wells were shut-in for 12 hours and then put back on production. The production increased from 20% to 280% for the 22 wells. Over the next 6 months the 22 wells averaged 5.66 m^3/day.

Table 2.4 Permian Basin crude oil properties.

Test	Results
WAT (°C)	32–41
Maximum pour point (°C)	−1 to 16
°API gravity	32–38
Paraffin content (%)	3.50–6.20
Asphaltene content (%)	0.25–0.75
Asphaltic resin content (%)	0.70–1.20
Field paraffin deposition history	Moderate to severe
Asphaltene stability	Good

2.6.2 Well tubing deposition

2.6.2.1 Case history no. 1

A field in the Northern Niagaran Reef Trend located in Michigan, United States was experiencing severe paraffin deposition in the tubing string of rod-pumped wells [81]. Crude oil and well data from the field are summarized in Table 2.5.

Table 2.5 Niagaran Reef Trend crude oil characteristics.

Test	Range	Average
Maximum pour point (°C)	−18 to 21	10
WAT (°C)	21–38	28
Paraffin content (%)	2.50–25.0	12
Asphaltene content (%)	0.01–0.50	0.1
Asphaltic resin content (%)	0.10–1.0	0.6
°API gravity	41–57	49
Oil production (m^3/day)	2.06–38.25	11.90
Water production (m^3/day)	0–29.37	5.71
Gas production (MCFD)	2–561	119
Well depth (m)	1229–1845	1524
Completion interval (m)	2.13–53.96	15.24
Bottom-hole temperature (°C)	32–43	38

The paraffin deposited in the tubing had a carbon number range of C_{18}–C_{66} with a peak maximum carbon number of C_{38} and an average melting point of 74°C. The flowline paraffin deposits had a carbon number range of C_{18}–C_{60} with a peak maximum carbon number of C_{30} and a melting point of 68°C. Tubing and flowline deposition failures occurred on a monthly basis. To combat this problem the wells were treated once every 2 weeks with hot diesel or water. In 1979 this treating cost was $1409 per well per month. In an attempt to reduce treating cost, application of dispersants was evaluated. Flask tests were run at 2% surfactant/dispersant concentration in produced water with 2 hours of agitation to screen dispersant candidates. From the testing the best performing dispersant was chosen for application (Additive 8 in reference). The dispersant was successfully trialed in one well using a batch treatment of 2% Additive 8 with enough produced water to provide 2 hours of contact time on the selected well. Following the success of this trial, the program was expanded to 21 wells. The average chemical cost over the remainder of the year was $687 per well per month for a savings of 51.2%. Total savings were $77,000. During this initial period, other dispersants (Additives 9, 10, and 11) were developed as improvements over Additive 8 for specific wells in the 21 well program. Over the next year the cost per well was further reduced and the total savings was approximately $137,000.

2.6.2.2 Case history no. 2

Wells in a field Wyoming, United States were drilled in 1981–1988 time-period with the number of active pumping wells in operation varying from 129 to more than 200 wells. All the pumping wells were being treated with 70 barrels of hot oil heated to 138°C and pumped down the annulus on a monthly basis. All wells were thought to be experiencing tubing paraffin deposition from the surface down to about 610–976 m. The producer thought that the hot oil treatments were removing the tubing wax down to these levels. The cost of the program was greater than \$500,000 per year. In an attempt to reduce the cost, a trial was initiated with batch treating 19 L of a dispersant/paraffin inhibitor in 0.8 m^3 lease crude oil once a week on five randomly selected wells. As part of the trial program, another five wells were randomly selected to stop hot oiling. This was to establish a baseline for the well failure rate timeframe. This program continued for 18 months and none of the 10 wells exhibited a tubing deposition problem. At this time all hot oil and chemical treatments were stopped to determine which of the wells in the field actually had a problem. Over the next 7 years, it was determined that only 41 of the more than 200 wells had a problem with tubing deposition and the hot oil costs were reduced to \$150,000 per year. These 41 wells were hot oiled on a monthly basis, but it was still necessary to hot oil down the tubing during a pulling job.

In an effort to understand the effects of the hot oiling a thermal heat transfer program was developed [55]. This program indicated that the hot oil treatments could only melt the tubing wax deposits down to a depth of 152 m below the surface. To treat the wax deposition below 152 m in the tubing, a series of trials were conducted to identify the best chemical treatment program. Batch treatments of dispersant/paraffin inhibitor in crude oil and surfactant/dispersants in produced water were conducted. Samples were collected at the wellhead during the treatment returns to identify which application method removed the most paraffin. HTCG analysis of the treatment samples indicated the surfactant/dispersant treatment in produced water removed the most high-molecular-weight wax from the tubing deposits. A program was initiated on 40 wells over the next few months. A total of 17 wells were treated with an initial large batch clean-up of 27 m^3 cold/hot water. Nine wells were treated with 71 L of surfactant/dispersant in 9.5 m^3 produced water. The other 14 wells received no initial clean-up. The 40 wells were then placed on a monthly treating schedule using 71 L of dispersant and 9.5 m^3 of ambient temperature produced water. Adjustments were made to the program and the majority of the wells were finally treated with a batch treatment of 35 L of dispersant in 4.8 m^3 of cold produced water. Treating frequencies and batch treatment volumes were adjusted based on the wells' response. Over the next 18 months, none of the wells required hot oiling down the tubing during a pulling job. Some failures did occur which required hot oil treatments down the casing. Over the final eight months, 19 wells experienced no problems, 10 wells had one problem, 7 wells had two problems, 3 wells had three problems, and 1 well had six problems requiring hot oiling down the casing. The annualized cost for the dispersant treatments was \$100,000.

2.6.3 Flowline deposition

2.6.3.1 Case history no. 1

This case history is for a paraffin inhibitor application on a two-well subsea development [85]. The field had dual flow lines and pigging capabilities and

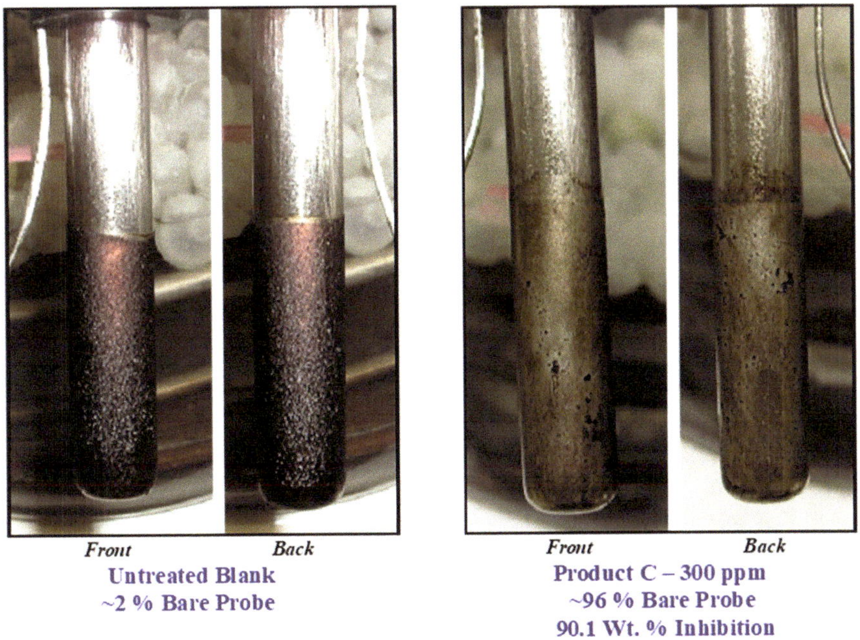

FIGURE 2.57

Cold finger results from flowline case history no. 1 testing.

applied paraffin inhibitor to reduce pigging frequency. The operator placed as much value in minimizing the risks associated with pigging (both with sticking the pig and safety concerns from the operation) as the revenue gained from minimizing deferred production. The injection pressure requirement of the paraffin inhibitor was 9000 psi. A paraffin inhibitor formulated with specifications for matching the injection pressure requirement of 9000 psi at 40°F was found to be effective at inhibiting deposition (through cold finger testing) at a relatively low dose rate; around 250–300 ppm. Fig. 2.57 shows results of the cold finger testing of the paraffin inhibitor compared to untreated crude oil from one well. Note the cold finger of the treated sample is mostly absent of deposition. Results with the other well were similar. In the cold finger testing the temperature differential between the crude oil WAT and the cold finger temperature was set between 20°F and 30°F in the various tests. These conditions were selected to simulate, (as best possible in a simple laboratory test) a section of flowline in which deposition first occurred since this region experienced the greatest amount of deposition.

2.6.3.2 Case history no. 2

This case history discusses a dispersant (in seawater) application to clean paraffin deposition from a severely restricted 17.4-m dual-flowline in the deepwater

Cottonwood Field in the GOM. In-depth discussion on this case history can be found in the literature [70,97,98]. Flow assurance concerns about paraffin deposition were identified in the initial field design [108] and a paraffin deposition management strategy was implemented at the start of production using pigging in conjunction with continuous paraffin inhibitor application to reduce the pigging frequency. However, in December 2007, an aggressively sized pig was stuck during a pig run. The flowlines to the platform were 6-in outside diameter and 4.8-in. internal diameter (ID) lines. The pig was a 5.1-in. elastomeric pig which got stuck approximately midway in the east flowline along with a significant accumulation of paraffin ahead of the pig.

Fig. 2.58 shows the field layout illustrating the location of the stuck pig. The obstruction eliminated production through half the dual flowline system. All production was subsequently diverted through the west flowline. This hindered both production output and the ability to carry-out the wax control strategy. Without pigging capabilities, complete control of wax deposition was not possible. The paraffin inhibitor itself was not able to completely prevent deposition and overtime the west flowline became severely restricted. Periodic solvent soak treatments were applied to keep the flowline operational. Replacement of the flowline with a cost of ~$30 million was too cost-prohibitive to consider.

Through persistent efforts over the following 2 years, the pig was dislodged in February 2010. Various procedures to dislodge the pig (alternating line pressurization, Micro Annular Pressure Pulse process, Subsea Umbilical Resonance Energy process) were attempted but not successful outright. The procedures were believed, however, to have contributed to the final pig release and breakthrough of the obstruction which came from maintaining a continual solvent-driven

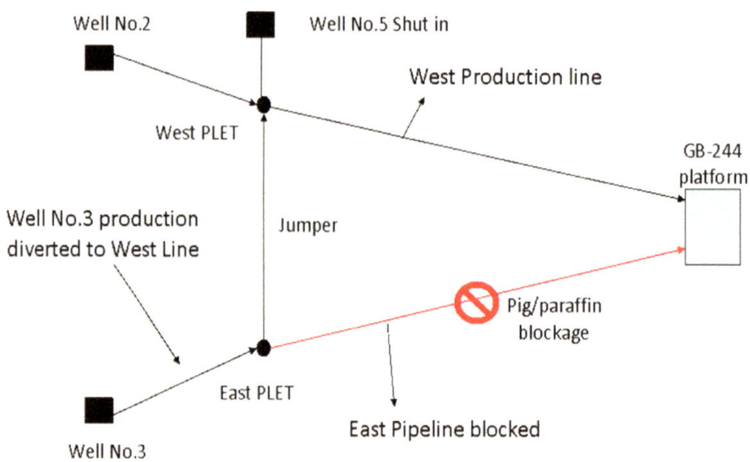

FIGURE 2.58

Cottonwood production system layout with stuck pig.

FIGURE 2.59

Dispersant performance in flask tests at 0°C for Cottonwood dispersant application.

controlled pressurization from the wellhead over the years. Although the removal of the pig opened the potential for resumption of production through the east flowline, the need to clean the wax deposition in both flowlines existed. The severe deposition present in the flowlines not only restricted production but also prohibited resumption of pigging operations since pigging in severely deposited flowlines is very risky toward sticking a pig. As a result, methods for cleaning the flowlines were sought following opening the east flowline. A paraffin dispersant treatment (in seawater) was chosen to clean the flowlines since in cold temperatures, dispersants can carry more paraffin than the best solvents can dissolve. Fig. 2.59 shows the flask test performance (with a quantitative measure of paraffin break-up) of the applied dispersant at varying concentrations in seawater performed with actual field paraffin deposit (HTGC shown in Fig. 2.11). As seen, the dispersant provided significant deposit break-up at dispersant concentrations above 5%.

A **four-stage** application was designed and executed. Two stages were related to the actual flowline cleaning—a dispersant application stage and a subsequent progressive pigging series stage to ensure the lines were completely clean after the application. The other two stages were for handling hydrate concerns before and after the flowline cleaning. Following the treatment, an additional phase for dewatering the flowlines prior to production restart was also needed.

For the application, two dynamically positioned marine vessels were required. A stimulation vessel was used to pump the dispersant/seawater application and a workboat equipped with storage tanks to capture/hold the paraffin deposition after separation and settling from the dispersant/seawater returns. The seawater/dispersant returns were recycled after paraffin removal. Fig. 2.60 illustrates the connection of the vessels to the production system along with a picture of the vessels

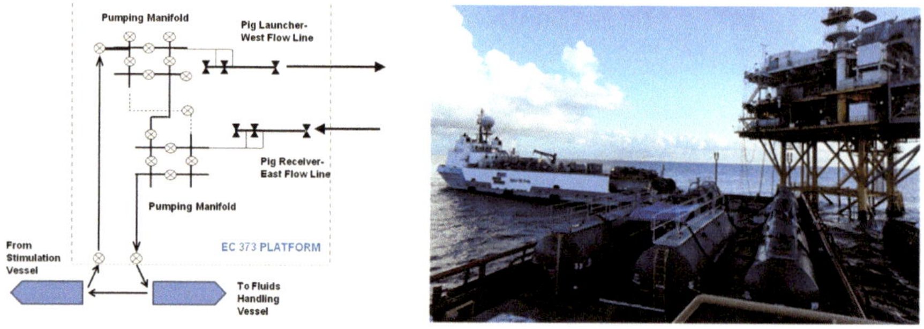

FIGURE 2.60

Connection of pumping and fluid handling vessels for the Cottonwood dispersant flowline cleaning application.

onsite. The application was pumped directionally from the more heavily restricted west line to the east line.

"Stage 1" consisted of hydrate control. The line was displaced with seawater and monoethylene glycol (MEG) to remove the crude oil and gas from the line. The treating mixture was captured and stored for reuse in Stage 4. No appreciable paraffin was recovered during this stage. "Stage 2" consisted of the dispersant cleaning application. A series of flowline displacements using 5%–7.5% dispersant concentrations were pumped at rates between 0.48 and 0.80 m^3/minute. The treating solution was separated from the removed wax and recycled with adjustments to the dispersant concentration. Throughout the dispersant treatment, hydrodynamic modeling showed a gradual increase in the ID of both flowlines providing indication that the dispersant treatment was working. Further in the returns, it could be observed that the dispersant was removing significant amounts of paraffin deposition from the flowlines. In the end, it was estimated that approximately 143 m^3 of paraffin sludge was captured. "Stage 3" consisted of a series of six pig runs of increasing diameter and strength pushed by primarily by seawater except the last portion of runs 3–5 included 5% dispersant such that some dispersant would be present ahead of the succeeding pig run. The sixth pig run was pushed by the seawater/MEG solution retained from Stage 1. Only relatively small amounts of paraffin ($\sim$0.1 m^3 total) were captured in the pig catcher during the six pig runs indicating most of the paraffin deposition had been removed during the dispersant stage. The sixth pig run with MEG was the "Stage 4" hydrate control phase. An additional slug of MEG was also later placed in the line prior to compressed gas flow from the platform to dewater and condition the line for well restart.

Following the cleaning operation, both lines were returned to their initially clean values. As mentioned, hydrodynamic modeling was used to assess the success of the application. Initial calculations indicated the west flowline was down from approximately 4.8-in ID to 2.2-in ID. The east flowline was reduced from 2.7-in. ID. In restoring the lines the dispersant application provided significant return for the Cottonwood development. It allowed production from wells to be brought on-line at a higher capacity, thereby generating increased revenue. It also allowed resumption of routine pigging operations. Cost of the application was ~12% of replacing the flowline.

It should be noted that most of the expense for this application was due to being a deepwater operation; related to the required vessel costs for pumping and fluids handling. In addition, extra expense was required for the hydrate control measures in dealing with deepwater lines with gas present. The dispersant chemical costs were extremely economically favorable as 90+ % of the treatment was from seawater. For dispersant operations, onshore or in shallow water application costs would be significantly lower. Still even with the deepwater challenges and costs, the Cottonwood dispersant application was a tremendous economic success.

2.6.4 Tank bottoms

2.6.4.1 Case history no. 1

A 12,700-m^3 tank containing crude oil in a refinery had a paraffinic tank bottom problem. Measurement of the bottom indicated a depth of 133 cm at the front sliding down to 18 cm at the back. Twenty potential dispersants were tested and the best dispersant selected for application. The tank was emptied down to the bottom and 0.62 m^3 of dispersant was spread across the tanks bottom surface. The tank was then filled with crude oil over an 88-hour period. During this time the combination of agitation from filling the tank and internal mixers dispersed the bottoms into the tank crude oil. Testing of the crude oil in the tank indicated it could be fed into the pipeline at a rate of 5%–10% of the loading volume. The stirring agitation was maintained during the tank drawdown. No upset occurred during this process. Measurement of the bottom following this process indicated a remaining bottom depth of 10.2 cm across the tank. The cost of the treatment was ~15% of the cost to mechanically clean the tank.

2.6.4.2 Case history no. 2

A 19,048-m^3 storage tank containing crude oil had a paraffin bottom. The bottom measured 68.6 cm sliding down to 18 cm across the tank. The tank was emptied and 0.62 m^3 of dispersant was distributed across the surface of the tank. The tank was refilled with crude oil with internal mixing. When the tank was full, the tank bottom deposit measured 18 cm down to 10 cm across the tank. This treatment cost was ~10% of the standard mechanical cleaning cost for the tank.

2.7 Summary

This chapter provided a summary of the flow assurance problems that paraffin (wax) can cause in petroleum production systems. The main problems that operators encounter are with paraffin deposition or improving the flow characteristics of high wax-content, high pour-point crude oils. In this chapter, the chemical and physical reasons why paraffin cause problems were discussed along with various management strategies. Implementation of effective paraffin management programs includes assessing potential problems, designing management strategies, and applying the programs. Thermal, mechanical, and chemical means (or combinations thereof) are all used for prevention and remediation of problems. Each was discussed in this chapter. The main focus of the chapter was on chemical treatment methods though. As such, different chemical treatments were discussed in detail. Many examples and case histories were included in the discussion to illustrate typical field problems and performance of different chemical treatments.

The optimal and/or most cost-effective paraffin management program for an operation depends on the specifics of the individual production systems. It is the hope of the authors that this chapter will provide guidance in better understanding and addressing paraffin issues. And with improved understand and knowledge, better selection and implementation of paraffin management programs and applications.

Nomenclature

ASTM American Society for Testing and Materials
BS&W basic sediment & water
CAPEX capital expenditures
CPM cross-polarized microscopy
DEH direct electrical heating
$\mathbf{D_f}$ diameter of flowline with gelled crude oil plug
DSC differential scanning calorimetry
ESP electric submersible pump
ETH-PIP electrically trace heated pipe in pipe
FP filter plugging
FTIR Fourier transform infrared spectroscopy
FWHT flowing wellhead temperature
GOM Gulf of Mexico
GOR gas-to-oil ratio
HTGC high-temperature gas chromatography
ID internal diameter
IP Institute of petroleum
L_{plug} length of gelled crude oil plug in flowline
MEC minimum effective concentration
MEG monoethylene glycol
OD outside diameter
OPEX operating expenditures
PIP pipe-in-pipe
PPD pour point depressant

SARA	saturates, aromatics, resins, and asphaltenes
TDS	total dissolved solids
VIT	vacuum insulated tubing
WAT	wax appearance temperature
$\Delta P_{restart}$	pressure differential required to restart a flowline with gelled crude oil
ΔT	cold finger temperature differential:

$$\Delta T = T_{crude\ oil} - T_{cold\ finger} \quad \text{when } T_{crude\ oil} \leq WAT.$$

$$\Delta T = WAT - T_{cold\ finger} \quad \text{when } T_{crude\ oil} > WAT.$$

τ	yield stress of gelled crude oil
ω	cold finger stirring rate (rpm)

References

[1] C.E. Reistle, Paraffin and Congealing Oil Problems, US Bureau of Mines Bulletin, Washington DC, 1932.

[2] P.A. Bern, V.E. Withers, J.R. Cairns, Wax deposition in crude oil pipelines, SPE-206−1980-MS, in: European Offshore Petroleum Conference and Exhibition, London, UK, 21−24 October, 1980.

[3] E.D. Burger, T.K. Perkins, J.H. Striegler, Studies of wax deposition in the trans Alaska pipeline, Journal of Petroleum Technology 33 (1981) 1075−1086.

[4] T. Fan, J. Wang, J.S. Buckley, Evaluating crude oils by SARA analysis, SPE-75228-MS, in: SPE/DOE Improved Oil Recovery Symposium, Tulsa, Oklahoma, 13−17 April, 2002.

[5] D.W. Jennings, K. Weispfennig, Experimental solubility data of various n-alkane waxes: effects of alkane chain length, alkane odd vs even carbon number structures, and solvent chemistry on solubility, Fluid Phase Equilibria 227 (2005) 27−35.

[6] UOP 46, Paraffin Wax Content of Petroleum Oils and Asphalts, Universal Oil Products LLC, Des Plaines, 1985.

[7] ASTM D5442-93, Standard Test Method for Analysis of Petroleum Waxes by Gas Chromatography, ASTM International, West Conshohocken, 2008.

[8] T.G. Monger-McClure, J.E. Tackett, L.S. Merrill, DeepStar comparisons of cloud point measurements & paraffin prediction methods, SPE-38774-MS, in: SPE Annual Technical Conference and Exhibition, San Antonio, TX, 5−8 October, 1997.

[9] ASTM D2500-17a, Standard Test Method for Cloud Point of Petroleum Products, ASTM International, West Conshohocken, 2011.

[10] IP 219, Petroleum Products − Determination of Cloud Point, Energy Institute, London, 2013.

[11] J.M. Howell, F.W. Jessen, Determination of the viscosity-temperature relationship for crude oils with the ultra-viscoson, Journal of Petroleum Technology 8 (1956) 95−97.

[12] A.B. Hansen, E. Larsen, W.B. Pedersen, A.B. Nielsen, Wax precipitation from North Sea crude oils. 3. Precipitation and dissolution of wax studied by differential scanning calorimetry, Energy & Fuels 5 (1991) 914−923.

[13] M.V. Kok, J.M. Letoffe, P. Claudy, D. Martin, M. Garcin, J.L. Volle, Comparison of wax appearance temperatures of crude oils by differential scanning calorimetry, thermomicroscopy and viscometry, Fuel 75 (1996) 787–790.

[14] P. Juyal, T. Cao, A. Yen, R. Venkatesan, Study of live oil wax precipitation with high-pressure micro-differential scanning calorimetry, Energy & Fuels 25 (2011) 568–572.

[15] D.W. Jennings, H. Hager, K. Weispfennig, Wax in crude oil: analytical methods and their uses for flow assurance purposes, in: Proceedings of Chemistry in the Oil Industry VIII, Manchester, UK, 3–5 November, 2003.

[16] ASTM D97-17b, Standard Test Method for Pour Point of Petroleum Products, ASTM International, West Conshohocken, 2008.

[17] ASTM D5853-17a, Standard Test Method for Pour Point of Crude Oils, ASTM International, West Conshohocken, 2017.

[18] K.M. Barker, M.E. Newberry, ASTM pour point what does it really mean, in: 5th International Conference on Petroleum Phase Behavior and Fouling, Banff, Alberta, Canada, 13–17 June, 2004.

[19] M.E. Newberry, Chemical effects on crude oil pipeline pressure problems, Journal of Petroleum Technology 36 (1984) 779–786.

[20] T.R. Bott, J.S. Gudmundsson, Deposition of paraffin wax from flowing systems, IP-77-007, Institute of Petroleum Technical Paper, 1977.

[21] C.C. Patton, B.M. Casad, Paraffin deposition from refined wax-solvent systems, SPE Journal 10 (1970) 17–24.

[22] P.E. Eaton, G.Y. Weeter, Paraffin deposition in flowlines, Paper 76-CSME/CSChE-22, 16th National Heat Transfer Conference, 8–11 August, 1976, St. Louis, Missouri, USA.

[23] Z. Huang, S. Zheng, H.S. Fogler, WAX DEPOSITION Experimental Characterizations, Theoretical Modeling, and Field Practices, CRC Press, 2015.

[24] M.A.L. Goncalves, S.P.G. Pinho, J.R.T. Montesanti, W. Shang, C. Sarica, A case study of scale-up of wax deposition model predictions using flow loop wax deposition data for pipeline design, in: 15th International Conference on Multiphase Production Technology, Cannes, France, 15–17 June, 2011.

[25] C. Carniani, M. Merlin, Basic design of waxy oil transportation through improved lab testing, SPE-36836-MS, in: SPE European Petroleum Conference, Milan, Italy, 22–24 October, 1996.

[26] P. Singh, R. Venkatesan, H.S. Fogler, N.R. Nagarajan, Formation and aging of incipient thin film wax-oil gels, AIChE Journal 46 (2000) 1059–1074.

[27] J. Hsu, M. Santamaria, J. Brubaker, Wax deposition of waxy live crudes under turbulent flow conditions, SPE-28480-MS, in: SPE Annual Technical Conference and Exhibition, New Orleans, LA, 25–28 September, 1994.

[28] R. Venkatesan, R. Tanti, V. Subramani, D. Vedapuri, E. Akparu, K. Johnson, Wax deposition testing in a large-scale flow loop, in: 11th North American Conference on Multiphase Production Technology, Banff, Alberta, Canada, 6–8 June, 2018.

[29] O.C. Hernandez, C. Sarica, J.P. Brill, M. Volk, E. Delle-Case, J. Creek, Effect of flow regime, temperature gradient and shera stripping in dingle-phase paraffin deposition, in: Proceedings of BHRG Multiphase Technology Conference, San Remo, Italy, 11–13 June, 2003.

[30] A. Benallal, P. Maurel, J.F. Agassant, M. Darbouret, G. Avril, E. Peuriere, Wax deposition in pipelines: flow-loop experiments and investigations on a novel

approach, SPE-115293-MS, in: SPE Annual Technical Conference and Exhibition, Denver, Colorado, 21–24 September, 2008.
[31] A. Matzain, Single phase liquid paraffin deposition modeling (MS thesis), The University of Tulsa, 1996.
[32] A. Matzain, Multiphase flow paraffin deposition modeling (PhD thesis), The University of Tulsa, 1999.
[33] P. Dwivedi, C. Sarica, W. Shang, Experimental study on wax deposition characteristics of a waxy crude oil under single phase turbulent flow conditions, Oil and Gas Facilities 2 (2013) 61–73.
[34] A. Rittirong, E. Panacharoensawad, C. Sarica, Experimental study of paraffin deposition under two-phase gas/oil slug flow in horizontal pipes, SPE Production & Operations 32 (2017) 99–117.
[35] Y. Chi, S. Zhou, N. Daraboina, C. Sarica, Experimental study of wax deposition under two-phase gas-oil stratified flow, in: 11th North American Conference on Multiphase Technology, Banff, Alberta, Canada, 6–8 June, 2018.
[36] C. Lira-Galeana, A. Hammami, Wax precipitation from petroleum fluids: A review. in asphaltenes and asphalts, in: T.F. Yen, G.V. Chilingarian (Eds.), Developments in Petroleum Science, Elsevier, Amsterdam, 1978, pp. 557–608.
[37] J. Coutinho, B. Edmonds, T. Moorwood, R. Szczepanski, X. Zhang, Reliable wax predictions for flow assurance, Energy & Fuels 20 (2006) 1081–1088.
[38] O.C. Hernandez, H. Hensley, C. Sarica, Improvements in single phase deposition modelling, SPE-84502-MS, in: SPE Annual Technical Conference and Exhibition, Denver, Colorado, 5–8 October, 2003.
[39] O.B. Rygg, A.K. Rydahl, H.P. Ronningsen, Wax deposition in offshore pipeline systems, in: BHRG North American Multiphase Conference, Banff, Alberta, Canada, 9–11 June, 1998.
[40] I.L. Purdy, A.J. Cheyne, Evaluation of vacuum-insulated tubing for paraffin control at Norman wells, SPE-22102-MS, in: SPE International Artic Technology Conference, Anchorage, Alaska, 29–31 May, 1991.
[41] S. Feeney, Project case histories and future applications of vacuum insulated tubing, Paper #60d, in: AIChE Spring National Meeting, Houston, Texas, 14–18 March, 1999.
[42] P. Singh, J. Walker, H.S. Lee, S. Ghareh, B. Thomason, D. Blumer, An application of vacuum-insulated tubing (VIT) for wax control in an artic environment, SPE Drilling & Completion 18 (2007) 127–136.
[43] R.G. Ezell, S. Fontenot, E. Robinson, L. Cunningham, A. Patrickis, High performance aqueous insulating packer fluid improved flow assurance and reduced annular pressure buildup in ultra deepwater wells, SPE-135177-MS, in: SPE Deepwater Drilling and Completions Conference, Galveston, Texas, 5–6 October, 2010.
[44] G. Lively, Flow assurance begins with downhole insulation OTC-14118, in: Offshore Technology Conference, Houston, TX, 6–9 May, 2002.
[45] F. Fleyfel, W. Meng, O. Hernandez, Production of waxy low temperature wells with hot gas lift, SPE-89853-MS, in: SPE Annual Technical Conference, Houston, TX, 26–29 September, 2004.
[46] X. Wang, Q. Qu, P. Javora, R. Pearcy, New trend in oilfield flow assurance management: a review of thermal insulating fluids, SPE-103829-MS, in: SPE International Oil & Gas Conference and Exhibition, Beijing, China, 5–7 December, 2006.

[47] R. Hunter, Wet insulation evolves to meet subsea flowline performance demands, *Offshore 68* (2008) 1–3.

[48] C. Candelier, S. Durica, F. Beys, Subsea pipeline electrical heat trace (EHT) – "Active" heating – Application for a deep water brown field development, in: 12th Offshore Mediterranean Conference and Exhibition, Ravenna, Italy, 25–27 March, 2015.

[49] R.C. Fisher, S. Hall, J.F. Cam, D. Delaporte, Field deployment of the world's first electrically trace heated pipe in pipe, OTC-23108, in: Offshore Technical Conference, Houston, Texas, 30 April–3 May, 2012.

[50] D.G. Parman, S.S. Ojeda, A. Marron, Case study: Heated offshore production tubing electrically for crudes that have wax issues, OTC 24137, in: Offshore Technology Conference, Houston, Texas, 6–9 May, 2003.

[51] L.D. Brown, J. Clapham, C. Belmear, R. Harris, A. Loudon, S. Maxwell, et al., Design of Britannia's subsea heated bundle for a 25 year service life, OTC-11017, in: Offshore Technology Conference, Houston, Texas, 3–6 May, 1999.

[52] K.A. Esakul, G. Fung, G. Harrison, R. Perego, Active heating for flow assurance control in deepwater flowlines, OTC-15188, in: Offshore Technology Conference, Houston, Texas, 5–8 May, 2003.

[53] F. Fleyfel, O. Hernandez, R. Sturgis, W. Meng, Evaluation of pipeline configurations with active heating for export of waxy crude oil, SPE-90054-MS, in: SPE Annual Technical Conference, Houston, Texas, 26–29 September, 2004.

[54] T.J. Straub, S.W. Autry, G.E. King, An investigation into practical removal of downhole paraffin by thermal methods and chemical solvents, SPE-18889-MS, in: SPE Production Operations Symposium, Oklahoma City, Oklahoma, 13–14 March, 1989.

[55] A.J. Mansure, K.M. Barker, Insights into good hot oil practices, SPE-25484-MS, in: SPE Production Operations Symposium, Oklahoma City, Oklahoma, 21–23 March, 1993.

[56] G.E. Addison, Paraffin control more cost-effective, SPE-13391-MS, in: SPE Eastern Regional Meeting, Charleston, West Virginia, 31 October - 2 November, 1984.

[57] K.M. Barker, G.E. Addison, J.A. Cunningham, Disadvantages of hot oil for downhole paraffin in rod pump systems, in: Proceedings of Annual Southwestern Petroleum Short Course, Lubbock, Texas, April, 1983.

[58] K.M. Barker, P.F. Lambert, M.E. Newberry, Paraffin removal using cold produced water, SPE-55800-MS, in: SPE Rocky Mountain Regional Meeting, Gillette, Wyoming, 15–19 May, 1999.

[59] G.E. Addison, Paraffin problems and solutions in Michigan, in: Proceedings of SPE-SPWLA Well Completions Symposium, Mt. Pleasant, Michigan, 21–22 March, 1984.

[60] K.M. Barker, Formation damage relating to hot oiling, SPE-16320-MS, in: SPE Production Operations Symposium, Oklahoma City, Oklahoma, 8–10 March, 1987.

[61] T.I. Mitchell, S.C. Donovan, J.B. Collesi, H.W. McSpadden, Field Application of a Chemical Heat and Nitrogen Generating System, SPE 12776, in: SPE California Regional Meeting, Long Beach, CA, 11–13 April, 1984.

[62] H.W. McSpadden, M.L. Tyler, T.T. Velasco, In-situ heat and paraffin inhibitor combination prove cost effective in NPR#3, Casper, Wyoming, SPE-15098-MS, in: 56th California Regional Meeting of SPE, Oakland, California, 2–4 April, 1986.

[63] C.N. Khalil, L.F. Neumann, C.A. Linard, I.G. Santos, Thermochemical process to remove paraffin deposits in subsea production lines, OTC-7575, in: Offshore Technology Conference, Houston, Texas, 2–5 May, 1994.

[64] N.O. Rocha, C.N. Khalil, L.C.F. Leite, R.M. Bastos, A thermochemical process for wax damage removal, SPE-80266-MS, in: SPE International Symposium on Oilfield Chemistry, Houston, TX, 5–7 February, 2003.

[65] J.M. Brown, J.B. Dobbs, Novel exothermic process for the removal of paraffin deposits in hydrocarbon production, NACE-98233, in: Corrosion Conference & Expo, San Diego, California, 22–27 March, 1998.

[66] D.W. Jennings, S. Asomaning, M.E. Newberry, Heat generation process for treating oilfield deposits, Patent US8691731, 2009.

[67] J.R. Illeman, Pipeline blockage remediation can reach five miles, *Offshore* (*March*) (1998) 106–107.

[68] B.F. Baugh, J. Illeman, J.B. Crawford, D. McStravick, W. Warren, Extended reach pipeline blockage remediation, OTC-8675, in: Offshore Technical Conference, Houston, Texas, 4–7 May, 1998.

[69] L. Abney, M. Kalman, J. Hooerhuis, C. Headworth, Flow remediation solutions for pipelines, OTC-15258, in: Offshore Technology Conference, Houston, Texas, 5–8 May, 2003.

[70] D. Jennings, O. Pogoson, J. White, S. Szymczak, M. Newberry, Paraffin dispersant applications for cleaning deepwater flowlines, in: 16th Annual Gulf of Mexico Deepwater Technical Symposium, New Orleans, Louisiana, 15–16 August, 2012.

[71] G.L. Poole, G. Brock, S. Szymczak, G. Casey, Successful pipeline cleanout: Lessons learned from cleaning paraffin blockage from a deepwater pipeline, SPE-115658-MS, in: SPE Annual Technical Conference and Exhibition, Denver, Colorado, 21–24 September, 2008.

[72] K.A. Ferworn, A. Hammami, H. Ellis, Control of wax deposition: An experimental investigation of crystal morphology and an evaluation of various chemical solvents, SPE-37240-MS, in: SPE International Symposium on Oilfield Chemistry, Houston, Texas, 18–21 February, 1997.

[73] S.P. Srivastava, A.K. Saxena, R.S. Tandon, V. Shekher, Measurement and prediction of solubility of petroleum waxes in organic solvents, Fuel 76 (1997) 625–630.

[74] K.M. Barker, M.E. Newberry, Y.R. Yin, Paraffin solvation in the oilfield, SPE-64995-MS, in: SPE International Symposium on Oilfield Chemistry, Houston, Texas, 13–16 February, 2001.

[75] D.W. Jennings, J. Breitigam, Paraffin inhibitor formulations for different application environments – from heated injection in the desert to extreme cold artic temperatures, Energy & Fuels 24 (2010) 2337–2349.

[76] D.W. Jennings, K. Weispfennig, Effect of temperature & shear on wax deposition: Cold finger investigation with GOM crude oil, Energy & Fuels 19 (2005) 1376–1386.

[77] K. Weispfennig, Advancements in paraffin testing methodology, SPE-64997-MS, in: SPE International Symposium on Oilfield Chemistry, Houston, Texas, 13–16 February, 2001.

[78] K. Weispfennig, D.W. Jennings, Paraffin deposition modeling using benchtop deposition tests, Paper #46b, in: AIChE Spring National Meeting, New Orleans, Louisiana, 10–14 March, 2002.

[79] M.E. Newberry, Chemical treatments for paraffin control in the oilfield, in: Proceedings of Annual Southwestern Petroleum Short Course, Lubbock, Texas, April, 1982.
[80] M.E. Newberry, Crude oil production and flowline pressure problems, SPE-11561-MS, in: SPE Production Operations Symposium, Oklahoma City, Oklahoma, 27 February - 1 March, 1983.
[81] M.E. Newberry, G.E. Addison, K.M. Barker, Paraffin control in the Northern Michigan Niagaran reef trend, SPE Production & Operations 1 (1986) 213–220.
[82] W.J. Matlach, M.E. Newberry, Paraffin deposition and rheological evaluation of high wax content altamont crude oils, SPE-11851-MS, in: SPE Rocky Mountain Regional Meeting, Salt Lake City, Utah, 23–25 May, 1983.
[83] D. Jennings, R. Yin, K. Weispfennig, M. Newberry, Tratamento de Problemas Parafinicos com Agentes Quimicos na Producao de Petroleo, IBP-18904, in: Rio Oil & Gas Expo and Conference, Rio de Janeiro, Brazil, 4–7 October, 2004.
[84] D.W. Jennings, K. Weispfennig, K. Effect of shear on performance of paraffin inhibitors: Cold finger investigation with GOM crude oils, Energy & Fuels 20 (2006) 2457–2464.
[85] D.W. Jennings, M.E. Newberry, Application of paraffin inhibitor treatment programs in offshore developments, OTC-19154, in: Offshore Technology Conference, Houston, Texas, 5–8 May, 2008.
[86] D.W. Jennings, M.E. Newberry, Paraffin inhibitor applications in deepwater offshore developments, IPTC-12127, in: International Petroleum Technology Conference, Kuala Lumpur, Malaysia, 3–5 December, 2008.
[87] C.H. Wu, K.S. Wang, P.J. Shuler, Y. Tang, J. Creek, R.M. Carlson, et al., Measurement of wax deposition in paraffin solutions, AIChE Journal 48 (2002) 2107–2110.
[88] M. Zougari, S. Jacobs, J. Ratulowski, A. Hammami, G. Broze, M. Flannery, et al., Novel organic solids deposition and control device for live-oils: Design and applications, Energy & Fuels 20 (2006) 1656–1663.
[89] M.E. Newberry, K.M. Barker, Formation damage prevention through the control of paraffin and asphaltene damage, SPE-13796-MS, in: SPE Production Operations Symposium, Oklahoma City, Oklahoma, 10–12 March, 1985.
[90] M.E. Newberry, K.M., Barker, K.P. Flynn, Identification and remediation of organic skin damage, Paper #60a, in: AIChE Spring National Meeting, Houston, Texas, 14–18 March, 1999.
[91] M.E. Newberry, K.M., Barker, Organic formation damage control and remediation, SPE-58723-MS, in: SPE International Symposium on Formation Damage, Lafayette, Louisiana, 23–24 February, 2000.
[92] L. Quintero, M. Newberry, R. Morales, C. Davis, Organic and inorganic formation damage removal using customized formulations, SPE-185735-MS, in: SPE Western Regional Meeting, Bakersfield, California, 23 April, 2017.
[93] K.M. Barker, M.E. Newberry, J.K. Johnson, Production stimulation by removal of paraffin and asphaltene deposition, in: Proceedings of Annual Southwestern Petroleum Short Course, Lubbock, Texas, April, 2002.
[94] H. Aslanov, A. Novruzov, A. Harun, Managing wax-deposition risks in oil subsea pipelines by integrating wax modeling and pigging performance, SPE Production & Operations 34 (2019) 625–634.

[95] C.J. Stevenson, S.R. Davies, I. Gasanov, P. Hawkins, M. Demiroglu, A.P. Marwood, Development and execution of a wax remediation pigging program for a subsea oil export pipeline, OTC-25889, in: Offshore Technology Conference, Houston, Texas, 4–7 May, 2015.

[96] G. Fung, W.P. Backhaus, S. McDaniel, M. Erdogmus, To pig or not to pig: The Marlin experience with stuck pig, OTC 18387, in: Offshore Technical Conference, Houston, Texas, 1–4 May, 2006.

[97] D. Jennings, J. White, O. Pogoson, D. Barros, K. Ramachandran, G. Bonin, et al., Paraffin dispersant application for cleaning subsea flowlines in the deepwater Gulf of Mexico Cottonwood development, IBP-1345-12, in: Rio Oil & Gas Expo and Conference, Rio de Janeiro, Brazil, 17–20 September, 2012.

[98] F. Shecaira, D. Barros, S. Ramachandran, G. Bonin, P. Waltrich, D. Jennings, et al., The Cottonwood field case history: The pig/paraffin obstruction of a long subsea, deepwater tie-back and its successful remediation, SPE 146156, in: SPE Annual Technical Conference & Exhibition, Denver, Colorado, 30 October–2 November, 2011.

[99] ASTM D4007-08, Standard Test Method for Water and Sediment in Crude Oil by the Centrifuge Method (Laboratory Procedure), ASTM International, West Conshohocken, 2011.

[100] IP 359, Determination of Water and Sediment Content of Crude Oils - Centrifuge Method, Energy Institute, London, 1982.

[101] G.P. Borghi, S. Correra, M. Merlini, C. Carniani, Prediction and scaleup of waxy oil restart behavior, SPE-80259-MS, in: SPE International Symposium on Oilfield Chemistry, Houston, Texas, 5–8 February, 2003.

[102] H.P. Ronningsen, Rheological behavior of gelled waxy North Sea crude oils, Journal of Petroleum Science and Engineering 7 (1992) 177–213.

[103] W.H. Thomason, Start-up and shut-in issues for subsea production of high paraffinic crudes, OTC-11967, in: Offshore Technology Conference, Houston, Texas, 1–4 May, 2000.

[104] A. Uhde, G. Kopp, Pipeline problems resulting from handling of waxy crude, Journal of The Institute of Petroleum 57 (1971) 63–73.

[105] S.A. Putra, E. Waspodo, Transportation of high pour-point oil through long hilly-terrain pipeline, SPE-862929-MS, in: SPE International Thermal Operations and Heavy oil Symposium and Western Regional Meeting, Bakersfield, California, 16–18 March, 2004.

[106] T.M. Williams, J.J. Hsu, H.L. Patterson, Measurement of break away yield stress of waxy crude oil and pipeline restart system design, in: Offshore Technology Conference, Houston, Texas, 6–9 May, 1996.

[107] K.M. Barker, D.B. Sharum, D. Brewer, Paraffin damage in high temperature formations, removal and inhibition, SPE-52156-MS, in: SPE Mid-Continent Operations Symposium, Oklahoma City, Oklahoma, 28–31 March, 1999.

[108] A. Alwazzan, M. Utgard, D. Barros, Design challenges due to wax on a fast-track deepwater project, OTC-19160, in: Offshore Technology Conference, Houston, Texas, 5–8 May, 2008.

CHAPTER

Asphaltene management

3

Priyanka Juyal[1] and Andrew T Yen[2]
[1]*ChampionX LLC, Sugar Land, TX, United States*
[2]*ENNOVA LLC, Stafford, TX, United States*

Chapter outline

Flow Assurance. DOI: https://doi.org/10.1016/B978-0-12-822010-8.00007-6

3.1 Introduction

Asphaltene deposition is a significant flow assurance problem thatfor many years has been causing problems in both upstream and downstream operations in the oil and gas industry [1–16]. This flow assurance challenge occurs in different geographical locations and under different operating conditions. During production, the variations in the pressure, temperature, or composition of the crude oil in the well may cause asphaltenes to destabilize, which will lead to precipitation and formation of aggregates that contribute to deposition and plugging of the well. The build-up of the deposits in the production tubing can cause a reduction in oil production, or in extreme cases, a complete shutdown of the production line can occur (Fig. 3.1). Gas injection and enhanced oil recovery projects of the brown fields involving carbon dioxide (CO_2) floods and water alternate gas injection can worsen the asphaltene deposition and it has been well documented. Solvent wash has been applied to mitigate the problem in some cases. For deep water, offshore fields frequent solvent washes are determined to be uneconomical, therefore many operators inject asphaltenes inhibitors to prevent asphaltene deposition on oil fields. The selection of asphaltene inhibitors requires a lab screening before field implementation.

This chapter discusses the chemistry of asphaltenes, prediction of asphaltene-related issues for green field projects, modeling efforts and experimental measurements to predict field issues, and chemical methods for prevention and remediation.

3.2 Chemistry of asphaltenes

3.2.1 Composition and structure

The molecular characterization of asphaltenes is as intricate as the challenges the petroleum fraction imposes in upstream (and downstream) petroleum industry.

FIGURE 3.1

Asphaltene deposition for offshore Alaska field.

Asphaltene-related problems during upstream production such as formation damage [17], deposition in pipelines [18,19], stabilization of water-in-oil emulsions [18–22] are the result of asphaltene composition and structure. A comprehensive understanding of the chemistry of asphaltenes is needed for the prediction of its function. If the structure is known, it is possible to make basic and coherent predictions about the function and reactivity [8,16,18].

Operationally, asphaltenes are defined as a vague solubility class. They constitute the most complex petroleum fraction that is insoluble in alkanes (usually heptane or pentane) and soluble in aromatic hydrocarbon solvents (usually toluene or xylene). This "vague" solubility definition of asphaltenes generates a wide distribution of chemical species and structures with enormous complexity, whose molecular level characteristics can dramatically change from one crude oil to another.

Asphaltenes are a polydisperse distribution of molecules with significant differences in the chemical and physical properties of the precipitated asphaltene fraction, which was observed in the laboratory using different *n*-alkanes as precipitants [8,16,18]. Shorter *n*-alkanes yield sticky asphaltenes whereas longer *n*-paraffins lead to the precipitation of powdery asphaltenes that are more aromatic relative to those precipitated by shorter *n*-alkanes [8,16,18]. Tavakkoli et al. showed that lighter hydrocarbons are stronger precipitants, and the amount of asphaltene mass increases with a decrease in the precipitant carbon number [16,23].

Compositionally, asphaltenes are highly polar, aromatic, and heteroatom-rich fraction comprising primarily of polycyclic aromatic clusters with alkyl side chains along with heteroatoms (O, N, S) and trace amounts of metals such as Ni, V, Fe. Many analytical methods have been used to study asphaltenes over the last 50 years, including X-ray, electron spin resonance (ESR), nuclear magnetic resonance (NMR), Fourier-transform infrared spectroscopy (FTIR), and mass spectrometry. Fourier transform ion cyclotron resonance mass spectrometry (FT-ICR Ms) has revealed the ultra-complexity of asphaltenes. Electrospray Ionization FT-ICR MS has identified polydisperse combinations of O_x, N_xO_x, S_xO_x, N_xS_x, and $N_xO_xS_x$, heteroatom classes (where x is number of the heteroatom) [24–28]. Highly condensed polycyclic aromatic hydrocarbons and nonpolar heteroatom containing compounds with O_x, N_x, S_x, and vanadium or nickel are observed in asphaltenes analyzed by atmospheric pressure photoionization (APPI) [24,29–33]. It is observed that the asphaltenes precipitated by heptane is different from asphaltenes generated asphaltenes generated from depressurization in the lab [25,26] In a recent study, Rowland et al. [34] found ~ 47,000 unique monoisotopic elemental compositions, by positive-ion APPI 21T FT-ICR MS, for an asphaltene sample [34,35]. The number of possible structural isomers of the organic compounds rises rapidly with the number of carbon atoms, and the carbon number in asphaltenes can range from ~15 to ~100, and they can exhibit millions of different components that are in many ways dissimilar yet close enough to stay within a relatively narrow solvent cut [36]. This extreme complexity of the asphaltenes is demonstrated by the detection, resolution, and unique molecular formula assignment of 462 mass spectral peaks >6σ at *m/z* 677 in a volcano asphalt sample with (+) APPI FT-ICR MS as shown in Fig. 3.2.

3.2.2 Solubility and aggregation

Solubility of asphaltenes in the crude oil has been approached variously, with the thermodynamic approach that favors complete solubility, and the colloidal approach that describes asphaltenes as a dispersion/suspension in petroleum fluid [16,18,37–41]. Various thermodynamic models derive from asphaltenes to be completely soluble, and asphaltene flocculation to be completely reversible well-understood thermodynamic process, whereas the colloidal model attributes a more complex irreversible mechanism to asphaltene flocculation [16,18,37–45]. The polydispersity associated with the asphaltene fraction may be indicative of asphaltene to be partly dissolved and partly suspended in the crude oil.

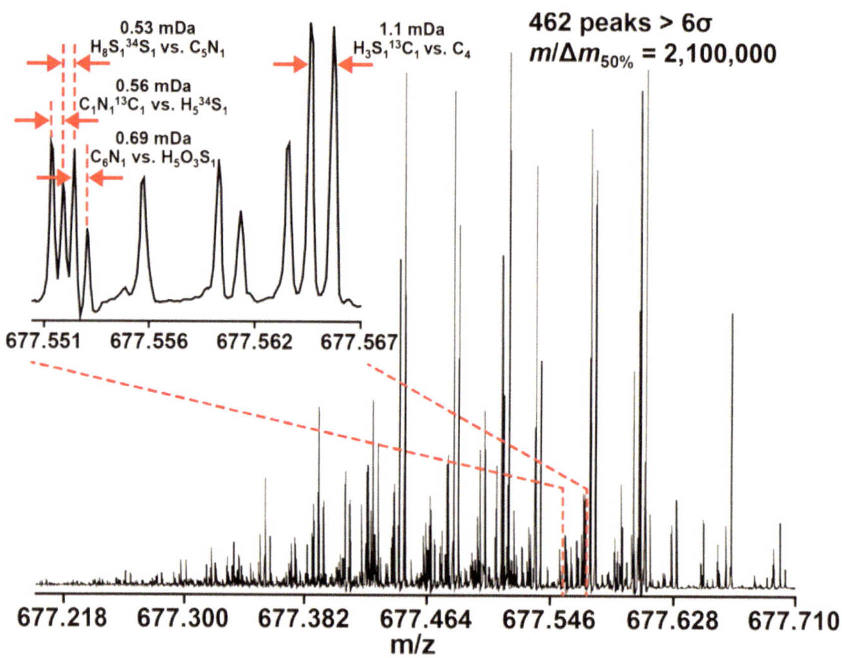

FIGURE 3.2

Mass scale expanded segment of the (+) APPI FT-ICR mass spectrum of an asphalt volcano sample at *m/z* 677.

Reproduced with permission from D.F. Smith, D.C. Podgorski, R.P. Rodgers, et al., 21 Tesla FT-ICR mass spectrometer for ultrahigh-resolution analysis of complex organic mixtures, Analytical Chemistry 90 (2018) 2041–2047 [24].

The asphaltene community has described asphaltenes as a set of nanoaggregates comprised of clusters of aromatic cores in a surrounding media of aliphatic moieties. This view developed in the 1960s, when X-ray scattering, NMR, ESR, and FTIR first emerged. Yen and co-workers were able to study the diffraction pattern of asphaltenes by using mixtures of model substances such as graphite, polyethylene, and polycyclic aromatic hydrocarbons, and concluded that the proportion of the XRD bands of real asphaltene samples are the result of the ratio aromatic/aliphatic moieties that compose asphaltene clusters [36,46]. With that premise, Yen was able to quantify average structural parameters related to the number of aromatic cores and their size in asphaltene aggregates as shown in Fig. 3.3, Yen suggested the structure for asphaltene aggregates, in which single core alkyl-substituted structures (*island* motifs) coexist with multiple core (*more than one*) alkyl-bridged compounds (*archipelago* motifs). In Yen's model, $\pi-\pi$ stacking is the force at molecular level that drives asphaltene aggregation. Yen suggested asphaltene self-association as the cause of divergent results for molecular weight, which at that time ranged between 1000 and 500,000 amu. By correlating mass spectrometry with XRD results, Dickie and Yen proposed, for the first time, a low molecular weight nature for petroleum asphaltenes. Dickie and Yen demonstrated that the weights of the aromatic cores in asphaltene molecules, known as disk weight in the X-ray scattering studies, were estimated to be between 260 and 850 amu [46].

The center of Yen's model is the concept of structures in which size progresses through aggregation and clustering, which is a colloidal view. Yen also

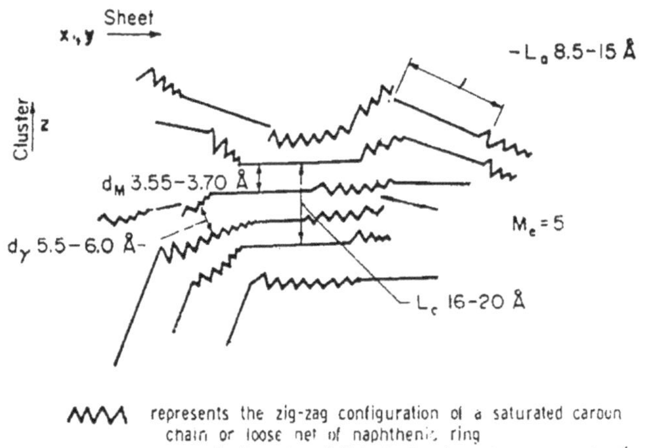

FIGURE 3.3

Molecular representation of asphaltene structure.

Reproduced with permission from T.F. Yen;.E J...G. Grdman, S.S. Pollack, . Investigation of the Structure of Petroleum Asphaltenes by X-Ray Diffraction. Analytical Chemistry 33 (1961), 1587–1594 [36].

focused his efforts on porphyrins and the structure of vanadium complexes using mass spectrometry, X-ray diffraction, NMR, and ESR spectroscopy [22,25,26,34].

The physical model by Yen inspired generations of researchers to study and refine the potential aggregation mechanism for asphaltene aggregation. Yen's model was modified by Mullins to account for nanoaggregates at low concentration. The modified Yen model suggests that the predominant asphaltene molecular architecture consists of a single, moderate-sized aromatic core with peripheral alkane chains known as island, whose representative structure is shown in Fig. 3.4. In the modified Yen model, the predominant intermolecular interaction is $\pi-\pi$-stacking. Therefore, the nanoaggregates consist of a single stack of PAHs surrounded by the alkane substituents, which average aggregation number is ~6. This type of nanoaggregates can produce clusters, with average aggregation number of ~8 [13].

Petroleum is an ultra-complex mixture of monofunctional and polyfunctional compounds. Asphaltenes are enriched with polyfunctional species whereas resins are enriched with monofunctional molecules. Yen's model missed the role of heteroatoms such as O, N, and S in asphaltene behavior; especially if those are a part of functionalities such as carboxylic acid, phenol, pyridine, pyrrole, sulfide, which can participate in strong intermolecular forces such as hydrogen bonding and acid/base interactions. Analyses on various field deposits and asphaltenes indicates there is an oxygen continuum as related to hard calcium-naphthenate deposit, asphaltenes, interfacial material and sodium-naphthenate emulsion [47]. This supports the importance of the heteroatoms on asphaltene aggregation and deposition and also challenges the theory that $\pi-\pi$-stacking is the only dominant intermolecular force for aggregation.

To determine the role of heteroatoms in asphaltene solubility, Chacón-Patiño et al. separated asphaltene fractions enriched in single-core (island) and multicore (archipelago) according to their partitioning in *n*-heptane by a proprietary fractionation method [48]. The results showed that unstable asphaltene fractions are enriched with polyfunctional compounds. IRMPD results demonstrate that the most unstable fractions from the Wyoming acetone fraction (island dominant) are enriched with archipelago motifs that exhibit poor ionization efficiency.

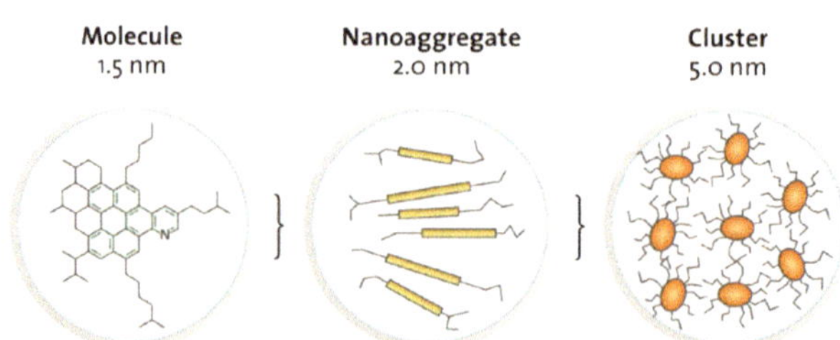

FIGURE 3.4

Yen-Mullins model compose of three regimes.

Collectively, the results of this study and the author's other works stress on the need for an asphaltene molecular model based on the existence of polyfunctional species capable of interacting with neighboring molecules via several intermolecular forces including London dispersion forces between aliphatic moeities, π-stacking of aromatic cores, hydrogen bonding between N- and O-containing functionalities, and acid–base interactions between the carboxylic acids and pyridinic rings.

Giraldo-Davila et al. fractionated the asphaltene sample known as Petrophase 2017 asphaltenes by thin-layer chromatography and performed FT-ICR MS analysis [35]. The authors found a direct correlation between adsorption degree on the stationary phase (silica gel), aggregation and emulsion stability. The fraction with the strongest adsorption on silica exhibited decreased the content of O-containing compound classes whereas S- and N-containing compound classes appeared concentrated. This fraction was shown to be highly unstable in Hep/Tol and produced the most stable petroleum emulsions. This work is consistent with microfluidic work done by Lin el al. [49] where the aggregation behavior of island-rich and archipelago-rich fractions were studied and showed different aggregation behavior. In a recent paper by Zhang et al. [50], a new model was proposed based on pancake bonding of free radicals to be complementary of π–π bonding. Further validation and studies are needed to fully understand the interaction of different molecular motifs (island and archipelago) within the asphaltene fractions and the aggregation mechanism of asphaltenes. The roles of heteroatoms are thus an area that require more attention and focus to understand their contribution to asphaltene aggregation.

3.3 Experimental techniques for asphaltene stability prediction

For a new developmental project, experimental measurements are often conducted to assess the stability of oils with respect to asphaltene stability on live oil samples. However, the measurement of asphaltene stability at reservoir conditions is rather expensive and may not always be feasible due to difficulties in obtaining such samples.

This section introduces experimental methods on live oil and dead oil that can be used to predict the stability of asphaltenes.

3.3.1 Solids detection system

For live oil testing, the asphaltene onset pressure (AOP) can be measured directly using a near infrared (NIR) laser or high-pressure microscopy (HPM) [51,52]. The solid detection system (SDS) has been used widely by different laboratories to detect asphaltene precipitation from live oil samples at reservoir conditions (high pressure and high temperature). In this method, the asphaltenes precipitate due to the changes in pressure, temperature, and composition of the live oil sample can be detected.

These experiments are often done iso-thermally. In case of iso-thermal depressurization tests, the asphaltenes are destabilized by gradually depressurizing the live crude sample from reservoir pressure, while the light (NIR) transmittance is being monitored continuously through the PVT cell. Prior to asphaltene precipitation, light transmittance increases with decreasing pressure due to oil expansion. Once the cell's pressure reaches the AOP, a sharp reduction of the light transmittance as detected with an inflection point. In the case that the oil does not exhibit an AOP, the transmittance continues to increase until the pressure reaches the bubble point. At this pressure, a gas phase in the form of bubbles appears, which causes the light transmittance to drop sharply due to the scattering of the light by the gas bubbles. In addition to light scattering, the HPM has been introduced and has the potential to distinguish different types of incipient phases and verify whether precipitated asphaltene particles cause deviation on transmitted light.

3.3.2 DeBoer plot

De Boer plot is a widely used quick screening method to predict the potential for asphaltene precipitation under reservoirs conditions [52]. De Boer correlated different samples from different fields to construct this diagram and compared to field experiences. The basis for the De Boer plot is Hirschberg's thermodynamic model of asphaltene solubility [53]. The risk of asphaltenes precipitation is evaluated by plotting the difference between the reservoir pressure and bubble point pressure versus reservoir fluid pressure. The unstable (severe problems, $SS > 2$) and stable (no problems, $0 < SS < 1$) regions are outlined based on the calculations of asphaltene supersaturation (SS) derived from the Hirschberg model [53]. An intermediate area or mild problems lies between the extremes with the calculated supersaturation values of 1 and 2 ($1 < SS < 2$). De Boer screen is considered an extremely conservative screen because of the assumption that the fluid is saturated with respect to asphaltene at reservoir conditions [54].

De Boer plot stability assessment of three different scenarios of asphaltene instability (severe, mild and no problem) is shown in Fig. 3.5.

3.3.3 Dead oil tests

Potential asphaltene problems in oil production systems can also be predicted through an analysis of the stability of the asphaltenes in dead crude. The methods for determining asphaltene stability are the saturates-, aromatics-, resins-, and asphaltenes (SARA)-based methods such as the asphaltene-resin ratio and the Colloidal Instability Index (CII) [55] and the precipitation-based method that

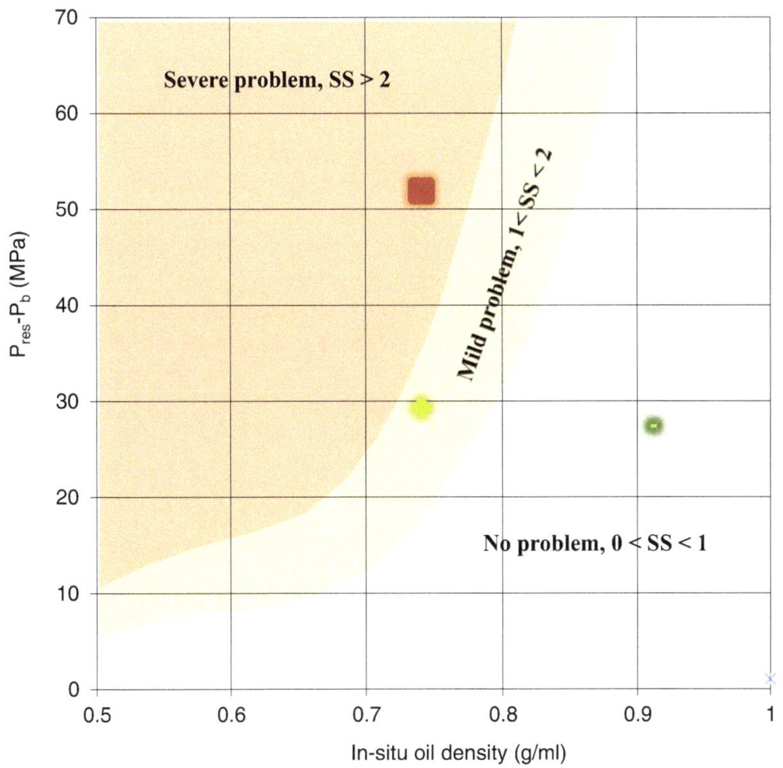

FIGURE 3.5

De Boer asphaltene stability screen.

includes Oliensis spot test [41], microscopy methods, and flocculation point analyzer (FPA) and the/asphaltene deposition detection unit [56,57].

3.3.3.1 Colloidal instability index

The crude oil CII was proposed to determine the stability of the asphaltenes for upstream field prediction. CII is an empirical method that correlated the composition of dead oil to a database with different fields with known asphaltene issues.

CII is defined by the following equation based on wt.% of different fractions in the crude oil sample.

$$\text{CII} = \{(\text{Saturates} + \text{Asphaltenes})/(\text{Resins} + \text{Aromatics})\} \tag{3.1}$$

A CII value below 0.7 usually indicates an oil is stable while values above 0.9 are normally obtained for unstable oils. Note that CII is an empirical method based on a particular SARA method and the database correlation with the method was based on a modified IP-143 method.

3.3.3.2 The flocculation point analyzer/asphaltene precipitation detection unit

The FPA/Asphaltene precipitation detection unit, which measures the onset of asphaltenes flocculation has also been used to predict the stability of asphaltenes in the dead crude oil [56,58]. These bench-top solids detection systems use an optical source to determine the onset of asphaltene flocculation during titration of an aliphatic solvent. In the initial stages of titration, the transmittance of the laser increases, due to a decrease in the density of the oil resulting from the addition of heptane. When the asphaltenes begin to aggregate, the laser transmittance decreases, resulting in an inflection in a plot of transmittance vs. volume of added heptane. The point of inflection corresponds to the point of asphaltene precipitation onset and provides a relative measure of how prone the asphaltenes are to precipitation based on field experiences. Both the AOP and the flocculation point are precipitation measurements, while precipitation is considered a precursor for deposition, it does not predict the deposition tendency of the fluid studied.

3.4 Asphaltene stability modeling

In the last 10 years, some commercial packages (PVT-Sim, Multi-Flash and Flash Pro) were developed for asphaltene precipitation. These modeling software packages are based on thermodynamics with different equation of states (Cubic, Peng-Robinson etc.). One promising approach was the introduction of Perturbed Chain version of the Statistical Associating Fluid Theory (PC-SAFT) to study asphaltenes precipitation behavior [59]. This modeling technique can simulate standard PVT experiments as well as the AOP for live oils and its blends with hydrocarbon gas (lean and rich gases) and CO_2. Furthermore, the effect of oil-based mud contamination, commingling of oils, and even the compositional grading produced in reservoirs that can lead to the formation of tar-mats can be quantitatively predicted with this modeling approach.

The modeling method using the PC-SAFT EOS was successful in correlating the data at different conditions and predicting asphaltenes precipitation behavior under different conditions. With the modeling approach, it will allow the operator to reduce the number of experiments that is needed. Results from this model for a Gulf of Mexico oil are displayed in Fig. 3.6.

3.4.1 Asphaltene instability trend modeling

Asphaltene instability trend (ASIST) is a modeling tool that can predict the AOPs during oil production [60–63]. A series of titration experiments using stock tank oil was used to determine the onset of asphaltenes at ambient conditions. The refractive index at onset conditions were measured and extrapolated to live oil

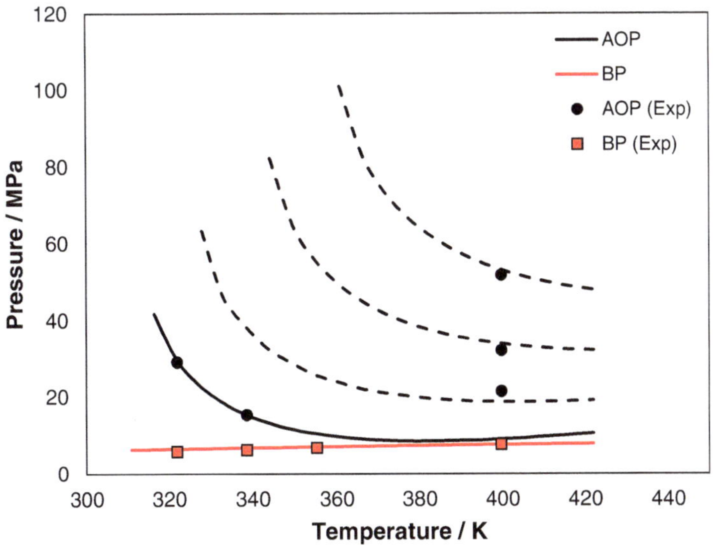

FIGURE 3.6

PC-SAFT prediction of different asphaltenes under gas injection scenarios for a Gulf of Mexico production.

conditions based on the linear relationship between refractive index at onset condition and the square root of molar volume of alkane.

The ASIST modeling approach has been applied successfully to many projects in the Gulf of Mexico. A couple of ASIST examples and comparison of results with SDS isothermal depressurization are highlighted in Figs. 3.7 and 3.8.

3.4.2 Asphaltene deposition model

Asphaltene precipitation is a necessary but not a sufficient condition for deposition. Additional work has been conducted to develop experimental and modeling tools to understand and predict asphaltene deposition in the production tubing [64–68].

Fig. 3.9 shows a schematic representation of the multistep mechanism for asphaltene precipitation, aggregation and deposition that is believed to occur in the wellbore. According to Vargas, when asphaltenes first precipitate, they can form small aggregates called primary particles, which, in turn, can further aggregate or travel to the inner surface of the production tubing and deposit. The rate of asphaltene precipitation depends on the solubility of asphaltenes at a given pressure, temperature, and composition, which can be calculated using the PC-SAFT equation of state or ASIST. Additional parameters are required to establish

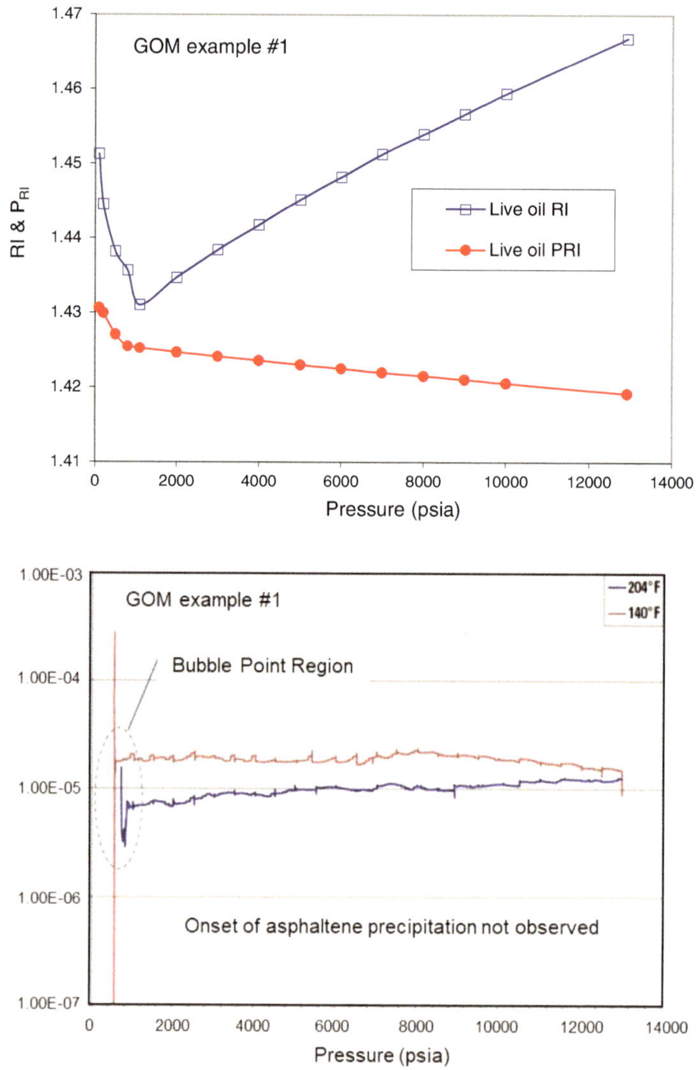

FIGURE 3.7

Comparison of ASIST prediction (top) versus SDS measurements (bottom) – GOM example #1.

the magnitude of asphaltene precipitation, aggregation, and deposition rates, which are usually tuned to available data, either from the laboratory or the field.

Kurup et al. [65] further developed the concepts proposed by Vargas [64] to establish a one-dimensional simulation tool called Asphaltene Deposition Tool

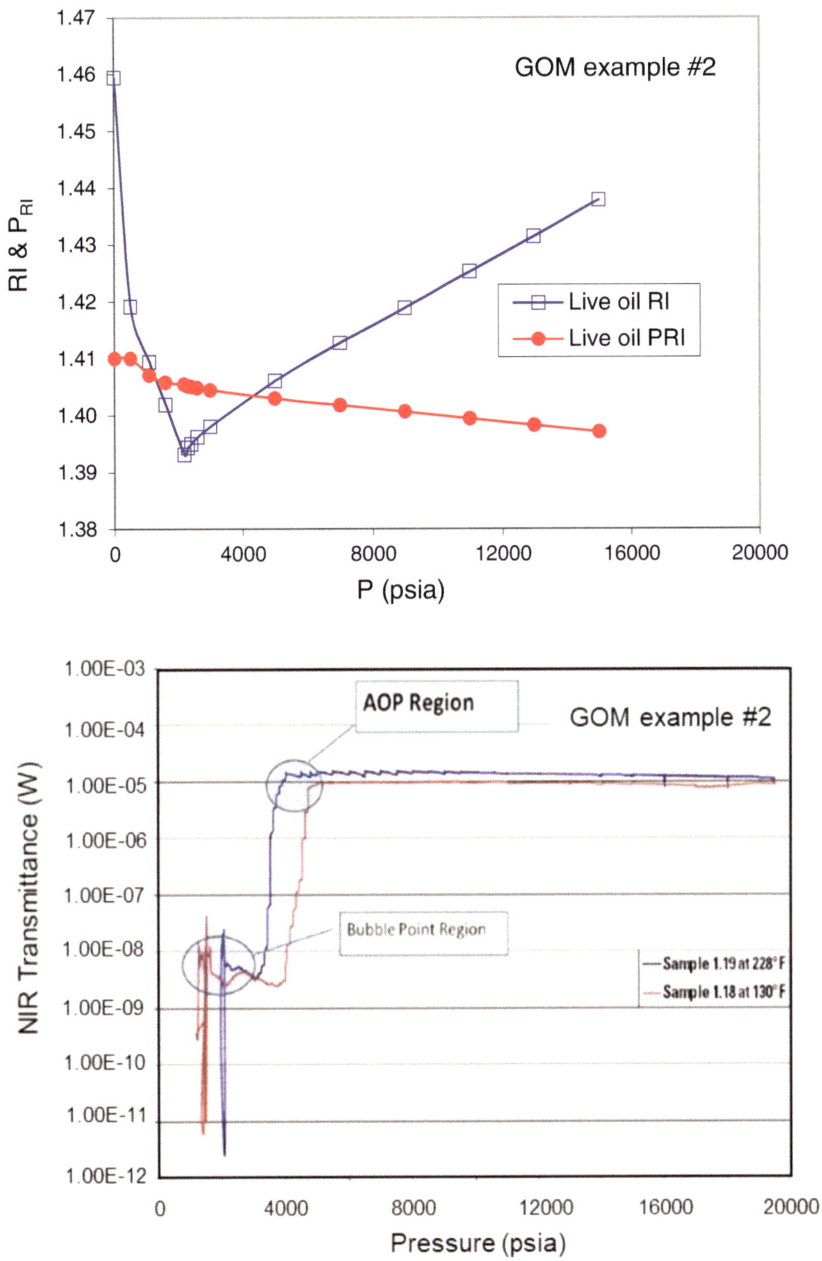

FIGURE 3.8

Comparison of ASIST prediction (top) versus SDS measurements (bottom) – GOM example #2.

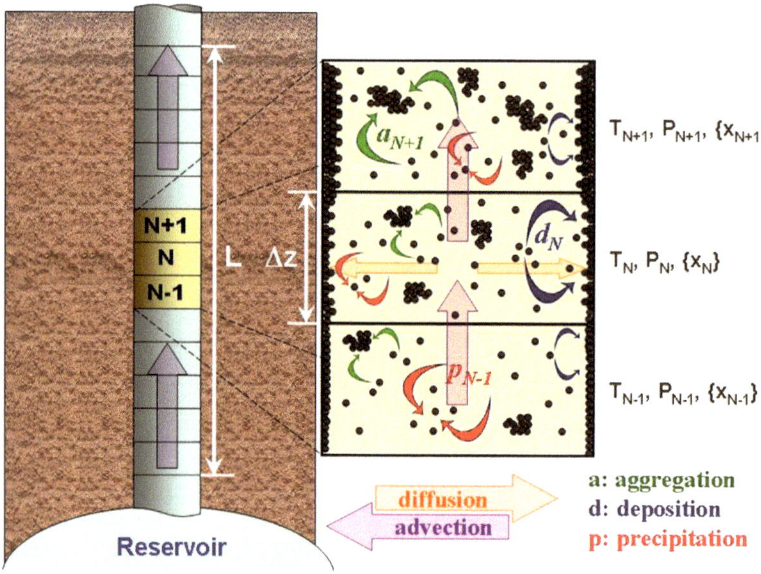

FIGURE 3.9

Schematic for the multistep mechanism for asphaltene precipitation, aggregation, and deposition in a production well.

Reproduced with permission from F.M. Vargas, J.L. Creek, W.G. Chapman, On the development of an asphaltene deposition simulator, Energy & Fuels 24 (2010) 2294–2299 [64].

(ADEPT). The modeling approach is showed in Fig. 3.10 and has been successfully used to forecast the occurrence and calculate the magnitude and profile of asphaltene deposition in some case studies. ADEPT can be an excellent simulator to enable operators to perform quick sensitivity studies and map out operating conditions to better identify potential asphaltene deposition risks.

3.5 Asphaltene inhibitor lab tests

The selection of asphaltene inhibitors requires a lab screening method before field implementation. This section summarizes commonly used lab evaluation techniques of asphaltenes inhibitors. The evolution of these test methods, inherent concepts, merits, and demerits can be found in a recent review paper [69]. For the purpose of this chapter, we use the operational definition for asphaltene inhibitor. The chemicals that applied downhole via continuous injection are defined as asphaltene inhibitors. We touch upon different precipitation and deposition-based tests utilized to screen and select asphaltene inhibitors.

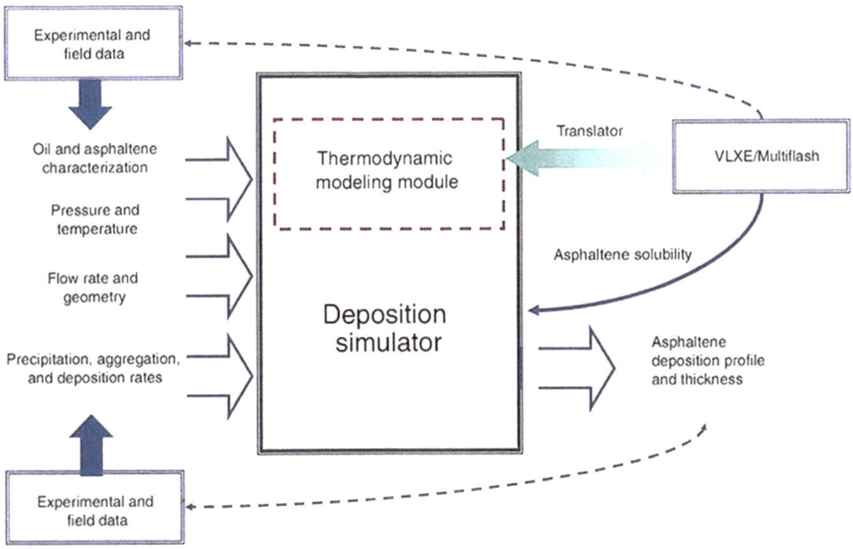

FIGURE 3.10

Modeling approach for ADEPT modeling.

3.5.1 Precipitation tests

In this category of tests, the basic premise is that the bulk precipitated species is responsible for deposition. Any favorable effect on precipitation by a chemical with respect to blank indicates activity. In most cases, the tests are done at ambient condition. But in few instances, tests at high temperature and high pressure (HTHP) are also available.

3.5.1.1 Asphaltene dispersion test

The asphaltene dispersion test (ADT) is the most common approach taken by the chemical vendors to evaluate the performance of chemicals on the dispersion efficiency of asphaltenes [70–79]. The ADT is commonly used as a primary screening method to test dispersants at different dosages quickly. There are different variations of this test, in general, the dispersant with a known concentration is added to the oil sample then an excess amount of asphaltene precipitant (for example, n-C_7) is added to the mixture to induce precipitation. The samples are left undisturbed to let flocculated asphaltenes settle. After a certain period, the amount of asphaltene sediments is compared among different samples by inspecting the level of sedimentation. A better dispersant will keep the precipitated asphaltene particles suspended in the solution and produce less amount of sediments

Visual detection is the simplest technique where tubes with and without dispersants are kept undisturbed for several hours. After that the amount of precipitate is measured. Melendez-Alvarez et al. [15] conducted such tests on various inhibitors at different dosages. Fig. 3.11 shows results of one such test where the chemical hardly performed at 70 ppm. But at 500 ppm the precipitate got diminished and the supernatant fluid took a dark color showing dispersion effect of the chemical.

3.5.1.2 Turbidity tests

In these tests, turbidity and light transmittance of the supernatant fluid is measured [78–85]. A high turbidity value (low transmittance) indicated the presence of a high level of dispersed particles in the mixture. Kraiwattanawong et al. [85] tested different asphaltene inhibitors on an Alaskan crude using Turbiscan, a modified ADT. The asphaltene precipitation was induced by the addition of *n*-alkanes (e.g., n-pentane, n-heptane, and *n*-decane) and the mean transmittance of the oil/precipitant mixtures was monitored and recorded every minute over an hour. Effective chemicals showed low transmittance value, which remained relatively constant throughout the experiment. This method does not require a long time for settling, therefore, is much rapid than other ADT versions.

- Indirect method

  Melendez-Alvarez et al. [15] employed NIR spectroscopy to evaluate asphaltenes dispersant. Crude oil as such does not absorb NIR frequency. But any precipitate in the crude oil-precipitant matrix is able to scatter NIR waves.

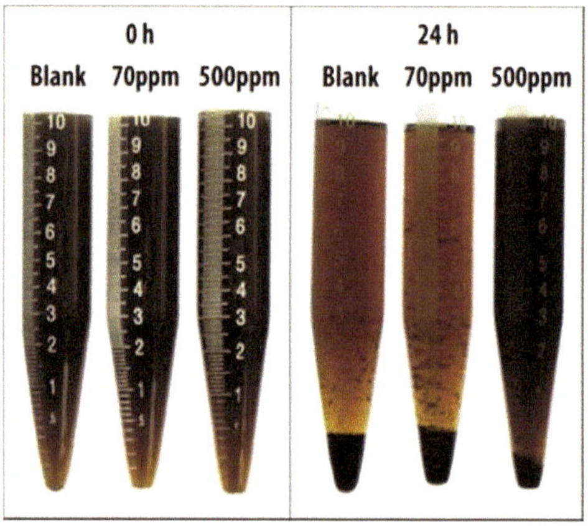

FIGURE 3.11

Asphaltene dispersion test results for a crude oil treated with a dispersant at 70 and 500 ppm with aging times of 0 and 24 h.

Reproduced with permission from A.A. Melendez-Alvarez, M. Garcia-Bermudes, M. Tavakkoli, et al. On the evaluation of the performance of asphaltene dispersants, Fuel 179 (2016) 210–220 [15].

Thus, loss of transmittance is proportional to precipitated species. They prepared different mixtures of crude oil (with or without dispersant) and precipitants (i.e., n-C_7) from 0 to 90 vol.%. Then the samples were left undisturbed for a specific time to age. After aging, the NIR light transmittance of each sample is measured. During the spectroscopy measurements, each sample is stirred by a magnetic stir bar at 1000 rpm to avoid the sedimentation of precipitated asphaltenes.

- Analytical centrifuged stability analysis for asphaltenes (ACSAA)

Jennings et al. [86–88] employed simultaneous centrifugation and NIR measurement in their method ACSAA. In this method, asphaltenes in dead oil samples were destabilized by the addition of precipitant in a vial. The oil/precipitant mixtures at different ratios were centrifuged in a rotor, which allowed for accommodating up to 12 samples at the same time. The changes in the light transmittance of each sample along the entire length of test vial was monitored during the centrifugation process. The transmittance data were then converted into an instability index ranging from 0 to 1. The instability index is defined as the change in integral transmission between time 0 and time t normalized by the maximum theoretical value. An instability index of zero implies a more stable solution with fully dispersed/dissolved asphaltenes that could not be disturbed with centrifugation.

Although precipitation type test such as ADT has been widely used as a quick screening method for asphaltene dispersants, there is no proven correlation between the performance of screened chemicals. According to Montesi et al. [74], most of the asphaltene dispersants selected based on the ADT method have failed to mitigate the asphaltene deposition problem in deepwater fields effectively. It is reasonable to speculate that certain chemicals that fail the ADT test may have the potential capability to mitigate or minimize asphaltene deposition but unfortunately overlooked. Kuang et al. [89] studied the relationship between the dispersive performance of three conventional asphaltene inhibitor and their ability to prevent asphaltene deposition in a packed bed column deposition test. The performance of the chemicals by these two methods did not necessarily correlate.

3.5.2 Deposition tests

Having smaller asphaltene particles does not guarantee less deposition on the metallic surfaces. According to Vargas et al. [64], the aggregation and deposition of particles to the surface are two competing phenomena, thus reducing the aggregation of asphaltene particles might facilitate the deposition process under certain conditions. Therefore, asphaltene inhibitor selection based on some conventional chemical screening methods, such as ADT, may not provide accurate assessments of inhibitor performance in the field. This calls for the development of more representative asphaltene deposition test methods.

3.5.2.1 Capillary deposition unit

Broseta et al. [90] first developed the method of using a capillary unit for asphaltene deposition tests. In this method, crude oil and asphaltene precipitant (e.g., n-C_7) were co-injected into a flow-through metallic capillary loop to induce asphaltene precipitation and subsequent deposition on the metallic surface of the tube [90–98]. The capillary deposition tests can be performed both at ambient and HPHT conditions. The effect of different variables such as temperature, pressure, flow rate, type of precipitant, and asphaltene inhibitors can be studied in this system.

The magnitude of asphaltene deposition is monitored via pressure drop across the capillary. A higher pressure drop indicates more deposition. The Hagen Poiseulle equation can be applied to translate the pressure drop data into the average deposition thickness in the capillary. At the end of the deposition test, the deposited asphaltene can be washed out by an effective solvent such as toluene for quantification. However, deposition may not necessarily uniformly distribute along the capillary. For example, Hoepfner et al. [92] and Lawal et al. [94] used scanning electron microscopy (SEM) to detect a thicker asphaltene deposit at the inlet of the capillary compared to the outlet end. Therefore, relying on only the pressure drop data might not necessarily provide accurate results. A nondestructive in-situ deposition profiling method was developed by Wang et al. [98] to determine deposition distribution along a capillary. After the deposition test, the remaining liquid inside the capillary was flushed out by nitrogen. A viscous fluid that is immiscible with hydrocarbon and cannot dissolve asphaltene was pumped from one end of capillary under constant flow rate while the other end of the capillary was opened to air. The incremental pressure buildup at the capillary inlet was recorded continuously as a function of injection time. This incremental pressure was determined by local diameter along the tubing which can then be used to back-calculate deposit thickness using Hagen Poiseuille equation, with the actual location determined by injection volume. The deposition profile was further validated by comparison with the total amount of deposit recovered from solvent wash.

An alternative approach was recommended by Gon et al. [95] to cut the capillary into segments before solvent wash and quantification. This approach was also used in the study by Kuang et al. [89] in which after the deposition experiment, the capillary tube was cut into ten similarly sized (1-foot) segments, and the amount of asphaltene was quantified for each section. Fig. 3.12 presents the asphaltene deposition profile along the capillary. The pattern indicates that the deposition of asphaltenes inside the capillary is not uniform in this particular case. The majority of asphaltene deposition appeared to occur at both ends (first and last segment) of the capillary.

3.5.2.2 Coupon deposition test

In a typical test, a piece of a metal coupon was submerged in a mixture of crude oil and precipitant at a certain mixing ratio [99,100]. The coupon was

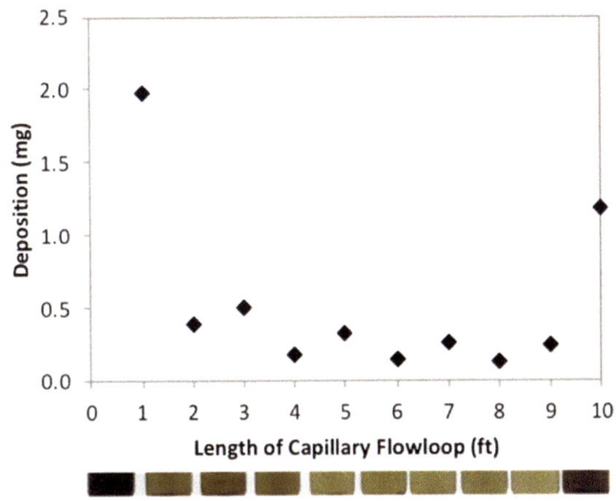

FIGURE 3.12

Asphaltene deposition profile in the capillary flow loop.

Reproduced with permission from J. Kuang, M. Tavakkoli, J. Yarbrough. et. al., Investigation of asphaltene deposition at high temperature and under dynamic conditions, Energy & Fuels 32 (2018) 12405–12415 [89].

continuously moved back and forth to provide a shearing environment. As asphaltenes precipitated, they started to deposit on the metal surface. After a certain period (typically seven days), the amount of asphaltene deposition could be quantified by weight change of the coupon before and after the test. The system can be utilized to evaluate the effects of different precipitants with varying mixing ratios, or the addition of brine and other chemicals. Although a straightforward approach is used for assessing the performance of chemicals, the tests were performed with dead oil samples at low shear and under ambient temperature and pressure that do not represent the actual conditions in the field.

3.5.2.3 Packed bed column

The packed bed column was initially developed by Vilas Boas Favero et al. [101] and later modified by Kuang et al. [102,103]. It had been used recently to assess asphaltene deposition on metallic surfaces for dead oil samples [101–105]. The column was packed with metallic spheres, which provided the primary surface area for asphaltene deposition. As illustrated in Fig. 3.13, in this technique, asphaltene precipitation was induced by continuous co-injection of oil and precipitant into the column. To ensure the presence of a well-mixed fluid, the T-junction at which the two liquids were mixed was placed inside a water bath equipped with ultra-sonication. After each test, the amount of asphaltene deposition could be directly quantified by measuring the weight of spheres before and after the experiment.

Moreover, the deposited material could be easily obtained from solvent wash and characterized directly. Therefore, the interactions between the deposited asphaltenes

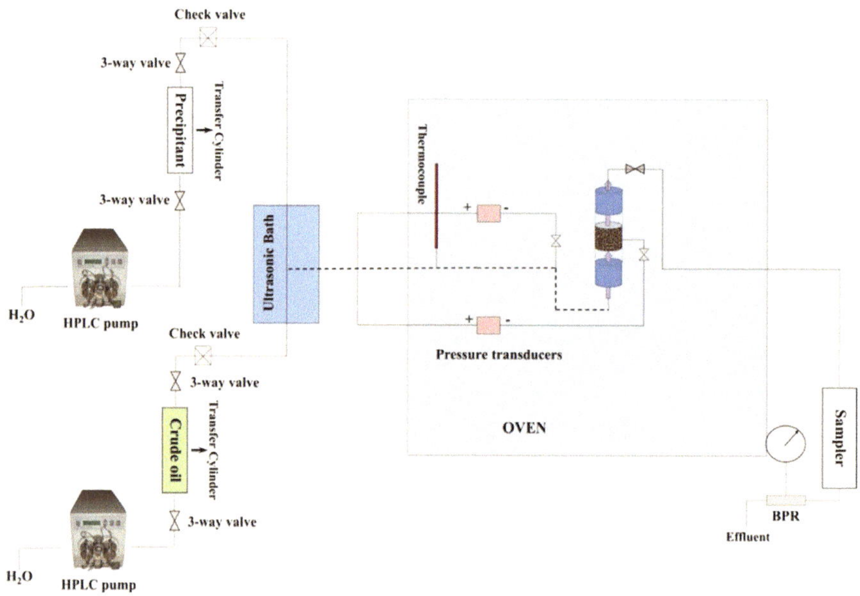

FIGURE 3.13

Schematic of packed bed column deposition test.

Reproduced with permission from J. Kuang, M. Tavakkoli, J. Yarbrough. et. al., Investigation of asphaltene deposition at high temperature and under dynamic conditions, Energy & Fuels 32 (2018) 12405–12415 [89].

and the sphere surface and the impact from asphaltene inhibitor can be studied. This experimental setup was capable of conducting deposition tests at both ambient and HPHT under dynamic flowing conditions. Compared to other techniques, the packed bed column requires less amount of oil samples. Besides, by changing the surface material of the spheres, the effect of surface coating on asphaltene deposition with and without the presence of asphaltene inhibitors can be evaluated.

Kuang et al. investigated the effect of a commercial asphaltene inhibitor on the deposition in the packed-bed column [89,102,103]. The deposition experiments were conducted with a dead oil sample at 80°C and 50 psi backpressure. As shown in Table 3.1, an increase in the dosage of the inhibitor caused a delay in detecting the initial spike in the pressure drop across the packed-bed column. However, the inhibitor did not reduce the total mass of deposited asphaltenes. This observation highlights the limitation of using pressure drop data to evaluate the performance of chemical additives and how it could lead to misleading assessments regarding the effectiveness of a chemical in reducing asphaltene deposition.

3.5.2.4 Microfluidic device

Recently, microfluidic devices have been employed to simulate the asphaltene deposition phenomenon inside porous media in a small benchtop micro-scale

Table 3.1 Results of packed-bed column deposition experiments with a dead oil sample with and without an asphaltene inhibitor.

Dosage of inhibitor (ppm)	Asphaltene deposits (mg)	Lag time (min)	Deposition flux (g/m^2/day)
0	197.6	406	29.9
50	197.7	436	30.9
200	218.2	634	35.6
400	200.3	667	30.7

system, equipped with visualization [104,106–110]. Fig. 3.14 illustrates the schematic of one of the micromodel deposition setups. In this method, asphaltenes were precipitated by the co-injection of crude oil and precipitant into the device by two syringe pumps. The two streams were mixed at a T-junction, before entering the porous media. An optical microscope was used to visualize and record the process of asphaltene deposition inside pore spaces [106]. The acquired images during the experiment were used to analyze deposition morphology and quantify the amount of deposited asphaltene with the aid of an image processing software (e.g., ImageJ). Besides, the pressure drop across the porous media was measured by a differential pressure transducer to evaluate overall permeability impairment caused by the deposition.

Lin et al. [109] examined the effect of four different alkylphenols with varying lengths of alkyl tail on asphaltene deposition using a microfluidic device. Fig. 3.15 demonstrates asphaltene deposition profiles inside that porous media. No asphaltene deposited when a model oil (0.5 wt.% asphaltene in toluene solution) flew through the system (Fig. 3.15). When n-C_7 was co-injected with the model oil into the porous media, cone-shaped asphaltene deposition occurred on the posts. Asphaltene deposition increased when the oil was dosed by 100 ppm of *p*-hexylphenol (Fig. 3.15C) or *p*-octylphenol (Fig. 3.15D). In comparison, 100 ppm of p-dodecylphenol (5e) or iso-dodecylphenol (Fig. 3.15F) decreased the deposition.

3.5.2.5 Core-flooding test

Besides production and transportation facilities, asphaltene deposition can also occur inside reservoirs, causing significant damage to the formation and production reduction as a result of wettability alternation and permeability impairment. Several core-flooding studies have been reported to investigate the deposition of asphaltenes in porous media. Crude oil and an asphaltene precipitant are co-injected into a core plug to induce asphaltene precipitation and deposition inside the porous media. Differential pressures along the core plug was continuously measured with multiple taps, which was then used to interpret the deposition inside each segment. According to Tavakkoli et al. [111], an important aspect for

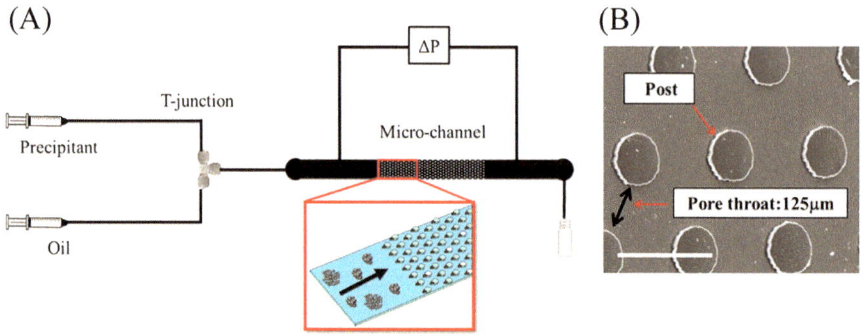

FIGURE 3.14

(A). Schematic of the micromodel deposition experiment setup. (B). SEM image of the porous media used in the deposition experiments. The scale bar is 200 μm.

Reproduced with permission from Y. Zhuang, A. Goharzadeh, Y.J. Lin, et al., Three dimensional measurements of asphaltene deposition in a transparent micro-channel, Journal of Petroleum Science and Engineering 145 (2016) 77–82 [107].

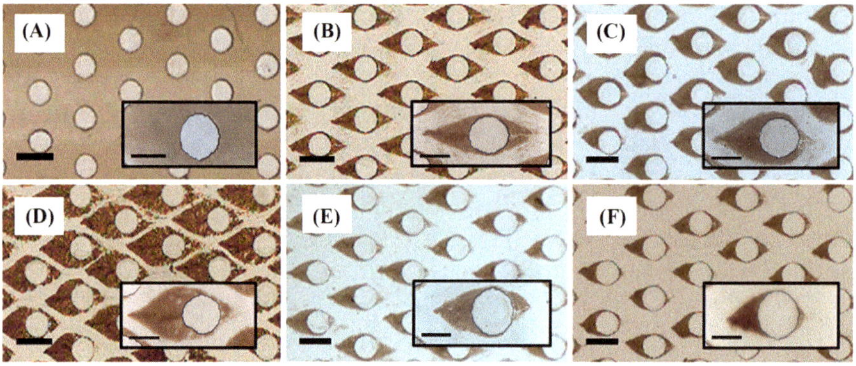

FIGURE 3.15

Asphaltene deposition profile of a crude oil sample with and without additive chemicals in a microfluidic device. (A) Oil without precipitant or chemicals. (B) Oil with precipitant; oil with precipitant plus 100 ppm of (C) *p*-hexylphenol, (D) *p*-octylphenol, (E) *p*-dodecylphenol, or (F) iso-dodecylphenol.

Reproduced with permission from Y.-J. Lin, P. He, M. Tavakkoli, et al. Characterizing asphaltene deposition in the presence of chemical dispersants in porous media micromodels, Energy & Fuels 31(2017) 11660–11668 [109].

a successful core flooding test is to maintain fluid solubility conditions slightly above the onset of asphaltene precipitation before the fluid enters the core sample to avoid sand face plugging at the core inlet.

Romero et al. evaluated the effectiveness of a nanofluid containing alumina nanoparticles for permeability restoration and inhibition of asphaltene deposition in the porous media from core flooding [112]. In that study, a series of core flooding tests were conducted. First, asphaltene deposition in the porous media was created by injecting heptane to a core saturated with crude oil. A 99% permeability reduction was observed compared to oil permeability before the damage. The permeability was restored to 63% of its original value after being subject to sequential solvent wash with diesel, alcohol, and xylene. A follow-up injection of nanofluid containing alumina nanoparticles increased permeability significantly (Table 3.2).

3.5.3 Live oil tests

Direct tests were performed using both live and dead oil. In this discussion we will first take up some tests performed on live oil. Tests involving live oil were thought to be more applicable as true conditions of asphaltenes precipitation were created.

3.5.3.1 Solid detection system

The SDS is a modified PVT apparatus where the PVT cell has some functionalities attached to it to facilitate detection of asphaltenes precipitation [18]. Such system with different variants had been used widely by different laboratories for the detection of asphaltene precipitation from live crude oil samples at reservoir conditions (HPHT). AOP can be detected at several temperatures by iso-thermal depressurization. Asphaltenes onset can be detected by using HTHP microscopy and/or light (NIR) transmittance. Before asphaltene precipitation, light transmittance increases with decreasing pressure due to oil expansion. Once the pressure of the cell reaches the AOP, a sharp reduction of the light transmittance with an inflection point occurs.

SDS can also be used to evaluate the performance of asphaltene inhibitors in their effect on shifting the apparent AOP to lower pressure values.

Table 3.2 The changes in effective permeability after each flooding test stage.

Stage	Effective permeability to oil (mD)
After asphaltene precipitation with n-heptane (induced skin without inhibition)	0.09
After a solvent wash with DAX (diesel, alcohol, xylene)	5.79
After treatment with nanoparticles	10.78
After second asphaltene precipitation with n-heptane (induced skin after inhibition)	2.54

Asomaning et al. designed their SDS experiments in such a way to quantify the amount of precipitated and deposited asphaltenes [113]. Upon the completion of each SDS test, the oil sample was displaced under pressure through a filter (0.2–0.45 μm pore size) to remove the precipitated asphaltenes. Some asphaltenes remained adhered to the inner side of the PVT cell cylinder were quantified by washing with toluene followed by evaporation. These experiments were carried out using 12 crude oil samples from the GOM area and other places. Around 10% of the total asphaltenes in the crude were found adhering to cell wall. The three tested chemicals did not affect the AOP values significantly but noticeable reduction (ranging from 35% to 92%) in the amount of collected asphaltenes was seen at 2000 ppm.

3.5.3.2 Organic solids deposition and control and realview deposition cell

Zougari et al. [114,115] developed a high temperature and high-pressure flow assurance device called organic solids deposition and control (OSDC) to study the deposition process of organic solids such as wax and/or asphaltene from live reservoir fluids. This method was based on the Taylor-Couette flow principles used in closed-cell with a concentric cylinder geometry that could operate under a wide range of production parameters such as temperature, pressure, composition, shear rate, surface type, and roughness. There were two testing modes, the batch mode, and the flow-through mode. The flow-through may be preferred as it reduces the particle depletion issue but requires large quantity of oil. It is suitable for oils with low asphaltene contents as enough deposit can be accumulated. The inner cylinder rotates, which provides desired shear force for the fluid present in the annulus. The deposition takes place on the inner side of outer stationary cylinder. The cell was initially filled with live crude sample at the desired temperature and pressure (above the AOP). After reaching equilibrium, the cell was depressurized to ~100 psi above Pb to provide the maximum precipitation. After completion of the test, the oil inside the cell was drained at the test pressure by nitrogen displacement, and the residue, which included the deposited material and the occluded oil, was removed by solvent wash. Finally, the n-C_7 asphaltene content of the residue was quantified using a modified IP-143 method. Akbarzadeh et al. [66] reported a limited amount of asphaltene deposition from batch OSDC experiments, which can lead to high uncertainty in the measured data. For this reason, they modified the batch system into a flow-through deposition set up. The new system was called RealView, which has both batch mode and flow-through mode. In the flow-through mode, there is a constant influx of fresh oil sample to the cell; thus, more fluid (~1 L) is required to run the test. It generates more asphaltene deposits than the batch mode which provides more accurate deposition rate results and adequate deposits for other post-analysis if needed. A schematic of the RealView test set-up is shown in Fig. 3.16.

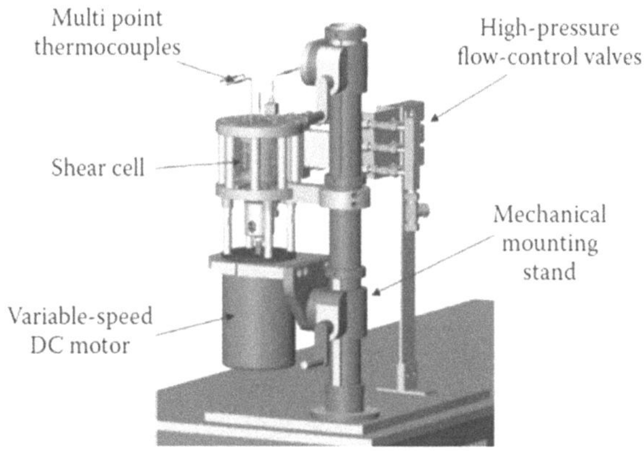

FIGURE 3.16

Schematic of the flow-through realview (aka. OSDC) deposition setup.

Reproduced with permission from M. Zougari, S. Jacobs, J. Ratulowski, et al. Novel organic solids deposition and control device for live-oils: design and applications, Energy & Fuels 20 (2006) 1656–1663 [115].

3.5.3.3 Asphaltene rocking cell test

The asphaltene rocking cell (ARC) test could be performed at high pressure, high temperature, different compositions (liquid and gas phases), and under dynamic conditions [99]. It consisted of a metallic pipe with a magnetic stainless-steel sphere moving freely inside the cell. The movement of the sphere was detected and monitored by the magnetic sensors located at both ends of the cell. The cell was charged with the appropriate fluid at the desired pressure and temperature, it is rocked up and down, causing movement of the ball inside the cell back and forth. A reduction in the ball movement indicated the precipitation and/or deposition of asphaltenes inside the cell. Fig. 3.17 shows example of a typical ARC test for a crude oil sample from the Gulf of Mexico without the addition of any chemical. In this experiment, the cell containing the dead crude oil sample was pressurized with synthetic field gas to 15,000 psi and heated to 115°C. After reaching the equilibrium, the cell was depressurized step by step while the movement of the ball inside the cell was monitored. As shown in Fig. 3.17A, at pressures between 15,000 and 11,000 psi, the ball inside the cell moves freely without any obstacles. However, once the pressure reaches 11,000 psi, the ball moving time increases dramatically, indicating the presence of asphaltenes preventing the ball from moving freely inside the cell. On the other hand, when the same test was repeated with the oil treated with 500 ppm of an asphaltene inhibitor, no blockage occurred throughout the whole pressure range (Fig. 3.17B).

The ARC is mainly used as a pass and fail test. The amount and magnitude of asphaltene precipitation and/or deposition cannot be determined, and it does not reveal any information about the nature of asphaltene deposits in the cell. There are several limitations to the method. (1). It is difficult to distinguish if the hindrance to

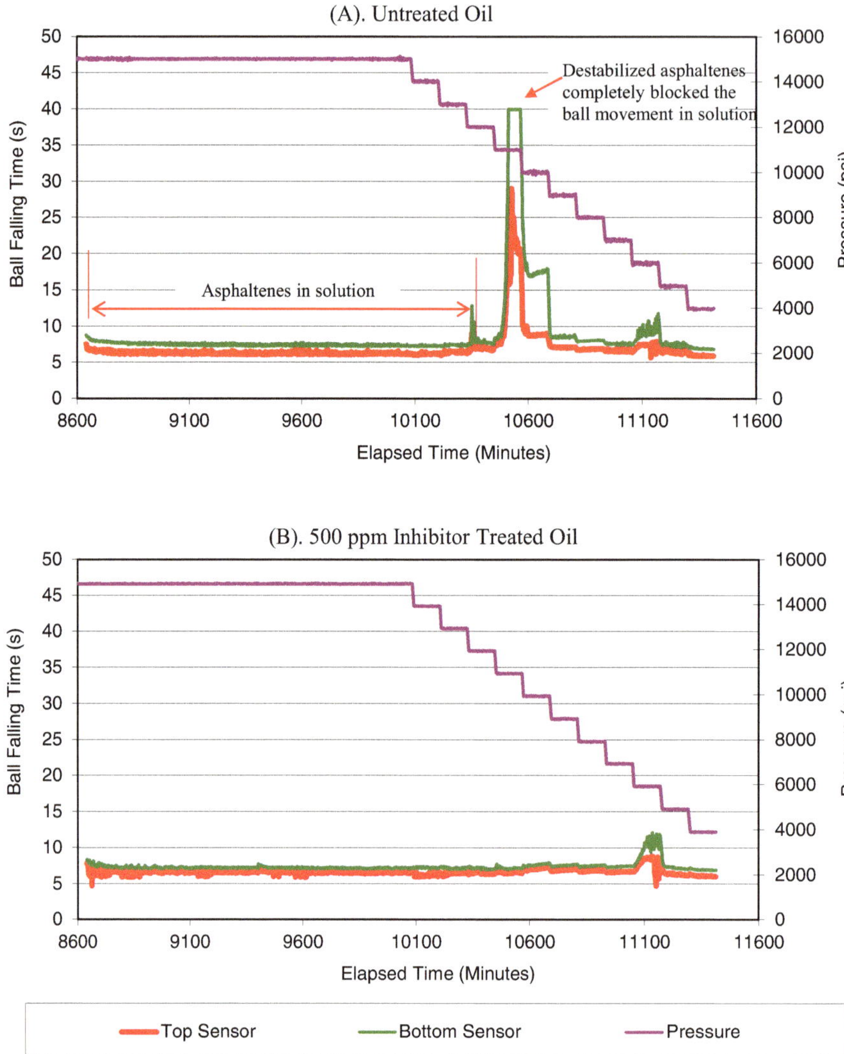

FIGURE 3.17

Example of asphaltene rocking cell test for a crude oil sample from the Gulf of Mexico. (A) Blank and (B) with 500 ppm of inhibitor.

ball movement is caused by sedimented or deposited asphaltenes; (2). the sedimented asphaltene could be smeared onto the lower part of the cell wall by the heavy rolling ball, causing artificial deposition; (3). the ball could get stuck at the two ends by asphaltenes accumulated there, giving a false positive signal of wall deposition. Moreover, the design is not sensitive enough; thus, the detection of asphaltene deposition for light crude oils with low asphaltene content can be difficult.

Asphaltene precipitation is a prerequisite but insufficient condition for asphaltene deposition. Although precipitation and aggregation tests focusing on dispersion mechanisms are still being widely used, the industry has slowly started to move towards representative deposition tests for asphaltene inhibitors selections. Several factors, such as sample size, cost, time, operational conditions, geometry, etc. can play an essential role in choosing the proper testing method. Furthermore, having access to more field performance data during production can help develop valuable correlations between the performance of chemicals in different lab tests and the actual fields.

3.6 Asphaltene control in oil production

Asphaltenes deposition can lead to formation damage, decline in productivity, and decreasing efficiency of equipment. Asphaltene deposition is a frequent and costly problem in upstream oil and gas industry with severe economic repercussions. Deposition in the wellbore and near well bore reservoir region and in the production-tubing causes restrictions that impact oil flow and productivity. Deposition may build leading to blockages and production shut-in necessitating intervention and remediation. Financial impacts of lost production and remediation increase manifold for offshore oil production. For example, for oil fields in the Gulf of Mexico, asphaltene deposition challenge may incur $70 million per well with the well shut-in and intervention [54] Fouling at the surface-controlled subsurface safety valve may escalate these costs to $100 million. Production downtime alone at $50/barrel for 10,000 bbl/day amount to $0.5 million per day. Oil industry expends significant efforts and resources in the risk assessment, monitoring and planning for management, control and mitigation strategies for optimum resource utilization and profitability. Therefore, asphaltene risk assessment, monitoring and asphaltene management, control and mitigation strategies for optimum resource utilization and profitability is an integral part of the field development strategy. Operators make decisions for both preventative and remediation tactics for the project during the planning stage. Preventative approach during upstream operations may include either control of operating conditions or chemical inhibitor applications or a combination of both. Remediation methods include mechanical removal or dissolution of the asphaltene deposits with asphaltene solvents and dispersants.

3.6.1 Prevention

3.6.1.1 Operating condition

Since, precipitation is prerequisite to field deposition, asphaltene deposition during oil production may be prevented by manipulation of the operating conditions outside the asphaltene precipitation envelope for the well [6,16,116]. For this purpose, field development requires robust predictions of the asphaltene stability boundaries with respect to pressure, temperature, and fluid composition. Accordingly, operating pressures could be manipulated so as to minimize pressure drops within the production system to either prevent precipitation or to move precipitation to a location where deposition is easier to handle. Asphaltene precipitation in the wellbore is minimal with higher temperature and increases with lowering of temperature. Therefore, increasing the wellbore temperature by insulation or heat injection such as steam may help curtail amount of asphaltene precipitation and consequently deposition.

As easy as it may sound, aligning the production conditions to stay out of the asphaltene precipitation region could be cost prohibitive and may not always be technically feasible since it involves targeting thermodynamic variables (pressure, temperature) and fluid composition (commingling, draw down) that impact asphaltene stability.

3.6.1.2 Chemical inhibitors and dispersants

Application of chemicals is the most frequently used method to control asphaltene deposition during production operations. Asphaltene inhibitors and dispersants are commonly deployed chemicals which are differentiated based on their mechanism of action [117,118]. Asphaltene inhibitors act by preventing asphaltene flocculation, thus they shift the onset of flocculation, moving the onset pressure from wellbore and near wellbore region to the location where it can be managed with relative ease [117]. Asphaltene dispersants do not impact initial flocculation of asphaltenes but act on pre-formed aggregates by keeping them suspended in the oil [73,117–123]. Mechanistically, inhibitors prevent aggregation of asphaltenes by strong colloidal interactions with the asphaltene [118]. The polar head groups (multiple polar sites) of the inhibitor attaches to the asphaltene and the alkyl chain keeps it in the oil phase. The aromatic cores of the inhibitors also interrupt pi-pi stacking of the asphaltene molecules [118,123]. The stronger these interactions for a particular inhibitor, the better performance its likely to have [118]. The effect of inhibition is not permanent, and there would be a point where the inhibitor would become ineffective due to continued stress such as long residence time in a system. In that respect, inhibitors can perhaps be better referred to as kinetic inhibitors that delay asphaltene aggregation [16]. As far as the chemical structure goes, generally asphaltene inhibitors are polymeric molecules, or resins that have multiple sites to interact with asphaltenes, thus stronger binding [118]. Asphaltene dispersants are typically nonpolymeric dispersant molecules. Additionally, meticulous formulation of inhibitor chemistry with carefully selected dispersants may provide excellent synergies and boosts the performance of the inhibitor.

Chemicals to prevent deposition of asphaltene must be added continuously to the production system before the onset of asphaltene flocculation. These chemistries are applied continuously at the wellhead either downhole through capillary tubing, slipstreams, or subsea umbilicals to help manage deposition in the tubing or the flowlines. However, protection of the wellbore and the near wellbore area requires squeezing the chemistry into the formation to mix with the crude oil before the onset of asphaltene precipitation. Asphaltene squeeze operations are less common compared to inorganic scale squeeze due to challenges associated with retention time and minimal effective dosage concerns.

3.6.2 Remediation

Upon the formation and the accumulation of the asphaltenic deposits, cleaning strategies should be applied. These deposits can be removed chemically or mechanically [6,16,118].

Mechanical methods consist of mechanical force to scrape of deposits inside the wells [124]. Wire line unit, milling operations and coiled tubing along with hydro blasting are often used. In some cases, pigging of flow lines are applied but there is a risk due to the possibility of pigs getting stuck in the flowline [125]. Although mechanical removal of asphaltene deposits is a quicker method to dislodge deposits, it cannot be applied to remove deposits from the formation face.

Solvent wash with aromatic solvent is one of the commonly used techniques to redissolve the deposited asphaltenes in the wells [126–129]. The solubility of asphaltenes in different hydrocarbon liquids has been examined with different techniques [102,130]. To select and develop the best solvents and the most appropriate solvent soaking conditions for asphaltene remediation, different lab screening tests have been developed. These tests varied from simple tea-bag tests where asphaltenes deposits are placed in a teabag to the utilization of microfluidics and packed bed column at HTHP. Different test conditions have been studied to find the best application such as soaking time, solubility of solvent and temperatures. In one example, pack bed test results show that the existence of the occluded oil during the aging and soaking process weakens the solubility power of the toluene. Solvents blends with dispersants can improve the solubility of aromatic solvents and help remediate tough deposits.

3.7 Case studies

3.7.1 Lab screening methods for field applications

Continuous injection of asphaltene inhibitors was proposed for wells in a Latin America offshore field with prolonged asphaltene deposition issues [131]. The performance of asphaltene inhibitors was evaluated on dead oil samples using heptane precipitation test and asphaltene precipitation detection test (APDT). In the heptane precipitation test, the chemical was added to heptane first, and then crude oil was mixed with the inhibitor-heptane solution. The mixture was shaken and left undisturbed to age for 1 hours and then centrifuged. After centrifugation, an aliquot of the supernatant liquid was diluted with toluene and subjected to absorbance measurements. According to the authors, a more effective asphaltene inhibitor keeps more asphaltenes in the solution resulting in a higher absorbance value. In this method, the relative performance of asphaltene inhibitors are calculated based on the following equation:

$$\text{Relative performance} = \frac{T_{\text{blank}} - T_{\text{inhibitor}}}{T_{\text{blank}}} \times 100 \tag{3.2}$$

where: T_{blank} and $T_{\text{inhibitor}}$ represent the transmittance values of untreated and treated samples, respectively. Among the tested inhibitors, the most effective chemical showed a relative performance of 51.7%.

In the APDT, the onset of asphaltene flocculation is measured based on the titration of crude oil samples with heptane. In the beginning, the transmittance increases because of a reduction in the mixture's density due to n-C_7 addition. After the onset point, asphaltenes start to aggregate, causing a drop in the transmittance value, creating an inflection in the plot of transmittance vs. volume of added n-C_7. The authors used the volume of added n-C_7 per mass of oil (mL/g) to reach this inflection point (expressed as APDT number) as an indication of asphaltene precipitation onset point. The same chemical with a dosage of 2000 ppm performed the best by showing the onset at 2.5 APDT number against 1.25 for the blank. This apparent shift in the onset was attributed to higher stability of asphaltenes in the crude oil. Additionally, an SDS test was conducted with this chemical using a live oil sample. Compared to a run with the blank sample, no significant shift in the AOP was observed. However, a 62.1% reduction in the amount of asphaltenes adhered to the PVT cell's inner surface was seen when the oil sample was treated with 500 ppm of the inhibitor. Moreover, an increase in the inhibitor's concentration did not significantly reduce the amount of adhered asphaltenes any further. Based on the laboratory test results, continuous injection of the inhibitor with an optimized concentration of 300 ppm at a depth of 14,743 ft. was started. After several months into the field trial, the production began to decrease gradually. As a result, the well was shut down, and a slickline tool equipped with a gauge ring was sent down the well. During this process, no obstruction was encountered above the chemical injection point; however, significant asphaltene deposition had occurred below the injection point, which was removed during the cleaning process The absence of asphaltene deposits above the injection point was attributed to the successful performance of the selected chemical in this particular well.

3.7.2 Evaluating asphaltene inhibitors for an offshore alaskan producer

In this work, Yen et al. performed laboratory tests on both live and dead oil samples to screen chemicals to inhibit asphaltene deposition in an offshore Alaskan producer [132]. This producer showed an asphaltene deposition problem in one of its gas-lift wells. In December 1998, a gauge ring was sent to the wellbore, and significant deposition was found below 7057 ft.

The heptane precipitation test was first used to study the chemicals' performance on the dead oil sample. The best performing chemical was then subjected to an SDS test with live oil at HPHT conditions. Asphaltene deposits, inside the cell, on the mixer, and the endcap were washed and collected with toluene. The toluene was then evaporated, and the solids were analyzed gravimetrically. Based on the promising high-pressure test results, the chemical was selected for the field trial.

A side stream filter loop was installed at the wellhead to monitor the field performance of the inhibitor. 250 ppm of the chemical could increase the filter's plugging time by approximately a factor of two (12 vs 26 days).

3.7.3 Development of multifunctional stabilizers of asphaltenes

Mena-Cervantes et al. investigated the performance of five different chemicals in preventing asphaltene deposition and selected one of them for a field trial based on laboratory observations [133]. The following tests were performed on these chemicals to evaluate their inhibiting and dispersing properties:

1. Ambient spectroscopy test to assess the dispersion efficiency of chemicals.
2. Atmospheric titration of dead oil samples via spectroscopy to evaluate the performance of chemicals in shifting the apparent onset of asphaltene precipitation by reducing the aggregation size of asphaltene particles at atmospheric conditions.
3. Electric field-induced deposition of asphaltene on metallic plates was performed to analyze if a given chemical can inhibit asphaltene deposition on metallic surfaces.
4. Asphaltene precipitation induced by pressure drop using SDS was done to analyze if a given chemical can diminish the aggregation size of asphaltene particles at reservoir pressure and temperature and shift the apparent AOP.

Based on the combined results of all four tests and economic criteria, such as the availability and production cost, the second-best performing chemical was selected for a field trial. During the chemical field deployment for 182 days, only two cleaning interventions were required against six performed before for the same period (Fig. 3.18), demonstrated the improved performance of the multifunctional asphaltene stabilizers.

3.8 Conclusion and path forward

Despite the enormous structural complexity associated with the asphaltene fraction of the crude oil, scientific community has made enormous advancements towards their structural elucidation at a molecular level. The structural information is important for the realization of structure-property function to predict behavior during various stages of crude oil production and processing. Knowledge of properties such as the molecular weight allows the flow assurance personnel to model the phase behavior of asphaltene, risk-assessment of the

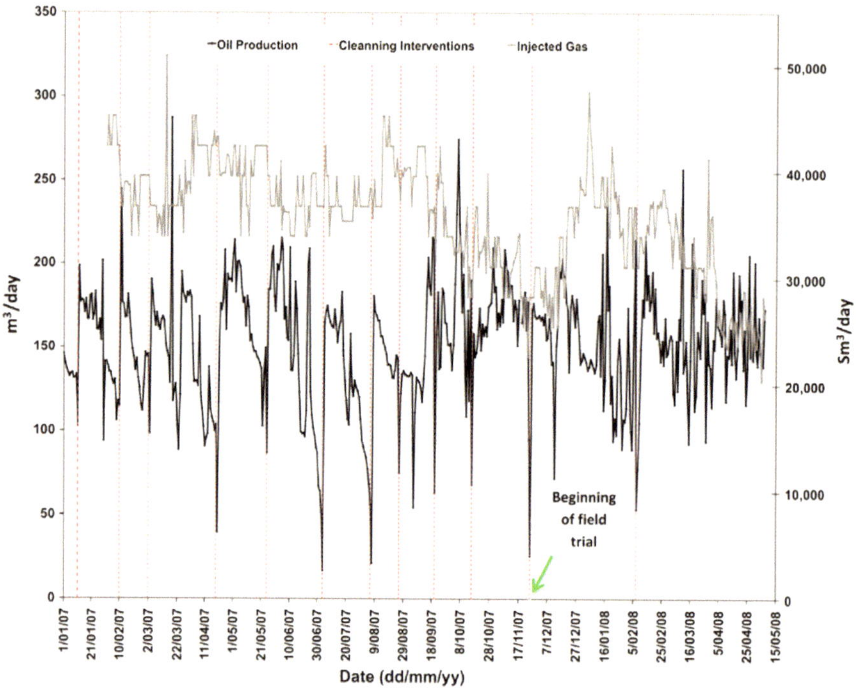

FIGURE 3.18

Average production rate, cleaning intervenes, and gas injection data for the production oil well W-1 during the year and period of a field trial for the chemical P1.

Reproduced with permission from V.Y. Mena-Cervantes, R. Hernández-Altamirano, E. Buenrostro-González, et al., Development of oxazolidines derived from polyisobutylene succinimides as multifunctional stabilizers of asphaltenes in oil industry, Fuel 110 (2013) 293–301 [133].

asphaltene associated challenge and proactive formulation of strategies towards management. However, with all the efforts expended by the industry and academia, a complete knowledge of the asphaltene deposition mechanism remains to be realized. Consequently, the problems associated with asphaltenes during oil production remain to be a major concern for the oil and gas industry.

A significant gap exists in the laboratory determination of onset of asphaltene precipitation at reservoir conditions, simulation of field deposition and extrapolation/correlation of laboratory observations to the field behavior. This absence of lab to field correlation is recognized to be intricately connected to the fact that the available laboratories tools and practices do not adequately mimic the field. The vague solubility definition of asphaltenes and lack of standard techniques for separation, quantification, and characterization of asphaltenes that are truly representative of the fraction that deposits in the field is one such impediment. Further complexity is introduced due to the disparity in results often observed from different techniques, and across different laboratories. Sensitivity constraints associated with the current tools used to assess asphaltene precipitation and deposition further removes laboratory from the field due to inaccuracy of data and resulting interpretations.

Modeling methods for prediction of asphaltene precipitation and deposition is another critical area where industry has continued to progress, but predictive tools to estimate the deposition mass and profile to capture deposition risks along the wellbore and flowlines with reasonable accuracy needs more work.

Dead oil is different in characteristics from that of live fluids. Further, the heterogeneity and complexity of asphaltenes is not addressed in the lab, especially when precipitants such as heptane are used for asphaltene content and instability analysis, adding uncertainty to asphaltene risk assessment as well as asphaltene management product recommendations. Despite extensive research towards development of chemical solutions to manage asphaltene deposition, products that would yield 100% management of this problem is still desired. The relative performance of chemicals tested on a given crude oil sample is dependent on the type of test used. In many cases exemplary chemical performance in laboratory does not translate to the field. Products selected based on dispersion-based methods do not necessarily manage deposition effectively, and industry is now leaning towards deposition testing towards screening and design of products. A test that would realistically capture field deposition accommodating the thermodynamics and asphaltene precipitation for various P and T conditions is the first step in the design of chemistries that are a step-change improvement in actual deposition control and allows for reconciliation of laboratory testing to the field.

With these considerations, flow assurance strategy to manage asphaltene deposition during oil production comprises of an integrated approach that relies on evaluation of several key elements such as the characterization of fluid properties, modeling, lab test results and field monitoring techniques to continuously reassess the risk, detect changes in the system as well as to confirm the effectiveness of control measures in place and continuous improvement.

Acknowledgment

The authors gratefully acknowledge our colleagues at ChampionX and ENNOVA for helpful discussions. Insights and suggestions from peers, specifically Jianxin Wang of Chevron, Qiwei Li of ChampionX, Amy McKenna and Ryan Rodgers of National High Magnetic Field Laboratory/FSU and Yunlong Zhang of ExxonMobil, are highly appreciated.

Nomenclature

ACSAA analytical centrifuged stability analysis for asphaltenes
ADEPT asphaltene deposition tool
ADT asphaltene dispersion test

amu	atomic mass unit
AOP	asphaltene onset pressure
APDT	asphaltene precipitation detection test
APDU	asphaltene precipitation detection unit
APPI	atmospheric pressure photoionization
ARC	asphaltene rocking cell
ASIST	asphaltene instability trend
CII	colloidal instability index
DAX	diesel, alcohol, xylene
EOS	equation of state
ESR	electron spin resonance
FPA	flocculation point analyzer
FT-ICR Ms	Fourier transform ion cyclotron resonance mass spectrometry
FTIR	Fourier-transform infrared spectroscopy
g	grams
GOM	Gulf of Mexico
h	hours
HPM	high-pressure microscopy
HPHT	high pressure and high temperature
HTHP	high temperature and high pressure
m/z	mass-to-charge ratio
mD	millidarcy
mg	milligrams
min	minutes
NIR	near-infrared
nm	nanometer
NMR	nuclear magnetic resonance
OSDC	organic solids deposition and control
PC-SAFT	perturbed chain version of the statistical associating fluid theory
ppm	parts per million
PVT	pressure, volume and temperature
rpm	revolutions per minute
SARA	saturates, aromatics, resins, and asphaltenes
SDS	solids detection system
SEM	scanning electron microscopy
SS	asphaltene supersaturation
XRD	X-ray diffraction
μm	micrometers

References

[1] C.E. Haskett, M. Tartera, A practical solution to the problem of asphaltene deposits - Hassi Messaoud field, Algeria, Journal of petroleum technology 17 (1965) 387–391.

[2] J. Dickie, M. Haller, T.F. Yen, Electron microscopic investigations on the nature of petroleum asphaltics, Journal of Colloid and Interface Science 29 (1969) 475–484.

[3] D.M. Clementz, Alteration of rock properties by adsorption of petroleum heavy ends: implications for enhanced oil recovery, in: SPE-10683-MS, SPE Enhanced Oil Recovery Symposium, Tulsa, OK, April 4–7, 1982.
[4] R.N. Tuttle, High-pour-point and asphaltic crude oils and condensates, Journal of petroleum technology 35 (1983) 1192–1196.
[5] R. Thawer, D. Nicoll, G. Dick, Asphaltene deposition in production facilities, SPE Production Engineering 5 (1990) 475–480.
[6] S.L. Kokal, S.G. Sayegh, Asphaltenes: the cholesterol of petroleum, in: SPE-29787-MS, Middle East Oil Show, Manama, Bahrain, March 11–14, 1995.
[7] S. Wang, F. Civan, Modeling formation damage by asphaltene deposition during primary oil recovery, Journal of Energy Resources Technology 127 (2005) 310–317.
[8] J.L. Creek, Freedom of action in the state of asphaltenes: escape from conventional wisdom, Energy & Fuels 19 (2005) 1212–1224.
[9] M. Jamshidnezhad, Prediction of asphaltene precipitation in an Iranian South oil field, in: PETSOC-2005-015, Canadian International Petroleum Conference, Calgary, Canada, June 7–9, 2005.
[10] S.F. Alkafeef, F. Al-Medhadi, A.D. Al-Shammari, A simplified method to predict and prevent asphaltene deposition in oilwell tubings: field case, SPE Production & Operations 20 (2005) 126–132.
[11] K. Indo, J. Ratulowski, B. Dindoruk, et al., Asphaltene nanoaggregates measured in a live crude oil by centrifugation, Energy & Fuels 23 (2009) 4460–4469.
[12] G.A. Mansoori, A unified perspective on the phase behavior of petroleum fluids, International Journal of Oil, Gas and Coal Technology 2 (2009) 141–167.
[13] O.C. Mullins, The modified Yen model, Energy & Fuels 24 (2010) 2179–2207.
[14] F.M. Vargas, M. Garcia-Bermudes, M. Boggara, et al. On the development of an enhanced method to predict asphaltene precipitation, in: OTC-25294-MS, Offshore Technology Conference, Houston, TX, May 5–8, 2014.
[15] A.A. Melendez-Alvarez, M. Garcia-Bermudes, M. Tavakkoli, et al., On the evaluation of the performance of asphaltene dispersants, Fuel 179 (2016) 210–220.
[16] F.P. Vargas, M. Tavakkoli, Asphaltene Deposition: Fundamentals, Prediction, Prevention, and Remediation, CRC Press, Boca Raton, 2018.
[17] D. Eskin, O. Mohammadzadeh, K. Akbarzadeh, et al., Reservoir impairment by asphaltenes: a critical review, The Canadian Journal of Chemical Engineering 94 (2016) 1202–1217.
[18] A. Hammami, J. Ratulowski, Precipitation and deposition of asphaltenes in production systems: a flow assurance overview, in: O.C. Mullins, E.Y. Sheu, A. Hammami, et al. (Eds.), Asphaltenes, Heavy Oils, and Petroleomics, Springer, New York, 2007, pp. 617–660.
[19] J.S. Buckley, Asphaltene deposition, Energy & Fuels 26 (2012) 4086–4090.
[20] H.W. Yarranton, H. Hussein, J.H. Masliyah, Water-in-hydrocarbon emulsions stabilized by asphaltenes at low concentrations, Journal of Colloid and Interface Science 228 (2000) 52–63.
[21] P. Qiao, D. Harbottle, P. Tchoukov, et al., Fractionation of asphaltenes in understanding their role in petroleum emulsion stability and fouling, Energy & Fuels 31 (2017) 3330–3337.
[22] A.C. Clingenpeel, S.M. Rowland, Y.E. Corilo, et al., Fractionation of interfacial material reveals a continuum of acidic species that contribute to stable emulsion formation, Energy & Fuels 31 (2017) 5933–5939.

[23] M. Tavakkoli, A. Chen, F.M. Vargas, Rethinking the modeling approach for asphaltene precipitation using the PC-SAFT equation of state, Fluid Phase Equilibria 416 (2016) 120–129.
[24] D.F. Smith, D.C. Podgorski, R.P. Rodgers, et al., 21 Tesla FT-ICR mass spectrometer for ultrahigh-resolution analysis of complex organic mixtures, Analytical Chemistry 90 (2018) 2041–2047.
[25] G.C. Klein, S. Kim, R.P. Rodgers, et al., Mass spectral analysis of asphaltenes. II. Detailed compositional comparison of asphaltenes deposit to its crude oil counterpart for two geographically different crude oils by ESI FT-ICR MS, Energy & Fuels 20 (2006) 1973–1979.
[26] G.C. Klein, S. Kim, R.P. Rodgers, et al., Mass spectral analysis of asphaltenes. I. Compositional differences between pressure-drop and solvent-drop asphaltenes determined by electrospray ionization fourier transform ion cyclotron resonance mass spectrometry, Energy & Fuels 20 (2006) 1965–1972.
[27] D.F. Smith, G.C. Klein, A.T. Yen, et al., Crude oil polar chemical composition derived from FT – ICR mass spectrometry accounts for asphaltene inhibitor specificity, Energy & Fuels 22 (2008) 3112–3117.
[28] S.S. Wang, C. Yang, C.M. Xu, et al., Separation and characterization of petroleum asphaltene fractions by ESI FT-ICR MS and UV-vis spectrometer, Science China Chemistry 56 (2013) 856–862.
[29] K. Qian, A.S. Mennito, K.E. Edwards, et al., Observation of vanadyl porphyrins and sulfur-containing vanadyl porphyrins in a petroleum asphaltene by atmospheric pressure photonionization Fourier transform ion cyclotron resonance mass spectrometry, Rapid Communications in Mass Spectrometry 22 (2008) 2153–2160.
[30] J.M. Purcell, I. Merdrignac, R.P. Rodgers, et al., Stepwise structural characterization of asphaltenes during deep hydroconversion processes determined by atmospheric pressure photoionization (APPI) fourier transform ion cyclotron resonance (FT-ICR) mass spectrometry, Energy & Fuels 24 (2010) 2257–2265.
[31] T.M.C. Pereira, G. Vanini, E.C.S. Oliveira, et al., An evaluation of the aromaticity of asphaltenes using atmospheric pressure photoionization fourier transform ion cyclotron resonance mass spectrometry - APPI (±) FT-ICR MS, Fuel 118 (2014) 348–357.
[32] X. Zhao, Q. Shi, M.R. Gray, et al., New vanadium compounds in venezuela heavy crude oil detected by positive-ion electrospray ionization Fourier transform ion cyclotron resonance mass spectrometry, Scientific Reports 4 (2014) 5373.
[33] M.L. Chacón-Patiño, C. Blanco-Tirado, J.A. Orrego-Ruiz, et al., Tracing the compositional changes of asphaltenes after hydroconversion and thermal cracking processes by high-resolution mass spectrometry, Energy & Fuels 29 (2015) 6330–6341.
[34] M.L. Chacón-Patiññ, S.M. Rowland, R.P. Rodgers, Advances in asphaltene petroleomics. Part 1: Asphaltenes are composed of abundant island and archipelago structural motifs, Energy & Fuels 31 (2017) 13509–13518.
[35] D. Giraldo-Dávila, M.L. Chacón-Patiño, A.M. McKenna, et al., Correlations between molecular composition and the adsorption, aggregation and emulsifying behavior of petrophase 2017 asphaltenes and their TLC fractions, Energy & Fuels 32 (2018) 2769–2780.
[36] T.F. Yen, J.G. Erdman, S.S. Pollack, Investigation of the structure of petroleum asphaltenes by X-ray diffraction, Analytical Chemistry 33 (1961) 1587–1594.

[37] R. Cimino, S. Correra, A. Del Bianco, et al., Solubility and phase behavior of asphaltenes in hydrocarbon media, in: E.Y. Sheu, O.C. Mullins (Eds.), Asphaltenes: Fundamentals and Applications, Plenum Press, New York, 1995, pp. 97–130.

[38] K.J. Leontaritis, G.A. Mansoori, Asphaltene flocculation during oil production and processing: a thermodynamic collodial model, in: SPE-16258-MS, SPE International Symposium on Oilfield Chemistry, San Antonio, TX, February 4 – 6, 1987.

[39] J. Beck, W.Y. Svrcek, H.W. Yarranton, Hysteresis in asphaltene precipitation and redissolution, Energy & Fuels 19 (2005) 944–947.

[40] S. Peramanu, C. Singh, M. Agrawala, et al., Investigation on the reversibility of asphaltene precipitation, Energy & Fuels 15 (2001) 910–917.

[41] H.F. Winterkorn, G.W. Eckert, Colloidal stability of asphalts, Industrial & Engineering Chemistry 33 (1941) 1285–1289.

[42] S.I. Andersen, Hysteresis in precipitation and dissolution of petroleum asphaltenes, Fuel Science and Technology International 10 (1992) 1743–1749.

[43] S.I. Andersen, E.H. Stenby, Thermodynamics of asphaltene precipitation and dissolution investigation of temperature and solvent effects, Fuel Science and Technology International 14 (1996) 261–287.

[44] J.S. Buckley, J. Wang, J.L. Creek, Solubility of the least-soluble asphaltenes, in: O. C. Mullins, E.Y. Sheu, A. Hammami, et al. (Eds.), Asphaltenes, Heavy Oils, and Petroleomics, Springer, New York, 2007, pp. 401–437.

[45] J.S. Buckley, Predicting the onset of asphaltene precipitation from refractive index measurements, Energy & Fuels 13 (1999) 328–332.

[46] J.P. Dickie, T.F. Yen, Macrostructures of the asphaltic fractions by various instrumental methods, Analytical Chemistry 39 (1967) 1847–1852.

[47] S.F. Niles, M.L. Chacón-Patiño, S.P. Putnam, et al., Characterization of an asphalt binder and photoproducts by fourier transform ion cyclotron resonance mass spectrometry reveals abundant water-soluble hydrocarbons, Environmental Science & Technology 54 (2020) 8830–8836.

[48] M.L. Chacón-Patiño, D.F. Smith, C.L. Hendrickson, et al., Advances in asphaltene petroleomics. Part 4. Compositional trends of solubility subfractions reveal that polyfunctional oxygen-containing compounds drive asphaltene chemistry, Energy & Fuels 34 (2020) 3013–3030.

[49] Y.J. Lin, T. Cao, M.L. Chacón-Patiño, et al., Microfluidic study of the deposition dynamics of asphaltene subfractions enriched with island and archipelago motifs, Energy & Fuels 33 (2019) 1882–1891.

[50] Y. Zhang, M. Siskin, M.R. Gray, et al., Mechanisms of asphaltene aggregation: puzzles and a new hypothesis, Energy & Fuels 34 (2020) 9094–9107.

[51] A. Hammami, C.H. Phelps, T. Monger-McClure, et al., Asphaltene precipitation from live oils: an experimental investigation of onset conditions and reversibility, Energy & Fuels 14 (2000) 14–18.

[52] R.B. De Boer, K. Leerlooyer, M.R.P. Eigner, et al., Screening of crude oils for asphalt precipitation: theory, practice, and the selection of inhibitors, SPE Production & Facilities 10 (1995) 55–61.

[53] A. Hirschberg, L.N.J. de Jong, B.A. Schipper, et al., Influence of temperature and pressure on asphaltene flocculation, SPE Journal 24 (1984) 283–293.

[54] D. Gonzalez, F. Gonzalez, M. Pietrobon, et al., Strategies to monitor and mitigate asphaltene issues in the production system of a Gulf of Mexico deepwater subsea

development, in: OTC-27123-MS, Offshore Technology Conference, Houston, TX, May 2–5, 2016.

[55] H. Laux, I. Rahimian, P. Schorling, The Colloidal stability of crude oils and crude oil residues, Petroleum science and technology 17 (1999) 349–368.

[56] S. Gharfeh, A. Yen, S. Asomaning, et al., Asphaltene flocculation onset determinations for heavy crude oil and its implications, Petroleum Science and Technology 22 (2004) 1055–1072.

[57] P. Juyal, V.N. Le, A.T. Yen, et al., Effect of crude oil aging on asphaltene inhibitor product recommendation, Journal of Dispersion Science and Technology 32 (2011) 1096–1104.

[58] S.I. Andersen, Flocculation onset titration of petroleum asphaltenes, Energy & Fuels 13 (1999) 315–322.

[59] F.M. Vargas, D.L. Gonzalez, G.J. Hirasaki, et al., Modeling asphaltene phase behavior in crude oil systems using the perturbed chain form of the statistical associating fluid theory (PC-SAFT) equation of state, Energy & Fuels 23 (2009) 1140–1146.

[60] J.X. Wang, J.S. Buckley, An experimental approach to prediction of asphaltene flocculation, in: SPE-64994-MS, SPE International Symposium on Oilfield Chemistry, Houston, TX, February 13–16, 2001.

[61] J.X. Wang, J.S. Buckley, N.E. Burke, et al., A practical method for anticipating asphaltene problems, SPE Production & Operations 19 (2004) 152–160.

[62] J.L. Creek, J.X. Wang, J.S. Buckley, Asphaltene instability induced by light hydrocarbons, in: OTC-19690-MS, Offshore Technology Conference, Houston, TX, May 5–8, 2008.

[63] J.L. Creek, J.X. Wang, J.S. Buckley, Verification of asphaltene-instability-trend (ASIST) predictions for low-molecular-weight alkanes, SPE Production & Operations 24 (2009) 360–368.

[64] F.M. Vargas, J.L. Creek, W.G. Chapman, On the development of an asphaltene deposition simulator, Energy & Fuels 24 (2010) 2294–2299.

[65] A.S. Kurup, F.M. Vargas, J. Wang, et al., Development and application of an asphaltene deposition tool (ADEPT) for well bores, Energy & Fuels 25 (2011) 4506–4516.

[66] K. Akbarzadeh, D. Eskin, J. Ratulowski, et al. Asphaltene deposition measurement and modeling for flow assurance of subsea tubings and pipelines, in: OTC-22316-MS, OTC Brasil, Rio de Janeiro, Brazil, 2011 October 4–6.

[67] D. Eskin, J. Ratulowski, K. Akbarzadeh, et al., Modelling asphaltene deposition in turbulent pipeline flows, The Canadian Journal of Chemical Engineering 89 (2011) 421–441.

[68] A. Khaleel, M. Abutaqiya, M. Tavakkoli, et al., On the prediction, prevention and remediation of asphaltene deposition, in: SPE-177941-MS, Abu Dhabi International Petroleum Exhibition and Conference, Abu Dhabi, UAE, November 9–12, 2015.

[69] S. Enayat, M. Tavakkoli, A. Yen, et al., Review of the current laboratory methods to select asphaltene inhibitors, Energy & Fuels 34 (2020) 15488–15501.

[70] J.I. Da Silva de Aguiar, C.P.P. Mazzeo, R. Garan, et al. What can we learn from analysis of field asphaltenes deposits? Enhancing product development through knowledge-based field-to-lab-to-field approach, in: SPE-193604-MS, SPE International Conference on Oilfield Chemistry, Galveston, TX, April 8–9, 2019.

[71] P. Juyal, V. Ho, A. Yen, et al., Reversibility of asphaltene flocculation with chemicals, Energy & Fuels 26 (2012) 2631–2640.

[72] M.B. Manek, Asphaltene dispersants as demulsification aids, in: SPE-28972-MS, SPE International Symposium on Oilfield Chemistry, San Antonio, TX, February 14–17, 1995.

[73] S.S. Schantz, W.K. Stephenson, Asphaltene deposition: development and application of polymeric asphaltene dispersants, in: SPE-22783-MS, SPE Annual Technical Conference and Exhibition, Dallas, TX, October 6–9, 1991.

[74] A. Montesi, R.A. Pinnick, S. Subramanian, et al., Asphaltene management in GOM DW subsea development, in: OTC-21587-MS, Offshore Technology Conference, Houston, TX, May 2–5, 2011.

[75] T. Kar, K. Naderi, A. Firoozabadi, Asphaltene deposition and removal in flowlines and mitigation by effective functional molecules, SPE Journal 25 (2020) 771–787.

[76] S.M. Hashmi, L.A. Quintiliano, A. Firoozabadi, Polymeric dispersants delay sedimentation in colloidal asphaltene suspensions, Langmuir 26 (2010) 8021–8029.

[77] D. Subramanian, A. Firoozabadi, Effect of surfactants and water on inhibition of asphaltene precipitation and deposition, in: SPE-177669-MS, Abu Dhabi International Petroleum Exhibition and Conference, Abu Dhabi, UAE, November 9–12, 2015.

[78] S.M. Hashmi, K.X. Zhong, A. Firoozabadi, Acid-base chemistry enables reversible colloid-to-solution transition of asphaltenes in non-polar systems, Soft Matter 8 (2012) 8778–8785.

[79] M.M. Shadman, M. Vafaie-Sefti, S. Ahmadi, et al., Effect of dispersants on the kinetics of asphaltene settling using turbidity measurement method, Petroleum Science and Technology 34 (2016) 1233–1239.

[80] C. Chen, J. Guo, N. An, et al., Study of asphaltene deposition from Tahe crude oil, Petroleum Science 10 (2013) 134–138.

[81] M. Curren, A. Kaiser, S. Adkins, et al., Effects of CEOR chemicals on asphaltene precipitation, in: SPE-191747-MS, SPE Annual Technical Conference and Exhibition, Dallas, TX, September 24–26, 2018.

[82] H. Firoozinia, K. Abad, A. Varamesh, A comprehensive experimental evaluation of asphaltene dispersants for injection under reservoir conditions, Petroleum Science 13 (2016) 280–291.

[83] S. Kashefi, A. Shahrabadi, M.N. Lotfollahi, et al., A new polymeric additive as asphaltene deposition inhibitor in CO_2 core flooding, Korean Journal of Chemical Engineering 33 (2016) 3273–3280.

[84] J. Al-Jabri, A. Sultan, E. Al-Mutairi, et al., Screening new formulations for asphaltene aggregation inhibition, in: SPE-192450-MS, SPE Kingdom of Saudi Arabia Annual Technical Symposium and Exhibition, Dammam, Saudi Arabia, April 23–26, 2018.

[85] H.K. Kraiwattanawong, S. Fogler, S.G. Gharfeh, et al., Effect of asphaltene dispersants on aggregate size distribution and growth, Energy & Fuels 23 (2009) 1575–1582.

[86] D.W. Jennings, R. Cable, G. Leonard, MS new dead-crude oil asphaltene inhibitor test method, in: OTC-25113-MS, Offshore Technology Conference, Houston, TX, May 5–8, 2014.

[87] D.W. Jennings, K.-P. Chao, R. Cable, et al., Method for field monitoring asphaltene treatment programs, in: OTC-27129-MS, Offshore Technology Conference, Houston, TX, May 2–5, 2016.

[88] D.W. Jennings, K.-P. Chao, J. Kim, Asphaltene inhibitor testing: comparison between a high pressure live-fluid deposition stability method, in: OTC-28650-MS, Offshore Technology Conference, Houston, TX, April 30–May 3, 2018.

[89] J. Kuang, M. Tavakkoli, J. Yarbrough, et al., Investigation of asphaltene deposition at high temperature and under dynamic conditions, Energy & Fuels 32 (2018) 12405–12415.

[90] D. Broseta, M. Robin, T. Savvidis, et al., Detection of asphaltene deposition by capillary flow measurements, in: SPE-59294-MS, SPE/DOE Improved Oil Recovery Symposium, Tulsa, OK, April 3–5, 2000.

[91] L. Nabzar, M.E. Aguiléra, The colloidal approach. A promising route for asphaltene deposition modelling, Oil & Gas Science and Technology–Revue d'IFP Energies nouvelles 63 (2008) 21–35.

[92] M.P. Hoepfner, V. Limsakoune, V. Chuenmeechao, et al., A fundamental study of asphaltene deposition, Energy & Fuels 27 (2013) 725–735.

[93] A.R. Farrell, B. Martin, D.M. Frigo, et al., Asphaltenes risk assessment and mitigation - designing appropriate laboratory test protocols, in: SPE-193560-MS, SPE International Conference on Oilfield Chemistry, Galveston, TX, April 8–9, 2019.

[94] K.A. Lawal, J.P. Crawshaw, E.S. Boek, et al., Experimental investigation of asphaltene deposition in capillary flow, Energy & Fuels 26 (2012) 2145–2153.

[95] S. Gon, D.M. Fouchard, Modified asphaltene capillary deposition unit: a novel approach to inhibitor screening, Energy & Fuels 30 (2016) 3687–3692.

[96] W. Chaisoontornyotin, N. Haji-Akbari, H.S. Fogler, et al., Combined asphaltene aggregation and deposition investigation, Energy & Fuels 30 (2016) 1979–1986.

[97] S.M. Hashmi, A. Firoozabadi, Effective removal of asphaltene deposition in metal-capillary tubes, SPE Journal 21 (2016) 1747–1754.

[98] J.X. Wang, J.L. Creek, T. Fan, et al., Asphaltene deposit distribution in capillary tests, in: Proceedings of the 11th International Conference on Petroleum Phase Behavior and Fouling, Jersey City, NJ, June 13–17, 2010.

[99] J. Bae, D. Fouchard, S. Garner, et al., Advantages of applying a multifaceted approach to asphaltene inhibitor selection, in: OTC-27171-MS, Offshore Technology Conference, Houston, TX, USA, May 2–5, 2016.

[100] D.M.D. Fouchard, J.E. Carmichael, Method of assessing asphaltene inhibitor efficiency, Patent WO2014149751A1, 2014.

[101] C. Vilas Bôas Fávero, A. Hanpan, P. Phichphimok, et al., Mechanistic investigation of asphaltene deposition, Energy & Fuels 30 (2016) 8915–8921.

[102] J. Kuang, J. Yarbrough, S. Enayat, et al., Evaluation of solvents for in-situ asphaltene deposition remediation, Fuel 241 (2019) 1076–1084.

[103] J. Kuang, N. Rajan Babu, J. Hu, et al., Experimental determination of asphaltene deposition, in: F.M. Vargas, M. Tavakkoli (Eds.), Asphaltene Deposition, CRC Press, Boca Raton, 2018, pp. 161–202.

[104] C. Hu, R.L. Hartman, High-throughput packed-bed microreactors with in-line analytics for the discovery of asphaltene deposition mechanisms, AIChE Journal 60 (2014) 14542.

[105] C.-A. Sung, M. Tavakkoli, A. Chen, et al., Prevention and control of corrosion-induced asphaltene deposition, in: OTC-27008-MS, Offshore Technology Conference, Houston, TX, May 2–5, 2016.

[106] Y.-J. Lin, P. He, M. Tavakkoli, et al., Examining asphaltene solubility on deposition in model porous media, Langmuir 32 (2016) 8729–8734.

[107] Y. Zhuang, A. Goharzadeh, Y.J. Lin, et al., Three dimensional measurements of asphaltene deposition in a transparent micro-channel, Journal of Petroleum Science and Engineering 145 (2016) 77–82.

[108] Y.-J. Lin, S. Barman, P. He, et al., Combined interfacial shear rheology and micro-structure visualization of asphaltenes at air-water and oil-water interfaces, Journal of Rheology 62 (2017) 1–10.

[109] Y.-J. Lin, P. He, M. Tavakkoli, et al., Characterizing asphaltene deposition in the presence of chemical dispersants in porous media micromodels, Energy & Fuels 31 (2017) 11660–11668.

[110] M. Tavakkoli, P. He, P.-H. Lin, M. et al. Asphaltene deposition and fouling in reservoirs, in: OTC-27933-MS, Offshore Technology Conference, Houston, TX, May 1–4, 2017.

[111] M. Tavakkoli, M.R. Grimes, X. Liu, et al., Indirect method: a novel technique for experimental determination of asphaltene precipitation, Energy & Fuels 29 (2015) 2890–2900.

[112] R.D.Z. Romero, H.M. Acuna, F. Cortes, et al., Application and evaluation of a nanofluid containing nanoparticles for asphaltenes inhibition in well CPSXL4, in: OTC-24310-MS, OTC Brasil, Rio de Janeiro, Brazil, October 29–31, 2013.

[113] S. Asomaning, Test methods for determining asphaltene stability in crude oils, Petroleum Science and Technology 21 (2003) 581–590.

[114] M. Zougari, J. Ratulowski, A. Hammami, et al., Live oils novel organic solid deposition and control device: wax and asphaltene deposition validation, in: SPE-93558-MS, SPE Middle East Oil and Gas Show and Conference, Manama, Bahrain, March 12–15, 2005.

[115] M. Zougari, S. Jacobs, J. Ratulowski, et al., Novel organic solids deposition and control device for live-oils: design and applications, Energy & Fuels 20 (2006) 1656–1663.

[116] J. Ali, J. Betancourt, C. Avila, Methodology for asphaltene control in production facilities in North of Monagas, Venezuela, in: SPE-56572-MS. SPE Annual Technical Conference and Exhibition, Houston, TX, October 3–6, 1999.

[117] S.J. Allenson, M.A. Walsh, A novel way to treat asphaltene deposition problems found in oil production, in: SPE-37286-MS, International Symposium on Oilfield Chemistry, Houston, TX, February 18–21, 1997.

[118] M.A. Kelland, Production Chemicals for the Oil and Gas Industry, first ed., CRC Press, Boca Raton, FL, 2009.

[119] K.J. Leontaritis, Asphaltene deposition: a comprehensive description of problem manifestations and modeling approaches, in: SPE-18892MS, SPE Production Operations Symposium, Oklahoma City, OK, March 13–14, 1989.

[120] M. Barcenas, P. Orea, E. Buenrostro-González, et al., Study of medium effect on asphaltene agglomeration inhibitor efficiency, Energy & Fuels 22 (2008) 1917–1922.

[121] C.-L. Chang, H.S. Fogler, Stabilization of asphaltenes in aliphatic solvents using alkylbenzene-derived amphiphiles. 1. Effect of the chemical structure of amphiphiles on asphaltene stabilization, Langmuir 10 (1994) 1749–1757.

[122] C.-L. Chang, H.S. Fogler, Stabilization of asphaltenes in aliphatic solvents using alkylbenzene-derived amphiphiles. 2. Study of the asphaltene-amphiphile interactions and structures using Fourier transform infrared spectroscopy and small-angle X-ray scattering techniques, Langmuir 10 (1994) 1758–1766.

[123] L. Goual, M. Sedghi, X. Wang, et al., Asphaltene aggregation and impact of alkylphenols, Langmuir 30 (2014) 5394–5403.

[124] S.H. Akbar, A.A. Saleh, A comprehensive approach to solve asphaltene deposition problem in some deep wells, in: SPE-17965-MS, Middle East Oil Show, Manama, Bahrain, March 11–14, 1989.

[125] G. Fung, W.P. Backhaus, S. McDaniel, et al., To pig or not to pig: the Marlin experience with stuck pig, in: OTC-18387-MS, Offshore Technology Conference, Houston, TX, May 1–4, 2006.

[126] A.I. Voloshin, V.V. Ragulin, A.G. Telin, Development and introduction of heavy organic compound deposition diagnostics, prevention and removing, in: SPE-93128-MS, SPE International Symposium on Oilfield Chemistry, The Woodlands, TX, February 2–4, 2005.

[127] K. Gharbi, K. Benyounes, M. Khodja, Removal and prevention of asphaltene deposition during oil production: a literature review, Journal of Petroleum Science and Engineering 158 (2017) 351–360.

[128] A. Del Bianco, F. Stroppa, L. Bertero, Tailoring hydrocarbon streams for asphaltene removal, SPE Production & Facilities 12 (1997) 80–83.

[129] M.H. Waxman, C.T. Deeds, P.J. Closmann, Thermal alterations of asphaltenes in peace river tars, in: SPE-9510-MS, SPE Annual Technical Conference and Exhibition, Dallas, TX, September 21–24, 1980.

[130] A. Miadonye, L. Evans, The solubility of asphaltenes in different hydrocarbon liquids, Petroleum Science and Technology 28 (2010) 1407–1414.

[131] A. Yen, Y.R. Yin, S. Asomaning, Evaluating asphaltene inhibitors: laboratory tests and field studies, in: SPE-65376-MS, SPE International Symposium on Oilfield Chemistry, Houston, TX, February 13–16, 2001.

[132] T. Balson, H.A. Craddock, J. Dunlop, et al., Prediction and solution of asphaltene related problems in the field, in: H. Frampton, H.A. Craddock, J. Dunlop, et al. (Eds.), Chemistry in the Oil Industry VII: Performance in a Challenging Environment, RSC Publishing, Cambridge, 2002, pp. 277–286.

[133] V.Y. Mena-Cervantes, R. Hernández-Altamirano, E. Buenrostro-González, et al., Development of oxazolidines derived from polyisobutylene succinimides as multifunctional stabilizers of asphaltenes in oil industry, Fuel 110 (2013) 293–301.

CHAPTER

Naphthenate and carboxylate soap treatment

4

Jonathan J. Wylde[1,2]

[1]*Heriot-Watt University, Edinburgh, Scotland, United Kingdom*

[2]*Clariant Oil Services, Clariant Corporation, Houston, TX, United States*

Chapter outline

Flow Assurance. DOI: https://doi.org/10.1016/B978-0-12-822010-8.00008-8

4.1 Introduction

4.1.1 Overview and chapter orientation

Carboxylate and naphthenate soap emulsion and scaling are relatively recent additions to the long list of production chemical challenges that impact oil recovery operations globally. These are salt species combining hydrocarbon components from the oil phase with metal cations sourced from the produced water phase. Its recognition as a serious threat to oil production operations has resulted in a particularly swift industry response, with many R&D collaborative efforts undertaken to better understand its detailed chemistry and devise appropriate and effective technical solutions and strategies for its mitigation and management.

The following chapter comprises three general sections: (1) an introduction to naphthenic and carboxylic acid chemistry and the development of understanding, (2) a broad review of soaps in the oil industry, including carboxylate emulsion and naphthenate scale impact on oilfield production operations, and last but certainly not least, and (3) a discussion regarding soap mitigation technologies, including acidic and nonacid chemical approaches and associated laboratory testing methods for naphthenate identification and inhibitor performance testing. This chapter will also feature an unambiguous naphthenate science lexicon to assist the reader as a terminologies reference guide. The lexicon of naphthenate terminologies was considered necessary as a significant number of scientific publications (to the author) expressed differing and sometimes conflicting interpretations for even the most rudimentary common oilfield naphthenate terminologies.

This chapter is primarily concerned with carboxylate and naphthenate related impact on acid crude production and this can be considered from two related oilfield perspectives:

1. Impact on dehydration processes, that is, development of persistent oil and water separation issues caused by the strong interfacial properties of the natural carboxylic (and naphthenic) acids present in the oil and particularly from their soap salt equivalents
2. The presence of naphthenate scale precipitate via complexation of higher molecular weight naphthenic acids with divalent metal cations, especially calcium, which can also lead to high, out of specification, calcium in crude content.

4.1.2 Naphthenates and recent history

The emergence of naphthenate-related fouling in oil production processes stimulated renewed interest in naphthenic acid chemistry toward the end of the 1990s. Naphthenate science and technology was (briefly) catapulted into R&D centerstage, and unsurprisingly there followed a flood of naphthenate publications in the technical literature. The industry demanded answers to the new, troublesome, and expensive threat posed by naphthenate scales and unresolvable emulsions to the production and refining operations, and the R&D community duly obliged, spurred on by an abundance of fresh new funding. Consequently, the number of groups, organizations, and research institutions involved in naphthenate R&D increased from c.1999 onwards. Fig. 4.1 presents a timeline showing (in the author's opinion) the most significant and relevant research group/operator involvement in advancing oilfield naphthenate science and technology across the period 1999–2020. A literature review focusing on the terms "naphthenic acids" and/or "naphthenates" from 1960 to the present date clearly shows that a distinct step change occurred in publications per annum that occurred from c.2000 onwards, which spiked in the year 2015, and which has seen an even more dramatic fall off since (Fig. 4.2).

The dramatic surge in technical publications correlated with not only conventional oil and gas soap research but also to the rise in commercial Athabasca oil sands mining, steam-assisted gravity drainage, and bitumen refining operations in Alberta, Canada [1]. Many other articles involved analytical technique development for detection, quantification, and elucidation of naphthenic acids—not only in the context of naphthenate deposits in the upstream oil and gas industry [2], but also with respect to the analysis of these species in oil sand process-affected water (OSPW)—an environmental consequence of extracting the vast bitumen deposits present in northern Alberta. Other publications focused on the technical challenges of naphthenates in the refining industry, where naphthenic acids can cause, among other challenges, significant corrosion of the overheads. Following the peak in publications around 2015, there followed a steep decline in naphthenate-related articles, almost like the industry suddenly lost interest. This was not really the case as many of the latter post-2015 publications were

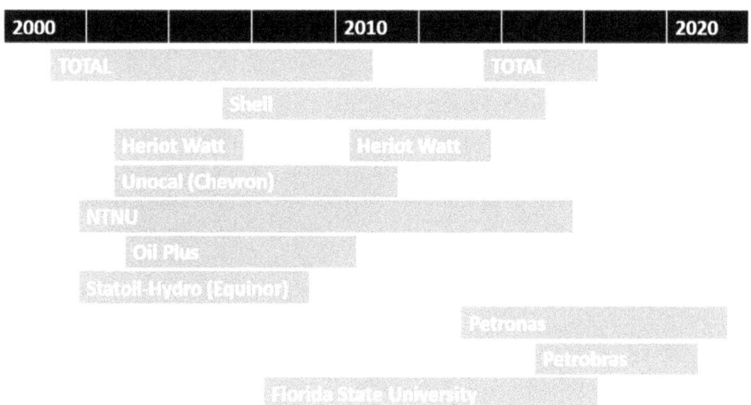

FIGURE 4.1

Gantt chart showing the prevalent R&D groups concerned with the study of naphthenic and carboxylate soaps from c.2000 to c.2020.

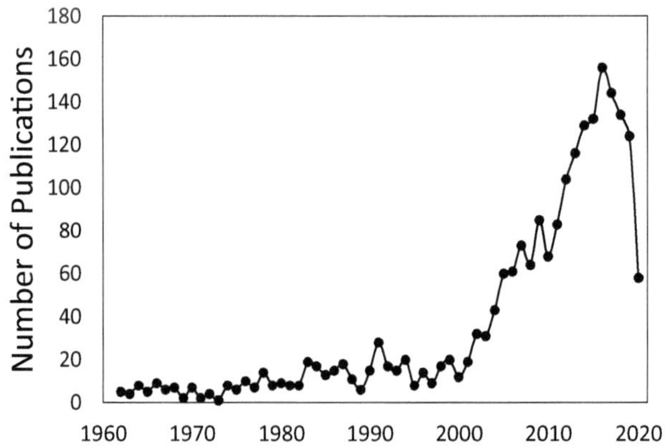

FIGURE 4.2

The number of publications per year containing the term "naphthenic acids" and/or "naphthenates" found in a search of the Engineering Village (Compendex and Inspec) search performed on September 01, 2020 (a Web of Science search also yields the same pattern).

concerned with oil sands naphthenate issues and enhanced oil recovery (EOR) applications. What were the reasons for the rapid decline in R&D interest in oilfield naphthenate and carboxylate soaps—especially after such a meteoric rise? A number of related factors are believed to have contributed to this 60% reduction in publication number in only 20 months. First and foremost, was timing. The oil industry is (unfortunately and exasperatingly) cyclical, and this is generally reflected in the oil price. The year 2015 witnessed one of the most significant crashes in the history of global oil production and this resulted unerringly in curtailed investment in oil industry R&D. Lower research budgets and reduced funding had a very direct impact on supporting R&D organizations and institutions, and this was further exacerbated by the fact that many key significant breakthroughs regarding the fundamental mechanistic understanding of carboxylate and naphthenate chemistry had just been published. With many of the immediate industry concerns addressed and adequate technologies for naphthenate mitigation now commercially available, it is of no real surprise that the appetite for the same level of R&D effort (and funding) simply was not there. Ultimately, the main knock-on effect of the oil downturn at this juncture was a flurry of discontinued and abandoned projects and prematurely dissolved naphthenate joint industry projects (JIPs). Several oil operating companies had by this time developed functioning naphthenate prediction models, which provide a significant theoretical understanding of when and where soaps are a challenge. In short, continued additional R&D effort would not have added significantly to the body of information allowing naphthenate scale-free oilfield operations. Lastly, the number of service company contributions to the naphthenate publication library had diminished considerably, which suggests a degree of "commoditization" had occurred in the chemistries and products used to control oilfield soaps, and therefore, in the world of limited R&D budgets, soap inhibitors do not often make the priority cut.

4.1.3 Definitions: acid crude oil and "naphthenates"

Biodegraded oils represent an enormously important component of the global oil inventory, with deposits found on every continent, with the greatest reserves located in Canada, Africa, South America, Gulf of Mexico, and Southeast Asia [3]. Crude oil typically comprises 85%–90% hydrocarbons and 10%–15% of polar and/or slightly polar species [4]. The polar species contain heteroatoms of nitrogen, oxygen, and sulfur, as well as traces of metals such as iron, nickel, and vanadium (e.g., associated with asphaltenes). A large proportion of biodegraded oils contain elevated levels of polar compounds, which bring associated processing challenges, including fouling and deposit issues, emulsion formation, catalytic fouling, corrosion, and environmental issues [5].

Crude oil polar species are majorly made up of the acid fraction. The first seminal publication regarding this topic appeared in 1874 and heralded a larger body of work regarding the pursuit of identification of these materials [6]. The acid fractions represent an incredibly diverse mixture of organic compounds with far-ranging molecular weight and structural differentiation. In one study alone, the research group identified in excess of 3000 different acid structures in a single heavy oil [7]. In the majority of oilfield crudes, the acid fraction is small, typically in the 2.5–4.0 wt.% range [8,9], yet their influence on the overall physicochemical behavior of the crude can be profound as a consequence of their interfacial activity, an inevitable trait of a polar compound [10].

Traditionally total acid number (TAN) is used to express the overall "acidity" of crude oil. This is represented as the quantity (in mg) of potassium hydroxide required to neutralize the acids in 1 gram of crude oil sample. Potentiometric and colorimetric titrations are used for TAN assay, and the most accepted and widely used method is the American Society for Testing and Materials (ASTM) D664-18e2 "Standard Test Method for Acid Number of Petroleum Products by Potentiometric Titration" [11]. Typically, crude oils have TAN values between 0.01 and 10 mg KOH/g crude oil, and when the TAN > 0.5 mg KOH/g, the crude oil is termed "acidic" [8,12]. Care must be exercised as TAN is not specific to naphthenic acids, and rather is an expression of all organic acid (carboxylic acids) and frankly all acid species in crude oil. It is, therefore, a common misinterpretation to assume a high TAN crude is high in naphthenic acids; this is just an indication. Another fallacy is that naphthenic acids specifically "reside" in the resin solubility fraction of crude oil; this has been shown to be untrue as they can also be present in the asphaltene fraction of a crude [13]. It is important to note that saturates, aromatics, resins, and asphaltenes (SARA) analysis of crude measures "solubility classes" and not just chemical families. All crudes contain acidic components; however, this does not necessarily mean that they will actually present significant flow assurance challenges, viz. "naphthenates" [14]. As will be seen later, there is a great deal of complexity to the exact speciation and analysis of acidic components in crude oil, and therefore, in the manner that production chemical challenges present themselves.

The American Petroleum Institute (API) definition of naphthenic acids includes structures with single and multiple fused cyclopentane and cyclohexane rings [15], as shown in Fig. 4.3. Here the carboxyl group is attached to the aliphatic side chain or the cycloaliphatic ring. In this definition, the R and R' group can essentially be any hydrocarbon group that is an aliphatic side chain or cycloaliphatic ring.

It is best practice to describe naphthenic acid homologues by the general formula $C_nH_{2n+z}O_2$, where n is the number of carbon atoms in the molecule and z signifies a homologous series. When the z term is "zero," then this indicates fully saturated aliphatic carboxylic acids and it decreases by 2 with increasing hydrogen deficiency from the formation of rings or double bonds in the molecule; sometimes referred to as double bond equivalents (DBE) [16]. There can be many different isomers present within the same homologous z series, and these can prove challenging to resolve via analytical chemistry, and typically requires a combination of techniques to resolve effectively. A further challenge is present insofar as homologues in different z series that can have the same nominal molecular weights but different formulas, for example, $C_{14}H_{26}O_2$ ($z = -2$) and $C_{15}H_{14}O_2$ ($z = -16$), are not always completely distinguishable under instrument resolution.

Table 4.1 lists several examples of major types of naphthenic acids classified by z number from 0 to -12 [17]. Highly aromatic acids, such as $z = -14$, -16, -18, etc., are not included. They are considered minor components of the naphthenic acid distribution in a crude [8]. The presence of other acids, such as diacids, hydroxy acids, carbonyl acids, and olefinic acids are also not considered in this definition for the same reason.

Naphthenic acid presence in crude oils has been known for a long time, can vary in quantity, and has been reported as high as 4 wt.% [18]. High naphthenic acid content crude oils have been reported globally, from California to Mexico, West Africa to the North Sea, and Indonesia to Russia [19].

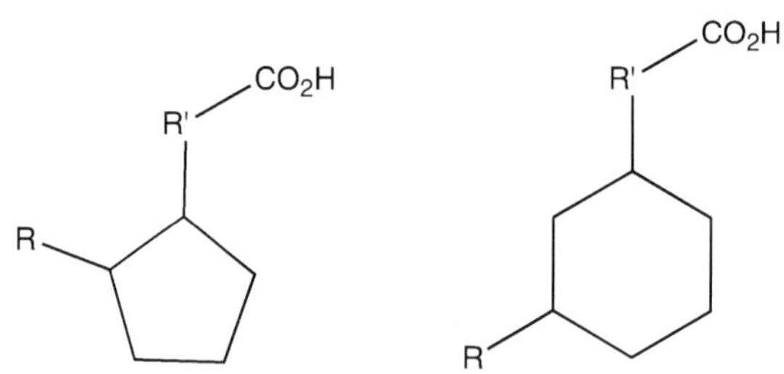

FIGURE 4.3

Structure of "true" naphthenic acids showing the required presence of at least one 5- or 6-membered cyclic ring structure.

Table 4.1 Examples of major naphthenic types and application of the general formula.

Z						
0 RCOOH	−2 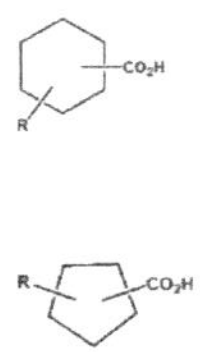 	−4 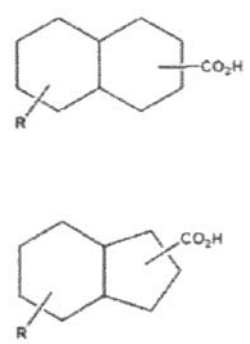	−6 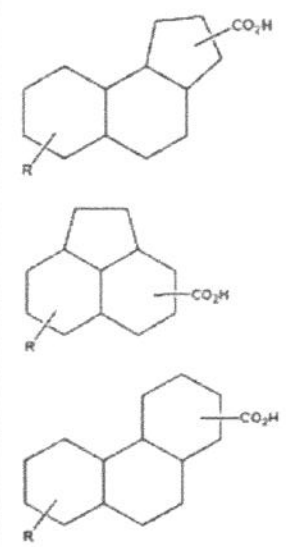	−8 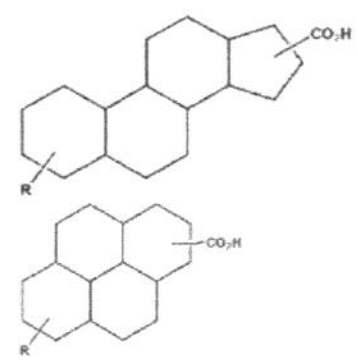	−10 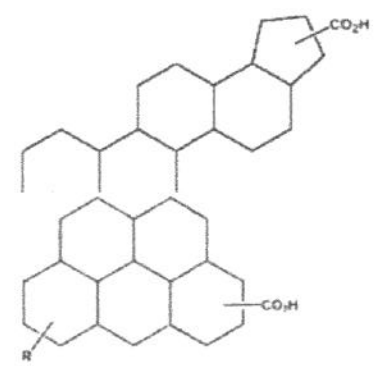	−12 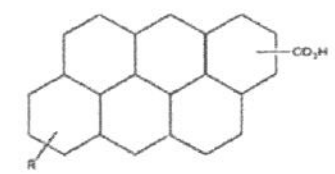

There is a great deal of ambiguity in the terminology used in the literature surrounding the subject of naphthenates, and therefore, it is important to level-set the definitions and terminologies used in this chapter. The ambiguity stems from the term "naphthenates" and "naphthenic acids" commonly being used interchangeably to describe all acidic components contained within crude oil. This is, of course, chemically speaking, incorrect and, as already identified, there is a large number of different chemistries that contribute to the total acidity of crude oil. Broadly, in the oil industry, the term "naphthenic acid" can represent all compounds with a carboxylic acid group, including straight-chained fatty acids as higher molecular weight species with one or more aliphatic or even aromatic structures.

Attention to detail is paramount when reading and interpreting data in the literature as there could be misdiagnoses due to ambiguity in the terms used and it is, therefore, important that analytical evidence is provided to support the claim(s). Although "naphthenic acids" originally implied that the acids contained naphthenic rings, today it is accepted that the term also includes cyclic, acyclic, and even aromatic acids in crude oil. In this review, "acid components" will be used to collectively describe the aliphatic, cyclic, and aromatic carboxylic acids present in crude oil. The term "carboxylic acids" generally imply aliphatic, straight-chained or branched organic acids, and "naphthenic acids" generally imply organic acids that contain one or more fully saturated 5- or 6-member alicyclic ring structures.

4.1.4 Origin of acidic crude

In-reservoir biodegradation processes occurring over many millions of years are thought primarily responsible for crude oil TAN values, however, considering the very long time periods involved, there will always be a question mark regarding, which TAN generation process actually dominated.

The relationship of TAN with biodegraded oils is the result of acid neoformation during biodegradation and/or the preferential removal of basic components from the oil [20–23]. Organic acids will also incorporate into the oil during source rock maturation, although the concentration of longer chain acids decreases rapidly with increasing chain length resulting in acetic acid being the major product [24]. Short-chain carboxylic acids, however, tend not to contribute appreciably to the crude TAN as they are water-soluble, and therefore, are readily depleted from crude relative to the higher chain length acids that remain in the oil via the extensive water washing that is associated with biodegradation.

Thermal degradation is also likely to contribute to crude oil TAN as linear and unsaturated acids are preferentially removed compared to other species [24]. A definitive study quantitatively analyzed 33 crude oils from the United Kingdom, Italy, and California and demonstrated that the carboxylic acid fraction was a major factor contributing to the acidity of these crude oils [25]. The graphs of TAN and carboxylic acid concentrations with a degree of biodegradation (Fig. 4.4) clearly show a significant correlation between TAN and the total carboxylic acid content.

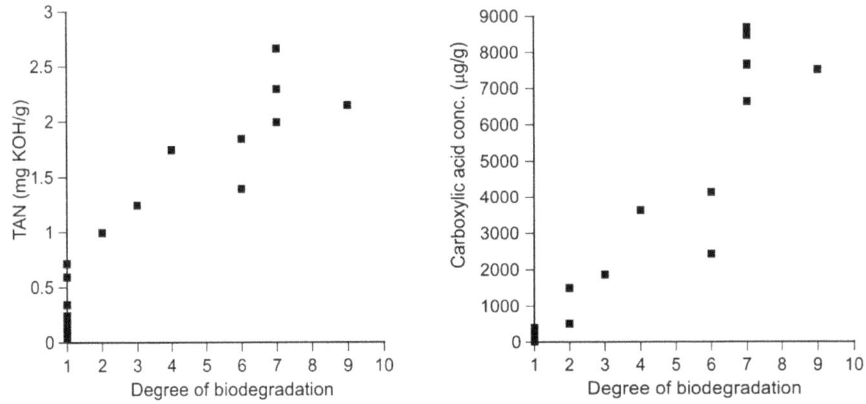

FIGURE 4.4

Degree of biodegradation versus TAN (left); degree of biodegradation versus total carboxylic acid content for a set of 33 crude oils (right).

Produced with data from W. Meredith, S.-J. Kelland, D.M. Jones, Influence of biodegradation on crude oil acidity and carboxylic acid composition, Organic Geochemistry, 31 (2000) 1059–1073 [25].

4.1.5 Carboxylate and naphthenate soap operational challenges

Having developed a common understanding of the acidic components and the idiosyncrasies in definition between carboxylic acids and naphthenic acids, we can now define further the operational manifestations of these acidic species in matters of flow assurance. Using these operational manifestations, this section will introduce field-based challenges as well as the classification system used throughout this chapter. Carboxylic and naphthenic acid soaps are acknowledged threats in the refinery industry and some early references describe the challenges they present, including; discoloration of kerosene [26] and influence on rust inhibition and detergency in lubricating oils [27]. Acid soaps were first implicated in upstream oil and gas operational challenges in the North Sea in the mid-1990s, a period when large-scale (basin-wide) water breakthrough [28] occurred. It is generally accepted that the native crude oil acid species are responsible for soap generation that impacts crude oil processing efficiency. The operational manifestations of acid soaps have two endmembers:

- Persistent emulsions and sludge pad build-up (see Fig. 4.5)
- Solid deposits are possible almost anywhere in the topside process (see Fig. 4.6)

There are additional challenges presented in downstream refining processing of high carboxylic–naphthenic acid crude, which include naphthenic acid corrosion [29], naphthenate fouling of heaters and exchangers [30], fouling and emulsions in desalting operations [31], high calcium in crude content and catalysis poisoning [32], as well as challenges in wastewater treatment [33].

FIGURE 4.5

Emulsion challenges brought about by carboxylates and naphthenates. Emulsion sludge in a skimmer vessel (left); wellhead sodium soap emulsion (right).

Reproduced from C. Smith, Chairman's introduction and aim of the workshop, in: Proceedings of the SPE ATW on Managing Naphthenates and Soap Emulsions, Pau, France, March 10–13, 2008 [2].

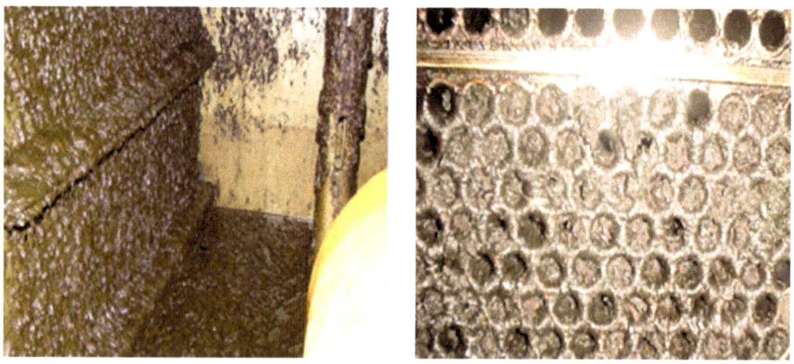

FIGURE 4.6

Examples of persistent solid deposition challenges brought about by carboxylate and naphthenate soap formation. North Sea separator plugged with calcium naphthenate (left); plugged Angolan process heater (right).

Reproduced from C. Smith, Chairman's introduction and aim of the workshop, in: Proceedings of the SPE ATW on Managing Naphthenates and Soap Emulsions, Pau, France, March 10–13, 2008 [2].

Almost 10% or more of North Sea crudes are at-risk of forming naphthenate scales, while 30% of southeast Asian crudes and 20% of West African crudes demonstrate soap emulsion or sludge problems [34]. There is a continuum between the two endmembers that will be described shortly in accordance with fluid composition and conditions that can result in more complex scenarios such as high calcium in crude combined with emulsion challenges.

4.1.6 The continuum model

During this chapter, the "Continuum Model", shown in Fig. 4.7, will be used to describe the operational challenges caused by soaps, linking them to the key chemical attributes of crude and brine. The model considers two major degrees of freedom to express the operational manifestations: First, the dominance of low molecular weight carboxylic acids compared to high molecular weight naphthenic acids; and second, the dominance of sodium to calcium in the formation water, and therefore, the resultant soap type.

These considerations lead to four endmember compositions; namely:

1. Calcium naphthenate (CAN)
2. Sodium naphthenate
3. Calcium carboxylate
4. Sodium carboxylate

The areas for these endmembers are deliberately small and consequently, a large unassigned area exists in the center of the model where most field cases likely reside. The point of course is that a continuum, or "solid-solution," exists for these two degrees of freedom and the field manifestations are often mixtures

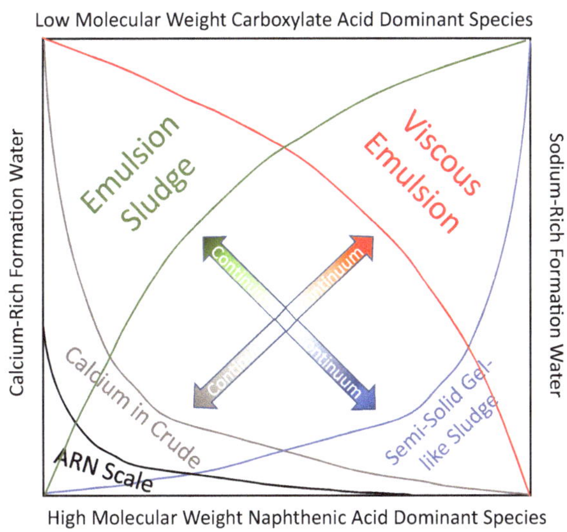

FIGURE 4.7

The *Continuum Model* representation of the operational challenges brought about by carboxylate and naphthenate soaps showing the four-endmember compositions.

of these endmembers, which also very often change over the life of a field. As will be appreciated, there are multiple equilibria at play that lead to the manifestation of field challenges; consequently and very often, each application is unique and requires a customized approach.

4.2 Fouling mechanisms of naphthenate and carboxylate soaps

Literature review, drawing from a few key references [2,34,35], suggests an approximate breakdown of global oilfield carboxylate and naphthenate related production challenges as:

70%–80%	Solid scale deposit related
10%–20%	Emulsion related
<10%	Calcium in crude related

The proportional R&D breakdown unsurprisingly reflected oil industry naphthenate concerns with CAN scale research accounting for the majority share of research activities; emulsion studies and calcium in crude come in as a poor second and third, respectively.

A good starting point to constructing any interpretative model for soap formation is to consider the intimate composition of oilfield crudes and the chemistry of the carboxylic and naphthenic acids therein, and then clearly identify the mechanistic aspects of acid-soap related fouling. The acids are polar and interfacially active in the produced fluids and consequently, one needs to understand their structure and physicochemical properties to construct an interpretive model.

4.2.1 Analytical techniques for acidic species in crude oils

4.2.1.1 Analytical chemistry basics of acid characterization

There has been a significant focus on the analytical characterization of acidic crude oil components, and a multitude of analytical techniques have been employed to isolate and characterize them. The subject matter itself has been extensively and comprehensively covered by other reviews [10,36], and therefore, the following section will select the most pertinent analytical techniques.

The most commonly used techniques for naphthenate speciation and quantification include elemental analysis, vapor pressure osmometry, differential scanning colorimetry, fluorescence emission spectroscopy, infrared spectroscopy, ultraviolet spectroscopy, ^{1}H and ^{13}C nuclear magnetic resonance (NMR) spectroscopy, gas chromatography (GC), and high-performance liquid chromatography (HPLC). Both GC and HPLC techniques performed in combination with mass spectrometry (Ms), have proven versatile and effective in naphthenate identification. Much of the early naphthenate characterization work tended to be concerned with their identification as crude oil biological markers for geochemical correlation and biodegradation studies and for feedstock hazard identification for refineries due to their corrosivity impact and wastewater fish toxicity potential. Fan reported on two complementary mass spectrometry methods: fluoride ion chemical ionization mass spectrometry, and negative ion fast atom bombardment-mass spectrometry when comparing naphthenic acid loads using a set of well-characterized crude oil samples [17]. These techniques opened the door to enabling specific high molecular weight naphthenic acid ionization and mass spectrometric elucidation for acid species identification and are referred to widely in the literature. Continuing with the mass spectrometry theme, Fourier transform ion cyclotron resonance mass spectrometry (FT-ICR-Ms) has been used in combination with IR and NMR for isolation and characterization of naphthenic acids from metal naphthenate deposits [37]. The ionization source, operational settings of the equipment, and the different solvents used for the samples can all have an impact on the quality of the analysis and a thorough understanding of these variables is critical in order to understand the analytical bias that may be introduced [38].

The second critical analytical chemistry requirement for acid component elucidation in crude is the method used for extracting the acids from oil or solid deposits or sludges. Historically, acid extraction was performed using liquid–liquid ion extraction with alkaline solutions [18,39,40] or other ion-exchange chromatography methods [41,42]. As will be detailed below, it was this area of research that was the rate-limiting step in the understanding of these substances as they simply were not able to isolate or even extract high molecular weight components (seemingly much above *m/z* of 600) despite the spectroscopic techniques existing to measure these species existing for many years.

4.2.1.2 The "ARN" acid discovery: an analytical breakthrough

Arguably the single most important discovery in the science of CAN deposition is the identification of a family of special polycyclic tetracarboxylic acids collectively known as ARN acids [43,44]. Isolated from the CAN deposits on the Heidrun field [28], the ARN acids first discovered contained 4 carboxylic acid groups (tetraprotic) with mole weights in the range of 1227–1235 g/mol, and 4–8 sites of unsaturation with a molecular formula of $C_{80}H_{142}O_8$. This same structure was confirmed in deposit analyses from other fields in the United Kingdom, China, and West Africa [44]. It is not clear from this work how much of the original solid deposit was composed of ARN acids, although most of the sample is implied; the limitations of the techniques employed are discussed [44]. From first discovery, ARN acids were, and are, intimately and directly involved in the deposition of the CAN soap scale. Etymologically the word "ARN" finds its origins in the Norwegian word for "eagle" (ørn (masculine) or ARN) as the structure and action of the ARN acid was postulated to resemble the four-talon claw of an eagle [44].

The enabling technology for the discovery of the ARN acid was undoubtedly rooted in innovations within the field of analytical chemistry, not just in the analytical devices themselves but also in the methodologies used to isolate the targeted naphthenic acid species. The most critical analytical chemistry development to the ARN acid discovery was the ion exchange resin (IER) method. The "Acid-IER" method is a sugar-based QAE Sephadex A-25 IER that is highly selective (>98%) to carboxylic acids from solids and nonorganic solvents as applied to CAN deposits [45].

Publication of the ARN discovery stimulated a fresh wave of naphthenate research focusing on ARN and other previously unextractable high molecular weight acid species. Complex organic acids could now be isolated for structural elucidation from oils and solid deposits, and physicochemical analyses along with precipitation experiments could be performed to develop a mechanistic understanding of this flow assurance challenge. Following the initial discovery, the structure of the ARN acids was quickly determined using a combination of NMR and FT-ICR-Ms [46]. This work was performed on a CAN deposit from West Africa and the main peak located at *m/z* 1230.0627 was consistent with the parent compound being $C_{80}H_{142}O_8$. The ^{1}H NMR spectrum was exceedingly complex, yet it was still possible to derive the structure (Fig. 4.8). The ARN structure appeared to lend itself to the very real possibility that it had an archaeal origin [47]. Archaea are one of the three main domains of life on earth along with bacteria and Eucarya and are known to inhabit harsh ecosystems, for example, volcanic springs, evaporitic settings, or deep-sea sediments. It is also known that archaea are also prevalent in the open ocean as picoplankton [48,49]. The ARN acids isolated from the CAN deposits showed a strong resemblance to structures isolated from core lipids found in membranes of archaea [50].

After initial discovery and reporting, ARN acids began to be identified more widely, for instance, TOTAL published multiple ARN finds in various oils from Africa and the North Sea [51,52]. Interestingly, ARN acids were found in several paraffinic oils with a very low TAN, including condensates that had no history of reported processing difficulties. Smith and coworkers measured the same C_{80} tetraprotic acids and trace amounts of other acids from crude oils sourced

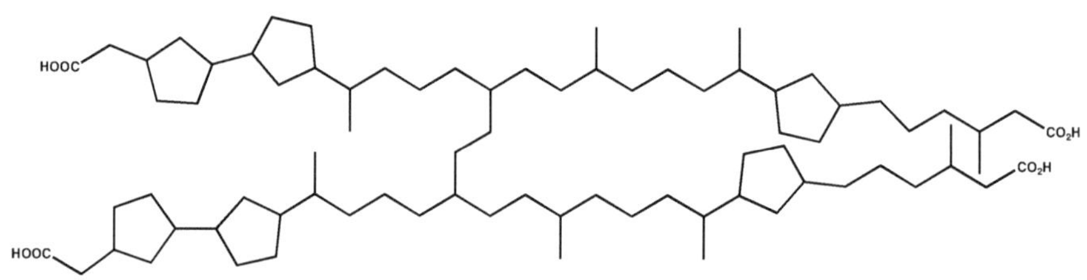

FIGURE 4.8

One of the four possible regioisomers of the tetraprotic acid with six rings as determined by NMR spectroscopy.

from the North Sea (the United Kingdom and Norway) and West Africa [53]. All three oils contained the same major acids; however, the proportions of each differed, possibly reflecting the growth temperatures from which the archaebacteria originated. In addition, the study revealed that mass spectrometry of individual acids isolated via preparative HPLC identified the presence of previously unreported C_{81} and C_{82}, 7 and 8-ring analog ARN acids (largest *m/z* of 1336). These same C_{81} and C_{82} acids were identified in a separate study with 7 and 8 membered rings determined using high-field NMR [54] along with some thought-provoking hypotheses regarding the origins of these high molecular weight acids. A separate study reported that the yield of tetraprotic acids in the extraction of the naphthenate deposit was very low at approximately 15%, however, optimization of the yield was not attempted (deposits originated from West Africa and the Heidrun field in offshore Norway) [46]. Three CAN deposits recovered from different fields were reported as containing between 28% and 41% (w/w) C_{80} tetraprotic acids [55]. Florida State University published work on CAN deposits, which indicated that each deposit was composed mainly of tetraprotic (ARN) acids with a C_{80} hydrocarbon skeleton. In one deposit, low molecular weight tetraprotic acids with C_{60-77} hydrocarbon skeletons were identified [56]. Heriot-Watt University investigated a large number of calcium and sodium naphthenate solids–sludges from the United Kingdom, Norway, and Asia. Careful extraction techniques were employed to preserve the make-up of the samples allowing a quantified compositional analysis of the solids deposits to be performed [57,58]. Analysis results varied significantly, and some samples indicated the presence of both higher and lower molecular weight naphthenic acids (acyclic, monocyclic, bicyclic, and alkylbenzoic) ranging from *m/z* 200–650, while others were characterized by their higher molecular weight naphthenic acids (ARN acids) content ranging from *m/z* 1230–1245, with only minor traces of lower molecular weight species. ARN acids were present throughout but at different concentrations. Reprecipitation experiments performed on the extracted acids confirmed the analytical dataset and showed that the experimental precipitates were spectroscopically similar to the indigenous precipitates. Scanning electron microscopy (SEM) and Energy Dispersive Analysis of X-rays confirmed the presence of calcium, carbon, and oxygen in various ratios and further corroborated these precipitates were CAN.

The initial flurry of post-ARN naphthenic acid analysis publications tapered off across the following three years, however, R&D efforts continued, reporting on technique refinements for identification and analysis of naphthenates (tetraprotic acids focus) in field deposits [59–62]. A separate but linked analytical initiative focused on naphthenic acid analysis in aqueous media that was spurred by the OSPW situation in Alberta, Canada [1,63]. This work is ongoing and illustrates that there are still improvement opportunities to be made in analytical technologies ranging from the extraction technique to the mechanism of carboxylic acid ionization as well as in the ultra-fine resolution of mass spectrometry.

4.2.1.3 ARN acid physicochemical properties

The physicochemistry of isolated crude oil acidic species was undergoing intensive investigation during the years leading up to the publication of the ARN discovery, as the interfacial activity and hence propensity of soaps to create challenging emulsions was well-known [39]. Original physical property measurement studies using extracted naphthenic acids combined with findings from subsequent ARN acid studies provide compelling evidence to explain carboxylate and naphthenate soap formation and inhibition mechanisms. Hemmingsen et al. evaluated the interfacial properties of acid extractions as a function of pH and concentration from successive alkali washes of acid crude oil [10]. For all extractions, there was strong interfacial activity with steep declines in interfacial tension (IFT) at low concentrations and a leveling out at higher concentrations. The same research group had previously documented detailed information on the crudes used, as well as the ability of their acid fractions to prevent hydrate formation via suspected emulsion control [10]. The significant interfacial activity of naphthenic acid was further explored with respect to inhibitory effects on asphaltene aggregates where the lowest IFT was measured for the highest pH extracted acid fraction [64]. The structure–activity relationship of these species in relation to the oil–water interface was highly complex, and the relative impact of different naphthenates on the IFT appeared to be influenced not only by the chemistry and size of the hydrophilic and hydrophobic groups, but also by their molecular structure and innate ability to self-assemble. The precise make-up and mixtures of acid species in a crude determines the interactions and the final IFT level. The results obtained, however, have to be viewed with a note of caution as the extraction techniques used tended to leave ~12% of the acids in the crude and these acids had molecular weights above 600 atomic mass units (AMU).

NTNU (Norges teknisk-naturvitenskapelige universitet - in English Norwegian University of Science and Technology) and Florida State University collaborated to measure physicochemical properties, including IFT and Langmuir monolayer properties for isolated ARN acids in water–toluene systems [37]. While not explicitly stated, it was strongly inferred that pure ARN acid was used as there was a distinct lack of other naphthenic acids reported and a referenced carboxylic acid extraction efficiency of ~98% [45]. The extracted acids, and therefore, implied C_{80} ARN acid, was highly oil–water interfacially active. As pH increased the interfacial activity became more pronounced as a consequence of increasing acid dissociation, and at pH 9, the IFT is lowered by ~40 nM/m at acid activity as low as 0.0050–0.010 mM. Undissociated, the acid had an area of ~160 Å^2/molecule; the high area reflecting an extended structural arrangement comprising the four carboxylic head groups separated by hydrocarbon chains. The quantity and purity of the ARN extract used were not explicitly detailed, however, it was stated that the original solid was "washed with toluene until the solution ran clear", which suggests removal of a substantial amount of organic material. A related study focused on isolated tetraprotic acid from a West African field source using acid-IER [65]. It is assumed that the 94% pure extract comprised C_{80} tetraprotic acid. This work is considered of particular note as the investigators added calcium to the acid in controlled experiments using various bulk and interfacial techniques. The extracted acid and corresponding calcium salt showed considerable surface activity, and both formed stable, elastic films at the oil–water interface. The IFT measured upon addition of the extracted acid is among the lowest published at 11.5 mN/m. The paper concludes that CAN forms a "plastic-like" film on a substrate, whereas the acid itself tended to crystallize during compression at low interfacial pressures.

Deposit extracts contain a multiplicity of naphthenic acid structures and are consequently not ideal for establishing fundamental structure–activity relationships. Tetraprotic acids "model compounds" on the other hand provide a convenient means of representing naphthenic acid and have been used in many studies to determine typical behavior and reaction trends [66]. The identification and subsequent study of model tetraacids may be considered potentially idealistic and over-simplistic, however, the insights gained from their use have proved invaluable with respect to understanding naphthenate reaction chemistry. Model naphthenates have provided excellent insight into potential mechanisms of soap formation and manifestation in the field [66–70]. The synthetic tetraprotic acid structure BP10, for example (Fig. 4.9), is believed to resemble ARN acids extracted from natural deposits [70]. Interestingly it is not hard to visualize the original ARN "eagle (claw talons)" visualization when looking at the structure for the ARN model compound BP10 in Fig. 4.9.

Nordgård et al. identified that model compounds form very similar cross-linked network arrangements to indigenous tetraprotic acids in the presence of calcium, and the ratio of these components are of fundamental importance to the film formation mechanism [66]. One important difference also highlighted in this work was that indigenous acids showed the characteristics of an amorphous, cross-linked polymer, whereas the model compound indicated more crystalline behavior with a liquid crystal mesophase. The crystalline structure was attributed to the regularity of the linear alkyl chains on the synthetic acids versus the presence of branching and more irregular structures of the indigenous acids. Sundman et al. published research regarding the complicated micelle formation chemistry of synthetic tetraprotic acids via their tensiometric behavior [69]. In the following year, the same author released the results of a study investigating the cationic preference of BP10 [70]. Cationic affinity

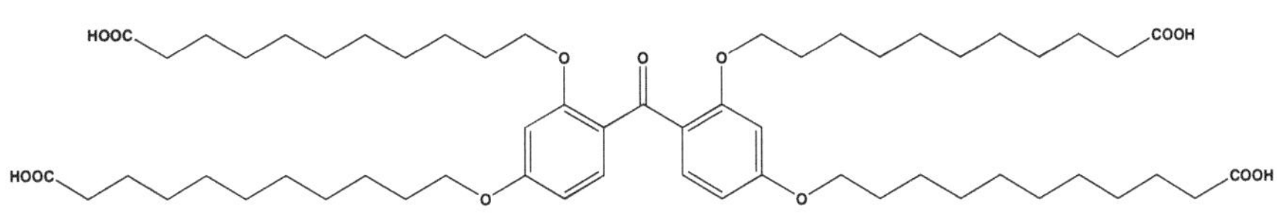

FIGURE 4.9

Structure of an example model tetraprotic acid BP10 used to study structure–activity relationships of ARN acids.

followed the order: $H^+ >> Ca^{2+} > Ba^{2+} \approx Sr^{2+} > Mg^{2+} >> Na^+$ and this sequence correlated with the hydration state and size of the cations. Significantly, the work revealed that at certain mixed cation and acid ratios, sodium became incorporated into the solid structure resulting in a mixed solid development of sodium and CAN (supporting the notion advanced in the chapter introduction of the existence of solid solutions).

4.2.2 Soap emulsions

4.2.2.1 Overview and field examples

As previously indicated, soap emulsions represent one extreme endmember of the continuum that exists for the manifestation of soap challenges in oilfield production chemistry (Fig. 4.7). The presence of carboxylic acids and their influence on emulsion stability is now well known, even though the presence of carboxylic soap was only recognized in the 1960s and R&D into their influence on emulsion stability really only began in earnest during the 1990s.

The operator Unocal experienced significant carboxylate soap emulsions during oil recovery operations in the Kutei Basin, offshore East Kalimantan in Indonesia in the early 2000s. A number of significant research papers regarding the occurrence were published across the proceeding 10 year period [71–77], and concerned the development of large quantities of greenish–brown toothpaste-like sludge (Fig. 4.10) occurring while producing from the Serang, Santan, Attaka, and West Seno fields. Approximately 100 barrels of oil per day (BOPD) of this very challenging slop-oil accumulated in onshore crude oil storage tanks and was problematic to remove due to the very high viscosity and pour point.

Kutei oil is light (38–41°API) with a low sulfur (<0.1 wt.%), wax (2.1–5.5 wt. %), and asphaltene (0.45–0.75 wt.%) content. Analysis of the sludge indicated that it comprised a mix of water, solids, entrained free oil, all stabilized by the presence of a "metallic soap." In the absence of effective in-situ sludge removal treatments, the sludge was manually recovered for expensive disposal. Intensive analysis of the filtered and dried (at 100°C) sludge via X-ray diffraction, NMR, FTIR, GCMS, and SEM identified that the greater proportion (~70 wt.%) comprised branched and straight-chain hydrocarbons in the range of C_{14}–C_{28}, and the remaining ~20 wt.% consisted of C_{28}–C_{30} carboxylate salts and <5 wt.% inorganics. Emulsification of oil, water droplets, and solids by carboxylate salts and bicarbonate ions was considered to be the most likely sludge formation mechanism [72]. The concentration of carboxylic acids salts (long-chain aliphatic acids salts) in the native sludge was estimated from the sodium analysis as 1.3 wt.%, and assuming an average mass of acids (associated with sodium) is 400 AMU then the estimated carboxylate concentration was approximately 23%. The formed emulsions were constantly undergoing shear, which only served to reemulsify any partially separated fluids, and the deepwater produced fluids themselves were cooled significantly during transmission, thereby losing the benefit of thermally induced demulsification. The prevalence of fine solid

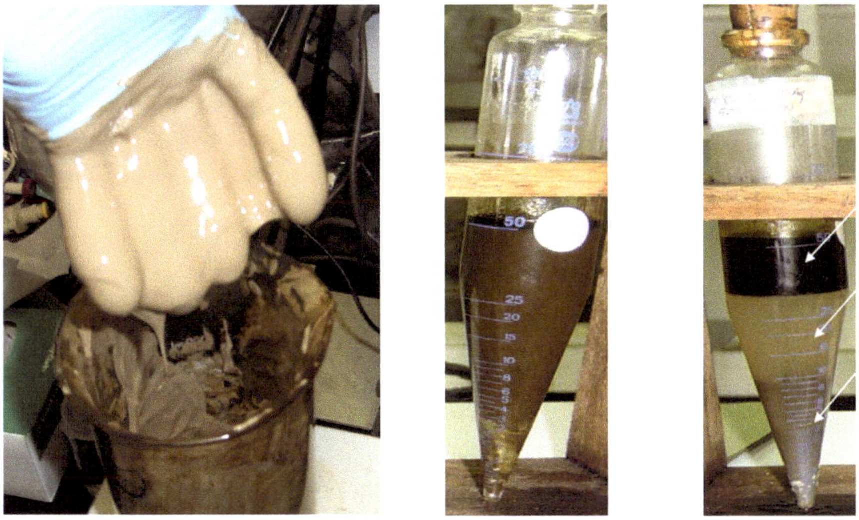

FIGURE 4.10

Pictures of the Kutei carboxylate stabilized emulsions. Fresh emulsion at worst case (left). Typical emulsion prior to breakage of soap sludge (middle). Sample treated with phosphoric acid-based demulsifier, heated above 140°F, then centrifuged (right).

Reproduced from D.L. Gallup, Processing Kutei Basin oils that generate soap emulsions upon contacting produced water, in: Proceedings of the SPE ATW on Managing Naphthenates and Soap Emulsions, Pau, France March 10–13, 2008 [75].

particulates (formation fines, residual drilling mud solids, wax particulates, etc.), routinely accompanied liquid production and contributed to the emulsion persistency [74]. ARN acids were latterly discovered in the Kutei Basin oils long after the original ARN paper publication, however, only trace levels of ARN-like soaps (diprotic and tetraprotic acids from C_{57}–C_{64} and tetraprotic acids at C_{83}–C_{84}) were detected in some samples and they did not appear to exert any significant bearing on the nature of the emulsion formed [71]. It was concluded that this class of soap was unique and certainly the closest representation of a pure carboxylate soap-induced emulsion identified in the literature. The bulk of acid species determined was simple straight-chained or slightly branched monocarboxylic acids (typically C_{16}–C_{34}) [75].

The geochemistry of the Kutei Basin crude oil is interesting, particularly the dominance of C_{20}–C_{30} acids, believed to be sourced from cuticular and epicuticular layers of leaves, part of the source rock for these oils. Concentrations of acids are significant when the leaves are tropical. Molecular changes may occur to the acids between the time they are in the leaf and then they become part of kerogen (type 3 kerogen of terrigenous origin) [71,76].

Gallup and coworkers considered the unusual nature of the Serang soap sludge compared to conventional CAN soap scales and revealed that "true" naphthenic acids (⌂-$(CH_2)_x$–COOH and σ-$(CH_2)_x$–COOH) were absent in the deposit analysis [78]. In addition, the Serang crude exhibited a low TAN (0.3–0.6 mg KOH/g oil) and low calcium levels, which are not conducive to typical ARN scale developments. A later work focusing on West Seno crude did identify some additional heavier molecular weight species *m/z* 829.8, which was assigned as C_{54}–C_{61} alkylbenzoic acids and C_{58} acyclic acid and deemed atypical of a Kutei Basin crude [77].

The Kutei soap emulsion is unique and occupies the top-right corner position on the *Continuum Model* (Fig. 4.7) as the soaps were almost exclusively sodium carboxylate type. There are examples of emulsion challenges for different emulsion types across the *Continuum Model* and it is particularly interesting when considering the gray area that lies between the extremes, and more so how does the relative presence of carboxylic and naphthenic soaps influence emulsions (beyond that of conventional oil water types?). From a mitigation perspective how can the *Continuum Model* be used to characterize an oilfield soap and recommend methods for its cost-effective mitigation? In the Kutei field, sodium carboxylate emulsion example is a combination of heat and acid demulsifier provided an effective means of resolving the emulsion development. It was also identified that production circumstances changed with time and a new emulsification issue developed, which then required the application of a combination of heat and nonacid inhibitor to resolve effectively [77].

A similar field example to Kutei was published by Salleh et al. [79] focusing on two field emulsion cases (A-52S and A-38S) from Petronas fields in Malaysia. Direct extraction followed by electrospray Ms identified an array of lower molecular weight naphthenic acids in the range of *m/z* 200–600, and the presence of acyclic C_{10}–C_{25}, monocyclic C_{20}–C_{32}, bicyclic C_{11}–C_{32}, and alkylbenzoic (C_{15}–C_{36}) naphthenic acid species was also confirmed. As per the Kutei deposit, there was little evidence of detectable levels of higher molecular weight ARNs acids [79]. The lower–medium molecule weight naphthenic acids were responsible for high viscosity and stable emulsion formation in the field after they were combined with sodium-containing brine at a pH >6.6.

When considering the Petronas examples with respect to the *Continuum Model*, we can see that it is located at the top right hand of the chart in Fig. 4.7, but is slightly more to the left and downwards compared to the Kutei Basin case. There are many other cases in the literature of emulsion formation caused by the presence of both low molecular weight carboxylic acids and medium molecular weight naphthenic acids and/or soaps [35,56,80–84]. None unfortunately are as well defined and detailed as the two highlighted above in terms of analytical characterization. In reading through the literature, one can get a good sense of the continuum described in Fig. 4.7 for emulsions and how there is variance caused by ion type and acid species molecular weight distribution. A key question to be answered in the next sections, therefore, is where does the continuum transition from solids to emulsions and how does the mix of acids and ionic character affect this and why?

Oil in water emulsion is also often reported as a challenge alongside the majorly troublesome water in oil emulsions [78,82]. This is often attributed to the large quantities of organic acids partitioned into the produced waters formed from the water-soluble salts of very low molecular weight acid species [78]. Any oil and water separation plant is a fine balance focused between export oil BS&W and water oil and grease qualities. The presence of interfacially active acidic species and their soaps understandably can have a profound effect on this and send either specification reeling very quickly due to a loss of interfacial control in the process plant.

4.2.2.2 Mechanism of soap emulsion formation and factors that influence stability

A mechanism for soap emulsion formation was first proposed in 2001 [78], with subsequent refinements [19,74,85,86], including the development of a thermodynamic model that described the formation of soaps under controlled laboratory conditions [87,88]. The best mechanism visualization was introduced by Brocart et al. [52], as shown in Fig. 4.11.

Short-chained acid species (carboxylic or naphthenic acids) in the crude oil encounter sodium and calcium ions from produced water and if system conditions permit, will saponify to form the corresponding acid soap. These acid soaps are highly interfacially active and preferentially migrate to the oil–water interface imparting stability to emulsions observed in the field.

The formation of soap molecules from parent acids occurs as produced oil, gas and water are transported from the reservoir, up-string to wellhead and then transmission through the process, experiencing temperature, and pressure changes

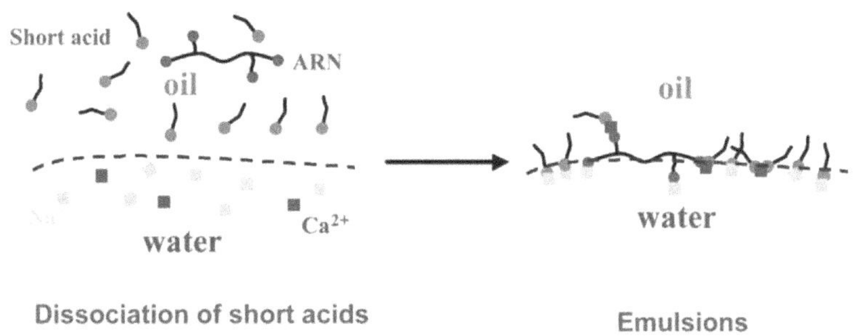

FIGURE 4.11

Basic mechanism of emulsion stabilization where the "short-chained" acids dominate and given the correct conditions saponify and become surface-active, preferentially clinging to the interface.

Reproduced with permission from B. Brocart, M. Bourrel, C. Hurtevent, et al., ARN-type naphthenic acids in crudes: analytical detection and physical properties, Journal of Dispersion Science and Technology 28 (2007) 331–337 [52].

along the way. Critical fluid and gas saturation equilibria are disturbed, which invariably impact produced brine pH. Of particular importance is the presence of acid gases (e.g., CO_2 and H_2S), which partition between the gas, liquid, and oil phases. Under reservoir conditions, the proportion of acid gas molecules dissolved (suspended) in the produced water is defined by temperature and pressure. Depressurization occurring during production will disturb the equilibrium and result in the dissolved gasses redistributing from the liquid to the gas phase. Solution kinetics will simultaneously redress dissolved acid gas loss and convert aqueous bicarbonate to dissolved CO_2 molecules, and in the process, a proton is consumed, and the brine becomes more basic as a whole. Depressurization below the production CO_2 bubble point, will drive the pH dramatically more alkaline [89–91]. The pH of the brine phase has a profound impact on soap formation and is often the deciding factor as to whether a soap will form; and essentially provides the basis for most, if not all, soap remediation methods.

The saponification reaction can be expressed simply (for linear carboxylic acids) as:

$$CH_3(CH_2)_nCOOH + M^+ \rightarrow CH_3(CH_2)_nCOO^-M^+ + H^+$$

As pH increases, the degree of dissociation of the carboxylic acid group increases, facilitating carboxylate metal salt formation. The process is the same irrespective of acid type or whether sodium and/or calcium ions are present in the brine. In the Kutei Basin case, sodium carboxylate formed preferentially. As system pressure and temperature decrease during production, the long-chain fatty acids (C_{16}–C_{32}) in the crude react with produced brine metal ions to form interfacially active metal carboxylate soaps at pH >6.2 [71].

Alkali extractions of crude oil acid components show a high degree of interfacial activity [92]. Pure carboxylic acid fractions isolated from crude oils have shown IFTs of >0.003 dynes/cm at 1400 mg/L [39], however, the presence of phenolic hydroxyl groups in addition within an oil can result in reduced activity of the carboxylic acids [93]. Carboxylic and naphthenic acids display very interesting phase behaviors as they can self-organize via the formation of aggregates such as dimers, micelles, and higher-order structures depending on their structural complexity [19]. The key to understanding emulsions is to study the influence different carboxylate and naphthenate salts impart on interfacial properties of known oil–water systems.

The number of literature accounts that clearly and demonstrably identify the exact combination of acidic structures contributing to a specific stabilized oilfield emulsion scenario is regrettably low. This is not surprising, however, considering the detailed crude analytics required, as well as correlating this data back to field observations, and all supported by tailored application test experiments. With CAN scales (where ARN acids provide the primary focus), the situation is reversed. It is for these reasons that only generalized models and directional understanding exist, supported by occasional isolated but significant field cases for the emulsion formation mechanisms of acid soaps. The Gallup study mentioned earlier employed a combination of bottle tests and dynamic separation tests to demonstrate the impact pH has on emulsification for different oils and acid fractions [78]. Model naphthenic acid solutions and acids extracted from crude oil were used in IFT tests and correlated the observed bulk phase experiments with observed interfacial behavior [85]. Brandal et al. developed on this theme and identified that the acids responsible for soap formation also expressed a very strong affinity for the oil–water interface resulting in a measurable lowering of IFT [94]. The study showed that the rate of soap formation was dependent on the stability of the positive monovalent carboxylate complex at the interface (considering steric conditions of the acid monomers at the interface as well as interfacial affinity of the cations). The IFT changes identified correlated closely with the size of divalent ions present in the aqueous phases; implicating both sodium and calcium ions in the observed interfacial activity of the system. A study published in the same year by Shepherd et al. arrived at the same conclusions using isolated and structurally simple carboxylic, naphthenic, and benzoic acids and corresponding salts and measuring the IFT between toluene and water [80]. The results revealed that all acids and salts studied were interfacially active to some degree, but in this system, the sodium salts were more interfacially active than the corresponding calcium salts. The experiments also identified stronger interfacial effects at higher pH, which corresponded with real field observations. The enrichment of these acidic species in emulsions (and therefore, interface) relative to the bulk crude in analytical studies is further an indication of their surface affinity. Mapolelo et al. noted selective enrichment of saturated fatty acids in field emulsion samples and as a consequence of their DBE characteristics, which expressed a more pronounced facility for binding monovalent cations compared to the parent crude [56]. The reader is directed to further additional literature regarding model acids and soap systems, which have contributed to the understanding of the influence that specific components have on given systems [95–100].

How significant are ARN acids when compared to the lower molecular weight, more polar acid species (in terms of hydrophilicity and lipophilicity, that is, surfactancy) in forming stable emulsions? They are interfacially active [65,96], but exhibit poorer interfacial activity potentials in troublesome emulsifying crude oils, and it is conceivable that they contribute in some way to the interfacial properties of the final emulsion. For other process-related parameters, conventional emulsion behavior rules apply, which are peculiar to the field and process facility. Beyond the water cut inversion point, oil and water separation improve; however, the presence of significant soaps will cause the inversion point to increase. Increased temperature promotes emulsion resolution, with nothing special being noted in the literature that specifically links temperature to carboxylate or naphthenate emulsion stabilization. Shear of course is a critical parameter required for emulsion formation and the more shear a given oil–water system experiences the smaller the droplets and the greater the degree of emulsification due to the larger interfacial area for soaps to cling to [83].

Several summaries making generalized observations on produced fluids that are prone to emulsion soap formation and their properties have been published [35,71,76,83]. In the carboxylate soap emulsion case (top right in the model of Fig. 4.7), the emulsion developed from lighter oils with an API typically >32 to 41° and with a relatively low TAN (0.05–0.6 mg KOH/g oil) and with variable organic acid contents. The oils tend to have low sulfur contents (normally <0.15 wt.%) and typically a moderate to high wax contents (4–15 wt.%), low asphaltene (<0.05–0.5 wt.%), and low resin contents (1–6 wt.%). These oils can have wax deposition problems in flowlines and export oil pipelines as well as having a low natural solvency for their asphaltene fractions, expressed as an increased facility for asphaltene deposition in well tubing and surface flowlines. The oils are typically not, or only slightly biodegraded and contain significant fatty carboxylic acids (C_{20}–C_{37}) that are involved in the formation of metal carboxylate soaps. Heavier acid components, including tetraprotic ARN acids, also feature, but usually in trace amounts. The produced water tends to be high in bicarbonate ion content (500+ to 1000 s of mg/L) and also contains very high volatile fatty acid (VFA) concentrations, which provide forewarning that oils may contain significant carboxylic acids. The overall salinity of the produced water is usually low, typically 7000 to <25,000 mg/L total dissolved solids (TDS). The CO_2 in the gas phase is low to moderate (0.8–3 mol.%), which results in a relatively low pH shift to neutral on depressurization (0.2–0.5 pH units typically). The soaps are salts of monovalent cations typically (e.g., Na, K) and rarely Ca, Fe, and Mg. The view that both sodium and calcium carboxylate emulsions coexist is reported in the literature with calcium carboxylate emulsion soaps mentioned as being rarer than their sodic analogues and are concentrated in West Africa, Southeast Asia, and the North Sea [83].

It is strongly suspected that many oilfield naphthenate and carboxylate soap emulsion events have gone undiagnosed and unreported, often being potentially masked or subsumed by partially correct interpretations of ongoing oil–water separation issues.

In summary and relating this newfound structure–activity knowledge back to the *Continuum Model*, we can conclude that a dominance of lower molecular weight, polar carboxylic acid species combined with sodium tends toward the more challenging emulsion scenarios caused by the stronger interfacial activity of these species. The higher polarity species in this continuum are low molecular weight, more linear, and less complex monocarboxylic acids. Increasing the calcium content still results in emulsions, however, they tend to be less challenging as dicarboxylic acid species have generally lower interfacial activity. Moving toward higher more-complex naphthenic acid species content also results in less challenging emulsions due to lower interfacial activities of the higher molecular weight, highly branched, acyclic acid species. The interfacial activity is still significant; however, they are less likely to result in sludge-like emulsions that are typical of the low molecular weight carboxylic acid species. High molecular weight polycyclic tetraprotic ARN acids have marginal influence on emulsion formation and do not provide any additional challenge beyond that of the other, more "normal" emulsifying species found in crude oil.

4.2.2.3 Impact of emulsion soap formation

Emulsion soaps can form downhole and upstream of the wellhead choke, however, given the importance of pH and depressurization on carboxylate soap formation, it is more usual to see their development commence in upper string or wellhead locations. Emulsions can make up a large proportion of overall production and upwards of 10% volume of fluid streams has been reported [72,74]. Their inherent viscosity and presence can dramatically and negatively impact production rates in some fields due to increased pressure drops across flow lines, etc.

The presence of emulsion soaps is critical to emulsion development and addition of other common oilfield emulsifying agents such as formation fines, asphaltenes, waxes, and inorganic scale particles, can further stabilize the formed emulsion. Emulsion soaps accumulate at the separator oil–water interface and impair droplet coalescence leading to "out of specification" BS&W in export crude (in extreme cases, this oil is processed as lower value "slop oil"). Emulsion soaps can plug water-processing equipment, including hydrocyclones (usually when combined with water phase suspended solids and oil remnants) and can reduce vessel and storage tank volumes resulting in nonoptimal residence times. Sludge removal from vessels can be challenging due to its high viscosity. Reservoirs that produce a great deal of sand and reservoir fines to surface can do so because of the production of soaps, which entrain these solids. Once emulsions are resolved, the particulates release and deposit, forming new flow restrictions within the process [74]. If unresolved, the presence of emulsion soaps can also impact oil in water quality, which will have a knock-on effect if the water is used in produced water reinjection (PWRI) operations or if it is discharged to the environment. The excess oil content will present an environmental burden, which in most parts of the world is now considered unacceptable and can result in unwanted plant shutdown and lost production and financial penalties.

Toxicity of a soap sludge can be a concern as they may accumulate heavy metal contaminants such as mercury, arsenic, and cadmium as well as the potential for naturally occurring radiative material (NORM) scales (e.g., $BaSO_4$). Furthermore, oil-wet soap particles can become suspended in the produced water and cause excessive oil in water, which can itself cause surface "sheening" and further environmental issues [74,101]. Reinjection of this water can result in the injection well formation damage due to near-wellbore formation wettability modification and/or accumulation at low porosity injection faces, which then proceed to progressively pack-off via a build-up of filter cake comprising emulsified oil remnants and entrained suspended particulate solids [102]. The phenomena were demonstrated synthetically using idealized conditions of 1 wt.% myristic acid (saturated C_{14} fatty acid) in toluene mixed with a 1 wt.% $CaCl_2$ brine at a pH >10, which generated solids, which when experimentally flowed through a core sample caused formation damage (as measured by conductivity impairment)

[103]. Idealized conditions and solids aside, it is feasible that emulsions soaps (not solid particulates) do end up getting reinjected and that they could cause adverse effects by, for example, wettability modification causing formation damage.

4.2.3 Naphthenate soap solids

4.2.3.1 Overview and field examples

CAN scale typically presents as thick organic, brown, asphaltenic, or wax-like solid material and can occur in high quantity in oilfield systems. On the *Continuum Model*, it is positioned in the bottom left-hand area of the plot, diametrically opposite to the pure carboxylate emulsion soaps discussed earlier.

CAN has been the subject of numerous reports across the last 20-years, far more than both soap emulsion and calcium in crude issues combined (approximately four times as many!) with papers focusing on oilfield CAN scale chemistry, occurrence, and technology for its remediation.

The first identification of oilfield ARNs from naphthenate deposits was from the Heidrun field in the Norwegian Sector of the North Sea. Papers include the original 2003 Vinstad et al. publication detailing Heidrun Naphthenate challenges [28], and in the same year, Mediås et al. reported a selective isolation method for carboxylic acids in crudes [71]. The seminal Baugh et al. paper in 2004 introduced ARN for the first time [43] and again a year later regarding higher molecular weight ARNs [44]. Rounding things off are two similar 2008 review papers looking at Heidrun and the management of the naphthenate soap challenges [104,105]. In 1998, the StatoilHydro (now Equinor) operated Heidrun TLP (Tension Leg Platform) and field was producing at plateau of 250,000 BOPD and 38,000 barrels of water per day (BWPD) (water breakthrough-commenced mid-1996). Heidrun crude was reasonably complex with contributions from multiple reservoirs and layers resulting in a 29° API crude with a TAN of 2.5–2.7 mg KOH/g oil. The "poorly maturated" oil presented ~2 wt.% naphthenic acids with an average molecular weight of 430 g/mol. Heidrun, PW typically contained approximately 1000 mg/L Ca^{2+} and 630 mg/L HCO_3^- with a reservoir pH of 5.8, which rose to pH 6.4 at the first stage separator. Process operational issues were identified in September 1996 in the form of solid scale deposits within the first and second stage separators, and by October 1996, the test separator also experienced fouling. Fouling continued throughout 1997 and by year-end, naphthenate scaling had spread to all parts of the liquid handling systems and consequently represented the most significant threat and challenge to Heidrun production. The solid material was jetted free from contaminated process units and it was noted that during test separator jetting, the lumps of material would collect and then block the jet water outlets. This observation quickly became a marker indicating whether the well in the test separator was producing significant sand to surface as the sand entrained within the naphthenate engendering development of a more robust scale mass. Fig. 4.12 presents photographic images showing naphthenate deposits freshly recovered from the separator, and then the same deposits after manual vessel entry. The CAN scale was observed to visibly solidify after exposure to air and became almost rock-like in hardness [28,104,105]. The sand entrainment phenomenon has since been reported in other more recent naphthenate scale production scenarios [106].

A second oilfield naphthenate case history example is from the then operated Marathon Oil UK operated Brae field in the UK North Sea, more specifically in the West Brae field, a subsea tie-back to the Brae Alpha platform via an 8.5 km 12″ flowline. West Brae came on-stream in the second half of 1997 [107] and was relatively cool at 63°C with a ~3000 psi reservoir pressure. The field well complements comprised six producers and a single injector. The 22° API field crude had a TAN of 2.5 mg KOH/g oil and produced water that showed 1950 mg/L Ca^{2+} and 1000 mg/L HCO_3^-. The first significant naphthenate scaling issue occurred within the South Brae platform heater located downstream of the slug-catcher, which was found to be severely fouled during the field 1998 planned shutdown.

Process naphthenate fouling occurred again in the following year necessitating an unplanned shutdown to clear large accumulations of scale from the coalescer, and there followed more naphthenate incidents as scaling spread through the South Brae process. Naphthenate fouling issue continued to dominate through 2002 and regularly impacted oil in water interface control and caused line blockages. From 2002 to 2005, the scale character slowly changed from a solid fouling precipitate to more of a modified slurry, which also altered its impact on the process. Production technicians were increasingly faced with naphthenate pad

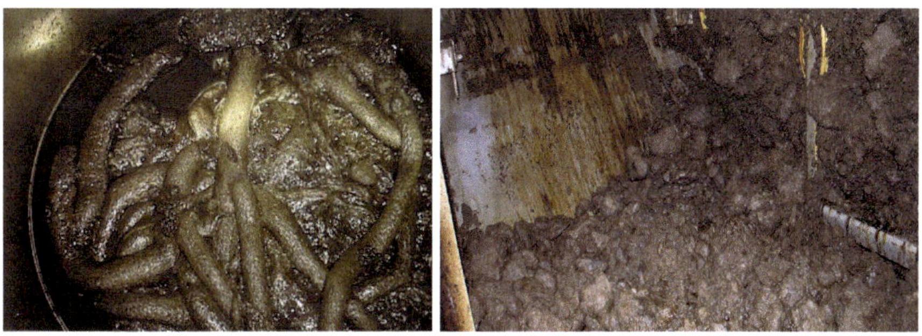

FIGURE 4.12

Freshly extracted calcium–naphthenate scale from the "tricock" valve on the first stage separator sampling the oil:water interface (left). Same scale built up in the same separator upon vessel entry (right).

Reproduced from L. Oduola, C. Igwebueze, O. Smith, et al., Calcium naphthenate solid deposit identification and control in offshore Nigerian fields, SPE-164055-MS, in: SPE International Symposium on Oilfield Chemistry, The Woodlands, TX, April 8–10, 2013 [106].

formation in the coalescer and other production vessels that had a stable oil–water interface. The event frequency declined, however, the PW degasser and hydrocyclone reject lines were regularly plugged, and the pads caused interface control issues and created operational challenges in the slop-oil system [107]. The CAN deposits removed from vessels on the Brae Alpha are presented in Fig. 4.13.

Other excellent naphthenate case histories are available from around the world, including the United States (California and Gulf of Mexico), Canada, Mexico, Venezuela, North Sea (United Kingdom and Norway), West Africa (Nigeria and Angola), Far East (Malaysia and Indonesia), and Russia [34,35,78,82,83,106,108–123].

4.2.3.2 Mechanism of calcium naphthenate soap scale and factors that influence stability

CAN scale and its control has received vastly greater attention in the literature compared to other emulsion challenges brought about by the same families of acids, albeit different molecular weights and dispersity. We already know about the existence of polycyclic tetraprotic ARN acids and that they play an influential role, but what are the interdependencies and what governs ARN containing crude oil naphthenate scale precipitation? How does the *Continuum Model* accommodate CANs, and how then can it be used to provide answers to the key questions pertaining to which factors dictate whether CANs present as solids or not, and where does the continuum transition from solids to nonsolids? Do the two challenges coexist, and are there examples?

Significant advances have been made in the mechanistic understanding of CAN soap deposits with several seminal studies [19,52,78,80,86,94]. Notable is the "idealized model" visualization (Fig. 4.14) proposed by Brocart et al. [52]. A "complex" acid system is presented comprising a mix of short-chained (low

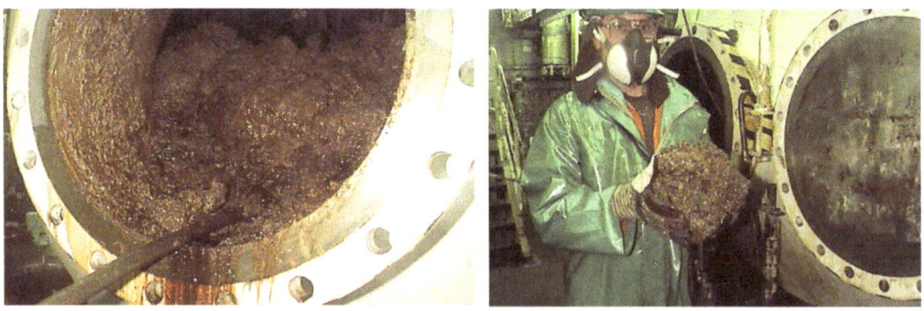

FIGURE 4.13

Manual removal of solid CAN from the Brae Alpha coalescer (left). The scale upon exposure to air turned concrete-like (right).

Reproduced from J. Hardy, West Brae naphthenate experience, in: Proceedings of the SPE ATW on Managing Naphthenates and Soap Emulsions, Pau, France, March 10–13, 2008 [107].

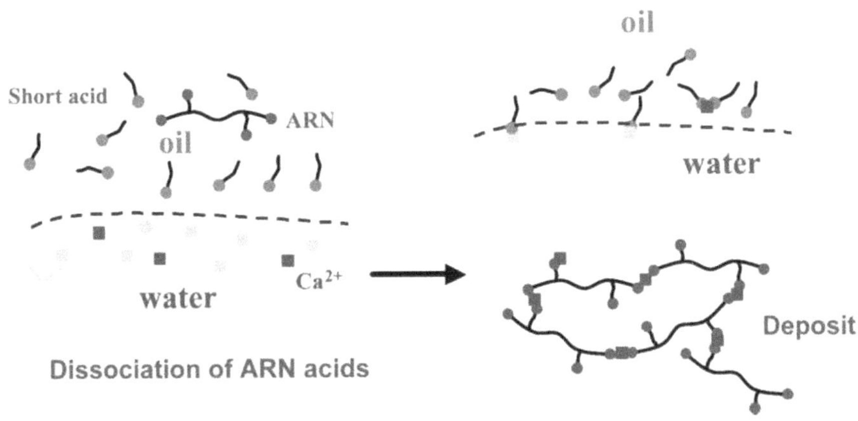

FIGURE 4.14

Basic mechanism of solid soap scale formation where the ARN acids under the correct conditions saponify and form a complex deposit. Short-chain acids are depicted to coexist with the ARN acids and become highly surface-active after saponification, preferentially clinging to the interface.

Reproduced with permission from B. Brocart, M. Bourrel, C. Hurtevent, et al., ARN-type naphthenic acids in crudes: analytical detection and physical properties, Journal of Dispersion Science and Technology 28 (2007) 331–337 [52].

molecular weight) acid species and higher molecular weight ARN acids. The acid species encounter brine rich in sodium and calcium ions and, given the correct saponification conditions, form the corresponding acid soaps. In this case, ARN acids complex with calcium to generate precipitating calcium-ARN solid deposits and the shorter-chained acid soaps preferentially migrate to the oil–water interface due to their enhanced interfacial activity. Refinements to this ideal will be proposed latterly.

The many naphthenate scale analyses published in the literature indicate that the ARN acid composition can vary greatly. Some scales contain high levels of high molecular weight C_{80+} tetraprotic acid salts of calcium [52,56], while other scales present much lower levels (<15%) of tetraprotic acids combined with an assortment of other lower molecular weight species [111]. This observation aligns with and supports the *Continuum Model*. The proof that ARN acids are critical to naphthenate precipitate formation is to be found in published physicochemical measurements of formation of cross-linked networks of ARN acids with calcium [66,70]. The ratio of tetraprotic acids to calcium dictates the nature of the film formed and the source of the naphthenates themselves also plays a significant role; naphthenate solids formed using field indigenous acids tend to adopt an amorphous morphology, while synthetic model acid species tend to create more crystalline CAN films. A refinement to the mechanism of Brocart [52] to incorporate these findings is presented in Fig. 4.15. In the refined model, the pure ARN

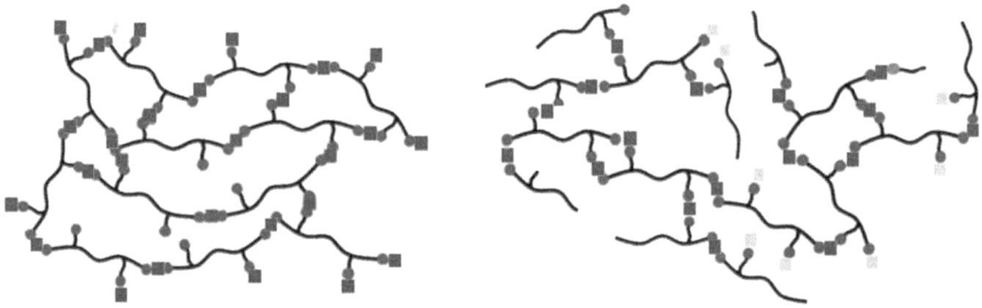

FIGURE 4.15

Mechanism proposal for the difference observed in scale types. One extreme scales that are composed of exclusively tetraprotic ARN acids and calcium in a cross-linked crystalline network (left). In another scenario, the ARN acids account for a small proportion (10–15 wt.%) of the scale, and lower molecular weight acids and sodium are incorporated into the structure leading to a more irregular amorphous arrangement (right). Consequently, the scale is softer and more sludge-like.

CAN scale is presented (left-hand image) and a less ARN pure CAN scale is also postulated (right-hand image) to account for observed field changes in the form of the solids encountered over time. The less pure ARN scale can include sodium as well as calcium atoms, and also incorporates lower molecular weight nontetraprotic carboxylate acid species. The purer CAN scale results in well-ordered deposits formed from extensive calcium cross-linked networks of tetraprotic acids, while addition of sodium and low molecular weight carboxylate species into the developing structure results in a more irregular arrangement and consequently forms a less crystalline, more amorphous film. This would support the field observations in the Brae case history where the scale form (hardness) changed over time.

Runham and Smith reported a successful management approach to naphthenate scale control on a North Sea facility where the naphthenate scale and carboxylate emulsion ranged in character from hard to slurry-like consistency [83]. Compositional assay indicated the deposits contained more oil and water as viscosity decreased, and naphthenate inhibitor use was proposed as a likely cause. Other explanations are worth exploring as it was noted (by the author) that the composition change occurred at constant inhibitor dose. One alternative considered the residence time of the material in the process; less time in the separator interface resulting in more oil and water-bound with the material [83]. The variability in deposit presentation was discussed by Shepherd et al. who reported on analysis of five different scale samples from distinct global locations and their differing sodium and calcium ion content, acid species, and thermal behavior [111]. The same work revealed a link between two of the sample's natures with the chemical treatment deployed onsite, which influenced their depositional character and composition. One study attempted a mass balance of the ARN acids extracted from a field deposit (c. 19 wt.%) and the same acids in the parent oil (<20 ppm). Mass balance calculation estimated a 500 kg/day deposition rate, which fell short of the 1000 kg/day actually observed; the discrepancy remains unresolved [124].

The major driving force for both CAN precipitation and emulsion formation is pH, and more specifically for saponification is deprotonation of the ARN and other carboxylic acids to allow complexation with sodium and calcium ions at higher pH. As discussed previously, pH change in oilfield production scenarios usually results from produced fluid depressurization causing dissolved CO_2 to move from the liquid to the gas phase and subsequent increase in the PW pH upwards as solution equilibrium is reestablished. Rousseau et al. reported the pH dependence of a naphthenate–calcite mixed scale in the Angola Block 17 deep-water field where the pH-driven CAN precipitation mechanism was likened to that of calcium carbonate scale [78]. In a mechanistic study of naphthenate scale formation, jar testing was performed on a series of model carboxylic, naphthenic, and benzoic acids to elucidate naphthenic acid precipitation mechanisms and discussed the effects of cations on bulk fluid and surface properties, including viscosity, emulsion stability, and IFT [80]. This work supports the hypothesis that a multitude of acid species in the crude oil contribute to naphthenate field scales.

The presence of ARN acids does not necessarily mean that a given crude will actually experience naphthenate-related operational issues [52]. Comparison of crude oil TAN, API, and CO_2 contents identified ARN acids in most crudes, and ARN was found across a broad range of those tested with TAN values in the 0.36–2.15 mg KOH/g oil range and API's from 20 to 36.5° [46,52]. These results, in conjunction with other unpublished data, were used to prepare a naphthenate prediction model, which was based on organic carboxylate species competition for available cationic species in produced water [52]. The perceived CAN scale risk is not simply dependent on ARN acid presence and its concentration but is in fact significantly influenced by the presence of other naphthenic acid species in the oil, which can act as natural inhibitors due to their competition with the ARNs for the available cations [115,125]. The prediction code devised was simple and effective and relied on pH measurements and the influence of phase equilibrium between oil and water of the acidic and basic species, and acid–base equilibrium in the water phase to model the complexing behavior of the carboxylate species. The model of Brocart et al. provided a semiquantitative risk assessment of CANs and incorporated an in-situ "inhibitive" property of the crude oil based on real field deposit correlations [115,125]. The work also suggested improvement opportunities with respect primarily to the oil phase behavior of tetraprotic acids in relation to other acidic species [115]. It is clear that the interrelationship between different acid species (ARN and otherwise) is complex and is determined by the dissociation constants of the acid species and the association constants of the associated soap species. Regardless of the competition factor, if ARN acids contact sufficient Ca^{2+} at the correct pH then precipitation will occur [126,127], although any carboxylate species present that express a higher affinity for Ca^{2+} will invariably complex but will present as nonadhering soft development rather than a solid CAN deposit. Literature suggests that CAN scale deposits must contain ARN acids. If one removes literature cases that do not feature advanced Ms Analytics, then all other studies have reported the presence of ARN acids in CAN scale deposits. It is not difficult to visualize that developing CAN proto nuclei solids can act as seeding antagonists to facilitate the precipitation of other species, including paraffin, asphaltenes, and other scales (and vice versa). Experimental evidence [64,85,128], as well as field observations, support this hypothesis with mixed scales being reported [119–123]. Decreasing the molecular weight from C_{80+} downwards eventually arrives at a critical point where non-tetraprotic acid species that are incapable of forming calcium cross-linked networks will instead form highly viscous oil dispersible species. In the *Continuum Model* of Fig. 4.7, this has been expressed as the transition from solid CAN scale to calcium in crude and the presence or absence of tetraprotic ARN acids is critical to determine, which mechanism is prevalent. Calcium in crude challenges will be discussed in more detail toward the end of this chapter.

The literature presents generalized observations and rules-of-thumb guidance on the produced fluids and their properties that lend themselves to CAN soap scale formation [71,76,83,129]. They tend to form in heavier oils with an API typically <18 to 27° and with a high TAN (0.8 to 8.0 mg KOH/g oil). Extreme outlier examples exist; the sanitized Field Z with 0.1 mg KOH/g oil [111] and another at 0.09 mg KOH/g oil [2]. These oils are significantly biodegraded, and therefore, considered naphthenic oils and are rich in naphthenic acids, including ARN tetraprotic acids. Typically, the wax content is low (<4 wt.%) and has a high asphaltene content (0.5 to >3 wt.%), high resin content (>6 to 17 wt.%), and moderate to high sulfur content. The oils rarely exhibit any nonnaphthenate flow assurance depositional challenges (wax or asphaltenes) as biodegradation has removed much wax and increased resin–aromatics and reduced saturates contents gives oil good solvency for their asphaltene loads. Produced waters typically are of a moderate salinity ($>25,000$ mg/L TDS) with a relatively high bicarbonate content (typically >450 to 1250 mg/L) and typically >300 ppm Ca^{2+}. The brines typically show moderate VFAs (<100 mg/L) and these do not appear to influence soap formation. Usually, there is significant CO_2 in the field gas (>3 to 12 mol.%) as these scales see a big water pH shift to alkali conditions during production depressurization (>0.5 to >1 pH unit shift).

In summary (and referring mechanistic structure-activity relationships to the *Continuum Model*), it can be concluded that a dominance of higher molecular weight naphthenic acid species in the oil along with a calcium dominance provides conditions for worst-case CAN scaling, that is, the bottom left corner of the *Continuum Model*. It does not have to be exclusively calcium-rich, and a continuum exists right across to sodium dominance, but just falling short of no calcium being present. The transition point location is at some point between the extremes for CAN precipitation and carboxylate emulsion formation (top right), however, it will be strongly influenced on a case-by-case basis by the calcium affinity and distribution of lower molecular weight naphthenate species. For the sake of the simplicity, the low molecular weight species such as pure carboxylic acid-containing oils will have far less influence on CAN solid formation and do not provide any additional challenge.

4.2.3.3 Impact of calcium naphthenate soap scale formation

Significant depressurization tends to occur in the upper reaches of producer wells and topside facilities, and therefore, pH change and naphthenate deprotonation for salt solid formation occur at those locations. It is rare for CAN scales to be encountered downhole, although some cases have been reported [117,119,120]. There are, however, question marks as to whether these are true CAN structures.

In the North Sea cases, naphthenate scale tended to form in surface units and then accumulate quickly and in high volume. Literature accounts describe scale deposits spread across the process, typically within separator internals (as plugging deposits), in valves, heaters, coolers, pumps, sample points and pipework, and interfering with normal production process operations requiring significant cost, time, and effort to resolve [28,83,101]. Images presented show CAN solids onsite and also the same precipitate after cooling and exposure to air, now in its hardened form. Cummine discussed operational issues due to CAN scaling on the Bleoholm FPSO from 2004 to 2005 and described the process to manually remove many tons of deposits from separator vessels and ship the material onshore for specialist disposal [113]. It is common for CAN solids to block-level transmission equipment in process vessels and impact safe operation and more often than not, upsetting oil and water quality due to loss of processing control, either through capacity (reduced residence time), poor operability (due to lack of level control), or blocked equipment (hydrocyclone reject lines and produced water skimmer vessels). Naphthenates can often be codeposited with other material such as sand, silt, and clays from reservoir fines, NORM scales, and sometimes with waxes and asphaltenes—these deposits can sometimes contain heavy metals, which gives them an intrinsic disposal challenge plus naphthenic acids themselves promote a known toxicity concern [101,130–132]. Produced water reinjection systems are at-risk of plugging and formation damage if solid particulates carry far enough through the system [102,103]. Exported oil can have a greater thermal fouling tendency across preheaters and exchanges and will have a greater risk for generating stable emulsions during refinery desalting operations [2,83].

4.2.4 High calcium in crude caused by oil-dispersible naphthenates

4.2.4.1 Overview and field examples

High calcium in crude developments occur in high density (14–24° API) dry export oils and present as well dispersed CAN salts [83]. In the *Continuum Model*, calcium in crude soaps represents a transitionary state where ARN acids are either absent or are present but remain dispersed throughout the crude oil rather than agglomerating and presenting as solid adherent scale. This is the least reported of the "upstream" naphthenate soaps [19,28,83,101], however, there are more accounts of "downstream" calcium in crude issues in crude oil desalting operations [31].

One such example occurred with the viscous (10 cP in the reservoir) Harding field (UK North Sea sector) export crude that had a TAN of 2.7 mg KOH/g, and was medium gravity (18–22° API). The Harding (then BP-operated) crude was used to make a high-quality refined needle grade coke (for making electrodes) and anode grade coke (used in the smelting industry). Anode grade coke feed crudes had strict metal contaminant content limitations, including a 40 mg/L limit for calcium, therefore, it was important that Harding crude contained as little entrained residual produced water as possible. Early Harding production was dogged by oil and water separation issues, even though inline heaters were employed to aid separation. Export water specification was achieved via a two-stage separation sequence combined with electrostatic coalescer use [133–135]. Feed crudes used in coke manufacturing were desalted prior to use. Harding crude desalting was frequently problematic as the operation failed to achieve the 40 mg/L calcium limit and consequently the coke quality was reduced. Investigation of Harding crude identified excess calcium was present that could not be explained by the residual produced water content alone. Investigation involving crude sample freeze-drying and subsequent FTIR and XRD identified the presence of carboxylate functionalities (assigned to naphthenate) and halite. The same techniques were then used to evaluate a remedial treatment program for use on Harding to address the high calcium in crude issue, and the resulting acid washing technique was adopted and implemented on the Harding platform to ensure the fluid pH was maintained below the critical CAN formation pH [133–135].

4.2.4.2 Mechanism of calcium in crude naphthenate soap and factors that influence stability

Data regarding the mechanistic aspects of high calcium in crude, and in particular, which carboxylic acid species are involved, has been challenging to unearth as the subject appears to have been poorly covered in the available literature. The scarcity can be understood (to an extent) when the Harding Field case detailed above is considered. If Harding crude had not been used as feed for specialized coke manufacture, then it is likely that the high calcium in crude issue might have simply been overlooked, or regarded as inconsequential, as it did not really impact Harding Field process operations too greatly. Piecing together the available evidence, it is reasonable to assume that the mechanism involves calcium complexing with crude oil carboxylate species in viscous high TAN crudes. The high calcium in crude presentation lies somewhere in the "middle ground" of the *Continuum Model*, between the extremes of ARN acid CAN solid scale and naphthenate soap emulsions. The calcium in crude development forms neither pure solid nor pure emulsion and comprises calcium, sodium, and carboxylate species.

The Brocart schematic [52] has been adapted to illustrate possible mechanistic explanation of calcium in crude (Fig. 4.16). A range of organic carboxy species is shown, from simple low molecular weight short-chain species up to the high molecular weight ARN acids. Calcium and sodium ions are complexed by carboxylate moieties. Three possible arrangements can form; high ARN content naphthenate deposits cross-linked, soap emulsions can present if the carboxylic acids are generally lower molecular weight types, or an intermediate state where a varied molecular weight range of calcium and sodium species are present in bulk crude and distributed between the crude and dispersed and entrained in water droplets and their interfaces with crude, which gives rise to calcium in crude.

4.2.4.3 Impact of calcium in crude salts

The calcium content in acidic crudes with a typical "dry" BS&W of $<0.5\%$ can range from 10 to 300 mg/L. High calcium crudes are known to cause issues in downstream refining and poisoning of expensive catalyst loads [32]. These salts have been known to take the form of viscous sludges and can enter the water plant thus reducing water quality and are not treatable via conventional means [101].

There are seemingly no true flow assurance challenges caused by the presence of these acid species and the presence of calcium in crude oil. The challenge is purely one of product quality specification and most, if not all, of the challenges actually occur at the refinery. Other unique scenarios, such as that described for

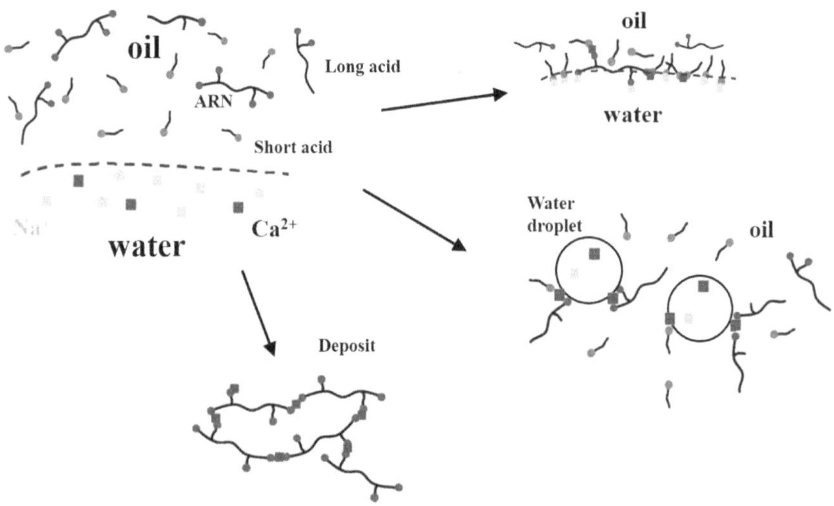

FIGURE 4.16

Basic mechanism of calcium in crude soap formation where medium to heavy molecular weight naphthenic acid species complex with calcium and remain oil-dispersible and/or interfacially active. Short-chain acids are depicted to coexist with the ARN acids although it is unclear as to their exact involvement in this process.

the Harding case study, could be present in examples around the world where equally unique processes may be required to provide adequate safeguards against the excessive accumulation of CAN salts in the exported crude oil.

4.2.5 Refinery challenges overview

The presence of acidic species in crude oil, especially carboxylic and naphthenic acids, causes a variety of concerns in downstream refining processes. Serious corrosion issues can be caused to refinery equipment, transportation piping systems, storage tanks, as well as the increased costs of building, maintaining, and operating these systems [29,136–140].

Under ambient conditions, naphthenic acids show little corrosivity, however, they do show increasing corrosivity to carbon steels from 180°C and worsening up to 395°C. Refinery distillation columns commonly operate from 245 to 395°C representing the worst possible thermal range for naphthenate corrosion.

The surface-active properties of carboxylic and naphthenic acids and their salts are well known to the refinery industry with respect to their potential to emulsify and also cause foam formation, and during desalting, the acids can leach cations (especially Ca^{2+} and Na^{+}). These acid crudes can also entrain metal species that can deleteriously impact catalyst efficiency [33]. The existence of high concentrations of naphthenic acids reduces crude oil quality and sales price, and can severely increase the toxicity of refinery wastewater effluent, even rendering it unsuitable for discharge due to toxicity limitations and incurring additional and significant disposal costs [33]. A summary of the issues has been provided in Fig. 4.17.

The processing, isolation, removal, and refining of acidic species from crude oil can occur in several ways with the most common and cost-effective being washing of acidic fractions using an aqueous solution of alkali or alkaline earth metals, followed by acidification with mineral acids to recover the insoluble naphthenic acid [33]. Other technique includes adsorption and solvent extraction [141–143], catalytic, or noncatalytic esterification [144,145]. Decarboxylation has also successfully been employed for the processing of high-acidity crude oils with both thermal and catalytic decarboxylation methodologies reported [146,147].

Acid removal techniques tend to have drawbacks; for instance, alkali washing is rarely 100% efficient as naphthenic acids exhibit poor water solubility, and emulsion formation with higher distillate cuts is another limitation associated with this removal method. Solvent extraction lacks environmental viability as it generates significant volumes of volatile organic solvent waste, and the adsorption technique is applicable to low-temperature lighter distillate fractions only. Decarboxylation requires higher temperatures (>250°C), which can lead to corrosion problems. Safe economic and environmentally acceptable methods for the processing of acid crudes, therefore, remain an active area of research [30,32,142].

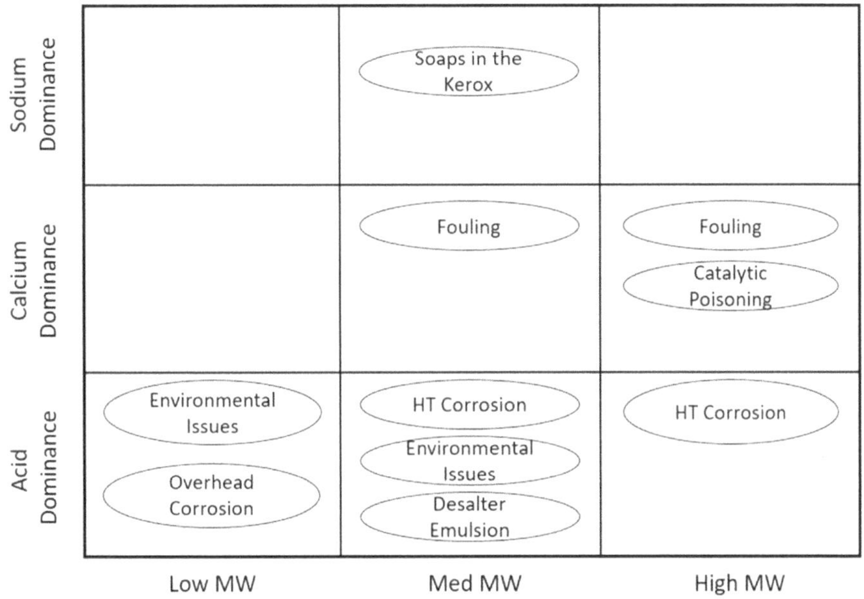

FIGURE 4.17

Summary of the refinery challenges brought about by the presence of acids and sodium and calcium salts of carboxylic and naphthenic acid soaps in refinery processes.

4.3 Chemical control methodologies and laboratory testing

4.3.1 Introduction

Carboxylate and naphthenate soap fouling is complicated by the number and distribution of acid species that could potentially contribute to the gross naphthenate scale or carboxylate emulsion development. The types and proportions of species, as well as the associated brine composition and system conditions all play roles in the final presentation of soaps. Mechanistic arguments and field case history examples detailed above provide support to the view that no two soap scenarios are the same, and that every system is governed by its own fine set of chemical equilibria. The complexity is unsurprising and is in turn reflected by the broad range of chemical solutions and mitigation strategies employed to treat individual oilfield soap scales, which usually required a combination of different technologies to be completely effective.

The impact of plant operation on naphthenate scale control cannot be overlooked and process plant parameter settings can play a major role in reducing the naphthenate risk. Considering the naphthenate formation mechanism then it is sensible and prudent to ensure where possible that large pressure drops are avoided, and that temperature increases are kept to a minimum. It is also important to minimize shear, reduce in-process fluid recycling, maintain level control, and control the slugging of wells. Plant optimization has been reported to achieve up to 90% reduction of the soap-related challenges, however, chemical technologies are nearly always required to assure complete control [13,129,148].

This section will focus on chemical methods for managing oilfield carboxylate and naphthenate soap control challenges and considers two distinct strategies of prevention (using inhibitors, dispersants, acids, nonacids, and demulsifiers) and that of remediation (of formed naphthenate deposits and slop oil). In addition, reviewed are the common test methods used to evaluate and select different control chemistries. The chemicals used have significant overlap for treatment of either soap emulsions or soap solids due to the commonality of the pH-driven mechanism for CAN scale and naphthenate emulsion formation. The interfacial activity of soap species allows the use of common interfacially active oilfield production chemicals such as emulsion breakers for their mitigation.

4.3.2 Preventive chemical strategies

4.3.2.1 Acids and acidic demulsifiers

The deployment of acid or acid-containing chemicals for control of soap challenges is the most widely used chemical mitigation technique employed by the industry from the earliest remedial applications to those used in the current day [78]. The saponification process is defined by pH increase; therefore, the addition of pH reducing chemistries will work if the pH is lowered below the naphthenate formation window value. The greater majority of literature case histories employ acids in one form or another to achieve soap suppression or remediation, and even in those applications where nonacids are deployed, they are introduced to augment and reduce the requirement of acid, rather than replacing it completely.

Early acid use for soap remediation tended to be direct acid dosing and only later did it become routine to codeploy acid and demulsifier, particularly for cases of soap-stabilized emulsions [73,77,129]. Acid is injected continuously or via batch application, and the quantity deployed is dictated by the buffering capacity of the system. Needless to say, the acid remediation operation requires to be very closely aligned with the facility corrosion management strategy. Continuous injection is "the norm," however; "one-shot" emulsion pad slop oil batch treatments are common in clean-up recovery operations. Besides pH adjustment, it has also been observed that continuous acid deployment can lead to dissolving and removal of historical solid naphthenate developments [129].

Acid acts by ensuring protonation of the carboxylate species in crude, and in doing so suppresses calcium and sodium complexation to form soaps. Operationally, the critical pH is $\sim 6.3 \pm 0.2$, above which the soap formation challenges begin. Given the range of different acid species, all with slightly different pK_a values, it stands to reason that the critical pH is not a definitive pH point but a range. There is field pH variance; however, the majority of indigenous organic acids pK_a values occur at or near the critical pH value. Many acids/specialist blends have been employed for naphthenate control over the years; however, the choice of acid should have little bearing upon the performance, if one simply considers protonation power [75]. Some patent literature suggests otherwise however; for example, claims exist regarding glyoxylic acid and formulations thereof preferentially working "better" for CAN control [149]. Evidence has also been presented that the acid type, while not discriminated via their carboxy protonation potential, is, however, differentiated by the emulsions that they influence, especially if these acids are "inhibited" via incorporation of interfacially active corrosion inhibitor-based surfactant molecules [113,114]. There are some generalizations that can be made for the different acid types and their specificity but again it is stressed that each case history remains unique.

Acetic acid is the most commonly used soap control agent and is typically deployed upstream of the primary separation unit [51,82,86,107,111,113,114,129]. Dosing is targeted at reducing produced water to approximately pH 6 as underdosing risks borderline naphthenate suppression and increased risk of side effects [83]. Acetic acid is preferred to mineral acid primarily due to its lower corrosivity potential, but it is not without its corrosivity challenges. Bretherton et al. reported injection point corrosion during its dosing for naphthenate control even though exotic metallurgy was used throughout the chemical injection system [150]. All acids used for naphthenate treatment can be configured with acid corrosion inhibitors to address their corrosion potentials.

Phosphoric acid is not as widely used as acetic acid for soap treatment but was deployed in the Kutei Basin case study as it was determined to perform more effectively than acetic acid products for carboxylate soap emulsion suppression [71,74,77]. It is more common to see phosphoric or phosphorous acid blended with emulsion breaker base chemistries for these types of emulsions. One point of note, however, is that special attention should be given to produced water compatibility as precipitation of phosphate scales (hydroxyapatite, vivianite, polyphosphates) can occur and has necessitated the use of additional scale inhibitors [43,77].

Other organic acids, including sulfonic acid, citric acid, glycolic acid, sulfamic acid, formic acid, hydroxy acetic acid, myristic acid, and lauric acid, have been deployed or considered for naphthenate scale control duty [73,83,129]. Subramaniyam reported that glyoxylic acid was preferred to glycolic acid and malic acid for removal of calcium in crude on Doba crude blend [149].

C_8 to C_{12} monocarboxylic acids have been shown to provide exemplary naphthenate inhibition capability compared to leading acids such as dodecyl benzene sulfonic acid (DDBSA) and phosphate ester-based inhibitors. It is believed that their performance is due to them expressing more surfactant-like properties in addition to protonation power [151]. The use of DDBSA is well reported [5] and this acid displays its own surfactant, and therefore, behaves in a unique fashion compared to other acidic species, especially if amine neutralized salts of DDBSA are used [77,152]. Formic acid has (anecdotally) been suggested as performing more effectively for naphthenate soaps compared to acetic acid. Literature review identified no evidence or reports to support this statement, and in fact, some cases state that for troublesome emulsion treatment, it is inferior to acetic acid, besides being generally more expensive and challenging to handle [73,129].

Mineral acids, and most commonly hydrochloric acid, have been used to address emulsions and naphthenate scaling [28,78,153]. The aggressive nature of these acids to materials of construction and product handling are the main drawbacks to their use. Naphthenate treatment applications employed mineral acid dosing [44,71], both in continuous and batch application modes, however, the bulk of more recent applications have switched to more benign organic acid formulation applications [115–120]. There are exceptions, and in circumstances where the target system has a high buffering capacity then mineral acid use may become unavoidable from a performance and economic perspective [78]. The corrosivity of HCl can be addressed via incorporation of an acid corrosion inhibitor to allow its use in naphthenate treatments; however, Turner et al. reported that CAN solubility in HCl is poor [129].

Acids are effective for carboxylate and naphthenate soap treatment; however, there are several drawbacks in their use. Corrosivity concerns have been discussed previously with respect to integrity of the injection system and processing facility materials of construction. A number of authors have highlighted the fact that acetic acid carries an additional integrity concern due to vaporization, which leads to top-of-line corrosion and the acid can also condense in other areas in the gas plant [129,130,154]. Health and safety concerns are at the forefront concerning handling, storage, and deployment of large quantities of acidic fluids with injection rates typically at 0.05%–0.20% of the produced water rate. The case studies were presented earlier for the Heidrun and West Brae naphthenate scale control scenarios in the North Sea with 38,000 BWPD and 30,000 BOPD water rates, respectively. It is, therefore, not uncommon for the acid injection rate to be in excess of 10,000 L/day [28,82,107,113]. A further issue caused by organic acid injection is that they could represent a nutrient source for unwanted bacteria and, for instance, if reinjected into the reservoir could cause souring [2].

Organic acid chelating agents such as ethylenediaminetetraacetic acid have been used for control of calcium in crude content in refinery applications [155,156]. The stoichiometric chelation of calcium ion in the refinery setting is achievable as the BS&W rates are much lower and crude washing chemical aids are used. There are no reported cases in the literature of this having been done in upstream higher water production environments and this is believed primarily due to the ready availability of cheaper naphthenate remediation options.

Acidic oils are notoriously challenging to dehydrate due to the high levels of surface acid components and demulsifier chemistries are routinely deployed to aid their separation. It would seem logical and sensible, therefore, in soap-prone production scenarios to combine both acid and demulsifier to address both emulsion separation and naphthenate scale formation. Most production processes require a delicate balance of production chemicals to ensure optimum dehydration and operation, which can easily upset to the detriment of oil and water quality [73,78]. Introduction of an acid treatment program for soap suppression will undoubtedly impact the incumbent process demulsifier program. Creation of a combined acid demulsifier product may not always be possible for high TAN crude processing, as process demulsifiers tend to be complex multicomponent formulations containing 20%–60% active content in a suitable solvent. Addition of acid into the formulation could be as high as 50% of the formulation, replacing the solvent and/or essential active demulsifier bases [73]. Junior et al. reported on a process scenario where a new demulsifier was required each time a new naphthenate mitigating chemistry or product was introduced [116].

The use of amine neutralized DDBSA is well-known for effective treatment of carboxylate soap emulsions and CAN scale control. The surfactancy and dispersancy characteristics of DDBSA and its salts present a combination of demulsification and dispersancy in one raw material and explains why its ubiquitous inclusion in many formulations. It should be noted, however, that the "acidity" contribution to soap control is negligible, certainly proven by the fact that DDBSA salts featuring simple amines such as ethanolamine all the way up to highly complex alkoxylated species (alkoxylated rosin amines) are described in the literature [157]. An acid demulsifier would traditionally comprise 20–30 wt.% salted DDBSA with demulsifier bases in an appropriate solvent system. One such variant was deployed on the Heidrun TLP to address the CAN scale deposition challenge and eventually replace the existing acid injection treatment [28]. The changeover to the naphthenate inhibitor was performed systematically and progressively, and keenly considered the impact on existing production chemical performance and BS&W and overboard oil in water. Ultimately, several optimization programs were undertaken before a final arrangement and product formula were identified [105].

The twofold mode of action for acidic demulsifiers is more complex than that of acids alone and involves interfacial crowding of surfactant molecules at the oil–water interface as well as pH suppression of acid functionalities [158]. The injected surfactant molecules surge toward the oil–water interface and physically prevent or hinder (inhibit) carboxylic and/or naphthenic acids from doing the same, resulting in a lowering of soap salt development. The acid components are maintained in their protonated states, encouraging them to remain in the crude oil (due to their reduced interfacial activity) [158]. This process has been represented schematically in Fig. 4.18. It is important that the demulsifier is configured to preferentially locate at the interface to the detriment of the indigenous acids and acid-soap species; hence the term "interfacial crowding."

It is beneficial that the acid demulsifier and process demulsifier are compatible and quasisynergistic and can complement each other's actions for best separation and naphthenic soap suppression results. It is also important that the crowding surfactant is not overly surface-active as this may create further emulsion challenges downstream. Ubbels and Turner reported that it is possible to

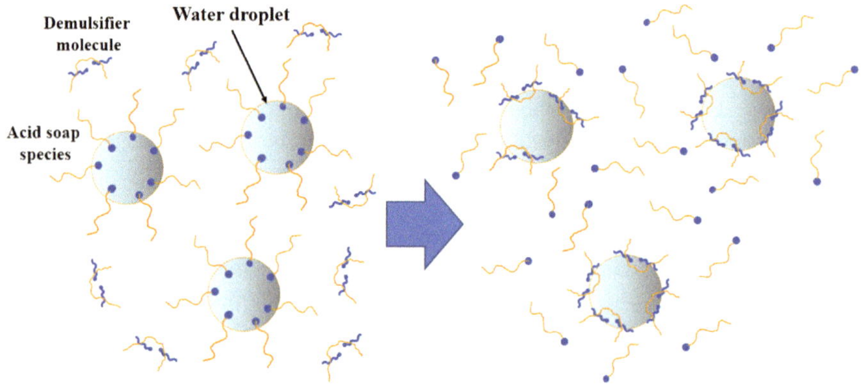

FIGURE 4.18

Schematic mechanism of interfacial crowding where injected chemistries preferentially occupy the oil–water interface effectively preventing problematic indigenous acid species from contacting ions in the produced water.

engineer the hydrophilic head group and oil-loving tail group to stop the reactions of the cations in the water with the naphthenic acids in the oil [81].

Observations regarding modification of field deposits recovered both pre and postchemical treatment provides evidence supporting the interfacial crowding hypothesis [111]. For CAN scale in "Field U", the deployment of a naphthenate inhibitor resulted in a deposit that was softer and depleted in calcium ions while still containing abundant high molecular weight acids. Crowding inhibition appears to have shielded the calcium ions in the water preventing their association and complexation with the high molecular weight ARN acids [111].

4.3.2.2 Nonacidic chemicals and low dose naphthenate inhibitors

Injection of large volumes of acid for pH modifications presents challenges for remote offshore scenarios where logistics and storage onsite can be limiting. An ideal solution, therefore, would be a nonacid "threshold-concentration" chemical solution, sometimes called low-dose naphthenate inhibitors. The term "low dose" inhibitor can be misleading as there are instances where products require to be injected at 100–200 ppm to achieve the same performance as acidification and sometimes even higher with up to 500 ppm being reported for several nonacid demulsifiers and naphthenate inhibitors [77].

There are examples of successful applications where prevention of CAN deposition or specific soap emulsion challenges was achieved at low dose (<10 ppm) [13]. It is typical to use nonacid products in combination with acids, or to lower the use of acid, and in some cases, it has been possible to reduce acid injection by up to 80% or even in some cases, complete cessation of acid injection has been possible [28,74,81,82,129].

In order to reduce acid consumption, the injection of dispersant chemicals has been performed to carry suspected naphthenate solids through a process by presumably preventing flocculation and/or agglomeration [129]. A common approach is to use DDBSA salted with complex amine chemistry. This type of product would likely provide a dispersant mode of action for growing agglomerations of tetraprotic CAN nuclei. It is also reported that nonacid formulations are specific to a given field or situation and may not be transferable to a different field, much in the same way demulsifiers very often are not transferable [101]. Dispersants have been described for naphthenate soaps in the water and oil phase and cause further problems downstream and just procrastinate the issue. Dispersants were not able to remove existing deposits of soap scale at 20–30 ppm dose rates [116]. Polymeric examples of soap control agents also exist with a 20,000 AMU copolymer of acrylic acid and 2-acryloylamido-2-methylpropanesulfonic acid (AMPS) described in the literature that presumable works via a dispersancy mechanism [159].

Amine chemistry has a general affinity for soap control and the patent literature teaches the use of primary amine-containing formulations for control of both emulsions and soap scales [152]. It is complex to extract from this work the exact chemistries because very broad claims are made for nearly all primary amine chemistry ranging from very basic structures like triethylenetetramine, to very complex structures such as polyisobutenyl succinimide. Blends of amines in organic solvents are claimed to control soaps effectively and supporting experimental data are provided [152]. The use of alkoxylated amines in addition to acid chemistries in alcohol solvents is described, further supporting the use of salted sulfonic acid formulation [157]. Latterly, the use of more complex rosin amine chemistry has been described in formulations on its own and in combination with demulsifier bases [160]. Formulations containing 6–12 wt.% of the rosin amine are described with any number of demulsifier bases also added in an alcohol or glycol solvent matrix. The inclusion of both organic and inorganic acid in the formulation is described and here again, the formulation examples look very much like acidic demulsifier formulations with added synergistic components, in this case, rosin amine ethoxylate [160].

The use of glyoxal has been claimed as effective at removing calcium from crude oil and a large body of test data are provided in the literature that shows improved performance on Doba crude over that of several commodity acids and other aldehyde chemistries [161]. The same researcher also claimed the use of various methyl, ethyl, propyl, isopropyl, mono, and/or diesters of maleic and fumaric acids and maleic anhydride for specific targeting of naphthenate calcium in crude occurrences [162].

Phosphate ester and its alkoxylated derivatives are low-dose naphthenate inhibitors used widely in the industry for mitigation of both soap emulsions and CAN scales [148,163]. Their use in combination with sulfonates (e.g., DDBSA) and ethoxylated fatty acids in combination with polymeric surfactants (polyglycol esters, polyimines, and epon resins) is further described along with their single-use or coinjection with acetic acid [148]. The phosphorous-containing surfactants employ interfacial crowding with the phosphate ester acting as a "booster" compound for an otherwise acid demulsifier formulation. Hellsten et al. describes the use of a more complex phosphorous-containing alkoxylate structures for the control of calcium carboxylate deposits [164]. Debord et al. and Venter et al. reported on nonacid CAN inhibitors comprising quaternary phosphonium compounds in combination with quaternary ammonium compounds and mixtures thereof that also optionally contain linear polymeric compounds with 20–50 repeated carboxylic and/or acrylic acid units [126,165]. The final phosphorous-containing polymers claimed are polymers of acrylic, carboxylic, or sulfonic acid end-capped with vinyl phosphonic acid and vinyl diphosphonic acid and likely also represent a dispersant type chemistry as they seem to be higher than normal molecular weight threshold scale inhibitor structures [166]. Multiple functionalities can be formulated into products to create products tailored for given systems. Poggeso et al. discussed naphthenate challenges in the West African deepwater fields Girassol and Dalia, and identified a product containing demulsifier functionality (via a blend of ethoxylated resins), naphthenate dispersancy (using an organic sulfonate), and asphaltene dispersancy (via amide and organic sulfonate chemistries) [108]. The same study also described the laboratory development of a further product incorporating added scale and corrosion inhibition properties [108]. Multifunctional chemical formulas are particularly relevant to deepwater chemical applications where injection of chemicals via umbilical often presents challenges, and costs per chemical line are excessive. The situation is further complicated when new and unforeseen production challenges emerge, requiring additional chemical delivered subsea for deployment. Very often integration–blending of the new production chemical into the existing chemical mix can be the only option as "spare capacity" is typically not engineered into the umbilical design [167–170].

4.3.3 Remediation and remedial chemical strategies

Oilfield soap removal refers to naphthenate scale deposition challenges. Methodology exists to remove emulsion pad build-up caused by low molecular weight carboxylate acid species in the oil, via batch treatment of acid or application of acid demulsifier.

Chemical CAN scale removal has not received much attention in the literature and remediation is usually performed using physical removal methods. The huge volume of scale that can be deposited is testament to this and the case histories that have been discussed detail physical remediation and removal methods used for CAN scales [25,104,107]. The preference is driven by many factors including rapid implementation for topside facility deposit removal; there are minimal access restrictions, and the physical removal process, although often unpleasant and time-consuming, is simple and almost guaranteed to succeed. The same cannot be claimed for chemical cleaning at surface, where additional challenges may be present.

There are some cases reported in the literature where continuous injection of acid (and acid demulsifiers) as well as expressing an inhibitory effect, can also serve to remove existing historical deposit build-up over time [35,107,129]. The formation of carboxylate and naphthenate soaps from acids is a pH-driven equilibrium reaction and is fully reversible, and therefore, acid application can impact the soap forming equilibrium in the reverse direction leading to carboxylate protonation and calcium–sodium ion ejection from its carboxylate complex. Protonation of existing soap deposits causes breakdown of the scale and drives toward solubilization of acid species into the oil. Berry et al. reported a glacial acetic acid-based naphthenate dissolver in an "environmentally friendly" solvent such as xylene or other based upon methyl ester and/or ethyl lactate [171]. It is strongly suspected that this dissolver formulation would also function equally well as a continuous injection inhibitor for soap control.

Field case histories of batch deployment of remedial chemical dissolvers do exist for deposition of naphthenate containing (although not dominant) organic scales downhole. The scales were remediated using a combination of dissolver technologies in conjunction with downhole tools [119,120,123]. In these cases, a customized chemical treatment was developed that is described as having multiple functionalities to not only dissolve the scale but also aid the separation of oil and water. Unfortunately, no detail is provided on the dissolver composition in these papers except that it was a microemulsion-based formulation consisting of solvents, surfactants, and inhibited acid. Yet again, this sounds not too dissimilar to an acid demulsifier. It is unlikely that the product described is specific to naphthenate scales as the situation described suggests it is questionable whether the organic deposits are wholly or even partly composed of ARN-containing CAN scale as no detailed compositional data is provided.

4.3.4 Laboratory and field testing

A variety of techniques has been introduced and discussed previously in this chapter for identification and elucidation of naphthenic and carboxylic acid soaps. In this section, CAN scale inhibition efficiency testing will take center stage. The two most widely used CAN soap scale inhibitor efficiency tests are static testing and dynamic testing. Static jar test involves contacting a brine phase, containing calcium ions, with a hydrocarbon phase, containing suitable extracted (or synthetic) tetraprotic ARN acids. When isolated from one another the fluids are stable; but when combined, will potentially form CAN soap solids. These solids can be isolated and gravimetrically compared to blank and control tests to determine the impact of naphthenate inhibitor efficiency [58,82,88,126,127,172].

The use of two different types of bespoke dynamic test rig is described in detail in the literature for both emulsions and solid deposits [114,118,153,173,174]. The two tests are similar in that they utilize a high-pressure cell to visualize emulsion formation thus simulating a production separator and oil and brine can be introduced at predetermined ratios both with and without gas. The amount of scale forming in the "separator" can be determined via differential pressure increase across a filter. The methods are claimed to closely mimic field operations and can be verified via lab to field correlations. The methods have been extensively commissioned by the industry and can be used to select and rank chemicals for control of soap-related issues.

A very simple field identification method has been adopted widely in the industry, the centrifuge test, and the filtration test. In the centrifuge test, a fresh crude or emulsion sample is put into a separatory funnel and the water is drained until the emulsion pad is encountered. The emulsion pad is then isolated and centrifuged, the "naphthenate pad" can be discerned at the interface of the centrifuge tube and the size of the pad can be related back as a percentage of the whole sample. The same sample is then filtered and the naphthenate pad isolated and transferred to a suitable vessel when it can be washed with hot 50% glacial acetic acid and solvent. CAN will disappears within minutes and typically other common deposits will remain insoluble. Different control chemicals are added to the test and compared visually to the blank and control tests yielding a percentage inhibition comparison.

4.4 Concluding remarks and remaining challenges

In summary, the term "naphthenate" is the general and rather loosely used term by the oil and gas industry to describe the operational challenges caused by the saponification of various carboxylic and naphthenic acid species present in some crude oils.

The resultant soaps (carboxylates and/or naphthenates) are highly interfacially active and regularly contribute to the formation of persistent emulsions and associated sludges. When specific high molecular weight tetraprotic acids (the so-called ARN acids) are present and encounter calcium ions, a persistent solid scale can form in notoriously high volumes. High calcium in crude is less well covered in the literature but would appear to be dominantly caused by the presence of higher molecular weight naphthenic acid species in the crude that are not tetraprotic ARN acids.

The formation of soaps is entirely pH-driven and injection of acids and acidic-based surfactants (demulsifiers) are successfully deployed around the world to prevent and/or reverse soap formation. A thorough review of the literature is provided detailing the formation mechanisms along with a summary of the state of the art in chemical control. A model, termed the *Continuum Model*, is proposed to describe the various operational manifestations and the dominant influencing factors to their formation.

After an intense 15 years of industry joint projects and research and development to understand and tackle the "naphthenate" challenge, the activity dropped off substantially and rapidly. The operational understanding of the problems encountered as well as how to deal with it was achieved and was likely "good enough" to not continue to fund further intense research. Consequently, there is still plenty of space for innovation and research to build a deeper understanding of the mechanisms at play. While seemingly all the influences and levers on the solid CAN scale deposition are directionally understood, a true and quantified structure-activity understanding still seems to be elusive and out of our clutches. The models, as described in Figs. 4.11 and 4.14, show that many generalizations are still present and while this is likely to be operationally robust, it does not truly explain why some crudes form scales, and some do not. This speaks to the petering out of literature works from 2010 to 2020 on this topic and the cessation of industry-supervised academic-driven studies.

Due to the complexity of the acid speciation in crude, with many thousands of discrete molecules existing, the application of big data analytics to the topic of why some crudes deposit and some crude do not could be an interesting area of discovery. The likelihood of this to occur out of industry interest is unfortunately low. Building large datasets of crude oil acid speciation and aligning this to the complementary operational and laboratory data could yield fresh insights and increase our understanding of how the different acids interplay with one another. We know, for instance, that ARN acids absolutely need to be present to form CAN scales, but we cannot explain why not all crudes that contain ARN acids will necessarily form these scales. Molecular modeling could also be usefully used to further complement a work of this nature; this would be, however, a laborious task that would likely take year of painstaking experimental work and field data gathering. Perhaps as a first step, what could be done is to gather and interpret all the data that is fragmented in the literature (and there is a lot of it when one considers collectively the few hundred articles available on this topic). The author remains convinced that putting this together in one place in a smart way would yield fresh insights. Several companies have clearly done this; however, these appear to be limited to their own oils (for example Ref. [52]) and never collectively as an industry. Furthermore, there is a wealth of data in other strands of research such as in the fields of oil sands and EOR.

For now though the industry has tactically and successfully managed to find several operationally reasonable solutions that have worked well for 20 years and likely will persevere for the next 20 years.

Nomenclature

AMU	atomic mass unit
API	American Petroleum Institute
ASTM	American Society for Testing and Materials
BOPD	barrels of oil per day
BS&W	base sediment and water
BWPD	barrels of water per day
CAN	calcium naphthenate
DBE	double bond equivalent

EDX	energy dispersive x-ray analysis
EOR	enhanced oil recovery
FAB-Ms	Fast atom bombardment-mass spectrometry
FT-ICR-Ms	Fourier transform-ion cyclotron resonance-mass spectrometry
FTIR	Fourier transform infrared spectroscopy
GC	gas chromatography
GCMS	gas chromatography-mass spectrometry
HPLC	high-performance liquid chromatography
IER	ion exchange resin
IFT	interfacial tension
JIP	joint industry project
LCMS	liquid chromatography-mass spectrometry
m/z	mass-to-charge ratio
mN/m	millinewton per meter
Ms	mass spectrometry
NORM	naturally occurring radiative material
NMR	nuclear magnetic resonance
OIW	oil in water
OSPW	oil sand process-affected water
R&D	research and development
SARA	saturates, aromatics, resins, and asphaltene
SEM	scanning electron microscopy
TAN	total acid number
TDS	total dissolved solids
TLP	tension leg platform
VFA	volatile fatty acid
XRD	X-ray diffraction

Acknowledgments

I would like to express my deepest gratitude to Alex for his editing skills; Sajjaat for helping with references; Emma for discussion on North Sea case studies and Clariant involvement and experience; Ken for multiple discussions about mechanisms, knowns, and unknowns; Anton and Grahame for chemistry matters; and finally Susan without whom none of this would likely be possible.

References

[1] D.M. Grewer, R.F. Young, R.M. Whittal, et al., Naphthenic acids and other acid-extractables in water samples from Alberta: what is being measured? Science of the Total Environment 408 (2010) 5997–6010.

[2] C. Smith, Chairman's introduction and aim of the workshop, in: Proceedings of the SPE ATW on Managing Naphthenates and Soap Emulsions, Pau, France, March 10–13, 2008.

[3] M.I. Head, M.D. Jones, R.S. Larter, Biological activity in the deep subsurface and the origin of heavy oil, Nature 426 (2003) 344–352.

[4] K.H. Altgelt, M.M. Boduszynski, Composition and Analysis of Heavy Petroleum Fractions, Marcel Dekker, New York, 1994.

[5] M.A. Kelland, Production Chemicals for the Oil and Gas Industry, second ed., CRC Press, Baton Rouge, LA, 2014.

[6] C. Hell, E. Medinger, Über das vorkommen und die zusammensetzung von säuren im rohpetroleum, Berichte der Deutschen Chemischen Gesellschaft 7 (1874) 1216–1223.

[7] K.N. Qian, W.K. Robbins, C.A. Hughes, et al., Resolution and identification of elemental compositions for more than 3000 crude acids in heavy petroleum by negative-ion microelectrospray high-field Fourier transform ion cyclotron resonance mass spectrometry, Energy & Fuels 15 (2001) 1505–1511.

[8] W.K. Seifert, Carboxylic acids in petroleum and sediments, in: L. Zechmeister, W. Herz, H. Grisebach, et al. (Eds.), Progress in the Chemistry of Organic Natural Products, vol. 32, Springer, Vienna, 1975.

[9] J.V. Headley, K.M. Peru, M.P. Barrow, et al., Characterization of naphthenic acids from Athabasca oil sands using electrospray ionization: the significant influence of solvents, Analytical Chemistry 34 (1999) 125–131.

[10] P.V. Hemmingsen, S. Kim, H.E. Pettersen, et al., Structural characterization and interfacial behavior of acidic compounds extracted from a North Sea oil, Energy & Fuels 20 (2006) 1980–1987.

[11] ASTM, Standard Test Method for Acid Number of Petroleum Products by Potentiometric Titration, ASTM International, West Conshohocken, 2018.

[12] G. Sartori, C.W. Savage, B.H. Ballinger, et al., Process for extraction of naphthenic acids from crudes, Patent US6281328, 2001.

[13] A.G. Shepherd, M. van Dijk, W. Koot, et al., Flow assurance in oil systems: on the role and impact of naphthenic acids, SPE-157295-MS, in: SPE International Production and Operations Conference and Exhibition, Doha, Qatar, May 14–16, 2012.

[14] B. Brocart, C. Hurtevent, Flow assurance issues and control with naphthenic oils, Journal of Dispersion Science 29 (2008) 1496–1504.

[15] W.K. Robbins, Challenges in the characterization of naphthenic acids in petroleum, in: 215th National Meeting, American Chemical Society, Dallas, TX, 29 March–03 April, 1998.

[16] J.A. Brient, P.J. Wessner, M.N. Doyle, Naphthenic acids, in: J.I. Kroschwitz (Ed.), Encyclopedia of Chemical Technology, fourth ed., John Wiley & Sons, New York, 1995, pp. 1017–1029.

[17] T.-P. Fan, Characterization of naphthenic acids in petroleum by fast atom bombardment mass spectrometry, Energy & Fuels 5 (1991) 371–375.

[18] I. Dzidic, A.C. Somerville, J.C. Raia, Determination of naphthenic acids in California crudes and refinery wastewaters by fluoride-ion chemical ionization mass-spectrometry, Analytical Chemistry 60 (1988) 1318–1323.

[19] K.S. Sorbie, A. Shepherd, P.C. Smith, et al., Naphthenate formation in oil production: in: general theories and field observations, in: Proceedings of the RSC Chemistry in the Oil Industry IX, Manchester, UK, 31 October –2 November, 2005.

[20] F.H. Behar, P. Albrecht, Correlations between carboxylic acids and hydrocarbons in several crude oils, Alteration by biodegradation, Organic Geochemistry 6 (1984) 597–604.

[21] R. Jaffe, M.T. Gallardo, Application of carboxylic acid biomarkers as indicators of biodegradation and migration of crude oils from the Maracaibo Basin, Western Venezuela, Organic Geochemistry 20 (1993) 973–984.

[22] S.D. Olsen, The relationship between biodegradation, total acid number (TAN) and metals in oils, American Chemical Society, Division of Petroleum Chemistry, Preprints 3 (1998) 142–145.

[23] S. Kim, L.A. Stanford, R.P. Rodgers, et al., Microbial alteration of the acidic and neutral polar NSO compounds revealed by Fourier transform ion cyclotron resonance mass spectrometry, Organic Geochemistry 36 (2005) 1117–1134.

[24] A.S. Mackenzie, G.A. Wolff, J.R. Maxwell, Fatty acids in biodegraded petroleums: Possible origins and significance, in: M. Bjorøy (Ed.), Advances in Organic Geochemistry, John Wiley, Chichester, 1983, pp. 637–649.

[25] W. Meredith, S.-J. Kelland, D.M. Jones, Influence of biodegradation on crude oil acidity and carboxylic acid composition, Organic Geochemistry 31 (2000) 1059–1073.

[26] B.H. Moerkeek, Discoloration of kerosine, in: World Petroleum Congress, WPC-239, London, UK, July 18–24, 1933.

[27] W.D. Bascom, S. Kaufman, C.R. Singleterry, Colloid aspects of the performance of oil-soluble soaps as lubricant additives, in: World Petroleum Congress, WPC-8518, New York, 30 May–05 June, 1959.

[28] J.E. Vinstad, A.S. Bye, K.V. Grande, et al., Fighting naphthenate deposition at the Heidrun field, SPE-80375-MS, in: SPE International Symposium on Oilfield Scale, Aberdeen, UK, January 29–30, 2003.

[29] E. Babaian-Kibala, M.J. Nugent, Naphthenic acid corrosion literature survey, NACE-99378, in: Corrosion Conference & Expo, San Antonio, TX, April 25–30, 1999.

[30] M.M. Ramirez-Corredores, Acidity in crude oils: naphthenic acids and naphthenates, in: M.M. Ramirez-Corredores (Ed.), The Science and Technology of Unconventional Oils, Elsevier, New York, 2017, pp. 295–385.

[31] S.N. Shah, M. Ibrahim, A. Mutalib, Extraction of naphthenic acid from highly acidic oil using hydroxide-based ionic liquids, Energy & Fuels 29 (2015) 106–111.

[32] C. Wu, A.D. Visscher, I.D. Gates, On naphthenic acids removal from crude oil and oil sands process-affected water, Fuel 253 (2019) 1229–1246.

[33] B. Albinet, Managing naphthenic acid and naphthenates in refineries, in: Proceedings of the SPE ATW on Managing Naphthenates and Soap Emulsions, Pau, France, March 10–13, 2008.

[34] J.P.T. Staff, Techbits: naphthenate deposits, emulsions highlighted in technology workshop, Journal of Petroleum Technology 60 (2008) 30–31.

[35] C. Smith, Integrated approach to designing systems for handling severe naphthenates/soaps problems, Paper 9, in: TEKNA Oilfield Chemical Symposium, Geilo, Norway, March 23–26, 2014.

[36] A.G. Shepherd, V. van Mispelaar, J. Nowlin, et al., Analysis of naphthenic acids and derivatization agents using two-dimensional gas chromatography and mass spectrometry: Impact on flow assurance predictions, Energy & Fuels 24 (2010) 2300–2311.

[37] Ø. Brandal, A.-M.D. Hanneseth, P.V. Hemmingsen, et al., Isolation and characterization of naphthenic acids from a metal naphthenate deposit: Molecular properties at oil-water and air-water interfaces, Journal of Dispersion Science and Technology 27 (2006) 295–305.

[38] A.G. Shepherd, K.S. Sorbie, G.B. Thomson, et al., Sensitivity study of naphthenic acids from flow assurance deposits characterized by low-resolution mass spectrometry, Energy & Fuels 24 (2010) 4387–4395.
[39] W.K. Seifert, W.G. Howells, Interfacially active acids in a California crude oil: Isolation of carboxylic acids and phenols, Analytical Chemistry 41 (1969) 554–562.
[40] N.A. Tomczyk, R.E. Winans, J.H. Shinn, et al., On the nature and origin of acidic species in petroleum: Detailed acid type distribution in a California crude oil, Energy & Fuels 15 (2001) 1498–1504.
[41] V.J. Cambridge, J.M. Wolcott, W.D. Constant, An investigation of the factors influencing transient interfacial-tension behavior in crude oil alkaline water-system, Chemical Engineering Communications 84 (1989) 97–111.
[42] D.M. Jones, J.S. Watson, W. Meredith, et al., Determination of naphthenic acids in crude oils using nonaqueous ion exchange solid-phase extraction, Analytical Chemistry 73 (2001) 703–707.
[43] T.D. Baugh, N.O. Wolf, K.V. Grande, et al., Characterization of a calcium naphthenate deposit - The ARN acid discovery, American Chemical Society, Division of Petroleum Chemistry, Preprints 49 (2004) 274–276.
[44] T.D. Baugh, K.V. Grande, H. Mediaas, et al., The Discovery of high-molecular-weight acids (ARN Acid) responsible for calcium naphthenate deposits, SPE-93011-MS, in: SPE International Symposium on Oilfield Scale, Aberdeen, UK, May 11–12, 2005.
[45] H. Mediaas, K.V. Grande, B.M. Hustad, et al., The acid-IER method – A method for selective isolation of carboxylic acids from crude oils and other organic solvents, SPE-80404-MS, in: SPE International Symposium on Oilfield Scale, Aberdeen, UK, January 29–30, 2003.
[46] B.F. Lutnaes, Ø. Brandal, J. Sjöblom, et al., Archaeal C_{80} isoprenoid tetraacids responsible for naphthenate deposition in crude oil processing, Organic and Biomolecular Chemistry 4 (2006) 616–620.
[47] J.J. Brocks, R.E. Summons, Sedimentary hydrocarbons, biomarkers for early life, in: H.D. Holland, K.K. Turekian (Eds.), Treatise on Geochemistry, Elsevier, Oxford, 2004, pp. 61–103.
[48] E.F. DeLong, L.L. King, R. Massana, et al., Dibiphytanyl ether lipids in nonthermophilic crenarchaeotes, Applied and Environmental Microbiology 64 (1998) 1133–1138.
[49] J.S. Sinninghe Damsté, S. Schouten, E.C. Hopmans, et al., Crenarchaeol the characteristic core glycerol dibiphytanyl glycerol tetraether membrane lipid of cosmopolitan pelagic crenarchaeota, Journal of Lipid Research 43 (2002) 1641–1651.
[50] M. De Rosa, A. Gambacorta, The lipids of archaebacteria, Progress in Lipid Research 27 (1998) 153–175.
[51] B. Brocart, C. Hurtevent, B. Escoffier, Analytical detection of ARN-type naphthenic acids in crude, in: Proceedings of the 6th International Conference on Petroleum Phase Behavior and Fouling, Amsterdam, The Netherlands, June 19–23, 2005.
[52] B. Brocart, M. Bourrel, C. Hurtevent, et al., ARN-type naphthenic acids in crudes: analytical detection and physical properties, Journal of Dispersion Science and Technology 28 (2007) 331–337.
[53] B.E. Smith, P.A. Sutton, C.A. Lewis, et al., Analysis of 'ARN' naphthenic acids by high temperature gas chromatography and high performance liquid chromatography, Journal of Separation Science 30 (2007) 375–380.

[54] B.F. Lutnaes, J. Krane, B.E. Smith, et al., Structure elucidation of C80, C81 and C82 isoprenoid tetraacids responsible for naphthenate deposition in crude oil production, Organic and Biomolecular Chemistry 5 (2007) 1873–1877.

[55] S. Simon, E. Nordgård, P. Bruheim, et al., Determination of C80 tetraacid content in calcium naphthenate deposits, Journal of Chromatography A 1200 (2008) 136–143.

[56] M.M. Mapolelo, L.A. Stanford, R.P. Rodgers, et al., Chemical speciation of calcium and sodium naphthenate deposits by electrospray ionization FT-ICR mass spectrometry, Energy & Fuels 23 (2009) 349–355.

[57] M.A. Mohammed, K.S. Sorbie, Naphthenic acid extraction and characterization from naphthenate field deposits and crude oils using ESMS and APCI-MS, Colloids and Surfaces A: Physicochemical and Engineering Aspects 349 (2009) 1–18.

[58] M.A. Mohammed, K.S. Sorbie, Spectroscopic Determination of naphthenic acid composition from various calcium naphthenate field deposits, SPE-121633-MS, in: SPE International Symposium on Oilfield Chemistry, Houston, TX, April 20–22, 2009.

[59] M.M. Mapolelo, R.P. Rodgers, G.T. Blakney, et al., Characterization of naphthenic acids in crude oils and naphthenates by electrospray ionization FT-ICR mass spectrometry, International Journal of Mass Spectrometry 300 (2011) 149–157.

[60] M.C.K. Oliveira, F.F. Roasário, J.N. Bertelli, et al., Flow assurance solutions to mitigate naphthenates problems in crude oil production, SPE-166235-MS, in: SPE Annual Technical Conference and Exhibition, New Orleans, LA, 30 Setember–2 October 2013.

[61] V. Molinier, M. Loriau, S. Lescoulié, et al., Quantification of the tetraprotic acids content in oils for the prediction of naphthenate deposits risks, IPTC-18704-MS, in: International Petroleum Technology Conference, Bangkok, Thailand, November 14–16, 2016.

[62] E.V. Barros, H.P. Dias, F.E. Pinto, et al., Characterization of naphthenic acids in thermally degraded petroleum by ESI(−)-FT-ICR MS and ^{1}H NMR after solid-phase extraction and liquid/liquid extraction, Energy & Fuels 32 (2018) 2878–2888.

[63] M.P. Barrow, K.M. Peru, B. Fahlman, et al., Beyond naphthenic acids: environmental screening of water from natural sources and the athabasca oil sands industry using atmospheric pressure photoionization Fourier transform ion cyclotron resonance mass spectrometry, Journal of the American Society for Mass Spectrometry 26 (2015) 1508–1521.

[64] I.H. Auflem, Influence of Asphaltene Aggregation and Pressure on Crude Oil Emulsion Stability (Ph.D. thesis), Norwegian University of Science and Technology, 2002.

[65] H. Magnusson, A.-M.O. Hanneseth, J. Sjöblom, Characterization of C80 naphthenic acid and its calcium naphthenate, Journal of Dispersion Science and Technology 29 (2008) 464–473.

[66] E.L. Nordgård, H. Magnusson, A.-M.D. Hanneseth, et al., Model Compounds for C80 isoprenoid tetraacids Part II. Interfacial reactions, physicochemical properties and comparison with indigenous tetraacids, Colloids and Surfaces A: Physicochemical Engineering Aspects 340 (2009) 99–108.

[67] E.L. Nordgård, J. Sjöblom, Model compounds for asphaltenes and c_{80} isoprenoid tetraacids: Part I- Synthesis and interfacial activities, Journal of Dispersion Science and Technology 29 (2008) 1114–1122.

[68] E.L. Nordgård, Model Compounds for Heavy Crude Oil Components and Tetrameric Acids: Characterization and Interfacial Behavior (Ph.D. thesis), Norwegian University of Science and Technology, 2009.

[69] O. Sundman, E.L. Nordgård, B. Grimes, et al., Potentiometric titrations of five synthetic tetraacids for indigenous C_{80} tetraacids, Langmuir 26 (2010) 1619–1629.
[70] O. Sundman, S. Simon, E.L. Nordgård, et al., Study of the aqueous chemical interactions between a synthetic tetraacid and divalent cations as a model for the formation of metal naphthenate deposits, Energy & Fuels 24 (2010) 6054–6060.
[71] D.L. Gallup, J.A. Curiale, P.C. Smith, Characterization of sodium emulsion soaps formed from production fluids of Kutei Basin, Indonesia, Energy & Fuels 21 (2007) 1741–1759.
[72] D.L. Gallup, P.C. Smith, J. Chpponeri, et al., Formation & mitigation of "metallic soap" sludge, Serang, Indonesia field, SPE-73960-MS, in: SPE International Conference on Heath, Safety and Environment in Oil and Gas Explorations, Kuala Lumpur, Malaysia, March 20–22, 2002.
[73] D.L. Gallup, J.F. Star, Soap sludges: Aggravating factors and mitigation measures, SPE-87471-MS, in: SPE International Symposium on Oilfield Scale, Aberdeen, UK, May 26–27, 2004.
[74] D.L. Gallup, P.C. Smith, J.F. Star, et al., West Seno deepwater development case history – Production chemistry, SPE-93969-MS, in: SPE International Symposium on Oilfield Chemistry, Houston, TX, February 2–4, 2005.
[75] D.L. Gallup, Processing Kutei Basin oils that generate soap emulsions upon contacting produced water, in: Proceedings of the SPE ATW on Managing Naphthenates and Soap Emulsions, Pau, France, March 10–13, 2008.
[76] D.L. Gallup, Geochemistry in Kutei Basin, in: Proceedings of the SPE ATW on Managing Naphthenates and Soap Emulsions, Pau, France, March 10–13, 2008.
[77] D.L. Gallup, V. Denny, C.Y. Khandekar, Inhibition of sodium soap emulsions, West Seno, Indonesia Field, SPE-130506-MS, in: SPE International Conference on Oilfield Scale, Aberdeen, UK, May 26–27, 2010.
[78] G. Rousseau, H. Zhou, C. Hurtevent, Calcium carbonate and naphthenate mixed scale in deep-offshore field, SPE-68307-MS, in: SPE International Symposium on Oilfield Scale, Aberdeen, UK, January 30–31, 2001.
[79] I.K. Salleh, M. Nadeem, K. Mustaqim, et al., Remediation of naphthenate stable emulsion for Malaysia oilfields, SPE-121630-MS, in: SPE International Symposium on Oilfield Chemistry, The Woodlands, TX, April 20–22, 2009.
[80] A.G. Shepherd, G.B. Thomson, R. Westacott, et al., A mechanistic study of naphthenate scale formation, SPE-93407-MS, in: SPE International Symposium on Oilfield Chemistry, Houston, TX, February 2–4, 2005.
[81] S.J. Ubbels, M. Turner, Diagnosing and preventing naphthenate stabilized emulsions and naphthenate deposits during crude oil processing, in: Proceedings of the 6th International Conference on Petroleum Phase Behavior and Fouling, Amsterdam, The Netherlands, June 19–23, 2005.
[82] C. Hurtevent, S. Ubbels, Preventing naphthenate stabilised emulsions and naphthenate deposits on fields producing acidic crude oils, SPE-100430-MS, in: SPE International Oilfield Scale Symposium, Aberdeen, UK, 30 May–01 June, 2006.
[83] G. Runham, C. Smith, Successful naphthenate scale and soap emulsion management, SPE-121522-MS, in: SPE International Symposium on Oilfield Chemistry, The Woodlands, TX, April 20–22, 2009.
[84] N. Borhan, I. Salleh, J.M.B.M. Ibrahim, et al., Investigation of naphthenates crude oil types and behavior to induce soaps-microemulsion and soap-fines foam in

Malaysian EOR fields, IPTC-18948-MS, in: International Petroleum Technology Conference, Bangkok, Thailand, November 14–16, 2016.

[85] T.E. Havre, Formation of Calcium Naphthenate in Water/Oil Systems, Naphthenic Acid Chemistry and Emulsion Study (Ph.D. thesis), Norwegian University of Science and Technology, 2002.

[86] M.H. Ese, K.P. Kilpatrick, Stabilization of water-in-oil emulsion by naphthenic acids and their salts: Model compounds, role of pH and soap/acid ratio, Journal of Dispersion Science and Technology 25 (2004) 253–261.

[87] M.A. Mohammed, K.S. Sorbie, A.G. Shepherd, Thermodynamic modelling of naphthenate formation and related pH change experiments, SPE Production and Operations 24 (2009) 466–472.

[88] M.A. Mohammed, K.S. Sorbie, Thermodynamic modelling of calcium naphthenate formation: Model predictions and experimental results, Colloids and Surfaces A: Physicochemical Engineering Aspects 369 (2010) 1–10.

[89] L.H. Allen, The importance of pH in controlling metal-soap deposition, Tappi Journal 71 (1988) 61–64.

[90] D. Benveneti, B. Carré, A. Gandini, Precipitation and solubility of calcium soaps in basic aqueous media, Journal of Colloid and Interface Science 237 (2001) 142–144.

[91] B. Lin, A.V. McCormick, H.T. Davis, et al., Solubility of sodium soaps in aqueous salt solutions, Journal of Colloid and Interface Science 291 (2005) 543–549.

[92] P.V. Hemmingsen, X. Li, J.-L. Peytavy, et al., Hydrate plugging potential of original and modified crude oils, Journal of Dispersion Science and Technology 28 (2007) 371–382.

[93] W.K. Seifert, Effect of phenols on the interfacial activity of crude oil (California): carboxylic acids and the identification of carbazoles and indoles, Analytical Chemistry 41 (1969) 562–568.

[94] Ø. Brandal, Interfacial Properties (O/W) of Naphthenic Acids and Metal Naphthenates, Naphthenic Acid Characterization and Metal Naphthenate Inhibition (Ph.D. thesis), Norwegian University of Science and Technology, 2005.

[95] G. Hováth-Szabó, J.H. Masliyah, J. Czarnecki, Phase behavior of sodium naphthenates, toluene and water, Journal of Colloid and Interface Science 242 (2001) 247–254.

[96] G. Hováth-Szabó, J.H. Masliyah, J. Czarnecki, Emulsion stability based on phase behavior in sodium naphthenates containing systems: Gels with a high organic solvent content, Journal of Colloid and Interface Science 257 (2003) 299–309.

[97] S.D. Taylor, J. Czarnecki, J.H. Masliyah, Stepwise thickening in aqueous foam films stabilized by sodium naphthenates, Journal of Colloid and Interface Science 282 (2005) 499–502.

[98] S.D. Taylor, J. Czarnecki, J.H. Masliyah, Aqueous foam films stabilized by sodium naphthenates, Journal of Colloid and Interface Science 299 (2006) 283–290.

[99] K. Moran, J. Czarnecki, Competitive absorption of sodium naphthenates and naturally occurring species at water-in-crude oil emulsion droplet surfaces, Colloids and Surfaces A: Physicochemical Engineering Aspects 292 (2007) 87–98.

[100] S. Quraishi, E. Acosta, S. Ng, The role of naphthenates and kaolinite in toluene and toluene-bitumen emulsions, Fuel 153 (2015) 336–345.

[101] P.C. Smith, M. Turner, Controls on soap scale formation including naphthenate soaps – drivers and mitigation, SPE-94339-MS, in: SPE International Symposium on Oilfield Scale, Aberdeen, UK, May 11–12, 2005.

[102] S. Sarac, F. Civan, Experimental investigation and modeling of naphthenate soap precipitation kinetics in petroleum reservoirs, SPE-106074-MS, in: SPE International Symposium on Oilfield Chemistry, Houston, TX, 28 February−02 March, 2007.

[103] S. Sarac, F. Civan, Mechanisms, parameters, and modeling of naphthenate-soap-induced formation damage, SPE Journal 14 (2009) 259−266.

[104] H. Mediaas, Operational challenges at heidrun field calcium naphthenate deposition, in: Proceedings of the SPE ATW on Managing Naphthenates and Soap Emulsions, Pau, France, March 10−13, 2008.

[105] H. Mediaas, Calcium naphthenate deposition at the heidrun field: the search for a solution, in: Proceedings of the SPE ATW on Managing Naphthenates and Soap Emulsions, Pau, France, March 10−13, 2008.

[106] L. Oduola, C. Igwebueze, O. Smith, et al., Calcium naphthenate solid deposit identification and control in offshore Nigerian fields, SPE-164055-MS, in: SPE International Symposium on Oilfield Chemistry, The Woodlands, TX, April 8−10, 2013.

[107] J. Hardy, West Brae naphthenate experience, in: Proceedings of the SPE ATW on Managing Naphthenates and Soap Emulsions, Pau, France, March 10−13, 2008.

[108] G. Poggeso, C. Hurtevent, D. Buchart, Multifunctional chemicals for West African deep offshore fields, SPE-74649-MS, in: SPE Oilfield Scale Symposium, Aberdeen, UK, January 30−31, 2002.

[109] A. Goldszal, C. Hurtevent, G. Rosseau, Scale and naphthenate inhibition in deep-offshore fields, SPE-74661-MS, in: SPE Oilfield Scale Symposium, Aberdeen, UK, January 30−31, 2002.

[110] P.C. Smith, M. Turner, Handling and management of calcium naphthenate and soap generation, in: Proceedings of the 6th International Conference on Petroleum Phase Behavior and Fouling, Amsterdam, The Netherlands, June 19−23, 2005.

[111] A.G. Shepherd, G.B. Thomson, R. Westacott, et al., Analysis of organic field deposits: New types of calcium naphthenate scale or the effect of chemical treatment? SPE-100517-MS, in: SPE International Oilfield Scale Symposium, Aberdeen, United Kingdom, 30 May−1 June 2006.

[112] J.L. Creek, N.A. Macleod, Can Tool, in: Proceedings of the SPE ATW on Managing Naphthenates and Soap Emulsions, Pau, France, March 10−13, 2008.

[113] C. Cummine, Operational issues due to CN deposition on BleoHolm FPSO – 2004 to 2005, in: Proceedings of the SPE ATW on Managing Naphthenates and Soap Emulsions, Pau, France, March 10−13, 2008.

[114] K. Melvin, C. Cummine, J. Youles, et al., Optimizing calcium naphthenate control in the Blake field, SPE-114123-MS, in: SPE International Oilfield Scale Conference, Aberdeen, UK, May 28−29, 2008.

[115] N. Passade-Boupat, M.R. Gonzalez, B. Brocart, et al., Risk assessment of calcium naphthenates and separation mechanisms of acidic crude oil, SPE-155229-MS, in: SPE International Conference and Exhibition on Oilfield Scale, Aberdeen, UK, May 30−31, 2012.

[116] J. Junior, L. Borges, C. Carmelino, et al., Calcium naphthenate mitigation at Sonangol's Gimboa field, SPE-164069-MS, in: SPE International Symposium on Oilfield Chemistry, The Woodlands, TX, April 8−10, 2013.

[117] S.S. Alian, K. Singh, A. Saidu, et al., Organic deposition: From detection and laboratory analysis to treatment and removal, SPE-165912-MS, in: SPE Asia Pacific Oil and Gas Conference and Exhibition, Jakarta, Indonesia, October 22−24, 2013.

[118] D.A. Nichols, F.F. Rosário, M.C.M. Bezerra, et al., Calcium naphthenate in complex production systems – Evaluation and chemical inhibition challenges, SPE-169756-MS, in: SPE International Oilfield Scale Conference and Exhibition, Aberdeen, UK, May 14–15, 2014.
[119] A.S. Mohamed, A. Goyal, M.D.M. Ismail, et al., Detection and removal of naphthenate deposit in oil well, OTC-25003-MS, in: Offshore Technology Conference Asia, Kuala Lumpur, Malaysia, March 25–28, 2014.
[120] A.S. Mohamed, S.S. Alian, J. Singh, et al., Remediation of well impaired by complex organic deposits embedded with naphthenate and contaminated with inorganics, OTC-26524-MS, in: Offshore Technology Conference Asia, Kuala Lumpur, Malaysia, March 22–25, 2016.
[121] L. Castro, G. Colmenares, The identification of calcium naphthenates in a heavy oil reservoir using laboratory techniques to assess solids and fluid samples, SPE-190145-MS, in: SPE Western Regional Meeting, Garden Grove, CA, April 22–27, 2018.
[122] S. Sidek, S.A.M. Hatta, K.G.H. Lian, et al., High molecular weight organic scale; a sticky journey in finding the solution, SPE-196028-MS, in: SPE Annual Technical Conference and Exhibition, Calgary, Alberta, Canada, 30 September–2 October, 2019.
[123] S.R.M. Shafian, A.I.A. Bakar, B. Abdullaev, et al., Restoring well productivity suffers from complex organic and inorganic deposition in malaysian oil producers: Case study and lessons learnt, SPE-199233-MS, in: SPE International Conference and Exhibition on Formation Damage control, Lafayette, Louisiana, USA, February 19–21, 2020.
[124] B.E. Smith, T. Whitfield, Calcium naphthenates: Our understanding, analysis, predictive methods, and results so far, in: Proceedings of the SPE ATW on Managing Naphthenates and Soap Emulsions, Pau, France, March 10–13, 2008.
[125] B. Brocart, C. Hurtevent, Naphthenates risk: TOTAL prediction tool, in: Proceedings of the SPE ATW on Managing Naphthenates and Soap Emulsions, Pau, France, March 10–13, 2008.
[126] J. Debord, P. Srivastava, Development and field application of a novel non-acid calcium naphthenate inhibitor, SPE-123660-MS, SPE Annual Technical Conference and Exhibition, New Orleans, LA, October 4–7, 2009.
[127] E.L. Nordgård, A. Hanneseth, J. Sjöblom, Inhibition of calcium naphthenate: Experimental methods to study the effect of commercially available naphthenate inhibitors, Journal of Dispersion Science and Technology 31 (2010) 668–675.
[128] J.-A. Östlund, M. Nydén, I.H. Auflem, et al., Interactions between asphaltenes and naphthenic acids, Energy & Fuels 17 (2003) 113–119.
[129] M.S. Turner, P.C. Smith, Controls on soap scale formation, including naphthenate soaps – drivers and mitigation, SPE-94339-MS, in: SPE International Symposium on Oilfield Scale, Aberdeen, UK, May 11–12, 2005.
[130] V.V. Rogers, K. Liber, M.D. MacKinnon, Isolation and characterization of naphthenic acids from Athabasca oil sands tailings pond water, Chemosphere 48 (2002) 519–527.
[131] N.S. Toor, E.D. Franz, P.M. Fedorak, et al., Degradation and aquatic toxicity of naphthenic acids in oil sands process-affected waters using simulated wetlands, Chemosphere 90 (2013) 449–458.

[132] J.R. Marentette, R.A. Frank, A.J. Bartlett, et al., Toxicity of naphthenic acid fraction components extracted from fresh and aged oil sands process-affected waters, and commercial naphthenic acid mixtures, to fathead minnow (*Pimephales promelas*) embryos, Aquatic Toxicology 164 (2015) 108–117.

[133] W. McLellan, M. Ferguson, G. McKay, et al., North Sea Harding project: A well engineering case history success, SPE-50676-MS, in: SPE European Petroleum Conference, The Hague, The Netherlands, October 20–22, 1998.

[134] P. Zhang, S.R. Green, Harding Central–Achieving 74% recovery, SPE-123923-MS, in: SPE Offshore Europe Oil and Gas Conference and Exhibition, Aberdeen, UK, September 8–11, 2009.

[135] J. Feith, J.P. Hoekstra, The Harding move: opening up new horizons, OTC-8162-MS, in: Offshore Technology Conference, Houston, TX, May 6–9, 1996.

[136] D. Johnson, G. McAteer, H. Zuk, The safe processing of high naphthenic acid content crude oils – refinery experience and mitigation studies, NACE-03645, in: Corrosion Conference & Expo, San Diego, CA, March 16–20, 2003.

[137] G. Bota, D. Qu, S. Nesic, et al., Naphthenic acid corrosion of mild steel in the presence of sulfide scales formed in crude oil fractions at high temperature, NACE-10353, in: Corrosion Conference & Expo, San Antonio, TX, March 14–18, 2010.

[138] G. Bota, P. Jin, F. Ferales, et al., Phosphate esters inhibitors in naphthenic acid corrosion – an experimental evaluation, NACE-2019–13491, in: Corrosion Conference & Expo, Nashville, TN, March 24–28, 2019.

[139] Y. Yoon, I. Kosacki, S. Srinivasan, Naphthenic acid and sulfur containing crude oil corrosion: A comparative review, NACE-2016–7598, in: Corrosion Conference & Expo, Vancouver, British Columbia, Canada, March 6–10, 2016.

[140] E.C. Rios, A.L. Oliveira, A.M. Zimer, et al., In situ characterization of naphthenic corrosion of API 5L X70 steel at room temperature, Fuel 184 (2016) 648–655.

[141] Y. Wang, Z. Chu, B. Qiu, et al., Removal of naphthenic acids from a vacuum fraction oil with an ammonia solution of ethylene glycol, Fuel 85 (2006) 2489–2493.

[142] K. Anderson, P. Goodrich, C. Hardacre, et al., Removal of naphthenic acids from crude oil using amino acid ionic liquids, Fuel 108 (2013) 715–722.

[143] K. Anderson, M.P. Atkins, P. Goodrich, et al., Naphthenic acid extraction and speciation from Doba crude oil using carbonate-based ionic liquids, Fuel 146 (2015) 60–68.

[144] Y.Z. Wang, X.Y. Sun, Y.P. Liu, et al., Removal of naphthenic acids from a diesel fuel by esterification, Energy & Fuels 21 (2007) 941–943.

[145] Y.Z. Wang, J.Y. Li, X.Y. Sun, et al., Removal of naphthenic acids from crude oils by fixed-bed catalytic esterification, Fuel 116 (2014) 723–728.

[146] A. Zhang, Q. Ma, K. Wang, et al., Naphthenic acid removal from crude oil through catalytic decarboxylation on magnesium oxide, Applied Catalysis A 303 (2006) 103–109.

[147] L. Ding, P. Rahimi, R. Hawkins, et al., Naphthenic acid removal from heavy oils on alkaline earth-metal oxides and ZnO catalysts, Applied Catalysis A 371 (2009) 121–130.

[148] S.J. Ubbels, J.P. Venter, M.V. Nace, Low dosage naphthenate inhibitors, Patent WO2006025912, 2006.

[149] M. Subramaniyam, Additive and method for removal of calcium from crude oils containing calcium naphthenate, Patent US9434889, 2016.

[150] N. Bretherton, R. Smith, G. Keilty, et al., Naphthenate control: Is acetic acid injection the answer? SPE-95115-MS, in: SPE International Symposium on Oilfield Corrosion, Aberdeen, United Kingdom, May 13, 2005.
[151] C.E. Reiners, E.C. Harvey, G.B. van Dijk, Naphthenate inhibition, Patent US20200199438, 2020.
[152] D. Smith, C. Smith, D. Watson, Method of treating hydrocarbon compositions, Patent GB2461611, 2010.
[153] H. Mediaas, K.V. Grande, B.M. Hustad, et al., A unique laboratory test rig reduced the need for offshore tests to combat calcium naphthenate deposition in oilfield process equipment, in: Proceedings of the TEKNA Oilfield Chemistry Symposium, Geilo, Norway, March 13–16, 2005.
[154] D. Larry, Y.M. Gunaltun, Correlation of cases of top of line corrosion with calculated water condensation rates, NACE-00071, in: Corrosion Conference and Expo, Orlando, Florida, March 26–31, 2000.
[155] J.J. Weers, S. Bieber, Calcium removal from high TAN crude, Petroleum Technology Quarterly 10 (2005) 21–25.
[156] S. Bieber, B. Fahey, J. Renbin, et al., Successful strategies for processing high calcium, high TAN crude oils, in: Proceedings of the Asian Refining Technology Conference, Kuala Lumpur, Malaysia, March 7–9, 2006.
[157] C. Khandekar, R. Wilson, J. Smith, Compositions and methods for inhibiting emulsion formation in hydrocarbon bodies, Patent WO2010017575, 2010.
[158] S.J. Ubbels, Naphthenate inhibitors: How do they work? in: Proceedings of the SPE ATW on Managing Naphthenates and Soap Emulsions, Pau, France, March 10–13, 2008.
[159] J.P. Vijn, P.I. Veld, Low dosage polymeric naphthenate inhibitors, Patent WO2012154378, 2012.
[160] C. Khandekar, S. Gopal, Compositions and methods for inhibiting naphthenate solids formation from liquid hydrocarbons, Patent WO2011063459, 2011.
[161] M. Subramaniyam, Additive and method for removal of calcium from crude oils containing calcium naphthenate, Patent US8840781, 2014.
[162] M. Subramaniyam, Compositions comprising combination of esters of carboxylic acids for removing metals from hydrocarbon feedstock, Patent US9080110, 2015.
[163] P.J. Venter, V.M. Nace, S.J. Ubbels, Low dosage naphthenate inhibitors, Patent US7776931, 2010.
[164] M. Hellsten, I. Uneback, A method for preventing the formation of calcium carboxylate deposits in the dewatering process for crude oil/water streams, Patent WO2008155333, 2008.
[165] J. Debord, P. Srivastava, C. Gallagher, et al., Inhibiting naphthenate solids and emulsions in crude oil, Patent US8003574, 2011.
[166] C.R. Jones, Soap control agent, Patent WO2005085392, 2005.
[167] M.A. Silvestri, S. Linares-Samaniego, R. Gingrich, Development of a combination corrosion-scale inhibitor for the subsea separation and boosting systems of the Parque das Conchas project in Brazil, OTC-20371-MS, in: Offshore Technology Conference, Houston, TX, May 3–6, 2010.
[168] M. Jordan, S. Linares-Samaniego, M. Alfonsa, Deployment, monitoring, and optimization of a scale inhibitor and associated corrosion inhibitor within a deepwater subsea facility offshore Brazil, SPE Projects, Facilities and Construction 6 (2001) 1–6.

[169] K. Spicka, C.J. Johnston, M.M. Jordan, et al., The impact of organic acid on scale inhibitor/corrosion inhibitor interaction, a case study from West Africa, SPE-141033-MS, in: SE International Symposium on Oilfield Chemistry, The Woodlands, TX, April 11–13, 2011.

[170] M. Ciolkowski, A. Neville, A systematic scale-corrosion study of a combined inhibitor in scale-corrosion environment via experimental design, NACE-2014–3960, in: Corrosion Conference & Expo, San Antonio, TX, March 9–13, 2014.

[171] S.L. Berry, H.L. Becker, J.L. Boles, et al., Naphthenic acid solids dissolver compositions and methods related thereto, Patent US20100029514, 2010.

[172] S.J. Dyer, G.M. Grahame, C. Arnott, Naphthenate scale formation – Examination of molecular controls in idealized systems, SPE-80395-MS, in: SPE International Symposium on Oilfield Scale, Aberdeen, UK, January 29–30, 2003.

[173] S.J. Dyer, H.L. Williams, G.M. Grahame, et al., Simulating calcium naphthenate formation and mitigation under laboratory conditions, SPE-100632-MS, in: SPE Oilfield Scale Symposium, Aberdeen, UK, 31 May–1 June, 2006.

[174] H. Williams, S. Dyer, G. Graham, Understanding the factors influencing the formation and control of calcium naphthenate solids and stabilised emulsions using a novel laboratory flow rig, SPE-106499-MS, in: SPE International Symposium on Oilfield Chemistry, Houston, TX, 28 February–02 March, 2007.

CHAPTER

Inorganic mineral scale mitigation

5

Gordon Michael Graham and Dario Marcello Frigo

Scaled Solutions Ltd., Livingston, Scotland, United Kingdom

Chapter outline

Flow Assurance. DOI: https://doi.org/10.1016/B978-0-12-822010-8.00010-6

5.1 Introduction

Inorganic mineral scaling is a common challenge experienced by almost any industrial or domestic process that involves water. It is commonly defined as a "hard, adherent mineral deposit that precipitates from brine/water solution." In the oilfield environment, it occurs because hydrocarbon production invariably involves the coproduction of water. The hydrocarbons are trapped in the porous medium of the reservoir sharing the pore space with water, and the entire accumulation almost always floats on an aquifer. This water has equilibrated over geological times to become saturated with the various minerals within the reservoir, and so these waters can become supersaturated and precipitate when they experience different conditions during passage through the production system. To estimate the magnitude of the issue, recent surveys have indicated that as a global average, for every barrel of oil equivalent produced, 3–4 barrels of water are also produced, which equates to more than 300 million barrels of associated water production per day, at current oil-production rates.

When left untreated, inorganic scaling leads to restrictions and eventually blockages of any conduit or equipment through which the produced fluids flow: reservoir pores, sand control screens, well tubulars, flowlines, and facilities. It can also lead to failure of safety-critical valves, malfunction of pumps, efficiency loss in heat exchangers, poorer fluid-separation efficiency, and enhanced corrosion. Its buildup results in production deferment, costly remediation activities, and, in some cases, if the scale cannot be cost-effectively managed, well or even early field abandonment. In their book first published in 2008, Frenier and Ziauddin [1] estimated that at that time, the economic impact of inorganic mineral scale was of the order of 1.4 billion USD, which Habibi et al. [2] clarified was to be attributed to prevention and remediation operations, that is, it did not include the associated cost of deferred or lost production. Frenier and Ziauddin also indicated that 28% of the decline in production in North Sea wells was directly related to scale [1]. As oilfields around the world continue to mature, and therefore produce more water, the risk to production and integrity and the associated costs of scale mitigation and remediation will continue to be very significant, for both conventional and unconventional wells.

There are thousands of publications in the literature on the subject of inorganic mineral scaling, and many books have also been published on the subject, of which key milestones include Cowan and Weintritt's 1976 seminal tome *Water Formed Scale Deposits* [3] and Frenier and Ziauddin's 2008 publication *Formation, Removal and Inhibition of Inorganic Scale in the Oilfield Environment* [1]. The 2009 book by Kelland, *Production Chemicals for the Oil and Gas Industry* [4], is a more general overview of production chemistry, with a chapter on the chemicals used in inorganic scale control. In the author's opinion, the current chapter together with these three books provide excellent reference material for production chemists and production engineers involved in the control and management of inorganic scale in the oilfield environment.

This chapter, therefore, aims to provide an overview of the main aspects of inorganic scaling in the oil and gas production environment, including why it occurs, the chemistry and physics governing its deposition, and the methods and challenges associated with the elements of its management: prediction, prevention, and remediation.

5.1.1 The role of water

Water is the transport medium for almost all inorganic scale deposits (perhaps the rare exception is the deposition of elemental sulfur that condenses from a gas phase containing hydrogen sulfide and oxygen). It is a versatile solvent because its polarity and capacity to hydrogen-bond enables it to dissolve many ionic and nonionic substances to a varying extent. At any given set of conditions, the solubility of the commonly found scales can vary over many orders of magnitude, from the highly soluble halite (sodium chloride, NaCl), moderately soluble gypsum (hydrated calcium sulfate, $CaSO_4{\cdot}2H_2O$), sparingly soluble barium sulfate ($BaSO_4$) through to scales that under certain conditions are among the least water-soluble substances known, such as lead sulfide (PbS). The differences in solubility arise from differences between the enthalpy and entropy of solution in water, which are largely related to the balance of lattice energy of the solid (the energy that must be added to disperse the ions from the crystal lattice) *vs* the hydration energy (the energy of interaction between the ions and the water in solution). Therefore, soluble ions, such as most of the chlorides and bromides of the alkali and alkaline-earth metals, have low lattice energies and high hydration energies; insoluble minerals tend to have the reverse.

The solubility of scale minerals changes with temperature, pressure, dielectric constant (primarily related to the ionic strength), and, for some minerals, the pH. Thus, any flowing stream subject to changes in these parameters will tend to dissolve, transport, and then reprecipitate minerals. In the oilfield environment, scale can arise from water that has dissolved minerals in the reservoir (usually at high temperature, high pressure, and relatively low pH due to the presence of various dissolved acid gases), which then experiences changes in temperature and pressure (and perhaps the influence of new phases appearing at the bubble or dew point of the hydrocarbon) as it flows through the production system, whereupon it reprecipitates and perhaps also deposits on the wetted surfaces (Figs. 5.1 and 5.2). This is the phenomenon of self-scaling that arises from environmental changes from a single stream or water and hydrocarbon (if present). It occurs in nature in Karst geology, which is characterized by dissolution and reprecipitation

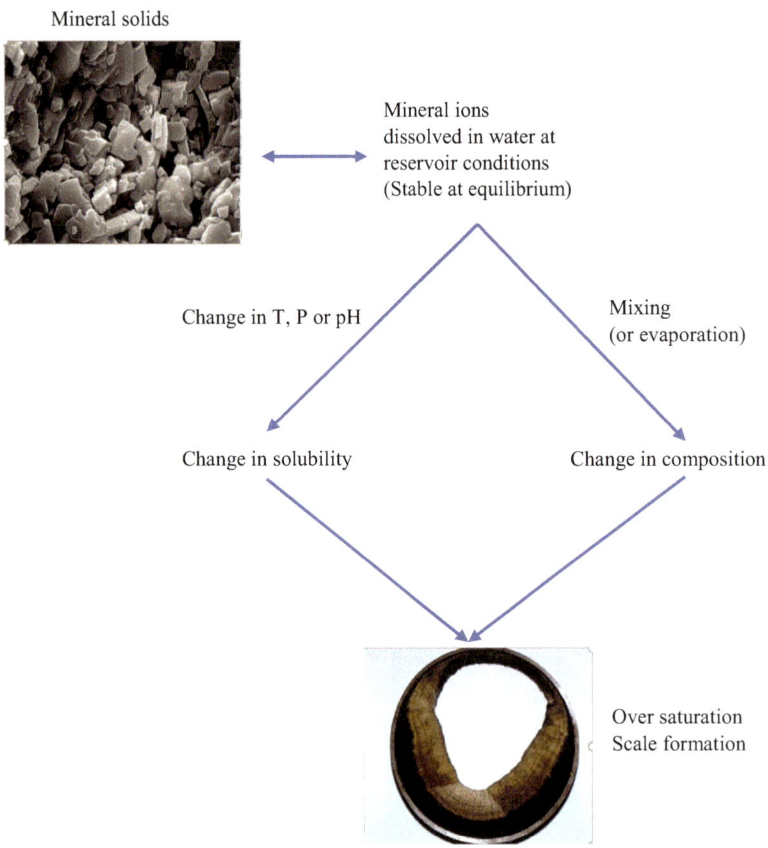

FIGURE 5.1

Illustration of how a stable aqueous solution can result in scale deposition.

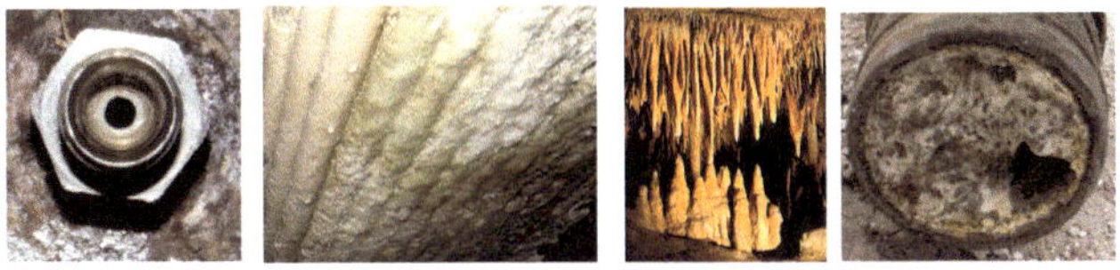

FIGURE 5.2

Scale deposits in industrial equipment, heat exchangers of desalination plant, naturally occurring and in an oil pipeline.

(including in caves) of calcium carbonate rocks, and via subsurface hydrothermal processes. It also accounts for the concentration in specific locations of much of the Earth's mineral wealth. It also occurs in many other industrial and domestic settings, for example, limescale depositing in household appliances, such as kettles, boilers, showers, and washing machines.

The other common process by which scaling occurs is the mixing of incompatible waters, where at least two streams transport the components of the scale, and each of the streams is undersaturated in the mineral but the mixture is supersaturated. In the oilfield, this could be waters from two different reservoirs mixing in the well or flowlines, or one of the streams could be water that is injected for secondary recovery. This also occurs in nature, for example, the tufa towers in Mono Lake, California, which arose from calcium-containing underground streams that entered the alkaline, sodium-carbonate-rich lake waters from below, leading to various forms of calcium carbonate.

5.1.2 Inorganic mineral scaling in the oil environment

In oil and gas production, the most significant aspect of inorganic mineral scaling relates to the adherence and buildup of that scale at critical points in the production process and the impact of this buildup on safe and effective production. These critical points include the subsurface production environment, such as the near wellbore formation area, wellbores, production tubulars, subsurface safety valves (SSSVs), flowlines, manifolds, trunklines, risers, chokes and flow control valves, separation facilities, filtration equipment, heat exchangers, water handling and processing equipment, and the reinjection system. In short, any wetted part of the production system that handles water can be at risk of inorganic scaling and the associated reductions in throughput, as a result of the buildup of hard, adherent deposits choking and restricting the flow paths.

The management and mitigation of such scales is therefore also related to the location of the deposit, with those forming further upstream in the production system being more difficult and costly to treat and/or remediate. In particular, when scaling occurs downhole in wells, especially in subsea installations with "wet trees," the costs associated with its mitigation and/or remediation can be orders of magnitude higher than those associated with scale formation in process facilities [5–10].

In addition to the natural formation waters and aquifer waters present in the oil and gas reservoirs, other waters are often injected into the production system and are usually applied to maintain reservoir pressures via secondary recovery or to enhance production, as in enhanced oil recovery (EOR) approaches. Moreover, even for individual production wells, there are often different formation zones producing different fluids, including different waters.

The types of scaling challenges in the oilfield can depend on the recovery method, with there often being a general increase in issues across the series primary production, waterflooding, and EOR.

- Primary production where the reservoir drive comes from fluid expansion (aquifer or gas cap), nearly always accompanied by a significant drop in reservoir pressure. From a scaling viewpoint, the only waters produced are those already present in the reservoir mainly the connate water, aquifer water, and that present in the transition zone.
- Secondary recovery which nearly always means injection of water and/or (immiscible) gas to maintain reservoir pressure (to a variable extent) and help sweep the oil to the production wells. The impact on scaling is mainly from water flooding because the injected water is often incompatible with the reservoir waters and can thus lead to additional scaling mechanisms.
- EOR where species are introduced into the reservoir significantly different in nature from those already present. This is sometimes loosely called tertiary recovery, but EOR can also be secondary in nature, as in performed after natural depletion, or even from Day One (e.g., steam flooding). EOR can lead to more significant scaling issues at production wells, of which the worst occurs when flooding with alkaline fluids, such as in alkali surfactant polymer flooding. Flooding with CO_2 either *per se* or as a CO_2 water-alternating gas flood can also enhance scaling. The inorganic scale challenges associated with EOR are described in more detail by Jordan et al. [11] and more recently by Graham et al. [12].

Furthermore, certain types of well completions or the use of artificial lift can greatly influence scaling issues, either because they introduce restrictions in the flow path (e.g., sand control screens, injection control devices and valves, in-well pumps, chokes), which are then easier to block, or because they enhance turbulence (e.g., the previous items plus subsurface safety valves, gas lift), or both. Similarly, sources of heating, for example, electrical submersible pumps (ESPs), and heat exchangers, can also enhance certain types of scaling. As a general rule, the simpler the well, flowline, or piece of facilities is, the lesser its tendency to enhance scaling and the lesser its susceptibility to a limited amount of scaling.

Over recent decades, scale-control challenges have increased as the offshore oil and gas industry has increasing amounts of new production from harsher environments. These have included subsea production in increasing water depths, production at the extremes of temperature (both high temperature high pressure in the reservoir and/or "Arctic" conditions at surface) and the increased prevalence and challenges associated with less conventional and more difficult to inhibit scales such as the Pb/Zn/Fe sulfides, halite and silicate, and so on. An increasing fraction of the well stock is of the more complex, multizone type, containing increased amounts of "jewelry" (largely, control systems involving moving parts and gauges for monitoring), which makes the wells more susceptible to more limited amounts of deposit. For subsea production environments, sulfate-reduction technology (to reduce the level of sulfate in injected seawater from thousands of mg/L to a few tens of mg/L) is progressively seen as a default in order to minimize the risk of the (typically more troublesome) sulfate scales [5,8–10,13,14].

In general, although the oil industry has been managing scale since the early days of oil and gas production, the nature of the challenges changes almost on a field-by-field basis, and, therefore, optimizing and improving treatment approaches remain an active area of research and development.

5.2 Basic principles of inorganic scale formation

5.2.1 Types of inorganic mineral scale

The American Society of Testing Materials lists more than 100 deposits found in water systems. In the oilfield environment, the most common of these scales are the carbonates and sulfates of the divalent alkaline-earth metal ions Ca^{2+}, Sr^{2+}, and Ba^{2+} (Group IIA). Various iron deposits (e.g., hydroxides, carbonates, and sulfides) may also form. They often relate to corrosion issues within the production system itself, either sweet, sour, oxygen-induced, microbially induced, or associated with the backproduction of acid-stimulation treatments in wells which have been insufficiently inhibited against corrosion.

Although these scales may represent the most common inorganic deposits found in oil and gas production, and indeed calcium carbonate is the most ubiquitous affecting, to some degree, almost every oil and gas production environment, they represent a relatively small number of the different minerals that can, and do, deposit. Table 5.1, therefore, provides only a selection of the types of deposits often encountered.

An important aspect to remember is that few scale deposits comprise only a single substance. A $CaCO_3$ deposit may typically consist of up around 95% of that mineral (as whichever polymorph) but will often contain substantial abundance of solids produced as solids from the reservoir (sand, fines, small lumps of carbonate rock) or corrosion scales (iron carbonate and/or iron sulfides), which perhaps initially deposited as a layer on the steel upstream and then stripped off and entrained in the produced fluids. These can then become incorporated into the growing scale layer giving it a composite composition. Also, there is the potential for more than a single mineral to be supersaturated, which can lead to the primary deposit being a mixed scale. Many $BaSO_4$ scales are actually (Ba, Sr) SO_4, where there is no fixed stoichiometry between the two metals forming the

Table 5.1 Selected list of oilfield scale deposits, indicating their solubility in common acids

Type of Scale	Formula	Acid Soluble
Calcium carbonate	Calcite $CaCO_3$	Yes
	Aragonite $CaCO_3$	Yes
	Vaterite $CaCO_3$	Yes
Calcium sulfate	Anhydrite $CaSO_4$	No
	Hemihydrate $CaSO_4 \cdot ½H_2O$	No
	Gypsum $CaSO_4.2H_2O$	Slightly
Strontium sulfate	Celestite $SrSO_4$	No
Barium sulfate	Barite $BaSO_4$	No
Iron oxo species	Fe_3O_4, FeOOH, $Fe(OH)_2$, and so on.	Yes
Iron sulfides	Fe_xS_y	Highly dependent on Fe:S ratio
Iron carbonate	Siderite $FeCO_3$	Yes
Other carbonates	e.g., Magnesite $MgCO_3$	Yes
Other hydroxides	e.g., Brucite $Mg(OH)_2$	Yes
Halite	NaCl	Soluble in fresh water
Zinc sulfide	Sphalerite, Wurtzite ZnS	(Yes)[a]
Lead sulfide	Galena PbS[b]	(Yes)[a]
Silica and metal silicates	Amorphous SiO_2 [c] A variety of metal silicates	No

[a]*See Section 5.2.1.1.*
[b]*Galena is cubic; an orthorhombic form also exists but only at pressure well beyond anything encountered in the oilfield.*
[c]*While waters in sandstone reservoirs tend to be equilibrated with quartz, this cannot precipitate as a scale over oilfield production timescales; therefore, the more soluble amorphous SiO_2 tends to form instead.*

solid solution. Mixtures of $CaCO_3$ and $CaSO_4$ (the latter, probably as gypsum) are commonly encountered, as are (Sr, Ca)SO_4. Indeed, in some cases, a (Sr, Ca) SO_4 scale arising from incompatible mixing under seawater flood can show compositional grading, with a $SrSO_4$-rich zone nearest the tubing, progressively becoming more $CaSO_4$-rich toward the center. This reflects the changing scaling tendencies for each mineral over time as the well produces a progressively higher seawater fraction in its produced water. Recognizing the composite and potentially graded nature of a scale deposit can have real operation consequences, such as, when selecting how it should optimally be removed.

5.2.1.1 *Acid-soluble versus acid-insoluble scales*

For the purpose of management of oilfield scales, the more common scales can be classified into acid-soluble and acid-insoluble minerals (see Table 5.1), as this can have a considerable impact on how easily they can be removed. The most important decisions in formulating the scale-control strategy are the optimal balance between upfront prevention (a risk-based approach because the severity may be uncertain) and remediation (only employed when the scaling event has actually occurred). The easier and cheaper it is to remediate, the more likely that the strategy will lean toward this approach, at least until the actual severity is borne out by field experience. Conversely, if the cost of remediation is prohibitive (e.g., if it risked losing entire wells!), the more the strategy will lean toward prevention.

Calcium carbonate, the most prevalent scale worldwide, is soluble in a variety of acids and nonacid chemicals, which means remediation can be more routine than many other (acid insoluble) scales. Other carbonates are also acid soluble, of which the commonest after $CaCO_3$ is $FeCO_3$. It is also fairly common that mixed carbonate scales can form, or at least that a $CaCO_3$ matrix also contain, smaller amounts of $MgCO_3$, $SrCO_3$, $BaCO_3$, and $FeCO_3$.

Halite is also soluble in many acids, though it is typically even more soluble in low-salinity water. As such, some halite deposits that have been found to contain significant traces of carbonate scales are sometimes removed with a low concentration of acid, which should dissolve both components [15].

The commonest metal oxide and hydroxides scales are those of iron and are associated with (oxygen) corrosion. Magnesium hydroxide is rarely seen but can form, such as when the pH of seawater is adjusted to values above 9; therefore, it can be used (e.g.,) for annular top-ups in wells. Most of these species are soluble in mineral acids, such as HCl.

Some metal sulfides are acid-soluble, at least to some extent. As described below, iron sulfides scale occurs in different forms and in general, the higher the ratio of sulfur to iron in the form, the lesser their acid-solubility. An important aspect of iron sulfide scales is that different forms may be present in the same system, for example, in different depths within a well, which means an acid remediation treatment may only be partially successful. Zinc sulfide (ZnS) and lead sulfide (PbS) have some degree of acid solubility, though the solubility of PbS is enhanced in the presence of chloride ion (even in neutral solutions) due to the stability of the $PbCl_4^{2-}$ ion, implying that HCl is likely to be a better solvent than alternatives.

Gypsum is considered by some to be acid-soluble, though its solubility is normally insufficient for it to be effectively used as a remediation method. Gypsum can, however, be converted into calcium carbonate or calcium hydroxide by the use of converting agents, resulting in a much more acid-soluble mineral, though requiring a two-step process to dissolve it, and the effectiveness of the conversion has often been found to reduce over repeated treatments [16]. Nowadays, there are better single-step solvents for this scale.

Barium and strontium sulfates are negligibly soluble in acid. The former represents a significant challenge in many offshore production environments in various regions including the North Sea, West Africa, Offshore Brazil, and Gulf of Mexico. For the Middle East region, calcium sulfate, and in some cases strontium sulfate, are the most likely acid-insoluble scales. Silica and silicate scales have some solubility in certain acids or strong alkali, but neither has been found to be a practical solvent.

5.2.1.2 Iron scales

These are specifically highlighted because they have two potential origins in an oilfield system. The difference being whether the iron component arises from the formation water equilibrated with iron minerals present in the reservoir (siderite, ankerite, ferroan calcite, ferroan dolomite, iron-rich chlorite, etc.) or from corrosion of the (mostly low-alloy carbon steel) tubulars.

Iron scales are common in oil and gas production because corrosion is common. Oxygen corrosion in a water-injection system produces oxides and oxyhydroxides, at least initially (these species can then convert to sulfide species if microbes are allowed to proliferate in the injection system, producing H_2S as a result of anaerobic respiration, which then sulfidizes the iron oxo species). Corrosion in the production system leads to the formation of carbonate and/or sulfide solids, depending on the ratio of CO_2 to H_2S within the produced fluids. These corrosion scales can either be adherent and hence largely self-limiting because they passivate the surface or can be stripped off by the wall shear stresses, allowing for further reaction and generating scale particulates as suspended solids in the production stream. Hence, the continuous supply of iron ions into the produced water system, due to uncontrolled corrosion of production pipework, can result in deposits of either iron carbonates or iron sulfides. Another perhaps underestimated source of iron sulfide occurs when sour wells with carbon-steel tubulars are stimulated with strong acids (up to 28 wt.% HCl), either with insufficient corrosion inhibitor or at temperatures where the corrosion inhibitor is ineffective. This can generate high concentrations of dissolved iron, which then form a variety of iron sulfide and oxide scales when the spent acid is backproduced, albeit only while the high-iron-content fluids are still present.

Iron scales can also be produced in the conventional sense, that is, as a result of dissolved iron in the formation water, which is then subject to changing environmental conditions. Iron scales produced thus are generally less common than calcium carbonates and sulfates of barium, calcium, and so on because the geochemical conditions do not often favor high concentrations of dissolved iron in the formation waters. When scales do form due to elevated levels of iron in the produced waters, they are typically dominated by iron carbonate, which forms in an analogous manner to calcium carbonate. In such cases, iron sulfides would be less likely because the presence of significant amounts of H_2S in the reservoir fluids (to generate iron sulfide) would significantly reduce the maximum dissolved-iron concentration that could co-exist with it. Where iron sulfide could predominate is in a condition of microbial reservoir souring [17], where the formation water contains significant amounts of dissolved iron, which persists in reservoir zones that have not soured, and other zones (those almost invariably well swept with injection water) that now produce H_2S. Mixing of the two streams could generate a considerable iron sulfide scaling tendency.

Iron sulfide scales are sufficiently varied to merit special mention. There are many forms of them [18]. From a scaling viewpoint, it is convenient to divide them into two groups:

- More oxidized forms with a Fe:S of (or around) 1:2, such as pyrite and marcasite (both in form of FeS_2), which are often formed over geological timeframes and are present in many reservoir rocks. They do not form as scales in the oilfield initially but can form from less oxidized forms of iron sulfide by interaction with H_2S and other sulfur species in the production stream. These forms tend to be almost entirely unaffected by most acids.
- Less oxidized versions, FeS_x (x is closer to or equal to unity), which are the ones that commonly form during production as scales. These include cubic FeS, mackinawite, pyrrhotite, troilite, and greigite [19]. The acid solubility of these less oxidized forms is variable, but many are fairly well dissolved by mineral acids [20].

Of significance also is that certain of the iron sulfide scales, commonly formed in oil and gas production, are pyrophoric and can self-combust when exposed to air, in particular in water injection systems, where the scale is produced as a result of sulfidizing a finely divided rust initially generated by oxygen corrosion [21]. Further, the nature of iron sulfides with their often "mixed-wet" surfaces has been a major contributor to the stabilization of emulsions (compromising separation performance) and the formation of mixed organic and inorganic deposits and sludges, often referred to as schmoo [22]. This not only forms sticky deposits but is extremely effective at plugging water-injection wells.

5.2.1.3 Solid solutions and mixed scales

It was mentioned above that scale deposits almost never comprise a single mineral and as such all scales are "mixed scales," that is, they result from the fortuitous deposition of at least one mineral that precipitated from solution plus other species, which may have been incorporated from solids suspended in the fluids. Solid solutions are different. They result from the simultaneous precipitation of more than one scale mineral within the same lattice. This obviously means that the resulting lattice must be such that it can incorporate both minerals without too much lattice strain that would make such co-precipitation energetically unfavorable. On the contrary, it is typically *more* favorable for at least one of the component minerals to become incorporated into the solid solution than into its own pure mineral lattice.

An important result of this for scaling is that it is not necessary for either mineral to be supersaturated in solution relative to its own pure mineral for precipitation to occur. An example of this is the formation of (Ba, Sr)SO_4. This is commonly found as a deposit in North Sea wells (e.g., from those in the Brent and Stafjord sequences, which covers a large fraction of the northern North Sea).

If the scaling tendency for $SrSO_4$ is calculated for the produced waters, there are no circumstances under which it can be supersaturated. This is because the degree of supersaturation is calculated relative to the standard state of crystalline celestite ($SrSO_4$), whereas the lattice energy of (Ba, Sr)SO_4 in the solid solution must be greater, which means the ions in solution are being compared against a solid with a lower energy state: the $SrSO_4$ is supersaturated relative to $Ba_xSr_{1-x}SO_4$ at the relevant value of x observed in the field.

Other examples of solid solutions include (Ba, Ca)SO_4 (where up to 18% Ca can be present without fundamentally changing the morphology of the barite lattice, though the lattice dimensions do change) or (Sr, Ca)SO_4. Both of these form even when $CaSO_4$ (all three polymorphs) are undersaturated in the solution. In each case, if the primary supersaturated scale were not to precipitate, then neither would the secondary mineral.

An important example of solid solutions common in the oilfield is (Ba,Ra)SO_4, which results in the formation of radioactive LSA (low specific activity) barium sulfate scale. This is essentially pure $BaSO_4$ because the radium (Ra) content is vanishingly small, but still enough to make the material sufficiently radioactive to be a hazard. The waters are still grossly undersaturated to $RaSO_4$ in such cases, showing that the mineral would never precipitate on its own and that the incorporation into the $BaSO_4$ lattice is essential to generate the LSA material [1,23–25].

5.2.2 Inorganic mineral scale formation

Inorganic scales are formed as a result of exceeding the solubility of minerals in water. However, since the common scales relate to the precipitation of sparingly soluble minerals, one key parameter is that there needs to be sufficient throughput of scaling ions at a particular location to allow the scale to build up to operationally significant levels. For example, a severe scaling regime where a maximum of 500 mg/L can potentially precipitate (if the system reaches thermodynamic equilibrium, which, as explored later it almost never does), equates to a volume fraction of the mineral relative to the solution of less than 0.02% (assuming a mineral density of 2–4 g/cm^3). Clearly, therefore, for scale to become a significant operational issue, continuous replenishment of the scaling ions at the same location is required. In oil and gas situations, this occurs in the production system from the near wellbore area and throughout the production and processing facilities, as a result of the very large volumes of fluids being produced.

Although the relationship controlling mineral solubility is very complex, and relates to a very large range of factors, the tendency for a single stream to scale—known as self-scaling—is largely determined by the influence on solubility of temperature, pressure, and pH (normally, a secondary effect of pressure). Table 5.2 shows how mineral solubility changes from the reservoir to the surface, with the associated decrease in temperature and pressure and increase in pH.

Table 5.2 Changes in mineral solubility due to reduction in temperature and pressure and rise in pH as fluids moves from reservoir to surface.

Condition change from reservoir to surface		**Temperature Decreases**[a]	**Pressure Decreases**	**pH Increases**
Scale Type	**Mineral Name**	**Change in mineral solubility as moving from reservoir to surface**		
$CaCO_3$	Calcite, aragonite, vaterite	↑	↓[d]	↓
$CaSO_4\ 2H_2O$	Gypsum	(↑)	↓	(↓)
Caso$_4$ ½H_2O	Hemihydrate	↑	↓	–
$CaSO_4$	Anhydrite	↑	↓	–
$BaSO_4$	Barite[b]	↓	↓	–
$SrSO_4$	Celestite	(↑)	(↓)	–
$FeCO_3$	Siderite	↓	?[d]	↓
$Fe_{1-x}S$	Pyrrhotite	↑	↓[e]	↓
FeS_2	Pyrite	↓	↓[e]	↓
ZnS	Sphalerite	↓[c]	↓[e]	↓
PbS	Galena	↓	↓[e]	↓
Silicas	Amorphous Silica	↓	↑ $<$ 200 bar ↓ $>$ 200 bar	–
Silicates	Metal (M) Silicates	Depends on M	May depend on M	(↓)

[a]Assumes normal Joule-Thomson (JT) cooling effect with heat losses to the surroundings.
[b]Maximum solubility around 150 – 200°C, then reverses.
[c]Maximum solubility around 35 – 40°C; inverse solubility for most oilfield cases.
[d]The absolute dependence of carbonate mineral solubility on pressure in an oilfield system is likely to be much less than the effect of pressure on CO_2 partitioning and hence pH of the aqueous phase.
[e]Based on behaviour at 200°C.

For several minerals the absolute pressure dependence is not known, though, in general, most minerals become more soluble as the pressure increases because the total volume of the mineral dissolved in the water is lower than that of the water plus the undissolved mineral. However, for minerals where pH has a significant influence, the secondary effect of pressure on pH (how it controls the phase partitioning of acid gases, CO_2, and H_2S) can be much greater than the absolute effect on the mineral itself, as discussed in Section 5.2.2.1. Similarly, many instances of halite (NaCl) scaling in gas and gassy oil systems arise due to pressure drops, even though these have very little influence on the absolute solubility of halite in water; instead, the reduced pressure causes evaporation of water into the co-produced gas, which concentrates the remaining water stream to a point where the scale mineral is supersaturated.

From the consideration of Table 5.2, it becomes clear that the changes in environmental conditions associated with oil and gas production can lead to scaling, in particular, if the water starts out saturated in the minerals because they are present in the reservoir. It also explains how severe calcium carbonate and calcium sulfate scales are often associated with hot surfaces, such as heat exchangers or the motor housing of electrical submersible pumps (ESPs), where the skin temperature can be significantly higher than the bulk fluid temperature. Moreover, as heat exchangers are typically located in surface facilities, fluids have already depressurized to a considerable degree and pH may have increased, thus producing two drivers for calcium carbonate scaling.

Many self-scaling phenomena can be explained by the trends indicated in Table 5.2, but another main scaling mechanism must be considered: that of incompatible mixing, either of water streams with differing compositions or a stream with a production chemical that modifies its properties. Each mechanism is further explored below.

5.2.2.1 Self-scaling: carbonate scales

One mechanism of self-scaling is now illustrated with the example of the most prevalent scale in the oil industry. Essentially, all formation waters are in equilibrium with their surrounding rock minerals under reservoir temperature and pressure. For carbonate reservoirs or clastic sandstone reservoirs containing even a minute trace of calcium carbonate, the waters are saturated with respect to calcium carbonate. Therefore, almost all produced formation waters have the potential to become self-scaling with respect to this mineral. The key is whether the effect of reduced pressure (decreases solubility) is greater than that of reduced temperature (increases solubility). The pressure effect on the solubility of the mineral is actually very weak, but it exerts a much greater effect on the pH of the water and thus the solubility of the scale mineral. This is because dropping the pressure causes the hydrocarbon phase to expand, which tends to drive CO_2 out of the water phase and into the hydrocarbon. This effect seems to be greater when the hydrocarbon is a gas, which leads to the myth (still believed by many) that this scale can only occur downstream of the bubble point, where gas has broken out. This is not true because transfer of CO_2 from water to the oil above the bubble point can also lead to supersaturation. However, it is *most commonly* observed that calcium carbonate deposits downstream of the bubble point, perhaps because the enhanced turbulence caused by gas breakout also enhances the transfer of CO_2 between the phases: in other words, what was long interpreted to be a thermodynamic phenomenon is actually a kinetic one. In either case, the loss of this acid gas from the aqueous phase increases the pH, and water can become supersaturated to scale. Reductions in temperature increase the solubility of the CO_2 in the water phase, keep the pH lower, and mitigate the effect of the pressure drop.

Although the reactions and equilibria leading to the formation of calcium carbonate are numerous and complex, the process can be simplified by the two equilibria below.

Calcium carbonate formation:

$$Ca^{2+}_{(aq)} + 2HCO^{-}_{3(aq)} \leftrightarrow CaCO_{3(s)} + CO_{2(aq)} + H_2O \leftrightarrow CaCO_{3(s)} + CO_{2(g)} + H_2O \qquad (5.1)$$

A pressure drop (to bubble point) leads to the loss of CO_2 from the aqueous to the gas phase; therefore, the equilibrium shifts to the right and drives increased carbonate precipitation.

Carbon dioxide in water:

$$HCO^{-}_{3(aq)} + H^{+}_{(aq)} \leftrightarrow H_2CO_{3(aq)} \leftrightarrow CO_{2(aq)} + H_2O \leftrightarrow CO_{2(g)} + H_2O \qquad (5.2)$$

Loss of CO_2 as brine passes through the bubble point (or as described above as CO_2 partitions into an expanding hydrocarbon phase prior to the bubble point) leads to increased pH, at which point the solubility declines rapidly, leading to increased precipitation.

In summary, both calcium carbonate and other carbonate scales are mostly formed due to the interplay between changes in pressure and temperature during production. Considering a particular case of the near-wellbore reservoir of a production well, this area is subject to a pressure drop, the drawdown. Associated with this drawdown will be a temperature change as the relatively rapid change occurs irreversibly from a thermodynamic viewpoint. This is often approximated by assuming it to be isenthalpic (the total enthalpy of all fluids remains about the same during the transition) or, in more loose terms, it will be subject to a "Joule–Thomson (JT) effect" (strictly speaking, only valid for a single phase). For most systems, the JT effect cools the fluids; therefore, whether a calcium carbonate tendency is expected in the near-wellbore is determined by whether the effect of the drawdown exceeds that of the JT cooling. A calculation must be determined on a case-by-case basis via scale prediction. However, at very high pressures, the JT effect can invert, leading to heating as the pressure drops. In such a case, the drops in pressure and temperature reinforce one another and high scaling tendencies can result.

A similar situation occurs when the scaling tendency arises from evaporation of water, such as typically occurs with halite. If the pressure drop in the near-wellbore also causes heating, then this will significantly exacerbate the evaporation (and hence the scaling) tendency relative to a situation where the pressure drop is counterbalanced by a concomitant temperature drop.

5.2.2.2 Incompatible mixing: sulfate scales

As described earlier, during secondary production, water may be injected into the reservoir to maintain pressure and produce oil to the surface. This can then lead to the mixing of incompatible waters, which may include formation waters high in calcium, strontium, and barium mixing with injection waters high in sulfate, resulting in the formation of calcium, strontium and barium sulfate scales. The reaction for formation of strontium sulfate is shown in Eq. (5.3). This type of scaling has been a particular issue in many offshore oilfields for decades, with seawater being the natural choice of injection water, due to its abundance and availability.

$$Sr^{2+}_{(aq)} + SO^{2-}_{4(aq)} \rightarrow SrSO_{4(s)} \tag{5.3}$$

This type of scaling has also been observed in wells where more than one reservoir is being produced and where one water is rich in alkaline-earth metal ions and the other in sulfate ions. An example is in the Varg field [26], illustrated in Fig. 5.3, with the formation waters from the various wells (located in the different reservoir units) shown in the accompanying table. The field was seawater flooded, but barium sulfate scaling was observed before it broke through because of the variations in barium and sulfate concentrations in different parts of the field.

In the Varg case, the differences in formation water compositions were ascribed to differing geochemistry associated with proximity to a salt dome (note that all of the waters are of high salinity). In other cases, areal variations can occur either because of natural baffles within the reservoir (which may or may not be visible using seismic methods) or because there has been geologically recent ingress of water of a different composition, and the system has not yet equilibrated.

Sulfate scaling tends to be relatively pH independent, with precipitation controlled by the concentration of the scaling ions in each water stream, as well as the temperature and (to some extent) pressure.

While this scaling mechanism is illustrated here for sulfate scales, incompatible mixing can also occur with many other minerals, provided sufficient concentrations of each component are present separately in each stream. Calcium carbonate scaling can occur when a calcium-based completion brine interacts

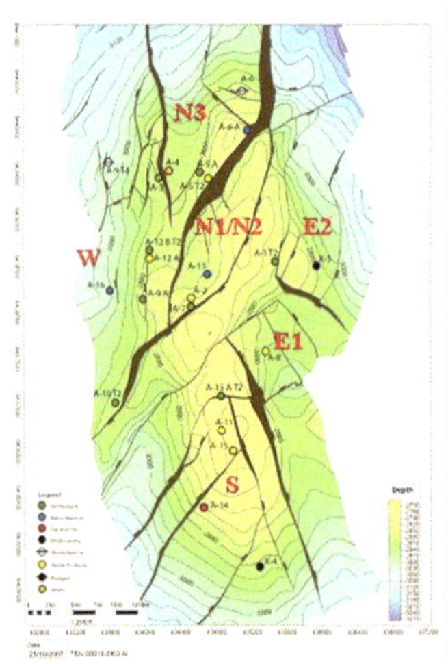

	A-03	A-01	A-06A	A-10	A-12	A-15
Segment	West – W	East - E2	North – N3	West – W	West – W	South – S
	10.07.2005	13.02.2005	10.07.2005	20.05.2005	10.07.2005	03.06.2005
	[mg/l]	[mg/l]	[mg/l]	[mg/l]	[mg/l]	[mg/l]
Sodium	46 000	40 400	39 400	44 000	48 200	42 900
Calcium	20 000	15 800	15 100	22 000	21 900	16 800
Magnesium	2 780	2 560	2 410	2 530	3 510	2 310
Barium	115	39	30	140	89	113
Iron	67	68	10	109	108	83
Strontium	701	764	549	850	673	839
Potassium	1 700	1 650	1 530	1 620	1 920	1 910
Chloride	116 000	100 000	94 200	116 000	124 000	107 000
Sulfate	42	65	100	13	27	53
Bicarbonate	60	59	365	16	68	97
TDS	187 000	162 000	154 000	187 000	201 000	172 000

FIGURE 5.3

Reservoir units of the Varg field with formation water variation between wells shown in the table.

with a bicarbonate-rich formation water, for example, during or after "losses" arising from a well operation. Sulfide scales can occur when a reservoir sours as a result of waterflooding, and some zones produce an H_2S-rich injection water while others produce formation water containing Fe^{2+}, Zn^{2+}, or Pb^{2+}. Calcium carbonate also occurs when triazine-based H_2S scavengers are deployed into a multiphase production stream because the chemicals are highly alkaline and increase the pH.

5.2.2.2.1 Formation water chemistries

As discussed in Section 5.3, the composition of the water streams, including the formation waters, gives the first indication of the likely scaling issues. This is illustrated in Table 5.3 for a small selection of North Sea formation waters and North Sea water compositions [27]. The typical compositions of formation water for the Middle East fields and its local Arabian Gulf seawater are listed in Table 5.4. Note that the seawater composition generally does not change much in

Table 5.3 Selected water chemistries for North Sea.

	Formation water (FW)			
Ion (mg/L)	**Brent**	**Forties**	**Miller**	**Seawater (SW)**
Na^+	9,000	28,370	25,591	10,890
Ca^{2+}	250	2,809	500	428
Mg^{2+}	50	400	91	1,368
Ba^{2+}	60	252	650	—
Sr^{2+}	25	574	60	—
Cl^-	14,300	52,360	39,879	19,700
HCO_3^-	1,050	250	2,050	124
$SO4^{2-}$	—	—	—	2,960

Table 5.4 Typical compositions of formation water in the Middle East and Arabian Gulf seawater.

Ion (mg/L)	**Formation water (FW)**			**Seawater (SW)**	
Na^+	29,680	51,187	50039	13,004	18,043
Ca^{2+}	13,574	29,760	11,055	474	652
Mg^{2+}	1,575	4,264	1,639	1,648	2,159
Ba^{2+}	8	10	—	—	—
Sr^{2+}	557	1,035	706	4.4	11
Cl^-	73,833	143,285	101,613	23,264	31,808
HCO_3^-	369	351	386	134	173
$SO4^{2-}$	404	700	325	3,200	4,450

different parts of the world and is generally similar to that identified in Table 5.3. This is not the case for those identified in Table 5.4 from the Arabian Gulf where higher salinity seawater is identified. This relates to the Arabian Gulf being shallow with evaporation leading to increased salinity, before exiting into the Arabian sea.

Regarding sulfate scales, those encountered in the North Sea, Gulf of Mexico, offshore Brazil, and other areas tend to be dominated by barium sulfate. Other regions, such as many Middle East fields, are dominated more by calcium sulfate and, in some cases, strontium sulfate. This is illustrated in the above tables with the selected Middle East field formation waters containing high calcium and/or strontium but very low barium concentrations [28,29], as shown in Table 5.4, whereas the identified North Sea examples in Table 5.3 clearly have high barium in the formation waters.

5.2.2.2.2 Unconventional reservoir production

In unconventional production, where completions involve multiple fractures, there is the potential for produced waters to have somewhat different compositions, at least in key aspects. Moreover, flowback waters may either be geochemically altered fracture water or this in combination with native FW, which then makes understanding the nature of the scaling risk more challenging. For example, Table 5.5 provides some example brine compositions produced from unconventional reservoirs in the Marcellus and Bakken formations [30,31].

The result is often initial flowback of water that is similar to that of the fracture water (generally, of low salinity) followed by much higher-salinity waters, unusually rich in either dissolved iron or barium or both. The inorganic scales commonly encountered in associated production from unconventional reservoirs include calcium carbonate, calcium sulfate, iron carbonate, iron sulfides, and iron oxo-hydroxides (the last, at least partially derived from dissolved oxygen in the

Table 5.5 Examples of high TDS produced water from Marcellus and Bakken.

Ion (mg/L)	Example #1	Example #2	Example #3
Na^+	64,000	80,000	79,000
K^+	314	—	900
Ca^{2+}	16,200	40,000	6,700
Mg^{2+}	1,200	3,400	900
Ba^{2+}	1,000	3,500	2
Sr^{2+}	500	—	386
Fe^{2+}	0–200	0–200	55
HCO_3^-	100	100	151
SO_4^{2-}	0	0	531

fracture water), some (Ba, Sr)SO_4 (in particular, if the fracture water contained significant sulfate ions), and halite [30,31].

5.2.3 Scale nucleation and growth

Scale formation occurs via two main stages involving nucleation (formation of small crystal embryos) followed by growth, where successive ions are added to the scale crystal in an ordered, "stepwise" fashion. This is broadly illustrated in Fig. 5.4. The process however is more complicated with the individual hydrated ions, initially forming hydrated ion pairs, which agglomerate together as potentially up to thousands of individual ion pairs. These pairs form ordered crystal nuclei (crystal embryos), and in doing so displace their waters of hydration to form small discrete crystals. From this point on, the crystals grow in a more ordered manner as individual ions are added to the surface.

5.2.3.1 Nucleation

Nucleation involves bringing together various ions in solution to create a small proto-solid as a separate phase, and in doing so, it creates a surface. Free energy is associated with this surface, which means this is initially energetically disfavored: the energy input associated with creating the surface exceeds that liberated by releasing the supersaturation in solution. However, the magnitude of this surface energy is proportional to the surface area of the nucleus, which in turn is

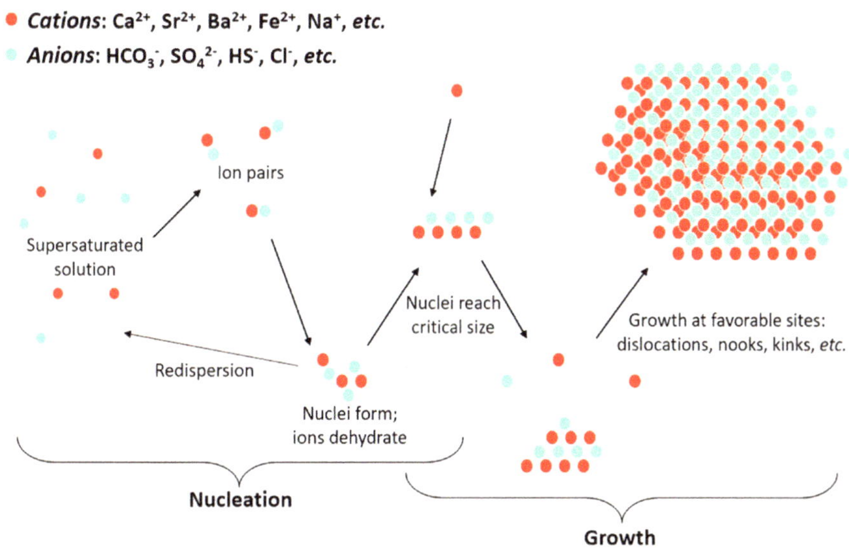

FIGURE 5.4

Processes associated with nucleation and growth of scale.

Image courtesy of Scaled Solutions Ltd.

proportional to r^2 if the nucleus is spherical (r is the radius of nucleus), whereas the magnitude of the released free energy of precipitation is proportional to the mass precipitated (which is proportional to r^3 for the spherical nucleus). Hence, precipitation is initially disfavored for small r but favored for large r, which essentially means precipitation is an activated process, as with almost all other chemical processes. Up to a critical size, nuclei form and break up stochastically, but beyond that size, there is no longer any energetic driving force to break up and the nucleus grows with a monotonic release of free energy. Nucleation is the process that occurs on the energetically disfavored side of the activation energy "hill." Once a nucleus has reached the critical size, then the system enters the crystal growth stage (Fig. 5.4). As with other activated processes, nucleation shows a Boltzmann-type dependence on temperature (proportional to $\exp(-E_a/kT)$, where E_a is the activation energy and T is temperature), which means it often has a fairly steep dependence on temperature.

Homogeneous nucleation is spontaneous formation of precipitates in bulk solution, in the absence of other surfaces. However, it is common to observe apparent homogeneous nucleation, which is really just heterogeneous nucleation onto suspended solids in the production stream. Moreover, a multiphase stream contains many surfaces other than the conduit walls, that is, the oil/water/gas interfaces.

Heterogeneous nucleation is nucleation on a surface, such as pre-existing scale, an added crystal, other foreign matter (e.g., sand, reservoir fines), or a pipe wall. This is observed by rubbing, tapping, or scratching a glass rod in a beaker full of a supersaturated, but metastable, solution and instantaneous nucleation and precipitation occur. Heterogeneous nucleation can occur at lower degrees of supersaturation because there is no requirement to create an initial surface (thermodynamically disfavored), and so heterogeneous nucleation is energetically the preferred route, if possible.

5.2.3.2 Crystal growth

A number of theories have been proposed to explain the mechanism of crystal growth, but in essence, the following processes are involved, upon which a variety of models have been based:

1. transport/diffusion of the reactive species through the solution,
2. adsorption at the crystal–solution interface,
3. surface diffusion and reaction at the interface, and
4. incorporation of the reactive products into the crystal lattice.

Theoretical models have been proposed using Burton, Cabrera and Frank (BCF) theory (spiral growth theory) [32], or the Periodic Bond Chain (PBC) theory [33]. These theories illustrate how crystal growth follows ordered specific growth patterns, with the crystal morphology formed by a series of identical unit cells added to one another to create the crystal. In both theories, crystal "imperfections," such as step edges, crevices, and dislocations, provide energetically favorable sites, for example, because growing at a "kink" in the surface can enable an ion to incorporate into the lattice with greater initial interaction from multiple sides. This is of particular significance in the use of threshold scale inhibitors because, if only a small fraction of a crystal is actively growing (the kinks, screws, and other dislocations), then retarding such growth only requires a similarly low concentration of chemical inhibitor, provided it preferentially adsorbs onto those growth sites.

Crystal growth also occurs as a result of transport of scaling ions to the surface and, therefore, diffusion is a critical process in crystal growth. Transport processes also depend on temperature, though diffusion tends to have a lesser steep dependence than the Boltzmann distribution-type dependence of nucleation. Another important aspect of crystal growth in aqueous solution is dehydration of the ions at the surface, which is obviously necessary before the ion can be incorporated into the growing lattice. The overall dependence of crystal growth on temperature (as well as other important aspects) depends on whether transport through solution or any reaction at the surface is the rate-limiting step. Given the difference in temperature dependence between these processes, the rate-limiting step could be different at different temperatures.

The main factors governing crystal growth are as follows:

- The supersaturation, which is the thermodynamic driving force for precipitation. This is quantified by the saturation ratio (SR) and is described further in Section 5.3.
- The absolute concentration of scaling ions in solution. The precipitation kinetics of highly soluble minerals tends to be facile (halite being by far the greatest of the commoner scales, though relatively soluble gypsum is also facile) relative to less soluble species because there are more ions in solution; therefore, the collision frequency between the scaling ions is greater.
- The temperature will either increase or decrease the supersaturation but will always give more rapid precipitation kinetics for a given supersaturation. It will also increase collision frequency (leading to enhanced probability of nucleation) and increase transport. The influence of pressure on scaling kinetics is generally weak or very weak.
- The viscosity of the aqueous medium, which is related to the diffusion coefficients of the ions, and hence also to the transport of ions to one another and to the growing crystal surface.
- The ease with which ions are dehydrated and either ions or ions pairs are incorporated into the crystal lattice, for example, the water activity is lower in high-salinity solutions; therefore, the degree of hydration of the individual scaling ions may be lower.
- The solution pH, which can change the surface charge of the growing crystal faces. Moreover, the growth of carbonate scales at oilfield pH values involves the incorporation of the bicarbonate ion, HCO_3^-, which must then deprotonate to the carbonate ion, CO_3^{2-} as it is incorporated into the lattice; predictably, a deprotonation step will be influenced by the concentration of protons in the medium by the law of mass action.

- The presence of inhibitors in solution. These inhibitors can either be purposely added (to control scaling) or they may be present naturally, and can be organic or inorganic. Organic inhibitors could be water-soluble organics from the oil, while inorganic ones could simply be ions present in the FW. An example of the latter is the presence of high concentrations of Mg^{2+} in seawater, which weakly inhibits the crystallization of calcite but not aragonite (both polymorphs of calcium carbonate).
- The hydrodynamics of the flow local to the part of the system of interest.

In the interests of brevity, it is not possible to describe all of the above aspects, though the last one is highlighted further in following section.

5.2.3.3 Influence of hydrodynamics on scaling kinetics

The importance of the flow regime and hydrodynamics in general on scaling kinetics has been noted by many authors, in particular, by the fact that systems with relatively mild scaling tendencies seem to experience relative low amounts of deposition except in key locations where the flow regime is disturbed, in particular, by "jewelry" in a well or by other important control devices within the gathering system or facilities. This gives scaling its own Murphy's Law, which is that the more susceptible a piece of equipment is, the more likely it is to induce scaling. And unlike many examples of this law, this is one with an increasingly understood explanation. The presence of susceptible equipment introduces additional turbulence and associated wall shear which then enhances the kinetics of nucleation and growth of the scale *per se*, and where the scaling arises from incompatible mixing of different waters, this also enhances mixing between the streams. Examples include what happens in gravel packs, through and within sand screens, within and just downstream of inflow-control devices, around gauges used for enhanced monitoring in "intelligent" wells, in and around valves of all kinds, including subsurface safety valves, chokes (more so if partially open), within artificial lift systems (around gas-lift valves, within ESPs, in and around the nozzle and diffusers of jet pumps), even at locations where a well's production tubing crosses over from one size to another. Similar phenomena are observed in gathering systems and facilities where high degrees of turbulence are generated, such as in pressure-control valves between separation stages, hydrocyclones, and so on.

The reasons for the effect of hydrodynamics are still only poorly understood but it is clear that two phenomena are important factors:

- The localization of areas of relatively high densities of turbulent kinetic energy, which not only helps to surmount the activation energy associated with scale nucleation but also enhances mass transport, which is required for both nucleation (increased collision frequency) and growth (transport from the bulk to the growing surfaces);
- Increased residence times as a result of "eddy currents" in the flow, which recycle the fluids within a particular location, giving extra time for nucleation and growth to occur.

A specific effect of turbulence-enhanced transport occurs when transport between phases is required for scaling, such as loss of CO_2 or water from the aqueous to the gas phase. It is surely no coincidence that deposition of carbonate and halite scales is often enhanced at locations of enhanced turbulence associated with pressures drops, such as around partially open chokes: the pressure drop provides both the thermodynamic driving force (via transfer of CO_2 or water to the gas phase) and the turbulent kinetic energy to enhance the kinetics of mass transport, rather akin to shaking a can of fizzy drink.

5.2.4 Thermodynamics and kinetics

From a thermodynamic point of view, a supersaturated solution will precipitate solid until it reaches an equilibrium state. However, knowing that an oilfield fluid is supersaturated in a mineral only indicates whether scale will or will not form given infinite time, and not whether scale will deposit over a relevant production timescale. It is possible for solutions to be supersaturated with several times the equilibrium level of solute dissolved but remain in a metastable state for a considerable period of time. Therefore, understanding the kinetics of oilfield scale formation, particularly for mildly supersaturated systems, is important. For such systems, the kinetics of precipitation and the local environment (availability of surfaces, flow paths and solutions tortuosity, turbulent flow regimes, etc.) control whether or not scale formation and field problems will be observed rather than the absolute thermodynamic saturation.

This is illustrated in Fig. 5.5 which summarizes the critical SR, or degree of oversaturation, that a solvent can hold in solution prior to spontaneous nucleation

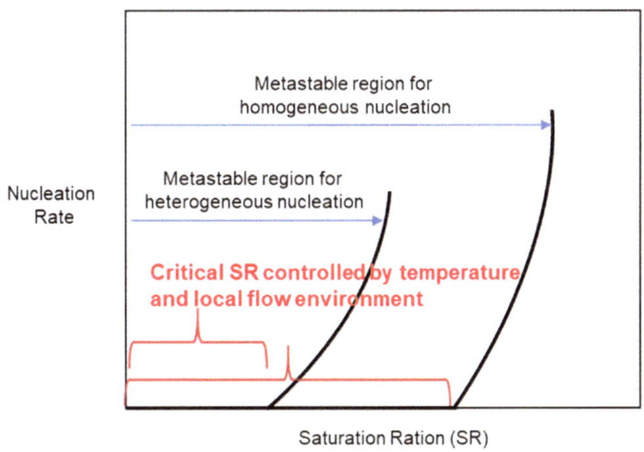

FIGURE 5.5

Illustration of the metastable region (critical SR) prior to spontaneous crystallization, under both homogeneous and heterogeneous conditions.

for both homogeneous and heterogeneous precipitation. As discussed in Section 5.2.3.3, this critical SR is controlled by the local flow environment and conditions and is therefore difficult to predict with accuracy.

At low supersaturation, field scale formation becomes more influenced by the local production environment. In particular, the flow regimes, the availability of surface nucleation sites and pressure drops during production, all impact on the kinetics of scale formation and therefore the propensity for scale to form and lead to production problems. There has been a considerable amount of work conducted in recent years to understand these phenomena.

Two important elements that influence scaling kinetics, and hence the critical SR value, are the system temperature and hydrodynamics. At higher temperatures, the chemical and physical processes that result in scale formation occur more rapidly and hence a supersaturated solution is less likely to remain in a metastable state over the expected production timescale. Conversely, at low temperatures, scaling kinetics can be slowed significantly and higher critical SR values are expected. For example, fluids being produced at a subsea pipeline temperature of 4°C can be significantly oversaturated with $BaSO_4$ and yet remain metastable.

For carbonate scaling, field experience indicates that wells producing at temperatures in the region of 85°C may be operated safely at SR up to 2.5, whereas for wells operating at higher reservoir temperatures, field scale formation can occur when the SR is lower (e.g., SR of 2) at 98°C. Other work suggests that an SR of 5 may be required for calcium carbonate precipitation at 80°C, whereas this drops to 1.2 at 150°C [34]. Related work, examining sulfate scales, has also indicated the importance of reaction kinetics and the local production flow environment, on the potential for oversaturated solutions to remain stable. This is illustrated in the Fig. 5.6.

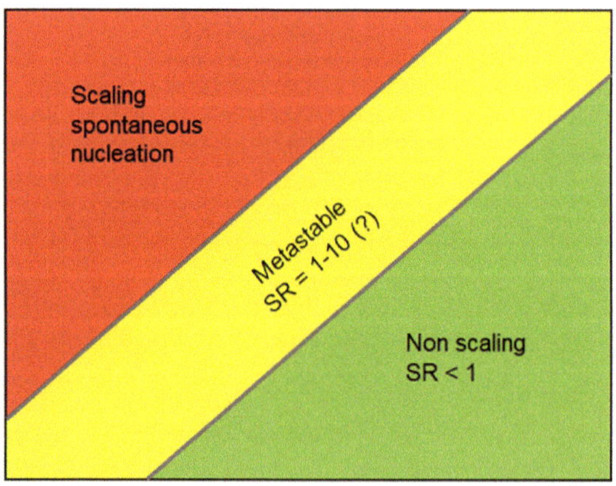

FIGURE 5.6

Scaling regions depending on saturation ratio.

It has long been recognized that scale tends to form in areas of the production system that experience high levels of turbulence and surface shear stress such as restrictions, ICDs, ESPs, sand screens, and so on [35]. However, the impact of hydrodynamics on scale formation has been increasingly reported and studied in recent years indicating that more severe hydrodynamic environments are expected to result in lower critical saturation ratios.

The risk of increased scale formation in and around ICVs and ICDs has been considered in publications with a focus on the theoretical potential for pressure drops across the nozzles, or mixing of incompatible waters around the devices, to result in precipitation [36–38]. Gustavsen et al. [39] provides a detailed report into the deposition of barium sulfate and strontium sulfate around ICD nozzles in a well in the Heidrun Field in the Norwegian Sea. It was proposed that high turbulence created by high velocity flow through the ICD nozzles was the cause of scale forming in these locations. The impact of turbulence caused by other devices that lead to irregularities in flow, such as downhole safety valves (DHSV), are also discussed. Jøsang et al. [40] have presented evidence of a direct relationship between the failure of DHSVs and field saturation ratio even in situations where scale inhibitors are present at concentration above those known to be effective in the bulk fluids). Scale deposition has also been noted as a failure mechanism for ESP which are another area of flow restriction and increased shear within production wells [41].

Several laboratory investigations aimed at increasing understanding of the impact of hydrodynamic factors (turbulence, wall shear stress and mass transport) on inorganic scale formation and deposition have been reported [36,41–47]. These have confirmed the effect that high shear, high turbulence production environments can have on scaling risk in terms of the lower critical SR values expected, higher scale inhibitor concentrations required and the likely locations of mineral deposits. Scale deposition may not always occur at the location of most extreme shear and turbulence due to the physical forces preventing surface adherence, but can often occur in neighboring regions upstream or downstream of a fluid mixing point such as that introduced by a gas lift valve or ICD. The upstream increase in scaling may initially seem counter-intuitive. However, it has been shown that this is caused by the incoming fluid at the mixing point disrupting the main flow and causing recirculating zones which can significantly increase the residence time of the fluid in the area immediately upstream providing increased potential for scale nucleation and growth.

5.3 Scale prediction

5.3.1 Scale prediction as a component of scale management

Scale prediction is the means by which the operational risks associated with scaling can be assessed, and as such, is a key input to developing a management strategy that minimizes the total cost of operations related to this risk.

Some control measures are optimally installed at the project stage, and so prediction has particular importance during the various stages of project development, including whether potentially high-cost capital items must be included (downhole injection in wells, a sulfate reduction unit for waterflood, etc.). However, it remains important throughout the field life cycle. First, the initial assumptions made during the project need to be verified by surveillance because the input parameters available at the project stage are always uncertain (variable quality of the formation water sample(s); uncertainty about production rates, system temperatures and pressures, transit times for injected fluids, etc.). Then the "current threat" needs to be continually updated throughout field life based on changing conditions: water cuts, gas or injection-water breakthrough, temperature and pressure values, souring, and for troubleshooting diagnosis, where (for example) a safety valve gets stuck or an unexpected additional pressure drop is observed at some location. Prediction can also play a role in understanding how the reservoir is behaving, elucidating geochemical effects that may influence the scaling prognosis. Hence, it is clear that effective scale management requires a feedback loop from the surveillance plan into scale prediction.

One of the main aims of the control strategy is to find the optimal balance between scale prevention and scale remediation. This optimum represents the one that leads to the lowest overall costs to the development from all of the threats associated with scale, production deferment, HSE risk (e.g., from compromised safety-critical equipment), and remediation costs, but other effects can also contribute (poorer separation, underdeposit corrosion, lower efficiency of heat exchangers, etc.). To approach this optimum, it is necessary to know with as much as confidence as practicable, the nature of the threat: how much scale, when (and how long), where (which can have a huge impact on control costs), and how is it likely to effect key parts of the system? All of this comes from scale prediction.

5.3.2 Scale prediction outputs

Scale prediction programs are based upon thermodynamics, calibrated against experimental solubility data, over a range of conditions (temperature, pressure, ionic strength, pH, etc.). There are a number of commercially available "thermodynamic" models on the market for inorganic scale predictions. These include, but are not limited to, ScaleChem (OLI Systems, USA), ScaleSoftPitzer (Brine Chemistry Consortium of Rice University, USA) and FLOW-iT$_{\text{SCALE}}$ (Scaled Solutions Ltd., UK). Broadly speaking, all of the common prediction models predict the thermodynamic oversaturation of a particular mineral scale at a certain set of conditions, along with the maximum mass of scale (or excess solute) that could precipitate if the system reaches equilibrium.

The main outputs are twofold:

- The saturation ratio (SR), which describes the thermodynamic driving force for scale to form and is essentially an indicator of the probability of scale forming. The SR is unity when the system is at equilibrium and supersaturated for values greater than unity; the higher the value, the greater is the driving force for precipitation and scaling. An alternative measure of the driving force is the Saturation Index, which is simply the base-10 logarithm of the SR, which means equilibrium is reached at Saturation Index being zero and negative values indicate undersaturation.

- The excess solute (ES), is the mass of the scale per unit volume of water that could form when any supersaturation is relieved by precipitation of that scale, that is, the maximum mass (in mg/L) that can precipitate for the system to reach equilibrium. The ES, together with the water production rate, can be seen as a general indicator of the severity of scale if the SR of the scale indicates that the scale is likely to form.

The theory behind how these parameters are calculated is explored further in Section 5.3.3.

There are currently no commercially available prediction programs that realistically describe the kinetics of scaling, that is, how quickly precipitation will occur, what fraction of that precipitate will form as deposited scale within a given time and how much will form as suspended solids within the produced fluids, though as discussed below there is ongoing active research in this area to improve this area of scale prediction (we note that work is progressing in this area, particularly with researchers at Rice University's Brine Chemistry Consortium group and also at Scaled Solutions Ltd.'s SWELL JIP, with new programs being developed in an attempt to capture the scaling kinetics). While certain aspects of the process are understood to some extent, as introduced in the previous section (e.g., the effect of temperature), attempts to create "operationally useable" models have been frustrated to date by the fact that these processes are influenced in a very complex manner by the effects of system hydrodynamics, as well as the presence of species in the system, often at trace quantities, that can influence scaling processes but that are not typically analyzed (or their nature is not even precisely known). Moreover, there is the additional complication that different parts of the system can be affected by scale deposits in different ways. Whereas 2 tonnes of scale deposited evenly over a kilometer of production tubing may cause negligible operational issues; at the other extreme end, 1 kg of scale deposited around the sliding component of a subsurface safety valve will likely cause it to fail, thereby being operationally significant. Hence, the industry uses general rules of thumb to relate the SR and ES, together with the water rates and the general "susceptibility" of the location to scale, to categorize the scaling tendency into scenarios such as the following:

- Where there is no or negligible risk: surveillance is still required to verify that the assumptions that led to this conclusion are correct, but otherwise, no specific measures are likely to be required to control scale.
- Where the risk is likely to be operationally significant but not necessarily severe. This means that measures should be put in place, such that either

prevention or remediation can be achieved cost-effectively, together with fit-for-purpose surveillance to verify the initial predictions and to determine the current scaling scenario, when this changes, and when treatments (scale removal or periodic squeeze treatments) may need to be repeated.

- Where the scaling tendency is severe enough that a high rate of scaling can be expected and preventive (as opposed to remedial) measures are likely to be the most cost-effective in terms of minimizing the overall total cost of operations related to scaling.

Broadly speaking, at low values of SR (e.g., SR $<$ 1.5–2 for most scales) and with production at low to moderate temperatures (up to $\sim$ 100°C), scaling is likely to be under kinetic control and therefore not expected to form sufficiently under normal production conditions (excluding high shear locations such as ESPs, chokes, etc.) to represent a significant operational risk. Such low SRs, especially with low levels of excess solute ($<$ 20 mg/L), often represent a low scaling risk.

5.3.2.1 Scale risk summary tables

Table 5.6 summarizes one such guideline with respect to scaling risk. It is noted that the values given in this table are indicative of those where field experience has shown scaling to fall broadly into the categories indicated. Nevertheless, the table enables a visual picture of where and under what conditions scaling is most likely to be an operational issue and where it will be most severe. In practice, scale prediction also depends on both water cut and absolute water rates, as well as the general susceptibility of the system to scale deposition and, like chemical kinetics, should also depend upon temperature (which is not included in Table 5.6). However, no simple set of scale prediction criteria exists to reflect this reality, and this should be borne in mind when interpreting scale-prediction data.

These guidelines, or rules of thumb, represent conventional production conditions, for example, moderate temperatures (up to $\sim$ 100°C). At high temperatures, the general SR levels at which significant scaling can occur, are reduced. It is further noted that these guidelines tend to be operator- and indeed experience-specific. Different operators and different scaling experts have their own in-house versions that perhaps better reflects their own operations and risk averseness.

Table 5.6 An example of a scaling risk assessment.

Key	Interpretation	$BaSO_4$	$CaCO_3$, $CaSO_4$ & $SrSO_4$
X	No scaling predicted	SR $<$ 1, ES = 0 mg/L	SR $<$ 1, ES = 0 mg/L
L	Low scaling risk	SR = 1–3, ES $<$ 10 mg/L	SR = 1–3, ES $<$ 50 mg/L
M	Moderate scaling risk	SR = 3–10, ES = 10–40 mg/L	SR = 3–10, ES = 50–250 mg/L
S	Severe scaling risk	SR $>$ 15, ES $>$ 40 mg/L	SR $>$ 15, ES $>$ 250 mg/L

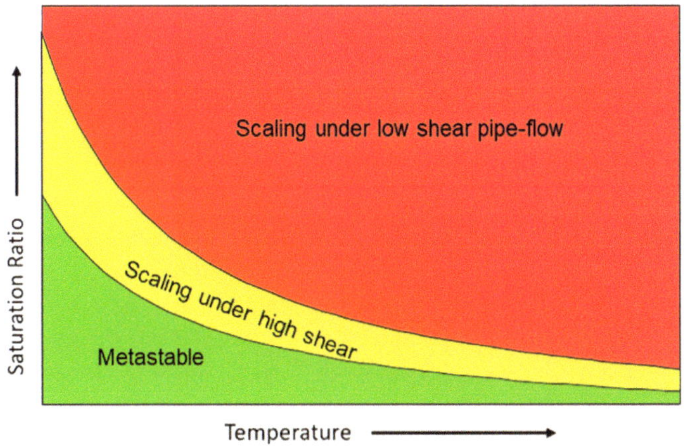

FIGURE 5.7

Illustration of interplay of shear/turbulence and temperature on the critical saturation ratios of inorganic scales.

Courtesy of Scaled Solutions Ltd.

The cost of being wrong about a scaling tendency can differ greatly depending on the setting.

In the work being undertaken by the SWELL JIP at Scaled Solutions, scaling risk as a function of fluid flow dynamics is being assessed, specifically, the influence of shear and turbulence on transport parameters and the localization of kinetic energy necessary to overcome the kinetic barriers to nucleation (an activated process). One aim has been to elucidate why scale deposition tends to concentrate at certain locations, often those where deposition will have the greatest operational impact. Another aim was to provide objective evidence to underpin the critical saturation ratios (such as those shown in Table 5.6) as function of mineral type, temperature and fluid flow dynamics, particularly, as a function of surface shear [44,45,48]. In this way, scale-prediction output can be interpreted with a better degree of granularity based on key parameters of individual systems, rather than the more broad-brush rules of thumb that have been applied to date. An illustration of the effects determined across a range of temperatures is presented in Fig. 5.7, which clearly indicates the limitations of the summary guideline tables above.

5.3.3 Theory of scale prediction

5.3.3.1 Solubility product and saturation ratio

When a chemical reaction is in equilibrium, the rate of the forward reaction equals the rate of the reverse reaction. There is no change in the concentration of reactants or products with time, and the concentrations of all are related by the equilibrium constant (K). The brackets indicate that molar concentrations are used.

$$A + B \rightleftharpoons C + D \qquad K = [C]\cdot[D]/[A]\cdot[B] \tag{5.4}$$

The use of concentrations to define equilibrium is an approximation that works well in dilute brines at low ionic strength. Real brines (including pretty much all oilfield waters) deviate from this approximation sufficiently that this simplistic treatment is inadequate. Instead of concentration, equilibrium is defined in terms of an effective "thermodynamic" concentration, known as the activity (a). This accounts for the fact that the participation of an ion (or any species) in solution reactions is influenced by the other species present: the hydration sphere around it, other ions that can interact with it (e.g., forming charge pairs, which can reduce its availability to react with other species), and the dielectric constant of the aqueous medium itself (which is largely related to its ionic strength). The way this is implemented in all electrolyte models, of which a scale-prediction program is one type, is to relate the concentration of a species to its activity using a multiplier, known as its activity coefficient (γ).

Hence, the more correct definition of equilibrium, which can then be directly related to thermodynamics parameters for the system, is

$$K = a_C \cdot a_D / a_A \cdot a_B \tag{5.5}$$

where $a_C = [C]\cdot\gamma^{\pm}{}_C$, and $\gamma^{\pm}{}_C$ is the activity coefficient of C.

The value of the equilibrium constant varies mainly with temperature, pressure and ionic strength. It is a valuable tool for determining the direction that a reaction will proceed and the amount of unreacted products that remain.

For scaling the relevant equilibrium process is the scale mineral reaching saturation in the oilfield water, for which the equilibrium constant is called the solubility product (K_{sp}). For example, for barium sulfate dissolved in water:

$$BaSO_{4(s)} \leftrightarrow Ba^{2+}_{(aq)} + SO^{2-}_{4(aq)} \tag{5.6}$$

$$K_{sp} = (a_{Ba})_{eqm} \cdot (a_{SO4})_{eqm} / (a_{BaSO4})_{eqm} \tag{5.7}$$

where the suffix *eqm* refers to the values at equilibrium. Although the expression for the equilibrium constant in Eq. (5.7) contains the term $(a_{BaSO4})_{eqm}$, the activity of a solid within the solid phase is conventionally assumed to be unity (though, there are some exceptions when considering solid solutions). The expression then reduces to the form:

$$K_{sp} = (a_{Ba})_{eqm} \cdot (a_{SO4})_{eqm} \tag{5.8}$$

If solid $BaSO_4$ is added to a solution that contains no Ba^{2+} or $SO_4{}^{2-}$ ions, some $BaSO_4$ will dissolve and the resulting solution will contain equimolar amounts of each ion; this is not normally the case in oilfield waters, and there is no need for it to be so as long as the entire solution is charge-balanced (equal moles of cations and anions accounting for different charges on each).

The general expression for the solubility product constant provides a convenient way to define the saturation state of $BaSO_4$ (or any other mineral) for a given solution and conditions. Therefore, bearing in mind that the K_{sp} is defined relative to the ion product at equilibrium $[(a_{Ba})_{eqm} \cdot (a_{SO4})_{eqm})]$, generalizing these values for any case, the saturation state can be defined thus:

$(a_{Ba}) \cdot (a_{SO4}) > K_{sp}$ - System oversaturated with $BaSO_4$ and might precipitate
$(a_{Ba}) \cdot (a_{SO4}) = K_{sp}$ - System is saturated with $BaSO_4$ at equilibrium and will not precipitate
$(a_{Ba}) \cdot (a_{SO4}) < K_{sp}$ - System is undersaturated with $BaSO_4$ and will not precipitate

Moreover, the driving force for precipitation (a necessary though not sufficient condition for scaling) can be quantified using this approach. The (a_{Ba}). (a_{SO4}) is known as the ionic product for $BaSO_4$, and so the saturation state can be quantified by the ratio of this to the solubility product, a value referred to as the *saturation ratio* (SR):

$$\text{Saturation Ratio}(BaSO_4) = (a_{Ba}) \cdot (a_{SO4}) / K_{sp} \tag{5.9}$$

This is the expression for the SR of $BaSO_4$, but of course, equivalent expressions exist for all scale minerals. As mentioned in Section 5.3.2, it is one of the two main outputs of scale prediction: it defines whether precipitation is possible, and if it is, the degree to which it is greater than unity is a measure of the thermodynamic driving force for precipitation. In broad terms, it can be seen as a measure of the probability that precipitation (and thence scaling) will occur.

5.3.3.2 Excess solute

For a system that is supersaturated, that is, the SR exceeds unity, there is more dissolved scale than the solution can support based on equilibrium criteria, and it will have a tendency to precipitate over some time scale. The excess solute (ES) is the quantity of that mineral that must precipitate for the system to achieve equilibrium (SR = 1). It has units of concentration because it refers to a mass of mineral (or number of moles for the purists) that will precipitate from unit volume of solution; mg/L is the commonest used, but lb/barrel is also occasionally seen. Excess solute is essentially the maximum mass of mineral scale which can precipitate if the system reaches equilibrium (SR = 1). The various components of the ES are shown in Fig. 5.8. The general proportions of which are for illustration purposes only: the actual values will depend on the scaling tendency, the mineral type, the residence time and a variety of other factors that determine the kinetics of nucleation and growth, both homogeneous and heterogeneous.

The ES is sometimes referred to as the "(potential) mass of scale," but this can be confusing. First, the ES is not a mass but a concentration; neither does it strictly refer to scale (a deposit) but is the potential for precipitation (generally, rather less deposits than precipitates). And perhaps worst of all, this nomenclature occasionally encourages engineers to calculate an expected "mass of scale" based

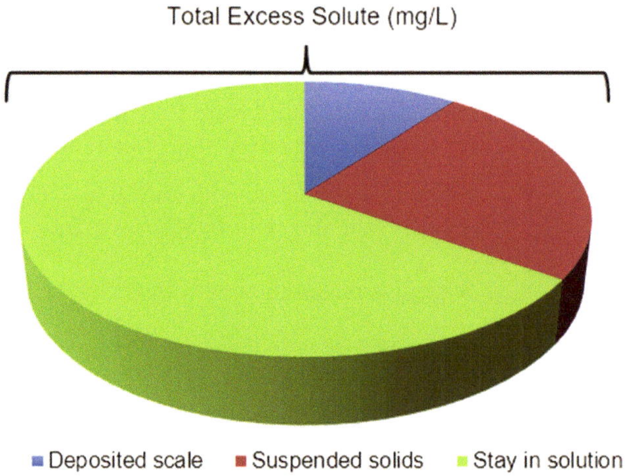

FIGURE 5.8

Component fractions of the excess solute (values are purely illustrative).

Image courtesy of Scaled Solutions Ltd.

on this number, typically overestimating what can be expected in the field by perhaps 2 orders of magnitude.

An important aspect of the ES is that it depends on how the user (and the prediction software) calculates it. For a system supersaturated in more than one scale with a common ion (*e.g.*, $BaSO_4$ and $SrSO_4$), the ES reported will depend on whether the user selects either or both scaling minerals to precipitate. If both are selected, most programs will precipitate the one with the highest SR, which, because this denudes the solution of the common SO_4^{2-} ion, means the ES for the other scale is greatly reduced or even zero. Generally, this does not reflect reality because only a fraction (often a small one, as illustrated in Fig. 5.8) of the ES for any mineral will normally precipitate during the residence times typical of most production systems. Therefore (in the case described here), the sulfate concentration is likely to remain high enough for both scales to be seen in the actual system. To account for this, and to ensure that the predictions are plausibly conservative, it is recommended to evaluate the ES for competing scales in separate calculations: selecting one and then the other; reporting both values but not selecting them in the model at the same time.

Given that some fraction of the ES will precipitate (the red and blue sectors in Fig. 5.8) and a smaller fraction will form a scale deposit (the blue sector in Fig. 5.8), the product of the ES and the water-production rate is likely to have an approximate relationship with the severity of scaling, other things being equal and subject to the following criteria:

- The SR is sufficiently higher than 1 that the driving force for precipitation means precipitation will occur significantly within the residence times of the system.

- The water cut is high enough for the produced water to contact the walls. Unfortunately, "high enough" is also a poorly constrained threshold, given that it depends on the nature of the oil (emulsifying tendency), the water (density, salinity) and the system hydrodynamics (the tendency for the denser brine phase to be blasted against the walls due to inertial forces even though it is emulsified. Some operators have reported scaling in "dry" systems, which is impossible, of course. The discrepancy arises because most operators define "dry" as less than 0.5 vol% water, which may be enough to allow scaling to occur, particularly in light-oil or gas systems. Moreover, a 1% produced-water fraction at bottom hole may have largely evaporated into the associated gas by the time it reaches the water-cut meter in the surface facilities.

5.3.4 Importance of quality input data

The number of oilfield developments that are currently designed with no information on the formation water composition is progressively decreasing, thankfully. Most new developments include sampling of the underlying aquifer, and the more thorough ones also attempt Elemental Residual Salt Analysis of the connate water from cores in the target reservoir, in case (i) this is expected to form a significant fraction of the produced water and (ii) it may be different from that of the aquifer or transition zone [49]. There have been cases where the aquifer and connate water have differed in compositions. In this case, it is important not only to recognize this but also understand the significance of producing one vs the other, which may be uncertain at the project stage and early in field life. Scale-control strategies have needed to be completely revised in cases where this was not recognized sufficiently early (Note: In the Elemental Residual Salt Analysis, cores are taken and their contents were washed out. The composition of the immobile, connate water is inferred from the washings).

Another potential source of error can be compositional grading across a field. A slightly tongue-in-cheek "rule" to remember is that all reservoirs are more heterogeneous and compartmentalized than you think, even when you've taken this rule into account. Hence, not only could there be vertical differences in the water composition (aquifer, transition zone, connate water). There could also be areal variations, not least because of reservoir compartmentalization. Something that is often not fully recognized early in field life or at the project stage. An illustration of this is the Varg field, shown in Fig. 5.3.

It has also been observed that in many cases nearly all samples obtained during the appraisal phase were contaminated to some extent, mainly due to drilling mud filtrate. Predictably, for water samples this particularly arises due to filtrate losses into the formation from water-based mud (though even oil-based mud typically has an inner phase containing brine). Inclusion of tracer in the drilling and completion fluids and analysis of the actual fluids are strongly advised; therefore, the extent of contamination can be quantified accurately and its impact on the sample can be assessed. A frequent issue is uncertainty in the barium content of a formation water because of traces of sulfate present in the drilling/completion fluids. Some "lower end" NaCl brines obtained from seawater evaporation have not had their sulfate content sufficiently reduced, while recycling of drilling fluids between fields without proper reanalysis and qualification can lead to nominally compatible well-construction and control fluids leading to loss of information that is critical to scale control.

The importance of accurate reservoir temperatures and pressures cannnot be understated. In many cases, knowing these values within a couple of °C and to the nearest 10–20 bar is quite adequate. Occasionally, however, we have encountered cases where taking a nominal value for reservoir temperature has led to incorrect conclusions, and it was necessary to account for geothermal gradients across the reservoir. A typical geothermal gradient is around 3°C/100 m TVD, and so there can be differences of 10°C–20°C, either between zones or even laterally within the same zone, if the reservoir is dipping significantly and of large lateral extent. This becomes particularly important when scale prediction is used to diagnose a potential issue within a well where the flowing bottom-hole pressure is known from a permanent downhole gauge. It has been found necessary in some cases to determine the reservoir temperature around that specific producing zone (rather than just some nominal value for the field) to model the scaling tendencies in the well accurately.

The same is true for reservoir pressure, though the effect is much less critical.

5.3.4.1 Reservoir equilibration

The important point with this parameter and indeed other parameters used in the modeling, including the brine composition itself, is that it is important to equilibrate any water within the reservoir under the prevailing conditions. Formation or aquifer waters for example cannot be oversaturated within the reservoir where they have come to equilibrium with the surrounding reservoir rocks and hydrocarbon phases over geological timescales. Thus, predictions into the future need to account for any changes in this parameter (e.g., changes in reservoir pressure), as it can significantly affect the "initial conditions" for the calculation. It is therefore critically important that waters are appropriately equilibrated at downhole reservoir conditions prior to commencing any scale prediction exercise.

Finally, we provide a note on surface samples, which are often the mainstay of scale surveillance in the operate phase. An element of formation water entering bottom hole will typically be subject to a degree of evaporation, which may alter its composition negligibly through to greatly, depending mainly on the evaporation tendency and the gas-water ratio. Rising up through the well, the tendency could be evaporative or condensing, depending on how the pressure and temperature change. It will be evaporative in an HTHP well, largely because the fluids are typically above the JT inversion point; therefore, the changes in pressure and temperature reinforce one another. Below this inversion point, they tend to counteract one another, and which "wins out" then often depends on the heat capacity of the fluids. Gas wells tend to condense water because the drop in temperature (low heat capacity) exceeds the effect of the drop in pressure (low hydrostatic head). These effects are not generally important if the well has a high water cut because the phase transfer of water per unit of water produced is very small. It can be important in very low-water oil wells, and is almost always very important in gas wells, where the produced water sampled at surface is often of low salinity for most of field life due to condensation from the gas phase. In most cases, these effects can be estimated and modeled and therefore accounted for with an understanding of the conditions in the various parts of the system. It should also be remembered that evaporation and condensation do not change the ratios of the ions, unless scaling has occurred. Therefore, analysis of these ratios can sometimes be used as a proxy for the absolute values, for example, a reduction in Na/Ca or Na/Mg ratios could indicate halite (NaCl) scaling in a gas well even if no reliable composition can be obtained for the water entering bottom hole.

5.3.5 Example of the utility of modern scale prediction packages

To illustrate the information derived from scale prediction packages, Table 5.7 provides brine chemistries for a series of North Sea formation waters and seawater [50]. Fig. 5.9 illustrates the information derived from the scale prediction codes, with respect to the severity of the scaling potential as a function of mixing of the formation water with injected seawater.

The scale prediction packages allow the profiling of the scaling challenge, which in this example is as a function of % SW. In this case, the maximum ES in these relatively severe scaling systems occurs between 8% and 20% SW, whereas the maximum SR occurs at approximately 50%–60% SW. The field engineers can then compare this information with the production functions (project engineers) or current analyses (ongoing operations) for the various wells in the field and infer the likely scaling risk from the composition of the produced water. In the above cases, it is clear that this scale will be a very significant problem with a

Table 5.7 Water compositions of selected North Sea formation waters and seawater.

Ion (mg/L)	Garn A FW	Miller FW	Forties FW	Seawater
Na^+	17,300	25,591	29,370	10,890
K^+	11,700	1,100	372	460
Ca^{2+}	1,711	500	2,809	428
Mg^{2+}	225	91	504	1,368
Sr^{2+}	199	60	574	8
Ba^{2+}	713	650	252	0
Cl^-	41,247	39,879	52,360	19,700
SO_4^{2-}	0	0	0	2,960

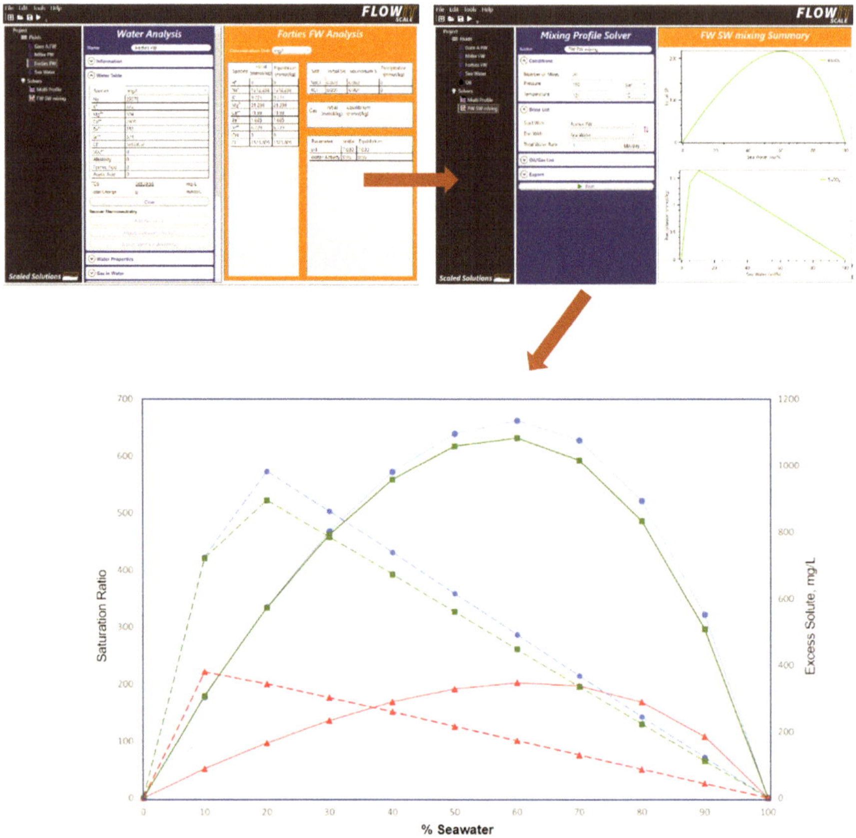

FIGURE 5.9

Barium sulfate scale prediction results for the North Sea brine systems (Flow-iT$_{\mathrm{SCALE}}$ model). SR is the solid line (left axis) and excess solute is the dashed line (right axis). Blue, Garn A; Green, Miller; and Red, Forties.

Courtesy of Scaled Solutions Ltd.

large impact on economics. Performing this prediction during the project phase would provide the impetus to install appropriate prevention methods, given that periodic remediation of severe $BaSO_4$ scale is very unlikely to be the most cost-effective strategy.

5.3.6 Limitations of scale prediction

The accuracy of scale prediction software is governed by the availability and accuracy of empirical (laboratory generated) thermodynamic data, primarily about the solubility of scale minerals under the conditions of interest. For the common carbonate and sulfate scales, the models are generally considered sufficiently accurate and reliable under most conventional oilfield production conditions (up to around 150°C/300 °F). Even within this range, many programs struggle in high-salinity systems where there is also a high concentration of calcium, which is frequently the case in high-salinity waters.

Under extreme HPHT/High Salinity conditions, the accuracy of the models is generally found to be poorer than at more conventional conditions, due to the more limited amount of thermodynamic data available and, in some cases, because the properties of the brine phase no longer conform to conventional models of a standard aqueous phase. Recent works have indicated that the accuracy of the models under many of these extreme conditions can be relied upon provided they have been properly calibrated [51–54]. Models that rely upon extrapolation from trends at lower temperature, pressure and salinity have been repeatedly found to go astray under these more extreme conditions, mainly because the actual solubility behavior does not follow those trends.

The solubilities of the less common minerals are less well known than those of the standard carbonate and sulfate scales, which means the reliability of predictions is often less certain *per se* and becomes even worse at more extreme conditions. The situation is made even worse when there are numerous forms of these minerals, such as sphalerite and wurtzite for zinc sulfide. Iron sulfides have several varieties and some of which are not even well defined (which means neither is their thermodynamic state functions). There are more varieties of silica and metal silicates than almost any other mineral group. Fortunately, there is only one known form of lead sulfide.

Although there are some limitations regarding the thermodynamics associated with scale prediction, the thermodynamics are at least broadly understood and reasonably reliable in most modern scale prediction packages. Perhaps the greatest limitation of scale prediction relates to the development of reliable kinetic models, though as indicated earlier, work is underway in this area as indicated earlier.

5.4 Scale control

5.4.1 Treatment options and scale control strategies

An review paper by Hinrichsen [55] gives a general overview of scale deposition and prevention in oil production facilities, interspersed with appropriate field examples. A more recent paper by Jordan et al. [56] suggests techniques for integrating scale control into the field life cycle management and how current scale prevention technology compares with that applied in fields that started production some decades ago. Further work by Graham et al. [5] summarizes the challenges associated with deepwater and subsea production and includes a review of the general scale control strategies adopted by the industry.

Once thermodynamic scale prediction models have been used to examine the likely impact of scale over the field life cycle, scale control strategies can be selected. These may be divided into four principal categories:

- **Fluid modification/selection of injection water for waterflood:** Seawater with reduced sulfate content; aquifer water injection; produced water reinjection (PWRI).
- **Flow modification of water:** Mechanical or chemical water shut-off, either in the wellbore or near-wellbore of production wells; chemical diversion of flood water deep in the reservoir; polymer flood to improve sweep efficiency and thus also retard early breakthrough of water.
- **Chemical inhibition:** Deployed by periodic scale squeeze treatments, continuous injection, progressive release from solid scale inhibitors (e.g., impregnated into fracture proppant, gravel-pack sand or dumped in the rat-hole of vertical and moderately deviated wells).
- **Scale removal:** Chemical dissolution, milling, jetting, pigging of flowlines, and periodic actuation of moving parts to shear off deposits (e.g., safety valves, sliding sleeves).
- **By-passing scale:** Fracturing and reperforating.

Most of this section will consider chemical inhibition, but fluid and flow modification will be briefly considered first. Scale removal is discussed in Section 5.6.

5.4.1.1 Judicious modification/selection of flood water

Techniques in this category apply primarily to prevention of sulfate scales, which are frequently caused by the mixing of incompatible waters. In principle, if at least one of the scaling ions is removed from one or other of the brine streams, then the problem will cease to exist. Most typically, the sulfate-ion concentration is reduced (not eliminated), which may eliminate the sulfate-scaling tendency but will very often reduce it to a level where it is either acceptably low not to require further control or it will reduce an unmanageable scaling issue into a manageable one.

5.4.1.1.1 Selection of compatible injection waters

One option to avoiding scaling from incompatible mixing in the production stream is to select a compatible water source for injection, for example, from a conveniently situated aquifer, such as the use of Utsira aquifer water (together with produced gas) to flood the Oseberg Sør field in Norway [57], or selecting of compatible low sulfate aquifer water for waterflooding in onshore Saudi Aramco fields [28]. An alternative is to reinject produced water. Other than certain limited changes that occur upon passage through the production system (rise in pH as the fluids degas, addition of low doses of production chemicals), this is typically the same water as already present in the reservoir and hence tends to be compatible with it *per se*. This option is the commonest method of water management in onshore fields given that it is the only practicable disposal method, and so it is obviously even more beneficial if this can be combined with secondary recovery by waterflooding. The main limitation with PWRI is that it can only start when the field in question (or adjacent fields, assuming their formation waters are compatible) produce sufficient water to achieve the required voidage replacement. Therefore, while it can represent a means to mitigate the overall scaling tendency, this often happens later in field life, and another source is required early on until rates of produced water are sufficient.

The following potential water sources for secondary recovery are available depending on the setting:

- Surface waters (lakes, rivers and the sea):
  - The composition of some can vary with the seasons (inland seas, effects of rivers and evaporation), weather and tides (estuaries).
  - Rivers and lakes are of low salinity and are potable; therefore, there could be competition for the resource.
  - Seawater has higher salinity and high concentrations of sulfate, potentially leading to sulfate scaling. On the other hand, for offshore operations seawater is plentiful and readily available.
- Subsurface waters (aquifer and produced waters):
  - Their composition is more varied than surface waters and can be of very high salinity
  - Use of aquifer water is more expensive than surface water because accessing it requires wells. In contrast, aquifer water may require less treatment before injection.

In offshore production environments, seawater is almost universally seen as the first water of choice for waterflood operations. However, in many areas, particularly those in the North Sea, Gulf of Mexico, South America, West Africa, and so on, the presence of moderate to high concentrations of barium ions in the formation waters leads to significant to severe barium sulfate scaling issues, the management of which is costly in both operating expenditure and production deferment. Experience has been similar in some fields in the Middle East but with strontium sulfate due to high concentrations of strontium in the formation waters. In extreme cases (e.g., high barium formation waters) or where the costs of control measures are excessive (such as subsea developments including those in Deepwater), engineering solutions using desulfation technology (see later in this section) may be employed to reduce the sulfate levels in the seawater prior to injection.

5.4.1.1.2 Produced water reinjection

The impact of produced water reinjection on scaling has also been covered in many publications and, just like selection of compatible aquifers, can have a significant impact on managing scaling in a production reservoir [58]. However, the produced water itself may already be scaling and therefore scale inhibition may be required on a continuous basis to protect injection lines and associated formation damage of the injection wells.

Further produced water volumes often need to be augmented by other injection water (seawater in an offshore setting) thereby increasing further any residual sulfate scaling risk and again requiring addition of inhibitors. Once injected into the formation, however, the scale inhibitor is rapidly denuded by adsorption onto rock minerals in the near injection well formation area. This means that the supersaturation within the injection water will be relieved by precipitation deeper within the reservoir, with the result that lower concentrations of scaling ions propagate through the formation, thereby potentially reducing the scaling risks subsequently observed at the production wells [59].

5.4.1.1.3 Geochemical reactions in the reservoir

While this is not purposeful modification of the flood water to make it less scaling, it is well documented that geochemical reactions within the reservoir can lead to modifications of the brine chemistry as waters interact with the reservoir mineral assemblage and mix with other waters within the reservoir. Some of these reactions also reduce the scaling tendency. Recent studies have shown that precipitation and deposition of sulfate scales deep within the formation, due to the in-situ mixing of injected "seawaters" and formation waters, can lead to the development of a large transient mixing zone deep within the reservoir. This, in turn, leads to the production of produced water containing reduced amounts of scaling ions and therefore reduced scaling in the near wellbore area. If the deposition occurs far enough from the production wellbore, then no loss of productivity is experienced, and the effect is beneficial. The total volume percent of scale forming when incompatible waters mix, or when other in-situ reactions lead to formation of inorganic precipitates, represents a very small percentage of the pore space. Therefore, precipitation in the deep reservoir has negligible effect on productivity and recovery efficiency, but can be beneficial by reducing the concentration of ions at the production wells. The interaction of a single water with composition different from that of the formation water with the reservoir minerals can also lead to similarly beneficial effects. This has been observed in high-temperature fields under seawater flood, where the Mg/Ca ratio is much higher in the seawater than the formation water. Ion exchange (or possibly also dolomitization reactions) occurs between these two (and other) ions and the reservoir minerals, increasing the calcium concentration in the flood water. At high temperatures the water becomes supersaturated in calcium sulfate, thus reducing the concentration of both these ions. More Mg/Ca exchange leads to more calcium sulfate precipitation such that the seawater can be significantly denuded in sulfate by the time it reaches the production wells, thus having a significant impact on sulfate scaling.

Examples showing the extent to which in-situ reservoir reactions can occur in oil and gas reservoirs have been described in many publications including conventional reservoirs, such as the Alba field [60] and in HT/high calcium reservoirs, such as the Clyde and Gyda fields [61]. Further, as highlighted above under PWRI, this was also the case whereby the excess solute associated with mixing seawater with produced water prior to reinjection precipitated during propagation through the reservoir, thereby minimizing any additional scaling risk at production wells [59].

Whether artificial or natural desulfation takes place, a significant concentration of scaling ions may still make their way through to the production well, and thus scale control in the production stream must still be expected, although the scaling tendency may be partially or in some cases significantly reduced relative to what would be predicted with breakthrough of unaltered seawater.

5.4.1.1.4 Desulfation of the injected water

In areas where using compatible water from an aquifer is economically unattractive or even unavailable, sulfate scaling can be mitigated by the partial removal of the sulfate ions from injected seawater [13,14]. This is achieved prior to injection by use of a desulfation plant, which will reduce the sulfate content from the levels encountered in seawater (typically in the range 2,700–3,000 mg/L) to a value in the range 20–100 mg/L. Desulfation involves nanofiltration to remove the relatively "large" divalent ions, of which the multiatom sulfate ion is the largest (including the hydration sphere of water molecules). The presence of a negative surface charge on the membrane can enhance the effectiveness of sulfate exclusion, though most membranes also reduce the divalent cations (Mg^{2+}, Ca^{2+}) to some extent. This is illustrated schematically in the Fig. 5.10.

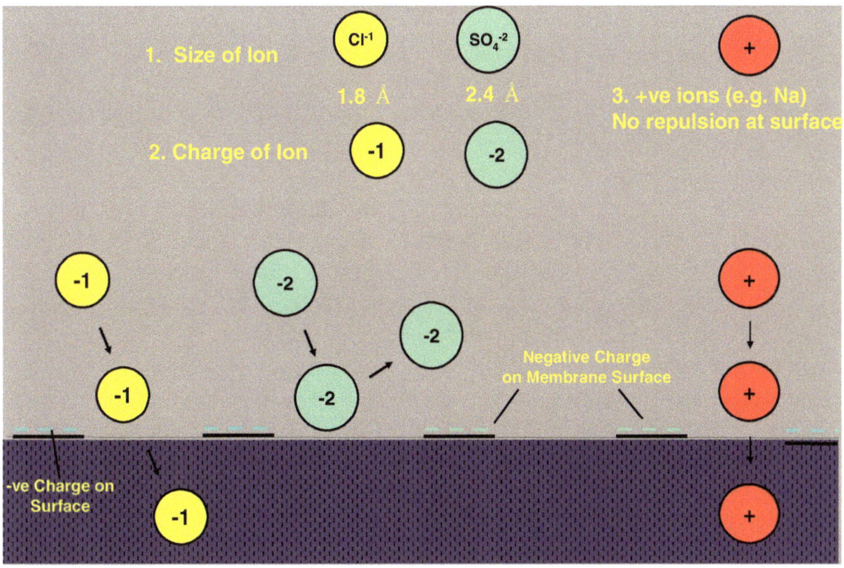

FIGURE 5.10

Basic mechanisms of nanofiltration: separation by size and charge.

Desulfation was originally developed by Marathon in the 1980s and first applied in the Brae field (North Sea), which had formation water containing in excess of 2,000 mg/L barium ions and would have resulted in an unmanageably high barium sulfate scaling potential if mixed with conventional seawater. Therefore, the concept was developed to minimize the sulfate scaling potential when seawater injection was used with high barium formation water by reducing the sulfate concentration. However, with today's more complex offshore subsea, and especially deepwater subsea fields, the increased costs and risks associated with squeeze treatments can often favor the high Capex, Opex, and operator attention associated with the installation and maintenance of a LSSW (low sulfate seawater) plant, rather than the much higher potential Opex costs and production deferment associated with unplanned interventions for scale removal. In general, however, owing to the high Capex costs that can be associated with LSSW systems, particularly those related to the weight requirements on an offshore platform, a rigorous technical and economic appraisal is required [6,8,9].

Important aspect to remember when selecting desulfation of the flood water include the following:

- With a few rare exceptions, all surface waters must be deoxygenated before injection into the reservoir, and this often involves the use of a physical removal system (stripping or a vacuum tower) plus chemical oxygen scavenger. By far the most commonly used scavenger is that based on bisulfite, which converts to sulfate upon reaction with the residual oxygen. Therefore, if oxygen removal is downstream of the desulfation package, this will reintroduce a small amount of sulfate back into the stream. Typically, a well-controlled deaeration system adds back less than 2 mg/L sulfate; a poorly controlled one may add significantly more because operators often compensate for a poorly maintained deaerator by adding more scavenger. This may or may not add significantly to the sulfate scaling risk.
- The efficacy of desulfation can be severely compromised if sulfate minerals are present in the reservoir. Barite is rarely found in significant quantities in reservoirs (though its presence at trace quantities can sometimes be inferred by the fact that the FW is saturated in it under reservoir conditions), but anhydrite is a common mineral, including as a cap rock for the reservoir. There is little point in removing sulfate concentrations down to less than 25 mg/L and then allowing them to rise again to perhaps several hundred mg/l because the now-undersaturated flood water dissolves significant amounts of the mineral from the reservoir.
- Desulfation is only practicable with relatively "clean" waters, such as seawater (ideally taken from a significant depth to minimize the need for significant prefiltration). Produced water cannot be treated because the oil content will plug up the nanofiltration units very quickly.

5.4.1.2 Flow modification

The ions that precipitate to form inorganic scales are only transported by water, and hence scale is a problem associated only with water production. Minimizing the volume of produced water reduces the potential for scale damage, apart from any other considerations (such as increasing oil production in some scenarios). Various techniques have been implemented to shut-off water production. These are easiest when the water is being produced from a cone underneath the well, or from a high permeability layer that is isolated from other, lower permeability, oil bearing strata (at least when the lower completion in the well facilitates the isolation of such a layer). However, these techniques usually only reduce water production. They do not eliminate it altogether, thus scale control remains an issue. Ultimately, except in unusual cases where the reservoir drive is exclusively expansion of a gas cap, flow of water (either expansion of an aquifer or injected water) nearly always has a large contribution to providing the energy to keep the oil flowing toward the production wells. This is "good water": the water we need to produce to produce oil. There is also "bad" water, which is water that does not significantly help produce oil, and this is (or should be) the target of water shut-off operations. Unfortunately, from a scaling viewpoint, the water doesn't really know whether it is "good" or "bad."

Most water shut-off occurs in the wellbore or near-wellbore of the production well. It can be mechanical, by placing a steel "patch" over the watered-out zone or (if it is "bottom water") by setting a plug above the zone. It can also be chemical, by injecting either a crosslinked gel, or sometimes, a gel combined with cement. Chemical treatments of this type must either be selectively placed in the target zone, or the whole zone is treated and then the dry zones are reperforated. There are also self-selective systems, which are chemicals that do not need to be selectively placed. They reduce the relative permeability to water to a much greater extent than their effect on the relative permeability to oil. Some success has been achieved with these, but it remains rather hit and miss. A key aspect is that they should only be used to shut off "bad" water, that is, where a zone is completely watered out. They will affect any zone producing both oil and water (which would then be "good" water) by greatly reducing the flow of both phases from that zone.

A final approach to flow modification is actually the basis of polymer flooding. This EOR method works by improving areal sweep relative to waterflooding, in particular, in cases where there is an unfavorable mobility ratio, that is, when the oil is viscous. Under such circumstances, polymer flooding will lead to much later breakthrough of the injected flood water, which will then translate to a later arrival and extent of scaling. This may not be the main reason for doing it, but it is often an additional benefit.

Field design and drainage strategy play an important role in flow modification, since this can significantly impact the production of injection water and the rate of increase in injection water at a particular well.

5.4.1.3 Nonchemical approaches: surface coatings and other "electromagnetic devices"

5.4.1.3.1 Surface coatings

The formation of hard adherent inorganic scales on metal surfaces leading to blocked pipelines and valves is clearly a surface phenomenon [62]. Over recent decades, the potential to engineer surfaces, by coating the surface with various "low-energy" coatings, has been the subject of much research. Indeed, the more inert a low surface charge/low energy or oil wet a surface, the greater potential for reduced scaling or at least reduced surface adhesion (meaning potentially that any attached scale can be more readily removed). To this end, many subsurface safety valves and other operationally critical components are often coated with specialized materials designed to either minimize the preference of scaling to occur at these locations and/or to aid removal of the surface deposit. This remains an active and ongoing area of research [63–66].

5.4.1.3.2 Electromagnetic devices

Other approaches include the use of electromagnetic and radiofrequency devices. They have shown promise for scale control in a number of industries. Their potential for application in the oil and gas production environment has therefore been the subject of much research and some considerable debate over many years. Various technologies have been tested in various laboratory studies and been evaluated in field trials with some success reported. In circumstances where electromagnetic or radiofrequency devices have been reported to provide benefits, the main mode of action appears to relate to the facilitation of bulk precipitation of crystals, thereby reducing the potential for scale to adhere and grow on the wall of pipes and vessels. Research and field application trials in this area continue [67–70].

5.4.2 Chemical inhibition

5.4.2.1 Scale inhibition versus chemical removal of scaling ions

No discussion of scale control by chemical addition would be complete without comparing two major chemical methods for preventing scale formation. The first of these is "chemical complexation/removal/sequestration" of the relevant cation (Ba^{2+}, Ca^{2+}, Sr^{2+}) or anion (SO_4^{2-}, CO_3^{2-}) from solution, using suitable chelating agents. The second is the addition of a chemical that prevents or inhibits crystal nucleation and growth.

5.4.2.1.1 Chemical complexation

The removal of scaling ions may be accomplished by complexation with chelating agents (often known as chelants), such as EDTA, by ion exchange or precipitation, or by the removal of the sulfate ions using nanofiltration to give LSSW. Reverse osmosis approaches, which are often applied to reduce overall salinity for injection of low-salinity water because it enhances microscopic sweep efficiency [12], have a similar effect on removing all ions from seawater, including sulfate. It must be remembered, though, that the salinity of low-salinity water normally has to be increased above that of reverse-osmosis-quality water prior to injection into sandstone reservoir to avoid formation damage. For offshore fields, this often means rediluting with seawater (or even the reject stream from the membrane array), such that an amount of sulfate is reintroduced, albeit at a much lower level than would be present in seawater.

Complexation using chemical chelants requires the addition of stoichiometric quantities (molar equivalent) of chemical, that is, most commonly, one molecule of complexing agent reacts with one scaling ion, to produce a soluble complex, thereby sequestering the scaling ion so that it can no longer react with other scaling ions. Chemical chelants are relatively high-molecular-weight species and are not generally specific to a single metal ion of choice, and, therefore, their use can rapidly become impractical in high salinity oilfield waters. They can be effectively used to control scaling in closed circuits (e.g., heating or cooling systems), in boiler-water applications, or in gas systems where rates of formation-water production are low (though even here, a threshold inhibitor would be preferred if it were effective; there are occasional cases where they are not).

The remainder of this section focuses on chemical inhibition, with chemical complexation described further in Section 5.6.

5.4.2.2 Introduction to scale inhibition

Using chemical scale inhibitors is one of the most common methods of preventing downhole and process mineral scale formation in oil fields. In most cases, their application proves effective, both from an economic and performance perspective, at managing inorganic scales throughout the lifecycle of a field or a production well. Periodic treatment is often not as attractive due to the continuous nature of the deposition process. Thus, a scale inhibitor must be continuously present in the produced brine, upstream of the point at which scale deposition starts to take place. This can be achieved by continuous injection, desorption or dissolution following scale inhibitor squeeze treatments or other slow-release techniques. The inhibitor may be applied by a variety of techniques, depending on where it is required. But in general, the further upstream in the production process the inhibitor is required, the more expensive and risky is its deployment.

If the near well formation and perforations need to be treated, then periodic squeeze application is the usual option. This is where a concentrated batch of the inhibitor is injected into the near-wellbore reservoir, where a significant fraction of it is retained (either by adsorption, precipitation or both), and the chemical continuously enters the stream of produced fluids passing through the treated zone until it is exhausted, whereupon the treatment must be repeated. Flowlines, manifolds and surface facilities can be protected by continuous injection of scale inhibitor. Subsea values, tubing and pumps can be protected by a range of options including continuous injection via capillary strings or dosed into the lift gas, such that everywhere should be protected downstream of the operating gas-lift valve. Other deployment options include solid inhibitors, prepacked (inhibitor impregnated packing) screens, or inhibitor-impregnated proppants in fractured completions.

The most important property any oilfield scale inhibitor must possess is the ability to prevent or inhibit crystal nucleation and/or growth at "threshold" concentration levels, or in other words, at concentrations much lower than would be required for a chemical to react stoichiometrically with the scaling ions. Oilfield scale inhibitors are therefore considered "threshold agents" and generally applied at dose rates ranging from < 1 mg/L to tens or (very occasionally) hundreds of mg/L, depending on the particular conditions. Many works have addressed the mechanism of threshold scale inhibition [71–88]. Moreover, the industry's drive to more challenging production environments over recent decades has led to a large range of new, more specialized scale inhibitor chemistries and formulations being developed for specific applications. This means that there is currently a very large number of scale inhibitor chemistries and treatment approaches (including hundreds of specific chemistries and many more formulations) available on the market (Note: Many of the commercially offered oilfield scale inhibitors include one or more active inhibitor chemistries, which are mixed, formulated (e.g., adding different solvents or adjusting pH) for different application by the chemical suppliers. This means that even with a specific chemistry manufactured by a specific chemical manufacturer, which the oilfield chemicals supplier procures, there may then be multiple different formulations using the same active inhibitor offered to oil producers for different applications). In part, this has been driven by the requirement to provide effective chemical inhibition in more challenging environments such as:

- Deepwater and subsea developments (where the cost of remedial treatments may be uneconomic);
- Expanding range of conditions (e.g., high downhole temperature and pressure, plus low seabed flowline temperatures);
- Requirement to keep equipment free of scale when it is either (safety) critical or very susceptible to even small amounts of scale or both (e.g., ESPs, SSSVs);
- More severe scaling regimes (e.g., high barium formation waters supported by SW injection in deepwater or other inaccessible regions, increased propensity for sulfide and iron scales in various conventional reservoirs and in unconventional production);
- Increased occurrence of less common scales, such as lead and zinc sulfides;
- HT environments where many conventional phosphonate- and polyacrylate-based inhibitor chemistries may be either thermally unstable or incompatible at these elevated temperatures, often due to high-calcium-content and high-salinity brines;
- Lower-pH environments, coupled with moderate or severe sulfate scaling regimes, under which conditions conventional phosphonate and polyacrylate inhibitors may be relatively ineffective;

- Changing environmental legislation which may preclude certain chemical types being applied in certain regions (e.g., an environmentally driven reduction in use of common phosphonate-based inhibitors in the North Sea has led to the development of effective, environmentally friendly polymers, etc.);
- Unconventional reservoirs with their novel scaling challenges.

The ability to select efficient scale inhibitors and increase operator confidence in their ability to prevent inorganic mineral scale formation under more challenging and changing production conditions can have a significant impact on the decisions taken at the design stage for new fields. For example, when considering the potential costs, benefits and risks associated with LSSW injection vs SW injection (in terms of the reduced risk of sulfate scaling when applying LSSW), the ability to effectively apply scale inhibitors continuously at the point of scaling and the economic risk associated, should the treatment be unsuccessful, needs to be taken into account.

When the scaling potential is low to moderate and chemical inhibitors are known to be effective, scale management via chemical inhibition will likely be selected to manage a sulfate scaling tendency under conventional production conditions (e.g., onshore or shallow water production) as the most appropriate and cost-effective solution (the answer may be different for a low carbonate tendency, at least initially). However, for a similar low-to-moderate scaling regime in a challenging environment and/or with very expensive production wells (such as those completed in deepwater offshore environments or other subsea complex "multi zone" wells), LSSW is often selected almost as the default position to control this sulfate tendency. This decision relates primarily to uncertainty in the effectiveness of treatment (chemical placement in all of the zones that need to be treated), the very high cost associated with such complex treatments, and the economic risk should the treatments fail in terms of loss of production (or even of entire wells) plus costly remediation treatments in expensive, high volume production wells. That said, chemical inhibition remains one of the most common methods for scale control and is applied in almost all oilfield production environments, even when LSSW is used to reduce sulfate scaling, as a residual sulfate scaling risk may still be anticipated in parts of the production system. Moreover, while the sulfate-reduction process also reduces calcium concentration to some extent, the overall effect on carbonate scaling risk is often minimal.

Scale inhibitor performance is often quantified in terms of the threshold concentration required to prevent scale, often referred to as the minimum effective dose (MED) or minimum inhibitor concentration (MIC). Besides depending on the nature of the scale mineral and the scaling tendency (see Section 5.3), different chemicals can be highly sensitive to the composition of the scale-prone water, such as the concentration of calcium and magnesium (both of which can affect the performance of inhibitors against barium sulfate scale), as well as the presence of transition metals, such as iron (often a bad actor) or zinc (which sometimes enhances performance). The pH of the water can also significantly affect performance within certain ranges. Most scale inhibitors are relatively weak acids that only work effectively when they are deprotonated into an anionic form, and this degree of ionization in solution can be strongly pH-dependent. Similarly, the pH can also affect the surface charge of the scale nuclei with which the anionic inhibitors must interact. So, for example, at high pH values (> 8–9 depending on the crystal face) the surface charge of calcium carbonate can change from positive to negative [89], making it much more difficult for an anionic inhibitor to interact with (and hence, inhibit) the scale.

5.4.3 Types of scale inhibitors commonly used in oil industry

These can be broadly classed as organic phosphonates/phosphate esters and relatively small polymers (with average molecular weights typically less than 20,000 g/mol). The polymers are therefore generally different from the very large polymers with molecular weights in the millions of g/mol used for other oilfield applications, such as EOR polymers or water-injection drag reducers. (Note: Some EOR polymers like partially hydrolyzed polyacrylamides also have some inhibitory effect on certain scales [12]). However, within each of these two broad classifications of scale inhibitors there is a large range of different individual species.

For the common carbonate and sulfate scales, the various inhibitors rely on the presence of anionic functional groups, such as phosphonates, acrylates, sulfonates (e.g., allyl and vinyl sulfonates, AMPS). For the polymers, the functional groups may be a single repeating monomer unit (homopolymers), but they may also be co-polymers containing two or even three repeating monomer units, with each containing different functional groups. The complexity and cost of production means that polymers very rarely go beyond three monomeric units (which are known as ter-polymers), except for very specific applications. The different chemicals are designed in order to maximize various aspects of the overall performance under specific conditions. Besides maximizing their inherent inhibition performance, changes in the molecular design may address compatibility with the produced waters, thermal stability, adsorption onto rock to optimize retention/release properties for squeeze treatments. Some monomeric units may even be incorporated mainly to facilitate the analysis of the inhibitor in produced water at suitably low concentrations.

5.4.3.1 Functional groups of scale inhibitors

Nearly all conventional inhibitors require anionic functional groups to be effective, which relates at least in part to the surface charge present on many scale minerals at the pH of typical oilfield systems. The surface charge of the commonest scale, calcium carbonate, is positive on most crystal faces up to a pH value of around 9; therefore, it is clear why anionic inhibitors would be effectively adsorbed to this mineral. The surface charge of barium sulfate depends on what other ions are present when it precipitates, becoming more positive in the presence of dissolved calcium (typically present in significantly greater molarity in most oilfield waters [90]), likely explaining why numerous inhibitors perform much better against this scale when significant concentrations of calcium are present. In contrast, the primary mineral present in sandstone, quartz, has a negative surface charge, which is consistent with the very poor adsorption of common inhibitors on this mineral. This obviously has a significant impact on the effect of deploying an inhibitor as a "squeeze treatment" (see Section 5.5) in a carbonate or sandstone formation: in the former, almost the entire rock matrix will adsorb the chemical, whereas in sandstone, it tends to be the nonquartz species (the detrital and diagenetic carbonates, and the high-surface-area clays) that provide the interaction that leads to chemical retention. It is worth mentioning that this model of simple ion charges is too simplistic because it ignores the effects of the ions in solution which (like calcium in the case of barium sulfate) can moderate or even flip the charge.

Coupled to this is that the performance of many scale inhibitors, particularly phosphonate and (poly)carboxylate-based inhibitors, relies on the formation of inhibitor–calcium complexes in solution (e.g., phosphonate-Ca^{2+} or carboxylate-Ca^{2+}) [72], which is known to play a significant role in both inhibition of the scales and chemical retention in the formation. Hence, inhibitors possessing functional groups with the ability to complex strongly with divalent cations, such as calcium, in solution, often provide the most effective chemical inhibitors.

Note that the same does not hold when inhibitors complex strongly with magnesium ions, which can be the case for many phosphonates. This complexation can lead to a significant negative impact on chemical performance against barium/strontium/calcium sulfate scales, primarily because, unlike calcium ions, magnesium ions do not interact strongly with the crystal surface and crystal lattice [72,91–93].

Similar negative effects are also seen for many inhibitos in the presence of high levels of dissolved iron [30,94,95]. Other dissolved ions (either already present or introduced with the inhibitor) can have a more positive effect, such as Zn^{2+} ions [96,97]. These initial comments serve to highlight how the complex nature of oilfield brines not only controls the scaling thermodynamics and the inherent risk of scaling as a function of production brine conditions (as described in Section 5.3) but also strongly influence the performance of inhibitors and does so in different ways.

More recently, newer chemistries and co-polymers have been introduced containing cationic species, such as amine groups that are protonated at typically oilfield pH values. Under certain conditions, these have been shown to enhance performance against common scales, either in terms of their inhibitory performance or in terms of their ability to retain more effectively in the reservoir and generate longer lifetimes when applied via a squeeze treatment [98]. One application of these novel species is in the inhibition of less common scales, such as lead and zinc sulfides, where their superior performance over conventional anionic species appears to be due, at least in part, to the surface charge of these minerals, which generally remain negative even in the presence of high salinity brines [99–102]. This is not to say that certain conventional products do not also offer performance against certain sulfide scales (e.g., HADMP is known to be reasonably effective against ZnS), or that certain other polymers cannot act as dispersants (as opposed to true inhibitors), which act by limiting the deposition of such scales rather than rigorously inhibiting their growth of nucleation.

5.4.3.2 Chemical structures of scale inhibitors

Figs. 5.11 and 5.12 provide some examples of the various inhibitor species commonly applied in oil and gas production systems. Fig. 5.11 covers a range of the discrete low-molecular-weight phosphorus-based molecules: inorganic phosphates, phosphate esters and (perhaps the most commonly used) phosphonates. Fig. 5.13 provides examples of some conventional polymeric scale inhibitors, including mono-polymers and multicomponent co-polymers. A large range of other phosphonates, phosphate esters, polymers, co-polymers, ter-polymers and other complex polyanions are available on the market.

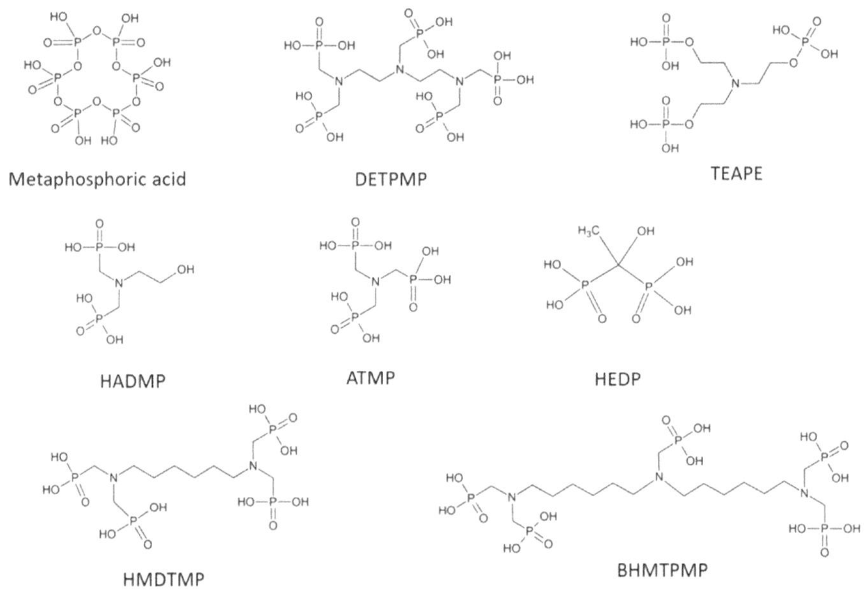

FIGURE 5.11

Common discrete phosphorus-based scale inhibitors: phosphate, phosphate ester and phosphonates (Note: The nondissociated forms are shown, whereas the active species will be anionic).

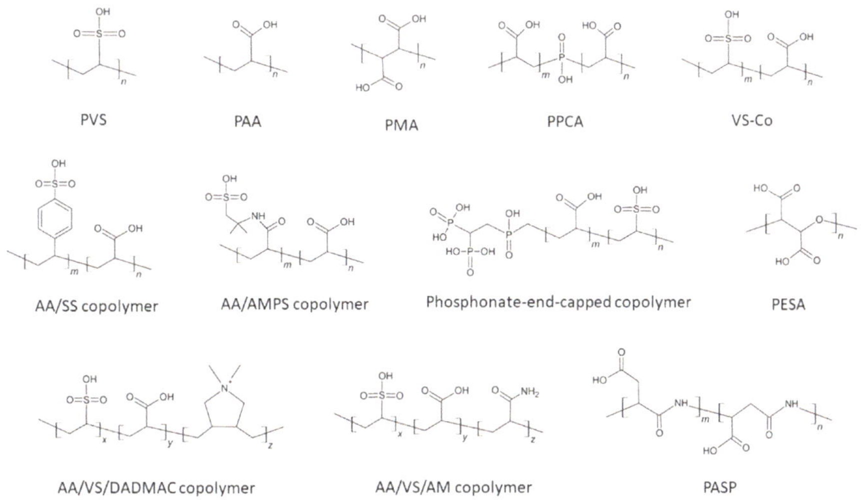

FIGURE 5.12

Chemical structures of selected polymeric scale inhibitors.

FIGURE 5.13

Illustration of compatibility issues associated with two scale inhibitor solutions prepared in seawater and heated to 160°C.

Image courtesy of Scaled Solutions Ltd.

5.4.3.3 Attributes of scale inhibitors

It cannot be emphasized enough that for oilfield chemical inhibitors (and indeed, all oilfield chemicals), it is not just the absolute performance of scale inhibitors that controls their effectiveness or applicability. Compatibility with produced fluids, in particular the high salinity produced oilfield waters which also often contain elevated concentrations of divalent cations, compatibility with other applied production chemicals, stability (particularly thermal stability), compatibility with wetted metallic and nonmetallic materials within the production system and within the chemical-delivery system, can all govern the selection and qualification of an inhibitor for a particular application which will be discussed later.

Chemicals have been developed over the years with improved stability and compatibility with higher salinity, complex brines, as well as striving to improve their inhibitory performance under oilfield applications (controlling a given scaling tendency at a lower dose rate), their retention in the reservoir under squeeze application, as well as finding chemistries that are effective against scales that have proven challenging, such as sulfides and silica/silicates. For high-temperature/high-pressure reservoirs (typically with reservoir T > 150°C and reservoir P > 9000 psi [103–105], especially those producing high-salinity waters rich in divalent cations (predominately, calcium), a combination of high thermal stability, high resistance to hydrolysis, and compatibility with the produced water has been critical for chemical selection. This has ultimately led to the requirement for sulfonated co-polymers which achieve the combination of high stability and high compatibility required.

5.4.3.3.1 Thermal stability

This varies both between categories of inhibitors but also within each category. Moreover, stability depends on time spent at the elevated temperature:

- A chemical continuously injected into a hot production stream may only need to survive at most a few hours.
- A chemical injected downhole via a capillary string may need to survive days or (when the well is shut) even weeks to months at temperature and as a concentrate in contact with a steel surface (both of which provide additional challenges to its stability.
- A chemical squeezed into a formation may need to survive months to years, albeit adsorbed onto rock or precipitated in-situ, both of which are likely to provide a limited degree of additional stability.

From the above, it is clear that the data presented here can only be a general guideline. Broadly speaking, phosphonates have thermal stabilities in the region up to 120°C–160°C under the slightly acidic conditions within a well (typically, pH 4.5–6). This compares with phosphate esters, which are generally hydrolyzed at less than 100°C (though some reports indicate slightly higher thermal stability). Polymers, in particular sulfonated polymers, have been shown to show prolonged stability under near neutral pH conditions at temperatures in excess of 200°C [103–105].

An important further consideration for squeeze treatment chemicals relates not only to their thermal stability at application conditions (which may be acidic conditions and therefore more prone to thermal degradation at the elevated downhole temperature), but their thermal stability when retained on the reservoir rock as for many chemicals including certain phosphonates this has been shown to impart improved thermal stability when attached to the rock [106,107]. Further, when considering scale inhibitor squeeze treatment, thermal stability of the chemical in produced waters (following release form the rock) can also be an important consideration, as opposed to the thermal stability at application conditions especially if the chemical is to be applied as an acidic form.

5.4.3.3.2 Brine compatibility

When considering brine compatibility, in addition to various sulfonated or other modified polymers, a range of modified "calcium tolerant" phosphonates have also been developed by the chemical manufacturers, where the structures of the common species are altered to enhance brine compatibility. Moreover, a degree of calcium intolerance can actually be useful for squeeze applications because precipitation of the calcium-inhibitor complex can be an additional chemical-retention mechanism which may prolong the squeeze treatment lifetimes.

5.4.3.3.3 Effect of chemical structure

The inclusion of an hydroxy and/or carboxylate group in the backbone increases the calcium tolerance of conventional phosphonate inhibitors. Adding sulfonate units into a carboxylate-based polymers does similar. An example of this is hydroxyamine diphosphonate (often referred to as HADMP), which has a very similar structure to the conventional triphosphonate (ATMP), but with one of the phosphonate groups essentially replaced with a hydroxyl group (see Fig. 5.11). This has a dramatic impact on the chemical's calcium and iron tolerance, as well as its absolute performance against different minerals. For example, whereas ATPMP is an effective inhibitor for calcium carbonate, it has limited applicability in cases where moderate to high dose rates are required in produced waters containing elevated calcium levels, due to its inherent incompatibility with calcium ions. HADMP, on the other hand, is one of the most calcium-tolerant and brine-compatible scale inhibitors. It not only retains reasonable performance against calcium carbonate but also has excellent iron tolerance and, unlike ATMP, can also inhibit (to some degree) iron carbonate and zinc sulfide scales.

These two chemicals illustrate how a relatively small change in structure can have a very significant impact on the overall applicability and performance of an inhibitor. Other phosphonate chemistries have also been modified to incorporate carboxylate and hydroxyl groupings, which again can have a significant impact on performance and applicability. The main difference between two similar phosphonates—the pentaphosphonates, DETPMP and BHMTPMP—is the length of the linkage between amine groups [$(CH_2)_2$ vs $(CH_2)_6$]. While this generally has only a small impact on their respective effectiveness as inhibitors, bis-HMTPMP is significantly more thermally stable [108].

Another general observation is that the smaller, discrete phosphonates such as HEDP, ATMP, and HADMP (and even the orthophosphate ion itself) are effective against calcium carbonate but have more limited performance against sulfate scales, whereas other phosphonates, such as DETPMP, HEDPMP, and bis-HMTPMP, offer performance against both carbonate and sulfate scales [76,86,108–110].

Although the above discussion highlights how modifications of phosphonate molecules have a significant impact on performance and applicability, it is in the area of polymer inhibitors, with the advent of not just co-polymers but various ter- and multifunctional polymers, where the ability to modify the components within the polymer backbone, has seen much activity over recent decades. For example, many of the co-polymers include sulfonated units as this increases their compatibility over that of polyacrylates in high-calcium brines and also improves thermal stability.

Fig. 5.13 shows an example of a combined compatibility and stability test for two scale inhibitor formulations prepared in seawater in which the active ingredient is of the same chemical type and which show almost identical performance as an inhibitor. However, it is clear that application of the second product could lead to considerable deployment challenges at the target temperature of 160°C.

Table 5.8 gives some general indications of the types of scales and conditions under which the various inhibitors or inhibitor types are typically effective. It must be stressed that these results only give a rough approximation of scale inhibitor performance and are not a substitute for rigorous qualification. Such qualification will then include other factors, such as the actual produced-water composition(s), pH values, operating conditions, and the exact nature of the chemical (note that this is even more significant for the inhibitors presented as a generic category. Even two nominally identical chemicals can differ in their solvent or surfactant packages. Polymers may vary in molecular weight). Scale inhibitor qualification under specific, representative conditions is always essential.

Table 5.8 General illustration of how chemical structure and active functional groups influences performance, stability and compatibility of inorganic mineral scale inhibitors.

	$SrSO_4$ / $BaSO_4$	$CaCO_3$	$FeCO_3$	ZnS	Ca Tolerant	< 80 °C	100 to 130 °C	> 150°C
Inorganic phosphonate	X	✓✓	X	X	X	✓✓	X	X
Phosphate Ester	✓✓	✓✓	X	X	✓✓ or ✓	✓✓	✓✓	X
ATMP	X	✓✓✓	X	X	X	✓✓	✓✓	X
HADMP	X	✓✓	✓✓	✓✓	✓✓✓	✓✓	✓✓	X
DETMP	✓✓	✓✓	X	X	✓	✓✓	✓✓	X
PAA	✓	✓✓	X	X	✓	✓✓	✓✓	✓ pH 5 ok
PMA	✓	✓✓	X	X	X	✓✓	X	X
PVS	✓✓	✓✓	X	X	✓✓✓	✓✓✓	✓✓✓	✓✓✓
Vs-Co	✓✓	✓✓	X	X	✓✓✓	✓✓✓	✓✓✓	✓✓✓

Keys:

X	Doesn't perform	✓	Poor	✓✓	Intermediate	✓✓✓	Good

5.4.4 Brief history of scale inhibitor development

Over100 years ago, it was discovered that some inorganic phosphorus compounds, called metaphosphates, could prevent calcium carbonate scaling in irrigation systems when only 2 ppm were added to the water. This small amount of metaphosphate could maintain a supersaturation of over 200 ppm of calcium carbonate in solution. A low concentration of metaphosphate could not only prevent scaling, but it also caused previously formed deposits of calcium carbonate to disappear. Clearly, this was not a stoichiometric phenomenon and so it was called a threshold treatment because it maintained the dissolved calcium carbonate on the threshold of precipitation.

Metaphosphates belong to a family of inorganic phosphates called condensed phosphates, dehydrated phosphates, or polyphosphates. They are inexpensive and are still used for treating municipal waters and cooling waters. The monomolecular phosphoric acid (orthophosphoric acid) is heated with an alkali to expel water and form sodium hexametaphosphate. The orthophosphates are generally regarded as having no scale-inhibiting properties for the sulfate scales, although they perform against carbonate scales [109,111]. In the early days of oil and gas production, polyphosphates were widely used in oil fields to inhibit scale, calcium carbonate in particular. In practice, it was found that solutions of sodium hexametaphosphate rapidly react with water, or hydrolyze, and revert to the orthophosphate.

A further disadvantage of the inorganic phosphates was overtreatment. If too much was added, which then decomposed to phosphate, resulting in the precipitation of calcium orthophosphate, $Ca_3(PO_4)_2$, from solution. The solubility of calcium phosphate is very low (20 mg/L, though numerous other calcium phosphate species can also precipitate from complex brines). It is therefore a less soluble scale than calcium carbonate at most typical oilfield conditions, and can be referred to as a pseudo-scale, that is, one formed as a result of inappropriate treatment with a chemical inhibitor. This issue remains a problem with organic amine phosphonates when applied under high temperature environments, whereby thermal degradation leads to the production of orthophosphate and the subsequent precipitation of calcium phosphate in produced waters. The disadvantages of the polyphosphates caused them to be replaced by the more expensive, but more reliable, organic compounds used today.

The organophosphorus-based scale inhibitors, including phosphate esters and amine phosphonate (ATMP, DETPMP, etc.) have almost entirely displaced the inorganic phosphorus compounds, and inorganic polyphosphates are now almost unknown in oil fields. For some applications, particularly at relatively mild temperatures (up to $\sim$ 100°C), phosphate esters are sometimes applied, as their thermal stability to hydrolysis lies between that of the inorganic phosphates and the more stable phosphonates. Depending on the particular structure, phosphate esters can inhibit both calcium carbonate and sulfate scales. Organic phosphonates are among the most common inhibitors applied, though their thermal stability and compatibility with more saline produced waters still often preclude their use in very HT/high calcium environments [103–105,111,112].

Phosphonates resist hydrolysis much better than do the inorganic condensed phosphorus compounds. However, hydrolysis is considerably more rapid at lower pH [113], reducing the potential use of phosphonates under extreme high temperature applications.

Phosphate esters cost less than amine phosphonates and should be considered for applications where stability is not a problem, in relatively mild temperature environments. This said, their propensity for hydrolysis and degradation may favor their application in regions of the world where amine phosphonates are not allowed due to environmental considerations (i.e., some phosphate esters are considered more environmentally friendly than amine phosphonates). In a phosphate ester, the phosphorous is connected to carbon through an oxygen atom while, in a phosphonate, the phosphorous is connected directly to carbon. The -O-P- bond is more readily broken by hydrolysis than the –C–P– bond is and hence phosphate esters are more readily degraded.

5.4.5 Chemical deployment

For **continuous injection** situations (e.g., flow lines and surface streams, downhole chemical continuous injection, DHCI, etc.), it is the chemical that protects at the lowest concentration that is considered most effective. Provided of course that the chemical is compatible under application and production conditions with the produced waters, thermally stable under the application conditions and does not interfere negatively with other production chemicals (e.g., corrosion inhibitors), interfere with other processes or is applied in a form that could cause additional problems (e.g., potential corrosiveness of neat acid-based scale inhibitors in carbon steel lines) and is compatible with the materials of the application system. For downhole application, such as via gas lift or capillary string application, other criteria like the resistance of the chemical to “gunking” or its ability to remain stable and “solid free” through the entire production system are also important [114]. Likewise, subsea chemical application has been examined and specifications for the system and chemical selection published by the BASICS JIP [115].

For downhole **squeeze** jobs, it is the chemical that protects for the longest time per squeeze that determines the most effective performance, rather than its absolute performance in terms of its ability to inhibit scale. Squeeze life is the return lifetime to its particular MED (minimum effective dose required to prevent scaling, MED; also called minimum inhibitor concentration, MIC) which determines the performance. This is an important concept for scale inhibitor selection for squeeze treatments as the interaction of the particular scale inhibitor with the reservoir rock substrate which governs its retention and release properties is often more critical than the absolute performance. A very effective (low MED) scale inhibitor may have very poor retention properties in the reservoir leading to a very short squeeze lifetime, compared with other products which have higher MEDs. This is illustrated in Fig. 5.14, indicating the potential return profiles of

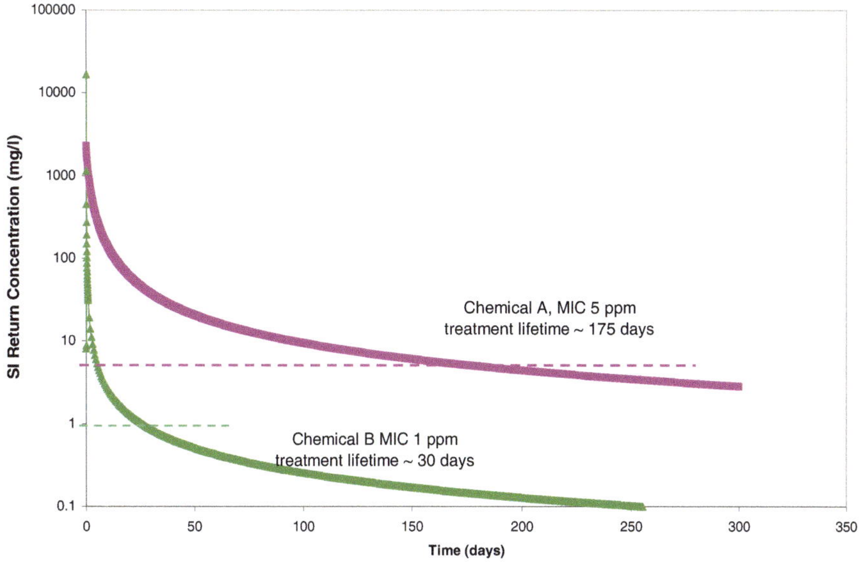

FIGURE 5.14

Comparison of return profiles for two scale inhibitors including chemical A (high MED) and chemical B (low MED). Modeled using Place-iT near wellbore simulator.

Courtesy of Scaled Solutions Ltd.

two chemicals A and B with different MEDs required to prevent scale. In this illustration, the poorer performing chemical in terms of inhibitor performance, chemical A, would lead to a longer treatment lifetime than the more effective chemical, chemical B. For the chemical squeeze process, in addition to performance, retention/release properties and other criteria such as stability and compatibility and so on, a further critical requirement relates to its ability to be applied without causing any serious "formation damage." This will be discussed further in Section 5.5.

Obviously, some level of expert laboratory assessment or extensive field experience with the chemicals is required to effectively qualify the products prior to deployment. In essence, the main rules associated with chemical application are:

1. It does no harm to the production system (i.e., does not have any negative impacts on production, or other production chemicals or degrade to a species that may either be hazardous or results in by-products which have a negative impact, etc.).
2. It does what it is designed to do.
3. It remains stable and effective (not forever but for as long as it is required to remain stable and effective in the production system it is being applied).

5.4.5.1 *Other approaches*

Other approaches for delivering scale inhibitor downhole in production wells include the use of scale inhibitor impregnated proppants [116–118]. Such treatments can be applied in prop and fractured wells, provided the integrity of the proppant is not compromised. Deep downhole chemical injection (DDHCI) has also been examined which is an (early) modification on the more conventional downhole chemical injection (DHCI) systems [114]. In DDHCI, the inhibitor is injected much lower in the production system than for conventional DHCI systems, though with similar issues in relation to providing continuous injection as opposed to "pulsed" injection [119]. Other approaches documented in certain fields has been to inject excess scale inhibitor at injection wells [120,121]. Although reported to be viable in some limited situations (such as those in fractured reservoirs with relatively short distances between injectors and producers), this should generally not be considered for most conventional reservoirs, due to the inherent nature of scale inhibitors being highly reactive with reservoir rock material. Therefore, in conventional reservoirs, the inhibitors are rapidly adsorbed in the near wellbore formation around the injector wells and hence do not generally propagate through the reservoir.

Other applications of solid and/or encapsulated inhibitors have also been developed for application downhole, examples of which are given by Bourne et al. [122] and Hsu et al. [123].

5.4.6 Chemical inhibition

In describing the threshold or inhibitor effect, it is generally regarded that the inhibitor molecules adsorb at the active growth sites, which may or may not be crystal defects (though crystal growth is often accelerated at any defects in the crystal structure). The adsorption of the inhibitor at these active sites thus prevents further crystal growth by interfering with the growth process and preventing new scaling ions from themselves adsorbing onto the crystal surface in the normal ordered pattern. Coupled with this, the morphology, the tendency to agglomerate, and the potential of the electric double layer (the zeta potential) of the growing nucleons are also altered by adsorption of inhibitor molecules at the growth sites.

In oilfield applications, it is recognized that precipitation is more likely to occur on a surface that is already present. Such surfaces may be existing scale deposits, metal surfaces that offer available sites for adsorption of lattice ions (production equipment, pipelines, etc.), or the rock formation. Heterogeneous nucleation on such surfaces, for most low scaling regimes, would be more likely than homogeneous because the free-energy barrier (for the reduction in supersaturation) may be reduced. This is particularly important as well as the local flow regime (e.g., surface shear and turbulence as described earlier), for mildly oversaturated systems. For highly oversaturated systems, scale precipitation is labile and therefore spontaneous nucleation occurs both homogeneously (in the bulk fluid) and heterogeneously (on surfaces).

5.4.6.1 Mechanisms in inhibition

The overall result of adding a scale inhibitor is a reduction in the crystallization tendency and the subsequent formation of scale, broadly by two pathways.

- Nucleation inhibition:
  - Disruption of the thermodynamic stability of the growing nucleons.
  - Endothermic adsorption of inhibitor species causing dissolution of crystal embryos.
  - When examined at inhibitor concentrations slightly below that required to completely prevent any crystallization and growth, this may lead to effective short residence time inhibition (delay in nucleation) with the inhibitor subsequently being consumed in lattice and, therefore, being less effective at longer residence times, once into the crystal growth phase.
  - Increased concentrations of inhibitor (higher dose rates) give complete inhibition.
- Crystal growth retardation:
  - Interfering/blocking of the growth process.
  - Irreversible adsorption of inhibitor at active growth sites.
  - When examined at inhibitor concentrations, slightly below that required to completely prevent any crystallization and growth, this may lead to effective long residence time growth inhibition following initial nucleation, but inhibitors may be less effective at preventing nucleation.

It is important to note that all common inhibitors perform effectively by both pathways, although their mode of operation may perform more effectively by one pathway over another.

This difference in recorded performance for different inhibitors performing through the different pathways, as a function of time, is illustrated further in Fig. 5.15 and the mechanisms involved are further illustrated schematically in Figs. 5.16 and 5.17.

In addition to preventing growth, it is recognized that crystals nucleated in the presence of threshold inhibitors have a different form (shape) than those nucleated under normal conditions. In extreme cases, where some scale formation occurs in inhibited waters, the scale is modified and is thereby less likely to adhere. It can potentially be more easily removed, or carried through the system, without adhering to metal surfaces, than scale formed in uninhibited waters. However, it is also been noted that when inhibitors are present at concentrations below the MED level (i.e., below the level required to prevent scaling), they can also serve to effectively reduce the activity of scaling ions in an otherwise highly oversaturated system, leading to increased surface nucleation and growth [124]. It can therefore be critically important not to under treat, and to carefully monitor production systems when under chemical inhibition.

This change in morphology is indicated below in Fig. 5.18 which shows SEM images taken for barite crystals grown on a metal surface from a Forties FW/SW mixture in the absence and presence of scale inhibitors applied below the MED to

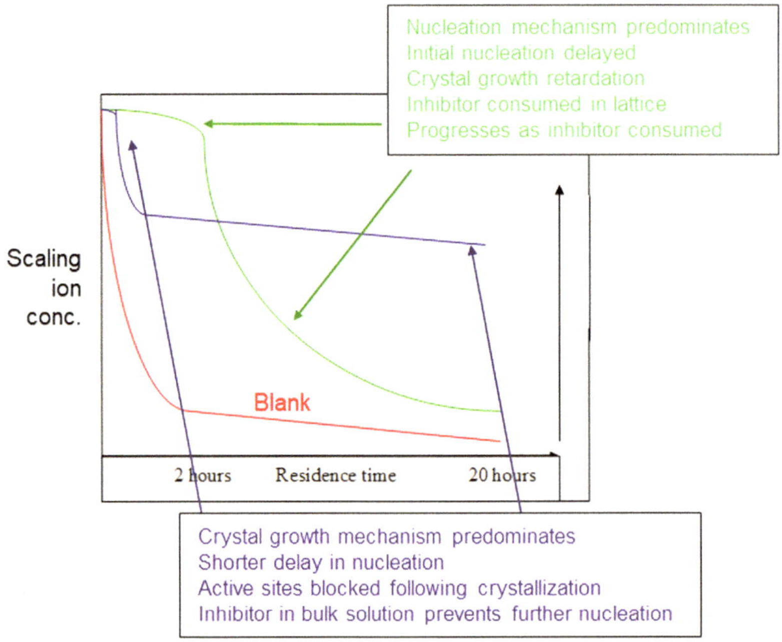

FIGURE 5.15

Schematic illustration of the process involved in nucleation inhibition. The concentration of Ba vs time for inhibited (Red line) and inhibited solutions.

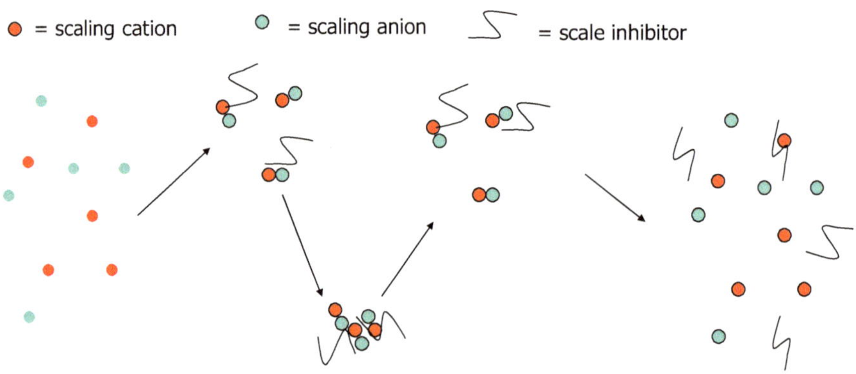

FIGURE 5.16

Schematic illustration of mode of nucleation inhibitor, leading to dispersion of nucleating ions.

Image courtesy of Scaled Solutions Ltd.

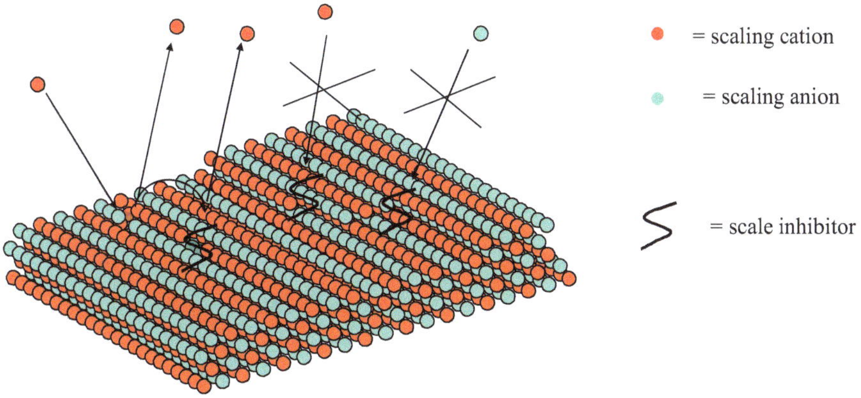

FIGURE 5.17

Schematic illustration of mode of crystal growth inhibition by blocking active growth sites.

Image courtesy of Scaled Solutions Ltd.

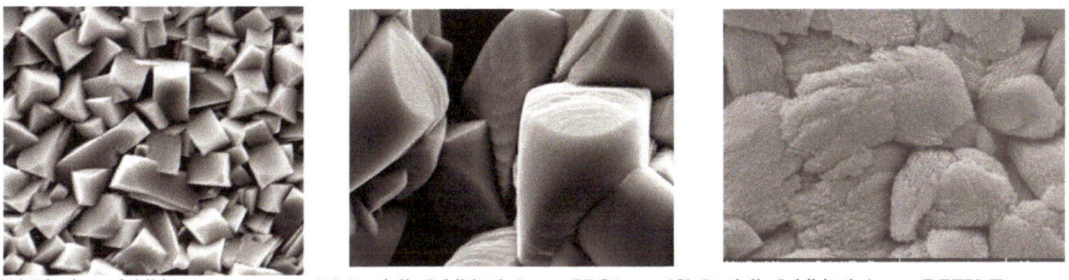

(A) Blank, no inhibitor (small rhombic crystals; < 5 micon) (B) Partially Inhibited, 1 ppm PPCA (large rounded crystals; > 10 micron) (C) Partially Inhibited, 1 ppm DETPMP (large "feathery" ppt; > 10 micron)

FIGURE 5.18

SEM images of barite crystals formed on metal surface from Forties FW/SW mixture. (A) Blank, no inhibitor leading to small rhombic crystals. (B) Inhibited barite growth, 1 ppm PPCA. (C) Inhibited barite growth, 1 ppm DETPMP.

slow the nucleation and growth processes. The effect of two common inhibitors, a polymeric PPCA and a phosphonate DETPMP, indicate the different modes of interaction of the different inhibitors [124]. The DETPMP inhibitor is effective at growth blocking, leading to a very "feathery" appearance where the crystal growth has been highly modified, whereas the PPCA is effective at preventing nucleation; hence, with the inhibitor present, fewer nucleation and growth sites are formed. But once crystallization occurs, the crystals grow larger, with the inhibitor being consumed and distorting the crystal shape.

Other terminologies: Scale inhibitors are also good **deflocculants or antiagglomerants**. In other words, they cause agglomerations of crystals to disperse. Scale inhibitor molecules contain ionic groups and these ionic groups impart the same electrical charge on solids to which they adsorb. The like charges repel one another, causing the solids to disperse. Very large polymeric inhibitor can adsorb on more than one particle and bind them together. They are used as flocculants to aid settling and filtration of solids.

5.4.7 Factors controlling the performance of generically different inhibitor chemistries

There are a number of structural features and influencing conditions directly controlling the effectiveness of inhibitors. Structural features comprise molecular weight, active functional groups, hydrophobic content, stereochemistry (as relating to lattice matching) steric restrictions and flexibility, pK_a values, and so on. Influencing conditions include reservoir and production attributes such as brine pH, concentration of scaling ions, type of scale (barite, carbonate, etc.), cation concentration and gradient (in particular Ca^{2+}/Mg^{2+}), and system temperature and pressure. Knowledge of these helps to select appropriate test procedures and to ensure that the laboratory assessments are conducted appropriately. The effectiveness of an inhibitor is influenced by the test methodology adopted and it is therefore imperative that experimental procedures mirror as closely as practically possible the exact nature of the scale control problem experienced in the field [71].

5.4.7.1 Structural features: active functional groups, pK_a, and pH

As described earlier, phosphonates and small polyelectrolytes (molecular weight <10,000) are most widely used as downhole scale inhibitors to control barium sulfate scale formation. These inhibitors tend to show good performance for a range of pH and temperature conditions. However, their effectiveness deteriorates markedly as the pH is lowered [71,72]. This is directly related to the pK_a values of the particular inhibitor species and the extent of dissociation (or ionic character) present at a particular pH. This behavior gives a strong insight into the manner in which these species interact with the scale crystals. Thus, barium sulfate scale prevention, with these common inhibitor species, is more difficult in low-pH environments, especially when the supersaturation is high. Alternatively, sulfonated polymers, such as poly vinyl sulfonate (PVS) (the functional groups of which are strongly acidic sulfonic acid units) provide good barium sulfate inhibition, even at low pH values (<4), because these functional groups remain dissociated (anionic form) at these lower pH values (i.e., they have low pK_a values). The relationship between pH and pK_a is illustrated in Eq. (5.10), showing that the solution pH has a considerable influence on the performance of different generic inhibitors, owing to the respective pK_a values of their functional groups.

$$\mathbf{SI}-\mathrm{OH} \rightleftharpoons \mathbf{SI}-\mathrm{O}^{-}+\mathrm{H}^{+}$$

$$pK_a \simeq -\log_{10}\left(\frac{[\mathrm{SI}-\mathrm{O}][\mathrm{H}^{+}]}{[\mathbf{SI}-\mathrm{OH}]}\right) \tag{5.10}$$

The effect of brine pH on different inhibitors can be summarized as: Common scale inhibitor (e.g., ATPMP, BHMTPMP, DETPMP, PAA, PPCA, etc.) are weak poly acids with pK_a values ~ 3.5–5.5, their performance is strongly affected by pH, with better performance at higher pH (i.e., pH $>$ 5). Polymeric inhibitors with sulfonate group (e.g., PVS, VS-Co, etc.) are strong acid, the functional group is dissociated at low pH, and thus they remain effective even at pH $<$ 5.

For sulfate scale inhibition, in relatively high supersaturation brine mixtures, the dissociated or anionic form of the inhibitor functional groups are required to be effective. The mode of attachment to the growing scale crystal is a strong electrostatic binding. In milder scaling environments, hydrogen bonding is also important in allowing scale inhibitors to interact with scale surfaces.

Finally, and although not directly relevant in terms of inhibitor performance, increased pH tends to result in less compatible scale inhibitors. For the phosphonate and polyacrylate-based inhibitors, application in higher pH fluids or even moderate pH fluids (pH 5–6, as often encountered in oilfield production), especially those containing high concentrations of calcium at elevated temperatures, can result in precipitation of a scale inhibitor–Ca^{2+} complex. Calcium tolerance of most phosphonate and polyacrylate based inhibitors is a function of pH, calcium (and other divalent cations) concentrations, and temperature, with reduced compatibility as temperature, calcium concentration, and pH increase. As discussed earlier, this can be addressed for the discrete phosphonates by the inclusion of other function groups, such as alcohol or carboxylate, and for polymers by inclusion of more compatible monomers into the backbone, such as the sulfonated monomers

5.4.7.2 Structural features: lattice matching and adsorption considerations

The ability of a SI to "lattice match" and dock onto growth sites is important for effective inhibition, as for a molecule to inhibit crystal growth it must adsorb onto the crystal surface. The binding groups of the inhibitor must therefore be able to match with bonding sites on the crystal surface. The ability to match with multiple binding sites increases the potential for irreversible adsorption at active growth sites. This is illustrated in Fig. 5.19.

Lattice matching considerations (i.e., the ability of the inhibitor's functional groups to line up with oppositely charged ions on the crystal surface and therefore bind strongly onto the surface) can be summarized as follows [76]:

- Small rigid phosphonates (e.g., ATPMP, HEDP) inhibitors: they are common carbonate scale inhibitors with minimal impact on sulfate scales.
- Higher-molecular-weight, flexible phosphonates (e.g., BHMDTMP, DETPMP): they are commonly used for both carbonate and sulfate scales inhibition.
- Flexible polymers (PAA, Vs-Co, PVS, and associated copolymers/multicomponent polymers): they adsorb via a multiplicity of binding sites and are less specific to adsorption at active growth sites, remain effective for variety of scales.

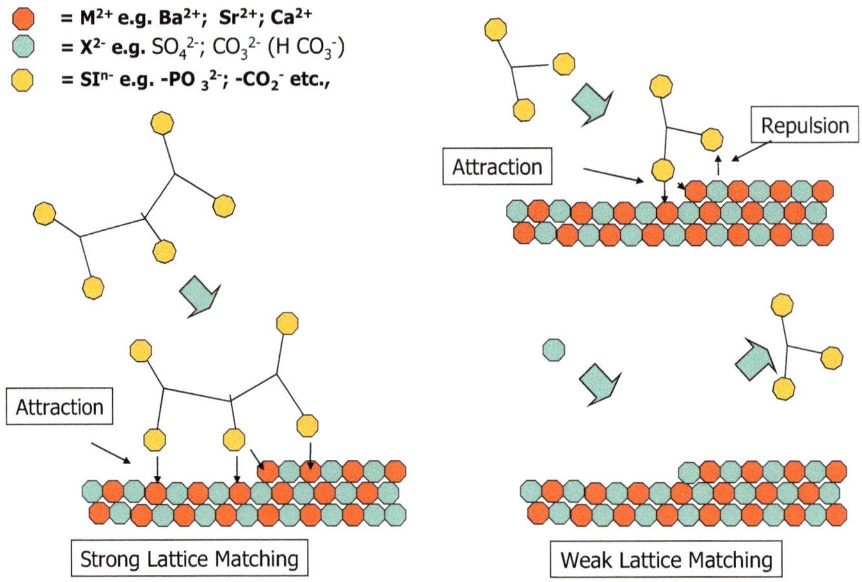

FIGURE 5.19

Illustration of the importance of lattice matching for small phosphonates in specific adsorption at active growth sites of particular crystal lattices.

In addition, the importance of dopants/impurities in crystal structure (i.e., solid solutions) on the lattice parameters must also be considered. The presence of other ions changes the lattice spacings (e.g., the presence of calcium in the barite lattice contracts the barite lattice structure) and can aid in more efficient lattice matching. The inhibitor species may bind more efficiently to such doped ions, thereby significantly improving their performance [91,93].

Lattice matching, as illustrated in Fig. 5.19, is useful in understanding the importance of the structural spacing of the functional groups within the inhibitor and the requirement for some degree of association and bonding to the growing lattice. However, as indicated it is only part of the story. Many conventional anionic inhibitors form complexes in brine solution with divalent cations, such as calcium (and certain other) ions and the resultant inhibitor–M^{2+} molecule can considerably change inhibition performance. In essence, the oppositely charged scaling ions on the scale surface may be the anionic species, with the anionic functional group of the inhibitors lined up via a cation bridge.

5.4.7.3 Structural features: molecular-weight considerations

It must be remembered that polymeric inhibitors, like all polymers, are not discrete molecules but are a range of different chemicals differentiated by the number of repeating monomer units making up the overall size of the individual chemical. All polymers, therefore, contain a large number of similar but different species within any commercially synthesized product. Inhibition efficiency against barium sulfate by polymeric species is controlled by molecular weight and polydispersity (PDI) considerations. In addition, incorporation of small levels of additives "to enhance" performance (or to provide a detectable tag) have been shown in some cases to have minimal impact on the performance of the bulk polymer (polyacrylic acid). For example, PPCA (and other tagged PAAs) can be treated as a "pure" PAA (low PDI) for general purposes [125–127].

To illustrate this point, Fig. 5.20 shows the molecular-weight profiles of a series of fractionated polyacrylates. The performance of the various fractions of the original inhibitor against barite scale are depicted in Fig. 5.21 for conventional bulk "bottle" tests conducted at 2 and 22 hours of residence time.

The effects of polymer molecular weight can be summarized as:

- Nucleation inhibition (short residence time): Maximum performance at low and medium molecular weight (molecular weight of order 3,000–5,000 g/mol) for polyacrylate-based polymers.

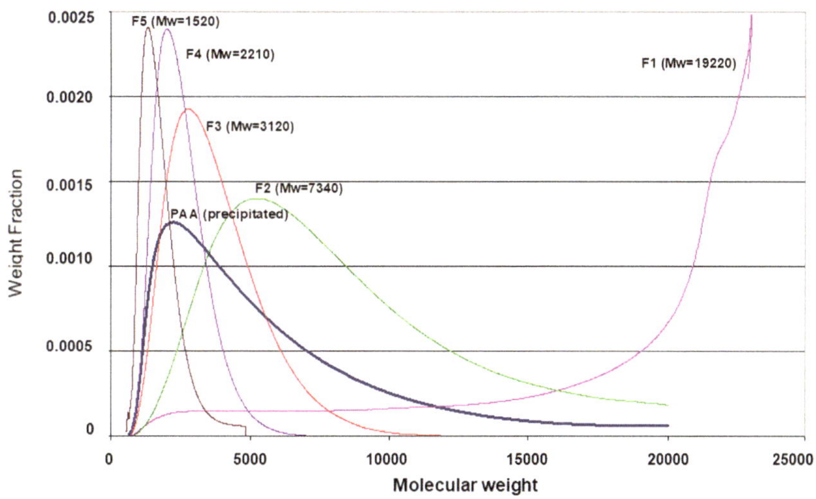

Fraction	Statistical		
	Mn	Mw	PDI
Precipitated, unfractionated	2,590	4,770	1.84
F1	11,740	19,220	1.64
F2	4,900	7,340	1.50
F3	2,540	3,120	1.22
F4	1,940	2,210	1.14
F5	1,350	1,520	1.13

FIGURE 5.20

Molecular weight analysis of bulk PAA and fractionated PAA samples.

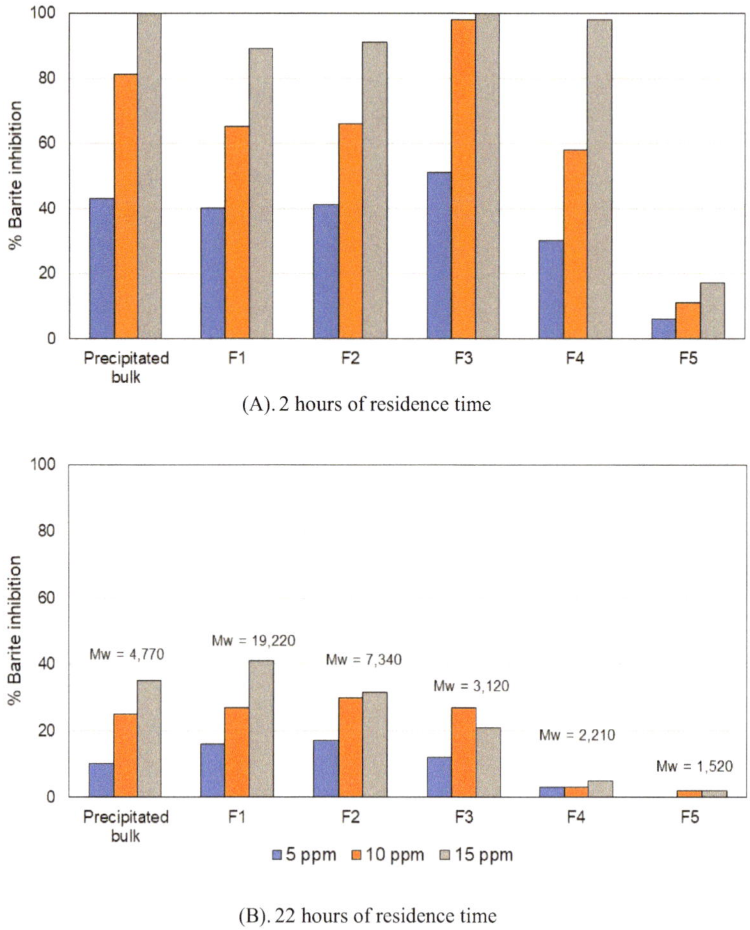

FIGURE 5.21

Static performance tests of PAA fractions in 50% FW (269 mg/L Ba^{2+})/50% SW (pH 5.5; T = 90°C). (A) 2 h of residence time. (B) 22 h of residence time.

- Crystal growth retardation (long residence time): Increased performance with increased molecular weight and multiplicity of binding sites.
- Low molecular weights generally give poor performance against barite scales with the sharp cut-off around 1,500–2,000 g/mol.
- Similar trends for different base polymers [86,127].

Scale inhibitor–brine compatibility and chemical retention are also functions of molecular weight, with higher-molecular-weight polymers, particularly polycarboxylate-based polymers, being less compatible in higher-salinity and higher-calcium waters.

5.4.7.4 Influencing conditions: brine composition, the controlling influence of Ca^{2+} and Mg^{2+}

The calcium concentration in a particular reservoir and in the inhibitor slug often determines the extent to which the inhibitor species is retained in the near-wellbore area during a squeeze treatment operation (i.e., on its adsorption or precipitation behavior). Divalent cations, such as calcium and magnesium, also control the inhibition process itself. Common ion effects are well known and influence the overall scaling tendency or level of supersaturation (as described in Section 5.3). However, for penta-phosphonate inhibitor species (e.g., DETPMP), significant improvements in inhibition efficiency have been reported by increasing the calcium concentration in the solution [72,92]. The effect of calcium and magnesium cation concentrations have been examined for a wide range of generically different inhibitor species, including penta-phosphonate, hexaphosphonate, phosphino-polycarboxylate, polyvinyl sulfonate, and sulfonated polyacrylate copolymers. This is shown in Figs. 5.22 and 5.23 which clearly indicate how different inhibitor species are affected by changes in [Ca^{2+}] and [Mg^{2+}] and how this difference relates to the cation affinity of the inhibitor's active functional groups. Results were obtained by comparing the barite inhibition efficiency of various species in a low/medium scaling brine (Brent type) and also in a more severe scaling brine (Forties type). The inhibition efficiencies were examined by static inhibition efficiency tests, with residence times ranging from 30 minutes to 24 hours.

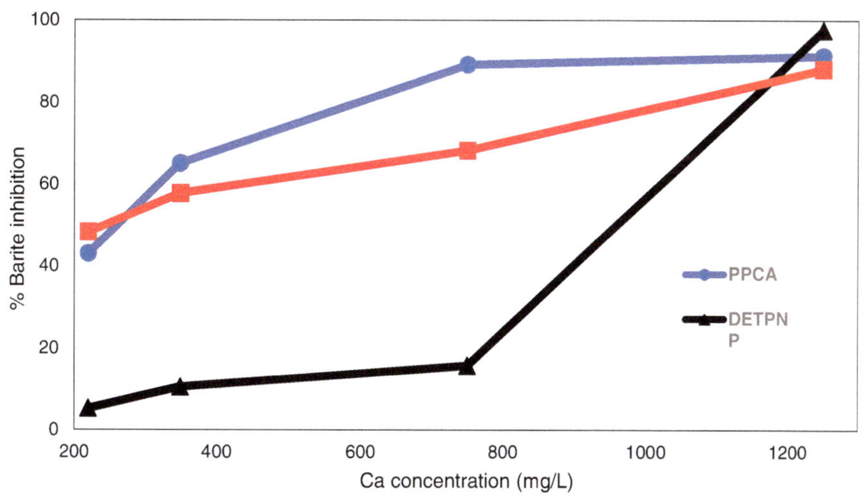

FIGURE 5.22

Effect of calcium on the scale inhibitor relative performance in 50% FW (60 mg/L Ba^{2+})/50% SW mixture.

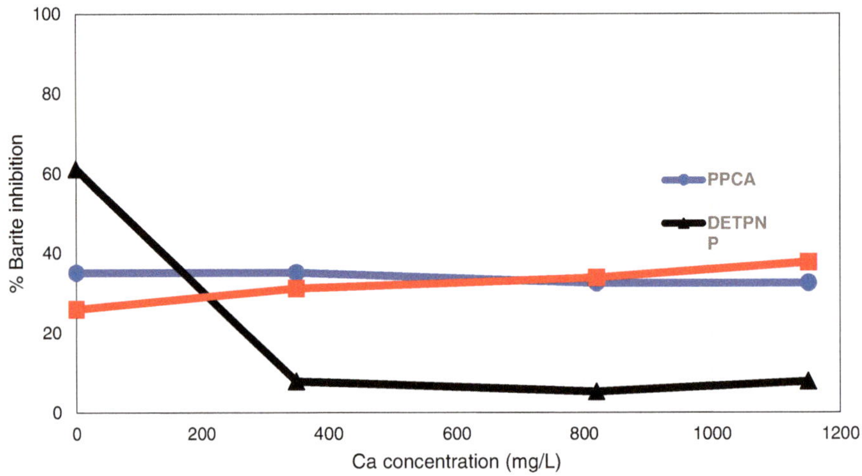

FIGURE 5.23

Effect of magnesium on scale inhibitor relative performance in 50% FW (60 mg/L Ba^{2+})/ 50% SW mixture.

Phosphonates are shown to be poor inhibitors at very low [Ca^{2+}], indicating that their effectiveness is controlled by the formation of Ca^{2+}−phosphonate inhibitor complexes. In contrast, polymeric polycarboxylate and polyvinyl sulfonate inhibitors are shown to be effective even at very low [Ca^{2+}], indicating that the formation of multiple bonds between the polymer and the crystal surface allows for stronger adsorption and, thereby, inhibition. While strong ionic bonds involving calcium cation bridging are required for the phosphonate-based species, the performance of the phosphonate is significantly reduced when the magnesium ion concentration is increased. The polymeric inhibitors are relatively unaffected by magnesium. Such trends, especially for the phosphonate species, were observed in a large number of studies [72,92].

The importance divalent cations present in oilfield waters cannot be overstated. For example, injection of seawater (high Mg^{2+}/low Ca^{2+}) into high Ca^{2+}/low Mg^{2+} formation waters can lead to the requirement to change the scale inhibitor in order to maintain effective scale control toward the later stages of a field or well lifetime, when producing at high percentage of SW in the produced water.

5.4.7.5 Influencing conditions: dissolved iron

In addition to the effect of calcium and magnesium, other divalent cations also play an important role, as any cation capable of complexing with the functional groups of the scale inhibitor can affect its activity.

The presence of dissolved iron can have a considerable impact on the performance of scale inhibitors, against both carbonate and sulfate scales. It also has a considerable impact on the complexity of the laboratory testing and qualification procedures, since all flow lines and test procedures must be conducted under fully anaerobic conditions. However, it is known that the presence of high levels of dissolved iron can render some scale inhibitor species ineffective [95]. In general, more specialized chemicals are often required when dealing with either conventional calcium carbonate or common Ca/Ba/Sr sulfate scales in brines containing high levels of dissolved iron or when dealing with iron bearing scales, such as iron carbonate.

The impact of dissolved iron is generally regarded as through one of three pathways [94]:

- Formation of complex: Iron react with iron to form strong complexes, effectively reducing the activity of the inhibitor and thereby increasing its MED.
- Compatibility issue: Precipitation of inhibitor–Fe complexes removes the inhibitor from solution and therefore renders the inhibitor inactive.
- Formation of different polymorphs: This is particularly relevant to calcium carbonate (and other carbonates) whereby iron incorporates into the lattice, forming new crystal polymorphs, which the inhibitors are no longer effective at inhibiting.

Although in general common scale inhibitors (such as DETPMP and PAA) are not designed to work against iron carbonate and are therefore relatively ineffective, other modified inhibitors have been shown to be effective against such scale [128]. In addition, when dealing with other iron bearing scales, such as iron sulfides, more specialized inhibitors may need to be assessed [129].

5.4.7.6 *Influencing conditions: effect of solution pH, bicarbonate and pressure*

The main effect on inhibitor performance due to pressure reduction relates to the associated changes in pH. Hence, for field wide protection, different inhibitors may be required for downhole, lower pH conditions than for topside, higher pH conditions. As pH has a significant impact on the calcium carbonate scaling potential, higher dose levels of inhibitor may be expected at higher pH levels. However, for the sulfate scales, pH has only a nominal impact on the supersaturation and therefore the effect of pH on sulfate scale inhibition is related to its control on the inhibitor effectiveness.

The effect of pH on the performance of scale inhibitors was summarized earlier (Section 5.4.7.1). It is worth stating that when testing or qualifying scale inhibitors, it is essential that the "in-situ test pH" and not the initially buffered pH (normally prepared and buffered at ambient conditions), is considered. Poor pH control and buffering in laboratory testing can lead to significant changes in the in-situ pH [71], potentially rendering the test results inappropriate at best, at worst, simply wrong. For example, pH buffered at ambient conditions can often give very low and nonfield representative in-situ test pH, resulting in the deselection of effective scale inhibitors.

Table 5.9 shows the impact of pH control on the recorded MEDs, for a phosphonate and polycarboxylate-based species. Under conventional bulk or bottle test conditions (conducted at ambient pressure), in the absence or presence of bicarbonate, the starting pH had minimal effect on the inhibitor performance (i.e., the MED changed little with starting pH). However, when similar tests were conducted via the dynamic tube blocking approach very different results could be obtained, depending on the accuracy with which the "in-situ" pH was adjusted and achieved. At pH 6 (with no bicarbonate and at the highest (unadjusted) pH (with bicarbonate present), dynamic performance results were similar to those in the bottle tests. However, at intermediate pH (with bicarbonate present - which required some degree of pH adjustment), very poor performance was observed. Unfortunately, in this case, the manner in which the system was pH buffered resulted in the in-situ pH being much lower than the target pH.

These tests reflect a situation whereby pH was buffered poorly under ambient conditions (i.e., in an open system) resulting in the consumption of bicarbonate ions as a result of the acid addition by the conventional reaction illustrated below in Eqs. (5.11)-(5.13). In these tests, the poor pH buffering in the presence of bicarbonate led to very low in-situ pH and hence very poor performance of the phosphonate and polycarboxylate-based inhibitors. It is also worth noting that the pH unadjusted system (with an ambient measured pH of ~ 7) resulted in an in-situ pH at the temperature and pressure of the dynamic test, in line with the

Table 5.9 Effect of pH buffering on in-situ pH and inhibitor performance.

Test	Ambient pH	$[HCO_3^-]$ ppm	Poly carboxylate MED ppm	Phosphonate MED ppm	In-situ pH[a]
Bottle Tests	5.5	0	10–15	4–5	–
	6.0	0	10–15	3–4	–
	7.0	0	10–15	3–4	–
	8.0	0	10–15	3–4	–
	6.0	250	10–15	3–4	–
	> 7.0	250	10–15	< 3	–
Dynamic Tests	2.5	0	> 200	> 100	–
	6	0	5–10	2–5	–
	6 (adjusted)	250	> 40	20–40	3.2–3.5
	7.3 (un-adjusted)	250	5–10	2–5	6.0

[a]*Modeled in-situ pH for dynamic tests.*

expected field pH. Therefore, in this example no pH adjustment at ambient conditions was actually required to achieve the target pH at in-situ at temperature and pressure.

$$HCO^{-}_{3(aq)} + H^{+}_{(aq)} \leftrightarrow H_2CO_{3(aq)} \tag{5.11}$$

$$H_2CO_{3(aq)} \leftrightarrow H_2O + CO_{2(aq)} \tag{5.12}$$

$$CO_{2(aq)} \leftrightarrow CO_{2(g)} \tag{5.13}$$

This example therefore serves to illustrate both the controlling influence of pH on the performance of different scale inhibitors, as well as highlighting the critical importance of pH adjustment in tests conducted at elevated temperature and pressure in sealed systems.

Given that field brines are often in the region of 4.5–5.5 at downhole (reservoir) conditions, and that these pH values are close to the pK_as of many conventional inhibitor functional groups, pH control in laboratory qualification is critically important. Some laboratories choose to control the ambient pH prior to introduction into the test apparatus rather than in-situ pH and so a degree of caution is needed when considering results [71,130].

In summary, as in every field of oilfield chemical qualification, laboratory test conditions must be determined in order to replicate as far as practically possible the conditions, including pH, that is likely to be experienced in the field at the point of scaling. This means ensuring that the in-situ pH (within the test apparatus) is matched to the field pH (at the point of scaling); not at ambient conditions

5.4.7.7 Influencing conditions: effect of temperature

Temperature has been shown to have a considerable impact on the performance and relative performance of generically different scale inhibitors. This has a particular significance when producing in subsea operations with long, unheated seabed flowlines, where temperatures can decrease to seabed temperatures of 4°C.

Fig. 5.24 shows the effect of low seabed temperatures on the performance of DETPMP, PVS and PPCA [75]. Other work has also shown that the performance of selected phosphonates, such as DETPMP, against barium sulfate scale can be considerably reduced at temperatures of around 50°C and below. It is therefore very important to ensure that the selected scale inhibitors perform across all temperature conditions experienced during production. It is also common that in subsea operations, one type of chemistry may be applied for downhole scale control at elevated temperatures, whereas a different chemistry (often a polymeric species) may be added at subsea wellheads, to ensure protection of the sea bed flowlines where lower temperatures may be encountered. Additionally, the use of different chemistries in seabed flowlines also assists in analysis of squeeze return

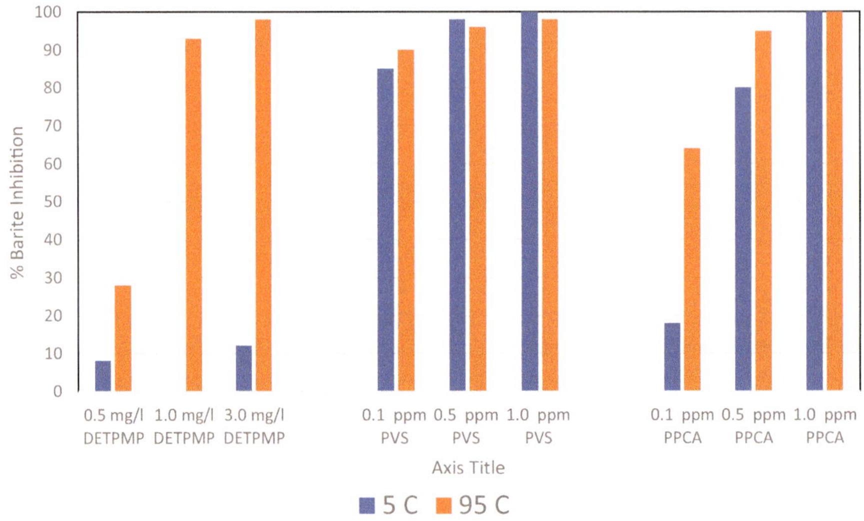

FIGURE 5.24

Effect of temperatures on the performance of scale inhibitors against barite under mild 50% FW (60 mg/L Ba^{2+})/50% SW (Blue bars; 5°C; Red bars: 90°C).

profiles, as the two inhibitors can be distinguished and so the concentration of returning squeeze chemical can be monitored.

5.4.7.8 Summary of factors controlling performance

This subsection has illustrated that the various factors relating to the complexity of the brine composition itself, changes which occur due to production (e.g., temperature, pressure and pH) or as a result of mixing of different waters (e.g., impact of dilution of FW with SW) control not only the level of oversaturation of different minerals but also have a controlling influence on the performance and relative performance of different inhibitors. Given the complex nature of oilfield brines and the significant diversity of brine compositions produced in oilfield environments, coupled with the changes in physical conditions experienced during production and that performance represents only one of the criteria required for effective chemical application, the authors hold that there is no substitute for effective laboratory qualification conducted under conditions representative of those pertaining in the field at the point of scaling. However, as also indicated, this is not always achieved and, in some cases (such as the case for pH control shown earlier), if the conditions within the laboratory test are not tuned effectively to those of the field conditions, incorrect chemical selection and qualification may occur.

The importance of effective laboratory qualification is therefore the focus of the next subsection of this chapter.

5.4.8 Laboratory assessment of scale inhibitors

Scale inhibitor testing is imperative if a large-scale treatment program is anticipated. Cost-effective treatment, of course, is the major goal, but other considerations must be made also. There are many thousands of scale inhibitor formulations on the market, many of which will not be suitable for the particular system to be treated. Furthermore, even for apparently similar products offered by different providers, the purity, formulation, molecular weight and in the case of polymers, polydispersity, of the inhibitor will vary among suppliers, all of which again can have an impact on the relative performance.

This section will concentrate primarily on the laboratory assessment of inhibitor performance tests by conventional static and dynamic test protocols, which represent the most common and routine approaches used within the oil industry, as well as other approaches such as the Kinetic Turbidity Test (KTT) and Jet Impingement test (JIT) which are becoming more commonly adopted for certain applications. It is also important to remember that it is critical to assess the other performance attributes outlined above (e.g., compatibility, thermal stability, etc.) prior to field deployment.

The main criteria for continuous injection inhibitors and squeeze treatment inhibitors for effective chemical selection are summarized as follows:

5.4.8.1 Continual injection inhibitors

As indicated previously, the main performance criteria required for continuous injection scale inhibitors can be summarized as follows:

- Inhibit scale formation at threshold (MED) levels, typically < 1 mg/L to tens of mg/L (though sometimes much higher);
- Other aspects which require to be addressed prior to field deployment:
  - Brine compatibility;
  - Physical and also performance compatibility with other production chemicals, such as corrosion inhibitors;
  - MeOH/Glycol compatibility, especially when applied through umbilical's and may come into contact with MeOH and/or glycol (mono-ethylene glycol, MEG) often used as hydrate inhibitors and for cold start operations in seabed flowlines;
  - Materials compatibility (with the materials of the injection system);
  - Impact on process conditions (e.g., stabilization of emulsions);
  - Thermal stability at application conditions (e.g., wellhead temperature or downhole injection temperature);
  - Chemical stability throughout the injection system. It is critically important for downhole continuous injection chemicals to prevent aspects, such as: gunking, instability leading to solids formation or high viscosities in the injection flowline, instability leading to the formation of components or phases with increased corrosivity, and so on.
  - Environmentally friendly

5.4.8.2 Squeeze treatment inhibitors

For downhole squeeze treatment inhibitors other critical aspects include:

- Minimal formation damage
- Appropriate interaction (retention mechanism) with reservoir (adsorption/precipitation) leading to long return profiles (typically 3–48 months) at inhibitor residual concentration > MED
- Minimal impact on process conditions (e.g., stabilization of emulsions) especially during the early return period where high concentrations of inhibitor chemical may enter the separators leading to poor separation and process upsets.
- Brine compatibility
- Compatible with other production chemicals
- Thermal stability at reservoir conditions over prolonged periods

For downhole squeeze inhibitors, their mode of interaction and the appropriate laboratory assessment of aspects relating both to formation damage and comparative inhibitor and subsequent field treatment modeling and application will be covered in Section 5.5.

Inhibitor performance in terms of the MED is one of the most important aspects for scale control additives, equaled only by the challenge of effective placement and deployment in today's ever more complex production environments.

The basic laboratory screening test protocols adopted throughout the industry for many years have been based upon static "bulk" inhibition performance tests (or "jar" tests) and dynamic "tube blocking" inhibitor performance tests [71]. However, even these common approaches which have been used throughout the industry for many decades are modified for various applications. For example, in mildly saturated systems conditions "bulk" or static tests will be modified to include stirred bulk tests, seeded bulk tests, seeded and stirred "bulk tests" to try to induce more rapid precipitation [131]. Similarly, dynamic tests are modified in terms of the type and length of the "scaling" coil, flow rates, introduction of in-line filters, prescaling of coils, and so on [71,129] in order to maximize scale formation within laboratory timescales. Although these tests are regarded as routine in the industry, they are not simple, given the somewhat complex nature of the systems under examination. Inappropriate application of the test, as in the case of poor pH buffering described earlier, can render the result meaningless and therefore care and knowledge cannot be underestimated when performing laboratory testing.

In addition, given the requirement to conduct tests and identify chemical inhibitors over normal laboratory timescales, in situations where kinetic factors control the propensity for scaling in the field, brine chemistries are often changed in order to exacerbate the scaling regime. This, however, can be a very risky approach to chemical qualification and the recommendation for such systems is to modify the test approach rather than the brine. To this end, more recently the importance of kinetics aspects (in particular turbulence and surface shear) has led to the development of alternative methods, particularly for mildly oversaturated systems, including KTT and JIT approaches, among others, as well as very large scale flow loops designed to more directly replicate the fluid flow dynamics [43,46].

5.4.8.3 Conventional bulk or "static" jar test

The conventional static "bulk" or "jar" test procedures commonly adopted are related to that described in the NACE standard TM 0197 [132]. Such tests have been described in many papers for both examination of the factors controlling inhibitor performance and for selecting scale inhibitor products prior to field applications [71]. These tests are used routinely throughout the industry for scale inhibitor selection and optimization studies. A general schematic of the static inhibitor performance test is illustrated in Fig. 5.25.

An example of the results typically recorded from static performance tests is illustrated in Fig. 5.26. This illustrates a number of aspects associated with chemical performance. First, the performance of the two polymers (PPCA and PVS), when tested at concentrations below that required to completely prevent scale, reduces with time over 24 hours, whereas that of the hexaphosphonate (hexa-P) remains stable over time. The results suggest a different mode of interaction, that is, the small polymers perform as nucleation inhibitors and crystal growth retarders, while the discrete phosphonates act more as growth blockers pinning active growth sites. All chemicals however are clearly working effectively at both nucleation and growth inhibitors and increasing the concentrations leads to complete cessation of scaling.

5.4.8.4 Dynamic tube blocking tests

Dynamic "tube blocking" performance tests are also routinely used for scale inhibitor selection in oilfield environments. However, these tests often give rise to different

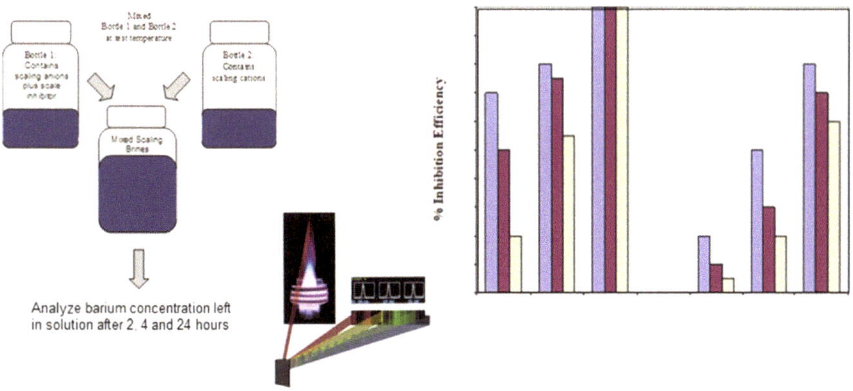

FIGURE 5.25

Schematic illustrating the steps involved in conventional static inhibitor performance tests.

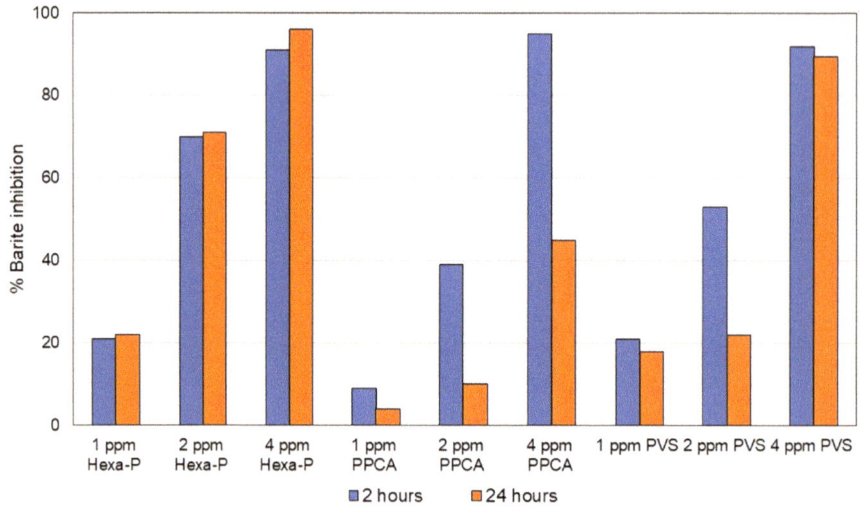

FIGURE 5.26

Comparison of inhibitor performance against barite scale in 50% FW (269 mg/L Ba^{2+})/ 50% SW (pH 5.5, 95°C).

selection and ranking of scale inhibitor products than that obtained through conventional "jar" tests. Such discrepancies have been explained in terms of: (i) short residence time (generally <10 s) compared with "jar" tests (2–24 hours), (ii) dispersant and/or antiagglomerant properties of certain inhibitor species, (iii) differences between nucleation and crystal growth inhibition effects, in addition to (iv) the impact of scale adhesion on the walls of the microbore tubing.

The dynamic "tube blocking" test is essentially measuring the ability of the inhibitor to prevent adherence and growth of scale on the walls of a (microbore) pipe over short in-situ time periods (although the individual test may run for several hours, the in-situ time within the coil is generally of the order < 10 seconds). The bulk or jar tests are measuring bulk or homogeneous precipitation over extended time periods.

There are a number of benefits associated with dynamic performance tests, which can be summarized as follows:

- Dynamic system, as in oilfield production;
- Examines growth and blockage of microbore metal coils;
- System under pressure, allowing testing under higher temperature conditions (> 100°C) and in the presence of bicarbonate ions without loss of pH control;
- Examination of systems in presence of dissolved iron is more readily achievable, provided feed brines are adjusted accordingly to prevent oxidation of ferrous Fe(II) to ferric Fe(III) [71,95,128,129];
- Systems can be automated to some degree, allowing improved throughput and determination of chemical MEDs and ranking of different products.

The main disadvantage however relates to the short residence times (generally < 10 seconds), and that the flow system is generally laminar. this means that for many mildly oversaturated systems, scale may not form in the timescales and the nonturbulent/low shear environment within the test coils.

A simple schematic of the dynamic inhibitor performance test is illustrated in Fig. 5.27. A HPHT dynamic tube blocking test rig typically used for inhibitor qualification is shown in Fig. 5.28. Fig. 5.29 provides an example of typical plots generally derived from tube blocking tests for scale inhibitor performance and qualification tests.

As indicated above, one of the main disadvantages of dynamic performance tests relates to the extremely short residence time (generally < 10 seconds) within the test coil (with fresh brine *inhibited* continuously injected). Although this may be appropriate for certain applications such as boiler systems, or when examining heat exchanger scaling, it is not appropriate to the longer residence time (often of the order of hours) in which oilfield scale inhibitors may be expected to perform. The second disadvantage relates to the in-situ flow conditions which are generally laminar in conventional tube blocking tests. Therefore, care must be taken when interpreting results and it is often recommended that both longer (more field specific) residence time static "jar" tests (in which field temperature and pressure and, potentially pH and bicarbonate content may be compromised) and dynamic "tube blocking" tests (in which residence times may be compromised) should be conducted where possible.

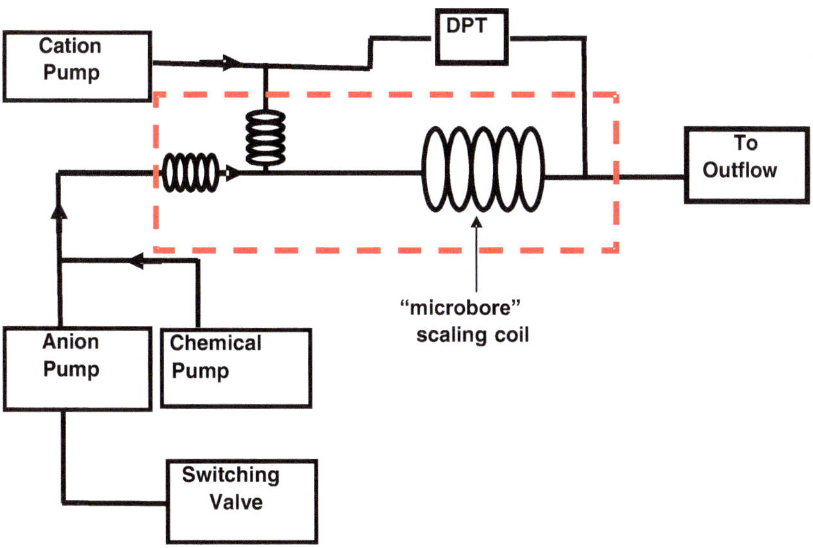

FIGURE 5.27

A simple schematic of the dynamic inhibitor performance test apparatus.

Courtesy of Scaled Solutions Ltd.

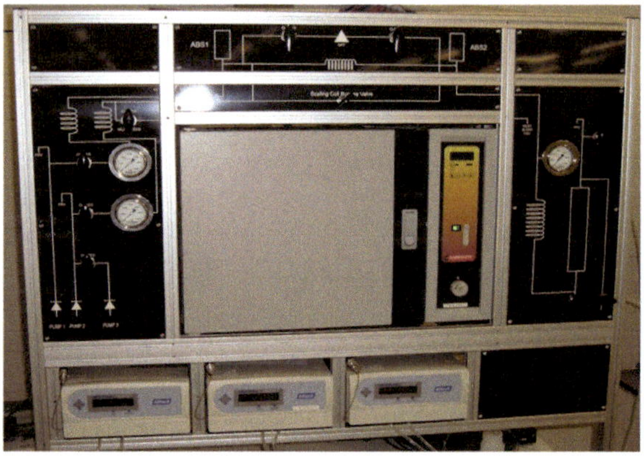

FIGURE 5.28

A HT/HP dynamic scale test rig.

Courtesy of Scaled Solutions Ltd.

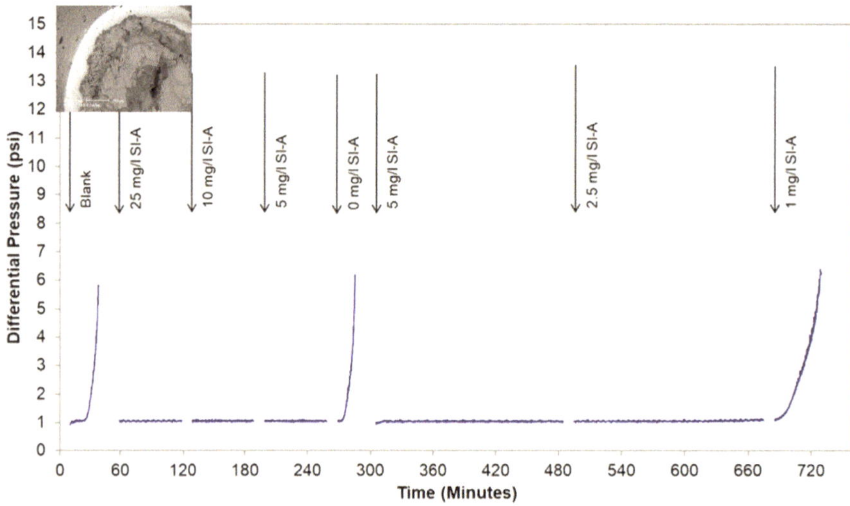

FIGURE 5.29

Typical differential pressure vs. time profiles recorded under moderate scaling conditions, including insert of SEM image of "scaled" microbore coil following a blank scale test.

Image Courtesy of Scaled Solutions ltd.

Finally, although the conventional dynamic "tube blocking" test probes (at least in part) the ability of scale inhibitors to prevent adherence and growth within microbore coils, the relationship between bulk homogeneous nucleation and growth and heterogeneous nucleation and growth at metal surfaces (adherence) is not well defined. This relationship is currently the subject of much ongoing research as it is recognized that heterogeneous growth at metal surfaces (scale adhesion) can be a much more serious problem to tackle than bulk precipitation [41,43,46,48]. For homogeneous (bulk) precipitation, suspended particles may remain in the bulk fluid and be transported to surface facilities rather than causing blockages to the flow lines.

The main aspect which cannot be well replicated in either conventional bulk or static jar tests and conventional tube blocking tests outlined above is to enable scale inhibitor ranking and qualification under mild scaling regimes where the scaling mineral remains metastable. In such cases scaling may be considered to be under kinetic control, or more importantly, controlled by the local flow environment with areas of increased surface shear and turbulence leading to increased risk of scaling. For such systems conventional test approaches have been modified and adjusted. Bulk or static jar tests are often modified to include stirring or seeding in low scaling environments [131]. Similarly, dynamic "tube blocking' tests are often prescaled or in-line filters introduced in order to exacerbate the propensity for scaling in mildly oversaturated conditions [129].

5.4.8.5 Kinetic turbidity testing

Kinetic turbidity testing (KTT) is a relatively new technique in the oil and gas industry, but is becoming more widely used as a method to evaluate scale inhibitor performance. With small volumes of brine required and short setup times, KTT is particularly useful for rapidly screening a large number of inhibitors or inhibitors at differing concentrations prior to further qualification using other methods. Use of the technique has been reported for the assessment of common oilfield scales (sulfates [133,134], calcium carbonate [135]) as well as halite [136,137], sulfides [138], and silicates [139]. A KTT apparatus is shown in Fig. 5.30.

The KTT approach involves measuring precipitation in a scaling brine using the absorbance of light [140]. A beam of monochromatic light, typically with a wavelength of 500 nm, is passed through a quartz cuvette containing a scaling brine (charged with a magnetic stirrer bar and placed in a heated cuvette holder) and the absorbance is measured at regular intervals. As precipitate forms in the test brine and the crystals increase in size over time, the proportion of light absorbed increases. A schematic absorbance profile is presented in Fig. 5.31. Region (A) shows no increase in absorbance over time as would be expected in a blank or fully inhibited test but could also represent the induction time prior to precipitate formation in a scaling brine. Region (B) shows an increase in absorbance due to precipitation and subsequent crystal growth. The crystals continue to grow and agglomerate until they reach a certain mass after which they will sink to the bottom of the cuvette due to gravity. This results in a decrease in the measured absorbance as shown in region (C). The maximum absorbance is dependent on the brine being tested but potentially also on temperature and pH (particularly for carbonate scale). KTT offers real time, in-situ monitoring of precipitation and is particularly useful for measuring scaling in mildly oversaturated systems where

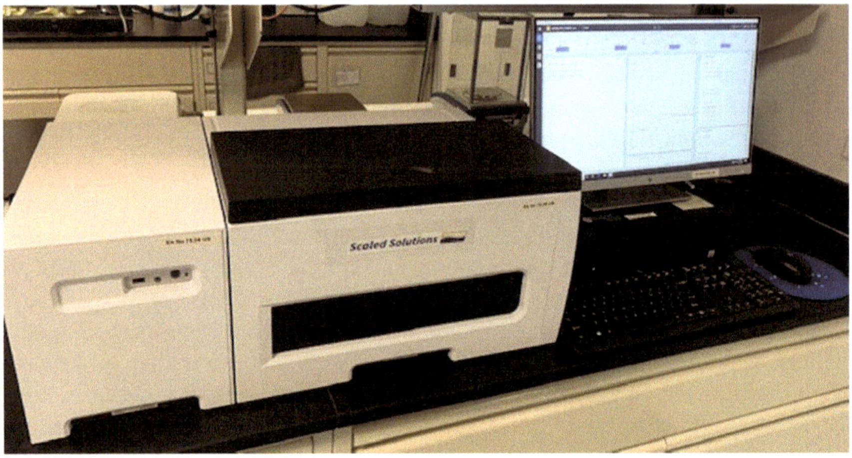

FIGURE 5.30

A kinetic turbulence test (KTT) apparatus.

Image courtesy of Scaled Solutions Ltd.

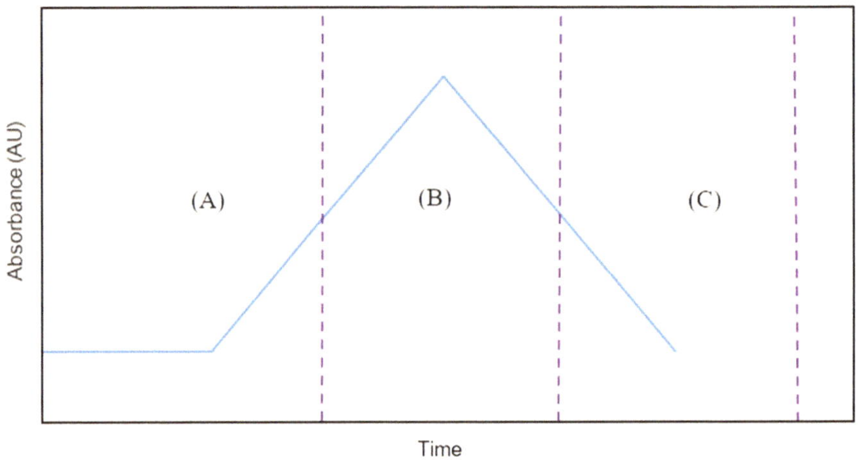

FIGURE 5.31

Schematic absorbance profile of KTT test.

changes in the scaling ion concentration cannot be measured by ICP (for example) or assessed by observation. However, it only measures precipitation and gives no information on how much of this will adhere to surfaces as scale.

KTT can also provide information on inhibitor mechanisms [133]. The absorbance profiles for a range of inhibitor mechanisms are shown in Fig. 5.32, along with profiles for a nonscaling test and scaling blank for reference. The absorbance

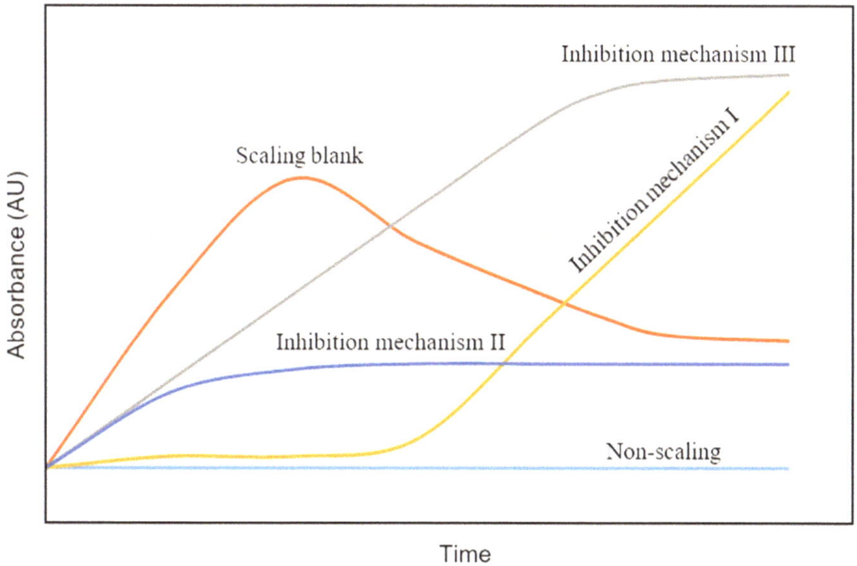

FIGURE 5.32

Absorbance profiles indicating different scale inhibitor mechanisms.

in the nonscaling test remains constant at 0 AU for the duration of the test. The absorbance in the scaling blank increases initially as precipitate forms; then decreases and levels off as some of the precipitate settles at the bottom of the cuvette. The inhibition mechanism profile shown as "I" in Fig. 5.32 represents a typical absorbance curve for inhibiting nucleation. Initially, the curve remains flat with an absorbance similar to the blank, indicating no precipitate has formed. However, the inhibitor may only provide protection for a limited period of time before precipitation occurs, as indicated by the increase in absorbance. The absorbance profile of an inhibitor acting as a dispersant and/or scale growth inhibitor is shown by the inhibition mechanism II curve. There is an initial increase in absorbance as nucleation occurs, however, this levels off as growth is inhibited and the precipitate remains suspended in solution, resistant either to further growth or Ostwald ripening. Either of which could lead to gravitational settling and would reduce the late-time absorbance. Inhibition mechanism III shows a curve representative of poor inhibition. While the rate of scaling is slower than the scaling blank, a larger number of small particles are formed, resulting in a greater maximum absorbance.

5.4.8.6 Jet impingement testing

Jet impingement testing (JIT) is a well-established method for examining corrosion processes under high shear flow conditions. In recent years, it has been developed for use in scale investigations [43–45,48]. It involves a high velocity jet of fluid from a small-bore nozzle being directed at a flat metal coupon held perpendicular to the jet. Advantages include the fact that shear stresses across the coupon surface as the fluid flows radially out can be calculated as a function of distance from the center of the coupon. The shear stress generated can be easily controlled by altering the nozzle diameter and flow rate. It is possible to achieve very high shears, representative of those expected in the field, which other laboratory techniques are unable to reach. In JIT testing, brine is circulated around the system using a peristaltic pump to control the flow rate through the nozzle and hence the shear stress generated on the surface of the coupon. A benchtop, ambient-pressure system is capable of testing up to 400 Pa shear stress at temperatures of up to 90°C (Fig. 5.33). Using more sophisticated equipment including high flow pumps and JIT autoclave cells, this approach has also been expanded to carry out tests at elevated temperature and pressure at shear stresses of up to 10,000 Pa [48]. Unlike most other test methods, JIT can be used to evaluate inhibitor performance by monitoring both mineral precipitation from solutions and surface scale deposition on the metal coupon.

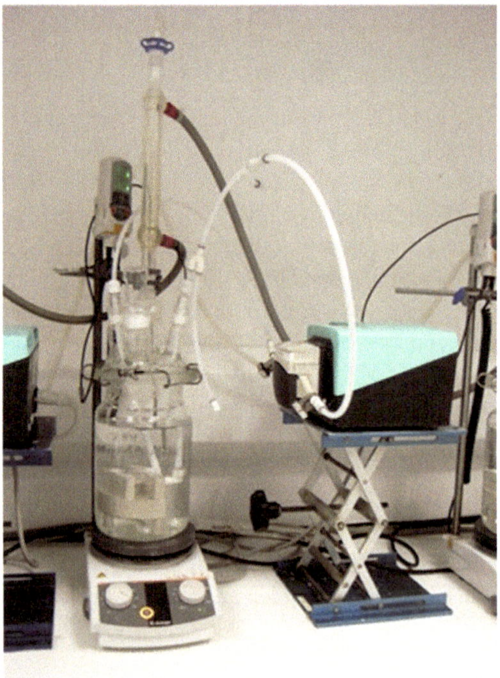

FIGURE 5.33

Laboratory jet impingement equipment.

Courtesy of Scaled Solutions, Ltd.

Due to the very high turbulence and shear stresses generated in this technique, the kinetic factors influencing scaling are enhanced and scaling can be induced in mildly oversaturated systems that remain may metastable with other test methods. It can also be used to indicate whether scale inhibitor performance will be maintained in environments that are expected to be subject to severe hydrodynamic conditions such as ICDs and ESPs where a higher dose rate may be required [43,44].

5.4.8.7 Summary: SI performance and qualification tests

The following table summarizes the main benefits and limitations of the main scale inhibitor testing approaches outlined above.

Test method and variations	Temperature and Pressure	Advantages	Disadvantages
Bulk "jar" or bottle tests • Static • Stirred • Seeded	Up to 90°C, ambient P (Can be conducted at higher T and P in more specialised autoclave equipment)	• Simple, rapid screening tool • Appropriate residence times can be examined • Allows interpretation of mechanism and impact of SI/ additive on scale formation and growth (i.e., different residence times can allow comparison of crystal growth and nucleation inhibitors)	• Normally limited in temperature $<95°C$ • Pressure limitations - 1 bar • Difficult to buffer if bicarbonate present • Difficult for quantitative assessment of carbonate scales and also calcium sulfate due to small percentage drop in cations which can often be below analytical detection limit; visual assessment may however be sufficient • Conventional unseeded "static" tests can miss scale formation under low SR conditions. *Note: stirred. seeded, stirred and seeded tests can sometimes overcome this*

(Continued)

Continued

Test method and variations	Temperature and Pressure	Advantages	Disadvantages
			• Difficult to perform tests conventionally in the presence of dissolved iron (more specialized anaerobic equipment required)
Dynamic tube-blocking tests (also dynamic scale loop, DSL) • with filter • prescaled	Up to 200°C, 10000 psi (generally limited to 5,000 psi)	• Dynamic, flowing system, as in oilfield production systems • Examines growth and blockage on metal surfaces (within micro-bore metal coils) • System under pressure • Higher temperature conditions readily achieved • Incorporation of bicarbonate ions without loss of pH control (caution required regarding in-situ pH) • Pre-scale tests overcome nucleation/dispersion effects • More routine for HT/HP than static/stirred autoclave tests • Examination of scaling and inhibition in presence of dissolved iron (anaerobic conditions) readily achieved	• Short residence time • Kinetics become significant factor at low temperature and/or low SR • Systems often made artificially severe (increased scaling ions) for low SR systems
Kinetic turbidity test (KTT)	0°C–95°C, ambient pressure	• Allows rapid screening • Absorbance profile allows interpretation of SI mechanism	• Limitations due to T and P are valid as with conventional static or bulk tests

(*Continued*)

Continued

Test method and variations	Temperature and Pressure	Advantages	Disadvantages
		• Introduces mixing with scaling and scale inhibition recorded directly as a function of turbidity • May be more appropriate than conventional statics for mildly oversaturated systems (i.e. stirred) • Can be readily conducted below ambient temperature with appropriate spectrometer	
Jet Impingement test (JIT)	Up to 90°C, ambient pressure; or autoclave up to 200°C and 2000 psi	• Allow tests to be conducted under very high shear stress and turbulence conditions. • Allows both deposition of scale on metal surfaces and precipitation from bulk fluid to be assessed in the same test. • Becoming more routine for mildly oversaturated systems	• Generally more time consuming than conventional bulk (jar) tests, KTT tests and dynamic tube blocking tests. • Allows SI qualification under kinetic control such as that experienced in very high shear locations (e.g., ICDs, etc.) • Useful for low SR conditions and avoids need to make brine more severe (aka conventional dynamic tube blocking tests)

5.4.9 Chemical qualification: final considerations

Chemical performance in terms of its inhibitory effect is only one criteria which a scale inhibitor must achieve in order for it to be applied safely and without causing negative impact on the production system. This means that chemical qualification for scale inhibitors (as with other production chemicals) requires a range of performance criteria to be met, depending on the application and conditions. The various requirements are summarized in Figs. 5.34–5.36 to illustrate the complexity of the chemical qualification process.

5.5 Scale inhibitor squeeze

5.5.1 Chemical squeeze process

Scale inhibitor squeeze treatments are one of the most common and effective methods for scale control in production wells because the location of chemical dosing typically protects everywhere scale can occur in the production system: the near-wellbore formation, perforation tunnels, gravel packs, sand screens, wellbore, or indeed anywhere downstream of this point. The technology is well understood and for scale inhibition is applied globally in multiple oilfields. It has therefore also been an active area of research and of much development of the process for certain applications (such as new chemical inhibitors to increase treatment, application of emulsified and invert emulsions scale inhibitor treatment packages, use of mutual solvents and other additive sin the reflush packages, development of oil miscible and oil soluble inhibitors, development of squeeze life enhancers both polymeric additives and particulate, etc.) have all been introduced over the years with a view to improving the base technology for certain applications. There are therefore multiple references in the open literature of which only a selection is indicated here [5,10,96–98,141–180].

As such, it protects all parts of the system upstream of the point where continuous downhole chemical injection is normally practicable, that is, just above the production packer in the well [114]. Continuous injection upstream of the packer is also possible, and has been achieved occasionally, but this requires a modified packer through which a chemical "macaroni string" can be passed, which significantly increases both the completion's complexity and (since the packer is primary well barrier) the risk to well integrity (see also references to DDHCI in Section 5.4.5). Even dip tubes that pass through the packer to deliver chemical within the producing interval do not protect the near-wellbore, gravel pack, perforations, and so on.

The downside of squeeze treatments is that they are periodic, discrete treatments that must be repeated throughout much of the life of the well, which means they require interruption of production (deferring hydrocarbon). It also requires that a fairly large amount of treatment fluid (both the chemical and the carrying media) be pumped into the well, thus introducing various risks, including that of formation damage and productivity impairment. Another aspect is that to protect against scaling associated with mixing of incompatible waters (most typically for sulfate scales), it is necessary to get sufficient coverage of the whole of the producing interval with a multistage treatment, which can be challenging in wells with long producing intervals and pressure differentials between zones. Finally, in certain production environments,

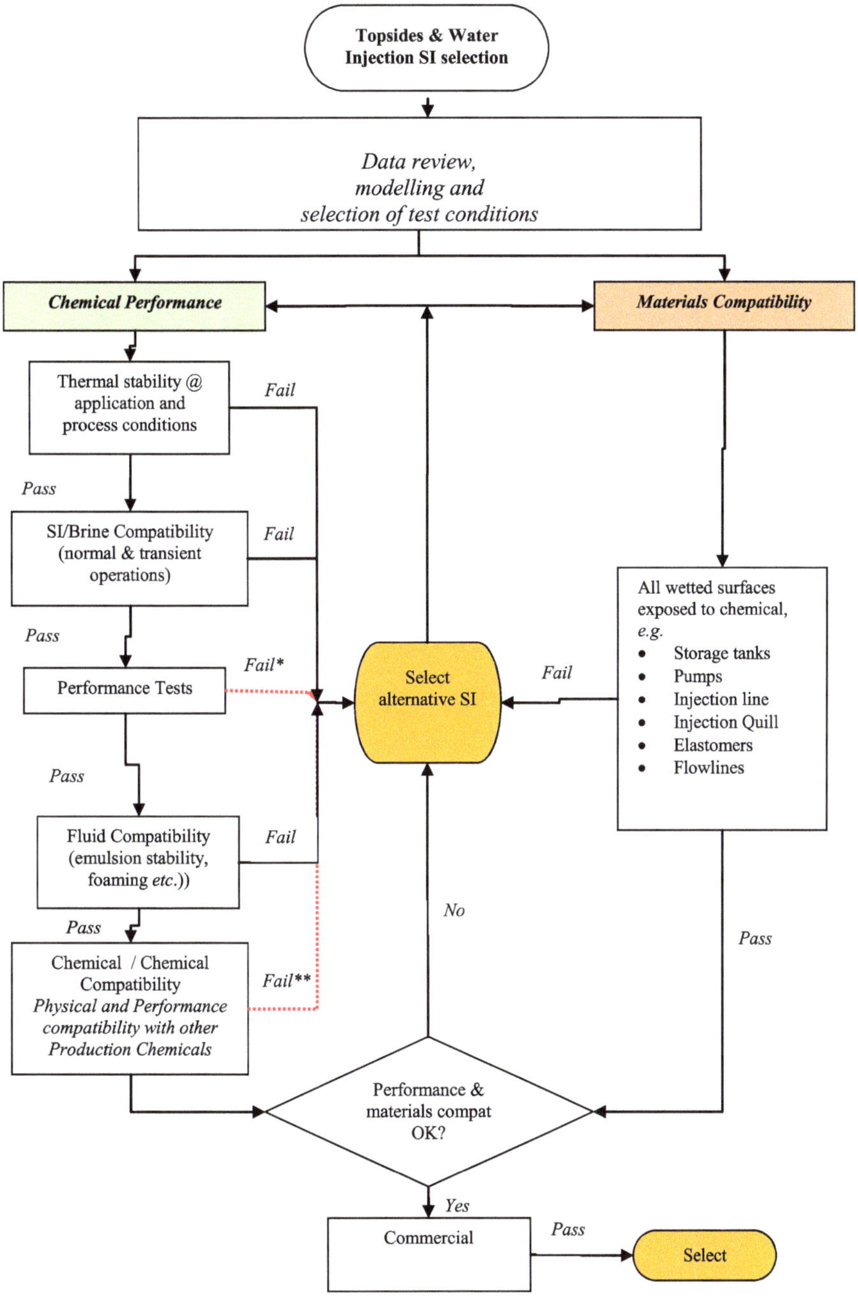

FIGURE 5.34

Overview of scale inhibitor qualification criteria for conventional topside applications.

Image courtesy of Scaled Solutions Ltd.

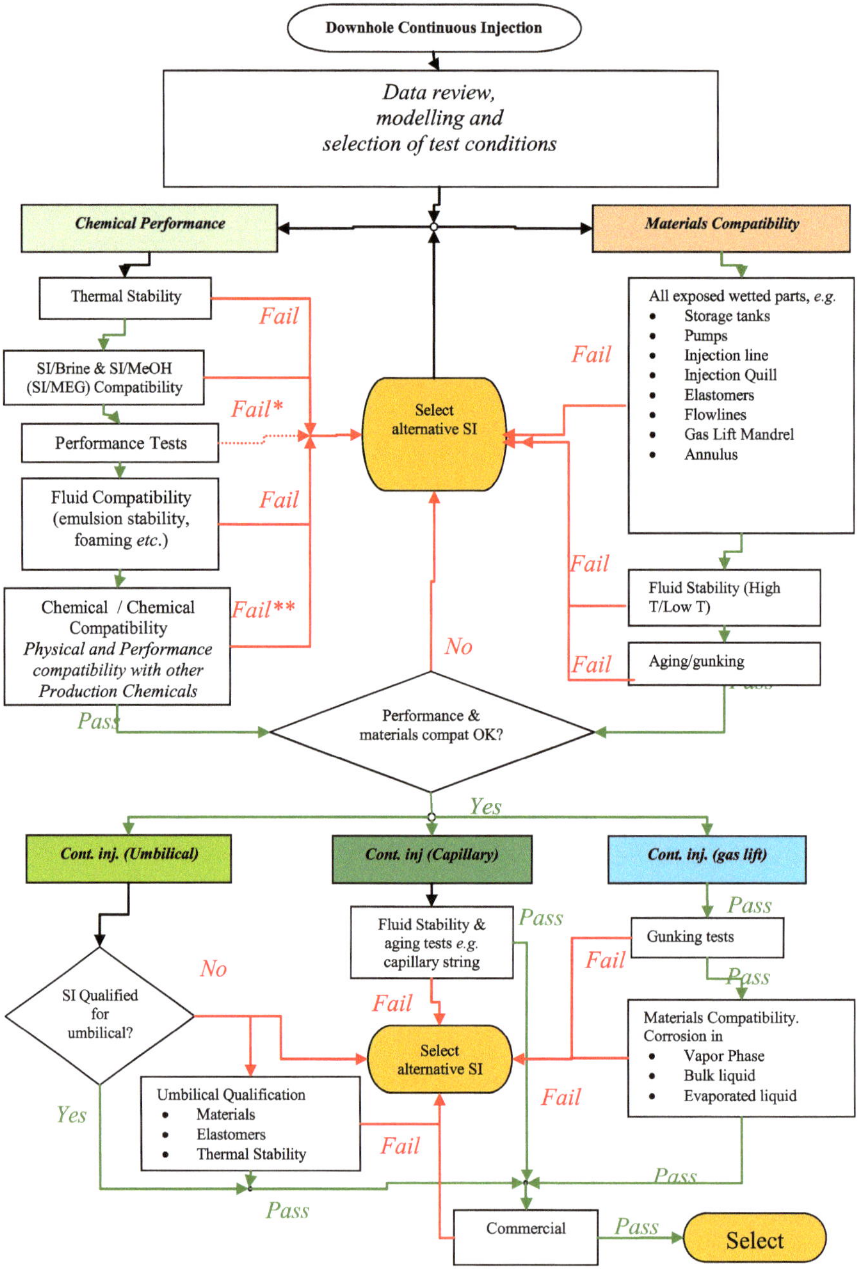

FIGURE 5.35

Overview of scale inhibitor qualification criteria for downhole continuous injection applications.

Image courtesy of Scaled Solutions Ltd.

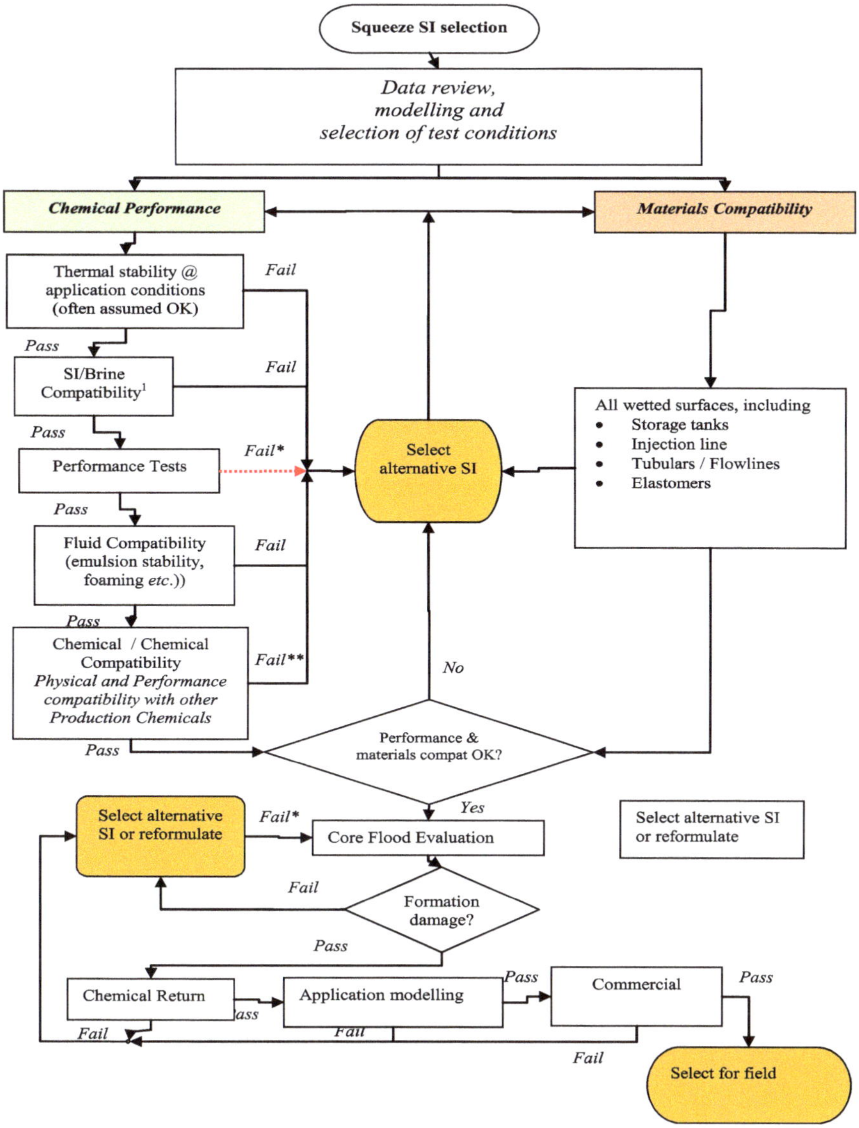

FIGURE 5.36

Overview of scale inhibitor qualification criteria for squeeze treatment applications.

Image courtesy of Scaled Solutions Ltd.

for example, remote facilities, subsea fields with wet trees, using scale squeezes as the primary means of scale management is committing to an expensive well treatment on a periodic basis. This has led to an increased use of sulfate reduction technology for deepwater subsea applications where barium sulfate is expected.

Squeeze treatments differ from other downhole "batch" treatments that may also be injected into the formation (such as dissolver treatments, commonly applied against inorganic scales and asphaltenes). The active chemical is designed to be retained within the formation so that it provides extended protection lifetimes as the chemical flows back, dose in the production water. Scale inhibitors work because they "stick" to both minerals: scale minerals and those within the formation; therefore, in a squeeze treatment, the inhibitor is designed specifically to react with the reservoir so that a proportion of the chemical is retained within the reservoir. On resumption of production, the retained chemical slowly releases back into the produced fluids. Ideally, the scale inhibitor will be released at a concentration just above its MED for an extended period (typically between 3 months and > 2 years) known as the "squeeze lifetime." Chemical squeeze treatments are also conducted for corrosion inhibitors and asphaltene inhibitors [181–183], though such treatments are much less common and technically much less well understood than for scale inhibitors.

5.5.1.1 General requirements of scale inhibitor squeeze treatments

Given that reactive chemicals (*e.g.*, scale inhibitors) will be applied directly into the producing reservoir in the vicinity of the production well, the two main criteria are as follows:

1. The avoidance of productivity impairment by formation damage, and
2. Maximization of the potential treatment lifetime to the predetermined MED.

Of these, the first is generally the more important criterion for this, and indeed, any chemical deployment, though the degree of formation damage depends strongly on the completion type and other constraints of the system, for example, drawdown limits. Either way, chemical qualification for squeeze treatments of any type generally requires some form of core flooding test designed to assess these basic criteria of formation damage and inhibitor retention/release properties. Guidance on this has been the subject of much work over recent years, resulting in various papers being published for core flood qualification of scale inhibitors as well as "core to field" guidelines [141–143,184]. Another important aspect of the laboratory core flood testing is the ability to reliably model treatment lifetimes based on core flood data, which is again an area where detailed understanding is required.

In essence, the squeeze process is to inject a concentrated SI solution and displace it into the formation (typically between 1 and 10 m distance from the wellbore), the treatment is shut in to allow either adsorption or precipitation of the chemical, and then, when production is resumed, the chemical is continuously released into the produced water. Eventually, the treatment is exhausted and must be repeated, which is most commonly indicated by the residual inhibitor concentration in the produced water dropping below the MED. More reliably, however, this surveillance techniques is augmented by others to ensure the continued effectiveness of the treatment. These include: (i) the abundance of suspended solids in the produced water (on the assumption that when control of deposition expires, this also allows scale particulates suspended in solution to restart; the solids analysis is obtained by filtering a fixed volume of produced water); (ii) monitoring scaling-ion concentrations in the produced water composition; (iii) productivity decline; (iv) changes in well pressures (where the gauge is either at the tree or downhole). Another method is both a surveillance and control method: actuating moving parts, such as safety valves, sliding sleeves may indicate build-up of scale (if they do not move, or move but do not seal, pass a leak-off test, etc.), but may also shear off limited amounts of scale deposited on the key surfaces. Of course, from a surveillance these are lagging indicators, which perhaps makes them less than ideal, in particular, if the surveillance signal is complete failure of some production-critical components, for example, complete failure of an ESP due to scale deposition.

A conventional aqueous-based scale inhibitor squeeze treatment typically comprises the following steps, as illustrated in Fig. 5.37:

- Preflush: A preflush, typically containing a dilute concentration of inhibitor and surfactant/demulsifier (around 0.1 vol.% each) dissolved in water that is compatible with the formation minerals (often either seawater or 3–5 wt.% KCl). This cools the near wellbore area, cleans up the rock surface to facilitate adsorption, and displaces the formation water away from the near wellbore in case it is incompatible with the high chemical concentration in the main scale inhibitor flush, in particular, at high temperature.
- Main Treatment Injection: The main flush, typically containing 5–20 vol.% active scale inhibitor dissolved in the same diluent medium as the preflush (seawater or KCl solution) is injected into the precooled near wellbore area. Besides potentially improving compatibility with the formation water, the precooling also reduces the rate of "reaction" (most commonly, adsorption) of inhibitor onto the formation rock, thus improving its propagation away from the wellbore.
- Overflush: The main flush medium (seawater or KCl solution, without the scale inhibitor) is then injected to displace the main treatment pill deeper into the formation, typically, to a distance of 1–10 m, though on ocassion up to 20 m. The main flush then penetrates beyond the cooled near-wellbore region and enters the hot region, close to original reservoir temperature because fluids travel faster than the temperature front.
- Shut-In: This normally lasts 6–24 hours to allow the inhibitor to react with the formation, either adsorbing to saturation and/or precipitating, depending on the retention mechanism(s) at play.
- Back production: The well is then put back onto production and the inhibitor is continuously dosed into the produced water flowing over at a progressively decreasing concentration until it reaches the MED (shown schematically in Fig. 5.37C), whereupon the treatment must normally be repeated.

Some stimulation may result from an inhibitor squeeze treatment due to the removal or sequestration of mineral scales or simply because the inhibitor

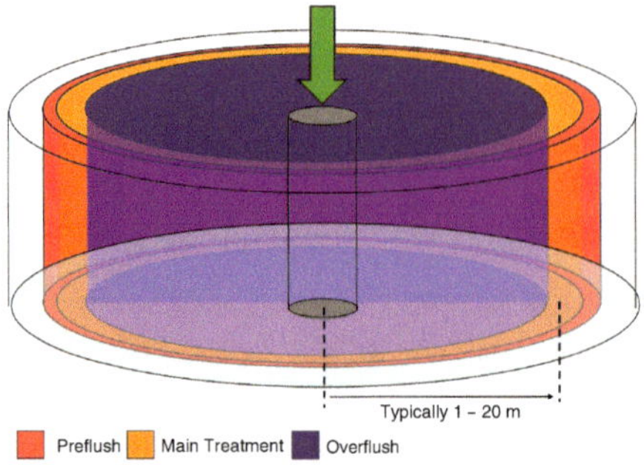

(A). Injection of the three stages

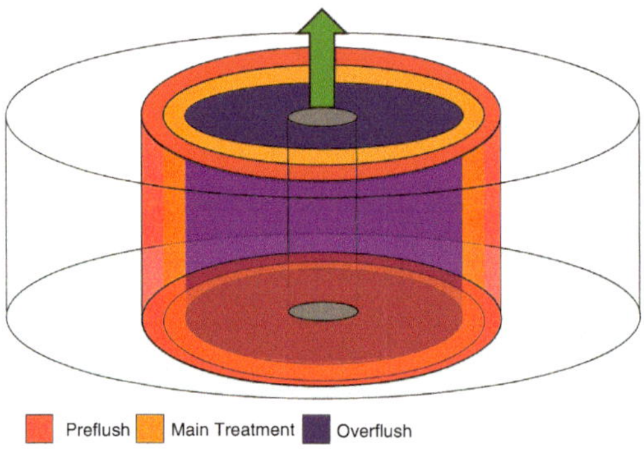

(B). Initial "flowback" stage

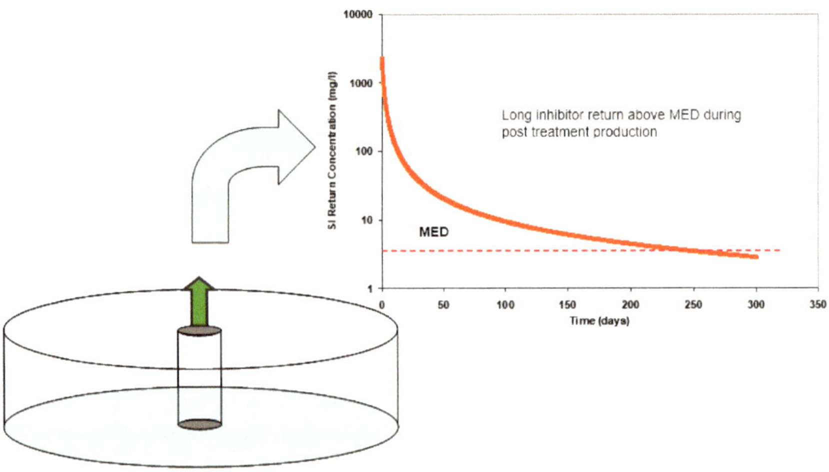

(C). Post treatment production phase

FIGURE 5.37

Schematic of a squeeze treatment in a single zone vertical well. (A) Injection of the three stages. (B) Initial "flowback" stage. (C) Posttreatment production phase.

Courtesy of Scaled Solutions Ltd.

chemical is acidic (this is particularly seen in carbonate cores). Also, scale dissolver treatments or stimulation treatments are often conducted immediately prior to a scale inhibitor squeeze treatment, and the effect is then both to stimulate and protect production rates.

In wells with poor lifting energy (low reservoir pressure/high hydrostatic head), it may be difficult and costly to lift the resulting column of water after a conventional aqueous-based squeeze treatment. Typical approaches adopted in such cases include the use of nonaqueous treatment fluids (which tend to have lower density than water and thus require less lifting energy) and/or temporary use of lift gas (perhaps via coiled tubing) to restart the well.

Conventional aqueous-based treatments can sometimes lead to unacceptable degrees of formation damage for a variety of reasons: fines migration and "water block" (more common with gas wells where not only capillary forces are greater and viscous drag forces by the hydrocarbon phase are lower but also in oil wells producing at low water cut from water sensitive formations). This should become apparent during the core-flood evaluation of a treatment in a field not previously squeezed provide the testing program was appropriately designed. For such cases, nonaqueous treatments or hybrid treatments may be preferred to aid posttreatment recovery, examples of which have been discussed extensively in the literature [5,166,179].

5.5.1.2 Squeeze optimization

There is a strong economic driver to obtain the longest treatment lifetime possible; therefore, near-wellbore simulators are typically used to model and optimize the treatment. This is normally done using a combination of data from core flooding and then actual field performance. In general, for an adsorption-based squeeze (that most commonly applied), the behavior in the core flood experiment is fitted to an isotherm, which then enables the posttreatment data (the inhibitor concentration as a function of pore volumes flowed through the core) to be compared with that predicted by the fitted isotherm, as shown in Fig. 5.38. This isotherm can then be used to predict the behavior in the field for a variety of different treatment dimensions (primarily the relative sizes and concentrations of the three different stages of the treatment) so that a preliminary optimal treatment can be developed for field application.

After the field data are obtained, these can then be fitted to actual field data, in case there is a significant discrepancy between that seen in the field and that seen in the core tests (e.g., as much arise if the core used for the tests was not sufficiently representative of the field). Further treatment optimization can occur with the new field data, and the treatments in any specific well or series of wells may evolve over time.

While squeezes are among the most effective means to control scale in wells, they are generally not very efficient in terms of the fraction of the applied chemical that is actually used to prevent scaling. Upon restarting production after the squeeze, typically a very high initial concentration of inhibitor is present in the

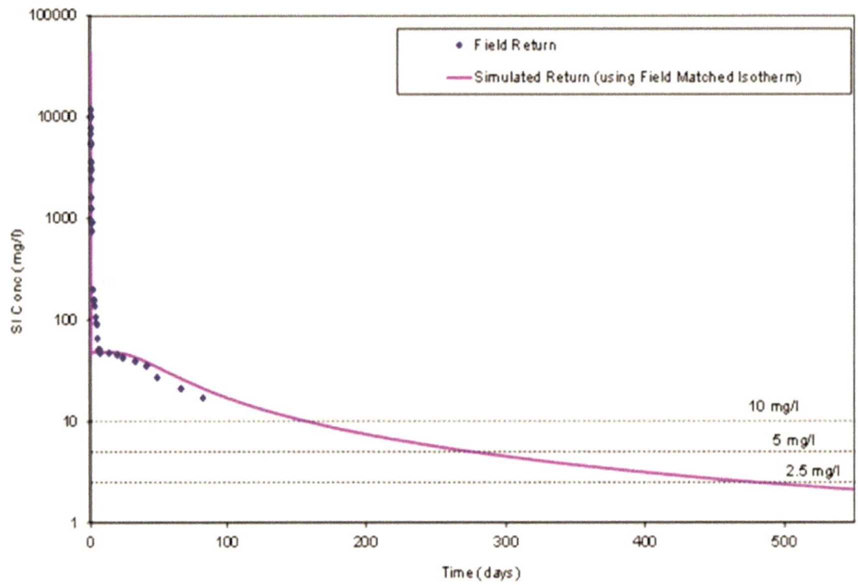

FIGURE 5.38

Field treatment lifetime modeling and optimization using computer near-wellbore simulation models.

Courtesy of Scaled Solutions Ltd.

initial flowback because there is invariably a significant amount of unreacted (neither adsorbed nor precipitated) chemical present, which then returns to production system. It may induce process upsets due to stable emulsions, and so on. One way to deal with this is to try to flow the well back slowly and dilute the stream with other wet wells that have not been squeezed. However, even if these flowback issues can be managed in the facilities, this large excess of chemical is wasted as it does not contribute to scale control. Hence, besides trying to prolong treatment lifetime in the most efficient way possible (rather than just applying a larger treatment), another challenge is to minimize this initial spike of wasted chemical (Fig. 5.39). This ongoing challenge is tackled both by design optimization using simulation models, application of alternative chemistries with more favorable retention/release properties, and adjusting the treatment strategy with modified treatment designs to maximize chemical retention properties by, for example, use of squeeze life enhancers (described in Section 5.5.2.5).

A key aspect of some squeezes is to ensure that the treatment covers as much of the producing interval as is necessary to protect the entire interval. Just treating the heel of a long-reach horizontal producing interval will not protect scaling closer to the toe. Certain specialized squeeze simulators also account for fluid placement in different reservoirs zones (see Section 5.5.7), with different

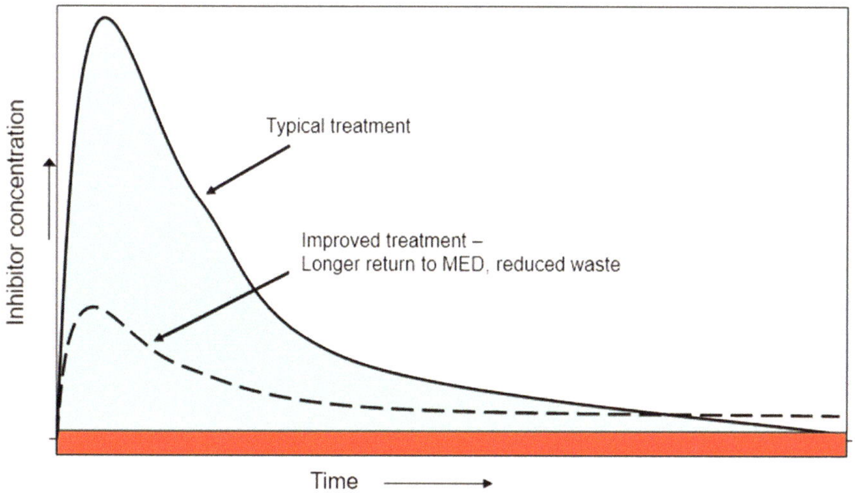

FIGURE 5.39

Schematic illustration of the ongoing challenge to optimize squeeze inhibitor treatments in terms of lower early-time flowback concentrations and extending treatment lifetime.

Image Courtesy of Scaled Solutions Ltd.

porosities and permeabilities, producing either water or oil, where pressure contrasts exist (those arising from differential depletion rather than just depth). They can also simulate fluids with different rheology.

5.5.2 Chemical retention mechanisms

There are two main mechanisms involved in the retention and release of scale inhibitor from the reservoir formation: adsorption/desorption and precipitation/re-dissolution. These will be discussed separately.

5.5.2.1 Adsorption/desorption

Chemical adsorption physicochemical interaction of the scale inhibitor molecule with the reservoir mineral surfaces. The level of adsorption and the subsequent release of the inhibitor (desorption) is a function of:

- Chemical type;
- Composition of the injected aqueous phase and the that of the produced water;
- The pH, which means both that of the applied chemical and any changes that occur during the application, for example, increased pH due to dissolution of the rock matrix by an acidic inhibitor;
- Temperature, either that of the reservoir or modified by the injection of cold fluids;
- Reservoir wettability;

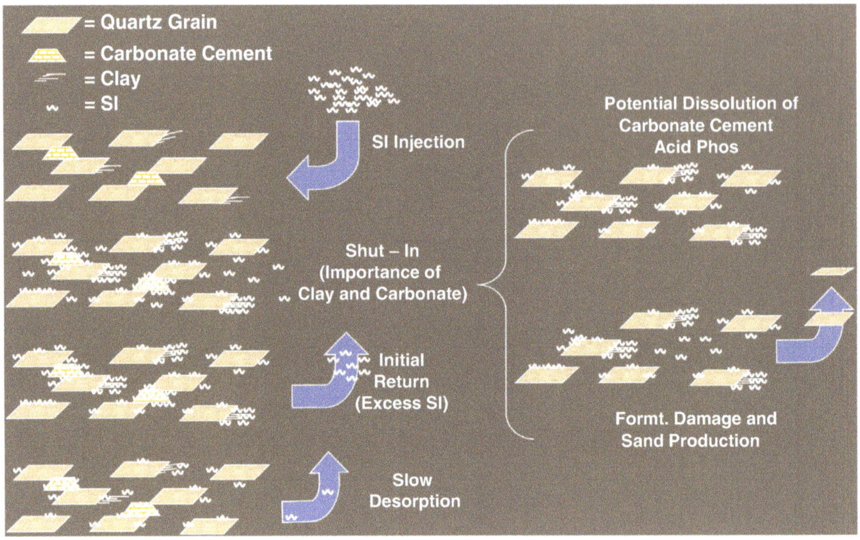

FIGURE 5.40

Schematic illustration of the adsorption/desorption process.

Image Courtesy of Scaled Solutions Ltd.

- Mineralogy, both the main matrix type (sandstone or carbonate), the type and abundance of secondary minerals (clays), other detrital and digenetic minerals, for example, carbonates in sandstone reservoirs.

The basic mechanism of adsorption and desorption squeeze treatments on a conventional sandstone reservoir is illustrated in Fig. 5.40. This figure illustrates the role of clays and carbonates in chemical retention in sandstone reservoirs. It also highlights a potential area of formation damage should an acidic scale inhibitor be used in sandstone reservoir where the carbonate is the main cementing mineral because dissolution of this cement may lead to movement of fines or sand production.

5.5.2.2 Precipitation /re-dissolution

The second main retention mechanism is precipitation, generally *via* a controlled precipitation squeeze which utilizes the fact that some inhibitors are chemically incompatible with divalent cations under certain conditions. This chemical incompatibility is a function of pH, divalent cation concentration and temperature. Incompatibility (and hence precipitation tendency) increases with increasing calcium concentration, pH and temperature. Unlike adsorption squeeze, chemical release in a precipitation squeeze is by inhibitor dissolution. The solubility and rate of dissolution are dependent on the temperature and composition of the produced water, including its salinity, pH, and divalent cation concentrations. This means that where the MED is low and the produced water has relatively low salinity and calcium content, dissolution is rapid, and the return profile strongly resembles that for an adsorption/desorption process.

For the most commonly applied precipitation squeeze treatments, the onset of precipitation is triggered primarily by the change in temperature as the treatment main flush is displaced into the hot region of the near wellbore during injection of the overflush. This gives potential risks of premature precipitation if the treatment is not effectively designed. In recent decades, however, additives have been used to more effectively control the precipitation process. For example, urea can be added to the main flush [176,177], which then hydrolyzes during shut-in with this solution in the hot region, deeper in the reservoir. The combination of released ammonia and bicarbonate causing an in-situ increase in pH, which triggers the precipitation of inhibitor–calcium complexes. Because the urea hydrolysis is relatively slow, it ensures the precipitation occurs almost entirely when the fluids are static in the reservoir, leading to a more controlled process that does not lead to significant formation damage. Fig. 5.41 provides a schematic illustration of a controlled precipitation squeeze and indicates the potential issue associated with premature or uncontrolled precipitation.

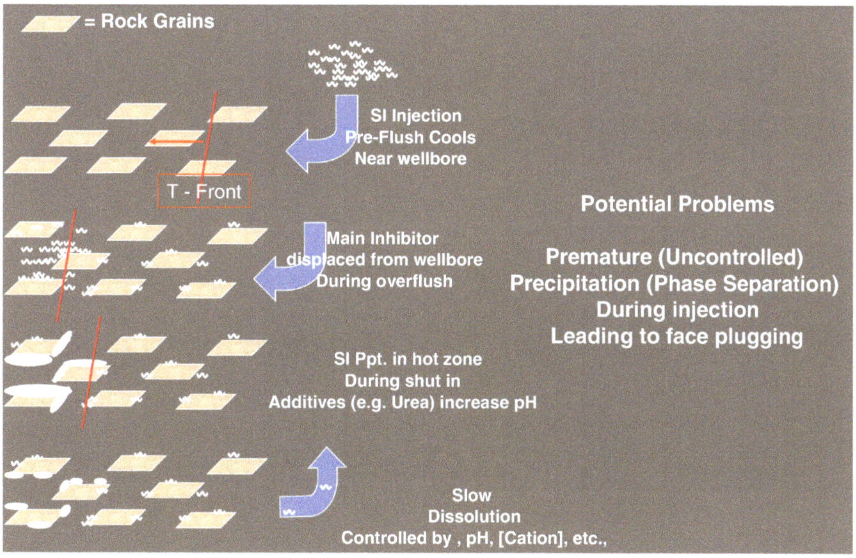

FIGURE 5.41

Schematic illustration of a well-controlled precipitation/dissolution squeeze treatment.

Image Courtesy of Scaled Solutions Ltd.

5.5.2.3 In-situ surface precipitation

A third retention mechanism occurs mostly when applying acidic phosphonate chemicals in reactive carbonate reservoirs or carbonate-rich sandstones. The low pH fluid induces the following sequential processes [180,185,186]:

1. Dissolution of the calcium carbonate, releasing locally high concentrations of Ca^{2+} ions and increasing the pH;
2. Precipitation of the calcium/scale inhibitor complex on the surface of the rock due to these changes.

For wells in carbonate reservoirs that are fractured, either naturally or due to acid-fracture stimulation treatments, this may be used to achieve enhanced chemical retention and therefore potentially increase squeeze treatment lifetimes (as long as the MED values are relatively low). But in cases where inflow is matrix dominated, such a precipitation-type treatment can lead to considerable formation damage and hence, loss of productivity. The effect of the resultant two-stage in-situ surface precipitation reaction is clearly evident in Fig. 5.42.

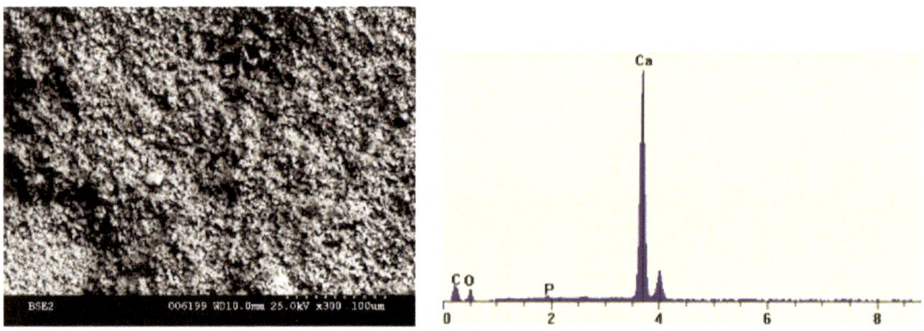

(A). Limestone core before treatment

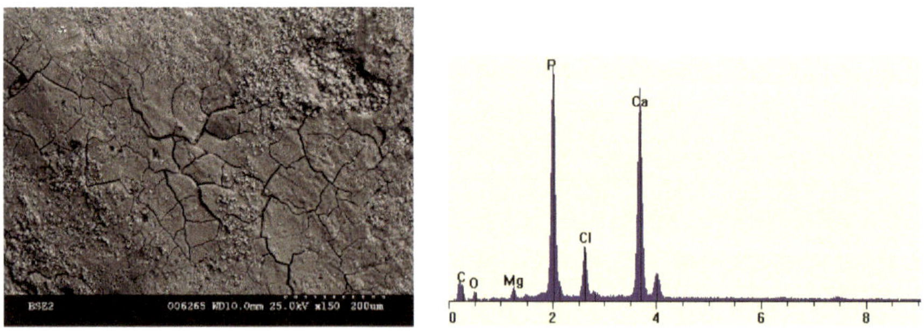

(B). Limestone core post treatment

FIGURE 5.42

SEM/EDX analyses of limestone core material treated with acid phosphonate: (A) before treatment with EDX spectrum showing calcium carbonate, (B) posttreatment with EDX spectrum confirming the surface precipitate as calcium phosphonate.

Images courtesy of Scaled Solutions Ltd.

5.5.2.4 Importance of temperature

All three retention mechanism processes discussed above are dependent on temperature, with both increased chemical adsorption and also increased precipitation occurring at elevated temperatures. To achieve maximum inhibitor retention, it is therefore important to ensure that the scale inhibitor slug is effectively placed into the hot region of the wellbore as a result of the overflush injection. This is based on the principle that as the injected fluids cool the near-wellbore, the temperature front advances into the reservoir at a slower rate than the injected fluid front. Therefore the overflush can be sized to effectively place the chemical into the hot zone of the near wellbore area. Such placement aspects are normally simulated using near wellbore treatment models. Fig. 5.43 illustrate the results from a typical simulation indicating the temperature profile and fluid front; they also show the expected chemical placement distance for an example simulation for a well with either a single zone (Fig. 5.43A) or two zones (Fig. 5.43B) in the producing interval. As a general rule of thumb, the overflush should constitute at around 2/3 of the total treatment volume (preflush + main flush + overflush) to ensure that the main treatment is displaced into the hot zone.

5.5.2.5 Squeeze life enhancers and extenders

Until around 15–20 years ago, it was considered impossible to "change" the properties of the near-wellbore reservoir to improve or enhance chemical retention, and chemical squeeze lifetimes were maximized by the choice of chemical, the pH, the chemical formulation, the treatment design itself and optimization thereof. The treatment design and volume optimization process were used to maximize surface contact of the chemical and enhance retention (see also squeeze treatment modeling), along with application of various solvent and preflush systems designed to enhance the chemical retention within the reservoir [145]. However, since then the ability to modify the near wellbore reservoir itself to enhance chemical retention has become more common, especially when dealing with highly quartzitic sandstones, which tend to be very poor at adsorbing chemicals. Two main approaches have been developed and deployed successfully in such cases:

- Polymeric squeeze-life enhancers: In these applications, large cationic polymers such as polydiallyldimethylammonium chloride and poly (ethyleneimine)polydadmac have been applied to create a more cationic surface on the formation, thereby allowing enhanced retention of anionic scale inhibitors). Field treatments with such chemicals have also demonstrated an additional benefit of a reduction in the potential for fines mobilization [187–190];
- Particulate squeeze life enhancers: This involves the injection of particles such as microcrystalline kaolinite, organosilane, or carbonate mineral into otherwise poorly adsorbing sandstone reservoirs, where they enhance chemical retention

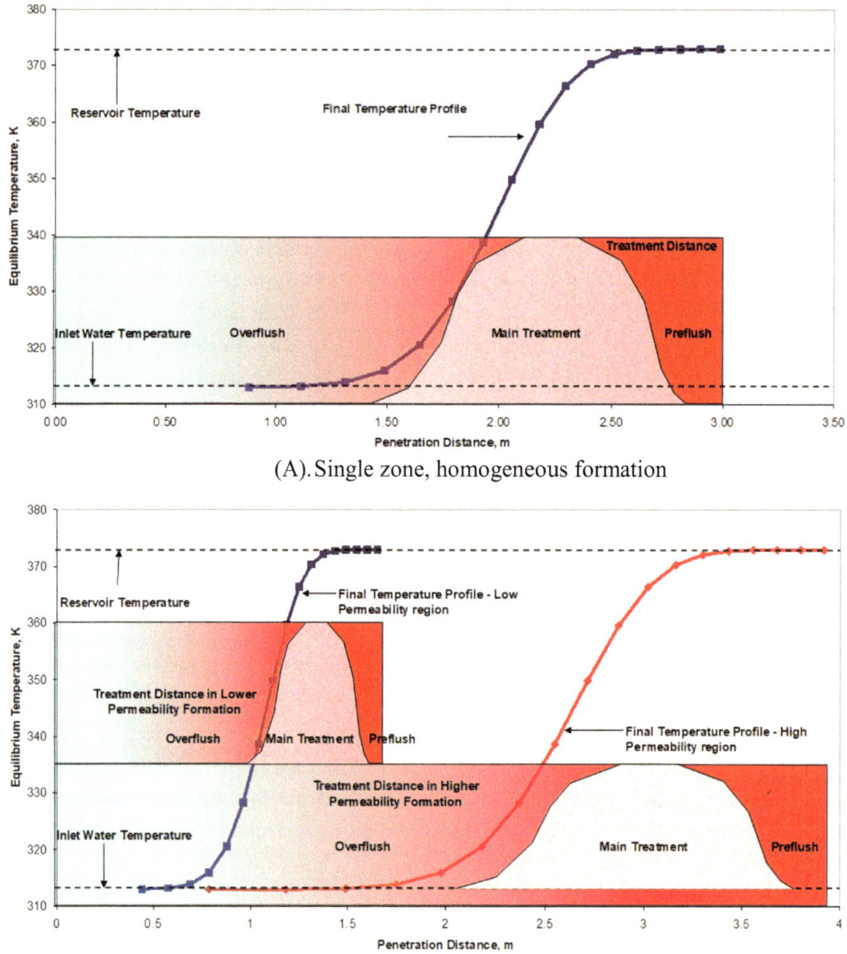

(A). Single zone, homogeneous formation

(B). 2-zone, heterogeneous formation

FIGURE 5.43

Simulated squeeze treatment distance and temperature profile using the Place *iT*™ Software.

Courtesy of Scaled Solutions Ltd.

and extend squeeze lifetimes by increasing both the surface area and adsorptive capacity of the availability surface. Benefits have been reported in high-permeability, highly quartzitic sandstones where there is a dearth of adsorption sites in the native rock [191,192]. More recently the concept of injecting particles has been further developed with the use of nanoparticles and modified nanoparticles [175,193].

5.5.3 Chemical testing of scale inhibitor squeeze treatments: reservoir condition core flooding

The two main aspects related to chemical selection are formation damage and inhibitor return profiles. Both aspects are examined via core flooding techniques with a range of techniques commonly used.

For formation damage testing, the main consideration is to compare pretreatment and posttreatment hydrocarbon (oil, gas) recovery in terms of their effective permeabilities immediately following chemical treatment and return. Performing the tests under conditions as close as practically possible represent the very near wellbore area. Formation damage to water flow can also be problematic in zones where both oil and water are flowing. Although it may appear that damaging water production is a good thing, disproportionately hindering the flow of water but not oil will just lead to a progressive rise in water saturation in the near-wellbore reservoir, which will then also impede the flow of oil. Conversely, it may appear beneficial if a squeeze chemical stimulates the flow of oil in a core test. This may be the case as long as it is fairly modest, but a significant change in permeability generally means a chemical reaction has occurred that was not the aim of the application and may have detrimental consequences, such as deconsolidating the rock and changing the chemistry of the scale-squeeze package, for example, loading it up with Ca^{2+}, Mg^{2+}, or Fe^{2+} ions, which could be either deleterious or beneficial. Either way, the impact needs to be understood.

For the inhibitor return stages the main criterion examines the shape of the inhibitor return profile from the core flood, under conditions that represent as close as practically possible those in the field at the location where the chemical will be retained and released (typically, within the reservoir at least 0.5–1 m away from the wellbore).

Core flood testing is performed using a core flood "rig" in which reservoir conditions can be simulated to ensure that any test results obtained are as representative as possible of the field in question. At the simplest level, the rig consists of a core holder housed in an oven. Fluids can be heated up, pumped through the core and sampled at the outlet. Typical core flood apparatus is shown in Fig. 5.44, with schematics of the main flow systems shown in Figs. 5.45 and 5.46.

Fig. 5.45 shows a basic schematic of a core flood rig that may be used to qualify or select scale inhibitors for squeeze application. This also shows a picture of (i) a typical Hassler-type core holder, (ii) the core mounted in a conventional Viton sleeve, and (iii) a schematic detailing the main features of the core holders.

The rig should be able to handle the sequential injection of oil or gas and aqueous streams in addition to the treatment chemicals themselves (which may or may not all be water-based). For light oils, this is commonly achieved by injection *via* a pump. For heavier oils, waxy oils or gases, injection is generally performed *via* accumulators inside the oven (see Fig. 5.46) or using more specialized piston pumps. The equipment is able to flow fluids through the core at in both directions, that is, W → F (wellbore to formation) and F → W (formation to wellbore) so as to simulate treatment injection and return. All wetted parts should

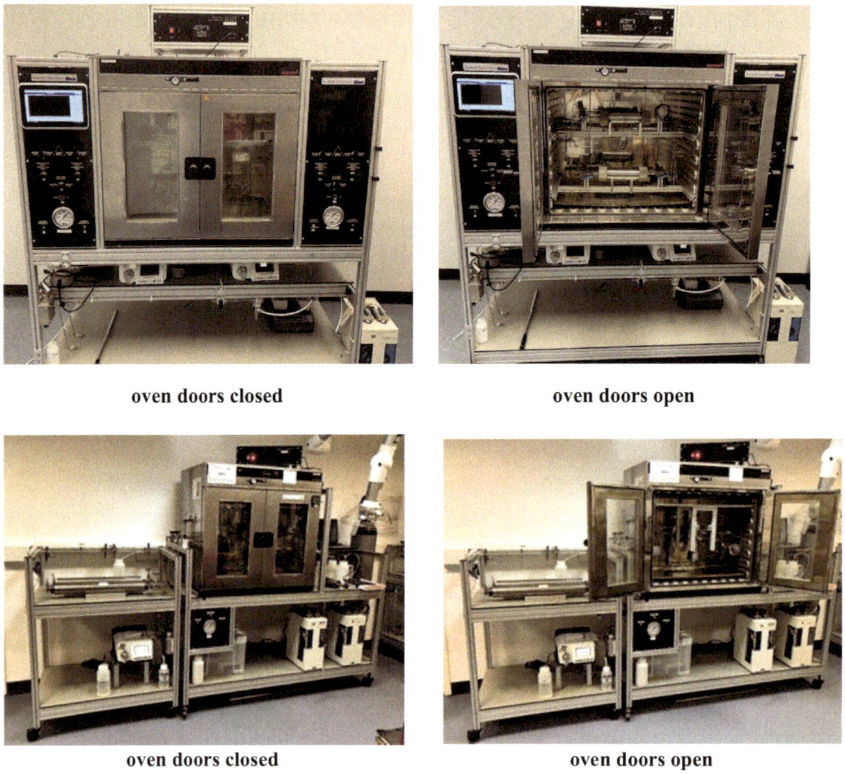

FIGURE 5.44

Core flood equipment typically used for scale inhibitor formation damage and inhibitor return evaluations; a) conventional 2 phase core flood rig b) more complex core flood rig (Left - oven doors closed; Right - oven doors open).

Images Courtesy of Scaled Solutions Ltd.

be manufactured from suitable corrosion resistant alloys to prevent corrosion in the flow line when injecting often acidic aqueous inhibitors packages and the high salinity brines during the pre and posttreatment periods.

Core plugs are typically 1 inch or 1.5 inches in diameter and typically 2–6 inches in length. The core is mounted inside the core holder, and (conventionally) is wrapped typically in high temperature resistant Viton sleeve (see Fig. 5.45), before applying a confining pressure to seal the system.

5.5.4 The importance of appropriate core flood testing protocols

Certain core flood protocols often adopted in some laboratories can have a significant effect on the relative performance of different chemicals. In some

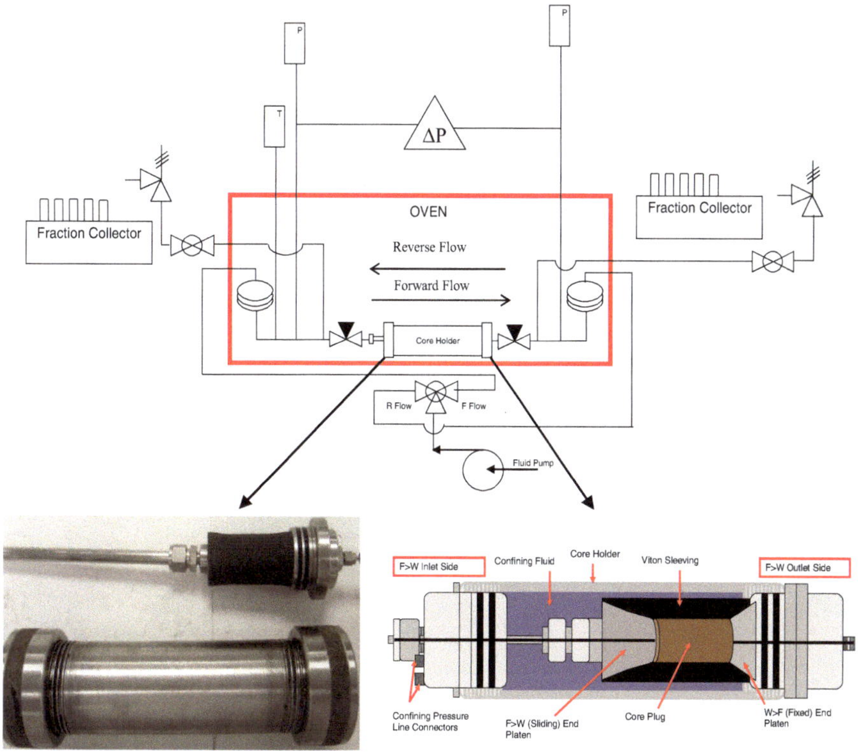

FIGURE 5.45

Simplified schematic of core flood equipment, fluid delivery (including oil) from precision pumps.

Images courtesy of Scaled Solutions Ltd.

cases, these can relate to artefacts of the protocol itself as opposed to differences in the performance of the chemicals. This can be a particular issue when tests attempt to combine a formation damage test and inhibitor retention/release test because the former requires sufficient chemical loading to detect flow impairment, whereas this can lead to chemical oversaturation of the core and thence to incorrect assessment of a chemical's (relative) performance [141]. Other aspects, such as the importance of appropriate chemical assay and the impact on chemical retention of returning preflush solvents, have also been explored. It is noted that formation damage and inhibitor return core floods often require some degree of compromise in terms of test design because certain protocols that favor enhanced chemical retention and extended return lifetime can also be detrimental to the formation (leading to increased *apparent* formation damage) and vice versa.

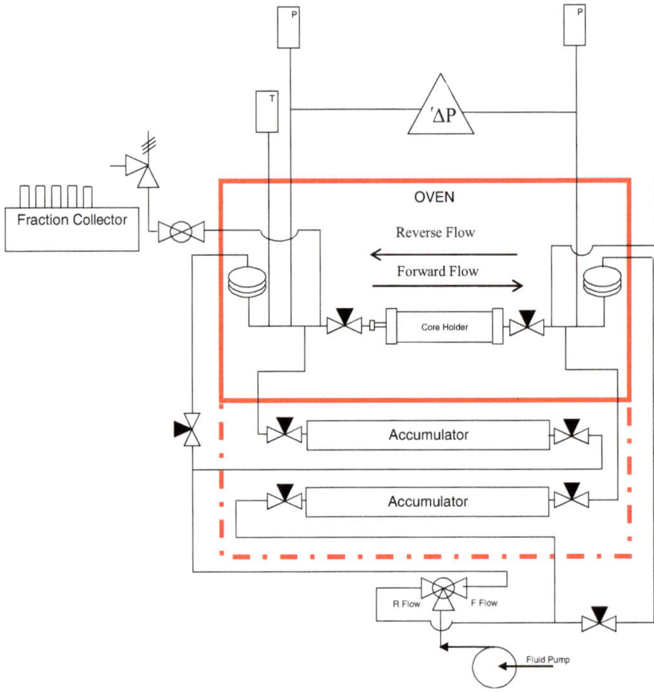

FIGURE 5.46

Schematic of scale inhibitor core flood equipment used for viscous/waxy oils or gas phase injection (hydrocarbon delivery via accumulators, aqueous and inhibitor phases via precision pumps).

Courtesy of Scaled Solutions Ltd.

This means that certain core test approaches may ensure that formation damage is minimized, whereas others may favor enhanced chemical retention. Unfortunately, neither may necessarily represent the field treatment sufficiently well. For example, treatment return (often referred to as flowback) of excess volumes of neat preflush solvents can sometimes minimize the observed formation damage due to displacement of both excessively retained or precipitated scale inhibitor and of residual water from the core. The effect of this is artificially to enhance the return permeability to oil, while reducing chemical retention and return lifetimes by stripping inhibitor from the rock. In such tests, merely selecting a different preflush solution can have a considerable impact on chemical qualification and selection [141–143].

Therefore, considerable care and understanding are required when qualifying and selecting chemicals for squeeze treatment via core flood testing. The main areas that require consideration from both a "reservoir" (geochemical) and a "chemical" perspective are outlined in the following sections.

5.5.4.1 Scale inhibitor/rock reactivity

Nearly all commonly applied scale inhibitors are either polymers or low-molecular weight organic phosphorus chemicals. They can be applied at a range of pH values ranging from strongly acid (pH ~ 1) to near-neutral (most typical pH 4.5–6.5) to slightly basic forms (pH ~ 8), depending on the conditions. Greater acidity (lower pH) makes the inhibitor more reactive with certain minerals in the core. Cores too (and the reservoirs whence they come) can vary in their reactivity, from highly quartz-rich geochemically "clean" sandstone to limestone or chalk. Reactivity also depends on temperature, as would be expected. Therefore, when considering the possible range of chemical formulations, reservoir mineralogy and temperatures, it is clear that the level of chemical reactivity during a squeeze can vary enormously. The following is the reactivity sequence from a mineralogical viewpoint, where < indicates "is less reactive to applied chemicals than":

$$\text{Quartzitic sandstone} < \text{Clay/carbonate-rich sandstone} < \text{dolomite}\left[CaMg(CO_3)_2\right] < \text{limestone}(CaCO_3)$$

This is illustrated in Fig. 5.47 with some example combinations of application chemicals and reservoir. Dolomite is considerably less reactive than limestone ($CaCO_3$), such that a "pure" dolomite reservoir is typically less reactive than (e.g.,) a calcite-containing sandstone.

5.5.4.2 Cautionary notes on combined formation damage and core flooding and core flood artefacts

It is important to appreciate the degree of reactivity between the core minerals and the applied inhibitor. Certain phenomena observed in core flooding, including artefacts such as chemical oversaturation, can lead to inappropriate chemical

Lowest reactivity formation → Highest reactivity formation

Reservoir Formation	*Applied Scale Inhibitor*
Clean sandstone	Neutral VS-Co
Clay-bearing sandstone	Other neutral polymers and co-polymers
Dolomitic sandstone	Neutral phosphonates
Carbonate-bearing sandstone	Acid VS-Co (acid dissolution)
Dolomite	Other acidic chemicals (dissolution and risk of precipitation)
Limestone Chalk	Acid Phosphonates (acid dissolution and precipitation)

Lowest reactivity inhibitor → Highest reactivity inhibitor

FIGURE 5.47

Illustration of relative reactivity of different inhibitor/rock systems.

Courtesy of Scaled Solutions Ltd.

selection. This is because when injecting multiple pore volumes of chemical to assess the critical formation damage aspect in the near wellbore area, reactive systems (e.g., acid phosphonates in carbonate containing material) will result in very high levels of retention - far in excess of that which can be achieved in the field treatment, resulting in nonrepresentative high return lifetimes and, thereby, incorrect chemical selection. Similarly, when assessing chemical retention and return, much smaller volumes of chemical would ideally be applied to assess chemical retention properties away from the very near wellbore area. Whereas in many cases, both aspects (formation damage and chemical retention and release properties) can, and often are, combined in a single test. This is not always the case and can lead to poor chemical qualification and selection. Graham et al. highlight a number of common artefacts of core flood testing that can lead to nonrepresentative results [141–143,184].

The main cautionary notes which if not properly accounted for in the experimental design and effectively interpreted when considering the results are as follows:

- core selection and core conditioning,
- ensuring the appropriateness of the residual assay techniques adopted (i.e., to ensure that the assay technique adopted is analyzing the active scale inhibitor species and not some analyzable moiety which may have been added or incorporated into the formulation),
- prevention of chemical oversaturation,
- avoidance of flowback saturation of neat (undiluted) preflush solvents, and
- main-treatment application temperature.

To illustrate the first point on core selection, Fig. 5.48 shows images of some typical core plugs prepared for core testing, giving examples of homogeneous plugs as well as typical reservoir plugs containing obvious heterogeneity. These figures serve to illustrate the importance of good core selection, since the impact of micro- or pore-scale heterogeneity within "homogeneous" intervals can have a significant impact on the resulting properties. Therefore, it is critical that the core plugs should not only be as representative as possible of the main producing intervals in terms of their basic porosity and permeability properties but also as comparable as possible in terms of their mineralogy (given the strong dependence on mineralogy within the core in terms of chemical reactivity and hence the potential for both formation damage and treatment lifetimes), for the comparative tests designed to qualify and select different products for field treatment.

5.5.5 SI Application considerations: formation damage and inhibitor retention/release properties

When selecting an appropriate scale inhibitor for squeeze application, a number of factors should be considered. Whether the inhibitor chemistry selected is

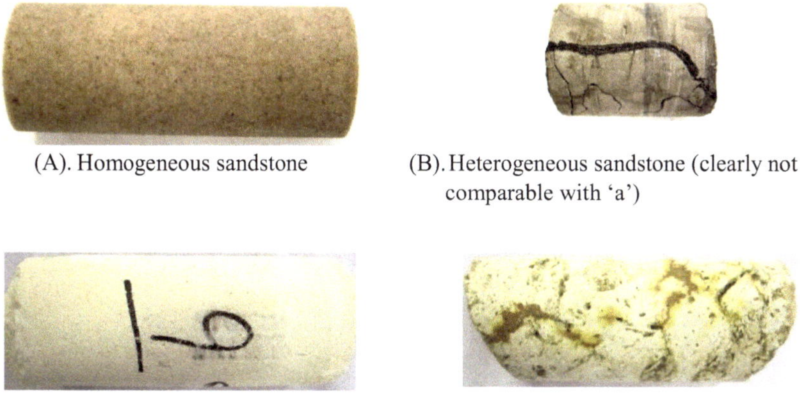

(A). Homogeneous sandstone

(B). Heterogeneous sandstone (clearly not comparable with 'a')

(C). Homogeneous carbonate

(D). Heterogeneous Carbonate (clearly not comparable with "c", despite both plugs coming from the same formation interval in this example)

FIGURE 5.48

Examples of variability in core plugs. (A) Homogeneous sandstone. (B) Heterogeneous sandstone (clearly not comparable with "a"). (C) Homogeneous carbonate. (D) Heterogeneous Carbonate (clearly not comparable to "c," despite both plugs coming from the same formation interval in this example).

Images courtesy of Scaled Solutions Ltd.

effective against the predominant scale forming in the well is of course of high importance, and compatibility with reservoir fluids (e.g., calcium tolerance in high salinity waters) and thermal stability, and so on have already been mentioned. In addition, three other aspects of inhibitor interaction with the formation must be considered:

- The likelihood of formation damage occurring which would reduce the overall productivity of the well.
- Scale inhibitor retention on the reservoir rock during squeeze application.
- The rate of scale inhibitor release during later production, and

The scale inhibitor must demonstrate acceptable behavior in relation to all three aspects to be suitable for squeeze deployment. We can illustrate this with a number of case histories from both clastic (sandstone) and carbonate reservoirs.

5.5.5.1 Formation damage vs. inhibitor return profiles: Sandstone pre-emptive squeeze

Fig. 5.49 shows comparative return profiles obtained for two products screened for pre-emptive application (treatment applied into dry oil-bearing formations) planned for the Mungo reservoir [179]. The results showed that chemical C had

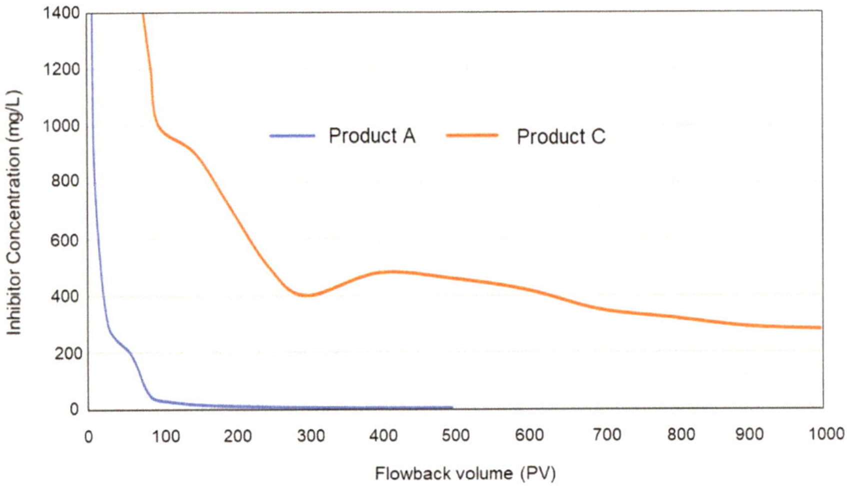

FIGURE 5.49

Comparative core flood return profiles recorded for two products (A and C) for pre-emptive squeeze in a sandstone reservoir. Y-axis is scale inhibitor in effluent (mg/L) and X-axis is pore volumes of brine injected through core.

Courtesy of Scaled Solutions Ltd.

the potential for significantly greater retention and release properties than chemical A. Both of these treatments were proposed by the vendors as "non-damaging and non-aqueous products" for potential application in the dry oil reservoir. However, when examining the permeability data recorded in the core flood tests, significant formation damage potential was observed for product C with a 10-fold reduction in flowback permeability (Table 5.10), highlighting the critical importance of understanding both the formation damage risk as well as the chemical retention/release properties.

In this case, product A was ultimately selected for field application. This chemical represented a nonaqueous/oil soluble product prepared by dissolving a conventional aqueous-based sulfonated co-polymer into mutual solvents. This allowed the otherwise aqueous-based scale inhibitor to be solubilized in an oleic phase therefore preventing or minimizing the potential for formation damage due to water blockages. When this product was applied in the field, full and immediate recovery to pretreatment productivity index (PI) was recorded. In addition, owing to the mutual solvent nature of the package, additional residual aqueous fluids were removed from the near wellbore area resulting in an increase in productivity of around 5%–10% which lasted for several months following the treatment. Thus, this treatment was effective at further stimulating a dry (<0.5% water) oil production well. We further note that since the product was being applied in a pre-emptive manner into dry or very low water cut wells, long

Table 5.10 Comparative formation damage (reduced flowback oil permeability) of for two products (A and C).

	Product A		Product C	
Formation damage evaluation	**W → F**	**F → W**	**W → F**	**F → W**
Initial K (mD)	150	144	87	85
Flowback (return oil permeability)	152 (0%)	146 (0%)	9	8
Final permeability at S_{wr}	108 (28%)	108 (28%)	7	7

treatment lifetimes in terms of bbls water protected was not a critical requirement. At low water cuts, even short lifetimes (in terms of total bbls protected) would lead to reasonably long treatment lifetimes.

5.5.5.2 *Formation damage vs. inhibitor return profiles: implications on limestone formation*

A best-in-class selection study was undertaken to select a suitable scale inhibitor for squeeze deployment in a limestone formation [186]. After successfully passing initial screening tests for fluid compatibility, stability, and inhibitor performance, two products were subjected to core flood tests to assess scale inhibitor retention and release/return properties (i.e., comparative treatment lifetimes) and formation damage potential as a final stage in the selection process.

5.5.5.2.1 Inhibitor retention and release properties, treatment lifetimes

Fig. 5.50 illustrates the inhibitor retention properties of the two products when injected into reservoir limestone core at downhole conditions. The figure shows the concentration of scale inhibitor in brine after passing through the core, normalized to the injection concentration. Any decrease in scale inhibitor concentration when compared to the inert lithium tracer indicates retention of inhibitor in the core. Fig. 5.50 shows that SI B has considerably more retention than SI A.

Fig. 5.51 however illustrates that although SI B has the higher retention in the core, this does not lead to the most effective release profile. In this case, the chemicals were of similar effectiveness as inhibitors having MEDs of ~ 2.5 mg/l. The inhibitor return profile for SI B reached the MED of 2.5 mg/L within ~1,200 PV of postflush compared with SI A at 3,000 PV. While SI B retains significantly more material on the core than SI A, it appears that the inhibitor is held too tightly to the rock to be released at a fast enough rate to maintain an effective dose in the brine for long. Thus, while high retention is certainly desirable for a squeeze application chemical, if the inhibitor is held too tightly to the rock then the effective return lifetime will be reduced. This therefore illustrates that, when considering treatment lifetimes, core flood approaches need to determine the release properties and not just the absolute retention properties.

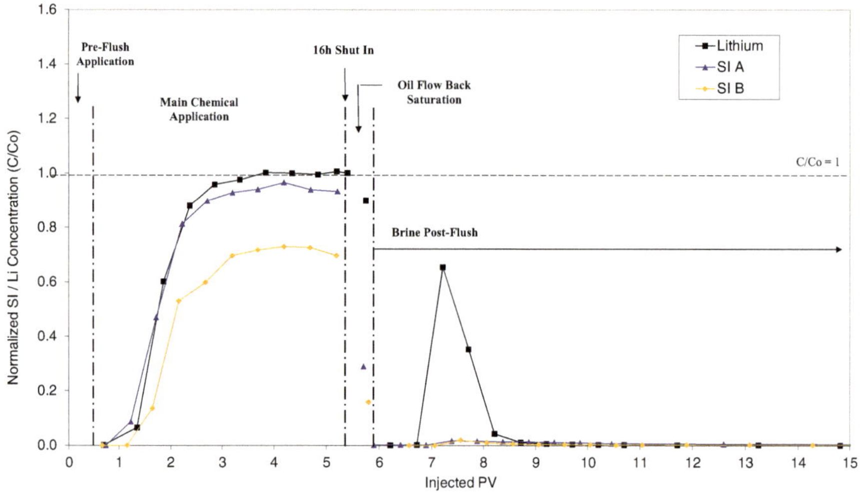

FIGURE 5.50

Comparative retention properties of two products in limestone cores. SI B is more reactive and retains significantly more than SI A.

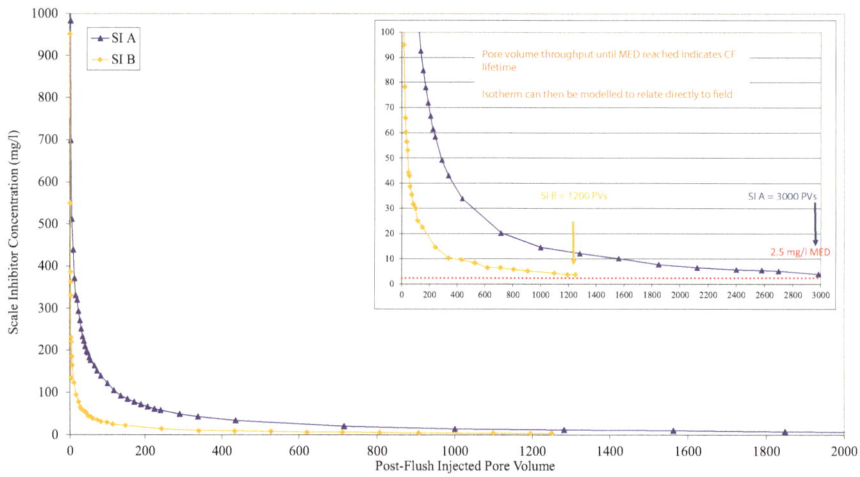

FIGURE 5.51

Comparative return profiles of two acidic products from limestone cores. SI B returns at much lower concentration than SI A.

Courtesy of Scaled Solutions Ltd.

From the inhibitor retention and release properties which have been demonstrated in Figs. 5.50 and 5.51, it would therefore appear that either SI A or SI B could be suitable for deployment *via* squeeze treatment, with a longer return lifetime expected for SI A. However, one other factor has not yet been discussed: the likelihood of formation damage if either of these products were applied in the field.

5.5.5.2.2 Formation damage aspects

As listed in Table 5.11, the application of both chemicals resulted in unacceptably high reductions in oil permeability following treatment, with the oil permeability reduced to 50% of the pretreatment value for SI A and only 32% for SI B. Neither product was therefore considered suitable for field application in this case.

Both of the low pH formulations tested gave an acceptable treatment return lifetime in the core flood test, but resulted in high levels of formation damage (reduced oil permeability). Further investigation showed that the likely cause of this formation damage was core dissolution resulting in excessive deposition of calcium-phosphonate complexes which blocked pore channels and reduced permeability (as described earlier in the two stage reaction leading to precipitation of calcium phosphonate when applying acid phosphonates into reactive limestone core material). Further testing was therefore carried out using similar products which were pH-adjusted to an application pH of 5.5. These partially-neutralized products were expected to be more benign to the formation (i.e., to reduce the initial dissolution reaction which occurred when the acidic chemicals contacted the limestone formation), but would this change in pH affect the retention/release properties of the scale inhibitors?

Fig. 5.52 shows the retention properties for the two partially-neutralized products SI C and SI D (pH adjusted to 5.5). When compared to Fig. 5.50, it is clear that the amount of retention is reduced compared to the more acidic versions.

Fig. 5.53 shows the core flood inhibitor return profiles for the partially-neutralized products. SI D would give a much longer treatment lifetime than SI C. The return concentration of SI C reached the MED of 2.5 mg/L after ~1,200 PV of postflush, while SI D remains above the MED even after 2,000 PV.

This would suggest that we could expect a significantly longer treatment lifetime from SI D in the field compared to SI C. However, as discussed in the next

Table 5.11 Formation damage results for two acidic inhibitors on highly reactive limestone substrate.

Inhibitor	pH	Pretreatment Oil Permeability (mD)	Flow Back Oil Permeability (mD)	Oil Permeability Recovery
SI A	3.1	1.01	0.51	50%
SI B	2.1	7.07	2.24	32%

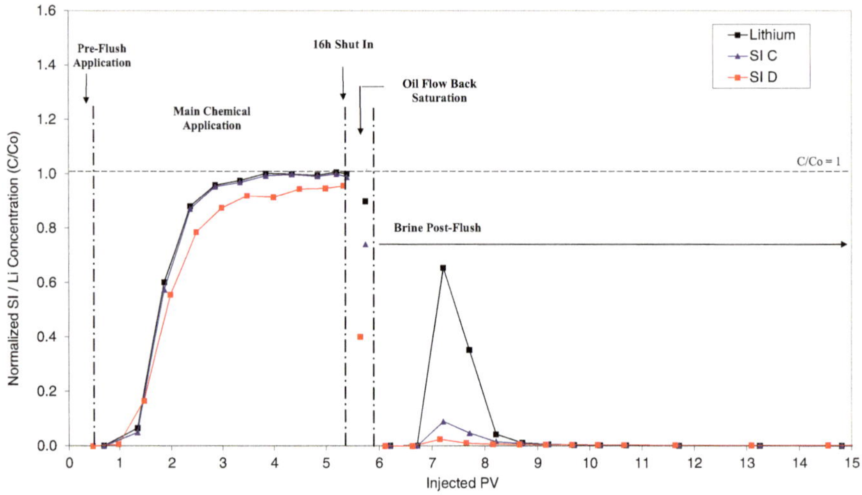

FIGURE 5.52

Comparative retention properties of two partially neutralized products (pH 5.5) in limestone cores.

Image Courtesy of Scaled Solutions Ltd.

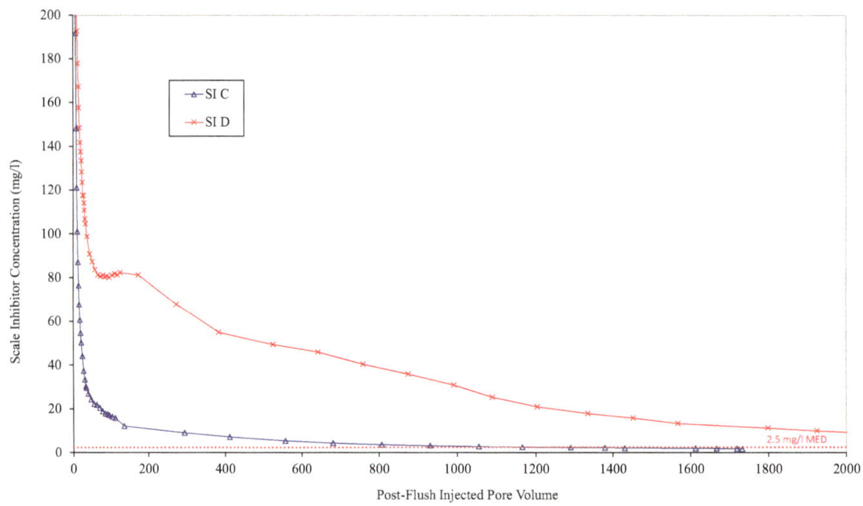

FIGURE 5.53

Comparative return profiles of two partially neutralized products (pH 5.5) from limestone cores.

Image Courtesy of Scaled Solutions Ltd.

Table 5.12 Formation damage results for two partially neutralized inhibitors on highly reactive limestone substrate.

Inhibitor	pH	Pretreatment oil permeability (mD)	Flow back oil permeability (mD)	Oil permeability recovery
SI C	5.5	4.04	3.65	90%
SI D	5.5	2.84	2.61	92%

subsection, final selection would then be made via upscaling the core flood results to the field condition.

Finally, as we saw in Case 1, we must also consider the likelihood of formation damage when using these products. Table 5.12 summarizes the measured oil recovery for the partially-neutralized products SI C and SI D. In comparison to the acidic products, in this case the formation damage has been minimized and both products show acceptable oil permeability recovery behavior, making them suitable for field application.

The increased retention of SI C on this reactive formation material did not impart any significant formation damage, when compared with SI D (Table 5.12). Acceptable lifetimes were predicted for field treatments based on the core flood return data outlined above. Treatments were subsequently designed and both products and applied in different wells in field trials resulting in minimal formation damage and acceptable lifetimes in the field.

This example illustrates that when applying scale inhibitors (or indeed other chemicals) *via* the squeeze process it is not only their absolute performance as inhibitors which determines their suitability for the field. Clearly, formation damage risk is also a significant factor. However, in the absence of formation damage, it is the return lifetime to the particular chemicals' MED. It also highlights that absolute retention is no substitute for determining the release properties under flowing conditions.

5.5.5.3 Examples of other factors controlling chemical retention

5.5.5.3.1 Effect of clay content in sandstone reservoirs

Jordan et al. [194] describe the importance of clay material (kaolinite) on the retention and release of a phosphonate scale inhibitor on sandstone cores. In their example, the same chemical was examined in core testing under identical conditions using two different clastic (sandstone) substrates. One contains ~ 10% kaolinite and the other representing outcrop core material is effectively clay free (94% quartz, 4% feldspar) with no kaolinite. This led to increased chemical retention on the kaolinite containing core and longer return lifetimes.

The reason for the enhanced adsorption and therefore improved return profiles relates primarily to the increased surface area of the clay material as indicated in the SEM image (Fig. 5.54). The image illustrates the much larger surface area within kaolinite clay material when compared with quartz crystals. The increased

FIGURE 5.54

Illustration of Surface area effects for kaolinite clay vs. quartz overgrowths (middle bottom).

Image Courtesy of Scaled Solutions Ltd.

surface area then allows for increased adsorption potential. The surface area appears to be the most significant factor when considering the impact of clay content in reservoir substrates, although different minerals still have different affinities for different scale inhibitors, due to their composition and surface charge. The importance of clay minerals in adsorption/desorption treatments has been further examined (e.g., [147]). The same general result is described in that increased chemical adsorption is achieved in clastic (sandstone) reservoirs containing high clay minerals when compared with clean (clay free) clastic reservoirs.

5.5.5.3.2 Effect of produced brine composition

Jordan et al. also determined the effect of produced brine composition on inhibitor return behavior [194]. They showed that when applying a phosphonate scale inhibitor (a hexaphosphonate) in core flood tests conducted under otherwise identical conditions, the nature of the postflush brine (SW at 428 ppm calcium vs. FW containing 5,000 ppm calcium) had a considerable impact on the return profile. The chemical returned at a much higher concentration in SW but effectively dropped below its MED in a shorter time, whereas when produced water was the high calcium formation water, the SI returned initially at a lower concentration due to more limited solubility of the inhibitor in the high calcium FW, but maintained this lower concentration for a much extended time. This therefore served to illustrate the extremely important impact of changing water chemistry. This is further observed in the case history from the Kittiwake field (North Sea) [170]. This case study shows that changes in brine chemistry, due to increased percentage of SW in the produced water, resulted in very poor and ineffective squeeze lifetimes for the originally qualified chemical (which gave good lifetimes when applied initially into wells producing high salinity (high calcium) formation waters). The loss of scale control in the Kittiwake field resulted in a change of scale inhibitor chemistry and a subsequent 10-fold increase in the field lifetime. In essence, the scale inhibitor was changed to a more conventional phosphonate which could not be applied when producing formation water due to significant compatibility in the FW which would have led to very poor return profiles at the early stage of the field lifecycle.

Although the above example related to adsorption/desorption treatments, the phenomena also has implicatons for precipitation teatments. In such cases, the nature of the produced water composition also has a controlling effect on the treatment lifetimes, since the inherent solubility of any precipitated inhibitor in the produced water controls the treatment lifetime and any potential benefit over a conventional adsorption/desorption treatment. This means that in some cases, especially when producing waters in which the precipitated chemcial is highly soluble, the precipitated scale inhibitor may be rapidly denuded in early postflush and then controlled by the adsorption/desorption properties.

5.5.5.4 Nonaqueous deployment packages for squeeze treatments

Wettability changes, water blocking and poor lift issues represent considerable challenges for scale inhibitor treatments into dry oil zones and also into fields with poor lift energy. An alternative method for scale inhibitor delivery into water sensitive wells or wells with poor lift capacity is to use a nonaqueous (or hybrid) scale inhibitor treatment package similar to that described in the example in section 5.5.5.1 [179]. Non aqueous packages allow oil continuity during treatment and reportedly also help to minimize relative permeability changes in the near wellbore area which may occur when applying conventional aqueous-based treatments. The use of hydrocarbon solvents and overflush solutions however aids recovery and lift due to the relative ease in lifting a column of hydrocarbon fluids when compared with water. These nonaqueous packages include: modified preflush packages (mutual solvents, etc.) and hybrid treatments, invert emulsion, oil soluble scale inhibitors, and oil miscible/oil flowable scale inhibitors.

5.5.5.4.1 Modified preflush and hybrid packages

To minimize near well bore damage and increase squeeze lifetimes, preflush solutions involving mutual solvents and other proprietary mixtures have been developed and deployed. These are often accompanied by conventional aqueous-based main treatments and with oleic (e.g., diesel) overflushes [195]. Such "hybrid" treatments were selected and applied in HT gas fields. Pre-emptive treatments in complex reservoirs are of particular concern. More recently, additional effort has been placed upon the selection and application of pre-emptive squeeze treatments in complex subsea reservoir systems, with the focus on minimizing the potential for formation damage, rather than extending squeeze lifetimes. These modified preflush/overflush treatments primarily use conventional aqueous-based main inhibitor pills and are therefore hybrid treatments. The mutual solvent can be very effective at maintaining oil continuity during back production, therefore, considerably reducing flowback recovery times and minimizing the potential for water blocks [178,195].

5.5.5.4.2 Invert emulsion

In invert emulsions, the water containing the scale inhibitor is in a water-in-oil emulsion [146,153,159]. The deployment of the scale inhibitor in these systems is as a nonaqueous package, which allows oil continuity during treatment. This prevents relative permeability changes in the near-wellbore area, which may occur when fluids are injected in conventional squeeze treatments.

The emulsions are injected downhole and then "break" as the reservoir returns to temperature during shut-in, thereby releasing the aqueous inhibitor to contact the pore structure. The oil phase prevents water from becoming trapped in oil-bearing zones and maintains continuity between the pore spaces, allowing rapid return of oil productivity. This method of chemical deployment allows the scale inhibitor to penetrate into parts of the reservoir that would not be accessible through aqueous injection, and hence improve the contact between the scale inhibitor and reservoir substrate

The technique of deploying scale inhibitors via emulsions is very versatile, as the emulsions can be used to carry dispersed or incompatible liquids or solids and oil soluble scale inhibitors can replace aqueous ones to further reduce formation damage. If the emulsion does not break down instantly, then the droplets can release the scale inhibitor slowly by osmotic expansion of the droplets, surfactant thermal degradation and droplet-droplet coalescence. The rate of release of the scale inhibitor is initially controlled by the breakdown of the emulsion. If the emulsion has not broken down completely, the release of inhibitor from the partially destabilized emulsion droplets is slower.

Inverse emulsion technology has several benefits, including rapid oil return post squeeze, reduction of deferred or lost oil costs, potential to improve contact between inhibitor and in-situ adsorption sites, and the ability to formulate packages for sensitive formations and low water cuts. Low pressure support reservoirs may not need artificial lift to restart the well when invert emulsions are used. Work is being undertaken to improve squeezes by using molecules with dual affinity for oil and water and deploying suspensions of solid scale inhibitors.

5.5.5.4.3 Oil soluble scale inhibitors

Another alternative method of scale inhibitor delivery in water sensitive wells or wells with poor lifting energy is to use an oil soluble inhibitor, rather than an oil-based carrier as above. In one particular field described by Wat et al. [165,166], there is a distance of up to 16 km (10 miles) from the platform to the templates, and therefore in a squeeze treatment, a large volume of fluid has to be pumped and back produced. This gives rise to high friction and low pump rates (reservoir pressure is 340–360 bar). There is also poor inhibitor placement due to the low injection rates because of the transition from the 2.5-inch methanol line (which the scale inhibitor is being pumped through) to the 5.5-inch tubing. The production template being considered by Wat et al. was connected to several producers and there was no separate test line. Therefore, after applying a squeeze treatment to a single well, other wells on the template had to be shut down so that the squeezed well could produce through the 9-inch production line.

Using an oil soluble scale inhibitor has advantages [166], included the ability to back produce a well without shutting down the others (as there is less difference in well head pressure), and less relative permeability effects for dry wells. The squeeze can be performed before water breakthrough without deferring production. They also expect extended squeeze lifetimes, since the oil soluble scale inhibitor can be placed deeper into the formation without water blocking, and possibly better placement with a higher squeeze rate into the formation.

Oil soluble scale inhibitors require suitable hydrocarbon solvents. The choice of hydrocarbon will depend on cost, handling, availability, compatibility with the oil soluble scale inhibitor (at platform, downhole and seabed temperatures). The effect of the hydrocarbon on the hydrolysis and partitioning kinetics of the inhibitor is particularly important, as this will determine the success of the squeeze. The disadvantage of using hydrocarbon solvents is that their heat capacity tends to be lower than that of water, and hence the near well cooling effect that benefits inhibitor propagation is reduced. In addition, hydrocarbon solvents will not displace connate water from the immediate near well area, as would an aqueous solution, and so care must be taken to ensure that the inhibitor at application concentration is compatible with the formation brine.

5.5.5.4.4 Oil-miscible/oil-flowable scale inhibitors

In addition to the oil soluble and invert emulsion systems discussed above, several service companies have developed oil miscible and oil flowable scale inhibitors. Their application is much the same, with the emphasis on reducing or preventing any damage in oil production zones by minimizing water injection.

5.5.6 Importance of accurate assay and monitoring

The main approach for monitoring the approach to the end of the squeeze lifetime (and therefore to prepare and plan for subsequent treatments) is by residual assay of the returning scale inhibitor. This has been an active area of research for many years and is recognized as an ongoing challenge especially for complex subsea wells where multiple wells co produce through a single flowline [107,196–198]. Inhibitor analysis can be difficult and sometimes near impossible when co-producing from satellite fields and particularly so when producing from subsea deepwater templates, due to the high costs associated with collecting individual samples. This is also one of the contributing factors that LSSW is often selected for pressure support as opposed to SW when deepwater subsea fields have an inherent sulfate scaling risk.

Analysis of the returning scale inhibitor, or some analyzable moiety within the scale inhibitor, is not analyzing the effectiveness of the treatment for inhibiting mineral scale and therefore additional monitoring should be conducted. This has been highlighted previously when an analyzable moiety is included within a scale inhibitor to be applied by squeeze. The moiety is not completely bound within the inhibitor. As the reservoir rock effectively acting as a chromatography column, the analyzable moiety returns at a different rate to the actual inhibitor [107,142]. Further, when analyzing phosphonate-based inhibitors, the most commonly applied approach is by total phosphorus (P) using ICP (inductively coupled plasma) spectroscopy. This, by its nature, suffers interference when other phosphorus containing species are present (such as other scale inhibitors or other phosphorous species). In addition, when applying phosphonates into high temperature reservoirs, thermal degradation leads to the release of ortho-phosphate. Care must be taken to ensure that the analyzed return concentration is assaying for the actual inhibitor species. In such cases, phosphonate speciation may be required in order to ensure that the assay is indeed detecting the active inhibitor [106,107].

To address some of the analytical challenges, collection of water samples and analysis for increased suspended solids and/or the presence of increasing amounts of scale particles represents an effective measure to determine the approach to the end of the squeeze lifetime [199,200]. This is because in most cases, bulk scale will always be present in the produced waters as the system approaches the end of the squeeze lifetime, and in most cases, before the onset of operationally significant scaling. Examples of this approach are described below.

Of note though is that accurate assay of the returning scale inhibitor concentration to determine the inhibitor concentration return profile is critical for deriving the field retention isotherm. The isotherm can be used to optimize subsequent treatments using near wellbore squeeze treatment simulators.

5.5.6.1 SEM analysis of produced scale particles

Sample collection, filtration and subsequent analysis by SEM or ESEM has become a routine approach to determining the effectiveness of a squeeze treatment at continuing to prevent mineral scale formation. The technique involves collecting a fixed volume of produced water, filtering on site, and then analyzing the filters for the presence and abundance of scale particles. Examples are shown below in Figs. 5.55 and 5.56. Fig. 5.55 shows an example of the increase in the amount of produced "scale fines," indicating the end of the squeeze lifetime. Fig. 5.56 are examples of partially inhibited scale formed as the treatment approaches the end of its lifetime.

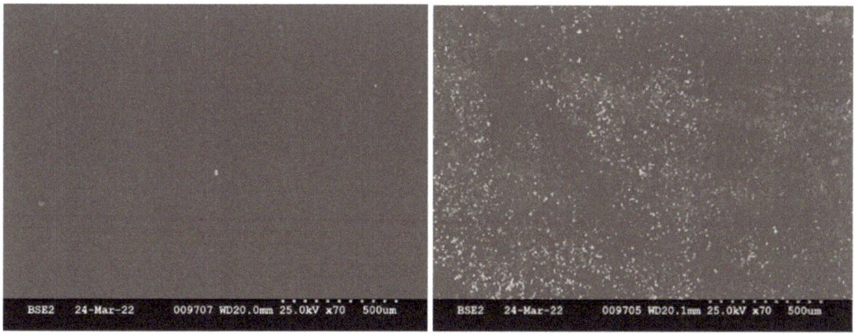

FIGURE 5.55

ESEM image of normal production with minimal "scale fines" with the system fully inhibited and not scaling (left) and increased scale fines when the system becomes uninhibited (right).

Courtesy of Scaled Solutions Ltd.

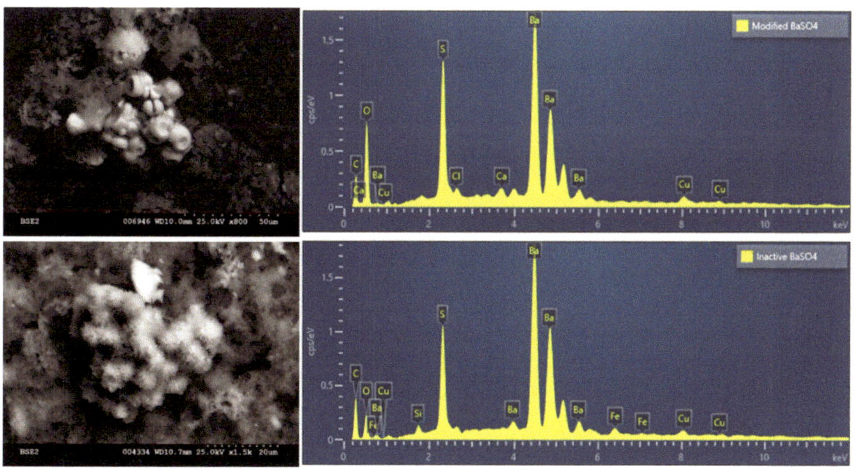

FIGURE 5.56

ESEM image of partially inhibited barite scale and associated EDX trace.

Courtesy of Scaled Solutions Ltd.

5.5.7 Isotherm derivation and near-wellbore simulation

A number of near wellbore "radial" treatment simulators are currently in common use for scale inhibitor squeeze treatment modeling and optimization. These include Squeeze (FAST JIP of Heriot Watt University) [147,157,160–162], Squeeze SoftPitzer (Brine Chemistry Consortium of Rice University) [96,156], and Place-iT and Place-iT Fractures (SWELL JIP of Scaled Solutions Ltd.) [26,141,142,148,149,163,180,186]. In addition, more sophisticated simulators including full reservoir models have also been used [151,152,172], though this requires a good deal of time-consuming near wellbore gridblock refinement of the reservoir model to produce useful results. In the majority of cases, the near wellbore radial models are used for treatment design and optimization. Of the three main near wellbore "radial" models mentioned above, the Place iT simulator differs in that it was designed primarily with a focus on simulating placement in complex multizone heterogeneous wells and minimized the reliance on more complex and time consuming "reservoir type" models, while allowing the treatment lifetimes form individual zones to be modeled and optimized based on the amount of chemical retained in each zone. Place-iT Fractures was designed as a further modification to specifically examine scale inhibitor squeeze treatment optimization in low permeability fractured wells such as those in unconventional reservoirs [149,163].

The general methodology however is to adopt a three-stage approach:

1. Design the initial treatment based on a derived core flood isotherm (or based on a similar treatment conducted in a closely analogous field).
2. Compare the field return data with the predicted inhibitor return, and if necessary rederive the field isotherm based on the initial treatment data.
3. Optimize future treatments using the matched field isotherm (in some cases this will be the original core flood isotherm).

Isotherm derivation and treatment modeling provides a more direct comparison of the expected return curves for chemical selection studies than that obtained from comparing the raw returns data (i.e., it removes artefacts such as chemical oversaturation). It also allows optimization of the planned field treatments. There are two key aspects of upscaling or treatment modeling that can significantly improve the potential for success in scale inhibitor squeeze treatments. These are modeling and optimization based on maximizing inhibitor returns, and modeling based on maximizing chemical placement. This second approach relates to designing the scale inhibitor treatment package to achieve maximum chemical injection into the water producing zones.

The modeling process allows simulation and upscaling of the core flood test data to a typical field application, allowing final selection of the best-performing chemical in terms of simulated field treatment lifetime. It also helps to overcome some of the artefacts associated with core flood testing (such as oversaturation for reactive systems) [141]. In most cases, the first field treatment is designed using the core flood derived isotherm. Once this first treatment has been applied in the field, the actual field return will be compared with that simulated using the core flood derived isotherm. Although this often records a reasonable match (eg as in Sorhaug et al. [26]), it is common that further optimization of the isotherm is beneficial to achieve an improved match to the field data. These "field isotherms" are often more accurate than core flood isotherms since they are generated using the real near-wellbore formation, and more accurately account for the gross effects of localized heterogeneities, differences in the near-wellbore fluid saturations and different flow regimes in the radial near-wellbore. This "field optimized" isotherm is then used for future treatments.

Reoptimization or further derivation of the isotherms is then normally only required when more significant changes occur during production, such as significant changes in water-cut or the zonal water production profile, and/or significant changes in produced water chemistry.

5.5.7.1 Isotherm derivation

The principle behind isotherm derivation from a scale inhibitor return core flood test is quite simple. It relies on mass balance, with the assumption that the core flood adsorption/release processes, or more specifically the release processes, occur at (or close to) equilibrium conditions. The process assumes injecting a known amount of scale inhibitor into the core and accurately measuring the amount of inhibitor eluted from the core during the injection, the amount of inhibitor retained in the core can be determined by simple mass balance. This allows us to determine an initial concentration of adsorbed inhibitor in the core at the start of brine postflush. By then measuring the eluted concentration of inhibitor during postflush, an isotherm plot of adsorbed inhibitor concentration (in mg/L or mg/kg of rock) vs. solution concentration can be created. This is then the adsorption isotherm for this chemical in the core test. This process can be done manually, but some simulators also include an option for automatic isotherm derivation. The scale inhibitor concentrations from field returns or a core flood postflush can be loaded into the simulator, which then iterates to find a Freundlich adsorption isotherm which generates an inhibitor return curve that matches the supplied data most closely. The comparison of the core flood data against the generated simulated inhibitor return profiles are presented for the user to select which isotherm is considered a best fit for the data (Fig. 5.57).

5.5.7.1.1 Types of isotherms

There are many forms of isotherm which have been derived for various adsorption processes, depending on the nature of the substrate and the adsorbing species. For scale inhibitor return modeling, two types of isotherms (Freundlich and Langmuir) are commonly used. Both are defined by two parameters, n and k for the Freundlich Isotherm, and a and b for the Langmuir Isotherm.

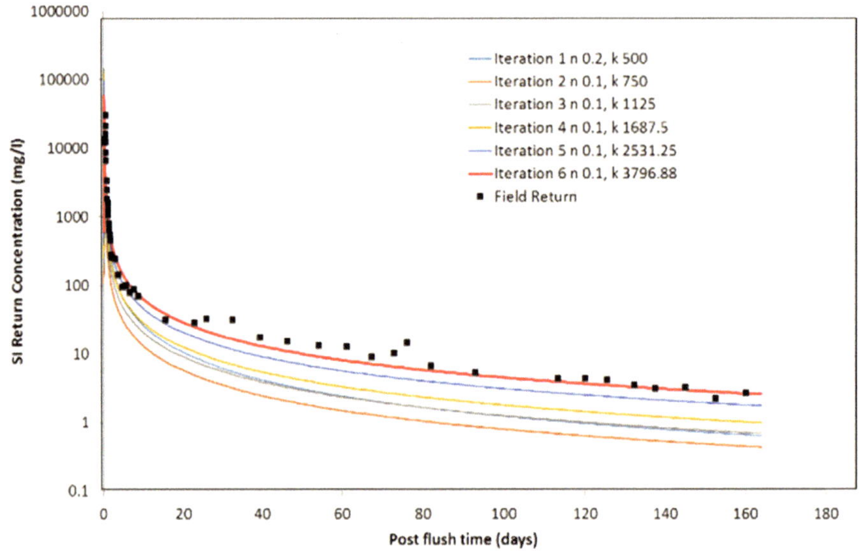

FIGURE 5.57

Automated isotherm simulation with field data (conducted using the Place-iT simulator).

Courtesy of Scaled Solutions Ltd.

Freundlich Isotherm:

$$\Gamma = kc^n \tag{5.14}$$

Langmuir isotherm:

$$\Gamma = \frac{abc}{1 + bc} \tag{5.15}$$

where: Γ is adsorption (retention), and c is equilibrium concentration in solution,

Note that the mathematical form of the two isotherms is different. The Langmuir Isotherm is characterized by a maximum adsorption limit *a*, whereas the basic Freundlich Isotherm has no upper adsorption limit. For this reason, the Freundlich Isotherm is often "capped" at a maximum adsorption level:

$$\Gamma = \begin{cases} kc^n, kc^n < \Gamma_{max} \\ \Gamma_{max}, kc^n \geq \Gamma_{max} \end{cases} \tag{5.16}$$

5.5.7.1.2 Matching the isotherm with core flood data

After obtaining an isotherm from laboratory core flood data, it is important to verify that the isotherm does, in fact, provide a good simulation of the core flood data. This can be done by using the software to model the core flood using the derived isotherm, and comparing the simulated return profile with that obtained in the laboratory. If a satisfactory match is obtained, the core flood isotherm can be used to simulate a field treatment.

5.5.7.2 Treatment modeling

Once a suitable adsorption isotherm has been obtained, it can be used to simulate a field treatment and examine how sensitive the simulated squeeze treatment lifetime is to a number of parameters. This process of field treatment modeling and optimization should take into account specific properties of the production well including length, heterogeneity, permeability contrasts, pressure contrasts, and so on.

5.5.7.2.1 Base case model: comparison of chemicals

Once core flood isotherms have been obtained, either manually or by automated isotherm derivation, a base case field model can be constructed and used to simulate a field treatment for the various scale inhibitors under consideration. At this stage, a generic treatment design can be used, typically representing a treatment to achieve propagation to between 3–5 m from the wellbore in a typical short reach well, and between 1–2 m in a long horizontal or deviated well completion. Base case treatments are often designed with 30% of the treatment volume as main-treatment and the remainder as overflush. The purpose of the base case modeling is to assess the likely treatment lifetimes obtainable for a field treatment based on the known parameters. This allows for the simulated treatment lifetimes for the different chemicals to be compared for a final selection.

5.5.7.2.2 Treatment optimization

Treatment design optimization can then be performed, using the near-wellbore simulator to investigate the sensitivity of the model to variations in treatment design (*e.g.*, increasing overflush volume, main-treatment volume, *etc*.). This optimization is often performed for two or more of the candidate scale inhibitor treatments, as the different adsorption/release properties of different chemicals may show different responses in the design optimization stage.

5.5.7.2.3 Cautionary note: core flood modeling

As described above, upscaling generally represents the most appropriate methodology for simulating field treatments. This should, however, be reviewed thoroughly, since small changes in the derived isotherm may have a significant impact on the predicted lifetime. This is particularly important if different inhibitor suppliers upscale their own core flood data and provide individual field treatment designs/lifetimes. Initial comparative modeling, especially for "best in class" chemical selection from different vendors should, if practically possible, be conducted by the test laboratory or by internal or independent experts as opposed individual vendors. Final treatment design modeling or field optimization is, however, often conducted by the selected vendor.

Modeled treatment lifetimes should be sense-checked to ensure that:

1. The same isotherm derived and used for the field treatment modeling has been used to rematch the core flood return treatment;
2. The core flood isotherm is based on conserving mass balance (i.e., isotherm "maximum adsorption" is similar to core flood "maximum adsorption");
3. Simulated core flood returns should be comparable to actual core flood returns, provided that oversaturation has not occurred. For core flood tests where similar retention and release properties are recorded, field simulated return profiles should also be broadly similar.

5.5.7.3 Example of isotherm derivation and treatment modeling

Fig. 5.58 provides a sequence of plots illustrating the methodology described above, and illustrates that the first point is to derive the core flood isotherm. Note: One important aspect of the isotherm is that it should tend toward the zero intercept. Due to analytical errors, especially when assaying the residual chemical released during the oil flowback stages, some manipulation of the maximum retention may be required to ensure that the isotherm tends toward zero.

If the isotherm is derived manually using mass balance calculations, the resulting isotherm is checked by simulating the core flood prior to field application modeling. In this example, a field treatment was subsequently conducted, and the

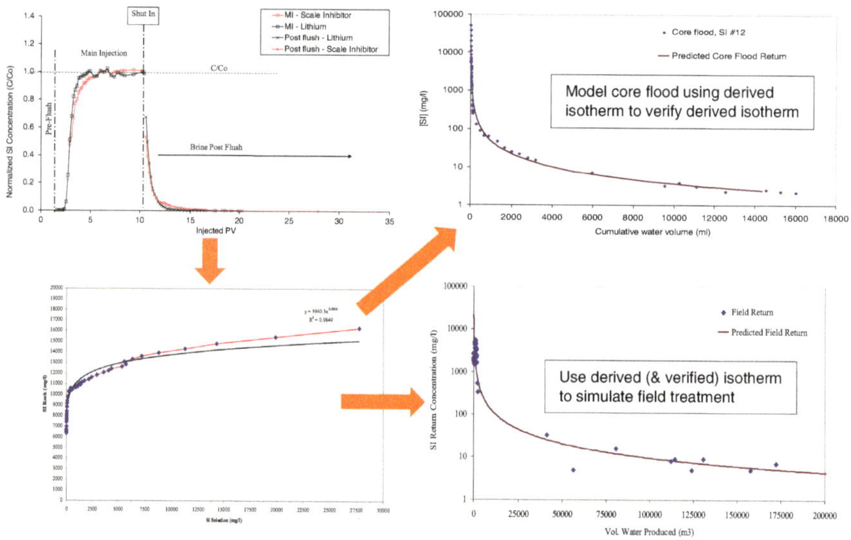

FIGURE 5.58

Derivation of isotherm, verification with core flood returns and application for field treatment modeling.

Courtesy of Scaled Solutions Ltd.

recorded field data is also shown, illustrating a very good match to that simulated from the core flood data.

5.5.7.4 Simulating treatments in tight fractured reservoirs

Low permeability reservoirs are often fractured to increase the connectivity of flow channels and to allow for increased hydrocarbon production. However, treatment of such reservoirs to prevent scale-induced flow assurance issues can be a challenge owing to the uncertainty inherent in these reservoirs, particularly in regard to water production and treatment placement. Fractured reservoirs can be categorized as: low permeability reservoirs with a typical matrix permeability of 1–5 mD, and tight shales with a typical matrix permeability of considerably less than 0.1 mD.

The pore throats of fractured shales are considered so small that there is no practical possibility of bulk fluid movement. If (advective) "Darcy" flow of injection fluids is assumed in shales, the calculated injection pressure required to induce flow is typically beyond the fracturing pressure. However, they may still be large enough for molecular diffusion. In low permeability reservoirs (typical permeability of 1–5 mD), a combination of diffusive and advective fluid transport mechanisms are suspected. Understanding the mechanisms of fluid flow into fractured reservoirs is essential to ensure that correct modeling and optimizations techniques are implemented.

In the case of fractured shales, some models (e.g., Place-iT Fracture model [149,163]) have the capability to model a shale reservoir system with multiple fractures explicitly, assuming diffusive transport of injected chemicals. The model requires fracture dimensions, aperture, length and width, which can be back-calculated from the treatment volume, given that the percentage of the fracture network volume suspected to be filled by the treatment is known. Furthermore, the model can be used to select which fractures produce water and which are most likely to be treated (Fig. 5.59). This increased capability allows for sensitivity modeling of field data to allow for better history matching (Fig. 5.60).

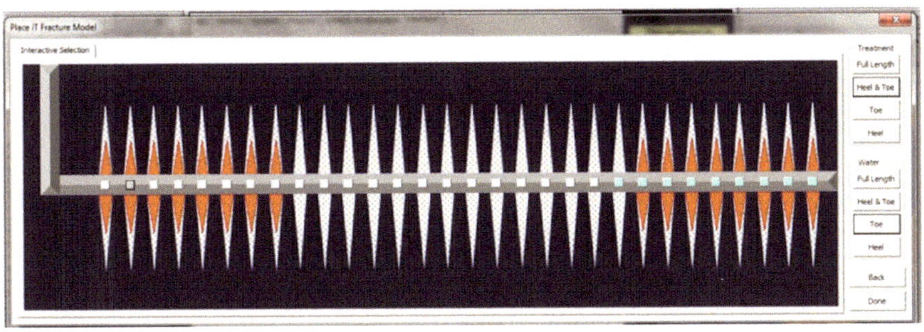

FIGURE 5.59

Software interface allowing the selection of treated and water production fractures.

Courtesy of Scaled Solutions Ltd.

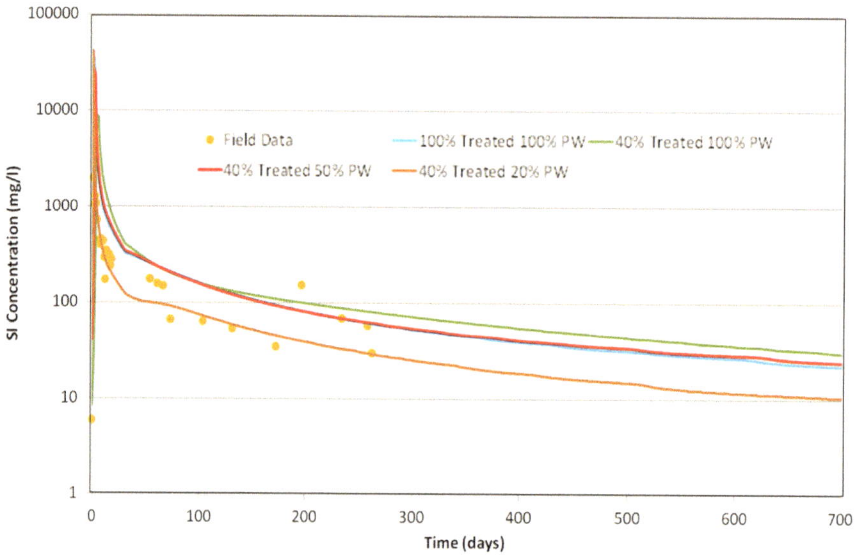

FIGURE 5.60

History-matching field data by varying the ratio of treated to producing fractures.

Courtesy of Scaled Solutions Ltd.

Such explicit squeeze simulators for fractured formations have been increasingly used as a design tool to optimize field treatments and history match squeeze-treated fractured shale producers in order to determine the regions that might be producing water vs those that are being treated. This relatively quick simulator allows for better understanding of these systems, for which there is no treatment protocol, and hence allows for informed decision making when designing scale inhibitor squeeze treatments.

5.6 Scale remediation

Prevention is better than cure, which can also be true in scale management. However, this is not always the case because a system may possess a scaling tendency for calcium carbonate that is marginal, and so it may not be clear that scaling will be operationally significant over field life. Even if it turns out to be an issue, this might mean having to clean the well out with acid perhaps twice during its lifetime, which may still compare favorably with applying periodic scale inhibitor squeeze treatments to prevent scale over many years of water production. Moreover, if the cost of "being wrong" about a scaling tendency is acceptable (i.e., the deposit can be readily and cost-effectively remediated), then there will be greater tendency to "wait and see," perhaps adopting a preventative strategy when operationally significant scaling has been confirmed. The other reason why scale may need to be remediated is that the preventative strategy has not been properly implemented: production was continued when continuous injection was interrupted for too long, or the scale squeeze was delayed too long. In either case, the system has experienced operationally significant amounts of scale. Finally, not all scales are readily inhibitable. So, while there are some chemical approaches to prevent halite, silica, silicates and certain hydroxides, they tend to be less effective and often more difficult to implement than for more conventional carbonate and sulfate scales. In such cases, remediation may be the only choice.

The removal of mineral scale deposits by mechanical means is often difficult, expensive, and risky, for example, when it involves physical intervention in a well. Therefore, chemical treatments are generally preferred. The ease of mechanical scale removal is related to the hardness of the deposit, which can vary greatly between different minerals, as well as depending on whether it is a single mineral or a mixture. Moreover, deposit hardness also depends on how quickly the deposit was formed (often related to the scaling tendency) and how long it has been in the system because weaker or more porous scales may regrow to harder deposits by Ostwald ripening over time. The Mohs Hardness Scale gives an indication of the hardness of the pure minerals, but it tends not to be a reliable indicator of the ease of mechanical scale removal (note that Mohs Hardness Scale for minerals ranges from 1 for talc to 10 for diamond). Gypsum is a relatively soft mineral (Mohs hardness ≤ 2) and is indeed normally one of the easier scales to mill or jet, but the values for the other common scale minerals (anhydrite, celestite, barite, calcite, and aragonite) all fall within the range 3–4, despite the fact that the harder sulfates (barite in particular) tend to be significantly more resistant than the carbonates (even though aragonite has the highest Moh's hardness of all of these). Another complication is that many scales are mixtures, and the mechanical properties of the composite deposit can be very different from those of the component parts. Often, the best way to find out how hard a deposit is is to sample and measure it.

5.6.1 Acid soluble vs acid insoluble scales

The acid solubility of scale minerals largely determines how readily the scale can be dissolved chemically, which is nearly always cheaper if it can be achieved practicably. In reality, the real determinant is whether a scale can be readily dissolved in a readily available, practical-to-use solvent, both in terms of dissolution capacity and dissolution rate. It just happens that acids often fit that descriptor best for a variety of scales. That is not to say that acid is always the optimal fluid even for the most readily acid-soluble scales, such as calcium carbonate. Acid is corrosive, and a strong mineral acid, such as HCl, can often be unacceptably corrosive to well tubulars at high temperature, even when a corrosion inhibitor package is included in the formulation. Hence, there is a tendency to use other less corrosive fluids to dissolve the scale, for example, it would be common to traverse the following series as the system temperature increases:

$$\text{High concentration HCl} \rightarrow \text{HCl/organic acid mixture} \rightarrow \text{organic acid mixture} \rightarrow \text{chelating agent}$$

Chelating agents are the main nonacid-based scale solvents, though they may be applied at slightly acidic pH values. They are essentially the primary (moderately) effective solvent against sulfate scales, and will also dissolve carbonate scales and some sulfides. The main reason a distinction is made between acid-soluble and acid-insoluble scales is because, although the latter can indeed be dissolved, their solubility and their rate dissolution tend to be much slower than that for acid-soluble scales dissolving in acid. This means that it is often only practicable to dissolve quite limited amounts of scale, in particular, where small amounts of scale are having a large impact on fluid flow or integrity. In other words, it is generally impracticable to dissolve a significant build-up of an acid-insoluble scale that has significantly reduced the diameter of a flowline or production tubing; physical removal tends to be the main option for this. However, chemical dissolution can be cost-effective and practicable even with acid-insoluble scales if, for example, the scale is in the near-wellbore or causing the malfunction of a valve or gauge. In such instances, the "culprit" could be only a few kilograms to a few tens of kilograms of scale, which means, conversely, removing that little scale could remediate the main scaling issue.

Each method will now be described in more detail.

5.6.2 Mechanical Remediation/Physical Methods

Mechanical or physical methods include the following:

- Milling or reaming on drill pipe;
- Jetting (e.g., with coiled tubing in a well or in facilities), either with fluid alone or augmented with particulates (such as beads that are designed to erode the scale much more than the steel tubulars), or sometimes combined with a scale dissolver that is thought to help soften the scale;
- Re-perforating to bypass scale deposited in the near-wellbore reservoir, or use of small explosive charges in the vicinity of the deposit to try to shake the scale loose;
- Replace the scaled tubing.

In considering jetting vs reaming in a wellbore, this may depend on how thick the deposit is. Many jetting tools work most effectively with the jets subtending a specific angle to the deposit, which may be unachievable for very thick deposits. Milling can be slower, but it may be the only option in some cases, particularly for thick deposits of the hardest scales, such as fairly pure barium sulfate.

Re-perforating with deep-penetrating charges can be very effective at re-establishing connectivity with the reservoir that was lost by scale deposition. If there is also significant deposition in the wellbore itself, then the guns may not be able to access the producing interval without removing that deposit first, most typically with another physical-removal method listed above. For wellbore scaling in long, highly deviated wells, especially those which follow snakelike profiles (*i.e.*, traversing different depths), coil tubing operations may not be able to reach the location of the scale because of friction.

The main drawback with mechanical methods, particularly when used offshore (in particular, in subsea operations), is that they can be expensive and time consuming. For example, the time required for coil tubing scale removal operations for a well can range from 7 days for a dry tree on a host platform or local jacket to 19 days for subsea. Milling on drill pipe is only possible if a rig is either available on the host or can be brought in over a wet tree. These are high-cost operations and can defer significant amount of production. In addition, tools can become stuck, especially if the thickness and/or hardness of the scale is not accurately known prior to the treatment. Moreover, this is a mechanical intervention in a well, with all of the well-control risks that entails, together with the risks of formation damage from applying kill fluids, kill pills, comminuted scale fragments that may end up in the producing interval. Physical removal also produces a solid mass of scale that needs to be disposed of, which may be more complex and costly if it is naturally radioactive.

As with downhole scale inhibitor squeeze treatments (or any other well treatment, for that matter), the costs associated with the mechanical and physical removal method can depend largely on the type of production environment, ranked as follows where "<" indicates "is less expensive than":

land-based operations ($) < offshore dry tree ($$)

< offshore subsea ($$$) < subsea deepwater ($$$$)

The costs of scale removal in subsea deepwater fields are so high that many wells are abandoned when scaling has been allowed to progress too far with a nonacid-soluble scale, such as barite.

The location of the scale also has a bearing because the further upstream in the production that deposition occurs, the more expensive it tends to be to remove, primarily due to more difficult accessibility. The wellbore is generally more expensive overall, and a production tubing tends to be significantly lower diameter than a subsea flowline, which means it may require less scale to restrict flow significantly. However, if scaling in a subsea flow leads to a stuck pig, then production from an entire field could be interrupted.

Reports from different operating companies indicate that for North Sea fields, where many fields undergo seawater flooding and hence are subject to barite scaling, an average of one coiled-tubing scale-removal operation is conducted per well over field life.

5.6.3 Chemical dissolution

Chemical dissolution treatments are generally preferred if they are practicable, in particular, if they require no more equipment than some tanks and pumps to deliver the chemical (compared with the costs of a rig or coiled-tubing unit for physical removal). As with all chemical treatments, the critical success factors include:

- Chemical effectiveness: how much and how quickly the solvent dissolves the scale
- Chemical delivery: how easy is it to place the chemical adjacent to the scale and keep it there for sufficient time for sufficient dissolution to take place

In some cases, remedial treatments are conducted immediately prior to scale inhibitor squeeze treatments or as combined treatments.

The carbonate scales are generally the easiest to remove chemically, due to their ready dissolution by a variety of acids and chelating agents. Sulfate scales normally require the use chelating agents, which are much slower-acting than acids on acid-soluble scales, and often require catalysts to accelerate dissolution. Pure barite is typically the most difficult to remove because the dissolution reaction is so slow, often requiring soak times of 24–48 hours. Moreover, dissolution of barite requires a trade-off between using a high concentration of chelating agent, which gives the highest dissolving capacity vs a more dilute form, which tends to dissolve the scale at a faster rate. For this reason, it is common to dilute the as-supplied barite solvent by a factor of two for application, though this must not be done with diluent containing significant amounts of calcium and magnesium (such as seawater), as this will significantly "spend" the solvent's dissolving power. Overall, this means that sulfate scales in general, and barite in particular, chemical dissolution of thick deposits is impracticable. Most success has been demonstrated where small amounts of scale are causing a large issue, such as in the near-wellbore or where a valve has become seized.

Mixed scales are often easier to remove by dissolution, as often the more soluble species is more rapidly dissolved, e.g., in a mixed scale containing both barite and calcium carbonate, the calcium carbonate would likely dissolve first, increasing the surface area for attack on the remaining barite. Also, deposits can sometimes detach even if only a fraction of the solid has been dissolved, simply by being mechanically undermined. If the calcium carbonate content of the scale were significant, the success can sometimes be enhanced by pretreating the deposit with an acid, such as HCl, which is suspected to make the scale more porous, increase its surface area and hence its reactivity to a subsequent barite chelating agent. However, it is essential to ensure that the HCl (pH around 0) does not mix with the barium sulfate solvent (pH of 10.5–12), or the reaction could be violent.

A solid solution of (Ba, Ca)SO_4 may be significantly more readily dissolved than pure $BaSO_4$, which means it may be very useful to understand the precise nature of the scale before attempting a scale-dissolution treatment. An actual sample of the scale (assuming it is representative of all of the scales present) not only gives the precise composition but can also be used in the laboratory to assess reaction times and rank the performance of different solvents.

Mutual solvents are applied either with the scale dissolver or as a small preflush prior to the addition of the chemical dissolver, as these can clean the surface and/or make the scale more water-wet [201].

5.6.4 Chemical deployment in scale dissolution

As with many oilfield treatment operations, effective chemical deployment can a significant challenge in scale dissolution treatments. For example, certain scale dissolvers require soak times of up to 48 hours; therefore, when the deposit is located in the middle or upper sections of a vertical production tubing, methods must be found to maintain the solvent pill opposite the deposit for that period, whereas the density of the pill (exceeding that of most produced waters and all produced oils) means it will tend to settle under gravity. Various methods have been used to address this, including the use of a thixotropic polymer pill underneath the solvent. Deposits at the very top of the well can be supported by closing the safety valve during the treatment. Scale at bottom hole obviously presents less of an issue in this regard because that is where the solvent will tend to end up due to gravity effects, though even at this location, there may be tendency for the pill to leak away too rapidly into the formation if the hydrostatic pressure from the fluid column in the tubing is too high. This can be adjusted by overdisplacing the treatment with a less-dense hydrocarbon, which may also be gas (from a donor well or oxygen-free nitrogen).

The challenges with keeping the scale dissolver in place during the soak period can be even greater with gas systems, as illustrated in Fig. 5.61. In the former, the scale is located lower in the well in a water producing or water saturated zone. In this case, the chemical can relatively easily be held in place during the soak time. Fig. 5.61b illustrates the case where the chemical is located higher up in the production tubing, surrounded by produced gas (or lighter oil in oil production systems). In such cases the ability to hold the scale dissolver in place can prove difficult. Treatment would be ineffective if the chemical drops away from the location of the scale.

The challenge of chemical deployment is of placing sufficient chemical across the entire interval, in this case, to dissolve the scale. Placement and diversion are a well-developed technology for well stimulation treatments, and similar techniques can be applied with scale dissolution in and around the producing interval of a well. Bullheading of fluids is nearly always the cheapest and simplest approach because it requires only minimal equipment (chemicals tanks, injection conduits to connect to the tree and pumps). It is rarely the most effective or efficient method, though it may be good enough to treat relatively short intervals where the contrasts between permeability and pressure between zones are not too great.

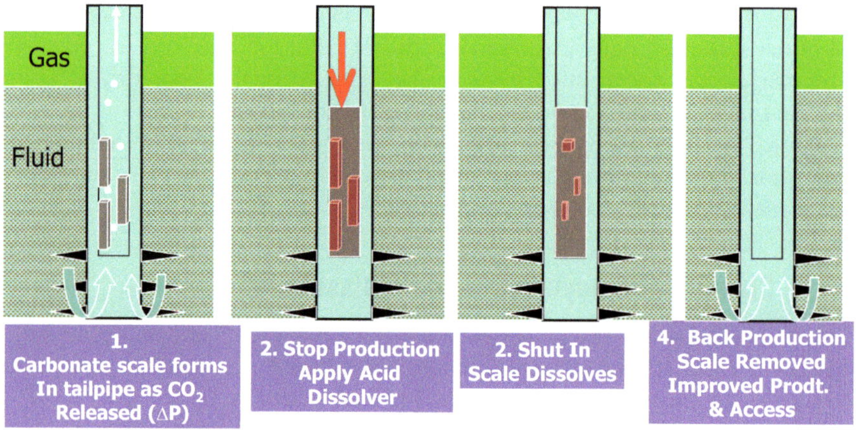

(A). Scale located in a liquid (water saturated location)

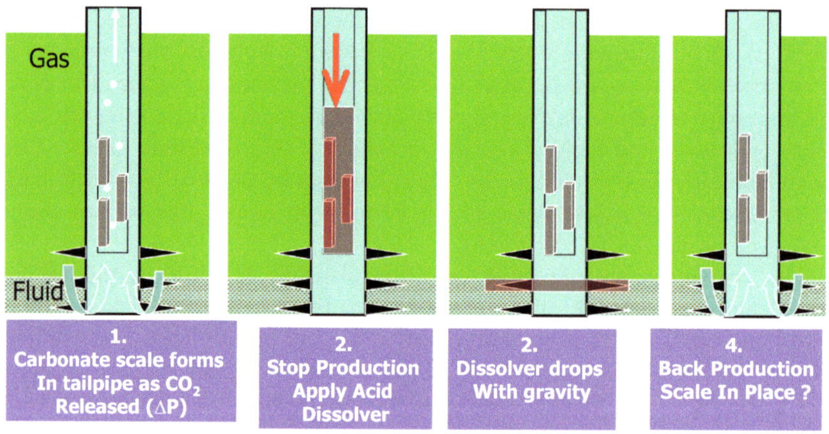

(B). Scale located in a gas (or possibly a light oil) region

FIGURE 5.61

Illustration of scale dissolver operations. (A) Scale located in a liquid (water saturated location). (B) Scale located in a gas (or possibly a light oil) region.

Image courtesy of Scaled Solutions Ltd.

In longer intervals, a highly reactive chemical (such as HCl reacting with calcium carbonate scale) could clean up the zone it first encounters during pumping (e.g., the "heel" of a horizontal well) so well that most of the treatment then enters this stimulated zone rather than the fluid front progressing along the wellbore, which would require displacing the wellbore fluids into nonstimulated zones.

The corollary of this is that scale solvents are likely to be more effective in such scenarios if their reaction rate is lower.

Viscosifying the solvent with a nondamaging polymer that will subsequently "break" (lose its viscosity due to degradation) can further assist this because it is likely to be easier to displace the wells fluids into the formation than for the stimulation fluid to leak off, which has the effect of advancing the stimulation-fluid front further along the wellbore. Other standard diversion techniques may also be applicable, depending on the completion type. Use of particulates or oil-soluble resins can temporarily shut off a stimulated zone and divert the fluids to other parts of the producing interval. Ballsealers in perforated completions may not seat well on perforations decorated with scale deposits.

Most effective approach is to use mechanical methods to improve placement effectiveness, such as placing the fluid with coiled tubing (CT) while taking fluid returns to surface. This will ensure complete coverage of the entire interval with the scale solvent. The fluid can then be bullheaded simultaneously via the CT and down the annulus to overdisplace the fluid into the near-wellbore to remove scale at this location too. If CT is going to be used, the additional costs of jetting may be marginal. In this case, jetting with scale solvent could be considered. This approach combines the advantages of jetting with the ability to treat the near-wellbore, as well as breaking up the scale into smaller, more reactive pieces, potentially avoiding the need to handle solids at surface (as would be the case when jetting with an inert solvent). Of course, the CT unit needs to be chemically tolerant to the scale solvent, including at the very high shear rates encountered in the CT and the jets.

Scale dissolution in facilities is generally more straightforward, and its effectiveness can be enhanced by judicious design. Circulating the solvent in a closed loop (through the affected pipe, vessel, heat exchanger, and so on and then back again (e.g.,) using chiksan pipe) can be an effective way to dissolve deposits in surface facilities. Rapidly recirculating the solvent for a number of hours at high shear can also increase its temperature, perhaps making the dissolution operation faster, though other methods may be available on site to heat the solvent to a suitable temperature.

5.6.5 Acids for Scale Dissolution

5.6.5.1 Common acids used for scale removal

Acidizing is the most common chemical method used in the petroleum production industry, for removing scale and for stimulating oil wells. Acids are used to remove carbonate, sulfide and oxide/hydroxide scales. Technical grade hydrochloric acid, also called muriatic acid, has been the preferred acid for many years because of its capacity for dissolving scale, low cost, and availability. It is generally applied at a concentration of 5–28 wt.%, though 15 wt.% is perhaps the most common. However, it is very corrosive to steel and therefore effective corrosion inhibitors must be added to limit damage of steel equipment. This means that in many regions, HCl has been replaced with less aggressive organic acids, such as acetic acid and formic acid and blends thereof, for example, 5% HCl and 9% formic or 9% formic and 10% acetic. In very high-temperature systems, chelating agents may need to be used because acids are too corrosive.

Calcium carbonate is the most common water-deposited scale that is soluble in acids. It dissolves very rapidly in HCl, with the liberation of carbon dioxide:

$$CaCO_3 + 2HCl \rightarrow CaCl_2 + CO_2 + H_2O \tag{5.17}$$

A comparison of the theoretical dissolving performance of various acids is shown in Table 5.13. However, it is important to note that these are equilibrium reactions. For HCl, the equilibrium sits very far to the right, which means the theoretical dissolving power is often approached (provided there is optimal contact between solvent and scale). This is not the case for the weaker acids because the CO_2 generated from the dissolution reaction largely remains in solution (unlike what happens on the laboratory bench top) and prevents the reaction from going to completion. Therefore, the actual dissolving power in the field is somewhat lower for formic acid and significantly lower for acetic acid, than the values indicated in Table 5.13. Being "weak" acids (i.e., largely nondissociated at application conditions), this effect cannot be compensated for by increasing the concentration, while it appears from the table that increasing the concentration of formic from 15 to 20 wt.% increases its dissolving power by 1/3, that is not the case in reality (though it does work, more or less for HCl). Moreover, many operators limit the concentration of formic acid to around 9–10 wt.% to avoid issues arising from reprecipitation of calcium formate, which could coat the scale and hinder further reaction. This is one reason why organic acids are often deployed as blends, in order to increase dissolving power without leading to reprecipitation.

Table 5.13 Capacities of some acids for dissolving $CaCO_3$ scale.

Acid	Concentration (%)	$CaCO_3$ dissolved (lb/gal)	CO_2 formed (Std. ft^3/gal)
Hydrochloric	15	1.84	6.99
Acetic	15	1.08	4.09
Formic	15	1.42	5.38
Hydrochloric	20	2.50	9.47
Acetic	20	1.43	5.41
Formic	20	1.90	7.20
Hydrochloric	25	3.22	12.20
Acetic	25	1.80	6.82
Formic	25	2.40	9.09

Acids other than HCl are generally more expensive. Therefore, HCl remains the solvent of choice for many moderate temperature regimes. In fact, the cost differential can become *much* smaller when the cost of corrosion inhibitor, iron-control additives, reducing agent, antisludging agents, and so on are included (which tend to be either unnecessary or much less of an issue with organic acids). Even in moderate temperature applications, it is important to ensure that the corrosion inhibitors are included in the formulation and their ability to prevent corrosion is assessed in the laboratory prior to treatment. Sulfuric or phosphoric acids are not generally used because their anion (sulfate and phosphate ion) can reprecipitate with calcium ions to form $CaSO_4$ or $Ca_3(PO_4)_2$.

Sulfamic acid is often used to treat carbonate scale within facilities because it is probably the easiest acid to handle from a safety perspective.

5.6.5.2 Carbonate scale removal

In practice, the volume of calcium carbonate to be dissolved is estimated from its thickness, additional pressure drops across the affected area (which can give significantly different values depending on what assumptions are made about its extent), calipers in wells, fluid residence times obtained using tracers, and so on. The amount of solvent (such as HCl) required to dissolve the total amount of calcium carbonate is then determined.

The reaction of calcium carbonate with HCl is very fast if the scale is clean. If it is oil coated, the reaction is slower. If the oil coating is heavy, a solvent rinse is often required to remove this prior to acidizing. A light contamination by oil is usually overcome by including a surfactant or mutual solvent either in the acid or as a preflush.

The volume of CO_2 liberated by acidizing carbonate scales can be very large and it can be generated rapidly. Provision should be made to vent the CO_2 formed if the pressures generated are likely to be unacceptable. If there is any risk of sulfide scales being present (often the case in sour systems), further precautions to prevent uncontrolled release of this toxic and corrosive gas are required. H_2S scavenger specifically formulated for acidizing, often based on aldehydes, should be incorporated into the package (conventional triazinane-based scavengers should not be used).

5.6.5.3 Iron scales

Iron carbonate, some iron sulfides and some iron oxide/hydroxides are also acid soluble. Additional precautions and care must be taken to ensure that effective sulfide scavengers are either included in the dissolver formulation and/or applied to the produced fluid streams. What steps are necessary depends on the likely extent of the issue, and whether it relates to corrosion (e.g., sulfide stress corrosion cracking), safety of personnel, or product specification. H_2S gas is also released when other sulfide scales are dissolved, such as zinc sulfide, lead sulfide, or (the occasionally encountered scale) mercury sulfide.

The generated dissolved iron can also be an issue because it is normally impracticable to deoxygenate acids, which means any dissolved Fe(II) can convert to Fe(III). Fe(III) will precipitate as a sludge when the pH of spent dissolver exceeds ~2. Iron-control agents are necessary to control the reprecipitation. For iron oxide/hydroxide scales, part of the iron is typically already in the Fe(III) form.

5.6.5.4 Scale conversion

This method does not strictly relate to acid-soluble scales, but to one barely acid-soluble one, namely, calcium sulfate. It is a largely superseded technology because modern chelating agents work relatively well with this scale, but conversion may still be used occasionally. Scale converters are chemical agents that convert a nonacid-soluble scale to another mineral that is readily acid soluble. Conversion agents are alkaline solutions, nearly always soluble carbonates or hydroxides, that react with calcium sulfate and convert it either to calcium carbonate ($CaCO_3$) or calcium hydroxide [$Ca(OH)_2$]. For example:

$$CaSO_{4(s)} + (NH_4)_2CO_{3(aq)} \rightarrow CaCO_{3(s)} + (NH_4)_2SO_{4(aq)} \quad (5.18)$$

Therefore, removal of calcium sulfate with a converting agent is a two-step process, which is the main reason it has lost ground relative to chelating agents. Treatment with the converting agent is slow, generally requiring a soak time of 24 hours or longer. The reactant (the $CaSO_4$ scale) and the product ($CaCO_3$) are both solids, which means the reaction is somewhat hindered by the product coating the reactant. The subsequent acid step is fast. If the scale also requires a preflush to clean up the scale and make it more water wet, then the dissolution process becomes even more complex because another stage is added in the sequence [202].

5.6.6 Chelating agents for scale dissolution

Chelating (or complexing) agents are anionic species that form water-soluble complexes with metal cations, including those that form scales. They are the main chemical agents applied to dissolve scales that are insoluble in acids.

If calcium sulfate is dissolved in water, the resulting solution is saturated with Ca^{2+} and SO_4^{2-} ions. By adding a chelant, the Ca^{2+} ions are complexed and the activity of Ca^{2+} in the water is reduced; therefore, the solution is able to dissolve a greater amount of calcium sulfate before it becomes saturated. As with other processes, a thermodynamically determined "stability constant" defines the amount of a particular cation that can be held in solution at equilibrium as a function of temperature, pressure, pH, salinity, and so on. If there is "spare capacity" in solution, then application of a chelant to a scale mineral could lead to its dissolution.

Scale dissolution by chelating agents is typically slow relative to acid reactions with acid-soluble scales, and the rate determining step is nearly always a surface reaction, often the diffusion of the metal-chelant complex away from the surface and into solution. (In contrast, with acids the rate-determining step can also be a surface reaction, but the fastest-acting systems are limited by transport of reactants and products in the bulk fluid.)

Some common complexing agents are listed in Fig. 5.62. EDTA is not very soluble in water alone but is very soluble in NaOH or KOH solutions. EDTA has four acidic carboxylate -OH groups, all of which can be replaced by the Na^+ or K^+ ions, giving a range of resulting pH values as more H^+ are replaced by the respective M^+ (either Na^+ or K^+). Typically, the application pH of EDTA-based solutions is in the range 5–11, where the lower range tends to be used for carbonate scales and the higher range for sulfate scales.

Most of the chelating agents have very high molecular weights, which contributes to their cost. For example, one mole of EDTA (molecular weight, 292.2 g/mol) complexes one mole of Ca^{2+} (atomic weight, 40.08). Therefore, it requires 7.3 mg of EDTA to complex 1 mg of Ca^{2+}. Nitrilotriacetic acid (NTA) has a lower molecular weight than EDTA. It requires 4.8 mg of NTA to complex 1 mg of Ca^{2+}.

Chelating agents will dissolve calcium carbonate, though they only tend to be used for this (instead of the more usual HCl or organic acid) in certain cases, viz.:

- where the carbonate also contains significant amounts of nonacid-soluble minerals,
- where the temperature is so high that even corrosion-inhibited acid may damage the tubulars, and
- where the formation water is very rich in either bicarbonate or sulfate ions because then flowback of the formation water with the calcium-rich spent acid

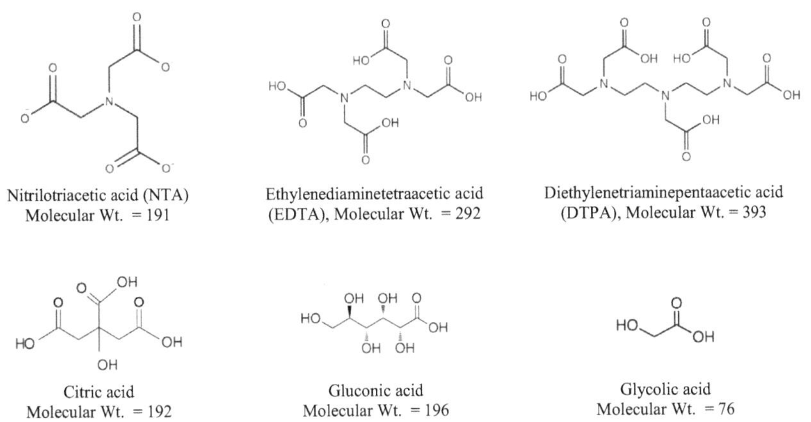

FIGURE 5.62

Common chelating agents, some of which are used as scale dissolvers.

Image courtesy of Scaled Solutions Ltd.

can lead to reprecipitation of calcium carbonate, resulting in a failed or compromised treatment. Cases are known where HCl-based scale dissolution failed in a well but was successful when replaced with (e.g.,) an EDTA-based treatment, which is not subject to reprecipitation.

5.6.6.1 Barium sulfate dissolution

Barium sulfate is difficult to remove chemically because of the limited dissolving capacity and slow dissolution rate. The most significant deposits must be mechanically removed. Most proprietary solvents comprise concentrated solutions of EDTA or DTPA raised to high pH by addition of potassium hydroxide (KOH) [203]. Potassium has a lower stability constant with DTPA than sodium does and therefore allows more scale to be dissolved at equilibrium, due to less competition. Others have suggested that the similarity in the ionic radius between Ba^{2+} (0.136 nm) and K^{+} (0.133 nm) facilitates exchange of these two metal ions between the barite surface and the K^{+} ion attached to the EDTA or DTPA moiety. Either way, the potassium salts of these chelants is very much more effective than the corresponding sodium salts.

Catalytic additives are generally included in the formulation to improve the rate of dissolution, which is very slow in their absence. A typical treatment requires 24–48 hours (sometimes up to 72 hours) of contact time with the solvent. Various factors control the performance of barium sulfate scale dissolvers including pH, temperature and the precise nature of these additives. Figs. 5.63

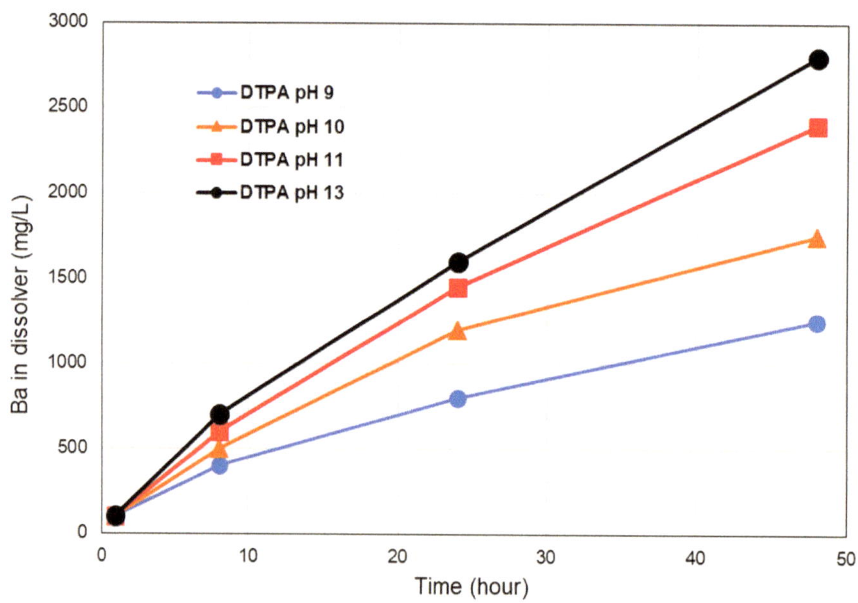

FIGURE 5.63

Effect of application pH on the dissolution of barite with 2.5 wt.% DTPA at 50°C.

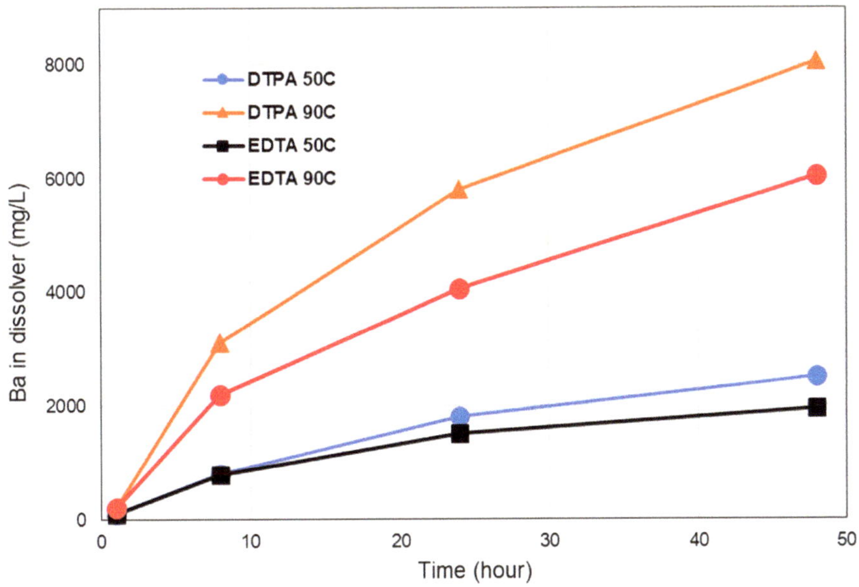

FIGURE 5.64

Effect of temperature on barite dissolution with 10 wt.% DTPA and EDTA (pH 12).

and 5.64 illustrate several such aspects controlling both the rate and dissolution capacity for barite in the absence of catalyst.

Fig. 5.63 shows the important effect of high pH to ensure that all acid groups on the chelating agent are fully deprotonated and allow for stronger complexation reactions with the barium ions. Fig. 5.64 shows improved dissolution at higher temperature and the superiority of DTPA over EDTA at the same chelant concentration. The results also indicate the importance of high temperatures during the scale dissolution process. This means that solvents are unlikely to be effective against barite if applied at temperatures closer to ambient.

5.6.6.2 Chelating agents for sulfide scales

Tetrakis(hydroxymethyl) phosphonium sulfate (THPS) can also be used to dissolve some sulfide scales [204–206]. This chemical is primarily used in the oilfield as an effective biocide, but it can also dissolve certain forms of iron sulfide, zinc sulfide and lead sulfide in particular, when ammonium chloride is added to the solution. Strictly speaking, THPS is not a chelating agent per se, but it interacts with ammonium ions in solution to generate a multidentate organophosphine-based ligand which can chelate the iron, zinc or lead in the scale, thus leading to scale dissolution. The ability to dissolve iron sulfide most likely assists the effectiveness of THPS in breaking down biofilms in which this mineral has been generally formed by microbially induced corrosion. THPS is significantly more expensive than HCl, but it does not release H_2S upon dissolution of the scale and neither is it very corrosive. The combination of THPS with a "salt" (perhaps NH_4Cl also) has been reported to dissolve iron oxide-hydroxides (such as akageneite, FeOOH) and siderite ($FeCO_3$), suggesting this chelating cocktail is a versatile solvent for various scales based on iron, zinc, and lead [207].

5.7 Summary

Inorganic scaling represents one of the most common production chemistry challenges experienced in the oil and gas production industry, representing a threat to production, system integrity and operability, as well as being a health, safety and environment (HSE) hazard. This chapter presents a general overview of this threat: the mechanisms of why and how it forms, and the role of scale prediction in assessing the threat, both initially (at the project phase) and throughout a well and field's lifetime and, therefore, how prediction is used to support creation of an optimal scale-management strategy that minimizes the total cost of operations associated with this threat.

The chapter discusses the various types of inorganic mineral scales encountered in oil and gas production including the carbonate and sulfate scales common to many oil and gas production environments, as well as the less common inorganic deposits. It covers the main mitigation approaches, including both scale inhibition and scale remediation approaches. It describes the types of chemical inhibitors employed and the factors which affect their performance across a range of different inorganic scaling and production conditions. Chemical deployment approaches are summarized, including both continuous injection and squeeze treatments, as well as the use of near-wellbore modeling to optimize such treatments.

In summary, the chapter has aimed to provide an understanding of challenges associated with inorganic mineral scaling in the oil and gas environment and provide an understanding of the principles and mechanisms involved and the common approaches employed for scale control and management.

Nomenclature

AM	acrylamide
AMPS	2-acrylamido-2-methylpropane sulfonic acid
ATPMP	aminotri(methylene phosphonic acid)
bbls	barrels
BHMTPMP	bis-hexamethylenetriamine penta(methylenephosphonic acid)
CT	coiled tubing
DADMAC	diallyldimethylammonium chloride

DDHCI	deep downhole chemical injection
DHCI	downhole chemical injection
DETPMP	diethylenetriamine penta(methylene phosphonic acid)
DHSV	downhole safety valve
DSL	dynamic scale loop
DTPA	diethylenetriamine penta-acetic acid
EDTA	ethylenediamine tetraacetic acid
EDX	energy dispersive x-ray analyzer
EOR	enhanced oil recovery
ES	excess solute
ESEM	environmental scanning electron microscopy
ESP	electrical submersible pump
FW	formation water
HADMP	hydroxyethylamino di(methylenephosphonic acid)
HEDP	hydroxyethylidene diphosphonic acid, etidronic acid
HMDT	hexamethylenediaminetetra(methylenephosphonic acid)
HMDTPA	hexamethylenediamine tetra(methylenephosphonic acid)
HSE	health, safety and environment
HTHP	high temperature/high pressure
ICD	inflow control device
ICP	inductively coupled plasma
ICV	inflow control valve
JIT	jet impingement test
JIP	joint industry project
JT	joule–Thomson
KTT	kinetic turbidity test
LSA	low specific activity
LSSW	low sulfate seawater
m	meters
M	metal ions
mD	milldarcies
Mn	number averaged molecuar weight
Mw	weight averaged molecuar weight
MED	minimum effective doase
MIC	minimum inhibitor concentration
NTA	nitrilotriacetic acid
PAA	poly(acrylic acid)
PASP	poly(aspartic acid)
PDI	polydispersity
PESA	poly(epoxysuccinic acid)
PMA	poly(maleic acid)
ppm	parts per milloin
PPCA	phosphino polycarboxylic acid
PV	pore volume
PVS	poly vinyl sulfonate
PWRI	produced water reinjection
SEM	scanning electron microscopy
SI	scale inhibitor

SR	saturation Ratio
SS	styrene sulfonic acid
SSSV	subsurface safety valve
SW	seawater
TEAPE	triethanolamine phosphate ester
TDS	total dissolved solids
THPS	tetrakis(hydroxymethyl) phosphonium sulfate
TVD	true vertical depth
VS-Co	sulfonated copolymer

References

[1] W.W. Frenier, M. Ziauddin, Formation, Removal and Inhibition of Inorganic Scale in the Oilfield Environment, Society of Petroleum Engineers, Richardson, TX, 2008.

[2] A. Habibi, C.E. Fensky, M. Roostaei, A laboratory workflow for characterisation of scaling deposits in thermal wells, Energies 13 (2020) 3184–3197.

[3] J.C. Cowan, D.J. Weintritt, Water Formed Scale Deposits, Gulf Publication Co., Houston, TX, 1976.

[4] M.A. Kelland, Production Chemicals for the Oil and Gas Industry, CRC Press, Boca Raton, FL, 2009.

[5] G.M. Graham, E.M. Mackay, S.J. Dyer, et al., The challenges of scale control in deepwater production systems, NACE-02316, CORROSION 2002, 2002, April 7–11, 2002, Houston, TX, USA.

[6] I.R. Collins, Scale management and risk assessment for deepwater developments, World Oil 224 (2003) 62–69.

[7] E.J. Mackay, M.M. Jordan, N.D. Feasey, et al., Integrated risk analysis for scale management in deepwater developments, SPE Production & Operations 20 (2005) 138–154.

[8] G.M. Graham, I.R. Collins, T.L. Johnson, Technical and economic analysis of the scale risks and uncertainties for subsea deepwater field developments, SPE-95049-MS, SPE International Symposium on Oilfield Scale, May 11–12, 2005, Aberdeen, UK.

[9] I.R. Collins, R. Stalker, G.M. Graham, Sulphate removal for barium sulphate scale mitigation a deepwater subsea production system, SPE-87465-MS, SPE International Symposium on Oilfield Scale, May 26–27, 2004, Aberdeen, UK.

[10] P.J. Bogaert, M.C. Berredo, C. Toschi, et al., Scale inhibitor squeeze treatments deployed from an FPSO in a deepwater, subsea field in the Campos Basin, SPE-102503-MS, SPE Annual Technical Conference and Exhibition, September 24–27, 2006, San Antonio, TX, USA.

[11] M. Jordan, E. MacKay, Predicting and managing inorganic scale associated with produced water from EOR projects, SPE-177144-MS, SPE Latin American and Caribbean Petroleum Engineering Conference, November 18–20, 2015, Quito, Ecuador.

[12] G.M. Graham, D.F. Frigo, Production chemistry issues and solutions associated with chemical EOR, SPE-193568-MS, April 8–9, 2019, Galveston, TX, USA.

[13] R.A. Davis, J.E. McElhiney, The advancement of sulfate removal from seawater in offshore waterflood operations, NACE-02314, CORROSION 2002, April 7–11, 2002, Denver, CO, USA.

[14] J.E. McElhiney, Deepwater project economics demand sulfate removal to ensure scale-free operation, Offshore Magazine, May 2003.

[15] R. Wat, A.M.B. Iversen, Y.H. Belsvik, The diagnosis of decline and successful recovery of PI in a "dry" HPHT gas well that has been affected by in-situ $CaCO_3$ and salt deposition, SPE-130521-MS, SPE International Conference on Oilfield Scale, May 26–27, 2010, Aberdeen, UK.
[16] C.F. Smith, T.J. Nolan, P.L. Crenshaw, Removal and inhibition of calcium sulphate scale in waterflood projects, Journal of Petroleum Technology 20 (1968) 1249–1256.
[17] S. Maxwell, I. Spark, Souring of reservoirs by bacterial activity during seawater flooding, SPE-92231-MS, SPE International Symposium on Oilfield Chemistry, February 2–4, 2005, The Woodlands, TX, USA.
[18] D. Rickard, D.W. Luther, Chemistry of iron sulfides, Chemical Reviews 107 (2007) 514–562.
[19] Q. Wang, J. Leal, I. Syafii, et al., Iron sulfide scale formation and removal in sour gas wells. SPE-179869-MS, SPE International Oilfield Scale Conference, May 11–12, 2016, Aberdeen, UK.
[20] H.A. Nasr-El-Din, A.Y. Al-Humaidan, Iron sulfide scale: Formation, removal and prevention, SPE-68315-MS, in International Symposium on Oilfield Scale, January 30–31, 2001, Aberdeen, UK.
[21] R. Walker, A.D. Steele, D.T.B. Morgan, Deactivation of pyrophoric iron sulfides, Industrial & Engineering Chemistry Research 36 (1997) 3662–3667.
[22] K.T. Ly, D.J. Blumer, W.M. Bohon, et al., Novel chemical dispersant for removal of organic/inorganic "schmoo" scale in produced water injection systems, NACE-98073, CORROSION 98, March 22–27, 1998, San Diego, CA, USA.
[23] A.F. Clemmit, D.C. Ballance, A.G. Hunton, The dissolution of scales in oilfield systems, SPE-14010-MS, SPE Offshore Europe, September 10–13, 1985, Aberdeen, UK, SPE-14010 (1985).
[24] M. Cowie, K. Mously, NORM management in the oil & gas industry, SPE-111842-MS, SPE International Conference on Health, Safety, and Environment in Oil and Gas Exploration and Production, April 15–17, 2008, Nice, France, SPE-111842 (2008).
[25] K.A. Mously, J.A. Campbell, M. Cowie, The International Association of Oil & Gas Producers (OGP) naturally occurring radioactive material (norm) management guideline, SPE-123482-MS, Asia Pacific Health, Safety, Security and Environment Conference, August 4-6, 2009, Jakarta, Indonesia, SPE-123482 (2009).
[26] E. Sorhaug, G.M. Graham, R. Stalker, et al., Scale challenges in Talisman's Varg Field, SPE-114077-MS, SPE International Oilfield Scale Conference, May 28–29, 2008, Aberdeen, UK.
[27] E.A. Warren, P.C. Smalley, North Sea Formation Waters Atlas: No. 15, Geological Society Memoir, Geological Society, London, 1994.
[28] K. Raju, H.A. Nasr-El-Din, Calcium sulfate scale: Field tests and model predictions, NACE-04397, CORROSION 2004, March 28-April 1, 2004, New Orleans, LA, USA.
[29] A.H. Aouda, A.A. Monem, T.A. Daghar, et al., Identifying the cause of unexpected solids deposition from a sodium bromide completion brine in a giant field, Offshore Abu Dhabi, SPE-183073-MS, Abu Dhabi International Petroleum Exhibition & Conference, November 7–10, 2016, Abu Dhabi, UAE.
[30] D. Shen, D. Shcolnik, R. Perkins, et al., Evaluation of scale inhibitors in Marcellus high-iron waters, Oil and Gas Facilities 1 (2012) 34–42.
[31] J.J. Wilde, J.L. Slayerm B. Frehlick, An exhaustive study of scaling in the Canadian Bakken: Failure mechanisms and innovative mitigation strategies from over 400

wells, SPE-153005-MS, SPE International Conference on Oilfield Scale, May 30–31, 2012, Aberdeen, UK.

[32] W.K. Burton, N. Cabrera, F.C. Frank, The growth of crystals and the equilibrium structure of their surfaces, Philosophical Transactions of the Royal Society of London. Series A 243 (1951) 299–358.

[33] P. Hartman, W.G. Perdok, On the relations between structure and morphology of crystals, ActaCrystallographica 8 (1955) 521–529.

[34] T. Ostvold, P. Rhandhol, Kinetics of $CaCO_3$ scale formation. The influence of temperature, supersaturation and ionic composition, SPE-68302-MS, SPE International Symposium on Oilfield Scale, January 30–31, 2001, Aberdeen, UK.

[35] L.R. Collins, Predicting the location of barium sulfate scale formation in production systems, SPE-94366-MS, SPE Oilfield Chemistry Symposium, May 11–12, 2005, Aberdeen, UK.

[36] L. Sutherland, C. Johnson, W. Taylor, et al., The influence of turbulence (or hydrodynamic effects of barium sulphate scale formation and inhibitor performance, SPE-164070-MS, SPE International Symposium on Oilfield Chemisry, April 8–10, 2013, The Woodlands, TX, USA.

[37] N. Akram, S. Hicking, P. Blythe, et al., Intelligent well technology in mature assets, SPE-71822-MS, SPE Offshore Europe Oil and Gas Exhibition and Conference, September 4–7, 2001, Aberdeen, UK.

[38] V.M. Kavle, S.M. Elmsallati, E.J. Mackay, et al., Impact of intelligent wells on scale management, SPE-100112-MS, SPE Europec/EAGE Annual Conference and Exhibition, June 12–15, 2006, Vienna, Austria.

[39] Ø. Gustavsen, O.M. Selle, F.H. Fadnes, et al., Inflow-control-device completion in a scaling environment: Findings and experiences obtained during production logging in the Heidrun Field, PE-134994-MS, SPE Annual Technical Conference and Exhibition, September 19–22, 2010, Florence, Italy.

[40] L.O. Jøsang, M.N. Psarrou, T. Østvold, et al., Scale under turbulent flow, Proceedings of the Tekna Oil Field Chemistry Symposium, March 18–21, 2012, Geilo, Norway.

[41] J.J. Wylde, D. Fell, Scale inhibitor solutions for high temperature ESP lifted wells in Northern California: A case history of failure followed by success, SPE-113826-MS, SPE Oilfield Scale Symposium, May 28–29, 2008, Aberdeen, UK.

[42] A. Fyfe, D. Nichols, M. Jordan, Implication of turbulent flow induced by gas lift on strontium sulphate scale formation and control within production tubing, SPE-204342-MS, SPE International Conference on Oilfield Chemistry, December 6–7, 2021, The Woodlands, TX, USA.

[43] N. Goodwin, M. May, D. Nichols, et al., Scale deposition and hydrodynamics - Benchtop to pilot rig, SPE-190756-MS, SPE International Oilfield Scale Conference, June 20–21, 2018, Aberdeen, UK.

[44] D.A. Nichols, D.M. Frigo, G.M. Graham, Advances in understanding effects of shear and turbulence on scale formation, adhesion and growth on surfaces, SPE-179903-MS, SPE International Oilfield Scale Conference and Exhibition, May 11–12, 2016, Aberdeen, UK.

[45] D.A. Nichols, H. Thomson, C. Simpson, et al., Scale and corrosion inhibitor evaluation under high shear flow conditions, SPE-184592-MS, SPE International Conference on Oilfield Chemistry, April 3–5, 2017, Montgomery, TX, USA.

[46] N. Goodwin, M. May, A. Kaur, et al., Pilot rig scale testing: Hydrodynamics and scale deposition, NACE-2021–16699, CORROSION 2021, April 19–30, 2021, Virtual.

[47] C. Simpson, G.M. Graham, I.R. Collins, et al., Sulphate removal for barium sulphate mitigation - Kinetic vs. thermodynamic controls in mildly oversaturated conditions, SPE-95082-MS, SPE International Symposium on Oilfield Scale, 2 May 11–12, 2005, Aberdeen, UK.

[48] G.M. Graham, H. Thomson, D. Bowering, et al., Correlation of shear and turbulence on scale formation and inhibition, SPE-169761-MS, SPE International Oilfield Scale Conference, May 14–15 (2014).

[49] P.J. Webb, O. Kuhn, Enhanced scale management through the application of inorganic geochemistry, SPE-87458, SPE International Symposium on Oilfield Scale, May 26–27, 2004, Aberdeen, UK.

[50] G.M. Graham, I.R. Collins, Assessing scale risks and uncertainties for subsea marginal field developments, SPE 84760, SPE International Symposium on Oilfield Scale, May 26–27, 2004, Aberdeen, UK.

[51] D. Nichols, N. Goodwin, G.M. Graham, et al., Scale predictions and mineral solubility under HPHT conditions, SPE-193564-MS, SPE International Conference on Oilfield Chemistry, April 8–9, 2019, Galveston, TX, USA.

[52] N. Goodwin, G.M. Graham, D. M. Frigo, Ultra HPHT solubility of calcium sulphate and other minerals, SPE-184568-MS, SPE International Conference on Oilfield Chemistry, April 3–5, 2017, Montgomery, TX, USA.

[53] Z. Dai, W. Shi, A.T. Kan, et al., Improvement of thermodynamic modelling of calcium carbonate and calcium sulfates at high temperature and high pressure in mixed electrolytes, SPE-169786-MS, SPE International Oilfield Scale Conference, May 14–15, 2014, Aberdeen, UK.

[54] G. Deng, A.T. Kan, Z. Dai, et al., Impact of high calcium concentration on sulfate scale prediction at high temperature from 120°C to 220°C, SPE-190744-MS, SPE International Oilfield Scale Conference, June 20–21, 2018, Aberdeen, UK.

[55] C.J. Hinrischen, Preventing scale deposition in oil production facilities: An industry review, NACE-98061, CORROSION 98, March 22–27, 1998, San Diego, CA, USA.

[56] M.M. Jordan, K. Sjuraether, I.R. Collins, et al., Life cycle management of scale control within subsea fields and its impact on flow assurance, Gulf of Mexico and the North Sea Basin, SPE-71557-MS, SPE Annual Technical Conference and Exhibition, September 30-October 3, 2001, New Orleans, LA, USA.

[57] N. Fleming, K. Ramstad, S.H. Eriksen, et al., Development and implementation of a scale management strategy for Oseberg Sør, SPE Production & Operations 22 (2008) 307–317.

[58] E.J. Mackay, I.R. Collins, M.M. Jordan et al., PWRI: Scale formation risk assessment and management, SPE-80385-MS, International Symposium on Oilfield Scale, January 29–30, 2003, Aberdeen, UK.

[59] N. Ostojic, S. Dyer, G.M. Graham, et al., Commingled PW/SW injection in Janice field - Overcoming late life scaling challenges and maximising recovery, Proceedings of the Tekna Oil Field Chemistry Symposium, March 23–26, 2014, Geilo, Norway.

[60] J. Paulo, E.J. Mackay, N. Menzies, et al., Implications of brine mixing in the reservoir for scale management in the Alba field, SPE-68310-MS, International Symposium on Oilfield Scale, January 30–31, 2001, Aberdeen, UK.

[61] R. Wright, R.A. McCartney, E. Sorhaug, Predicting the sulphate content of produced water in reservoirs under seawater flood and containing calcium-rich formation water, SPE-113974-MS, SPE International Oilfield Scale Conference, May 28–29, 2008, Aberdeen, UK.

[62] I.R. Collins, A new model for mineral scale adhesion, SPE-74655-MS, SPE International Symposium on Oilfield Scale, January 30–31, 2002, Aberdeen, UK.

[63] S. Baraka-Lokmane, T.V.J. Charpentier, A. Neville, et al., Comparison of characteristic of anti-scaling coating for subsurface safety valve for use in oil and gas industry, IPTC-17953-MS, International Petroleum Technology Conference, December 10–12, 2014, Kuala Lumpur, Malaysia.

[64] G.M. Graham, C.P. McMahon, The effect of scale inhibitor performance against bulk (homogeneous) and surface (heterogeneous) scale nucleation and growth by the addition of film forming corrosion inhibitors, NACE-02315, CORROSION 2002, April 7–11, 2002, Denver, CO, USA.

[65] W.C. Cheong, A. Neville, P. Gaskell, et al., Using nature to provide solutions to calcareous scale deposition, SPE-114082-MS, SPE International Oilfield Scale Conference, May 28–29 (2008).

[66] G.M. Graham, N. Goodwin, E. Albino, et al., Minimizing scale deposition through surface enhancement in downhole tools, SPE-169781-MS, International Oilfield Scale Conference, May 14–15, 2014, Aberdeen, UK.

[67] C. Gabrielli, R. Jaouhari, G. Maurin, et al., Magnetic treatment for scale prevention, Water Research 35 (2001) 3249–3259.

[68] C. Piyadasa, H.F. Ridgway, T.R. Yeager, et al., The application of electromagnetic fields to the control of the scaling and biofouling of reverse osmosis membranes - A review, Desalination 418 (2017) 19–34.

[69] L. Lin, W. Jiang, X. Xu, et al., A critical review of the application of electromagnetic fields for scaling control in water systems: Mechanisms, characterization and operation, npj Clean Water 3 (2020) 25.

[70] M.A. Singleton, I.R. Davis, Experimental investigation of the impact of electromagnetic devices on barium sulphate scaling, SPE-200664-MS, SPE International Oilfield Scale Conference, June 24–25, 2020, Virtual.

[71] G.M. Graham, I.C. Collins, R. Stalker, et al., The importance of appropriate laboratory procedures for the determination of scale inhibitor performance, SPE-74679-MS, SPE International Symposium on Oilfield Scale, January 30–31 (2002).

[72] G.M. Graham, L.S. Boak, K.S. Sorbie, The influence of formation calcium and magnesium on the effectiveness of generically different barium sulphate oilfield scale inhibitors, SPE Production & Operations 18 (2003) 28–44.

[73] O.J. Vetter, An evaluation of scale inhibitors, Journal of Petroleum Technology 24 (1972) 997–1006.

[74] G.R. Chesnut, G.D. Chappell, D.H. Emmons, The development of scale inhibitors and inhibitor evaluation techniques for carbon dioxide EOR floods, SPE-16260-MS, SPE International Symposium on Oilfield Chemistry, February 4–6, 1987, San Antonio, TX, USA.

[75] K.S. Sorbie, N. Laing, How scale inhibitors work: Mechanisms of selected barium sulphate scale inhibitors across a wide temperature range, SPE-87470-MS, SPE International Symposium on Oilfield Scale, May 26–27, 2004, Aberdeen, UK.

[76] J.S. Gill, R.G. Varsanik, Computer modeling of the specific matching between scale inhibitors and crystal structure of scale forming minerals, Journal of Crystal Growth 76 (1986) 57–62.

[77] W.H. Leung, G.H. Nancollas, Nitrilo (methylene phosphonic acid) adsorption on barium sulphate crystals and its influence on crystal growth, Journal of Crystal Growth 44 (1978) 163–167.

[78] M.C. Van der Leeden, J. ReedijK, G.M. Van Rosmalen, The influence of various phosphonates on the growth rate of barium sulphate crystals in suspension, Stu Geol 38 (1982) 279–287.

[79] J.S. Gill, C.D. Anderson, R.G. Varsanik, Mechanism of scale inhibition by phosphonates, IWC-83-4, 44th International Water Conference, October 24-26, 1983, Pittsburgh, PA, USA, IWC-83-4, 44th International Water Conference, (1983).

[80] G.H. Nancollas, Oilfield scale: Physical chemical studies of its formation and prevention, March 26-27, 1985, London, UK, Proceedings of the RSC Chemistry in The Oil Industry (1985).

[81] Z. Berkovitch-Yellin, J.V. Mil, L. Addadi, et al., Crystal morphology engineering by "tailor-made" inhibitors: A new probe to fine intermolecular interactions, J Am Chem Soc, 107 (1985) 3111–3122.

[82] M.P.C. Weijnen, G.M. Van Rosmalen, M.C. Van der Leeden, Influence of the molecular structure of phosphonate inhibitors on various aspects of barite and gypsum crystallization, March 16-22, 1986, Granada, Spain, Proceedings of the International Meeting on Geochemistry of the Earth Surface and Processes of Mineral Formation (1986).

[83] P.J. Breen, B.N. Diel, H.H. Downs, Correlation of scale inhibitor with adsorption thermodynamics and performance in inhibition of barium sulphate in low-pH environments SPE-20688-MS, SPE Annual Technical Conference and Exhibition, September 23–26, 1990, New Orleans, LA, USA, SPE-20688 (1980).

[84] M.C. Van der Leeden, G.M. Van Rosmalen, Inhibition of barium sulphate scale deposition by polycarboxylates of various molecular structure, SPE Production Engineering 5 (1990) 70–76.

[85] P.J. Breen, H.H. Downs, B.N. Diel, The use of adsorption thermodynamics in the development of scale inhibitors for high Barium oilfield brines, April 8-11, London, UK, Proceedings of the RSC Chemistry in The Oil Industry, (1991).

[86] G.M. Graham, A Mechanistic Examination of the Factors Influencing Downhole $BaSO_4$ Oilfield Scale Inhibitors and the Design of New Species, PhD Dissertation, Heriot-Watt University, 1994.

[87] M.B. Tomson, G. Fu, M.A. Watson, et al., Mechanisms Of mineral scale inhibition, SPE Production and Operations 18 (2003) 192–199.

[88] A.Y. Lu, W. Shi, J. Wang, et al., The mechanism of barium sulfate deposition inhibition and the prediction of inhibitor dosage, Journal of Chemical & Engineering Data 64 (2019) 4968–4976.

[89] A. Hiorth, L.M. Cathles, M.V. Madland, The impact of pore water chemistry on carbonate surface charge and oil wettability, Transport in Porous Media 85 (2010) 1–21.

[90] I.R. Collins, Surface electrical properties of barium sulphate modified by adsorption of poly alpha, beta aspartic acid, Journal of Colloid and Interface Science 212 (1999) 535–544.

[91] A.J.B. Hennessy, G.M. Graham, The effect of additives on the co-crystallisation of calcium with barium sulphate, Journal of Crystal Growth, 237– 239 (2002) 2153–2159.

[92] S.S. Shaw, K.S. Sorbie, L.S. Boak, The effects of barium sulfate saturation raio, calcium and magnesium on the inhibition efficiency-Part1: Phosphonate scale inhibitors, SPE Production & Operations 27 (2012) 306–317.
[93] Fazrie Wahid, Assessing the Controlling Effect of Divalent Cations on DETPMP Inhibition of Barite Scale via Molecular Modelling, PhD Thesis, Heriot Watt University, 2004.
[94] B.G. Kriel, C.A. Lacey, R.H. Lane, The performance of scale inhibitors in the inhibition of iron carbonate scale, SPE-27390-MS, SPE Formation Damage Control Symposium, February 7–10, 1994, Lafayette, LA, USA.
[95] G.M. Graham, R. Stalker, R. McIntosh, The impact of dissolved iron on the performance of scale inhibitors under carbonate scaling conditions, SPE-80254-MS, SPE International Symposium on Oilfield Chemistry, February 5–7, 2003, Houston, TX, USA.
[96] A.T. Tan, Z. Dai, M.B. Tomson, The state of the art in scale inhibitor squeeze treatments, Petroleum Science 17 (2020) 1579–1601.
[97] A.T. Kan, G. Fu, H.A. Al-Saiari, et al., Enhanced scale-inhibitor treatments with the addition of zinc, SPE Journal 14 (2009) 617–626.
[98] H. Montgomerie, P. Chen, T. Hagen, et al., Development of a new polymer inhibitor chemistry for downhole squeeze applications, SPE Production & Operations 24 (2009) 459–564.
[99] S. Baraka-Lokmane, C. Hurtevent, C. Tillement, et al., Development and qualification of new zinc and lead sulphide scale inhibitors for application under harsh conditions, SPE-173761-MS, SPE International Symposium on Oilfield Chemistry, April 13–15, 2015, The Woodlands, TX, USA.
[100] S. Baraka-Lokmane, C. Hurtevent, M. Rossiter, et al., Design and performance of novel sulphide nanoparticle scale inhibitors for North Sea HP/HT fields, SPE-179866-MS, SPE International Oilfield Scale Conference and Exhibition, May 11–12, 2016, Aberdeen, UK.
[101] B.G. Al-Harbi, A.J. Graham, K.S. Sorbie, Laboratory investigation of zinc and lead sulphide inhibition, SPE-189572-MS, SPE International Conference and Exhibition on Formation Damage Control, February 7–9, 2018, Lafayette, LA, USA.
[102] C. Tortolano, T. Chen, P. Chen, et al., Mechanisms, new test methodology and environmentally acceptable inhibitors for co-deposition of zinc sulfide and calcium carbonate scales for HTHP application, SPE-169810-MS, SPE International Oilfield Scale Conference and Exhibition, May 14–15, 2014, Aberdeen, UK.
[103] G.M. Graham, S.J. Dyer, K.S. Sorbie, et al., Scale inhibitor selection for continious and downhole squeeze application in HP/HT conditions, SPE-47197-MS, SPE Annual Technical Conference and Exhibition, September 27–30, 1998, New Orleans, LA, USA.
[104] G.M. Graham, S.J. Dyer, P. Shone, Potential application of amine methylene phosphonate based inhibitor species in HP/HT environments for improved carbonate scale inhibitor performance, International Symposium on Oilfield Scale 17 (2002) 212–220.
[105] G.M. Graham, S.J. Dyer, P. Shone, et al., Comparative core flooding and field application implications for scale inhibitor downhole squeeze treatments in high temperature (374F) reservoirs, SPE-65008-MS, SPE International Symposium on Oilfield Chemistry, February 13–16, 2001, Houston, TX, USA.

[106] R. Wat, L. Hauge, K. Solbakken, et al., Squeeze chemical for HT applications-Have we discarded promising products by performing unrepresentative thermal-aging tests? SPE Production & Operations 23 (2008) 334–342.

[107] G.M. Graham, I. Munro, N. Harvison, et al., Improved scale inhibitor assay for sulphonated polymers in oilfield brines, SPE-131100-MS, SPE International Conference on Oilfield Scale, May 26–27, 2010, Aberdeen, UK.

[108] S.J. Dyer, C.F. Anderson, G.M. Graham, Thermal stability of amine methyl phosphonate scale inhibitors, Journal of Petroleum Science and Engineering 43 (2004) 259–270.

[109] G.M. Graham, M.M. Jordan, G.C. Graham, et al., The implication of HP/HT reservoir conditions on the Selection and Application of Conventional Scale Inhibitors: Thermal stability studies, SPE-37274-MS, International Symposium on Oilfield Chemistry, February 18–21, Houston, TX, USA.

[110] W. Wang, A.T. Kan, M.B. Tomson, A novel and comprehensive study of polymeric and traditional phosphonate inhibitors for high temperature scale control, SPE-155108-MS, SPE International Conference on Oilfield Scale, May 30–31, 2012, Aberdeen, UK.

[111] G.M. Graham, S.J. Dyer, P. Shone, High temperature core flooding experiments for the selection of appropriate potential application of amine methylene phosphonate based inhibitor species in HP/HT environments for improved carbonate scale inhibitor performance, SPE-60217-MS, SPE International Symposium on Oilfield Chemistry, January 26–27, 2000, Aberdeen, UK.

[112] G.M. Graham, S.J. Dyer, K.S. Sorbie, et al., Scale inhibitor selection for continuous and downhole squeeze application in HP/HT conditions, SPE-49197-MS, SPE Annual Technical Conference and Exhibition, September 27–30, 1998, New Orleans, LA, USA.

[113] S.J. Dyer, G.M. Graham, K.S. Sorbie, Factors affecting the thermal stability of conventional scale inhibitors for application in high pressure/high temperature reservoirs SPE-50717-MS, SPE International Symposium on Oilfield Chemistry, February 16–19, 1999, The Woodlands, TX, USA.

[114] N.J. Goodwin, O.G. Svela, J.H. Olsen, et al., Qualification procedure for continuous injection of chemicals in the well -method development, SPE-154934-MS, SPE International Conference on Oilfield Scale, May 30–31, 2012, Aberdeen, UK.

[115] API, Technical Report 17TR6. Attributes of Production Chemicals in Subsea Production Systems, American Petroleum Institute, Washington, DC, 2012.

[116] A.M. Fitzgerald, L.G. Cowie, A history of frac-pack scale-inhibitor deployment, SPE-112474-MS, SPE International Symposium and Exhibition on Formation Damage Control, February 13–15, 2008, Lafayette, LA, USA.

[117] R.J. Powell, R.D. Gdanski, M.A. McCabe, et al., Controlled-release scale inhibitor for use in fracturing treatments, SPE28999-MS, SPE International Symposium on Oilfield Chemistry, February 14–17, 1995, San Antonio, TX, USA.

[118] J. Leasure, T. Lieng, B. Hicks, et al., Multi-year scale inhibition using a ceramic proppant based particulate chemical carrier-Learning from the first application Offshore Congo, SPE-200686-MS, SPE International Oilfield Scale Conference, June 24–25, 2020, Virtual.

[119] J.J. Wylde, G.D.M. Williams, F. Careil, et al., Deep downhole chemical injection on BP-operated miller: Experience and learning, SPE Production & Operations 21 (2006) 188–193.

[120] E.J. Mackay, Scale inhibitor application in injection wells to protect against damage to production wells: When does it work?, SPE-95022-MS, in SPE European Formation Damage Conference, May 25–27, 2005, Sheveningen, The Netherlands.

[121] V. Ragulin, A. Mikhailov, O. Latipov, et al., Scale management of production wells via inhibitor application in supporting injection wells, SPE-87461-MS, SPE International Symposium on Oilfield Scale, May 26–27, 2004, Aberdeen, UK.

[122] H.M. Bourne, S.M. Heath, S. McKay, et al., Effective treatment of subsea wells with a solid scale inhibitor system, SPE International Symposium on Oilfield Scale, January 26–27, 2000, Aberdeen, UK.

[123] J.F. Hsu, A.K. Al-Zain, K.U. Raju, et al., Encapsulated scale inhibitor treatments experience in the Ghawar Field, Saudi Arabia, SPE-60209-MS, SPE International Symposium on Oilfield Scale, January 26–27, 2000, Aberdeen, UK.

[124] G.M. Graham, L.S. Boak, C.M. Hobden, Examination of the effect of generically different scale inhibitor species (PPCA and DETPMP) on the adherence and growth of barium sulphate scale on metal surfaces, SPE-68298-MS, SPE International Symposium on Oilfield Scale, January 30–31, 2001, Aberdeen, UK.

[125] E.J. Hills, G.M. Graham, B. Langlois, et al., Molecular weight and compositional controls on the performance of polymeric scale, Proceedings of the TEKNA International Oil Field Chemistry Symposium, March 13–16, 2005, Geilo, Norway.

[126] G.M. Graham, K.S. Sorbie, Examination of the change in returning molecular weight obtained during inhibitor squeeze treatments using polyacrylate based inhibitors, SPE-29000-MS, SPE International Symposium on Oilfield Chemistry, February 14–17, 1995, San Antonio, TX, USA.

[127] G.M. Graham, K.S. Sorbie, The effect of molecular weight on the adsorption/desorption characteristics of polymeric scale inhibitors on silica sand and sandstone core, NACE-94050, Corrosion/94, February 28 - March 4, 1994, Baltimore, MD, USA.

[128] T. Johnson, C. Roggelin, C.M. Simpson, et al., Phosphonate based scale inhibitors for high iron and high salinity environments, Proceedings of the RSC Chemistry in The Oil Industry, October 31 – November 2, 2005, Manchester, UK.

[129] G.M. Graham, C.S.A. MacPherson, C. Simpson, et al., Development of appropriate test methodologies for the selection and application of lead and zinc sulphide inhibitors for the Elgin/Franklin fields, SPE-100627-MS, SPE International Oilfield Scale Symposium, May 31-June 1, 2006, Aberdeen, UK.

[130] G.M. Graham, H.L. Williams, R. Wat, et al., Design and application of a novel HT/HP “stirred reactor” test rig to study scale formation and control, SPE-93425-MS, SPE International Symposium on Oilfield Chemistry, February 2–4, 2005, The Woodlands, TX, USA.

[131] K. Sjursaether, S. Bernsten, J. Davidson, et al., The influence of entrained solides, turbulence and scale inhibitonr concentrations on the kinetics of barium sulfate scale formation in moderate scaling regims, Proceedings of the Tekna Oilfield Chemistry Symposium, March 13–15, 2010, Geilo, Norway.

[132] NACE, TM0197: Laboratory screening test to determine the ability of scale inhibitors to prevent the precipitation of barium sulfate or strontium sulfate, or both, from solution (for oil and gas production systems), NACE, Houston, TX, 1997.

[133] S. Heath, A novel laboratory approach to test scale inhibitor efficiency and correlate mechanistic actions with functional chemistry in brines with low to mild barium

sulphate scaling tendency, Proceedings of the RSC Chemistry in the Oil Industry, November 4–6, 2019, Manchester, UK.

[134] M. Callum, Mechanistic investigations into synergistic interactions of polymeric and phosphorous containing scale inhibitors for inorganic scale control within oil-field process equipment, Proceedings of the RSC Chemistry in the Oil Industry, November 4–6, 2019, Manchester, UK.

[135] I. Aarag, K.F.S. Alnes, Testing of scale inhibitor efficiency in a pre-scaled environment, Proceedings of the Tekna Oil Field Chemistry Symposium, March 22–24, 2015, Geilo, Norway.

[136] H. Lu, J. Penkala, J. Brooks, et al., Novel laboratory and field approaches to control halite and other problematic scales in high salinity brine from Bakken, SPE-174788-MS, SPE Annual Technical Conference and Exhibition, September 28–30, 2015, Houston, TX, USA.

[137] S. Paudyal, G. Ruan, J. Lee, et al., Development of dynamic tube blocking test method to study halite scale deposition and inhibition, SPE-204389-MS, SPE International Conference on Oilfield Chemistry, December 6–7, 2021, The Woodlands, TX, USA.

[138] J. Kerr, The evaluation and development of exotic scale dispersants using kinetic turbidity testing, Proceedings of the Tekna Oil Field Chemistry Symposium, March 26–29, 2017, Geilo, Norway.

[139] H. Lu, J. Brooks, R. Legan, et al., Novel laboratory test method and field applications for silica/silicate and other problematic scale control, SPE-190705-MS, SPE International Oilfield Scale Conference, June 20–21, 2018, Aberdeen, UK.

[140] T.D. Baugh, J. Lee, K. Winters, et al., A fast and information-rich test method for scale inhibitor performance, OTC-23150-MS, Offshore Technology Conference, April 30 - May 3, 2012, Houston, TX, USA.

[141] G.M. Graham, S. Kidd, R. Fitzpatrick, et al., Appropriate simulation of field treatments using core flood data, SPE-131131-MS, SPE International Conference on Oilfield Scale, May 26–27, 2010, Aberdeen, UK.

[142] G.M. Graham, R. Stalker, L.W. Hoeth, et al., Effective laboratory simulation of scale inhibitor squeeze treatments - Core to field II, SPE-157105-MS, SPE International Conference on Oilfield Scale, May 30–31, 2012, Aberdeen, UK.

[143] I. Mokogwu, S. Wilson, G.M. Graham, Formation damage vs chemical performance - Important aspects for upscaling scale inhibitor squeeze core floods for carbonate systems, SPE-189560-MS, SPE International Conference and Exhibition on Formation Damage Control, February 7–9, 2018, Lafayette, LA, USA.

[144] F.H. Browning, H.S. Fogler, Precipiation and dissolution of calcium phosphonates for the enhancement of squeeze lifetimes, SPE Production & Operations 10 (1995) 144–150.

[145] I.R. Collins, L.G. Cowie, M. Nicol, et al., The field application of a scale inhibitor squeeze enhancing additive, SPE-38765-MS, SPE Annual Technical Conference, October 5–8, 1997, San Antonio, TX, USA.

[146] I.R. Collins, J.A. Hewartson, Extending squeeze lifetimes using miscible displacement, SPE-74650-MS, SPE International Symposium on Oilfield Scale, January 30–31, 2002, Aberdeen, UK.

[147] K.S. Sorbie, R.D. Gdanski, A complete theory of scale-inhibitor transport and adsorption/desorption in squeeze treatments, SPE-95088-MS, SPE International Symposium on Oilfield Scale, May 11–12, 2005, Aberdeen, UK.

[148] J.S. James, D. Frigo, M.M. Townsend, et al., Application of a fully viscosified scale squeeze for improved placement in horizontal wells, SPE-94593-MS, International Symposium on Oilfield Scale, May 11–12, 2005, Aberdeen, UK.

[149] A. Kuar, V.E. Spooner, D. Frigo, et al., Modelling squeeze treatments in fractured systems - A case history, SPE-179887-MS, SPE International Oilfield Scale Conference, May 11–12, 2016, Aberdeen, UK.

[150] O.M. Selle, M. Springer, I.H. Auflem, et al., Gelled scale inhibitor treatment for improved placement in long, horizontal wells at Norne & Heidrun fields, SPE Production & Operations 24 (2009) 425–438.

[151] E.J. Mackay, K.S. Sorbie, Modelling scale inhibitor squeeze treatments in high crossflow horizontal wells, SPE-50418-MS, SPE International Conference on Horizontal Well Technology, November 1–4, 1998, Calgary, Canada.

[152] E.M. Mackay, A.P. Matharu, K.S. Sorbie, et al., Modelling of scale inhibitor treatments in horizontal wells: Application to the Alba Field, SPE International Symposium on Formation Damage 15 (2000) 107–114.

[153] H.A. Nasr-EL-Din, M.K. Saiari, G. Hashem, Optimization of a novel emulsified scale inhibitor system to mitigate calcium carbonate scale in a sandstone reservoir in Saudi Arabia, SPE-80389-MS, International Symposium on Oilfield Scale, January 29–30, 2003, Aberdeen, UK.

[154] O.M. Selle, O.K. Nygard, E. Storas, et al., Foamed scale inhibitor and scale dissolver treatments for improved placement in a subsea Norne well, SPE-155407-MS, SPE International Conference on Oilfield Scale, May 30–31, 2012, Aberdeen, UK.

[155] R. Wat, O.M. Selle, H. Borstad, et al., Scale inhibitor squeeze treatments strategy on Heidrun, SPE-68944-MS, SPE European Formation Damage Conference, May 21–22, 2001, The Hague, Netherlands.

[156] M.A. Tomsom, A.T. Tan, G. Fu, Control of inhibitor squeeze via mechanistic understanding of inhibitor chemistry, SPE-87450-MS, SPE International Symposium on Oilfield Scale, May 26–27, 2004, Aberdeen, UK.

[157] M.J. Todd, A.J. Savin, K.S. Sorbie, Dual chelant mechanism for the deployment of scale inhibitors in controlled solubility/precipitation treatments, SPE-152870-MS, SPE international Conference on Oilfield Scale, May 30–31, 2012, Aberdeen, UK.

[158] C. Sitz, J. Hardy, J. Merryman, et al., A new class of non-damaging, environmentally responsible scale squeeze enhancement aids, SPE-155254-MS, SPE International Conference on Oilfield Scale, May 30–31, 2012, Aberdeen, UK.

[159] R.N. Smith, T.A. Lawless, H.M. Bouren, et al., Planning and execution of a field trial utilising new invert emulsion squeeze technology, SPE-60210-MS, International Symposium on Oilfield Scale, January 26–27, 2000, Aberdeen, UK.

[160] K.S. Sorbie, M. Yuan, M.M. Jordan, Application of a scale inhibitor squeeze model to improve field squeeze treatment design, SPE-28885-MS, European Petroleum Conference, October 25–27, 1994, London, UK.

[161] K.S. Sorbie, A general coupled kinetic adsorption/precipitation transport model for scale inhibitor retention in porous media: I. Model formulation, SPE-130702-MS, SPE International Conference on Oilfield Scale, May 26–27, 2010, Aberdeen, UK.

[162] K.S. Sorbie, A simple model of precipitation squeeze treatments, SPE-155111-MS, SPE International Conference on Oilfield Scale, May 30–31, 2012, Aberdeen, UK.

[163] V.E. Spooner, R. Stalker, R. Wright, et al., Improving scale inhibitor squeeze design for naturally fractured reservoirs, SPE-1697555-MS, SPE international oilfield scale

conference and exhibition, SPE International Conference on Oilfield Scale, May 14–15, 2014, Aberdeen, UK.

[164] B.L. Calberg, N.O. Wolf, K.W. Pober, et al., Inhibition of reservoir scale, Patent US4860829, 1989.

[165] P. Chen, T. Hagen, H. Bourne, et al., The challenge of squeezing a water sensitive hp/ht reservoir - Lab and field experiences with a novel non-aqueous inhibitor/ squeeze enhancer package, SPE-87435-MS, SPE International Symposium on Oilfield Scale, May 26–27, 2004, Aberdeen, UK.

[166] R. Wat, H. Montgomerie, T. Hagen, et al., Development of an oil-soluble scale inhibitor for a subsea satellite field, SPE-50706-MS, SPE International Symposium on Oilfield Chemistry, February 16–19, 1999, The Woodlands, TX, USA.

[167] G.M. Graham, S.J. Dyer, P. Shone, et al., High temperature coreflooding experiments for the selection of appropriate scale inhibitor products for potential application as downhole squeeze treatments in high temperature reservoir environments. SPE-68314-MS, SPE International Symposium on Oilfield Scale, January 30–31, 2001, Aberdeen, UK.

[168] E. Sorhaug, M.M. Jordan, D. Marlow, et al., Achieving extending squeeze life in challenging wells - A non-conventional approach, SPE-131093-MS, SPE International Conference on Oilfield Scale, May 26–27, 2010, Aberdeen, UK.

[169] M. Falla, K. Oeski, D. Frigo, et al., Design and laboratory testing of scale-inhibitor squeeze packages for HP/HT gas-condensate reservoirs, SPE-94865-MS, SPE International Symposium on Oilfield Scale, May 11–12, 2005, Aberdeen, UK.

[170] G.M. Graham, D. Frigo, N. Poynton, Calcium sulphate (anhydrite) scale control in a mature asset - The Kittiwake and Fulmar reservoirs, SPE-80386-MS, SPE International Symposium on Oilfield Scale, January 29–30, 2003, Aberdeen, UK.

[171] G.M. Graham, A. Juhasz, H. Guan, Modification of surface wettability following chemical (scale inhibitor) application in chalk and dolomite cores - Stimulation vs. damage, SPE-73716-MS, SPE International Symposium of Formation Damage, February 20–21, 2002 Lafayette, LA, USA.

[172] F. Wahid, G.M. Graham, J.M. Jordan, et al., Enhanced scale inhibitor placement by gravity slumping in high permeability horizontal wells - Draugen Field North Sea, SPE-95067-MS, SPE International Symposium on Oilfield Scale, SPE International Symposium on Oilfield Scale, May 11–12, 2005, Aberdeen, UK.

[173] M.M. Jordan, K. Sjusaether, I.R. Collins, Scale control within North Sea chalk/limestone reservoirs: The challenge of understanding and optimizing chemical placement methods and retention mechanism - Laboratory to field, SPE Production & Facilities 20 (2005) 262–273.

[174] G.M. Graham, P. Hill, K.S. Sorbie, et al., The mechanism of formation damage and recovery following chemical squeeze application in low permeability water sensitive reservoirs, Proceedings of the 10th NIF International Oilfield Chemical Symposium, February 28-March 3, 1999, Fagernes, Norway.

[175] N. Ghorbani, M. Wilson, N. Fleming, et al., Using nanoscale dispersed particles to assist in the retention of polyphosphinocarboxylic acid (PPCA) scale inhibitor on rock, SPE-156200-MS, SPE International Oilfield Nanotechnology Conference and Exhibition, June 12–14, 2012, Noordwijk, The Netherlands.

[176] J. Hen, A. Bruncter, B.K. Peterson, et al., A novel scale inhibitor chemistry for downhole squeeze application in high water producing North Sea wells, SPE-30410-MS, SPE Offshore Europe, September 5–8, 1995, Aberdeen, UK.

[177] H.M. Bourne, G.D.M. Williams, J. Ray, et al., Extending squeeze lifetime through in-situ pH modification - Laboratory and field experience, Proceedings of the International Oil Field Chemistry Symposium, March 2–5, 1997, Geilo, Norway.

[178] N. Poynton, R. Tidswell, J. Steel, et al., Squeezing aqueous based scale inhibitors into a water sensitive reservoir - Development of a squeeze strategy, SPE-60219-MS, SPE International Symposium on Oilfield Scale, January 26–27, 2000, Aberdeen, UK.

[179] G.M. Graham, A. Gyani, J.J. Jordan, et al., Selection and application of a non-damaging scale inhibitor package for pre-emptive squeeze in mungo production wells, SPE-74665-MS, in International Symposium on Oilfield Scale, January 30–31, 2002, Aberdeen, UK.

[180] J. Al-Ashhab, H. Al-Matar, G.M. Graham, et al., Scale inhibitor squeeze operations in low permeability carbonate reservoirs, SPE-114080-MS, SPE International Oilfield Scale Conference, June 20–21, 2008, Aberdeen, UK, SPE-114080 (2008).

[181] S. Punnapala, D.S. Abdallah, M. Grutters, et al., Preventing asphaltene deposition by inhibitor squeeze injection: case studies from giant carbonate Abu Dhabi fields, SPE-206319-MS, SPE Annual Technical Conference and Exhibition, September 21–23, 2021, Abu Dubai, UAE.

[182] M. Grutters, S. Punnapala, D.S. Abdallah, et al., Asphaltene Mitigation in Giant Carbonate Abu Dhabi Fields: A Techno-Economic Comparison between Continuous Injection and Formation Squeeze. SPE-207993-MS, Abu Dhabi International Petroleum Exhibition & Conference, November 15–18, 2021, Abu Dhabi, UAE.

[183] G.M. Graham, D. Bowering, K. MacKinnon, et al., Corrosion inhibitor squeeze treatments - Misconceptions, concepts and potential benefits, SPE-169604-MS, SPE International Oilfield Corrosion Conference, May 12–13, 2014, Aberdeen, UK.

[184] G.M. Graham, S. Kidd, R. Stalker, et al., Effect of coreflood test methodology on appropriate simulation of field treatments, SPE-151779-MS, International Symposium and Exhibition on Formation Damage Control, February 15–17, 2012, Lafayette, LA, USA.

[185] G.M. Graham, I. Wattie, E.J. Mackay, et al., Scale inhibitor selection criteria for downhole (squeeze) application in a chalk reservoir, SPE-65025-MS, SPE International Symposium on Oilfield Chemistry, February 13–16, 2001, Houston, TX, USA.

[186] D. Abdallah, C. Elechi, G. Graham, et al., Inorganic scale challenges and their mitigation in an onshore Abu Dhabi field, SPE-177881-MS, Abu Dhabi International Petroleum Exhibition and Conference, November 9–12, 2015, Abu Dhabi, UAE.

[187] N. Bt Alwi, I. Bt Salleh, J.M. Ibrahim, et al., Development of improved chemical formulations for scale control for alkaline surfactant polymer flood in a seawater flooded reservoir, SPE-179883-MS, SPE International Oilfield Scale Conference, May 11–12, 2016, Aberdeen, UK.

[188] L. Sutherland, M. Jordan, N. Champion, Enhancing scale inhibitor squeeze retention in carbonate reservoirs, SPE-190715-MS, SPE International Oilfield Scale Conference, June 20–21, 2018, Aberdeen, UK.

[189] P.J. Shuler, Method for prolonging the useful life of scale inhibitors injected within a formation, Patent US 5038861, 1991.

[190] P. Chen, H. Montgomerie, S. Heath, et al., Method of increasing scale inhibitor retention, Patent US2017/0002258, 2016.

[191] N. Flemming, K. Ramstad, A. Mathisen, et al., Innovative use of kaolinite in downhole scale management: squeeze-life enhancement and water shutoff, SPE Production & Operations 24 (2009) 439–449.

[192] N. Flemming, K. Ramstad, L. Wichers, et al., Mechanical alteration of near-wellbore mineralogy for improved squeeze performance, SPE Production & Operations 25 (2010) 524–532.

[193] P. Guraieb, R. Tomson, I. Littlehales, Development of scale squeeze enhancement technology via application of metal nanoparticles coupled with polymer scale inhibitors, SPE-193541-MS, SPE International Conference on Oilfield Chemistry, April 8–9, 2019, Galveston, Texas, USA.

[194] M.M. Jordan, K.S. Sorbie, M. Marulier, et al., Implication of produced water chemistry and formation mineralogy on inhibitor adsorption/desoprtion in reservoir sandstone and their importance in the correct selection of scale squeeze chemicals, Proceedings of the RSC Chemistry in the Oil Industry, April 14–17, 1997, Ambleside, UK.

[195] D. Frigo, G.M. Graham, I.J. Littlehales, et al., Design and laboratory testing of a hybrid scale-inhibitor package for an HTHP gas-condensate reservoir, SPE-94865-MS, SPE International Symposium on Oilfield Scale, May 11–12, 2005, Aberdeen, UK.

[196] G.M. Graham, K.S. Sorbie, L.S. Boak, et al., Development and application of accurate detection and assay techniques for oilfield scale inhibitors in produced water samples, SPE-28997-MS, SPE International Symposium on Oilfield Chemistry, February 14–17, 1995, San Antonio, TX, USA.

[197] S. Heath, S. Green, A. Badawy, et al., On-site analysis of scale squeeze inhibitors in remote environments usingfluorescence and time resolved fluorescence, SPE-200682-MS SPE International Oilfield Scale Conference, June 24–25, 2020, Virtual.

[198] S. Murugesan, D. Ahoor, L. Souza, et al., Next generation on-site scale inhibitor analysis - Surface enhanced nanotechnology detection, SPE-200702-MS, SPE International Oilfield Scale Conference, June 24–25, 2020, Virtual.

[199] M.M. Jordan, C.J. Johnson, M. Robb, Evaluation methods for suspended solids and produced water as an aid in determining effectiveness of scale control both downhole and topside, SPE-92663-MS, SPE International Symposium on Oilfield Chemistry, February 2–4, 2005, The Woodlands, TX, USA.

[200] M.M. Jordan, J. Buckman, C.J. Johnson, Direct monitoring of the performance of scale control programs across the produced water life cycle via suspended solids analysis, SPE-93892-MS, SPE European Formation Damage Conference, May 25–27, 2005, Sheveningen, The Netherlands.

[201] M.M. Jordan, I. Hiscox, J.F. Dalton, et al., The design and deployment of enhanced scale dissolver/squeeze treatment in subsea horizontal production wells, North Sea Basin, SPE-73717-MS, International Symposium and Exhibition on Formation Damage Control, February 20–21, 2002, Lafayette, LA, USA.

[202] R.C. Hoerauf, New two-step xylene scale treatment removes calcium sulphate, Oil Gas Journal 86 (1988) 56–59.

[203] A. Mendoza, G.M. Graham, K.S. Sorbie, Sulphate scale dissolution: Examination of the factors controlling the effectiveness of EDTA and DTPA based on scale dissolvers and their comparative effectiveness against different sulphate minerals, Proceedings of the 10th NIF International Oilfield Chemical Symposium, February 28 - March 3, 1999, Fagernes, Norway.

[204] R.E. Talbot, P.D. Gilbert, M.A. Veale, et al., Tetrakishydroxymethylphosphonium sulfate (THPS) for dissolving iron sulfides downhole and topsides - a study of the chemistry influencing dissolution, NACE-02030, CORROSION 2002, April 7–11, 2002, Denver, CO, USA, NACE (2002).

[205] M.W. Jawish, R. Suleiman, Q. Wang, et al., New formulations for the control of iron sulfide deposits in oil production facilities, SPE-192228-MS, SPE Kingdom of Saudi Arabia Annual Technical Symposium and Exhibition, April 23–26, 2018, Dammam, Saudi Arabia, SPE-192228 (2018).

[206] D. Patel, R. Ramanathan, H.A. Nasr-El-Din, Optimization and thermal stability of the THPS and NH4Cl blend to dissolve iron sulfide FeS scale at HPHT conditions, Abu Dhabi International Petroleum Exhibition & Conference, November 11-14, 2019, Abu Dhabi, UAE, SPE-197632-MS (2019).

[207] J.J. Wylde, I.G. Winning, Challenges and solutions associated with the development of an iron scale dissolver chemistry, NACE 04730, CORROSION 2004, March 28 - April 1, 2004, New Orleans, Louisiana, USA.

CHAPTER

6 Sand control completion using in-situ resin consolidation

Philip Nguyen and Mike Sanders
Halliburton, Houston, TX, United States

Chapter outline

6.1 Sand control

6.1.1 Mechanisms and causes of sand production

Sand production has been a dilemma for oil and gas operators worldwide for years. Formations that are considered soft or poorly consolidated are often plagued by sand production. These formations are typically shallow, geologically young, with little or no natural cementation to hold the individual sand grains

Flow Assurance. DOI: https://doi.org/10.1016/B978-0-12-822010-8.00002-7

together [1,2]. Approximately 70% of the world's oil and gas reserves are contained in poorly consolidated reservoirs where solids production is likely to become a problem during the life of the field [3].

Wellbore instability often results from a combination of mechanical and chemical instability; however, mechanical factors play a dominant role in wellbore instability during the drilling phase of operations [4,5]. Drilling a well in a formation changes the initial stress state and causes stress redistribution in the vicinity of the wellbore. The redistributed stress state may exceed the rock strength, and cause formation failure [6]. Further contributing factors to sand production is a combination of high drawdown coupled with reservoir depletion [6,7]. Pressure disturbance affects fluid in the pores of the formation and creates additional near-wellbore mechanical stress. The effective stress on the formation increases more rapidly near the wellbore, and this increase results in a higher risk of formation failure during production. Formation failure is more likely to occur later as the reservoir is depleted through production. As stresses increase due to depletion of reservoir pore pressure, this increases grain-to-grain stresses to the failure point, which causes sand production.

6.1.2 Problems/issues of sand production

Production fluid flow rates affect sand production. High flow rates result in high drag forces on the sand grains. If these drag forces exceed the formation's restraining forces, sand will be produced [1]. Field evidences have shown that the onset of water production in these unconsolidated, edge water-drive reservoirs would result in sand production. Water encroachment reduces capillary pressures and promotes intergrain weakening, as well as dissolution of the cementing material, and these phenomena result in sand production [8,9]. Wu et al. [10] showed that the effect of watercut on perforation failure and sand production is most significant for sandstones with high clay content.

Formation sand production results in plugged gravel packs, screens, perforations, tubulars, surface flow lines, and surface equipment as well as lost production [7]. In addition to damage caused by plugging, surface equipment can erode because of abrasive, sand-laden fluid flowing from the well. In the worst case, sand production problems can cause total well failure or the need for recompletion resulting from casing collapse, openhole collapse, or both. Operation costs multiply when sand production wears out equipment and forces expensive workovers, which often require production shutdown of the well. For example, sand control failure could become a severe problem when it occurs on an offshore platform.

6.1.3 Sand control methods

Sand-control methods comprise of two main categories: mechanical sand bridging and chemical in-situ sand consolidation.

6.1.3.1 Mechanical sand bridging

The sand bridging methods for controlling sand production includes using sand screens [11,12], gravel packing [13,14], or frac-packing [15–18].

6.1.3.1.1 Sand screens

The first attempts to control sand were made with screens. The term screen means a pipe with openings, regardless of type, intended to exclude the formation particles by causing them to bridge on the openings. The slotted screens may be of many forms from torch-cut slots (large in size and totally incapable of stopping the fine-grain sands which usually cause trouble) to horizontal or vertical saw-cut slots. In more recent times, slots have been cut with lasers. The slot size of wire-wrapped screens is determined by the proximity of each wrap of a continuous wire around the circumference of a slotted or perforated pipe. Saw-cut slots are cheaper than wire-wrapped slots. The wire-wrapped screen usually permits the use of harder and more corrosion-resistant metal. In addition, saw cuts have a limit as to how small the spacing can be due to damage to the saw blades when the thickness becomes too small.

6.1.3.1.2 Gravel packing

The need for gravel packing arose when screens proved unsatisfactory in controlling very fine sands. The function of the gravel-pack material is to act as a filter, allowing fluid movement but preventing formation particles from entering the well bore. Effective sand control requires a good gravel-pack design and execution, including obtaining a representative sample of formation sand, analyzing the formation grain-size distribution, selecting an optimum gravel size in relation to formation sand size to control formation sand movement, and using screen slot width to retain the gravel [19]. The primary function of screens is to provide mechanical support to prevent sand or gravel placed in the perforation tunnel and formation from flowing back into the wellbore. Keeping gravel in place will support the perforation tunnel and fracture and thus maintain a conductive path and production.

Gravel pack treatments rely on the stable bridges of gravel grains at the formation interface to minimize embedment and penetration of formation particulates into the pack. The filtering mechanism provided by the gravel pack is effective for controlling sand production while still maintaining adequate production rates, but permeability of the pack decreases over time. This decreased permeability results from a continuous migration of formation fines toward the wellbore which causes progressive plugging and results in high skin damage and reduced productivity over time.

6.1.3.1.3 Frac-packing

A frac-pack completion is designed to access reservoirs with high-permeability formations by combining propped fractures to bypass near-wellbore damage and a gravel pack to retain and filter formation sand. Reducing drawdown is one of the major objectives when performing a frac-pack procedure. Operators can reduce pressure drop across the formation face and perforations by distributing the pressure drawdown in a larger area around the wellbore. Although the reservoir pressure gradient is diverted into the fracture and away from the wellbore, substantial flow remains near the wellbore, and fluid flow converges into any remaining perforations [16].

As part of frac-pack treatment procedure, a tip screenout is desirable to help generate short and thick proppant-packed fractures toward the wellbore. A tip screenout is achieved when proppant reaches the leading edge of the fracture and prevents further fracture extension. The annulus between the wellbore wall and the sand screen is tightly packed to maximize connectivity with propped fractures and to prevent development of void spaces. The width of the fracture is further increased by continued injection of the fracture fluid, providing increased proppant packing inside the fracture [16]. Propped-fracture operations that use sand-control screens have been successful. However, screens employed in these applications increase well completion costs and sometimes fail with time [20,21]. In case a refracturing treatment is required, the screen might need to be removed or the producing intervals might require reperforating before the recompletion can be performed. Screenless frac packs using resin coated sand or proppant have been proposed as an alternative solution.

6.1.3.1.4 Shunt tubes

Shunt-tube technology was introduced in the early 1990s [22,23]. This technology continues to evolve to enhance and ensure complete annular packing in gravel pack and frac-pack treatments as the interval lengths in horizontal wells are getting longer. Basically, shunt tubes are perforated rectangular tubes or tubes with nozzles which are attached to sand screens and function in both transporting and packing of gravel slurry. The shunt tube provides an alternative flow path for the gravel slurry to bypass sand bridges or sand dune areas formed unexpectedly in the annulus so that gravel placement can continue. The exit ports located along the length of the shunt tube enable the gravel to fill up all the voids in the screen-wellbore wall annulus to ensure its complete packing.

6.1.3.1.5 Packing voids behind perforated casing

During production of fluids from subterranean wells, formation particulates may be carried from the formation into the production flow stream and into the production tubing, creating voids behind the casing with time. These voids may undesirably impact stability of the well, potentially impacting production [19]. The production of these particulates may cause problems both downhole and at the surface.

6.1.4 Other wellbore stabilization methods for sand control

6.1.4.1 Perforating

Perforations play a critical role in achieving a successful fracturing treatment performed in vertical and horizontal wells. Perforations, and ultimately fracture initiation, determine the communication path between the wellbore and the fracture plane extending from the wellbore into the formation. Good perforation design and implementation can help prevent non-planar fracture geometries [24]. To minimize these problems and their effects on a successful fracturing stimulation, a perforation design should be in phase with the anticipated fracture direction to help ensure the following:

- A fracture initiates perpendicular to the minimum horizontal stress, which gives a maximum fracture width.
- A single fracture propagates.
- Fracture initiation and extension pressures are reduced.

Perforating should be performed without excessive underbalance or in a near-balanced situation to minimize the possibility of perforation collapse and/or formation material sloughing into the wellbore [24]. A primary goal of perforating is to generate and maintain open perforations that can later be mechanically supported with the completion.

6.1.4.2 Oriented perforating

Using proper perforation orientation minimizes stress placed on the proppant pack and perforation tunnels. The perforations should be aligned with the maximum stress to simplify the fracture geometry and minimize the possibility of any restrictions or changes in direction in the near-wellbore region [24].

In horizontal and deviated wells, perforations should be oriented in the vertical plane with the wellbore drilled in the direction of maximum horizontal stress; however, perforations in vertical wells should be oriented in the direction of the maximum horizontal stress. Magnitudes of the horizontal stresses may not be significantly different in many soft formations. Under these conditions, the deviated and horizontal wells should still be perforated in the vertical plane since this will be in the direction of the overburden, which should be the maximum stress. In this situation, the wellbore orientation or direction is not critical. A vertical well should be perforated using 180 degrees phasing when there is little or no difference between the horizontal stresses, since the fracture will tend to stay in the direction of initiation. If there is a likelihood of significant horizontal stress contrasts, then 60 degrees phasing will result in the best connection to the fracture. In this case, however, all perforations may not be packed.

6.1.5 Perforating techniques for completions using sand-consolidation treatments

This section provides the design and perforating techniques for completions using sand-consolidation treatments to:

- provide a good flow connection from the reservoir to the wellbore, and
- promote uniform and strong consolidation of the near-wellbore region.

6.1.5.1 New intervals

For perforating new intervals, observe the following guidelines:

Selection of the perforating gun carrier size and type should depend on the type of well operation and the tubular sizes.

Perforation charges should be selected based on the basis of the formation's native consolidation or compressive strength characteristics. Generally, for any formation, a higher shot density will reduce drawdown during production and promote both sustained production rates and lower stress across the sand consolidation, resulting in a longer sand-free life for the consolidated interval.

Underbalanced perforating methods are preferred to provide perforations that are free-flowing and open to injection of the consolidation fluids or gravel pack fluids

For best results, select the perforated interval for good sand quality. Low clay content and higher permeability are important considerations for choosing the perforation interval for successful, long- lived sand-consolidation completions. The perforated interval should coincide with the zones that have higher permeability and low clay content as indicated by the log interpretation. Clean, high-permeability formation sand can be consolidated to yield both higher consolidation strength and higher retention of the native permeability.

6.1.5.2 Deep penetrating perforating charges

Clean reservoir sands having sufficient consolidation strength to support open perforation tunnels after underbalanced perforating should be perforated with deep-penetration charges. The deeper open perforation tunnel will enhance productivity by bypassing near-wellbore damage. After it is consolidated, the tunnel should remain open to enhance reservoir inflow.

6.1.5.3 Big-hole perforating charges

Big-hole charges should be selected for zones that will be prepacked. Reservoirs with marginal or dubious sand quality can benefit from forced placement of gravel or proppant in the near-wellbore region (prepacking) to form a high-strength, high permeability filter between the formation sand and the wellbore.

Note: if sand quality is dubious, and prepacking is not a practical option, completion methods other than sand consolidation may be better choices.

Perforations that are prepacked and consolidated will be filled with gravel when the well is producing; therefore, produced oil and gas must flow through the gravel-pack proppant in the perforation tunnels. Because production flow pressure drop in packed perforations is highly dependent on the cross-sectional flow area, the perforation charge selection should be strongly influenced by entry-hole diameter. Large-diameter holes give greater cross-sectional flow area and thus reduce the pressure drop through the packed tunnels.

6.1.5.4 Shot density

Various sources indicated that a shot density of 4–8 shots/ft of perforations is recommended for resin treatments unless the perforations will be prepacked prior to the resin treatment.

The higher shot density will reduce the velocity through the perforated interval and thus would provide for less stress on the resin coated formation sand.

It has been hypothesized that resin entering the perforations will flow vertically when it is out in the formation. In that case, it is suspected that treatment 100% of the perforation tunnels may not be a requirement for success of the treatment as the formation outside the casing will be consolidated due to the vertical movement of the fluids.

6.1.5.5 Hole size

The hole size is not considered as important in resin treatments except when a prepack is to be performed prior to the resin treatment. In that case, a shot density of 8–12 SPF with Big Hole charge are recommended.

Note: ensuring that the perforation tunnels actually are able to perforate through the tubulars and connect with the formation is of paramount importance for any treatment but especially so with a resin treatment.

6.1.5.6 Cleaning perforations

A compressed zone which may have as little as 20% of the native formation permeability is typically found after jet perforating a well. Underbalanced perforating is one way of breaking down that compressed zone and bringing it into the wellbore where it can be circulated out of the well.

Perforating using a jetting tool will not result in the compressed zone and should be investigated for instances where perforating underbalance due to fear of sticking the perforating guns is a consideration.

Before prepacking the formation, flush the perforations free of perforation debris, formation sand, and drilling-mud residue; then, if needed, stimulate to promote good injectivity in each perforation.

Note: an acid treatment will not substantially remove the compressed zone in normal sandstone formations as the compressed zone will be comprised of crushed quartz which is not readily soluble in HCl or even HF acid.

In some cases, a well-designed stimulation treatment can open the area around the perforations and increase injectivity. This increase can enhance consolidation treatment placement by enabling more complete injection fluid along the interval and into the area surrounding the wellbore.

6.1.5.7 Prepacking perforation guidelines

To determine whether prepacking is needed, apply the following guidelines:

- If the interval to be consolidated is a clean, high permeability interval that has not produced sand, it does not need to be prepacked. Otherwise, as a general rule, perforations should be prepacked before the consolidation treatment is pumped.
- Prepacking is recommended both for existing perforated intervals that have already produced sand and for new completions of reservoir sands with low consolidation strength that may not support open tunnels after perforating. This step will help ensure the existence of open-flow pathways between the wellbore and the reservoir.
- Existing perforations that have produced sand should be prepacked with gravel or synthetic proppant to provide a good flow connection between the reservoir and the wellbore and to ensure that there are no void areas in the near wellbore region.
- Previous production of formation sand from existing perforations can leave the near-wellbore region in poor condition for achieving good consolidation strength and for supporting a low restriction path for reservoir inflow to the wellbore.
- If an existing interval that has made formation sand is to be reperforated, the recommendation for prepacking applies as well. The near-wellbore region may be in a condition of poor consolidation and may need stimulation and prepacking to establish a good connection between the reservoir and the wellbore before consolidation.
- Prepacking and stimulation are necessary for repairing zones that have produced sand of sufficient quantities to alter the integrity of the near-wellbore layering and permeability.
- The prepack gravel used should be sized to the formation grain size per standard industry practices.

6.1.5.8 Prepack procedure

Prepacking the perforation tunnels can be performed using the completion brine and 40/60 sand (which is much more permeable than the formation sand) using low sand concentrations such as 0.25–0.50 lbs of sand per gallon of carrier fluid. The carrier fluid must be extruded into the formation to allow for packing the perforations tunnels with the 40/60 sand thus a viscosified fluid is not typically chosen for the prepack as it will be more difficult to inject into the formation matrix. Once a sandout has occurred, reverse any excess sand from the wellbore.

Cavities or unstressed formation can affect uniform distribution of the plastic. Shroeder and Tucker [25] found that prepacking increased longevity four times and brought the level of success of the remedial jobs to the same as that for new completions.

6.1.5.9 Prepacking correlation with success

The correlation between prepacking and consolidation success has been firmly established by several field-lifetime studies. All were "old" intervals that had produced sand before treatment. The results clearly show that prepacking old intervals before consolidation has significant benefits.

Sand-control success for old intervals after prepacking is equivalent to success in "new" intervals that have not produced sand. The prepacking operation apparently can restore the sand distribution around a sand-producing well to its undisturbed state [19].

6.1.6 Chemical sand consolidation

In addition to applying mechanical techniques to control sand production, chemical means that provide consolidation, or cementation, of the formation material, have been used since the early 1940s to solve sand production problem [26]. More often, resin consolidation treatment is performed in wells where completions using gravel packing and sand control screens have been deemed to be economically unattractive. Basically, this sand consolidation process consists of injecting curable resins into the target production intervals where they harden and transform the unconsolidated or weakly consolidated formation surrounding the wellbore into a competent mass, while retaining most of its original permeability. The consolidation treatment enables the wellbore to (1) withstand high drawdown pressure and stress cycling during production and shut-in of the well, and (2) immobilize near-wellbore formation particulates.

6.1.6.1 Preferred resin characteristics

The resins used for sand consolidation should have the following characteristics:

- Low viscosity – It should have a low enough viscosity to be pumpable through surface lines and into the formation, especially in tight, low-permeability formations. High-viscosity consolidation fluids can cause fracturing of the sand formation, particularly in low-permeability sands. It has also been noted that the fluid used to displace the resin treatment further into the formation can form channels instead of evenly displacing the viscous consolidating fluid.
- Adherence to sand grains – The resin should stick to the sand. This is accomplished by using a resin which can be drawn into the pendular region at the sand-grain contacts. Substances which furnish a chemical bond (in addition to any physical bond) from sand-to-resin, called coupling agents, are added to make the resin adhere better to the sand surfaces [27].

- Strength – The resin itself should be strong and provide for an adequate increase in compressive strength of the sand to allow for higher drawdown pressures during production of the well.
- Long storage life – A resin that goes bad quickly would not be dependable enough for field use. It should have long shelf life to overcome long duration often required as part of logistic, shipping and transport, project plans, after being mixed/prepared, and delivered between continents. Typically, the individual components are kept separated, until they are delivered and mixed at wellsite.
- Resistant to chemical attack – The hardened (polymerized) resin should be resistant to attack by normal oil-field fluids, including oil, water, gas, and stimulation fluids.

6.1.6.2 General resin descriptions

Curable resins are polymers that become fully crosslinked and hardened with time once activated with an activator. The curable resins are commonly used in consolidating near-wellbore formations for sand control, coating fracturing sand or proppant in primary hydraulically fracturing treatments, screenless frac-packing treatments, or remedial treatments, for proppant flowback control.

Various categories of curable resins are presented below:

- Internally catalyzed or preactivated systems contain both a curable resin and an activator. The resin is mixed with the curing agent or hardener at the surface to form a single resin mixture before injecting downhole. Polymerization time is tailored for the bottomhole temperature of the well being treated. An obvious disadvantage is apparent in that delays after activation could result in plugging tubing or pipe in which the resin mixture is being delivered if it is not completely removed from this fluid-delivery equipment.
- Externally catalyzed systems are resin systems that can be activated by a post-flush containing a catalyst to activate the resin in-situ. The resin is first pumped into the formation followed by a post-flush. This post-flush fluid can be either hydrocarbon or aqueous solution containing a catalyst. Its main advantage is the unlimited placement time of the resin, especially in long target interval, or interval with low injectivity. An obvious disadvantage is the activator solution has to follow the same flowpath as the resin. The low-viscosity activator solutions may finger through a high-viscosity resin so a low viscosity resin system is preferred.
- Solvent-based resins: These conventional systems have been used for sand consolidation in the petroleum industry since the start of resin consolidation as a sand control method in the mid to late 1940s. The systems in use today can either be internally or externally catalyzed.
- Aqueous-based resins: Unlike the solvent-based resins, the aqueous-based resins have high flash points and low viscosity. They can be foamed with

nitrogen to enhance their uniform placement in intervals with highly contrast permeabilities.

Note: A good wetting surfactant is needed to ensure good wetting of resin on the sand surface ensuring it remains on the treated surface without completely being displaced from the contact points between grains during injection of over-flush or post-flush fluid stage [28]. Ideally, this often requires a balance between a desirable consolidation strength and an optimal retained permeability for a resin-treated sand pack.

6.1.6.3 Curable resins

Current commercially available curable resin systems are based on the polymerization reactions of a phenolic resin, an epoxy, or a furan resin. Resin polymerization is activated by a catalyst or a curing agent. In a catalyst-activated polymerization, a small amount of catalyst causes the resin molecules to join together into long chains. In a curing polymerization, the curing agent reacts directly with the resin and becomes an integral part of the resin structure.

6.1.6.3.1 Phenolic resins

Phenolic resins are the first commercial synthetic resins (plastics). They can be divided into two different types: novolacs and resoles. The phenolic resins are prepared by either acid-catalyzed (novolacs) or base-catalyzed (resoles) addition polymerization of phenol with formaldehyde. Phenolic resins have been widely used for the production of molded products including billiard balls, laboratory countertops, and as coatings and adhesives [29].

This resin material has typically been used in preparing resin pre-coated sand or resin pre-coated proppant for improving its crush resistance in minimizing fines generation when placed in high-closure fractures, or in preparing curable, resin pre-coated proppant for controlling its flowback during fracturing fluid recovery and well production as part of hydraulic fracturing treatments, or for filling and packing voids behind casing.

6.1.6.3.2 Epoxy resins

Epoxy resins are thermosetting polymers that provide unique mechanical and resistance properties. They are well known for their bonding adhesive and surface-coating performances. An epoxy resin is a class of thermoset polymer made from a monomer that contains at least two epoxide groups. Epoxy resins can be cross-linked or cured into a three-dimensional network using catalysts, curing agents, or hardeners, including polyfunctional acids, amines, phenols, thiols, alcohols and anhydrides [29].

Epoxy resin systems typically are mixtures of both a curable resin and an activator to form one-component consolidating systems when injected downhole to treat a short formation interval. Combining the resin with the activator into a single component helps ensure that wherever the formation matrix is treated, the consolidation will take place without any uncertainty as often encountered with externally catalyzed consolidation systems. However, the pot-life of an epoxy resin system is relatively short when compared to that of a furan resin, which can only be activated with an external activator or extremely high temperatures. Because of its short pot-life, epoxy systems should only be applied in consolidation treatments of short intervals, preferably shorter than 10 ft in length, that can accept treatment fluids readily. As the resin mixture is being injected downhole, its viscosity increases with time as it is exposed to downhole temperatures. As its viscosity increases, the resin mixture becomes more difficult to penetrate the formation sand matrix, and is less effective in coating onto formation sand surfaces, or becomes less effective in being uniformly displaced by an over-flush fluid.

The application of epoxy consolidation system mainly involves the following procedure [28]:

1. Injecting a volume of preflush fluid, which is simply a common brine and/or a hydrocarbon such as diesel or base oil, for conditioning the formation matrix,
2. Injecting a volume of epoxy resin mixture followed by a hydrocarbon spacer, and
3. Injecting a volume of over-flush fluid, which is simply a common brine or a hydrocarbon such as diesel or base oil, to displace the resin from the wellbore and to displace the resin from occupying the pore spaces inside the matrix of the formation,
4. Shutting in the well for complete curing of the resin.

6.1.6.3.3 Furan resins

Furan resins are thermosetting oligomers obtained by crosslinking reactions of furfuryl alcohol and its derivatives when heated and mixed with catalysts [29]. These resins have been used as general adhesives and binding agents for a variety of applications, such as chemical resistant cements, and binders for explosives, wood adhesives, fiber composites, and other impregnating and coating compositions.

Furan or furfuryl alcohol-based resins have commonly been used to provide consolidation in unconsolidated formations to help prevent formation sand production [30–35]. This resin system is self-activated at temperatures of 300°F and above. For wells with bottomhole temperatures less than 300°F, an external activator, typically an acid catalyst, is applied as a post-flush fluid to activate the resin after it has been placed into the formation.

A typical procedure of a furan resin treatment for low-temperature range involves [27]:

1. Injecting a preflush treatment fluid to condition the formation matrix,
2. Injecting the resin to coat the weakly consolidated formation matrix,
3. Injecting a spacer fluid to separate the resin from the activator, displace all resin into the formation, and ensure complete removal of resin from treatment fluid delivery equipment,

4. Allowing a period of time for completing resin treatment and spacer fluid displacement in all target intervals before injecting a post-flush activator,
5. Injecting a post-flush activator solution to activate the curing/polymerization of the resin,
6. Shutting in the well for complete curing of the resin.

Note: When treating single intervals the resin is followed by a small volume of spacer followed immediately by the post-flush activator (typically HCl acid) which is then displaced to the perforations.

It is necessary to separate the resin and activator to prevent premature mixing that would result in premature curing of the resin in the wellbore or in the pore space in the formation near the wellbore.

The spacer fluid and the post-flush activator serve as displacement fluids to help displace the resin from occupying the pore spaces between the sand grains. The retained permeability of the formation matrix depends strongly on how well the consolidating fluid is displaced from the formation porosity or how well the pore spaces are cleared of resin. Ideally, it is preferred to have the pore spaces relatively empty and only the regions at the contact points between sand grains filled or occupied with consolidating resin. Of course, the larger the regions of contact between grains, the greater the consolidation strength that can result for the treated formation matrix.

6.1.6.4 Design parameters for consolidation treatments

To prevent formation failure, the ultimate goal of resin placement for near-wellbore consolidation is to uniformly treat a formation over the entire exposed interval, regardless of any existing permeability contrasts. The main drawback of this method is the uncertainty of resin penetration and its distribution in the formation during a treatment.

6.1.6.4.1 Short-interval consolidation

Because a formation has the potential to vary sharply in permeability, curable materials have usually been applied to consolidate only short intervals. Depending on the interval lengths in a wellbore, resin treatments can be divided into two categories: short and long interval consolidations. For intervals shorter than 10 ft, resin can be distributed into the interval by bull-heading or using coiled tubing (optionally coupled with opposing wash cups or inflatable packers). Depending on the injectivity rate (at a pressure which is less than the frac pressure) and the bottomhole temperature of the target interval, a relatively short pot-life, one-component resin, such as epoxy resin, can be used in treating wells with short intervals.

6.1.6.4.2 Long-interval consolidation

Resin consolidation treatment in longer production intervals typically requires installation of isolation devices along the wellbore length to divide it into shorter segments to ensure the success of fluid placement. Installation of isolation devices greatly increase the complexity of the resin treatment process when considering the number of fluids, the volumes, and their repeating sequence required to be placed in each short interval. This is discussed in detail in paper SPE 36888 [36].

A resin system that can be activated with a post-flush fluid stage (i.e., containing an activator) is most preferable for long-interval consolidation treatment.

6.1.6.4.3 Consolidation depth

Achieving ideal consolidation in the field is impossible because strata of different permeabilities accept injected fluids at different rates, and because all perforations may not be open. Consolidation of the formation surrounding the wellbore by penetrating the resin two to three times the wellbore diameter should be sufficient to stabilize the wellbore in most cases. This consolidation depth is based on stress concentration profiles around the wellbore [37]. Ideally, a consolidated zone a few inches thick around the wellbore is all that is required [19,31,38]. The consolidated radius must be increased to withstand the drawdown stress imposed by fluid production. Increasing the radius of the consolidated sand mass can also help stabilize a formation around plugged perforations that would not accept resin injection.

Originally, consolidation treatments were designed for a radial penetration of 1.0 m. It has been noted that that a radial penetration depth of 0.5–0.75 m may be sufficient [19]. This would reduce the cost of the treatment and reduce the spacing of the near wellbore porosity affected by the resin treatment. The skin calculation shows that a thinner zone with regained permeabilities as low as 25% can have a relatively small effect on productivity. The authors recommended to design treatments for 1 m penetration depth in dirty sands and in new areas. After successful usage has been established, field trials should be carried out to establish whether this value can be reduced to 0.75 or maybe even 0.5 m.

The typical resin volume recommended is currently 100 gal/ft. The original recommendations from the 1940s and 1950s was 1 bbl/ft of perforations. As the intervals being treated became longer with a higher shot density more failures were observed. Increasing the treatment volume of 2 bbl/ft of perforations was the answer. Shroeder and Tucker [25] noted that a resin treatment volume should be at least 30–50 gal/ft but increasing the treatment volume to 100–150 gal/ft increased the productive life of the resin treatment.

Some people recommend adding 50% excess to the calculated amount to allow for vertical migration and using a minimum volume of 200 gallons.

Shroeder and Tucker showed that increased resin volumes significantly improved producing life [25]. From 50 to 100 gal/ft, productivity was increased six times. It was doubled again from 100 to 150 gal/ft, to 200,000 bbls of fluid produced before sanding up. A resin volume of 150 gal/ft gives approximately 4.5–5 ft of coverage with 30% porosity sands.

The typical radial penetration utilized in a resin treatment is approximately 3-ft. Depending on the wellbore diameter and the porosity this requires a volume of

75 gal/ft for 7″ pipe with 30% porosity or 50 gal/ft for 7″ pipe with 20% porosity. This is a theoretical calculation and is used as a starting point. The resin chemicals come in various package sizes so another aspect of the design is to ensure that full packages are used.

6.1.6.4.4 Zonal isolation

The major well completion prerequisite to satisfy prior to using sand consolidation treatments is that a good cementation along the zone to be treated exists. If a Cement Bond Log (CBL) indicates a poor bond behind the casing in or near the zone to be consolidated, the well should be cement squeezed. Preferably another CBL should be run to check the results of the squeeze before the sand consolidation treatment.

6.1.6.4.5 Mineralogy

A review of the X-Ray Diffraction data for the perforated interval can help identify problem areas. Higher clay content sands require a higher concentration of resin due to the high surface area of the clays. In some cases, a preflush containing a permanent clay stabilizer may be recommended. Also, in some cases, an acid treatment to shrink (HCl) or dissolve (HF) excess clays may be recommended. A clay stabilizing treatment before the consolidation treatment can be used to ensure that water sensitive clays are stabilized before contact with the dehydrating preflush required for the consolidation.

6.1.6.4.6 Effect of clay content

Clay prefers to be water wet. Therefore, a strong cationic surfactant has been used in the current resin system to allow the resin to overcome the wetting incompatibility and adhere to the clay surface. Fig. 6.1 shows the consolidation strengths obtained for various sand packs treated with an epoxy resin corresponding to different amounts of clays [28]. It seems that the consolidation values vary with the amount and the type of clay mixed in the sand pack. In addition, the selected surfactant enhances the capillary action and helps pull the resin into the crevices between sand grains and particulate. Despite being displaced from the pore spaces between sand grains, the resin adheres and continues to remain at the areas close to the contact points between sand grains, resulting in good consolidation strengths. The top portions of sand packs tend to have lower UCS values than those of the bottom portions. This is a direct result of resin displacement by the overflush fluid from the top portion down toward the lower portion (Fig. 6.4).

As the concentrations of fines and small particulate are increased in the sand pack, its permeability decreases (Fig. 6.5) [28]. Despite of some loss in permeability after being treated with resin, the degree of damage strongly depends on the initial permeability of the sand pack. The sand packs with high initial permeabilities tend to retain better permeability than those with low initial permeabilities. Some sand packs have higher regained permeability than the initial permeability. One possible explanation for this occurrence is the ability of the

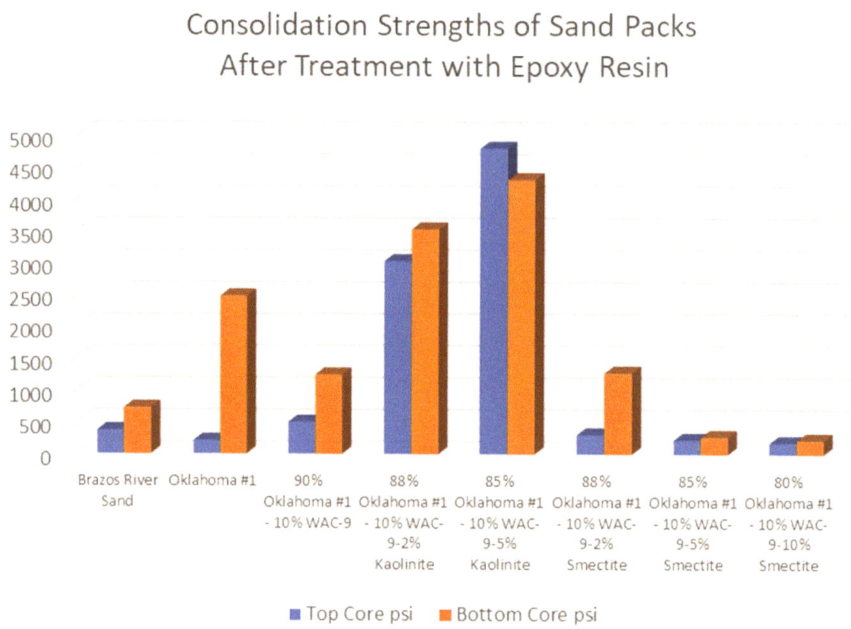

FIGURE 6.1

Consolidation strength of sand packs after treatment with epoxy resin.

consolidated sand pack to trap the fines and sand particles in place, preventing them from moving and blocking the flow channels during fluid movement. Additionally, as often observed in the past with resin treatments, water-swollen clays were found to shrink when exposed to resin, allowing the regained permeability of the resin-treated formation to be improved [31] (Fig. 6.2).

6.1.6.4.7 Effects of over-flush volumes

Even after being displaced with several pore volumes of overflush fluid, the resin still coats the solids to provide good consolidation strengths for the treated sand packs (Fig. 6.3) [28]. The thin resin adheres to the sand and clay surfaces as it is displaced in the matrix; it does not plug the pore spaces and helps retain excellent regained permeability (Fig. 6.4) [28].

In a typical core flow or sand pack type test with resin the resin coating on the sand and fines will gradually become thinner as more pore volumes of overflush are applied, although SEM photos shows that they are typically not opened wider.

In contrast to Brazos River sand, the SEM micrographs show that the weakly consolidated formation cores obtained from a West Africa well are filled with cementation materials covering the pore spaces within the matrix and the pore throats were not visible to any great extent.

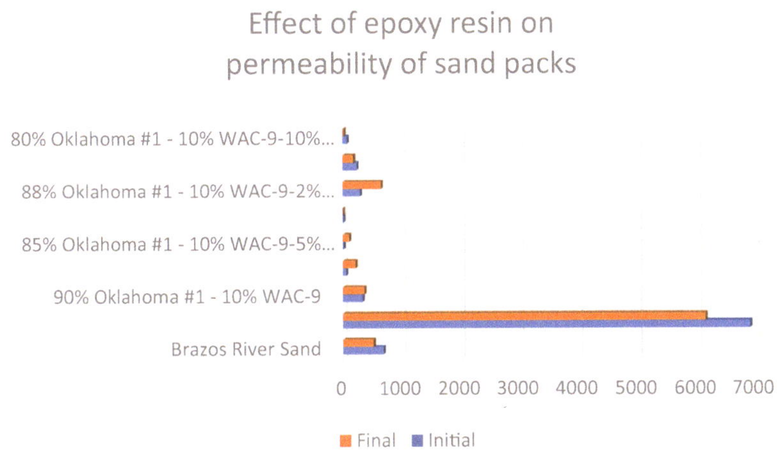

FIGURE 6.2

Effect of epoxy resin treatment on the permeability of sand packs.

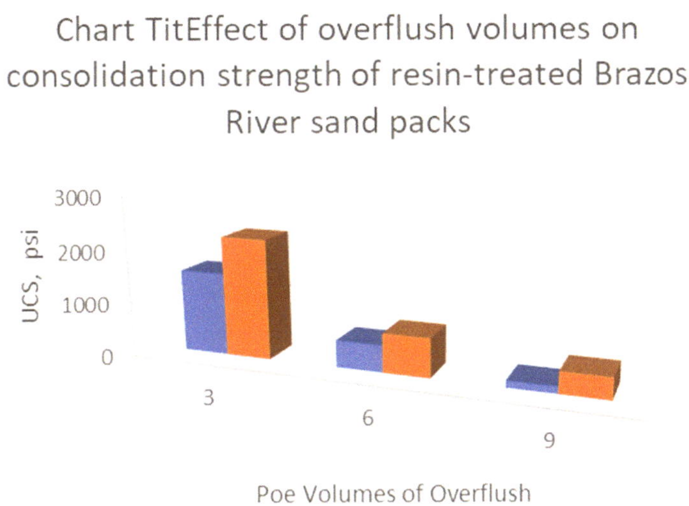

FIGURE 6.3

Effect of overflush volumes on the consolidation strengths of resin-treated Brazos River sand packs.

6.1.6.4.8 Consolidation strengths

Unconfined compressive strength (UCS) is used in the laboratory to measure how well sand grains are cemented together. UCS has often been used as a guideline within the industry to provide some basic insight into the formation strength and

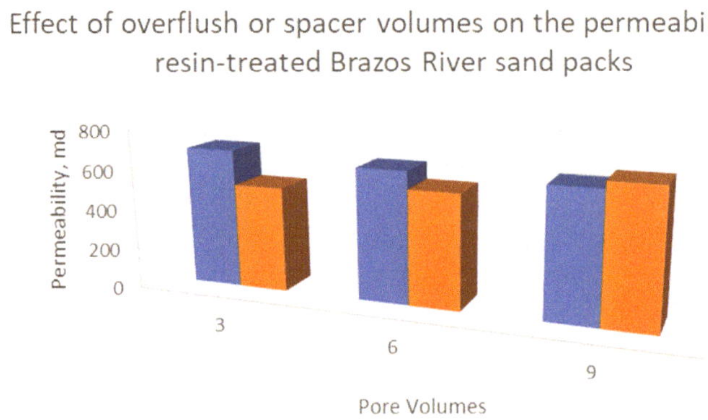

FIGURE 6.4

Effect of overflush or spacer volumes on the permeability of resin-treated Brazos River sand packs.

whether or not a formation will produce sand. In a producing well, however, the stress load on the consolidated sand comes in the form of a drawdown stress imposed by fluid flowing through the sand. Flowing fluid causes the consolidation to fail in a shear mode (tension plus compression).

A series of laboratory experiments was conducted to relate compressive strength to drawdown stress at failure. Fluid (oil) was flowed through a series of consolidated samples of known compressive strength until they failed. The results showed that each 1000 psi in compressive strength corresponds to 1700 psi in drawdown stress [19].

Higher strengths are desirable to offset the natural weakening of the resins with time and exposure to wellbore fluids. Strength degradation can be caused by exposure to HF acid. Overnight exposure can cause 50% or more compressive strength loss. Exposure to HCl over several months could cause a significant strength loss. Use of HF is usually only recommended before the consolidation treatment. It is spent in the formation before being produced back through the consolidation zone. HCl can be used after the consolidation treatment, with limited exposure times.

Nguyen et al. [28] described a series of triaxial tests performed on real core samples obtained from a weakly consolidated formation in West Africa at different confining pressures to develop an elastic/plastic material model to evaluate formation failure. This study showed an increase in UCS was developed when the consolidation material was applied to the formation material. The treated core samples showed no apparent change in the physical characteristics of porosity distribution, grain to grain contacts and such. The treated samples did, however, exhibit a more gradual transition from elastic to plastic properties that typically occurred at a higher stress magnitude than the untreated cores. The significant change in the plastic behavior of the consolidated core samples suggest that these cores are capable of sustaining much higher levels of strain before failing than the untreated samples. This data shows that the introduction of the consolidating resin has a significant effect on the mechanical behavior of the formation, providing an increase in the mechanical strength and introducing a much larger area of plastic deformation. The benefits of a consolidation treatment are summarized as follows:

Clays and formation fines were encapsulated by the resin material, essentially eliminating one significant damage mechanism normally associated with formation yielding or plastic failure.

The very thin coating of consolidating material between the formation sand grains greatly increase cohesive strength of the rock.

The nature of the consolidation material suggested that the plastic properties of the rock may be changed significantly.

The combination of controlling fines movement and altering the plastic characteristics of the rock may provide a means of relieving high stress concentrations around the wellbore without completely failing the rock or introducing a sand production problem.

6.1.6.4.9 Regained permeability

The regained permeability as determined using a core flow test in the laboratory is offset by the increase in strength of the sand pack. A high regained permeability along with unconfined compressive strengths of 500–1000 psi are desirable.

Note:A reduction in permeability over the consolidated zone of about 25% to a reduction in the productivity index of about 5%–10%, which is acceptable for sands with high unconsolidated permeabilities but may be unacceptable for low permeability sands [39].

Permeability loss should not be confused with loss in productivity. For example, radial flow analyses show that a retained permeability of 70% allows about 90% of the original productivity [40].

6.1.6.4.10 Formation water salinity

Residual formation water at the sand grain surface can interfere with the chemical reaction which forms the binding agent. In particular, the salinity of the connate water affects the final compressive strength of the consolidated rock. Most systems will require a low salinity brine preflush prior to injecting the resin into formations containing high salinity formation water.

6.1.6.4.11 Wellbore integrity

Ensuring that wellbore has been pressure tested prior to the resin treatment is critical. A leak in the tubulars (when using a bullhead treatment) can result in the resin being lost to that leak and not being available for its intended purpose and potentially causing unpredictable wellbore problems

A leak at the surface can allow some of the treatment fluids to follow the path of least resistance and u-tube up into the coiled tubing/wellbore ID annulus above the perforations. This means that not all of the treatment fluids are available for its intended purpose. It also can also mean that the location of the various stages of the treatment are unknown. If some of the resin is allowed to move up into the annulus and is later pushed back downhole by pumping into the annulus at the surface it may have the effect of placing resin into the perforations tunnels without the required overflush thus causing an unintended reduction in the permeability of the formation near the wellbore. If the resin is left in the annulus it can cause the tubulars to become stuck.

6.1.6.4.12 Cleaning the wellbore tubulars

The use of coiled tubing can help prevent damage being caused by material from the tubulars/wellbore ID such as scale, rust, paraffin, asphaltene from breaking loose and being injected into the perforations ahead of the resin treatment.

If a treatment is to be performed using a bullhead treatment down the production tubulars or if the resin treatment will be performed using a workover rig and packer it is important to ensure that the wellbore has been cleaned prior to pumping the resin treatment and thus locking any materials lost to the formation in place. Scrape the casing while circulating. Repeat scraping if needed and continue circulating until the returns are clean. If need be, pump a viscous pill to remove solids.

6.1.6.4.13 Determining the bottomhole treating temperature

Consolidation systems which use an internal catalyst results in a consolidation fluid that has a limited pump time, much like the cement in a cement squeeze job. The following guidelines should be observed when designing a treatment that will include a sand-consolidation treatment:

Ensure that a wellbore temperature simulation has been run before choosing a consolidation system; once the consolidation fluid begins curing, the reaction will proceed rapidly. Sufficient time must be available for pumping before the consolidation fluid begins curing.

Close attention should be paid to the pre-job injection rate. If the injection rate has to be decreased because of high pressures, the job may need to be aborted.

If the treatment is being pumped down the production tubulars, have a contingency procedure available. Reversing the treatment back out of the well is not an option.

6.1.6.4.14 Wellbore cleanout/washout

Before performing the consolidation treatment, ensure that the wellbore is clean of debris, sand fill, or other obstructions that could prevent uniform injection of the treatment fluids into the entire interval. Procedures and materials used to perform operations for removing debris or sand should not reduce or damage injectivity into the interval. If the wellbore is deviated, carefully remove all materials that could block perforations on the bottom side of the hole. Techniques that can be used include viscous sweeps, turbulent flow washing, radial jet nozzles, and nitrified or foamed fluids. If any sand fill is left adjacent to the perforations it will not allow those perforations to be treated with the resin treatment and is a likely cause for failure of the treatment.

6.1.6.4.15 Gas lift valves and landing nipples

With regard to gas lift valves and landing nipples, when applying a sand-consolidation treatment, the following options are recommended:

- Run coiled tubing (CT) through the production tubing to protect gas lift mandrels and landing nipple profiles. As additional benefits, the CT can be used to wash down and can also serve as a clean conduit for the sand-consolidation treatment.
- The gas lift valves should be pulled, and dummy valves should be installed in the side pocket mandrels to protect the side-pocket mandrel gas injection ports.
- Bullheading the treatment down the production tubing may result in the resin filling or partially filling the landing nipple profile. Due to the change in ID, the lock mandrel may not be able to seat properly in the landing nipple.

6.1.6.4.16 Fluid control

"When resins that harden into insoluble plastics are used, the placement technique must guarantee that all the consolidating fluids are injected only into the formation. There must be no possibility that these fluids can accumulate in the rathole or in the workstring/casing annulus. Provided that there are no leaks, wellbore fluid flow can be controlled by correct selection of the densities of the fluids in the annulus and rathole" [19].

Because a plastic consolidation resin will cure to form a hard plastic. adequate control must be maintained over fluid flow in the wellbore. Two areas of immediate concern are the rathole and the annular space between the workstring and casing.

6.1.6.4.17 Rathole

The area below the perforations should be limited to 2 to 3 feet to help prevent treatment fluids from swapping with the rathole fluids. If possible, fill the rathole with a compatible fluid having a higher density than that of any treating fluid being used.

6.1.6.4.18 Annulus fluid

It is imperative that the annulus fluid level be kept full during a resin consolidation treatment. A low annulus fluid level can allow the resin treatment fluids to u-tube up into the annulus as that becomes the path of least resistance rather than the perforations. If the resin is allowed to u-tube up into the wellbore/coiled tubing annulus it will (1) coat the wellbore and CT with the resin thus potentially causing a sticking problem and (2) if this error is compounded by pumping down the annulus to flush the treatment fluids out of the annulus prior to pulling the CT it can potentially allow for the resin to be pumped into the near wellbore area without the required overflush or afterflush thus resulting in an area of reduced permeability near the wellbore. If acid was allowed to u-tube up into the annulus and it was not neutralized properly prior to pumping the resin treatment then it can invade the near wellbore area and contact the resin. A low pH (4–5) can prevent an epoxy resin from curing properly.

Filling the annulus with a lighter weight fluid such as diesel or a light weight is an option that can be investigated.

When the bottomhole pressure (BHP) is so low that the annulus will not stand full of diesel, a retrievable packer should be run on the workstring. Customarily, about 30 ft of tailpipe is used below the packer. Having a full annulus or running a packer eliminates fluid-level rise in the annulus.

6.1.6.5 Acid stimulation treatment

Acidizing is a method for removing clay, scale, and other particulate materials from the formation sand. Plugging of perforations by particles suspended in the workover fluid or loosened from the walls of the casing or tubing is common in workover operations. The particles impair injectivity (and subsequent productivity) and must be removed before consolidation. The presence of these damaging solids can usually be identified during an injectivity test. In addition to improving injectivity, acidizing benefits a consolidation treatment by cleaning the sand-grain surfaces. Fresh silica is exposed when clay, feldspar, and other minerals are removed from the pore space. Maximum bond strength is developed between the consolidation resin and a freshly etched silica surface. Thus. even when no injectivity problem exists, an acid treatment is frequently performed to improve the strength of the consolidated sand [19].

Before the formation is prepacked or consolidated with a sand-consolidation service, an acid stimulation treatment is recommended.

The acid treatment can perform the following functions:

Remove scale: The acid treatment can remove scale, rust, and other contaminates from the tubing.

Improved injectivity: The acid treatment can improve injectivity, especially if effective acid diversion methods are used to make the acid stimulation more uniform across the zone. This helps ensure that the sand-consolidation treatment is injected into all perforations.

Higher compressive strengths: The acid treatment can remove clays and other materials and allow the consolidation fluid to adhere directly to quartz grains in the formation.

Higher initial permeability: The acid treatment should improve the permeability of the formation before the consolidation fluid treatment, and

thus, should account for the minor losses of permeability due to the consolidation fluid taking up space in the pore throats of the formation.
Less post-treatment skin: The lower the skin after the treatment, the lower the overall drawdown and the less reason for the formation sand grains to move with the produced fluids.
Fines removal: The acid treatment will dissolve and/or displace the fines in the near wellbore area rather than consolidate them into place, which should improve the regained permeability.

6.1.6.5.1 Acid volume

Typical volumes for an acid treatment are 50 gal/ft for 15% HCl and 100 gal/ft for HCl/HF Acid.

6.1.6.5.2 Acid additives

Corrosion inhibitors are added to the acid to protect the surface equipment as well as the downhole equipment from the well being treated.

Emulsions can occur when a low-pH fluid (acid) is mixed with an oil (crude or diesel). These emulsions are readily stabilized by the fines released or produced during the acidizing process. A mutual solvent can be used to offset the formation of emulsions. Surfactants can also be used but this may cause a wettability problem with the resins. The resins will typically oil wet a sand.

When a highly asphaltic crude is encountered, a spearhead of an aromatic solvent can be used effectively ahead of the acid treatment. Typical solvents include toluene and xylene.

Note: A spacer should be placed between any mutual solvent or solvent and the resin system as they can negatively affect the ability of the resin to provide ultimate compressive strength.

6.1.6.5.3 Pickling or cleaning the workstring

Before pumping an acid treatment into the reservoir to remove damage from the near-wellbore area in the perforated interval, the workstring/conduit to the perforations should be "pickled" (cleaned) and a flush pumped down the workstring to remove residual pickling treatment fluids.

6.1.6.5.4 Neutralizing the acid

After an acid treatment is pumped, the entire system must be cleaned and acid residue must be neutralized before the sand-consolidation service fluids are pumped.

The acid can cause the cure of an epoxy resin to be extremely slow or stop it altogether. It is thought that this is due to the hardener being a base. The reaction between the acid and the base hardener may deactivate some or all of the hardener.

If it suspected that the pH of the fluid in the pore spaces is low as a result of an acid treatment, or in the case of a CO_2 flood field where the water in the pore spaces might have a low pH, the acid residue must be neutralized before the sand-consolidation service fluids are pumped.

Even though large volumes of fluids may be flushed through the system ahead of the sand-consolidation treatment fluids, some acid may remain in low spots in the pump, lines, and tubing. The sand-consolidation fluid and acid may be the highest-density fluids pumped during the job; this may result in the mixing of these two fluids in the system's void areas or unswept volumes. This mixture may lead to undesirable results.

Note: Acid will cause furan resins to harden and epoxy resins not to harden.

Before injecting the consolidation fluid, we recommend flushing an acid neutralizer pill through the pump and discharge lines, manifold, tubing, and near-wellbore area.

The neutralizing solution for flushing the pump and discharge lines should consist of the following items, mixed thoroughly:

- 10 bbl fresh water
- 100 lbm of NH_4Cl
- 50 lbm of NH_4HCO_3

In addition, a neutralizing spacer should be pumped into the tubing ahead of the preflush stage.

- 500 gal of 3% ammonium chloride (minimum)
- 100 lbm of ammonium bicarbonate

Another potential neutralizer pill is as follows:

- 10 bbl fresh water.
- 250 lbm of KCl ($\pm$7%).
- 100 lbm K_2CO_3.

6.1.6.6 Equipment Cleanup

Good well cleanup procedures are necessary to ensure the success of consolidation treatments.

Thoroughly flush and completely drain pumps, tanks, and manifolds before mixing chemicals. Ensure that no contaminants are present. The suction hoses to outside tanks should be of small diameter and as short as possible to minimize dead pockets of fluid when pumping at low rates.

6.1.6.7 Field consolidation treatment procedure/method

When good injectivity into all perforations is established and prepacking of the treatment zones is accomplished, the actual consolidation process can begin.

Note: If a gelled fluid is used as the carrier fluid, a post-prepack acid job is necessary to remove the gel before the consolidation fluid treatment is pumped.

- Completely flush and drain pumps, tanks, and manifolds before mixing chemicals. Ensure that no contaminants are present.

- Ensure that suction hoses to outside tanks are of small diameter to minimize dead pockets of fluid when pumping at low rates.
- Avoid fracturing pressures.
- Filter all aqueous fluids going to the well.

6.1.6.7.1 Placement volumes

Placement volumes are typically calculated during a radial penetration depth of 1 m. More information about this topic can be found on pages 10 and 11.

6.1.6.7.2 Injectivity testing

Performing an injection test prior to the resin treatment is a critical step. It helps to ensure that the injection pressure and the subsequent pump rate allows for placement of the resin treatment and the afterflush/displacement to the perforations prior to the resin curing to the point where it is too difficult to inject into the formation matrix.

Pumping at below the frac pressure is critical. "Exceeding the fracture pressure creates a channel that leads the treating fluids away from the area immediately surrounding the perforations. The consolidation treatment probably will not succeed because much of the consolidation chemicals are at substantial distances from the perforations where sand control is needed [19]."

The pump pressure/pump rate will typically be calculated using the radial form of Darcy's Law and will depend on the permeability of the formation and the viscosity of the fluid being injected.

$$Q = \frac{7.08 \times k \times h \times \Delta P}{\beta_0 \times \mu \times \ln(r_e/r_w)}$$

Q = flow rate (bbls/day), k = permeability (darcies), h = height or thickness of formation (ft), ΔP = pressure drop across length (psi), μ = viscosity (cp), r_e = drainage radius (ft), r_w = wellbore radius (ft), β_o = formation volume factor.

A typical pump rate will be in the 0.50–0.75 bpm range.

6.1.6.7.3 Pumping rates

Adjust injection rates according to the design used for each individual well. The objective is to coat the near-wellbore formation sand with the consolidation fluid chemicals, not to bypass some perforations by creating a fracture that may initiate at only a few perforations. Limit rates so that pressures do not exceed 80% of estimated fracturing pressure while injecting the treatment fluids.

Keep injection rates as low as practical throughout the job. The objective is to saturate the near- wellbore area with the consolidation chemicals, not to create a fracture. Injection rates should be held to no more than 2–3 gallons per minute per perforation if possible [41].

Note: In certain cases, it might be necessary to increase the rate to pump the resin into the perforations before it becomes too viscous to pump into the formation.

6.1.6.7.4 Preflush

Begin by placing the preflush in the tubing, annular area, and any space where fluid might be injected, even inadvertently, during the treatment. If the annulus will not "stand full," fill it with filtered preflush. If the fluid level falls during the job, refill the annulus as needed.

Note: A recommended enhancement to the standard matrix treatment is the use of nitrogen commingled with the treating fluids. The nitrified treatment method is recommended for improved uniformity of consolidation placement throughout the treated interval. However, the time required to place the job must be compared to the placement time allocated for the curable resin which is becoming more and more viscous as the time increases.

6.1.6.7.5 Post-flush fluid

The volume of post-flush is based on testing to determine retained permeability and consolidation strength as a function of overflush volume.

6.1.6.7.6 Catalyst (if Applicable)

The catalyst causes the resin to harden. The recommended overflush volume per foot of net pay is a compromise between regained permeability and adequate consolidation strength.

6.1.6.8 Placement techniques

Various methods have been used to place the consolidation treatments, including:

- Direct (matrix) injection
- Through coiled tubing
- Coiled tubing and jetting or jetting/pulsing tools
- Using nitrified sand-consolidation service for gravel-pack screen repairs
- Using selective injection packers or straddle packers

6.1.6.8.1 Direct (matrix) injection

The direct (matrix) injection method does not use tools, coiled tubing, or any other mechanism for aiding in the placement of the consolidation fluid treatment. The consolidation fluid treatment is pumped from the surface, generally through production tubing, and into the perforations.

When nitrogen is used to enhance placement efficiency, rates of 300–800 scf/bbl are generally used in all fluids. The actual rate selected can vary depending on factors such as design nitrogen fluid quality (gas phase volume percent), bottomhole treating pressure, treating liquid pumping rate, and equipment limitations.

The major disadvantage of pumping the treatment through the production tubing is that treating pressures could begin to increase and approach or exceed the fracturing pressure. If the fracturing pressure is exceeded, the only way to remove the treating fluid from the production tubing is by either continuing pumping or attempting to use reservoir pressure to flow the fluids back to surface. The risk of this action is that most of the afterflush may be pumped into the fracture rather than into the matrix of the formation and thus may allow the resin to set up in the pore spaces and possibly plug the near-wellbore reservoir.

6.1.6.8.2 Bullhead

Bullheading refers to pumping the consolidation chemicals directly down the permanent production tubing or casing. The well can be either a tubingless or a conventional completion with a pipe diameter of 2–3/8″, 2 7/8″, or 3–1/2″. The chemicals are pumped through the Christmas tree.

Because it is impossible to circulate the rathole full of brine without a rig, a high-density fluid injection is recommended to fill the rathole. To minimize the amount of high-density fluid needed, the rathole depth should be limited to 25 ft. Current safety information indicates that o nitrotoluene should not be used with consolidation systems that include organic acid chlorides as catalysts.

Note: a much smaller rathole (2–3-ft deep) is recommended whenever possible.

Alternativity, a denser brine than any of the treatment fluids can be used.

In conventional completions, the annulus between the tailpipe below the packer and the casing should be filled with diesel or a light brine if there is any concerns that the annulus will not stay full during the treatment.

6.1.6.8.3 Application through coiled tubing

Coiled tubing provides the following advantages:

- The option of moving the end of the coiled tubing past the interval being treated while fluids are pumped. This option is also possible when using conventional tubing for intervals less than 30 ft, although it can be time consuming.
- The ability to sweep the interval during treatment.
- Tools can be run on coiled tubing to improve fluid placement efficiency.
- Coiled-tubing movement rates of approximately 10 ft/min are desirable, provided this rate will allow a complete sweep of the perforated interval from bottom when starting a different fluid to top and back to bottom while injecting the fluid.
- Every perforation should be passed twice during each stage. Usually, the consolidation fluid will be the smallest volume chemical. Determine the minimum consolidation fluid liquid pumping rate and the consequent time to inject it. Then, adjust the coiled tubing movement rates to allow sweep from bottom to top and back to bottom.

6.1.6.8.4 Application with coiled tubing and jetting

A rotating/jetting tool can be used to blast scale off the walls of production tubing without pulling the string from the well. Without high-pressure nozzles, the tool can also be used to direct a spray of nitrified consolidation fluids into the perforations or into the gravel and formation sand behind a failed gravel-pack screen. Additionally, the tool rotates and sweeps the interval by pipe movement, which enhances the placement of the treatment into every perforation along the interval.

Reciprocate the coiled tubing at a sufficient rate so that each fluid is injected into every perforation. This method is particularly effective in repairing damaged gravel-pack screens. Using the side spray of the jetting tool directs the stream of treating fluids toward the failed gravel pack.

6.1.6.8.5 Application of nitrified sand consolidation and gravel-pack screen repairs

When using the jetting tool with commingled nitrogen for screen repairs, consolidation fluid volume is usually 15–30 gal for every ft of perforated interval. Generally, nitrogen rates of 300–800 scf/bbl are used in all fluids. These gravel-pack screen repairs are typically run without the aid of pre-consolidation acid stimulations or pre-packing. However, an acid treatment can be pumped before the consolidation treatment if stimulation of the well's productivity is needed or if scale must be removed to promote consolidation fluid adhesion to the sand and gravel.

6.1.6.8.6 Application using selective injection packer or straddle packer

Using a selective injection packer or straddle packer assembly provides treatment to every foot of the interval, even when formation permeability varies dramatically. The treatment fluids can be injected into a short interval using either an injection tool of opposed cups (similar to a perforation washing tool) or a short-interval straddle packer assembly. Either method limits the length of interval treated at one time. Consequently, permeability and injectivity into the shorter length of zone are likely to be more uniform.

A particular concern in working with selective injection techniques is that chemicals or sand can be routed out the injection port and through the perforation tunnels, then travel up to another perforation above the packer onto the top of the tool. This rerouting may result from a high-permeability fracture or cavity behind the casing. Pickup and slack-off weight for the injection tool should be checked frequently to confirm that sand or other sticking material is not accumulating above the tool. If sticking is indicated, reverse-out the treatment or bullhead with sand-consolidation service afterflush. Pull the tool above the interval to eliminate sticking concerns.

If acid preflushes are used ahead of the sand-consolidation treatment, remove incompatible fluids from the annulus before and after any consolidation fluid stage is injected.

6.1.6.9 Restoring well to production

A well should be brought back to production slowly after a consolidation treatment.

Swabbing and gas lifting are the two most common methods of restoring flow to the well. The gas-lift systems are designed to relieve the hydrostatic pressure slowly by unloading the kill fluid from the top down. By contrast, swabbing is potentially more detrimental. As the swab cup is pulled up, the hydrostatic head of fluid above the cup is instantaneously removed from the fluid below. If the well is swabbed too deeply or too rapidly, the instantaneous pressure from the reservoir into the well can become quite large. Presumably. the pressure drop could be sufficient to cause even a good consolidation to fail.

After consolidation, most wells will return to the productivity achieved before treatment. Wells that have been reperforated and effectively acidized before the consolidation treatment may produce at higher rates. Significant productivity loss caused by consolidation can be treated by either HCl acidizing or reperforating. However, if the problem is caused by poor reservoir quality, then little can be done to improve the production rate [19].

6.1.6.9.1 HCl acidizing

Occasionally, productivity is difficult to restore after consolidation. Many reasons have been proposed to explain why productivity is lower even when injectivity was good while the consolidation treatment was being pumped. Two such reasons are perforation plugging with debris loosened from the casing wall by the powerful solvents in the resin systems and partially catalyzed resin draining down the casing wall and accumulating in the perforations.

Whatever the cause, acidizing with HCl after the consolidation resin has cured effectively can enhance well productivity. HCl will not harm the consolidation strength once the resin has cured. The HCl can dissolve pieces of scale and rust in the perforations. It can also slightly attack the cement at each perforation, possibly enlarging the perforation tunnel and creating a new fluid flow path. Consolidations may be acidized with 50–100 gal/ft of 15% HCl. The acid is injected slowly and allowed to soak for a few hours before production is restored [19].

6.1.6.9.2 Reperforating

If efforts to restore productivity with HCl have been unsuccessful, reperforation of the consolidated interval can be attempted. The same interval that was consolidated must be reperforated to avoid reaching sand that was not treated with resin. This requirement calls for good depth control and close attention by service company personnel.

The consolidated radius around the wellbore ranges from 1 to 4 ft, depending on reservoir stratification and the ability of the formation to accept consolidating fluids uniformly. This 1- to 4-ft radius is beyond the penetration distance of typical through-tubing perforating guns. Field experience has confirmed that intervals can be perforated without causing sand production.

Jet perforators penetrate because of pressure, not because of heat; hence, the jet will not melt or degrade the plastic. The rock pores should be open and available for flow as in any other perforating operation [19].

6.1.6.10 Lessons learned and recommendations

6.1.6.10.1 Typical reasons for failure of a consolidation treatment

- Failure to treat the entire perforated interval due to mitigating factors such as:
- Mud plugged perforations or mud invasion of the formation.
- Sand fill covering the perforations.
- Lack of isolation in the primary cement job which can lead to inadequate coverage of the resin treatment (the treatment fluids will always follow the path of least resistance).
- Interval is too long (>3–4 meters)
- Interval has non-uniform permeability and/or non-uniform pressures
- Zone has a high percentage of water swelling clays.
- Zone has exposed shale stringers which can fracture more easily than the consolidation material can enter the fluid filled sand. For best results, avoid perforating dirty sections of clean sands.
- Dirty fluids.
- Poor quality control or substitutions of incompatible chemicals.
- Formation damage is extensive and is not removed resulting in a less than recommended pump rate and thus less control on the resin curing process.
- Fluid contamination or non-uniform treatment due to gravity settling of the resin into rathole fluids. This is discussed later.
- Annulus not full. There is a high risk of failure due to sticking the washpipe. Fluid and pump pressures should be monitored to be sure resin doesn't circulate up the back side. Use a packer if necessary.
- The rathole is not taken into account on the design of the treatment thus some of the resin is exchanged with the rathole fluid.
- The resin tends to be viscous which often results in low injectivity, especially during treatment in low permeability formations.
- Injection pressure that surpasses the fracture pressure generates channel or path of least resistance causing resin treatment fluids to treat the formation areas away from the intended areas around the perforations and near-wellbore.

6.2 Fines Migration control

6.2.1 Mechanisms and causes

Migration of formation fines are known to cause severe formation damage during production, limiting the potential production of the well. Hibbeler et al. [42] provided an excellent review of fines migration mechanisms. Many factors contribute to the migration or movement of formation fines [43,44]. They examined the impact of salinity changes, pH, and flow rate on fines migration and clay swelling that affect permeability decline. Formation sand and fines can be released during production as a result of shear failure if drawdown pressures exceed the yield strength of the rock [1].

The introduction of fine particulates into a propped fracture can have a major impact on fracture conductivity as demonstrated by McDaniel and Willingham [45] where proppant was mixed with formation sized fines. Table 6.1 shows an example where the addition of 100 mesh fines to a 20/40 proppant pack results in significant loss in conductivity to the proppant pack. The damage to conductivity will be more sever if the fines have a broad size distribution as is normally the case in reservoir rock. There are many opportunities for fines contamination of the proppant which is used to prop fractures. Many of these are associated with fracturing processes itself.

Numerous field case histories have demonstrated the severe impact of fines migration on well production potential. No matter how the wells were completed,

Table 6.1 This plot illustrated the significant loss of pack permeability that occurs when 20/40 mesh sand is mixed with 100-mesh fines [45].

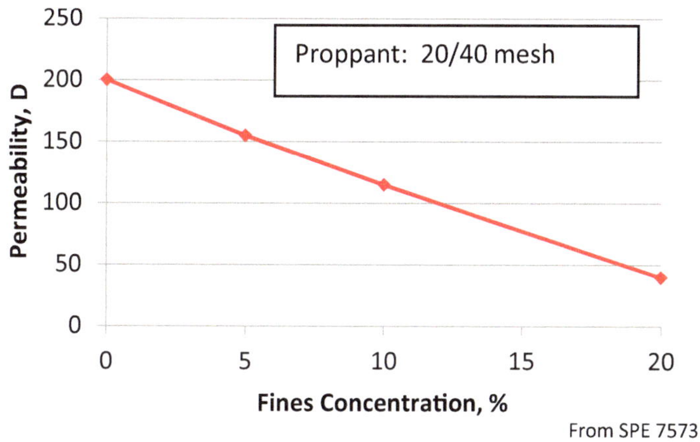

Fines Concentration	Permeability
%	Darcy
0	200
5	155
10	115
15	80
20	45

whether frac-packed or acid-fractured, fines migration could drastically reduce the production potential of these wells. If fines particulates were allowed to move freely with the production fluids, it was expected they would be the main culprit in choking off the well's early high production flow rates. All evidence supports the inevitable effect of fines migration on formation damage, which in turn severely impacts the production potential of producers [43,46,47]. Even in cases involving injector wells, the ability to prevent fines from migrating helped the injector wells maintain their injection capability [48–50].

6.2.2 Previous fines migration control methods

Various techniques have been developed in the industry over the years as fines-stabilizing solutions to overcome the effects of fines migration [43,46,47].

Other chemical treatments, such as the inclusion of: (1) clay-stabilizing surfactants as part of the completion fluids, or (2) polymers in remedial operations, have been applied in an effort to minimize fines migration and enhance well productivity [51–53]. These treatments commonly require that the treatment fluids be injected deep into the formation matrix, allowing the surfaces of the fines and pores to be in contact and to interact with the treating fluid. These treating fluids focus on flocculation or agglomeration of clay fines or involve changing the surface-wetting properties to reduce the tendency to migrate in aqueous fluids. Large volumes of treating fluid are often required to achieve the desired results. Most such treatment fluids provide only temporary solutions because they tend to desorb with time and with the production of fluids from the well.

Another solution also involves temporary clay stabilizers [54]. These stabilizers minimize the tendency to disperse or deflocculate naturally occurring fines within the formation matrix. Such chemical systems may include mono- and divalent salts or low-molecular-weight quaternary amines. This solution is considered temporary because the subsequent changes in water salinity can cause chemical dispersion or swelling of susceptible clays.

Gravel-pack completions provide mechanical approach designed to control formation-sand production. However, fines often migrates into the pack, resulting in its permeability damage. In addition to fines damage in the gravel, the sand screens used to retain the gravel can also experience a plugging problem caused by the fine particles. Alternately, if these fine particles pass through the screen, local erosion of the screen can become an additional concern. Various cleanup methods have been implemented to remove screen plugging and increase productivity [55,56]. Efforts are often focused on removal of the fines from the gravel pack, without consideration of the need to prevent fines production at its sources. An acid treatment, such as hydrofluoric acid acidizing, used in the workover to remove the bridging fines from sandstones, can often further weaken the formation and enhance the production of fines and exacerbate the problem [56]. Acidizing has often been used to dissolve fines by "unblocking" and enlarging pore throat geometry in the formation near the wellbore to increase the permeability of the formation. Production performance in wells that have been acidized or acid-fractured has often been disappointing. High production rates typically last only for a short time, followed by a drastic drop in production because of the damage caused by fines plugging.

6.2.3 Controlling fines migration into proppant pack

Major advancements have been made in providing chemical agents that can mitigate conductivity damage in propped fractures due to fines and the migration of fines. Since their introduction in 1996, surface modification agents (SMAs) continue to play an important role in controlling fines migration in oil and gas wells to enhance and maintain their productivity.

The SMA material has been described as a “nonhardening resin manufactured from renewable resources” [57]. A particularly useful SMA was made from “tall oil,” a byproduct of the paper and pulp industry; this SMA combined surface activity with a polymeric material. The polymer is insoluble in both water and oil, but is soluble in a few highly oxygenated solvents, which makes it particularly suited for treating proppant. It is described as thermally stable, with a polar polyamide backbone with long, pendent fatty chains. The polar polyamide portion of the polymer tends to adsorb strongly on mineral surfaces while the hydrophobic fatty chains tend to extend away from the mineral surface. This leaves the mineral surface highly hydrophobic and sticky to the touch [58].

Lehman et al. [59] provided field case histories to illustrate long-term benefits from applying SMA as a conductivity enhancement material in fracturing treatments compared to those without. Initially, when the conductivity enhancement material was introduced, it was believed that the benefit would only be temporary or short-term. Since then, long-term benefits illustrated by cumulative production volumes have been well-established and recognized by numerous operators.

Proppant pack stability is important in maintaining the propped fracture conductivity. High stress, coupled with stress cycling, can lead to proppant crushing and subsequent migration of fines formed by proppant crushing and bridging off, eventually plugging the pack’s permeability. The tendency for crushed proppant fines to migrate is mitigated by the application of SMA to the surface of proppant as it is used in the fracture treatment. The same application of SMA to proppant can also prevent formation fines from migrating into the proppant pack, thereby conserving the effective fracture width. This application has found widespread use in poorly consolidated formations using frac-pack completions. In these applications, greater than normal fracture conductivity can be achieved by using a proppant size larger than normally recommended to prevent formation fines invasion as the SMA-coated proppant traps the formation fines at the formation/fracture pack interface [57].

Application of SMA-coated proppant in coalbed methane producers was found to provide long-term coal fines control [60]. Without SMA, typical fracture stimulation treatments in these applications required frequent refracturing. In the SMA treatment's ability to affect fracture sustainability, it was observed that the load recovery from the frac treatment and the dewatering process were both significantly accelerated.

The applications of SMAs in controlling fines can be divided into primary and remedial types of treatments.

6.2.3.1 Primary fines control

In general, the liquid, nonhardening SMA is coated on the proppant as it moves though the proppant conveying equipment at a rate that is corresponded to the proppant delivery rate dictated by the fracturing treatment injection rate. Activation of this coating to make the proppant tacky can be immediate or delayed depending on the need. Once activated, the thin, hydrophobic coating binds any proppant fragments that may arise from the pumping operation. This thin coating remains very viscous and tacky which protects the proppant pack indefinitely against movement of fines and stabilizes both the proppant pack and the pack/formation interface. Even when closure stress increases and some of the proppant is crushed during pack rearrangement, the fragments are bound in place by the tacky coating, thus preserving the conductivity of the pack. Since the proppant is tacky, it is difficult for the proppant to form a tight pack, resulting in an increase in conductivity. Typically, about 30% increase in conductivity is observed when conductivity testing is performed using water [61–65]. This effect is lost once the closure stress reaches a value where proppant begins to crush and the pack does compress as can be seen in the Fig. 6.5 below.

Nguyen et al. [67] performed a study to examine the mechanisms in which the coatings of SMA materials on proppant can significantly impact the migration behaviors of formation particulates into the proppant pack and maintain its permeability. Soft, unconsolidated formation material was used in simulating formation faces. The mechanisms involved an on-the-fly, direct coating of proppant with surface-modification agents. An SMA does not cure or set over time but alters the cohesive properties on the surface of the proppant. The treatment renders the formation sand and fines immobile so that migration and plugging do not occur. Only minimum embedment or intrusion of formation fines into SMA-coated proppant pack was found, retaining all proppant width. A close inspection of the SMA-coated proppant pack indicated that formation fines stopped at the interface without migrating deep inside the pack. When this was repeated without the SMA coating embedment of fines was observed in the sand pack.

6.2.3.2 Remedial fines migration control—treating unconsolidated/weakly consolidated formations or propped fractures

Evaluation of treating of ultra-thin tackifying agent (UTTA), that has been prepared by diluting SMA either with a large volume of a solvent-based or an aqueous-based solution, on formation particulates was performed to demonstrate its effectiveness in controlling migration of formation fines into the pore spaces of the proppant pack and its impacts on the pack permeability [66,67].

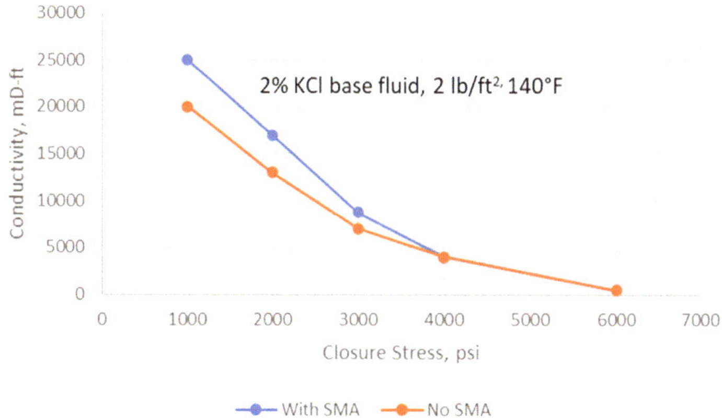

FIGURE 6.5

Top Left: SMA coated 20/40 mesh Ottawa sand after the API conductivity testing; results shown in the graph. Note that all the fines generated from the conductivity testing are adhering to the SMA coated grains unlike the uncoated sample shown in the bottom left photo [66].

6.2.3.2.1 Solvent-Based ultra-thin tackifying agent

This particulate SMA is soluble only in a few select oxygenated solvents and is not soluble in normal well fluids or well treating fluids. Since the oxygenated solvent is very water-soluble, it rapidly partitions to the aqueous phase and is extracted from the UTTA, leaving behind a thin SMA polymer coating on formation sand or mineral fines particles. The macromolecule is very stable and resists attack by acid and caustic except under extreme conditions.

The polymer spreads on the silicate surface, and the hydrophobic groups extend away from the polar mineral surface and appear as bushy appendages. The polymer was designed so that the proper spacing between hydrophobic branches would allow them to "just fit" together, resulting in a significantly increased surface area that is available for association. Although sufficient force can pull molecules apart without breaking carbon-carbon bonds, these associations can re-form repeatedly upon contact. This process can be likened to a molecular hook-and-latch system [68].

6.2.3.2.2 Aqueous-based ultra-thin tackifying agent

An aqueous-based SMA (ASMA) is a water external emulsion having an SMA contained within the internal phase [69,70]. New emulsion technology allows SMA release to be delayed so that the extreme tackiness and insolubility does not damage mixing equipment and can be easily cleaned.

This aqueous-based UTTA can be prepared by diluting ASMA with a large amount of brine for deep injection into the formation matrix or the existing proppant packs.

6.2.3.2.3 Applications of ultra-thin tackifying agent

One new application made possible by emulsion delivery is the inclusion of the ASMA in the first fluid injected into a fracture, which coats the fracture face with a hydrophobic film. This film provides stability to soft formations, inhibits water invasion, minimizes damage attributed to formation spalling, and improves proppant distribution by causing proppant to adhere to the frac face. This enables its use for remedial treatments to repair fines-damaged frac-packs where frequent acid treatments are required to maintain acceptable production.

Acid treatment dissolves formation fines in the perforations and the formation and displaces the remaining fines further away from the wellbore. The use of UTTA after acidizing helps stabilize formation particulates by locking them in place. This action could enhance the effectiveness of acid treatment because the generated wormholes, or channels, would remain open without collapsing or becoming plugged by fines migration.

Similarly, after gravel pack has been remediated with a "cleanup" treatment, UTTA can be applied to lock formation fines in place far away from the gravel pack and prevent them from migrating back. The use of UTTA in the prepack treatment or as part of the prepad of a gravel packing should stabilize any potential migrating fines in the formation, preventing them from moving into the gravel pack.

Propped fractures can also be remediated using UTTA to help displace formation fine particulates and anchoring them far away from the wellbore.

UTTA can be injected as a prepad or pad treatment fluid in front of the main fracturing treatment to treat and stabilize the formation faces of the fracture as they are generated during hydraulic fracturing. Recent study by Weaver et al. [71] has shown that coating proppant with a thin, highly dielectric, hydrophobic film can result in a long-term reduction in mineral dissolution rate, compaction, embedment, and polymer-anchoring sites often associated with proppant permeability damage. It is believed that the use of this UTTA system in the prepad fluid may allow the fracture faces to be coated with a thin hydrophobic film. This coating prevents polymer reactions with activated mineral surfaces, allowing water to be repelled (especially in gas wells) and minimizing the effect of proppant diagenesis in addition to stabilizing the formation sand and fines in the fracture faces.

Other fines migration studies were performed through unconsolidated, fine-laden sand pack columns using either solvent-based or aqueous-based SMA [68,70,72–74], or simulating post-acidizing treatment [68–70].

6.2.4 Fines migration field case histories

Field case histories have been reported in numerous SPE papers and articles since the introduction of SMAs in 1996 as these polymers continue to play an important role for fines migration control in various well conditions and environments [60,64,75–78].

6.3 Proppant flowback control

Flowback of fracturing sand or proppant from hydraulically fractured or screenless frac-packed wells has been known to cause catastrophic problems similar to the dilemma resulted from sand production. At considerable expense, fractures are packed with proppant to provide hydrocarbon conductivity to the wellbore by ensuring that the fractures remain open after the fracture stimulation treatment during production operations. Since proppant pack width has a direct correlation to fracture conductivity, it follows that if proppant is produced back into the wellbore, then the fracture conductivity is reduced. In addition to lost productivity, proppant flowback increases operational expense and results in production downtime required for cleaning out wellbores and replacing pumps and equipment damaged erosion, plugging, sticking and wear caused by produced proppant entering the wellbore.

Rapid proppant production is often observed during the fracturing fluid flowback following the stimulation treatment, particularly when immediate flow is used to induce rapid formation closure. Inadequately broken fracturing fluids at this stage can exacerbate proppant flowback. If this technique is used, very aggressive fracturing fluid breaker program should be employed [79]. Many operators will flow wells back at low rates during fluid recovery to minimize proppant production. Over displacement of a fracture treatment to mitigate proppant flowback is often used when using multi stage completions in horizontal wells to avoid mechanical issues in the well [80]. In this situation, displacing proppant away from the near wellbore may result in significant reductions in near wellbore fracture conductivity resulting in significant loss of production as the price for proppant free production.

Another type of proppant production is a slow, but persistent production of proppant that occurs in many reservoirs [81]. This is a nuisance which also leads to frequent well cleanouts, equipment erosion and pump replacements. Precautions can be made to treat proppant to prevent proppant flowback but in some areas, only a few wells will experience the problem and it is more economical to deal with the problem wells rather than applying proppant flowback control to all of the wells during the fracturing treatments as discussed by [82].

6.3.1 Primary proppant flowback control

Common primary treatment (a treatment performed in conjunction with the fracturing treatment) methods used to control proppant flowback include curable resin pre-coated proppants [81–84], intermixing proppant with fibers [85], intermixing proppant with deformable particles [86], on-site coating of proppant with curable resin [87–89] and on-site coating of proppant with a nonhardening resins to modify proppant surface properties [61,62,76].

6.3.1.1 Curable resin pre-coated proppant

Curable resin pre-coated proppants are often used as tail-in materials with the aim of filling the fracture volume near the perforations with a consolidated proppant pack that will act as a screen for the preceding volume of unconsolidated proppant. Proppants with a range of size and strength are coated with resin that has been partially cured so that at temperatures below about 120°F, the surface is glassy and the coated sand is free flowing. As the temperature increases, the surface begins to soften and become tacky. When the tacky grains are pressed together, the resin welds between the grains and eventually resulting in a consolidated pack. The hardening process requires heat, stress and time to complete. At low temperatures, long times are required and may require long shut-in time before full production can begin.

Certain additives (accelerators) can be included in the frac fluid which accelerates the softening of the resin coating, promoting the adhesion or tackiness between grains and permitting earlier well flowback. However, extended time is still required to achieve good grain-to-grain consolidation. At high temperatures, the resin may cure before formation closure has occurred resulting in a cured resin coated proppant with little consolidation. Rapid and thorough gel break is necessary to get adequate grain-to-grain contact to ensure that good consolidation takes place. Some fracturing fluid additives are known to accelerate the curing of resin pre-coated proppants so that hardening occurs before grain-to-grain contact is achieved. Several manufactures supply curable pre-coated proppants and use a variety of resins. Some of these can have a significant impact on fracturing fluid stability and requires on-site testing to determine what corrective fluid formulations are required.

Curable resin pre-coated proppants are generally useful in wells with low to moderate flowback rates (5–30 BPD/perforation) and generally provide weak consolidation strength (5–50 psi) [90]. The main advantage of RCP systems is their ready-for-use applications, like those of uncoated sand or proppant. However, RCPs have the following disadvantages:

- Short shelf-life during warehouse storage.
- Activator must be used to initiate curing of the resin if the well has low bottomhole temperature.
- Closure stress must be applied on the proppant grains to enhance the bonding between grains.
- Does not handle the effect of stress cycling as part of well operations.
- Only selected sizes of proppant are delivered to the wellsite.

6.3.1.2 Fibers and deformable particulate

Fibrous materials [85] and deformable particulate [86] have been used to control proppant flowback. These solid materials are mixed with the proppant and become an intimate part of the proppant pack. A network is created between the proppant and the fibrous strands. The main functions of the fibrous strands are to induce bridging at the perforations and allow solids-free fluid to flow through. The use of deformable particulate requires the application of closure stress on the proppant-deformable particulate so the particles will adhere to one another by means of inter-embedment. Without the closure stress, a stable pack cannot be established. Although these materials reduce proppant flowback, they do not eliminate the problem entirely and are usually accompanied by significant decreases in fracture conductivity [91].

6.3.1.3 On-site liquid resin coating

The use of on-site liquid resin coating technology provides a superior solution for primary treatment to mitigate proppant flowback. The same curable resin systems that have been used for sand consolidation treatments can be applied in proppant flowback control treatments. Products exist within the industry that include three resin systems formulated to cover bottom hole temperatures, ranging from less than 70°F to more than 550°F. Each of these results in high strength (>100 psi) proppant consolidation, but are designed to cure at significantly different rates so that the proper system can be selected to suit the reservoir properties of temperature and fracture closure rate. The ability to select ensures that good strength can be developed at low temperatures and yet provide adequate time before curing for the higher temperature regimes. The liquid resin process requires only proppant grain contact and does not require stress to ensure consolidation as the precoated curable resin coated proppants do. In addition, 100% of the resin goes to creating welds that consolidate the proppant. It is generally a good practice to perform fracturing fluid quality control rheology test with fracturing fluid that has been exposed to precoated proppant or liquid resin coated proppants to ensure compatibility [89,92,93].

Scanning electron micrographs of a consolidation of RCP 20/40 mesh sand after curing at low temperature and stress showed weak regions of bonding between grains and the small footprints left where the consolidated proppant was broken apart. Scanning electron micrographs of a proppant coated (20/40 mesh) with LRC show the stronger regions of bonding and the large footprints which is characteristic of a strong consolidation formed at the same temperature and stress conditions.

6.3.1.4 Liquid-curable resins

As with any technology, the development of the liquid-curable resins (LCRs) has undergone various phases that often involved laboratory and field testing. Laboratory testing represents ideal conditions, while field trials help expose the difficulties that a resin system must overcome to meet the requirements of field applications.

In the 1980s and 1990s, the preparation of an on-the-fly LCR literally involved mixing multiple components at the wellsite just before the fracturing treatment. All components were either premeasured or measured at the wellsite [87,94,95].

It was not simple to readjust the resin-treatment volume required for the fracture-treatment design (i.e., treatment size, amount of proppant). Excluding a single component could lead to failure of the treated proppant pack to consolidate. Because both resin and hardener components were premixed into a single component and stored in a container, the resultant resin mixture had a short shelf life and was required to be used in coating as soon as possible before the resin viscosity increased to the extent that it could not be pumped and metered. If a premature screenout occurred during the fracturing treatment, the remaining mixed resin had to be disposed of properly. This mixed resin could not be saved for another fracturing treatment because of safety requirements and its limited shelf life.

The coating of this early version of LCRs involved metering the resin mixture directly into the frac-blender tub in which the resin interacted with fracturing carrier fluid before coating onto the proppant. This coating process is known as "wet coating." Surface chemistry must be "just right" to ensure effective coating of the resin onto the proppant grains. Both the carrier fluid and coating agent are in competition to wet the proppant. If the balance is not correct, the following problems can develop:

- A less effective coating of the LCR is placed on the proppant as it competes with the carrier fluid, and/or poorer-than-predicted consolidation-performance properties are achieved.
- Direct interaction of the coating agent with the carrier fluid can affect transport properties of the carrier fluid, such as its crosslinking and breaking times.

In 2001, an LCR and a new coating process for this resin system were developed to overcome these problems [88]. This LCR comprises a liquid resin and a hardening agent, which are preblended and delivered to the wellsite in separate containers. As long as these two components remain separate and uncontaminated, they have a shelf-life of several months. The new coating process involves direct coating of proppant with the mixed resin during the transfer of the proppant from its storage container into the carrier fluid. Coating of the mixed resin onto the proppant is completed before the addition of the proppant to the gel-carrier fluid. Therefore, the gel cannot compete to coat the proppant because the resin coating is already in place. This leads to more consistent properties from the coating agent and helps eliminate the possibility that an excess amount of coating agent is carried in the well. While this LCR can still affect the breaker and crosslinker performance, the effects are far less severe than in the earlier "wet-coating" process.

The LCR was designed to handle the wide range of bottomhole static temperatures (BHST) in the wells. Current LCR can be selected from one of the following resin systems, based on the bottomhole temperature of the well:

- A low-temperature, epoxy-based resin designed for a temperature range of 70°F–225°F.

- A mid- to high-temperature, epoxy-based resin designed for a temperature range of 200°F–300°F.
- A high-temperature, furan-based resin designed for a temperature range of 275°F–550°F.

6.3.1.4.1 Advantages of liquid-curable resins treatments

In formations where controlling proppant flowback following fracture treatments is a primary consideration, LCR can help improve well productivity in several ways.

On-the-fly application. LCR is applied on-the-fly (i.e., it is metered directly onto the proppant where LCR is spread and coated evenly on the proppant grains before being mixed with the fracturing fluid as the fluid is being pumped downhole). The system eliminates problems inherent to storage and handling of resin-precoated proppants. LCR can be coated on any natural sands, either white or low-grade sands, or any types of man-made proppants at the wellsite, regardless of their mesh sizes.

Enhances or maintains proppant-pack conductivity. RCPs and fibrous flowback-control materials placed in the matrix of the proppant pack often reduce conductivity under high closure stresses. Conductivity testing performed in the laboratory using modified API linear flow cells shows that coating LCR on 20/40-mesh, high-strength ceramic proppant actually improves the conductivity of the proppant pack [93]. Several factors contributed to this increase in conductivity. The tackiness of LCR coating alters the proppant-pack density by increasing intergrain friction, thus providing higher porosity within the pack. LCR not only bonds the proppant grains to each other, but also bonds to the fracture faces. This bonding distributes the point-source load of the proppant across the formation face, thus reducing the spalling effect and the formation fines intruding into the proppant pack.

Consolidation strength. Consolidation strength of the consolidated proppant pack is critical to controlling proppant flowback, especially in wells with high production flow rates. RCP often cannot provide the necessary consolidation strength because most of the resin coated on the proppant is already cured. In addition, high closure stress is required to provide good grain-to-grain contact before the resin cures. In contrast, all resin of LCR coated on the proppant is curable. Even with no closure stress, a LCR-coated proppant pack has more than sufficient consolidation strength to reduce or stop proppant flowback under the most severe conditions, while providing and maintaining high fracture conductivity.

The LCR was formulated with a proprietary additive to help with the removal of any crosslinked gel coating on the proppant to enhance the contact between proppant grains, thus increasing the consolidation of the proppant pack even without applied closure stress. As a result, even under low or no closure-stress conditions, high consolidation strength of the coated proppant pack can still be developed. In addition to the ability to provide consolidation strength, this resin is also formulated to provide elasticity, which is beneficial to effectively handling the repeated stress-strain cycles that occur during normal production operations. [93]. The capillary pressure between grains pulls the liquid resin to the contact points, thus helping prevent resin from occupying the pore spaces. The bonding between grains, illustrated by the footprints at the contact points, helps establish the consolidation strength of the proppant pack to withstand stress load or high shear. This consolidation strength (i.e., unconfined compressive strength) is proportional to the concentration of LCR coated on the proppant (Fig. 6.6) [93]. The consolidation strength of the proppant pack is dependent on the fluid/proppant system, proppant size, curing temperature, curing time, and resin concentration coated onto the proppant.

6.3.1.4.2 Field application of liquid-curable resins

For wells with long or multiple perforated intervals, the best practice has been to coat all proppant in all concentration stages to minimize uncertainty and to ensure complete success of the resin treatment. However, based on the experiences gained from earlier fracturing treatments using RCPs, only the tail-in portions of the proppant were treated with LCR in most cases. Typically, about 15%–30% of proppant was treated with LCR at the tail-in stages for consolidating the proppant grains in most of the fracturing treatments. Concentration of LCR was typically applied at 3% (volume by weight of proppant). Based on production responses, the operators might choose to coat proppant with a concentrations ranging from 1.5% to 3.5%.

Note: Al-Ghurairi et al. [91] provided a case study of a potential issue of having the first proppant placing in the fractures to be the first proppant to be produced back as a result of fluid convection or proppant banking. As a result of this observation, it has been recommended to coat the first 5%–10% of the proppant with LCR, and the last 10%–15% of the proppant with LCR, prior to their

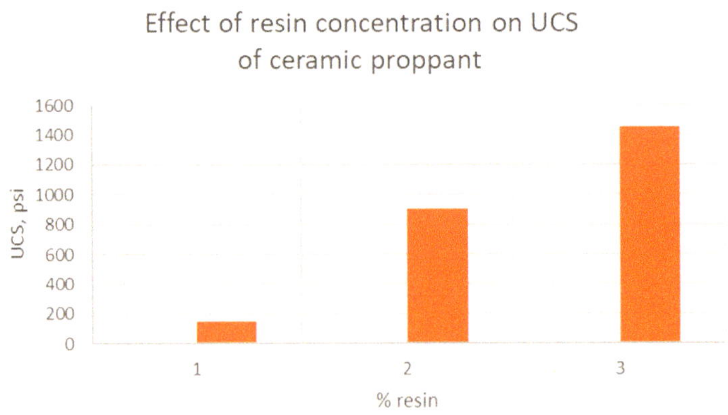

FIGURE 6.6

Effect of LCR concentration on consolidation strength of treated ceramic proppant.

placement in the fractures, to overcome the effect of fluid convection or proppant banking.

The liquid resin and the hardener were delivered to the well location in separate containers. They were metered in proportion with the desired fluid and proppant rate being pumped during the treatment. These individual components were separately metered into a single stream and then pumped through a static mixer, which provided sufficient mixing to create a homogeneous activated-resin blend. The mixed LCR blend was then injected to the bottom area of the sand screw, which had its lower end installed inside the sand hopper. The auger action of the sand screw helped distribute the resin liquid onto all the dry proppant as it moved the proppant from the sand hopper to the blender tub. Once dropped into the blender tub containing the fracturing fluid, the coated proppant was mixed into the fracturing fluid before this slurry mixture was pumped downhole.

An aggressive breaker schedule was applied to ensure that early gel breaking would allow the proppant grains to obtain grain-to-grain contact before the resin began to cure. The shut-in duration of the well after the fracturing treatment often depended on whether the operator wanted to apply forced closure. Without applying forced closure, the shut-in duration of the treated well was dependent on the well BHST, ranging from 4 to 48 hours to allow the coated resin to achieve near or complete curing before the well backflow back was started.

6.3.1.4.3 Case histories of liquid-curable resins

LCR systems covering all temperature ranges were introduced beginning in 2001. To date, more than 3500 hydraulic fracturing and screenless frac-packing treatments using LCR systems were performed worldwide [88,89,92,93,96–100].

A case history illustrating the impact of LCR in a high-temperature, high-pressure gas well application was described by Nguyen et al. [97]. This case compares four wells completed with LCR-coated proppant to four wells without LCR coating on the proppant. These wells were completed in the same reservoirs and used similar quality proppants. In this case, there is a definite increase in early production that can be attributed to increased fracture conductivity. There is also an increase in long-term productivity, suggesting that there is a much longer effective fracture length in the LCR-treated wells that is actually providing an increased drainage radius.

From late 2004 through 2008, more than 3200 hydraulic-fracturing treatments have been pumped by more than 70 operators using the LCR process in the Permian Basin [89]. More than 150 million pounds of proppant have been coated with the liquid resin. Fig. 6.7 provides a summary of the number of fracturing treatments and amounts of LCR-coated proppant used per year in the Permian Basin since 2005 [89]. More operators in the Permian Basin have seen the fracture conductivity increases that LCR coating gives them and want to coat a higher percentage of the total proppant pumped.

The wells were drilled in more than 40 formations ranging in depth from 1400 to 16,000 ft, with BHSTs from 90°F to 250°F. Depending on the operator and

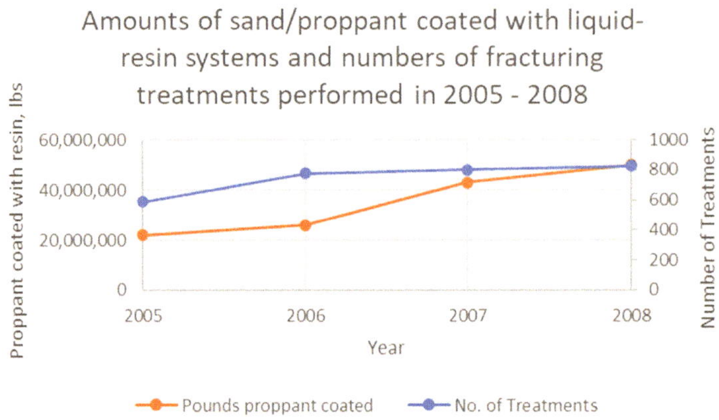

FIGURE 6.7

Amounts of sand/proppant coated with liquid-resin systems and numbers of fracturing treatments performed since late 2004 [89].

formations, well-drilling spacing ranged from 160- down to 10-acre infill drilling. Approximately 95% of the wellbores at perforated intervals of the fracture-treated wells were vertical or slightly deviated, and approximately 5% were horizontal. The majority of the fracturing treatments were performed in either sandstone or dolomite formations, but other formations, such as limestone and shale, were also included.

The fracturing carrier fluid was designed for BHSTs ranging from 90°F to 250 °F, including mostly borate crosslinked hydroxypropyl guar, CO_2 foam of zirconate crosslinked carboxymethyl-hydroxypropyl guar, and occasionally with just linear gel. The fracturing fluid was generally pumped down casing or tubing. The average pad size for the wells was between 15% and 50% of the total fluid volume. Pump rates applied in the fracturing treatments ranged from 15 bbl/min to more than 50 bbl/min; however, more than half of the fracturing treatments were performed with pump rates of 30–40 bbl/min. Various proppant materials were pumped, including mostly natural sands (such as Brady and Ottawa) and sometimes man-made proppants (intermediate- or high-strength ceramics), with mesh sizes ranging from 20/40 to 8/16. The maximum proppant concentration varied depending on formation lithology and permeability up to 8 lbm/gal, with an average proppant amount intended for placement into the fractures ranging from 50,000 to 200,000 lbm per fracturing treatment.

Except for a few premature screenouts, more than 95% of the fracturing treatments were successfully performed as per design. The treatment applications of LCR-coated proppant have provided an excellent solution for proppant-flowback problems compared to the use of RCPs or any other additives that have been applied in the area. LCR-coated proppant drastically reduces the number of wells with proppant flowback, allowing the operators to maintain the well production at high drawdown without losing production because of well shutdown and/or workovers. In fact, the number of workovers is significantly lowered in wells that have been treated with LCR compared to those treated with RCPs.

6.3.1.5 Non-curable surface-modification agents

For treating proppant surfaces, a highly viscous surface-modification agent (SMA) can be applied as a liquid additive to water-based fluids on-the-fly during hydraulic fracture-stimulation treatments [61,62,76]. This agent instantaneously coats the proppant with a thin, tacky coating that dramatically increases intergrain friction and the fluid velocity required to cause mobility of the proppant pack. This tackiness can also significantly reduce the proppant settling rate and the ultimate density of the settled pack, resulting in a better vertical distribution of proppant within a fracture. When used with resin-coated proppant, either precoated or coated on-the-fly, the SMA allows the well to be returned to production quickly; consolidation is not necessary before production is resumed. The SMA can allow more aggressive well-cleanup procedures after fracture stimulation. Because it does not require any shut-in time for activation or curing, the SMA is particularly useful in forced-closure operations. Under certain conditions, aggressive flowback techniques have improved stimulation performance, reportedly by producing damage-inducing fracturing fluids out of the fracture and allowing for forced closure. One of the major limitations of these procedures, however, is the rate at which fluid can be produced without proppant flowback. Uncontrolled proppant production leads to increased completion costs and reduced stimulation treatment performance.

For wells with expected low flow rate (< 5 BPD/perforation) production, low cost proppant flowback control can be achieved using a non-curable resin system (SMA) that coats the proppant leaving the surface hydrophobic and very tacky. The hydrophobic coating on the sand behaves like VELCRO, that is, the coated grains adhere to each other upon contact, but can be pulled apart only to adhere again upon contact with the same or different coated grains. SMA coating on fracturing sand for example, increases the critical flow velocity to produce proppant by 3–6 times [76]. Experimental data indicates that 20/40 mesh Brady sand has a critical flow rate of 0.6 BPD per ½-inch perforation. When coated with 1% SMA a fluid rate of 3.2 BPD/perforation is required to initiate proppant flowback. This figure also shows experimental data for 16/30 sand both uncoated and SMA coated and the impact on critical flow rate. Once the fluid rate exceeds the critical fluid rate, untreated proppant flows freely. The SMA treated proppant produces very slowly until a channel is created into the proppant pack that is stable. This slow production of proppant is why SMA has not marketed as a flowback control agent, however many operators use this material routinely to allow rapid fracture cleanup while minimizing proppant flowback. Often, the savings from the improved cleanup time alone will justify the SMA treatment cost.

Aqueous-based SMA (ASMA) is applicable for all of the same uses as solvent-based SMAs, such as on-the-fly coating of proppant during fracturing, frac-packing, or gravel-packing.

6.3.1.5.1 Field application of surface-modification agent

The SMA is coated on the proppant on-the-fly during the HRWP, hydraulic fracturing, or frac-pack treatment, preferably throughout the proppant stages. It can be coated in a dry-coat or a wet-coat process. In the dry-coat process, the SMA is added directly onto dry proppant using the liquid-metering systems as the proppant proceeds from the sand hopper and is augered with a sand screw into the blender tub. In the wet-coat process, SMA is added to the downstream side of blender tub where it becomes coated onto the proppant by means of mixing action as the proppant slurry is injected downhole.

The proppant was treated with SMA using a concentration of 1% (volume by weight of proppant) throughout proppant stages. The reasons for this 100%-coating were (1) uncertainty about where untreated proppant and treated gravel were placed near the wellbore and (2) to help ensure that an effective fines-migration control into proppant pack is achieved.

The liquid SMA solution was delivered to the well location and metered in proportion with the proppant rate pumped during the treatment. The SMA was injected to the sand at the bottom of the sand screw, where the bottom end of the sand auger was installed inside the sand hopper. The auger action of the sand screw helped distribute the SMA onto all the dry proppant as it moved the proppant from the sand hopper to the blender tub. Once dropped into the blender tub containing the carrier fluid, the coated proppant was mixed with the fracturing fluid before the slurry mixture was pumped downhole. This direct dry-coating maximizes the coating effectiveness of SMA onto the dry proppant and minimizes any chemical interaction between the SMA and the carrier fluid.

6.3.1.5.2 Case histories of surface-modification agents

Since the introduction of SMAs in 1996, more than 20,000 treatments of hydraulic fracturing, frac-packing, and high-rate water packing were performed in which proppants were coated with this type of polymer [61,62,64,91,97,101–103].

Lehman et al. [60] reported that the use of SMAs to coat proppants used in propping hydraulic fractures resulted in sustained and more uniform production from wells. From that publication, the production-decline curves from some of their data appear to show a significant change in decline rate compared to the use of untreated proppant.

A field study of SMA-treated proppant was reported to the Arkansas Oil and Gas Commission 2004 CBM Workshop that disclosed long-term results on gas production [102]. These were coalbed methane (CBM) wells in the San Juan Basin that typically required refracturing annually to produce at an economical rate. With the SMA-treated proppant, no refracs have been requiredproduction has remained essentially constant for five to six years. This longevity was initially attributed to prevention of fines invasion into the proppant pack; however, recent findings indicate that diagenesis protection has contributed to this fracture conductivity endurance.

It is evident that the production rate for the well stimulated without SMAs declines at a significantly greater rate than the well using SMAs. While a small increase in average daily-production rate was apparent after production stabilized, the major impact was the effect of the SMA to moderate the production decline. In addition, it was reported that the SMA-treated wells tolerated interruptions in production with little effect on the daily production rates, while the untreated wells showed a significant negative impact from well-production interruption.

6.3.1.6 Other proppant flowback control methods

6.3.1.6.1 Forced closure

Forced fracture closure is a technique used to close the fracture rapidly, trapping proppant in a uniform distribution [79]. This technique can potentially eliminate proppant flowback because the stresses exerted by the closing fracture walls help hold the proppant in place. However, closure stress does not always hold proppants in place. Test results indicate that closure stress might actually contribute to flowback. Fracture-closure stress is normally the mechanism relied on to hold proppant in place. Unfortunately, many times it is not effective and results in proppant flowback [104].

6.3.1.6.2 Mechanical screens

The primary function of screens is to provide mechanical support to prevent the proppant or gravel placed in the screen/wellbore annulus from flowing back into the wellbore. The gravel and proppant act as the primary filter to prevent formation sand and fines production. Screens help keep the gravel or proppant in place to support the perforation tunnel and fracture and maintain a conductive path for production. The screens can become damaged from plugging, scale buildup, corrosion, or erosion caused by producing formation fines and sand [20,21]. The installation of sand screens restricts the wellbore diameter.

6.3.1.6.3 Screenless frac-pack with near-wellbore consolidation

Combining resin consolidation of near-wellbore area with hydraulic fracturing (screenless frac-pack) in high-permeability formations provides an alternative solution in which negative or zero-skin completions can be achieved [105]. Frac-pack provides a highly permeable conduit to connect the untreated formation with the wellbore, bypassing the consolidated, permeability-reduced area near the wellbore region. All proppant that is applied in propping the fractures should be coated with a liquid-resin system to provide consolidation within the pack and prevent proppant flowback. A curing period is required to allow the resin-treated proppant to consolidate with the near-wellbore, resin-treated formation and form a hardened, permeable pack bed. This method increases the viability of screenless completions in unconsolidated formations. It can be applied either as a primary completion or a remedial treatment.

6.3.2 Remedial methods for proppant flowback

6.3.2.1 Reducing production rate

A common procedure for controlling proppant flowback or formation-solids production is to reduce well production flow rate until the well produces solids-free [106]. This method is probably the most convenient and economical means to solve the proppant-flowback problem in the short term to allow the operator time to assess the situation and implement alternative solutions. However, the obvious drawback of this approach is the loss in revenue that occurs when the well is not producing at its full potential.

6.3.2.2 Remedial proppant treatments

It is sometimes desired to remediate a proppant flowback problem from an earlier fracturing treatment. The resin systems used in remedial proppant treatments have much lower viscosity than those of LCR coated on proppant during a fracturing treatment [107–109]. A number of diluents were examined to reduce the viscosity of the resin mixture so it can easily be placed in the formation while still providing adequate consolidation strength. The current systems use diluents that contain water-soluble solvents that are soluble and compatible with the chosen curable resins. When exposed to water within a propped fracture or subterranean formation, the water-soluble solvent will have a higher affinity for the water in the pore spaces than the curable resin, and it will partition into the water the more viscous resin to be deposited on the proppant or formation-sand surfaces. Selection of the water-soluble solvent depends on the chosen curable resin. The consolidating treatment fluid provides cohesion at the contact points between the proppant grains to keep them in place without causing plugging in the pore spaces of the proppant matrix.

6.3.2.2.1 Curable resin systems

Current consolidating treatment fluids can be selected from one of the following liquid curable resin (LCR) systems, based on the bottomhole temperature of the well:

- A low-temperature, epoxy-based resin designed for a temperature range of 70°F–225°F.
- A mid- to high-temperature, epoxy-based resin designed for a temperature range of 200°F–300°F.
- A high-temperature, furan-based resin designed for a temperature range of 275°F–550°F.

Low temperature, furan-based resin can be designed for temperature range from 70°F to 250°F – required a post-flush of acidic fluid, or an internal catalyst.

Field treatment of curable resins. Resin consolidation treatments can be pumped into the proppant in a fracture to minimize or stop proppant flowback. It can be run where coated frac proppant cannot, for example, when fracturing down tubing, or when other concerns make a premature screenout and the resulting consolidated sand too expensive to drill out. This treatment is also preferred where multiple fractures have occurred and the tail-end proppant could not cover all the perforated interval due to earlier screenouts. A treatment is run after the initial cleanout of the interval so that treating chemical may contact all perforations and zones of the fracture. Ordinarily, an acid or filtercake cleanup pretreatment of the near-wellbore zone will occur before the sand-consolidation treatment. This pretreatment removes all near wellbore gel and filtercake residue so that these materials are not consolidated in place.

The consolidation treatment can be commingled with nitrogen (300–800 scf/bbl) to increase the working volumes of the treatment and provide the diverting effects of two-phase flow during the treatment.

When pumping the consolidation treatment into perforations, consolidation fluid pump rates should be kept adequately low to avoid exceeding fracture pressure or disturbing the near-wellbore proppant distribution.

In addition, coiled tubing with side-spraying nozzles can be used in this application. The workstring may then be swept across the interval during the treatment, thus treating every perforation over the zone of interest.

Unless it is a single-component system, the resin system with two components are metered together on-the-fly through a static mixer to form a homogeneous mixture just before injecting downhole by means of coiled tubing (CT) or jointed pipe before being placed into the perforated interval [109–111]. The application of this curable resin system mainly involves the following steps:

- Inject a volume of preflush fluid to remove fines and oil residue and prepare the surfaces of proppant particulates so that they accept and allow the resin to coat the surfaces.
- Inject a volume of low-viscosity consolidation fluid to coat the proppant and establish consolidation between grains.
- Inject a volume of post-flush fluid to displace the excess resin from occupying the pore spaces within the matrix of the proppant pack, thereby minimizing the damage to the pack permeability. Shut in the well to allow curing.

Note: For a resin system that depends on an external catalyst to activate curing, a spacer fluid is placed between the curable resin fluid and the post-flush fluid (i.e., containing a catalyst) to prevent a direct contact between the curable resin and the post-flush fluid that potentially may result in premature activation of the resin in the tubing.

Selecting a suitable resin formulation and employing an appropriate placement method must go hand-in-hand to ensure the success of the treatment. The consolidating treatment fluid provides cohesion at the contact points between the proppant grains to keep them in place without causing plugging in the pore spaces of the proppant matrix. [107]. The objective of this treatment is to treat the proppant near the wellbore, rather than trying to treat the entire proppant pack in the fractures.

The typical volume is based on treating the fracture out to 10-ft from the wellbore. The assumption is that there are two frac wings with 0.25″ wide, 42% porosity and a safety factor of two.

Since most of the wells are not completed with 180-degree phasing (oriented) perforations, not all perforations are aligned with the propped fractures. The non-aligned (or out-of-phase) perforations located in poorly consolidated formations are prone to producing proppant or formation sand during production [112]. The resin treatment described in this method can provide an effective means for sand control by consolidating the formation sand surrounding all the perforations.

Injecting a curable resin into the perforated interval of propped fractures is the most common practice for remediating proppant flowback. It is often believed that the larger the injected volume of curable-resin fluid, the higher the potential for success. However, field evidence has shown that this is not always the case [106]. Applying large volumes of curable resin with an ineffective placement technique does not guarantee a remedial solution to the proppant-flowback problem. The treatment can be expensive and might not produce the required results.

6.3.2.2.2 Aqueous-based curable resin

A new technique using an aqueous-based resin emulsion has been described by Villesca et al. [108] as a remedial treatment to mitigate proppant flowback. This resin system replaces earlier solutions which were based on solvent dispersed in resin systems as discussed by Nguyen et al. [107] which were more difficult to handle and place within the fracture. It is a method which provides good consolidation strength and allows for proppant-free, high rate production. One advantage of this remedial system is that only wells in which proppant flowback is a problem are treated as opposed to treating the proppant in all wells during the primary stimulation treatment. A preferred placement of this treatment is done by using coiled tubing and foaming with nitrogen to improve distribution of the treating fluids to cover all of the exposed perforations along the wellbore and effectively treat longer fractured intervals. The combination of low concentration aqueous resin emulsions and high quality foam also provides a very cost effective solution for proppant flow back remediation as presented by Nguyen and Rickman [109]. Scanning electron micrographs shows the resin contact points created between individual proppant particles that develop very high strength bonds and effectively stop proppant flow back via foamed aqueous based resin treatment.

6.3.2.2.3 Case histories

Numerous field treatments of both curable, solvent-based and aqueous-based resins were successfully performed as part of proppant flowback control treatments [89,92,93,107,108,110,113–117].

6.3.3 Lessons learned/recommendations

- Perform QA/QC testing of LCR-coated proppant with the fracturing fluid that will be used in the fracturing treatment to help ensure the crosslinking time and break time followed the fracturing treatment design and schedule.
- Before performing the fracturing treatment using LCR-coated proppant, a thorough clean-up of perforations and wellbore is performed to ensure that no restriction can prevent LCR-coated proppant from entering and packing all the perforations, including those not aligned with the propped fractures and/or void spaces that might be existing behind the casing. This simple procedure greatly enhanced the success of the entire treatment.
- The resin-coating process can raise some concerns should premature screenouts occur during the fracturing treatment and the proppant slurry settles inside the wellbore. In this case, the well is flowed back as soon as possible so that tubing will not be needed to clean out the wellbore. However, if the screenout occurs toward the end of the treatment, it is a desirable feature that helps ensure complete coverage of all perforations. When allowed to remain in the wellbore to semicure, the LCR-coated proppant pack can be cleaned out with tubing, preferably coupled with a drill bit or notched collar.
- In some cases, the operator might want to perform forced closure right after the fracturing treatment for a few hours to help enhance closure of the fractures and heat up the proppant pack. After this short shut-in period, the well is shut in again for several hours, ranging from 4 to 48 hours, depending on the bottomhole temperature of the well, to allow the LCR-coated proppant to achieve a complete cure.
- To minimize the amount of proppant producing back during forced closure, the concentration of breaker is increased by 50%–100% of the normal amount during the tail-in stage. This increase in breaker concentration promotes the breakdown of the crosslinked polymer fluid to enhance the grain-to-grain contact of the LCR-coated proppant and minimize the drag force during flowback of forced closure, especially with crosslinked fluid. The LCR-coated proppant grains are tacky in aqueous-based fluid, thus their grain-to-grain contacts should help minimize their movement and flowback with the fracturing fluid, even though the curing of the LCR on the proppant has not been well-established.
- Following treatment, cleanup solution should be circulated through equipment that might have had direct contact with the LCR, including sand screws, frac-blender tub, pumps, and hoses, to prevent resin residue from setting inside this equipment. This cleaning helps ensure proper equipment functioning for the next fracturing treatment.

Nomenclature

ASMA	Aqueous-based SMA
BHP	Bottomhole pressure
BHST	Bottomhole static temperature
BPD	Barrels per day
CBL	Cement Bond Log
CBM	Coalbed methane
CT	Coiled tubing
bbl	Barrels
ft	Feet
gal	Gallons
lbm	Pounds mass
LCR	Liquid-curable resins
RCP	Curable resin pre-coated proppant
scf	Standard cubic feet
SEM	Scanning electron microscope
SMA	Surface modification agents
SPF	Shots per foot
UCS	Unconfined compressive strength
UTTA	Ultra-thin tackifying agent

References

[1] E.B. Rogers, Sand control in oil and gas wells, Oil & Gas Journal (1971) 54–60.

[2] A.P. Kooijman, et al., Horizontal wellbore stability and sand production in weakly consolidated sandstones, SPE-36419, in: SPE Annual Technical Conference and Exhibition, Denver, Colorado, 6–9 October, 1996, SPE Drilling & Completion, 15 (2000), 274–283.

[3] P. Singh, R. van Petegem, A novel chemical sand and fines control using zeta potential altering chemistry and placement techniques, IPTC 17614, in: International Petroleum Technology Conference, Doha, Qatar, 19–22 January 2014.

[4] G.O. Suman Jr., Unconsolidated sand stabilization through wellbore stress state control, SPE 5717, in: SPE Annual Fall Meeting, Dallas, Texas, September 28–October 1 1975.

[5] X. Chen, C.P. Tan, C.M. Haberfield, Wellbore stability analysis guidelines for practical well design, SPE 36972, in: SPE Asia Pacific Oil and Gas Conference, Adelaide, Australia, 28–31 October, 1996.

[6] M.R. Mclean, M.A. Addis, Wellbore stability analysis: a review of current methods of analysis and their field application, IADC/SPE 19941, in: IADC/SPE Drilling Conferences, Houston, Texas, 27 February to 2 March 1990.

[7] H.H. Abass, H.A. Nasr-El-Din, M.H. Bataweel, Sand control: sand characterization, failure mechanisms, and completion methods, SPE 77686, in: SPE Annual Technical Conference and Exhibition, San Antonio, Texas, 29 September to 2 October 2002.

[8] C.D. Hall Jr., W.H. Harrisberger, Stability of sand arches: a key to sand control, Journal of Petroleum Technology 22 (7) (1970) 821–829.

[9] H. Vaziri, B. Barree, Y. Xiao, I. Palmer, M. Kutas, What is the magic of water in production sand? SPE 77683, in: Annual Technical Conference and Exhibition, San Antonio, Texas, 29 September to 2 October 2002.
[10] B. Wu, C.P. Tan, N. Lu, Effect of water cut on sand production–an experimental study, SPE 92715, SPE Production and Operations 21 (3) (2006) 349–356.
[11] G.F. Briggs, A.G. Fiedler, Ground water and wells: a reference book for the water-Well Industry, Johnson Division Universal Oil Products Company, Saint Paul, MN, 1975.
[12] C.J. Coberly, Selection of screen openings for unconsolidated sands, Drill and Production Practices (1937).
[13] C.J. Coberly, E.M. Wagner, Some considerations in the selection and installation of gravel packs for oil wells, Petroleum Technology (1938) 1–20.
[14] R.J. Saucier, Consideration in gravel pack design, JPT (1974) 205–212.
[15] B.J. Chapman, S. Vitthal, L.M. Hill, Prefracturing pump-in testing for high-permeability formations, SPE 31150, in: International symposium on formation damage control, Lafayette, Louisiana, 14–15 February 1996.
[16] R.C. Ellis, An overview of frac packs: a technical revolution (evolution) process, Journal of Petroleum Technology 50 (1998) 66–68.
[17] R.R. Hannah, E.I. Park, D.A. Porter, J.W. Black, Combination fracturing/gravel-packing completion technique on the Amberjack, Mississippi Canyon 109 Field, SPE Production & Facilities 9 (4) (1994) 262–266.
[18] B.V. Hainey, J.C. Troncoso, Frac-pack: an innovative stimulation and sand control technique. SPE 23777, in: SPE international symposium on formation damage control, Lafayette, Louisiana, 26–27 February 1992.
[19] W.L. Penberthy Jr., C.M. Shaughnessy, Sand Control, Series on Special Topics, Vol. 1, Richardson, Texas, 1992, pp. 65–83.
[20] M. Asadi, G.S. Penny, Sand control screen plugging and cleanup, in: Paper SPE 64413 presented at the Asia Pacific Oil & Gas Conference and Exhibition, Brisbane, Australia, 16–18 October 2000.
[21] F.O. Stanley, et al., Matrix acidizing horizontal gravel-packed wells for fines damage removal, SPE 65519, in: SPE/Petroleum Society of CIM International Conference on Horizontal Well Technology, Calgary, Alberta, Canada, 6–8 November 2000.
[22] D.W. Bryant, L.G. Jones, Completion and production results from alternate-path gravel-packed wells, SPE Drill Completion (1995).
[23] L.G. Jones, et al., Gravel packing horizontal wellbores with leak-off using shunts, SPE-38640, in: SPE Annual Technical Conference and Exhibition, San Antonio, Texas, 5–8 October 1997.
[24] Frac-pack Completion Services, Halliburton Energy Services, Inc., Houston, (1994), 5.
[25] R.H. Schroeder, M.J. Tucker, Evaluation of completion practices for improved sand control success and longevity, SPE 5027, in: 49th Annual Fall Meeting of the Society of Petroleum Engineers of AIME, Houston, Texas, 6–9 October 1974.
[26] W.H. Harrisberger, et al., Application of recent advances in the mechanics of sand consolidation, SPE 3589, in: Fall Meeting of the Society of Petroleum Engineers of AIME, New Orleans, Louisiana, 3–6 October 1971.
[27] B.M. Young, et al., Resin-sand pack–a successful sand-control technique, API-69–256, in: Drilling and Production Practice, Washington, DC 1 January 1969.
[28] P. Nguyen, R. Dusterhoft, Stabilizing wellbores in Unconsolidated, Clay-Laden Formations, SPE Production and Operations (2006) 339–348.

[29] S.H. Goodman, (Ed.), Handbook of Thermoset Plastics, second ed., William Andrew, New York, 1999, pp. 23–71.

[30] B.M. Young, An improved sand consolidation process with clay conditioning, SPE 1339, available from SPE, Richardson, Texas, 1965.

[31] B.M. Young, Consolidation technique in clayey sand boosts flow rate, Oil & Gas Journal (1967) 107–109.

[32] J.D. Weaver, J.A. Morgan, Furan resin process replaces workovers in gas storage reservoirs, SPE 17742, in: SPE Gas Technology Symposium, Dallas, Texas, 13–15 June 1998.

[33] P.D. Fader, et al., New low-cost resin system for sand and water control, SPE 24051, in: Western Regional Meeting, Bakersfield, California, 30 March to 1 April 1992.

[34] N.J. Shotts, B.W. Surles, P.D. Fader, Case histories of low-cost sand consolidation in thermal wells, SPE 24840, in: SPE Annual Technical Conference and Exhibition, Washington, DC 4–7 October 1992.

[35] M. Parlar, et al., New chemistry and improved placement practices enhance resin consolidation: case histories from the gulf of mexico, SPE 39435, in: SPE International Symposium on Formation Damage Control, Lafayette, Louisiana, 18–19 February 1998.

[36] M. Gunningham, et al., Selective sand consolidation technique using a pin-point injection packer, SPE 36888, in: European Petroleum Conference, Milan, Italy, 11–14 October 1996.

[37] J.C. Jeager, N.G.W. Cook, Fundamentals of rock mechanics, third ed., Chapman and Hall, New York City, 1979.

[38] W.F. Hower, W. Brown, Large-scale laboratory investigation of sand consolidation techniques, Journal of Petroleum Technology 13 (1961) 1221–1229.

[39] G.O. Suman, World Oil's Sand Control Handbook–Part 7-Completion, Gulf Publishing Company, Texas, 1975, pp. 47–53.

[40] J.L. Rike, Review of sand consolidation experience in South Louisiana, Journal of Petroleum Technology, 18 (1966) 545–550.

[41] Halliburton Internal SANFIX Tech bulletin.

[42] J. Hibbeler, T. Garcia, N. Chavez, An integrated long-term solution for migratory fines damage, SPE 81017, in: Latin American and Caribbean Petroleum Conference, Trinidad, West Indies, 27–30 April 2003.

[43] T.W. Muecke, Formation fines and factors controlling their movement in porous media, Journal of Petroleum Technology (1979) 144–150.

[44] C. Gruesbeck, R.E. Collins, Entrainment and deposition of fine particles in porous media, Society of Petroleum Engineers Journal 22 (1982) 847–856.

[45] R.R. McDaniel, J.R. Willingham, The effect of various proppants and proppant mixtures on fracture permeability, SPE 7573, in: SPE Annual Fall Technical Conference and Exhibition, Houston, Texas, 1–3 October 1978.

[46] A.K. Sarkar, M.M. Sharma, Fines migration in two-phase flow, JPT, May 1990.

[47] B.G. Sharma, M.M. Sharma, Polymerizable ultra-thin films: a new technique for fines stabilization, SPE 27345, in: SPE Formation Damage Control Symposium, 7–10 February 1994.

[48] E.V. Oort, et al., Impairment by suspended solids invasion: Testing and prediction, SPE Production & Facilities (1992) 178–184.

[49] S. Pang, M. Sharma, Predicting injectivity decline in water injection wells, SPE Formation Evaluation (1997) 194–201.

[50] D.A. Shumbera, et al., Improved water injector performance in a Gulf of Mexico deepwater development using an openhole frac pack completion and downhole filter system: case study, SPE 84416, in: SPE Annual Technical Conference and Exhibition, Denver, Colorado, 5–8 October 2003.

[51] L.J. Kalfayan, D.R. Watkins, A new method for stabilizing fines and controlling dissolution during sandstone acidizing, SPE 20076, in: SPE California Regional Meeting, Ventura, 4–6 April, 1990.

[52] H.C. McLaughlin, J.D. Weaver, Oil well treating method and composition, US Patent 4366071, 1982.

[53] H.C. McLaughlin, E.A. Elphingstone, B.Hall, Aqueous polymers for treating clays in oil and gas, SPE 6008, in: Annual Technical Conference and Exhibition of the SPE of AIME, New Orleans, Louisiana, 3–6 October 1976.

[54] R.E. Himes, et al., Environmentally safe temporary clay stabilizer for use in well service fluids, CIM/SPE 21604, in: PETSOC Technical Meeting, Calgary, 10–13 June 1990.

[55] M. Asadi, G.S. Penny, Screen performance and cleanup characterization in horizontal well drill-in-fluids using a large-scale radial flow-cell, SPE-56777, in: SPE Annual Technical Conference and Exhibition, Houston, Texas, 3–6 October 1999.

[56] F.O. Stanley, et al., An economic, field-proven method for removing fines damage from gravel packs, SPE 58790, in: SPE International Symposium on Formation Damage Control held in Lafayette, Louisiana, 23–24 February 2000.

[57] J.D. Weaver, J.D. Baker, S. Woolverton, M.A. Parker, Application of surface modification agent in wells with high flow rates, SPE 53923, in: Latin American and Caribbean Petroleum Engineering Conference, Caracas, Venezuela, 21–23 April 1999.

[58] J.D. Weaver, et al., Control of fine particulate flowback in subterranean wells, US Patent 5853048, 1998.

[59] L.V. Lehman, et al., Conductivity maintenance: long-term results from the use of conductivity enhancement material, SPE 82241, in: SPE European Formation Damage Conference, The Hague, The Netherlands, 13–14 May 2003.

[60] L.V. Lehman, M.E. Blauch, L.M. Robert, Desorption enhancement in fracture-stimulated coalbed methane wells, SPE 51063, in: Eastern Regional Conference, Pittsburgh, Pennsylvania, 9–11 November 1998.

[61] Nguyen, et al., Enhancing fracture conductivity through surface modification of proppant, SPE 39428, in: Formation Damage Control Conference, Lafayette, Louisiana, 18–19 February 1998.

[62] P. Nguyen, B. Dewprashad, J. Weaver, A new approach for enhancing fracture conductivity, SPE 50002, in: Asia Pacific Oil and Gas Conference and Exhibition, Perth, Australia, 12–14 October 1998.

[63] M. Blauch, et al., New insights into proppant-pack damage due to infiltration of formation fines, SPE 56833, in: Annual Technical Conference and Exhibition, Houston, Texas, 3–6 October 1999.

[64] B.T. Dewprashad, et al., Modifying the proppant surface to enhance fracture conductivity, SPE 50733, in: SPE International Symposium on Oilfield Chemistry, Houston, Texas, 16–19 February 1999.

[65] R. Dusterhoft, P. Nguyen, M. Conway, Maximizing effective proppant permeability under high-stress, high gas-rate conditions, SPE 90398, in: Annual Technical Conference and Exhibition, Houston, Texas, 26–29 September 2004.

[66] B. Dewprashad, et al., New insight into failure of resin-coated proppant consolidation, SPE 40017, in: SPE International Symposium on Oilfield Chemistry, Houston, Texas, 16–19 February, 1998.

[67] Nguyen, et al., Mitigating formation invasion to maintain fracture conductivity, IPTC 10571, in: International Petroleum Technology Conference held in Doha, Qatar, 21–23 November 2005.

[68] Nguyen, et al., Controlling formation fines at their sources to maintain well productivity, SPE Production and Operations (2007) 202–215.

[69] L.K. Vo, et al., Development and applications of an aqueous-based surface modification agent, SPE 165172, in: European Formation Damage Conference, Noordwijk, The Netherlands, 5–7 June 2013.

[70] P.D. Nguyen, J.D. Weaver, Water-based, frac-pack remedial treatment extends well life, SPE 144065, in: European Formation Damage Conference, Noordwijk, The Netherlands, 7–10 June 2011.

[71] J.D. Weaver, et al., Sustaining fracture conductivity, SPE 94666, in: European Formation Damage Conference, Scheveningen, The Netherlands, 25–27 May 2005.

[72] R. Rickman, et al., Application of curable thin film for sand control, SPE 98228, in: SPE International Symposium and Exhibition on Formation Damage Control, Lafayette, Louisiana, 15–17 February 2006.

[73] D.J. Haggerty, et al., Sand consolidation testing in an API RP 19B Section IV perforation flow laboratory, SPE-120901, in: European Formation Damage Conference, Scheveningen, The Netherlands, 27–29 May 2009.

[74] P.D. Nguyen, et al., Controlling solids flowback to maintain production of hydrocarbons: a review of successful chemical technologies in the last decade, IPTC 13725, in: International Petroleum Technology Conference, Doha, Qatar, 7–9 December 2009.

[75] S. Ingram, I. Paterniti, R. Pauls, et al., Enhancing and sustaining well production: granite wash, texas panhandle, SPE 106531, in: Production and Operations Symposium, Oklahoma City, Oklahoma, 3 March 2007.

[76] J.D. Weaver, J.D. Baker, S. Woolverton, M.A. Parker, Application of surface modification agent in wells with high flow rates, SPE 54645, in: SPE Western Regional Meeting, Anchorage, Alaska, 26–27 May 1999.

[77] H. Wei, et al., Long-term production results of high-rate water packs: case histories from nanhai west bay, South China sea, SPE 127914, in: SPE International Symposium and Exhibition on Formation Damage Control, Lafayette, Louisiana, 10–12 February 2010.

[78] F. Lemesnager, et al., Innovative Frac Stimulation of low perm oil zone in high deviation well offshore congo, SPE 115892, in: SPE Annual Technical Conference and Exhibition, Denver, Colorado, 21–24 September, 2008.

[79] J. Ely, Experience proves forced fracture closure works, World Oil (1996) 37–41.

[80] K. van Gijtenbeek, et al., Overdisplacing propped fracture treatments—good practice or asking for trouble?, SPE 154397, in: EAGE Annual Conference & Exhibition Incorporating SPE Europec, Copenhagen, Denmark, 4–7 June 2012.

[81] R.J. Vreeburg, et al., Production backproduction during hydraulic fracturing - a new failure mechanism for resin-coated proppants, Journal of Petroleum Technology (1994) 884–889.

[82] L.R. Norman, et al., Application of curable resin-coated proppants, SPE Production Engineering (1992) 343–349.

[83] B.L. Graham, J.E. Clifford, Sand control and selective completion technique Ojai (Silverthread) field, Ventura county, California, SPE-5375, in: SPE California Regional Meeting, Ventura, California, 2–4 April 1975.

[84] C.D. Pope, et al., Curable resin-coated sand controls proppant flowback, SPE 16209, in: SPE Production Operations Symposium, Oklahoma City, Oklahoma, 8–10 March 1987.

[85] R.J. Card, et al., A novel technology to control proppant back production, SPE Production & Facilities (1995) 271 (November).

[86] C. Stephenson, A. Rickards, H. Brannon, Increased resistance to proppant flowback by adding deformable particles to proppant packs tested in the laboratory, SPE 56593, in: Annual Technical Conference and Exhibition, Houston, Texas, 3–6 October 1999.

[87] J. Murphey, D. Dalrymple, P. Creel, et al., Proppant flowback control, SPE 19769, in: Annual Technical Conference and Exhibition, San Antonio, Texas, 8–11 October 1989.

[88] P.D. Nguyen, J.D. Weaver, Controlling proppant flowback in high-temperature, high-production wells, in: Paper SPE 82215 presented at the European Formation Damage Conference, The Hague, The Netherlands, 13–14 May 2003.

[89] J. Trela, P. Nguyen, B. Smith, Controlling proppant flowback to maintain fracture conductivity and minimize workovers—lessons learned from 1,500 fracturing treatments. Paper SPE 112461 presented at the International Symposium and Exhibition on Formation Damage Control, Lafayette, Louisiana, 13–15 February 2008.

[90] B. Dewprashad, et al., Epoxy resin consolidation system improves worker safety, chemical compatibility, and performance, SPE 38189, in: European Formation Damage Conference, The Hague, The Netherlands, 2–3 June 1997.

[91] F. Al-Ghurairi, R. Solares, K. Bartko, L. Sierra, Results from a field trial using new additives for fracture conductivity enhancement in a high-gas screenless completion in the jauf reservoir, Saudi Arabia, SPE 98088, in: International Symposium and Exhibition on Formation Damage Control, Lafayette, Louisiana, 15–17 February 2006.

[92] B. Krismartopo, et al., A fracture treatment design optimization process to increase production and control proppant flow-back for low-temperature, low pressured reservoirs, SPE 93168, in: Asia Pacific Oil and Gas Conference and Exhibition, Jakarta, Indonesia, 5–7 April 2005.

[93] X. Chang, C. Bao, W. Lubis, D. Hui, P. Nguyen, Screenless frac-pack completions—case studies from jidong fields, China, SPE 108909, in: Asia Pacific Oil and Gas Conference and Exhibition, Jakarta, Indonesia, 30 October–1 November 2007.

[94] B. Dewprashad, H. Abass, D. Meadows, J. Weaver, B. Bennett, A method to select resin-coated proppants, in: Paper SPE 26523 presented at the Annual Technical Conference and Exhibition, Houston, Texas, 3–6 October 1993.

[95] B. Dewprashad, P. Nguyen, R. Kuhlman, M. Besler, Consolidation for Screenless completions and proppant-flowback control in hot wells, in: Paper SPE 48874 presented at the International Oil and Gas Conference and Exhibition, Beijing, China, 2–6 November 1998.

[96] R.E. Patterson, et al., High-pressure, high-temperature consolidated completion in the continental shelf environment of the gulf of mexico: case history, OTC 18976, in: Offshore Technology Conference, Houston, Texas, 30 April–3 May 2007.

[97] P. Nguyen, J. Weaver, R. Rickman, Prevention of geochemical scaling in hydraulically created fractures— laboratory and field studies, SPE 118175, in: Eastern

Regional/AAPG Eastern Section Joint Meeting, Pittsburgh, Pennsylvania, 11–15 October 2008.

[98] J. Luna, E. Soriano, R. Garcia, J. Galvan, A. Barrera, Sustaining fracture conductivity increases cumulative production in tight-gas reservoir—acase history, in:. Paper SPE 111992 presented at the North Africa Technical Conference & Exhibition Marrakech, Morocco, 12–14 March 2008.

[99] R. Yassine, M. Amr, M. Salem, et al., Practical Methods to sustain production after hydraulic fracturing treatment in marginal wells—case historyfrom Western Desert Egypt, in: Paper SPE 112603 presented at the North Africa Technical Conference and Exhibition, Marakech, Morocco, 12–14 March 2008.

[100] J. Leal, R. Solares, F. Al-Ghurairi, L. Sierra, Long term evaluation of a new liquid resin additives used for fracture conductivity enhancement in a gas producer screenless completion, Jauf Reservoir, Saudi Arabia, in: Paper SPE 126154 presented at the Saudi Arabia Section Technical Symposium, AlKhobar, Saudi Arabia, 09–11 May 2009.

[101] S. Ali, S. Vitthal, J. Weaver, Improvements in high-rate water packing with surface-modification agent, SPE 58755, in: SPE International Symposium on Formation Damage Control, Lafayette, Louisiana, 23–24 February 2000.

[102] L. Robert, Arkansas Oil and Gas Commission, 2004 CBM Workshop.

[103] W. Whitley, et al., Conductivity endurance in the unconsolidated environment of the Gulf Of Mexico, in:. Paper OTC 17116 presented at the Offshore Technology Conference, Houston, Texas, 2–5 May 2005.

[104] J. Weaver, M. Blauch, M. Parker, B. Todd, Investigation of proppant-pack formation interface and relationship to particulate invasion, SPE 54771, in: European Formation Damage Conference, The Hague, The Netherlands, 31 May–1 June 1999.

[105] Nguyen, et al., Stabilizing wellbores in unconsolidated formations for fracture stimulation, SPE Production & Facilities (2000) 262–269.

[106] S.W. Cole, et al., Remedial proppant consolidation using resin technology, SPE 52192, in: SPE Mid-Continent Operations Symposium held in Oklahoma City, Oklahoma, 28–31 March 1999.

[107] P.D. Nguyen, et al., Application of diluted consolidation systems to improve effectiveness of proppant flowback remediation–laboratory and field results, SPE Production and Operations 24 (1) (2009) 50–59.

[108] J. Villesca, et al., Development and field applications of an aqueous-based consolidation system for proppant remedial treatments, SPE 128025, in: SPE International Symposium and Exhibition on Formation Damage Control, Lafayette, Louisiana, 10–12 February 2010.

[109] P. Nguyen, R. Rickman, Foaming aqueous-based curable treatment fluids enhances placement and consolidation performance, SPE 151002, in: SPE International Symposium and Exhibition on Formation Damage Control, Lafayette, Louisiana, 15–17 February 2012.

[110] S.R. Ingram, P.D. Nguyen, A.L. Hood, Remediation of production loss due to proppant flowback through coiled-tubing intervention, SPE 106532, in: SPE Production and Operations Symposium, Oklahoma City, Oklahoma, 31 March–3 April 2007.

[111] S.R. Ingram, P.D. Nguyen, Proppant flowback stopped with coiled tubing intervention, World Oil (2008) 99–106 (February).

[112] P.A. Fletcher, et al., Using fracturing as a technique for controlling formation failure, SPE Production & Facilities (1996) 117–121.
[113] D. Dapara, et al., Preventing proppant and formation-sand production in high water cut, heavy-oil wells: a field study from Argentina, IPTC 13174, in: International Petroleum Technology Conference, Doha, Qatar, 7–9 December 2009.
[114] P. Nguyen, et al., Remediation of production loss due to proppant flowback in existing wellbores, SPE 102626, in: SPE Russian Oil and Gas Technical Conference and Exhibition, Moscow, Russia, 3–6 October 2006.
[115] M. Durkee, et al., Coiled tubing deployed resin treatment remediates proppant and formation sand flowback in high-temperature wells: a field study in South Texas, SPE 126606, in: SPE Tight Gas Completions Conference, San Antonio, Texas, 15–17 June 2009.
[116] P. Nguyen, et al., Effectively controlling proppant flowback to maximize well production: lessons learned from Argentina, SPE 165174, in: SPE European Formation Damage Conference and Exhibition, Noordwijk, The Netherlands, 5–7 June 2013.
[117] E. Dagher, M. Tawakol, P. Nguyen, Proppant flowback remedial treatment brings dead well back to life, World Oil, 233 (1) (2012).

CHAPTER

7 Condensate and water blocking removal

Mashhad Fahes
The University of Oklahoma, Norman, OK, United States

Chapter outline

7.1 Introduction

Gas reservoirs hold a significant amount of hydrocarbon reserves around the world, with a total of 7019 trillion cubic feet of worldwide proven reserves by the end of 2019, equal to 1170 billion barrels of oil equivalent [1]. For comparison, the total worldwide oil proven reserves at the end of 2019 were reported to be 1734 billion barrels [1]. In fact, the global share of gas as a primary source of energy has been on a steady incline, reaching 24.2% in 2019 [1–3]. This resource

Flow Assurance. DOI: https://doi.org/10.1016/B978-0-12-822010-8.00005-2

is precious not just because it is one of the cheapest sources of energy, but also because it is one of the cleanest fossil fuels, producing 60% less CO_2 than coal in power generation and 20% less greenhouse gases than oil in cars [4–6]. The renewed focus on hydrogen production from natural gas also puts this fossil fuel resource at the center of renewable energy production [7,8]. However, limiting methane leak during the life cycle of natural gas production is critical to reduce its negative environmental impact [9–11].

Most gas reservoirs hold significant amounts of heavy hydrocarbon components that condense to liquid form at certain conditions of temperature and pressure [12–15]. This includes the majority of the giant gas reservoirs. One example is the South Pars/North Dome gas-condensate field located to the south of Iran and north of Qatar, which is by far the world's largest natural gas field. As of 2017, the field holds 1235 trillion cubic feet of recoverable gas reserves, while the second-largest field, Urengoy, located in Russia, holds 222 trillion cubic feet of recoverable gas reserves [16]. The gas from the South Pars/North Dome field yields 40 barrels of condensate per million cubic feet of gas production [17]. The condensate yield is the main parameter used to characterize the gas reservoirs, which ranges from dry gas, with very minimal condensate yield, to rich gas-condensate reservoirs, yielding over 100 barrels of condensate per million cubic feet [12]. At high pressure, this condensate is present as a part of the gas phase in the reservoir. It condenses along the production path, initially past the formation, resulting in the valuable production of liquid condensate. However, when the pressure in the reservoir drops, which initially occurs in the near-wellbore region of the formation, liquid condensate will form inside the formation. This condensed liquid introduces two-phase flow challenges and impedes the flow of gas causing a decline in well productivity [13,15,18–20]. While early in the life of most gas-condensate wells, it is possible to maintain a high enough pressure to avoid liquid condensation inside the formation, pressure drawdown eventually catches up, and liquid starts to condense and accumulate in the pore space.

Project planning for field development in gas-condensate reservoirs requires accurate fluid characterization, production monitoring, and carefully tailored solutions to maintain well deliverability. Capturing representative fluid samples before the on-set of liquid condensation is a critical step in this regard. This can include a downhole fluid sample or the more common method of recombining gas and condensate samples collected at the surface before any change occurs to the gas–oil ratio (GOR). Laboratory fluid tests conducted using this initial reservoir fluid include measuring fluid composition, constant composition expansion (CCE), and constant volume depletion (CVD). While collecting and testing fluid samples can be an optional scenario in some types of reservoirs, it is a critical component of managing a large gas-condensate reservoir in order to plan the stages of intervention and maintain well deliverability. The results from these tests allow for accurate predictions of condensation overtime and the expected recovery factors at various pressure values.

7.2 Background theory

In this section, the relevant theoretical background is briefly discussed to allow for a thorough discussion of the problems associated with liquid condensation and blocking. This also sets the stage for the discussion of the ways, in which these problems are addressed in the current industry practice as well as in the research and development of promising technologies in this domain.

7.2.1 Fluid phase behavior

Hydrocarbon fluids are mixtures of a large number of hydrocarbon components, in addition to mostly small amounts of some nonhydrocarbon components that can include helium, nitrogen, carbon dioxide, hydrogen sulfide, and other impurities. The hydrocarbon components range from the smallest molecule of methane to some of the largest and more complex hydrocarbon structures of asphaltenes [12]. These mixtures vary greatly in composition between various fields and sometimes between various wells within the same field [21].

Each fluid mixture can exist in either a liquid phase, a gas phase, or an equilibrium of both liquid and gas phases, depending on the conditions of temperature and pressure. A map of this phase behavior for a particular fluid at different values of pressure and temperature is called a phase diagram. Laboratory pressure–temperature–volume experiments on fluid samples can be used to generate these phase diagrams. More commonly, a limited number of laboratory measurements, which includes the fluid composition, is used to generate phase diagrams through equation of state modeling. The composition of the fluid greatly impacts the shape of the phase diagram. An example of a phase diagram for a gas-condensate fluid is shown in Fig. 7.1.

The combination of the initial conditions of temperature and pressure in the reservoir, the phase diagram, and the surface conditions of temperature and pressure are used to determine the type of reservoir fluid. In this example, this reservoir fluid is at a reservoir temperature of 222°F and an initial reservoir pressure of 7330 psia. According to the phase diagram shown in Fig. 7.1 [22], the reservoir fluid at these initial conditions is at a point that is above the dewpoint line, represented by the boundary of the two-phase envelope. This indicates that the whole fluid would be in the gas phase at these initial conditions of temperature a pressure, and that this fluid corresponds to a gas reservoir.

When the temperature and pressure conditions correspond to a point within the two-phase envelope, this fluid would exist in a two-phase state. One of the most important fluid properties that need to be determined for a gas-condensate fluid is the value of dewpoint pressure at reservoir temperature. Hydrocarbon liquid condensation starts to occur in the near-wellbore region at the moment the well bottomhole flowing pressure reaches this value. In this next section, we discuss the pressure profile in the reservoir that results in the condensation problem

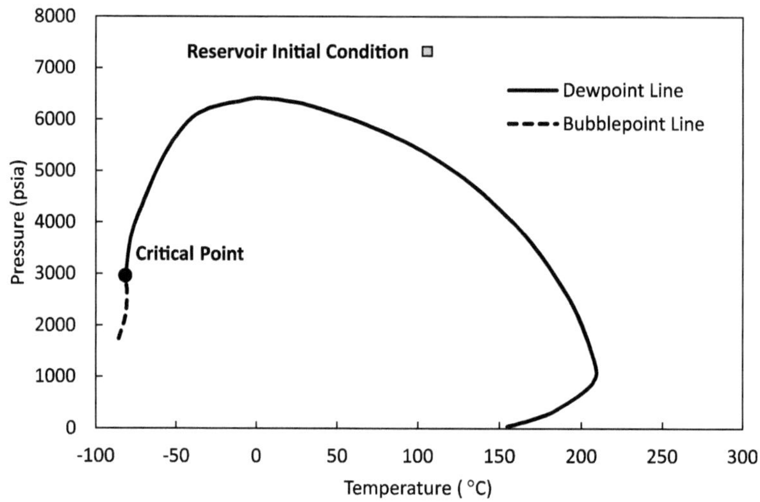

FIGURE 7.1

Phase behavior of a gas-condensate fluid from a well in the Tarim Basin in China. Reservoir temperature is 105.6°C and reservoir pressure is 50.56 MPa.

Produced with data from: X. Zhou, H. Wang Hongjun, Y. Chen, Z. Xu. The study on the contributing factors of phase state of gas condensate: examples from Tarim Basin, Marine and Petroleum Geology 15 (1998) 21–31.

being mainly concentrated in the near-wellbore region. For this fluid, at the temperature of 222°F, liquid starts to condense once the dewpoint pressure of 5360 psia is reached. While it could take a longer time for the reservoir pressure itself to reach this value, it is highly likely that a well in this reservoir would start producing at a wellbore bottomhole flowing pressure that is lower than the dewpoint pressure, revealing that the liquid condensation in the formation around the wellbore for this example is expected early in the life of the well.

7.2.2 Pressure profiles

Single-phase fluid flow in porous media can mostly be described by Darcy's law, which is represented in differential form by Eq. (7.1) [23]. In this equation, k is the value of the absolute permeability of the rock when the pore space is completely occupied by one fluid.

$$v = -\frac{k}{\mu}\frac{\partial p}{\partial x} \tag{7.1}$$

It is important to note that the fluid flow behavior deviates from the Darcy behavior under certain conditions. This includes flow at low pressure that causes gas slippage, commonly observed in laboratory measurements [24], and flow at a high velocity where forces of inertia cannot be ignored [25], such as the case of high-velocity gas flow near the wellbore in gas wells. Gas slippage can be avoided in laboratory measurements of gas flow in conventional rocks by applying some back-pressure and conducting the measurements at pressure values that exceed 60 psia throughout the rock length. The situation is different in the case of gas flow in shale rocks, where gas slippage needs to be taken into consideration [26]. The flow of gas at a high velocity is less efficient than what Darcy's law calculations would predict. In other words, additional pressure is needed to achieve a given fluid velocity as compared to the pressure calculated using Darcy's law. This is captured in the laboratory using Forchheimer's equation shared in Eq. (7.2) [25].

$$-\frac{\partial p}{\partial x} = \frac{\mu}{k}v + \beta\rho v^2 \tag{7.2}$$

However, the impact of near-wellbore non-Darcy flow in the reservoir is typically captured through the introduction of an associated skin factor that introduces into the flow equation the equivalent pressure drop as a result of deviation from Darcy behavior [27]. This allows for the use of the simplified Darcy model of fluid flow in reservoir simulators without significantly compromising the accuracy of the results.

The differential form of Darcy's law can be integrated according to the given geometry and boundary conditions of the system. For the case of single-phase gas flow to a vertical well and assuming pseudosteady-state conditions, gas well productivity in oilfield units can be described using Eq. (7.3), where the flow rate q is in Mscf/d [27].

$$\bar{p}^2 - p_{wf}^2 = \frac{1424q\bar{\mu}\bar{z}T}{kh}\left(\ln\frac{r_e}{r_w} - 3/4 + s\right) \tag{7.3}$$

This equation can be used to calculate the pressure profile in the reservoir as a function of distance from the wellbore when producing from a vertical well. A simple reservoir example case is shared in Table 7.1. Pressure as a function of distance from the wellbore is plotted in Fig. 7.2.

The observed pressure profile in Fig. 7.2 highlights the logarithmic relationship between pressure and radius. Over 2000 psi of pressure loss is observed in the first 10 ft of the formation, in the near-wellbore region, compared to less than 2000 psi of pressure lost in the remaining 1400 ft of the reservoir. This pressure drawdown behavior is common in radial flow as the fluid accelerates in the increasingly-smaller cross-sectional area available to flow as it approaches the wellbore. According to this example, the pressure reaches 5360 psia at a distance of 10 ft from the wellbore. If this fluid is described by the phase diagram in Fig. 7.1, then it is expected that the fluid would be flowing as a 100% gas phase through the reservoir until it reaches a distance of 10 ft away from the wellbore. At that point, liquid condensation starts throughout that near-wellbore region. One can see how this zone is critical for the gas to flow all the way to the wellbore.

Table 7.1 Reservoir properties for example case.

Reservoir/fluid property	Value
Reservoir temperature (°F)	222
Average reservoir pressure (psia)	7330
Well bottomhole flowing pressure (psia)	3330
Thickness (ft)	20
Permeability (md)	20
Wellbore radius (ft)	0.33
Average pressure boundary (ft)	1500
Skin factor	0
Average gas viscosity (cp)	0.022
Average compressibility factor	0.92

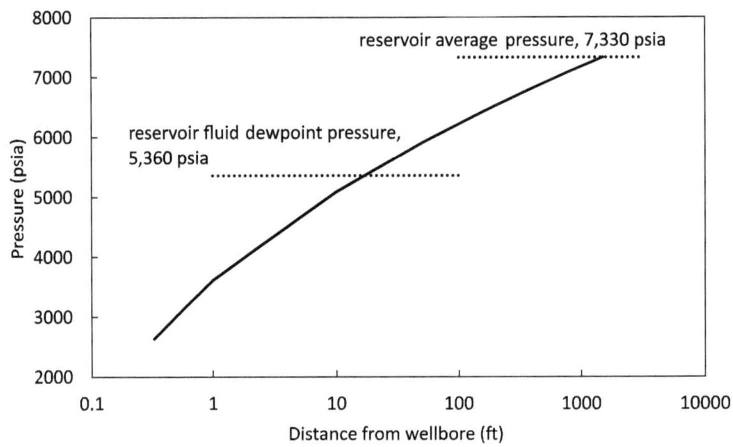

FIGURE 7.2

Pressure profile in the reservoir during production under pseudo-steady state conditions for the example reservoir case in Table 7.1.

Constricting the flow of gas in this region, through the accumulation of liquid condensate, can significantly impact productivity of the well.

Tracking the amount of liquid that condenses out of the gas at reservoir temperature requires measuring the liquid volume percent (or liquid dropout) that drops out of the gas-condensate fluid at various pressure values. This amount is zero when the pressure is higher than the dewpoint pressure. From the value of dewpoint pressure, and as pressure decreases, the liquid volume percent starts to increase until it reaches a maximum value. After that, further reduction in pressure results in the liquid volume percent decreasing as the liquid starts to evaporate again. This is shown in an example in Fig. 7.3 for a gas-condensate fluid

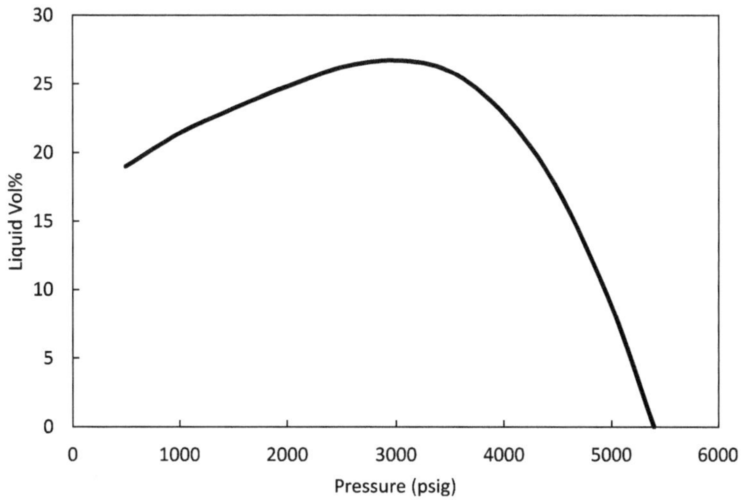

FIGURE 7.3

Example of liquid volume percent for a gas-condensate fluid as a function of pressure at the reservoir temperature of 260°F. The fluid is from the Cupiagua field in Columbia.

Produced with data from C. Franco, R. Zabala, O. Botero, E. Mora, C. Candela, A. Castillo, Inhibited gas stimulation to mitigate condensate banking and maximize recovery in Cupiagua field, SPE Production & Operations 28 (2013) 154–167.

from Cupiagua field in Columbia. This is a rich gas-condensate fluid where the maximum liquid volume percent is 27vol.% observed at a pressure of 3000 psig [28]. This however does not correspond to the liquid occupying only 27% of the pore space. Overtime, and as gas flows into the condensation region, more liquid condenses and accumulates until it builds up enough saturation for the condensate to start flowing. Liquid saturation in the pore space can reach values as high as 60% or 70%. With time, as reservoir pressure decreases, and depending on the value of the maintained gas rate, the boundary of the condensation zone starts expanding deeper into the reservoir and more liquid saturation buildsup.

So far, we discussed the mechanism through which liquid condenses out of gas-condensate fluids and how the pressure distribution in the reservoir impacts this condensation. Next, we discuss the implications of liquid condensation on gas and liquid mobility and on well deliverability.

7.2.3 Two-phase flow challenges

The presence of more than one phase inside the pores of the rock results in reduced flow capacity as each fluid constricts the space available for the other fluid to occupy. This is captured through the concept of relative permeability.

7.2.3.1 Relative permeability

As described in Eq. (7.1), the rock flow capacity is measured by a value for rock absolute permeability, k, when the pore space of the rock is occupied by one fluid. When more than one fluid is present in the pore space, Darcy's law is used to describe the flow of each of the fluids by assigning an "effective permeability" to that fluid at the given flow conditions. For a fixed rock/fluid system, this effective permeability is a function of fluid saturation, as well as saturation history, and can also be impacted by the flow rate, that is representated by the capillary number.

In the case of two-phase flow, it is common to describe the fluids as one being the "wetting phase," with higher affinity to the rock surface, and one being the "nonwetting phase," with lower affinity to the rock surface. In gas–liquid systems, and under the natural conditions of the reservoir rock, liquid is the wetting phase and gas is the nonwetting phase.

If we take an example of linear two-phase flow in a cylindrical core, and assuming for the time being the case of two liquid phases so that the assumption of fluid incompressibility can be utilized, the flow of the wetting and nonwetting phases can be described according to Eqs. (7.4) and (7.5), respectively, which represent the linear steady-state solution of Darcy's law for incompressible fluids. In this case, capillary pressure is also ignored, where the pressures in both phases at the inlet and outlet of the core are considered to be the same.

$$\frac{q_w}{A} = \frac{k_{w,eff}}{\mu_w}\frac{p_1 - p_2}{L} \tag{7.4}$$

$$\frac{q_{nw}}{A} = \frac{k_{nw,eff}}{\mu_{nw}}\frac{p_1 - p_2}{L} \tag{7.5}$$

The ratio of effective fluid permeability to the absolute rock permeability is termed as the relative permeability of the rock to that fluid and is given by Eqs. (7.6) and (7.7).

$$k_{rw} = \frac{k_{w,eff}}{k} \tag{7.6}$$

$$k_{rnw} = \frac{k_{nw,eff}}{k} \tag{7.7}$$

The value of the relative permeability for each fluid varies mostly between 0 and 1, where a value of 0 corresponds to the possible presence of an immobile amount of that fluid in the rock and the value of 1 corresponds to the case of the pore space being fully occupied by that fluid. The value of the relative permeability at the intermediary saturation values depends on many factors, including the wettability condition, saturation history, as well as flow rate [29–31]. An example of the variation in relative permeability as a function of fluid saturation is shown in Fig. 7.4.

One can observe from this relative permeability example that the wetting phase (liquid) has a lower mobility than the nonwetting phase (gas) at a given

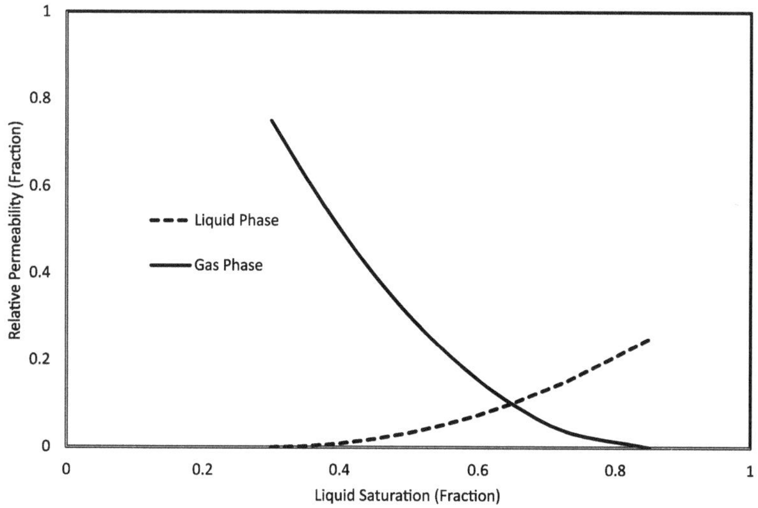

FIGURE 7.4

An example of two-phase relative permeability as a function of fluid saturation.

fluid saturation. For example, the relative permeability of the wetting phase at a wetting phase saturation of 60% is only 0.07 while the relative permeability of the nonwetting phase at a nonwetting phase saturation of 60% is 0.5. Another observation is that both fluids experience capillary trapping at low saturations. In this example, the wetting phase does not start experiencing mobility until it accumulates a saturation of 30% and the nonwetting phase during an imbibition process (typical for a production phase in the reservoir) loses mobility when the residual nonwetting phase saturation of 15% is reached. Consulting Fig. 7.4, one can start to see the impact, liquid condensation has on the mobility of the gas phase. It is not unheard of to have a condensate saturation of around 35%–40% in a near-wellbore area of a reservoir that has a relatively lean gas-condensate fluid. According to the relative permeability curve shared in the figure, gas mobility in that region is reduced by around 50%. The shape and values of the relative permeability curve are impacted by the conditions of wettability of the system and by capillary pressure, permeability, and pore structure. We address some of these dimensions in Section 7.4.1.

7.2.3.2 Condensate saturation regions in gas-condensate reservoirs

One of the common ways to characterize the impact of liquid condensate on gas flow is to divide the reservoir into regions that represent the different zones of condensate saturation [32,33]. The distribution of these regions can vary in five different phases throughout the life of the well.

During phase-1, early in the life of the well when the wellbore flowing bottomhole pressure is higher than the dewpoint pressure, the hydrocarbon fluid is present in one phase, and that is gas. Condensate saturation in this case is zero throughout the reservoir. The produced GOR during this phase is constant.

As the pressure at the wellbore drops below the dewpoint pressure, the condensation and accumulation of liquid start. However, this condensate is initially immobile and trapped inside the pores until enough condensate saturation has accumulated. During this phase-2 of production, two regions exist: one near the wellbore with a low and immobile condensate saturation and one away from the wellbore where pressure is still higher than dewpoint pressure and condensate saturation is zero. Depending on the richness of the gas-condensate fluid and the operating conditions of pressure and rate, this phase of production can last a significant amount of time or can be a very short-lived phase. With the loss of some condensate in the formation, an increase in the GOR is observed. The produced condensate is not flowing in the reservoir as a liquid but condensing out of the produced gas along the production path and in the separation facilities.

During phase-3 of production, enough condensate has accumulated near the wellbore to start generating a mobile liquid phase. As long as the reservoir pressure is still higher than the dewpoint pressure, three regions are observed in the reservoir: a region near the wellbore where a mobile liquid condensate saturation exists and both gas and liquid concurrent flow occur, a second region beyond that where the pressure is lower than dewpoint pressure and some immobile condensate saturation exists but gas is the only flowing phase, and an external region where pressure is still higher than the dewpoint pressure and no condensate saturation is present. With time, the outer borders of the two inner-most regions keep expanding into the reservoir. During the transition from phase-2 to phase-3, a sharp peak in GOR is possible from the sudden loss of condensate in phase-2, followed by a steady increase in GOR during phase-3 as the "condensate bank" expands, and more immobile condensate accumulates in the reservoir. The produced condensate during this phase is a combination of liquid condensate flowing from the formation and condensate dropping from the flowing gas phase along the production path and in the separation facilities. A depiction of condensate saturation regions during phase-3 is shown in Fig. 7.5.

The fourth phase of production occurs if the well bottomhole pressure drops beyond the peak in liquid dropout as discussed in Fig. 7.3. When the wellbore bottomhole flowing pressure drops beyond that peak, this initiates the evaporation of some of the condensate in the formation. A fourth inner-most region is then observed with a lower condensate saturation and an enhanced gas mobility. This increase in the gas velocity further contributes to the evaporation of the condensate, especially, that this gas is leaner for having lost a significant amount of its heavy components in the prior two regions. This may not result in a decrease in the GOR since condensate is still accumulating deeper in the reservoir, however, it could have the impact of reducing the slope of GOR increase.

The last phase of production occurs later in the life of the well when the reservoir pressure throughout the drainage area drops below the dewpoint pressure. While this results in a significant loss of heavy components throughout the

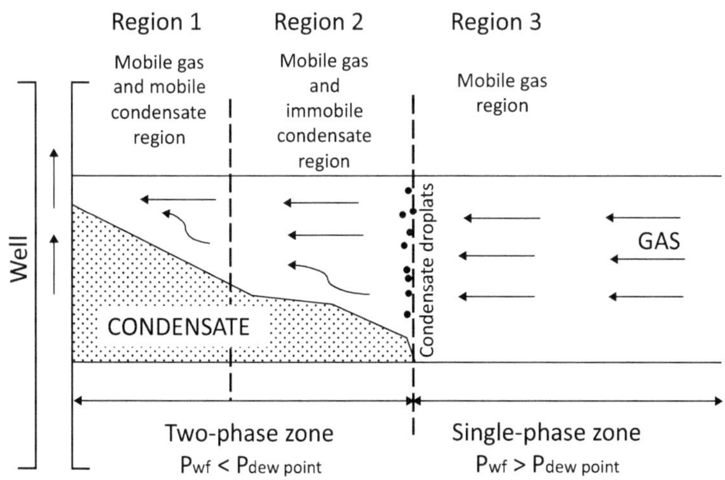

FIGURE 7.5

Condensate-saturation regions in the formation during phase three of production in a gas-condensate reservoir.

Reproduced with modifications from K. Ganie, A.K. Idris, D.F. Mohshim, W.R. Sulaiman, I.M. Saaid, A.A. Malik, A review on the wettability alteration mechanism in condensate banking removal, Journal of Petroleum Science and Engineering 183 (2019) 106431.

reservoir, it results in a leaner gas flowing into the wellbore and less problematic accumulation in that critical near-wellbore region.

7.2.3.3 Impact of two-phase flow on the productivity index

The accumulation of an oil (condensate) saturation in the near-wellbore region has a significant impact on the ability of gas to flow. Before the pressure drops below the dewpoint pressure and condensate starts accumulating, gas is flowing at its maximum capacity. In this case, one can also assume an immobile water saturation in the rock and treat it as a part of the rock system assuming a gas relative permeability of one. An increase in the oil saturation in the wellbore to 55% to allow some of that condensate to start flowing results in a 78% reduction in gas flow capacity in that region. We can consider this as a damage zone that has a radius of say, r_s, equal to 10 ft, and with a damage permeability, k_s, that is 22% of the permeability of the formation. Accordingly, a skin factor can be calculated using Hawkins' formula shown in Eq. (7.8) [27].

$$s = \left(\frac{k}{k_s} - 1\right) \ln \frac{r_s}{r_w} \tag{7.8}$$

The value of the skin in this case is 12.1. Damage skin has a significant impact on well deliverability, which can be assessed using a productivity index described in Eq. (7.9) [27]. Using this equation, the simple case of 55%

condensate saturation in 10-ft near the wellbore can reduce the productivity index of the well by 57%.

$$J = \frac{q}{\overline{p}^2 - p_{\mathrm{wf}}^2} = \frac{kh}{1424\overline{\mu}\overline{z}T\left(\ln\frac{r_e}{r_w} + 0.75 + s\right)} \tag{7.9}$$

Next, we share field cases of gas-condensate reservoirs where liquid blocking resulted in a loss in productivity and we review some of the mitigation strategies used in the industry to overcome this problem.

7.3 Field examples and industry practice

The impact of liquid blocking on gas-well deliverability has long been observed in the field. In this section, examples of such cases are presented and discussed. In each case, the reservoir fluid and rock properties are shared as available, the impact of liquid accumulation on production is summarized, and the mitigation strategies employed by the operating companies are shared.

7.3.1 Ichthys gas-condensate field in Australia

The Ichthys gas-condensate field was discovered in 2000 [34,35] and started producing in July 2018 [36]. The total gas in the place is estimated at 12 trillion cubic feet, with 500 million barrels of condensate [37]. The field is located 220 km offshore Western Australia with an average water depth of 250 m [38]. The sandstone formation in the Ichthys field is the unconsolidated Grebe sands that are over 500 m thick, which required significant work on engineering suitable drilling mud for overbalanced drilling [34]. The field covers an area of 800 km^2 where field development was planned for 50 wells expected to produce over 40 years [35]. Well spacing optimization was studied to maintain the longest gas rate plateau and minimize the well interference resulting in drainage area radii of 1500–2000 m [35].

Analysis of core plug samples reveals a permeability ranging from under 1 md to over 500 md and porosity values between 8% and 17% [39]. The results of CVD tests using a reservoir fluid sample, shown in Fig. 7.6, reveal a modest maximum liquid dropout of around 2.8vol.% at a pressure of 1500 psig with a dewpoint pressure of 5000 psia at reservoir temperature. While no reservoir pressure data have been published, the field is reported to be at an initial pressure that is close to the dewpoint pressure [36].

Production from this field was controlled to maintain the well-flowing bottomhole pressure above dewpoint pressure in order to avoid condensate blocking. The impact of condensate accumulation on the well deliverability was assessed in two wells by increasing the gas production rate allowing the pressure to drop below the dewpoint

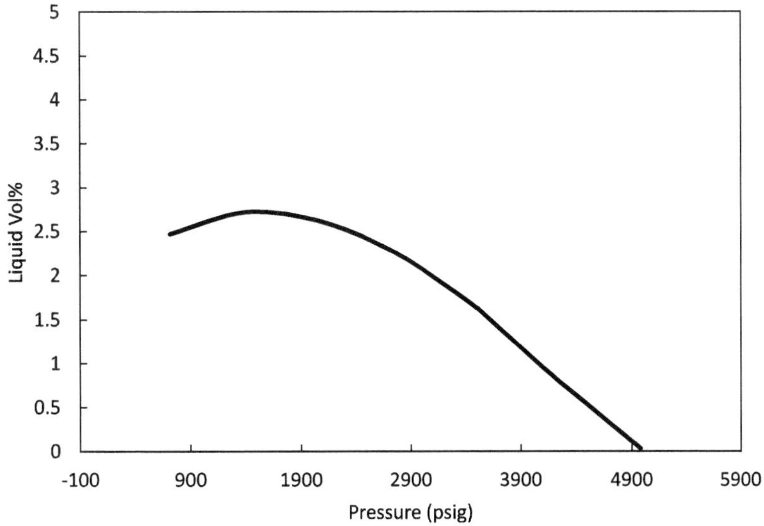

FIGURE 7.6

Liquid dropout as vol.% for the gas-condensate fluid in the Ichthys gas-condensate field.

Produced with data from R. Iizuka, Monitoring and modelling condensate banking effects with the Ichthys field production data, SPE-202440-MS, in: SPE Asia Pacific Oil & Gas Conference and Exhibition, Virtual, 17–19 November, 2020.

pressure [36]. Well A has an average permeability of 20 md while well B has an average permeability of 6 md. Production rates were not reported; however, it was noted that well A had a production potential of 200 MMscf/d. During the high-rate test, well A lost 25% of its productivity while well B lost 40% of its productivity. This was associated with the skin factors of 5.5 for well A and 10.5 for well B.

Lowering the production rate was successful in recovering most of the lost well productivity for these wells; lower production rates correspond to higher wellbore pressure values and that increase in pressure helps to reduce the amount of liquid accumulation since it results in the evaporation of the liquid back into the gas phase. Modeling work revealed an oil saturation accumulation of only around 16% near the wellbore for well A, which was responsible for the 25% loss in productivity. Lowering liquid accumulation and the corresponding skin factor for that well required maintaining a production rate well below 50 MMscf/d, less than a quarter of the potential for this well [36].

7.3.2 Cupiagua gas-condensate field in Columbia

The Cupiagua gas-condensate field is an associated gas field in Columbia. It was discovered in 1988 and production started in 1996 [40–42]. The field is relatively smaller than other giant gas-condensate fields, with an estimated original gas in place of 4.5 billion cubic feet. The field is at a depth greater than 10,000 ft subsea with a reservoir thickness that can reach 5000 ft. The initial reservoir pressure exceeds 6200 psia and the reservoir temperature exceeds 240°F [43]. The porosity is around 6% and the average permeability varies between 0.001 and 10 md [44]. This low range of permeability values makes this formation more prone to productivity loss as a result of liquid accumulation. At low permeability values, capillary trapping of the wetting phase is higher due to the smaller pore spaces and pore throats. As a result, condensate needs to accumulate to higher saturation values before significant condensate mobility can be recorded.

The fluid contained in the Cupiagua field is a very rich gas-condensate with a yield of 280 stb/MMscf [45]. The dewpoint pressure is around 5300 psia and the maximum liquid dropout can reach 27vol.% as shown in Fig. 7.3 [28]. This high yield, combined with water production and other damage mechanisms, posed significant problems to well productivity. The formation is also highly fractured, resulting in challenges related to the loss of near-wellbore permeability in low net-pressure zones [46]. The combined skin factor can reach values past 40 [46–48].

Dry gas injection is one of the potential methods of delaying the onset of condensate accumulation. When dry gas is injected, it helps to maintain a higher pressure in the reservoir and can support a longer period where the wellbore bottomhole pressure can be maintained above the dewpoint pressure. This strategy was implemented early in the life of the field starting in 1998 but the reservoir pressure had dropped to 5000 psi by 2005 resulting in condensate dropout throughout the formation. Without gas injection, only 18% of the hydrocarbons in place could be recoverable, but with gas injection the expected recovery factor increased to 60% [45]. It was later observed, however, that sweep efficiency is not uniform and ineffective in some areas, which resulted in a lower actual recovery factor.

Critical condensate saturation in this low permeability formation is reported to be 32%, meaning that condensate has to accumulate to occupy 32% of the pore volume before it can have any measurable mobility. Two wells tested between 1996 and 1998 revealed a 74%–93% reduction in productivity mostly due to condensate impact [46].

Condensate removal was attempted in some wells using inhibited diesel, which reached 5–10 ft around the wellbore. However, this reveals only temporary success [28]. Dry gas injection was supplemented with alcohol and surfactant blends in order to reach larger areas of the formation and enhance the sweep efficiency. This field is an example of where the combination of high condensate yield and low permeability results in a significant productivity loss and requires continuous field-wide management strategies to maintain any level of productivity.

7.3.3 Other examples

Here we briefly review a few other field examples and simulation studies where a multitude of removal and mitigation strategies are examined to tackle condensate and/or water blocking. Some of these strategies, like solvent injection, require a periodic application to remove the condensate that accumulates near the wellbore. Others, like gas injection, focus on recharging the reservoir with either dry gas or nonhydrocarbons to help either re-evaporate the condensate or increase reservoir pressure. Well stimulations, such as fracturing and acidizing, are used to improve the near-wellbore permeability and alter the pressure profile in order to delay the onset of liquid accumulation and lower its impact on well productivity. Newer technologies, discussed in Section 7.4, include the use of wettability modifiers to allow for long-term reduction of capillary trapping of the liquid phases, in addition to huff-n-puff techniques and other innovative methods.

7.3.3.1 Solvent injection

In some cases, where the condensation and accumulation of liquid is slow enough that it can be tolerated for certain periods of time, one common approach is to frequently inject solvents into the producing well to work on removing the liquid block. This has been applied for both condensate and water blocking cases. The most common solvent used is methanol [49,50] but in some cases isopropyl alcohol and inhibited diesel [28,51] have been used. The solvent acts on altering the properties of the fluid–fluid interface in addition to impacting the phase behavior of the system.

A popular field example of solvent inject is the one applied in the Hatter's Pond field in Alabama, published by Al-Anazi et al. [49]. The field produces from two deep formations, with depths larger than 18,000 ft and a reservoir temperature of 315°F. The permeability is low, ranging between 2 and 6 md, with a 12%–15% porosity. The reservoir fluid has a dewpoint pressure that is 3030 psi at the reservoir temperature, and is quite rich in liquids, having a maximum liquid dropout of 40vol.% [49]. The field was discovered in 1974 and by 1997 the reservoir pressure had dropped to 2700 psi. With the reservoir pressure resulting in condensate dropout throughout the reservoir, wells in this field experience gradual productivity loss. One well experiencing such loss was initially tested at 2.7 MMscf/d with 348 stb/d of condensate production. A couple of years later, its production had declined to 0.25 MMscf/d, a 91% reduction in gas productivity, with 87 stb/d of condensate production [49]. Solvent injection was used in order to remove the condensate and improve the production rate. A 1000 bbl of methanol was injected into the well. Methanol is soluble in both water and condensate and results in a reduction in the value of the dewpoint pressure. This treatment resulted in doubling the productivity of the well for a period of 4 months as a result of liquid block removal of both water and condensate. The reaccumulation of condensate after treatment reduced the productivity to a value that is still 50% higher than pretreatment productivity. The authors stipulate that the permanent removal of the water from the near-wellbore area and the remaining methanol present in the residual saturation impacted this extended productivity enhancement by enhancing condensate mobility and altering the pressure profile in that zone [49].

7.3.3.2 Gas cycling

The main purpose of gas cycling or dry gas injection is to maintain a higher reservoir pressure in order to delay the onset of condensation in the formation. This technique is a suitable choice when the operator associates more value and use for the liquid condensate than the dry gas. Multiple factors impact this decision, such as the richness of the condensate, the available market for liquid versus gaseous hydrocarbons, and the projected severity of well productivity loss as a result of condensation.

Burachok [52] reviews a list of commercial projects where gas cycling was implemented in order to improve the condensate recovery factor. One of these examples is Kaybob, the largest gas-condensate field in Canada [53]. The field was discovered in 1961 and production started in 1968. It is a high-permeability formation (100–300 md) that has a complex geology with a highly-fractured formation and experiences a weak water drive [53]. The fluid is a rich gas-condensate with a condensate yield of 190 stb/MMscf [54]. It has a dewpoint pressure of 4075 psia and a maximum liquid dropout of around 15vol.%. The initial reservoir pressure of 4315 psia is only 240 psi above the dewpoint pressure [54]. Gas cycling was considered and implemented early in the design of the development strategy for this field [55,56]. Without gas injection, 79% of the original gas-in-place could be produced, however, only 43% of the original liquid-in-place would be recovered [54]. A partial gas cycling scheme was chosen to be the most suitable for implementation in this field where around half of the produced gas is reinjected for around 10 years [53,54]; since larger amounts of gas injection could lead to early breakthrough of lean gas. Under this scheme, 83% of the original gas-in-place was expected to be recovered along with 58% of the original liquid-in-place. This scenario of gas injection was implemented between 1969 and 1983. In reality, only 50%–60% of the original gas-in-place was recovered due to sweet efficiency complications [53]. While gas cycling is the most established technique in the development of gas-condensate fields, the complexities around early lean-gas breakthrough as a result of complex reservoir geometry can limit the success of this approach. In addition, the renewed focus on the value of natural gas requires careful economic evaluation regarding how much gas is diverted toward reinjection and whether other well interventions can be more cost-effective at various stages of the life of the field.

7.3.3.3 Nonhydrocarbon gas injection

Many gas-condensate fields can be benefited from pressure maintenance to extend the plateau of gas production before well flowing bottomhole pressure drops below the dewpoint pressure and to delay the whole reservoir pressure falling below the dewpoint pressure. The injection of nonhydrocarbon gases has been considered by operators and implemented in some field cases to avoid the use of valuable hydrocarbon gases. In this section, we review the use of nitrogen and carbon dioxide in recharging reservoir pressure.

While nitrogen can be a great choice for injection due to its abundance and to the available and relatively cheap technology to compress it, studies have shown that the use of nitrogen results in increasing the dewpoint pressure of the hydrocarbon system, which can lead to a significant loss in liquid hydrocarbons throughout the reservoir [57]. On the other hand, nitrogen contributes to liquid swelling, allowing condensate to flow, and it can also result in a reduction in the value of the maximum liquid dropout [58]. A field in Abu Dhabi was subjected to 15 years of hydrocarbon dry gas injection and then was considered for transition to a 20% nitrogen/80% hydrocarbon gas injection before a planned move to 100% nitrogen injection [52,59]. The process of partial nitrogen injection resulted in nitrogen breakthrough and the move to 100% nitrogen injection was halted [59]. A lean gas-condensate field in Iran was also considered for nitrogen injection [58]. Its initial pressure was 4200 psia, with the dewpoint pressure at reservoir temperature being 4070 psia, a maximum liquid dropout of 4.7vol.%, and a condensate yield of 56 stb/MMscf. The study was able to identify scenarios of nitrogen injection that can result in improved gas and condensate production, however, the economic evaluation of the project showed that the financial investment would not be recovered even when a rate-of-return of only 5% was considered.

CO_2 injection toward enhanced gas recovery has the advantage of also contributing to CO_2 sequestration in geologic formations, a topic that is of great interest to many researchers and funding agencies [60,61]. It is more common to consider the injection of CO_2 in depleted gas reservoirs rather than utilizing it as a part of new field development. The utilization of CO_2 as a part of gas-condensate field development is still in early stages of investigation. Some laboratory work has shown CO_2 to be ineffective in impacting the depletion of a gas reservoir as compared to hydrocarbon gas injection [62]. The trapping of the CO_2 and the phase behavior complications can reduce its efficiency when applied in enhanced gas recovery [52,63,64].

7.3.3.4 Well stimulation

As discussed earlier, the use of gas cycling or gas injection targets the delay in the onset of condensation through pressure management and contributing to re-evaporation of condensate liquid. Solvent injection, as presented earlier, targets the removal of liquid condensate through the repeated injection of solvents into the producing well. The technique of well stimulation, however, is concerned with enhancing the permeability in the near-wellbore region of the producing well. This can contribute two outcomes. On one hand, the increased permeability through fracturing or acidizing alters the pressure profile in that critical region, resulting in a delay in the pressure dropping below dewpoint pressure. On the other hand, the region of enhanced permeability would experience more favorable two-phase flow conditions and lower amounts of liquid trapping and gas blocking.

The application of these techniques has multiple constraints and limitations, in addition to not solving the problem of condensation beyond the zone of enhanced permeability. Government regulations can significantly impact the investments in fracturing operations where fracturing has been outright banned in some locations. Matrix acidizing that targets the development of wormholes are the most effective carbonate formations that are almost purely made of calcium carbonate [27].

Hydraulic fracturing has shown in some cases to be more effective than solvent injection in enhancing well productivity [50], however, when the cost of fracturing and the economics of the operation are considered, especially in lean gas-condensate reservoirs and in reservoirs where the pressure has dropped below dew-point pressure, the cost of this approach is larger than the return [50,65]. The efficiency of hydraulic fracturing is highly dependent on the specific reservoir case, wherein rich gas-condensate reservoirs, liquid accumulation around the fracture face can completely block the flow of gas into the fracture. However, the use of horizontal well multistage fracturing, when economically justified, can result in an extended period of production before the wellbore pressure drops below the dew-point pressure and could be worth investigating [66]. In some simple cases of carbonate formations where simple acidizing systems can be used, matrix acidizing can be easier to justify from the economic point of view if it can result in a reasonable delay in the onset of condensation. Acidizing is the most commonly used technique in carbonate reservoirs since it is the go-to methodology used to overcome near-wellbore damage from drilling and completion fluids. The use of slow-reacting chemicals can also result in similar patterns of changes to the pore structure toward reducing capillary trapping [67]. However, both matrix acidizing and slow-acting chemicals do not solve the problem of condensate blocking in the long term and would require the future application of solvent injection or to be combined with gas cycling or with the technique of wettability alteration that is discussed in the next section.

7.4 Recent advances in research and development

Several new techniques to address the problem of condensate blocking are currently undergoing testing and development. The aim of most of these new techniques is to reduce the frequency of well interventions by introducing long-term solutions to the problem of liquid blocking. In this section, we review the state-of-the-art in this domain.

7.4.1 Wettability alteration

The process of altering the wettability of the rock in the near-wellbore area is concerned with affecting a long-term impact on the trapping of liquid in the pore space. While the injection of solvents and gases targets the fluids in the formation and wettability alteration targets the rock. This technique has been researched over the past two decades and has not yet been widely utilized in field applications. In this section, we cover the fundamentals upon which this technique is based, we then review some of the experimental and simulation studies in this domain, and finally we present a few examples of attempted field applications.

7.4.1.1 Impact of wettability on two-phase flow

As discussed earlier, when studying two-phase flow in porous media, the fluids in the rock are labeled as "wetting phase" and "nonwetting phase." The wettability of a rock-fluids system is determined by the value of the interfacial tensions at the three two-phase interfaces of this system. The interfacial tension represents an interfacial force per unit length, and the balance of these interfacial forces at the line where the three phases intersect determines the contact angle, at which the fluid–fluid interface meets the solid surface as seen in Fig. 7.7. This contact angle in such a simple system is determined according to Young's equation:

$$\sigma_{SG} - \sigma_{SL} = \sigma_{LG}\cos\theta \tag{7.10}$$

These surface forces have a significant impact on the movement and distribution of a fluid inside the rock pores in the presence of another fluid. Capillary pressure is a direct outcome of these forces. Consider a capillary tube inserted in a container where two fluids are present, as shown in Fig. 7.8. Assume fluid #1 is the fluid that wets the surface, and thus the contact angle through fluid #1 is less than 90 degrees. Inside the capillary tube, the fluids meet at a two-phase interface, while the perimeter of that interface represents the three-phase line of contact. This line is subjected to interfacial forces that result in an upward movement of the interface. The rise of fluid #1 inside the capillary tube stops when the hydrostatic forces balance the interfacial forces. At this equilibrium condition, the pressure at the fluid interface inside the capillary tube has two different values on each side of the interface, with the pressure in the nonwetting phase (fluid #2 in this case) having a larger value; the difference between these two values is the capillary pressure, determined by Eq. (7.11).

$$P_c = \frac{2\sigma_{LG}\cos\theta}{r} \tag{7.11}$$

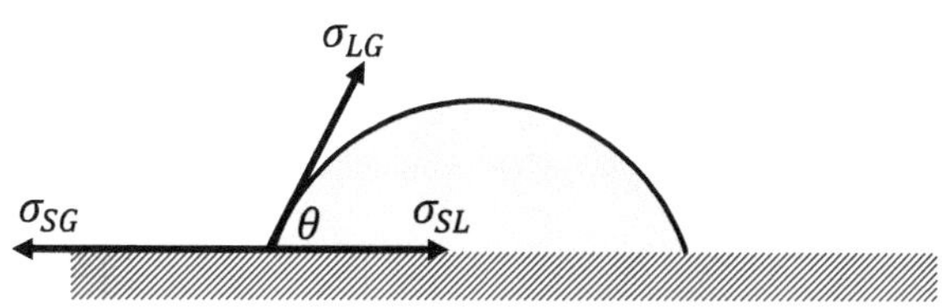

FIGURE 7.7

Contact angle defining the wettability of a fluid–fluid–solid system. The relative value of the interfacial tensions at the solid–fluid interfaces determines whether the angle within a given phase is acute or obtuse, thus defining the wettability of the system.

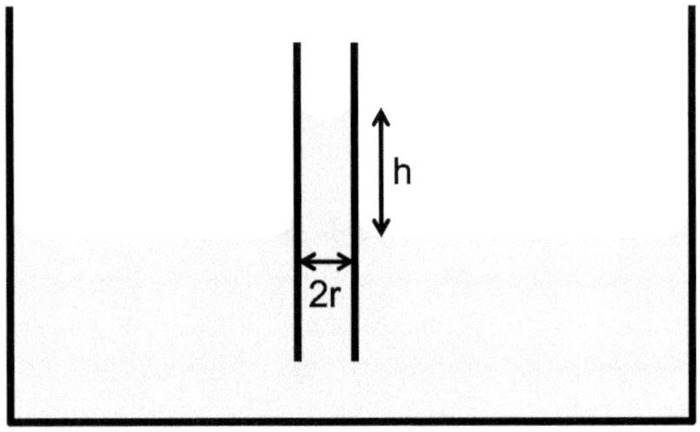

FIGURE 7.8

A schematic of capillary rise in a capillary tube. The pressure within the nonwetting phase inside the tube (in this case a gas) is higher than the pressure within the wetting phase at the curved interface between the two fluids.

As can be seen from this equation, the value of the capillary pressure is impacted by the size of the capillary, in addition to the properties of the solid surface and the two fluids. Inside the porous medium, capillary pressure contributes to the trapping of the wetting phase in an irreducible wetting-phase saturation and contributes to the trapping of the nonwetting phase in a residual nonwetting phase saturation.

Inside porous media, capillary pressure is simplified as a correlation that provides a value of that pressure as a function of fluid saturation. Since fluid distribution inside the rock at a given fluid saturation is not unique, capillary pressure can have a range of values at that saturation. This is addressed by determining the two capillary pressure curves, one with maximum values representing a drainage cycle (during a process through which the wetting phase saturation is decreasing), and one with minimum values representing an imbibition cycle (during a process through which the wetting phase saturation is increasing), as shown in an example in Fig. 7.9 [68]. This "hysteresis" effect on capillary pressure also impacts relative permeability curves where it is mostly observed in the relative permeability value of the nonwetting phase as shown in Fig. 7.10, where the nonwetting phase experiences more trapping through an imbibition cycle than it does during a drainage cycle.

It is typical to ignore the capillary pressure when modeling multiphase flow in porous media due to the complications it introduces to the system of equations. However, the impact of capillary pressure is captured through its direct impact on the relative permeability curve, and especially on the values of trapped fluid saturations. Accordingly, rock permeability, which is associated with the size distribution and structure of the pore space, has a significant impact on trapped fluid saturations and relative permeability parameters. The impact of wettability on capillary trapping is also captured in its impact on the relative permeability curve, which directly translates to an impact on the two-phase flow characteristics of a gas-condensate system. Accordingly, controlling the wettability in the accessible near-wellbore region can be utilized to modify the fluid distribution and two-phase flow in that region toward reducing the impact of liquid condensation on gas mobility [18].

7.4.1.2 Wettability alteration to mitigate condensate blocking; experimental studies

It was in the year 2000 when the first set of experiments were published on the impact of altering rock wettability on the flow of a liquid–gas two-phase system was conducted [69]. The research team at the Reservoir Engineering Research Institute conducted these initial studies as a proof-of-concept that altering the wettability of this system is possible and that it has the desired impact on the flow within porous media. It was determined that reducing the affinity between the liquid phase and the solid surface results in reducing the immobile liquid saturation and enhancing the mobility of the liquid at a given liquid saturation [70,71]. This was also shown to be effective for both water–gas systems and oil–gas systems at room temperature as well as at elevated temperature and pressure [18].

Challenges with altering rock wettability relate to identifying the suitable chemical that: (1) is environmentally suitable, (2) is stable at reservoir conditions and can adsorb to the rock for extended periods of time, (3) does not aggregate and result in blocking some of the rock pores, and (4) can be targeted to achieve various degrees of wettability alteration as desired. The initial type of chemicals was ideal for long-term alteration of wettability and resulted in permanent adsorption to the rock surface

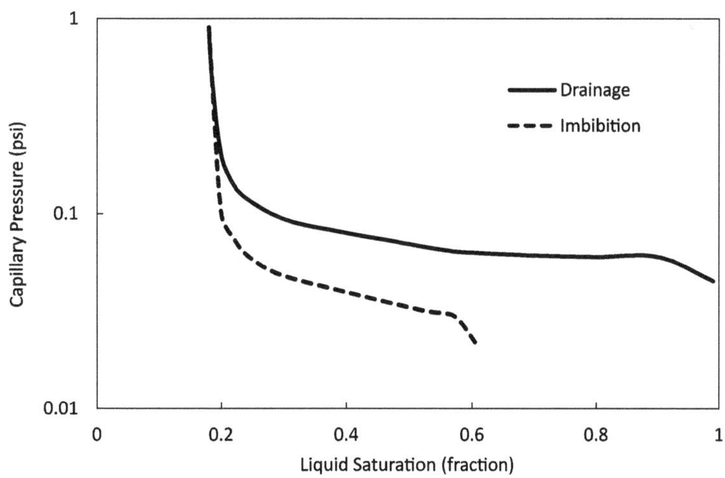

FIGURE 7.9

Capillary pressure in porous media, and the impact of hysteresis; or saturation history.

Produced with data from: G.R. Jerauld, S.J. Salter, The effect of pore-structure on hysteresis in relative permeability and capillary pressure: pore-level modeling, Transport in Porous Media 5 (1990) 103–151.

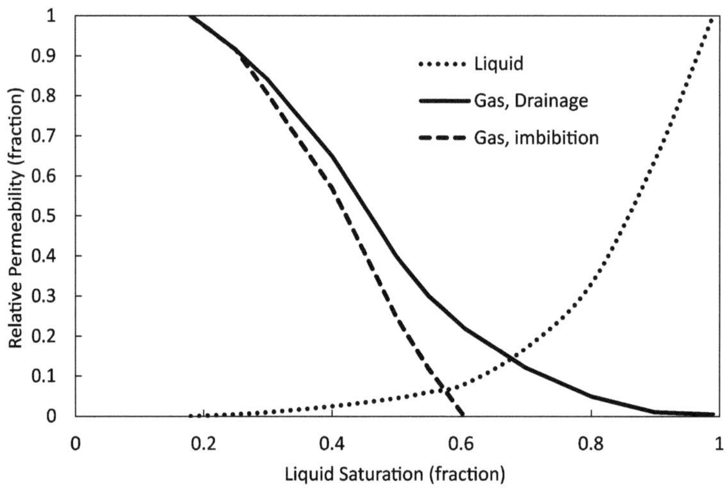

FIGURE 7.10

The impact of capillary trapping and fluid distribution on relative permeability; curve above corresponds to the capillary pressure curve in Fig. 7.9.

Produced with data from G.R. Jerauld, S.J. Salter, The effect of pore-structure on hysteresis in relative permeability and capillary pressure: pore-level modeling, Transport in Porous Media 5 (1990) 103–151.

[18], however, it was determined to be unfriendly to the environment. Challenges started to arise after that where newly designed chemicals would wash away during flow or result in very minimal changes to the oil–gas–rock system [72].

The use of fluorinated chemicals to achieve wettability alteration has been explored by multiple research groups. This includes the use of surfactants and polymers in both water-based and alcohol-based solvents. The application of surfactants has been explored to tackle both condensate blocking and water blocking [73]. While great success is reported in terms of changes to the wettability of the system [15,73–77], one of the main challenges has been the occasional observation of reduction in rock absolute permeability as a result of the chemical treatment [72,78]. Unfortunately, many of the published studies on these chemicals do not explicitly report the value of the absolute permeability of the rock before and after treatment. This requires characterizing the dry-core gas permeability before and after it was exposed to the chemical treatment. Fahes and Firoozabadi reported great success for chemical 11–12 P in altering rock wettability with no impact on the absolute permeability of the sandstone rocks used [18]. However, the company that synthesized that chemical reported that it was deemed unsuitable for field application citing environmental concerns. Replacement surfactants provided by the same company did not measure up in terms of ability to permanently adsorb onto the rock and resulted in a 25% reduction in rock permeability. Multiple publications by other research groups reported success in using surfactants without referencing the impact on the absolute permeability of the rock nor the durability of the treatment. Most of the publications that do report the value of the permeability note that reduction in the rock flow capacity mostly occurs in low-permeability rocks (close to 10 md). However, some publications also report permeability reduction even in 100 md high-permeability rocks. This reduction in permeability can vary between 5% and 30%. While the improvement in the production capacity due to alteration of wettability can exceed the negative impact of reduction in permeability, the risk of blockage is too great to make such surfactants a suitable choice for field application.

In the past decade, there was significant research done on the use of nanoparticles to reduce the liquid wetting for application in gas-condensate reservoirs. The advantage of these fluorinated particles is that they can be used at much lower concentrations. They work by introducing dimensions that impact wettability, lower surface energy, and produce a modified surface roughness that also contributes to reducing liquid wetting properties. Mousavi et al. present a good review of the use of nanoparticles to achieve modified wetting conditions. Their research team demonstrated for the first time the use of functionalized fluorinated silica nanoparticles for application in gas-condensate reservoirs [79,80]. While most surfactant research for wettability alteration in gas-condensate systems replied on surfactants developed by chemical companies, most of the work on utilizing nanoparticles relies on in-house synthesis of treatment mixtures. The process is described in most of the publications; however, a detailed explanation is shared by Villegas et al. [81]. Great success is reported in altering the wettability of both sandstone and carbonate rocks through functionalized silica, iron oxide, calcium carbonate, magnesia, and other nanoparticles [81–90]. The impact on wettability is reported to depend on the minerology of the rock. Recently, a state of super gas-wetting has been reported, with a contact angle of 145 degrees through the hydrocarbon phase [91], something that has not been achieved before using surfactants. The focus of many of the experimental studies in this domain is on maximizing the contact angle through the liquid phase, however, this has been shown through simulation work to not be the right target; an optimum state of intermediate-wetting can result in better well productivity than a super gas-wetting state near the wellbore [78,92,93]. This is addressed further in the next section. Some of the studies reference concerns about pore-throat plugging by the nanoparticles, while other studies reference a slight increase in the rock permeability when an acidic solvent is used. Consistent assessment and documentation of the impact of wettability alteration chemicals on rock absolute permeability is critical to allow for the identification of the chemical, solvent, and procedures that can achieve the desired wetting conditions without adverse effects on the flow capacity of the rock.

In the past couple of years, a whole array of other chemicals was utilized in altering rock wettability toward reduced liquid-wetting. Nowrouzi et al. succeeded in altering the rock wettability using supercritical gases, R134A and R404A, that are widely used in refrigeration [94]. The performance was dependent upon the pressure and aging time used. Contact angles through the liquid hydrocarbon phase that exceed 100 degrees were achieved using fluorochemical foam [95] andusing fluorochemical modified graphene oxide [96]. A nonfluorinated benzoxazine monomer was successfully used in reducing water-wetting toward application in fields where water blocking is a concern [97]. Similarly, a nonfluorinated $scCO_2$-based silylation was used to change the wettability of a water–CO_2 system [98]. In summary, a whole array of chemicals has been developed or utilized to alter the rock wettability toward reduced liquid-wetting. Field application should always start with experimenting to identify the suitable chemical that will achieve for the specific rock-fluid system the desired state of wettability. In some cases, reservoir simulation can be performed to identify the impact of wettability alteration in combination with well stimulation or gas cycling depending on the field development plans and the specific conditions of rock and fluid properties [99].

7.4.1.3 Wettability alteration to mitigate condensate blocking; simulation studies

While identifying an ideal chemical for use in wettability alteration has faced many challenges, the impact of successful alteration of wettability on condensate blocking and gas-condensate well deliverability has been studied by multiple research groups by performing reservoir simulations. In these studies, a well model is built that allows for a region around the wellbore to be given "altered" flow characteristics, namely a modified relative permeability curve, and then various scenarios are considered to assess whether altering wettability can lead to enhancement in well productivity.

The advantage of simulation work is that it can guide experimental work and field applications by addressing questions such as the degree to which the wettability should be altered, the impact of the value of the condensate yield for the reservoir fluid, the reservoir rock permeability, and changes in reservoir pressure.

Early work on simulating the impact of wettability on well productivity focused on revealing that it is possible to enhance gas production by altering the wettability of a small region around the well. Kumar et al. published the first study in 2006 that showed an 80% increase in the gas production rate when a region of 20 ft around the wellbore is modified to alter the irreducible oil saturation [100]. They modeled a lean gas-condensate system in a low-permeability reservoir and assumed that the modified wetting conditions result in an increase in the end-point gas relative permeability. Delavarmoghaddam et al. in 2009 modeled the impact of the gas-condensate fluid on the effectiveness of wettability alteration [101]. Their results suggest that the treatment is more effective in the case of lean gas-condensate systems, however, the relative permeability model used was not realistic. Zoghbi et al. suggested for the first time in 2010 the need to assess the impact of various degrees of wettability alteration [92]. This includes considering the fact that a reversal of wettability toward a strong gas-wetting condition, while enhancing condensate mobility, will reduce the mobility of the gas in the treated region. They modeled a case of a rich gas-condensate fluid in low-

and high-permeability reservoirs. Their results show that lowering the gas relative permeability curve in the treated region when a gas-wetting state is achieved can still provide superior productivity to the original case of strong liquid-wetting. However, optimizing the wetting conditions to achieve an intermediate state of wettability is the way to go when the target is to maximize the well productivity. Treatment in low-permeability reservoirs (1–10 md) showed to be more effective than in high-permeability reservoirs (100 md). Following that study, several research groups between 2014 and 2021 used similar relative permeability models and further investigated the impact on well productivity confirming improved well deliverability when optimum wetting conditions are targeted [102–104]. The results were consistent; however, it did not address a large-enough number of variables to provide guidance for when this approach is expected to be effective, economical, and suitable for consideration.

The assessment of the added value from wettability alteration needs to consider the economic dimension in order to be determined as viable. This requires observing not just the changes in gas recovery factor, but also the change in gas rate over time, which would be discounted to present value in order to compare the various scenarios. The introduction of the economic dimensions resulted in a new set of screening criteria that can help to assess the suitability of various wells as candidates for wettability alteration. The economic dimension that considers the cost of the treatment in addition to the incremental changes in the production of both the gas and the liquid streams through discounting was first considered by Weiss [105,106]. The results showed that treatment in lean gas-condensate reservoirs is not economical and should be avoided. High permeability reservoirs can benefit from treatment when the vertical well is allowed to drain a large area. On the other hand, targeting optimum wetting conditions seems to be ideal for low permeability reservoirs with medium to rich gas-condensate fluids. Further investigations by Ajagbe [93,107] to start establishing screening criteria for wettability alteration showed that a large treatment radius is needed in the case of rich gas-condensate fluids.

7.4.1.4 Wettability alteration to mitigate condensate blocking; field applications

Although the technique of wettability alteration is still undergoing research and development, there have been several field trials to assess the impact of chemical treatment toward reduced liquid wetting on gas-well deliverability. We review five such cases that have been published. The earliest being for a well in China in 2008 and the latest for a well in Columbia in 2021. One of these cases reports success in sustained improvement in well productivity [108]. Another reports partial success in regaining some of the well deliverability [109,110]. Two cases report only short-term temporary enhancement in productivity [99,111], and the last one reports a reduction in well productivity [112]. This variation in the range of results and the low success rate is further evidence that this technology is promising but requires more investment in further research and development before it can be a reliable go-to method for treating condensate and water blocking problems.

In 2009, a gas-condensate well in a sandstone reservoir in Oklahoma was subjected to a chemical treatment to alter the wettability in the near-wellbore region [108]. The reservoir rock had a permeability of 20 md, rendering it a low-permeability formation. The gas production rate is reported to be 0.2 MMscf/d when the well first started producing, which decreased to 0.1 MMscf/d 4 months into production. After treatment, the production rate increased around fourfolds to an average of 0.5 MMscf/d over the year following treatment.

In 2013, a gas-condensate well in Saudi Arabia was subjected to a chemical treatment to alter wettability [109,110]. The formation had a 14 md permeability, thus categorized as a low-permeability formation. The gas rate in this well had decreased by a factor of 10 as a result of condensate blocking. The treatment resulted in a sustained 83% increase in the gas production rate and a 313% increase in the condensate production rate. Although these improvements did not bring the production rate back to its original level before the decline occurred, it is a success due to the sustained improvement in rate compared to pretreatment conditions.

The three other cases with temporary improvement or a failed treatment were in tight reservoirs. In 2008, a well producing from a 0.1 md permeability sandstone formation in China was subjected to chemical treatment. The improved production rate was only sustained for 4 days before it went back to its pretreatment levels [99]. In 2012, a well producing from a sandstone formation in Columbia was treated and the improvement in gas production lasted only 25 days then declined to pretreatment levels [111]. Both cases seem to indicate a possible concern related to the durability of the chemical used and indicate a possible washing off of the chemical from the pores.

In 2021, a well producing from a 3 md tight sand reservoir in Columbia was treated with a chemical treatment dispersed in gas (GaStim) [112]. A 66% reduction in the gas production rate was observed after the treatment. Experimental investigation following the failed treatment showed that rock permeability was adversely impacted by the treatment. The liquid-to-gas ratio of the injected chemical seemed to be key in limiting the condensation of the dispersed chemical and the subsequent blocking of the pores [112].

7.4.2 CO_2 huff-n-puff

The concept of using CO_2 in the process of liquid-condensate removal from the near-wellbore region has been known for decades but has not been widely utilized in the field. A large volume of research is dedicated to the study of the physical properties of oil–CO_2 mixtures for the purpose of enhanced oil recovery. These mixtures form after exceeding a threshold of pressure that is designated as the minimum miscibility pressure, which is a property specific to the oil in question depending on the composition of that oil [113,114]. The miscibility of CO_2 in the oil results a significant reduction in oil viscosity and an increase (swelling) in oil volume. The density behavior of these oil–CO_2 mixtures, however, is counterintuitive; the density of the mixture can increase to values higher than the original oil density because the mass of the mixture can increase faster than the volume does as a result of the molecular structure of CO_2 [115,116].

When it comes to condensate removal, the injection of CO_2 into the production well at suitable pressure and rate conditions results in mixing between the CO_2 and the liquid condensate. The mixing is expected to result in an increase in the liquid volume and accordingly an increase in liquid saturation and relative permeability. The reduction in liquid viscosity also aids the ability of the condensate to flow toward the wellbore. The changes to the interfacial tension between the liquid and the gas result in reducing capillary trapping and this also contribute to an enhancement in flow characteristics. Similar to the case of solvent injection, this process of CO_2 injection would need to be applied frequently to remove the condensate that builds up over time. A given slug volume of CO_2 is injected (huff), an amount of soaking time is allowed, and then the well is put on production (puff).

A good review of work done on this technique is available in a paper by Ayub and Ramadan [117]. In this study, a three-dimensional-simulation model is utilized to study the parameters impacting the success of CO_2 huff-n-puff application in a hypothetical gas well. Four cases were considered with low (10 md) and medium (40 md) permeability values and with a maximum liquid dropout ranging from 22 to 35vol.%. The simulation considered injection of 100% CO_2 for 1 week in a supercritical state (temperature and pressure surpassing the critical point of CO_2). This was followed by 2 months of soaking time. The impact of this treatment on the gas production rate, condensate removal, and condensate re-accumulation was assessed. Two of the cases benefited from the CO_2 huff-n-puff process, and those were the cases that experienced mild condensate accumulation in the first place. These cases experienced improved productivity for 3 years following the treatment. For the cases where a high amount of condensate accumulation occurred due to a richer gas or a higher permeability that favors faster drop in pressure, the amount of condensate that was removed from the near-wellbore region was not significant enough to result in an improvement in well productivity. One of the factors that contribute to the success of this technique is the boundary condition that allows for pressure buildup during the soaking time. A field application of CO_2 huff-n-puff in a field in Oman also attributed the success to pressurization as a result of cross-flow from a lean gas formation [118]. These results shed some doubt on the significance of the contribution of CO_2 to the improvement in well productivity. It is possible that some scenarios could see a positive huff-n-puff effect, however, these should be carefully designed to assess the potential benefit of the treatment. The CO_2 huff-n-puff technique is definitely still an area that is suitable for further experimental and simulation studies to carefully design the screening criteria and parameters that can help operators to better utilize it.

7.4.3 Other new technologies

The injection of solvents toward the removal of condensate from the near-wellbore region has been discussed earlier. Such solvents are typically alcohol-based, which can include methanol or isopropyl alcohol. The focus in these chemicals is on the mixing between the solvent and the liquid condensate and on the subsequent evaporation as a result of altered phase behavior. A new approach toward the removal of liquid condensate targets increasing the temperature of the fluid in the near-wellbore region to allow for the evaporation and recovery of the condensate [119–121]. This is accomplished using thermochemical fluids where the chemical reaction between the injected fluids releases heat and results in an increase in both temperature and pressure. Hassan et al. [122] used ammonium chloride and sodium nitrate to study the removal of condensate from sandstone and carbonate rocks in the laboratory. Reservoir simulation was then utilized to evaluate the impact on well productivity where a 6.5 fold increase in the gas production plateau time was observed as a result of treatment.

Ultrasonic and microwave technology has also been shown to impact the fluid movement in porous media [123,124]. Research toward the application of this technology in mobilizing liquids that are experiencing capillary trapping is limited, however, it does show positive results. In some cases, microwave treatment was successful in the removal of 100% of the condensate from a core plug [125]. Its effectiveness, however, does depend on multiple factors that include the mechanical and electrical properties of the rock [125]. More research is required to identify the conditions that allow this technology to be efficient in removing the condensate bank. One advantage of this technique could be the one-time installation of such downhole instruments that would allow the frequent application of the treatment at lower costs.

7.5 Final remarks

With a wide range of options available to tackle production from gas-condensate reservoirs, it is important to recognize that the ideal approach is not a one size fits all. Gas-condensate reservoirs require careful characterization of the rock and fluid system early in development. Reservoir simulation and continuous monitoring can help identify the suitable approach to mitigate condensate blocking at each phase of production. Most of these techniques require investing in laboratory testing before field application in order to examine compatibility and optimize the injection parameters. Research and development in this area is still a rich domain where more experimental work and optimization models can be developed to guide field applications and streamline the process of identifying the right approach suitable for a particular field.

Nomenclature

CCE	constant composition expansion
CVD	constant volume depletion
GOR	gas oil ratio
md	millidarcies
MMscf/d	million standard cubic feet per day
stb/MMscf	stock tank barrels per million standard cubic feet
A	cylindrical core cross-sectional area, cm^2
k	rock absolute permeability, Darcy
k_{rw}	wetting phase relative permeability, dimensionless
k_{rnw}	nonwetting phase relative permeability, dimensionless
k_s	damage zone permeability, Darcy
$k_{w,eff}$	wetting phase effective permeability, Darcy
$k_{nw,eff}$	nonwetting phase effective permeability, Darcy
ft	feet
h	formation thickness, ft
J	well productivity index
L	cylindrical core length, cm
p	fluid pressure, atm
P_c	capillary pressure
p_{wf}	flowing bottomhole pressure, psia
p_1	inlet pressure in linear flow, atm
p_2	outlet pressure in linear flow, atm
$\overline{p}$	average reservoir pressure, psia
q	gas production rate, Mscf/D
q_{nw}	nonwetting phase volumetric floe rate, cc/s
q_w	wetting phase volumetric flow rate, cc/s
r	radius of capillary tube
r_e	radius of well drainage area, ft
r_s	damage zone radius, ft
r_w	wellbore radius, ft
s	skin factor, dimensionless
T	reservoir temperature, °R
v	fluid velocity, cc/s
x	distance in the direction of flow, cm
$\overline{z}$	average gas compressibility factor, dimensionless
β	high velocity coefficient
ρ	fluid density
μ	fluid viscosity, cp
$\overline{\mu}$	average gas viscosity, cp
μ_w	wetting phase viscosity, cp
μ_{nw}	nonwetting phase viscosity, cp
σ_{LG}	interfacial tension between the liquid and gas phases
σ_{SG}	interfacial tension between the solid and gas phases
σ_{SL}	interfacial tension between the solid and liquid phases
θ	contact angle between the fluid–fluid interface and the solid surface

References

[1] BP, Statistical review of world energy, 69th edition, London, 2020.

[2] A. Demirbas, The importance of natural gas as a world fuel, Energy Sources, Part B 1 (2006) 413–420.

[3] M. Balat, World natural gas (NG) reserves, NG production and consumption trends and future appearance, Energy Sources 27 (2005) 921–929.

[4] Union of Concerned Scientists, Environmental impacts of natural gas, 2014 <https://www.ucsusa.org/resources/environmental-impacts-natural-gas>.

[5] R.E. James III, D. Kearins, M. Turner, et al., Cost and Performance Baseline for Fossil Energy Plants Volume 1: Bituminous Coal and Natural Gas to Electricity, National Energy Technology Laboratory (NETL), Pittsburgh, 2019.

[6] L.W. Zeltmann, Use of natural gas for electricity generation, WPC-24265, in: 13th World Petroleum Congress, Buenos Aires, Argentina October 20–25, 1991.

[7] B. Balasubramanian, A.L. Ortiz, S. Kaytakoglu, et al., Hydrogen from methane in a single-step process, Chemical Engineering Science 54 (1999) 3543–3552.

[8] G. Ji, J.G. Yao, P.T. Clough, et al., Enhanced hydrogen production from thermochemical processes, Energy & Environmental Science 11 (2018) 2647–2672.

[9] J. Tollefson, Methane leaks erode green credentials of natural gas, Nature 493 (2013) 12.

[10] G. Pétron, G. Frost, B.R. Miller, et al., Hydrocarbon emissions characterization in the Colorado Front Range: A pilot study, Journal of Geophysical Research 117 (2012) 1–19.

[11] G. Myhre, D. Shindell, F.-M. Bréon, et al., Anthropogenic and natural radiative forcings, in: T.F. Stocker, D. Qin, G.-K. Plattner, et al. (Eds.), Climate Change 2013: The Physical Science Basis. Contribution of Working Group I to the Fifth Assessment Report of the Intergovernmental Panel on Climate Change, Cambridge University Press, Cambridge, 2013, pp. 659–740.

[12] J. McCain, Properties of Petroleum Fluids, second ed., PennWell Books, Nashville, 1990.

[13] M.A. Sayed, G.A. Al-Muntasheri, Mitigation of the effects of condensate banking: a critical review, SPE Production & Operations 31 (2016) 85–102.

[14] M.A. Sayed, G.A. Al-Muntasheri, Liquid bank removal in production wells drilled in gas-condensate reservoirs: a critical review, SPE-168153-MS, in: Proceedings of the SPE International Symposium and Exhibition on Formation Damage Control, Lafayette, Louisiana, February 26–28, 2014.

[15] K. Ganie, A.K. Idris, D.F. Mohshim, et al., A review on the wettability alteration mechanism in condensate banking removal, Journal of Petroleum Science and Engineering 183 (2019) 106431.

[16] R. Sandrea, Global natural gas reserves - A heuristic viewpoint, 2008 <https://web.archive.org/web/20080515152021/http://www.mees.com/postedarticles/oped/v49n12-5OD01.htm>.

[17] South Pars/North Dome gas-condensate field, Wikipedia, 2021 <https://en.wikipedia.org/w/index.php?title = South_Pars/North_Dome_Gas-Condensate_field&oldid = 1032078250>.

[18] M.M. Fahes, A. Firoozabadi, Wettability alteration to intermediate gas-wetting in gas-condensate reservoirs at high temperatures, SPE Journal 12 (2007) 397–407.

[19] D.B. Bennion, F.B. Thomas, R.F. Bietz, et al., Water and hydrocarbon phase trapping in porous media-diagnosis, prevention and treatment, Journal of Canadian Petroleum Technology 35 (1996) 29–36.

[20] D. Afidick, N.J. Kaczorowski, S. Bette, Production performance of a retrograde gas reservoir: a case study of the Arun field, SPE-28749-MS, in: Proceedings of the SPE Asia Pacific Oil and Gas Conference, Melbourne, Australia, November 7–10, 1994.

[21] H. Nasrabadi, A. Firoozabadi, R.O. Esposito, et al., Interpretation of an unusual bubble point pressure variation in an offshore field, SPE-113574-MS, in: Proceedings of the Europec/EAGE Conference and Exhibition, Rome, Italy, June 9–12, 2008.

[22] X. Zhou, H. Wang Hongjun, Y. Chen, et al., The study on the contributing factors of phase state of gas condensate: examples from Tarim Basin, Marine and Petroleum Geology 15 (1998) 21–31.

[23] J.W. Amyx, D.M. Bass, R.L. Whiting, Petroleum Reservoir Engineering: Physical Properties, McGraw-Hill Book Co, New York, 1960.

[24] L.J. Klinkenberg, The permeability of porous media to liquids and gases, API-41--200, in: Drilling and Production Practice, New York, New York, January 1, 1941.

[25] A. Firoozabadi, D.L. Katz, An analysis of high-velocity gas flow through porous media, Journal of Petroleum Technology 31 (1979) 211–216.

[26] V. Shabro, C. Torres-Verdin, F. Javadpour, Numerical simulation of shale-gas production: from pore-scale modeling of slip-flow, Knudsen diffusion, and Langmuir desorption to reservoir modeling of compressible fluid, SPE-144355-MS, in: Proceedings of the North American Unconventional Gas Conference and Exhibition, The Woodlands, Texas, June 14–16, 2011.

[27] M. Economides, D. Hill, C. Ehlig-Economides, et al., Petroleum Production Systems, second ed., Pearson Education, London, 2013.

[28] C. Franco, R. Zabala, O. Botero, et al., Inhibited gas stimulation to mitigate condensate banking and maximize recovery in Cupiagua field, SPE Production & Operations 28 (2013) 154–167.

[29] W.G. Anderson, Wettability literature survey - Part 5: the effects of wettability on relative permeability, Journal of Petroleum Technology 39 (1987) 1453–1468.

[30] F.M. Carlson, Simulation of relative permeability hysteresis to the nonwetting phase, SPE-10157-MS, in: Proceedings of the SPE Annual Technical Conference and Exhibition, San Antonio, Texas, October 4–7, 1981.

[31] R.A. Fulcher Jr., T. Ertekin, C.D. Stahl, Effect of capillary number and its constituents on two-phase relative permeability curves, Journal of Petroleum Technology 37 (1985) 249–260.

[32] Ø. Fevang, C.H. Whitson, Modeling gas-condensate well deliverability, SPE Reservoir Engineering 11 (1996) 221–230.

[33] D. Marokane, A.B. Logmo-Ngog, R. Sarkar, Applicability of timely gas injection in gas condensate fields to improve well productivity, SPE-75147-MS, in: Proceedings of the SPE/DOE Improved Oil Recovery Symposium, Tulsa, Oklahoma, April 13–17, 2002.

[34] A.J. Hinton, T. Nolan, V. Tilley, et al., Taming the grebe sand—Tophole drilling success in the Ichthys field, SPE-121439-MS, in: Proceedings of the SPE Asia Pacific Oil and Gas Conference & Exhibition, Jakarta, Indonesia, August 4–6, 2009.

[35] A. Kawasaki, S. Batiot, A. Ion, Ichthys field development - Implementation of a multi-disciplinary well optimisation workflow for a large gas-condensate reservoir, SPE-171469-MS, in: Proceedings of the SPE Asia Pacific Oil & Gas Conference and Exhibition, Adelaide, Australia, October 14–16, 2014.

[36] R. Iizuka, Monitoring and modelling condensate banking effects with the Ichthys field production data, SPE-202440-MS, in: Proceedings of the SPE Asia Pacific Oil & Gas Conference and Exhibition, Virtual, November 17–19, 2020.

[37] W. Townsend, Ichthys: from challenge to reality, SPE-161522-MS, in: Proceedings of the Abu Dhabi International Petroleum Conference and Exhibition, Abu Dhabi, UAE, November 11–14, 2012.

[38] N. Coone, Challenges in riser design - The Ichthys gas field development, OTC-26785-MS, in: Proceedings of the Offshore Technology Conference Asia, Kuala Lumpur, Malaysia, March 22–25, 2016.

[39] L. Knapp, T. Fukunari, K. Yoshimatsu, et al. Feasibility of reservoir monitoring using 4D seismic in the Ichthys gas-condensate field, NW Australia: velocity-geology-pressure relationships defined in core, SPWLA-JFES-2017-K, in: Proceedings of the SPWLA 23rd Formation Evaluation Symposium of Japan, Chiba, Japan, October 11–12, 2017.

[40] LADB Staff, Foreign oil companies discover huge natural gas deposits in colombia, University of New Mexico, Albuquerque, 1994.

[41] N.G. De'Ath, Discovery history of the giant Cusiana and Cupiagua oil fields, Columbia, Bulletin of the Geological Society of Malaysis 41 (1997) 31–39.

[42] S.-T. Lee, M. Chaverra, Modelling and interpretation of condensate banking for the near critical Cupiagua field, SPE-49265-MS, in: Proceedings of the SPE Annual Technical Conference and Exhibition, New Orleans, Louisiana, September 27–30, 1998.

[43] P.R. Ballin, P.J. Clifford, M.A. Christie, Cupiagua: modeling of a complex fractured reservoir using compositional upscaling, SPE Reservoir Evaluation & Engineering 5 (2002) 488–498.

[44] D.A. Castaño, N. Ruiz, J. Muñoz, et al., A new approach for CT interventions in high-permeability-contrast wells of Cupiagua field, Colombia, SPE-163919-MS, in: Proceedings of the SPE/ICoTA Coiled Tubing & Well Intervention Conference & Exhibition, The Woodlands, Texas, March 26–27, 2013.

[45] D.E. Tipping, N.F. Saavedra, Cupiagua uncertainties offer insight into piedemonte exploration prospects in Colombia, Journal of Petroleum Technology 57 (2005) 36–42.

[46] L.A. Giraldo, H.-Y. Chen, L.W. Teufel, Field case study of geomechanical impact of pressure depletion in the low-permeability Cupiagua gas-condensate reservoir, SPE-60297-MS, in: Proceedings of the SPE Annual Technical Conference and Exhibition, Denver, Colorado, March 12–15, 2000.

[47] C.A. Franco, A. Restrepo, L.G. Acosta, et al., SDLA: fighting skin damage in Colombian fields - A war story, SPE-98136-MS, in: Proceedings of the SPE International Symposium and Exhibition on Formation Damage Control, Lafayette, Louisiana, February 15–17, 2006.

[48] A. Restrepo, J. Duarte, B. Exploration, et al., A multi-parameter methodology for skin factor characterization: applying basic statistics to formation damage theory, SPE-107674-MS, in: Proceedings of the European Formation Damage Conference, Scheveningen, The Netherlands, May 30–June 1, 2007.

[49] H.A. Al-Anazi, G.A. Pope, M.M. Sharma, A Successful methanol treatment in a gas/condensate reservoir: field application, SPE Production & Operations 20 (2005) 60–69.

[50] D.H. Pham, S.K. Hoang, V.X. Trinh, et al., Condensate Banking characterization and quantification of improvement from different mitigations using pressure transient analysis: a case study in Hai Thach field offshore Vietnam, OTC-30142-MS, in: Proceedings of the Offshore Technology Conference Asia, Kuala Lumpur, Malaysia, November 2–August 19, 2020.

[51] F.O. Garzon, H. Anazi, J.A. Leal, et al., Laboratory and field trial results of condensate banking removal in retrograde gas reservoirs: case history, Journal of Petroleum Technology 56 (2007) 72–75.
[52] O. Burachok, Enhanced gas and condensate recovery: review of published pilot and commercial projects, Nafta-Gaz 77 (2021) 20–25.
[53] S.C. Atchley, D.M. Cleveland, L.W. West, A case for renewed development of a mature gas field: the Devonian Swan Hills Formation at Kaybob South field, Alberta, Canada, Bulletin of Canadian Petroleum Geology 56 (2008) 165–190.
[54] J. Flores, Compositional simulation to develop an optimum gas cycling scheme Kaybob Beaverhill Lake "C" Pool, Alberta, Journal of Canadian Petroleum Technology 18 (2013) 96–103.
[55] M.B. Field, I.M. Wytrychowski, J.K. Patterson, A numerical simulation of Kaybob South gas cycling projects, Journal of Petroleum Technology 23 (1971) 1253–1262.
[56] M.B. Field, J.W. Givens, D.S. Paxman, Kaybob South - Reservoir simulation of a gas cycling project with bottom water drive, Journal of Petroleum Technology 22 (1970) 481–492.
[57] C.A. Kossack, S.T. Opdal, Recovery of condensate from a heterogeneous reservoir by the injection of a slug of methane followed by nitrogen, SPE-18265-MS, in: Proceedings of the SPE Annual Technical Conference and Exhibition, Houston, Texas October 2–5, 1988.
[58] M. Sadooni, A. Zonnouri, The effect of nitrogen injection on production improvement in an Iranian rich gas condensate reservoir, Petroleum Science and Technology 33 (2015) 422–429.
[59] H.M. Hamza, M. Al Suwaidi, O. Al Jeelani, et al., A case history of nitrogen injection monitoring in rich recycled gas condensate reservoir, onshore gas field, SPE-177790-MS, in: Proceedings of the Abu Dhabi International Petroleum Exhibition and Conference, Abu Dhabi, UAE, November 9–12, 2015.
[60] S.M. Benson, D.R. Cole, CO_2 sequestration in deep sedimentary formations, Elements 4 (2008) 325–331.
[61] P. Folger, Funding for carbon capture and sequestration (CSS) at DOE: In brief, UNT Digital Library, University of North Texas, Denton, 2016.
[62] H. Gachuz-Muro, B. Gonzalez-Valtierra, E. Luna-Rojero, et al., Laboratory tests with CO2, N2 and lean natural gas in a naturally fractured gas-condensate reservoir under HP/HT conditions, Proceedings of the SPE-142855-MS, in: SPE Enhanced Oil Recovery Conference, Kuala Lumpur, Malaysia, July 19–20, 2011.
[63] E. Shtepani, F.B. Thomas, D.B. Bennion, New approach in gas injection miscible processes modelling in compositional simulation, Journal of Canadian Petroleum Technology 45 (2006) 35–40.
[64] J. Narinesingh, D. Alexander, Injection well placement analysis for optimizing CO2 enhanced gas recovery coupled with sequestration in condensate reservoirs, SPE-180866-MS, in: Proceedings of the SPE Trinidad and Tobago Section Energy Resources Conference, Port of Spain, Trinidad and Tobago, June 13–15, 2016.
[65] A. Kerunwa, P.O. Ariche, N.C. Izuwa, Economic evaluation of hydraulic fracturing in a gas condensate reservoir operating below dewpoint, Open Journal of Yangtze Gas and Oil 5 (2020) 73–86.
[66] A.E. Ignatyev, I.R. Mukminov, E.A. Vikulova, et al., Multistage hydraulic fracturing in horizontal wells as a method for the effective development of gas-condensate fields in the Arctic region, SPE-149925-MS, in: Proceedings of the SPE Arctic and

Extreme Environments Conference and Exhibition, Moscow, Russia, October 18–20, 2011.

[67] R.N. Moghaddam, M. Van Doorn, A. Dos Santos, An efficient treatment technique for remediation of phase trapping damage in tight carbonate gas reservoirs, SPE-202996-MS, in: Proceedings of the Abu Dhabi International Petroleum Exhibition & Conference, Abu Dhabi, UAE, November 9–12, 2020.

[68] G.R. Jerauld, S.J. Salter, The effect of pore-structure on hysteresis in relative permeability and capillary pressure: pore-level modeling, Transport in Porous Media 5 (1990) 103–151.

[69] L. Kewen, F. Abbas, Experimental study of wettability alteration to preferential gas-wetting in porous media and its effects, SPE Reservoir Evaluation & Engineering 3 (2000) 139–149.

[70] G.-Q. Tang, A. Firoozabadi, Relative permeability modification in gas/liquid systems through wettability alteration to intermediate gas wetting, SPE Reservoir Evaluation & Engineering 5 (2002) 427–436.

[71] G.-Q. Tang, A. Firoozabadi, Wettability Alteration to intermediate gas-wetting in porous media at elevated temperatures, Transport in Porous Media 52 (2003) 185–211.

[72] J. Fahimpour, M. Jamiolahmady, M. Sohrabi, A Combined experimental and theoretical investigation on application of wettability modifiers in gas-condensate reservoirs, SPE-159807-MS, in: Proceedings of the SPE Annual Technical Conference and Exhibition, San Antonio, Texas, October 8–10, 2012.

[73] Y. Zhong, H. Zhang, E. Kuru, et al., Mechanisms of how surfactants mitigate formation damage due to aqueous phase trapping in tight gas sandstone formations, Colloids and Surfaces A: Physicochemical and Engineering Aspects 573 (2019) 179–187.

[74] M. Sayed, G. Al-Muntasheri, A safer generation of wettability alteration chemical treatments, SPE-184566-MS, in: Proceedings of the SPE International Conference on Oilfield Chemistry, Montgomery, Texas, April 3–5, 2017.

[75] S.-A. Hoseinpour, M. Madhi, H. Norouzi, et al., Condensate blockage alleviation around gas-condensate producing wells using wettability alteration, Journal of Natural Gas Science and Engineering 62 (2019) 214–223.

[76] A. Hassan, M. Mahmoud, M.S. Kamal, et al., Novel treatment for mitigating condensate bank using a newly synthesized Gemini surfactant, Molecules 25 (2020) 3030.

[77] A. Hassan, M. Mahmoud, A. Al-Majed, et al., Gas condensate treatment: a critical review of materials, methods, field applications, and new solutions, Journal of Petroleum Science and Engineering 177 (2019) 602–613.

[78] N. El Cheikh Ali, B. Zoghbi, M. Fahes, et al., The impact of near-wellbore wettability on the production of gas and condensate: insights from experiments and simulations, Journal of Petroleum Science and Engineering 175 (2019) 215–223.

[79] S. Sharifzadeh, S. Hassanajili, M.R. Rahimpour, Wettability alteration of gas condensate reservoir rocks to gas wetness by sol-gel process using fluoroalkylsilane, Journal of Applied Polymer Science 128 (2013) 4077–4085.

[80] M.A. Mousavi, S. Hassanajili, M.R. Rahimpour, Synthesis of fluorinated nano-silica and its application in wettability alteration near-wellbore region in gas condensate reservoirs, Applied Surface Science 273 (2013) 205–214.

[81] J.P. Villegas, I. Moncayo-Riascos, D. Galeano-Caro, et al., Functionalization of γ-alumina and magnesia nanoparticles with a fluorocarbon surfactant to promote

ultra-gas-wet surfaces: experimental and theoretical approach, ACS Applied Materials & Interfaces 12 (2020) 13510−13520.

[82] A. Safaei, F. Esmaeilzadeh, A. Sardarian, et al., Experimental investigation of wettability alteration of carbonate gas-condensate reservoirs from oil-wetting to gas-wetting using Fe_3O_4 nanoparticles coated with Poly (vinyl alcohol), (PVA) or Hydroxyapatite (HAp), Journal of Petroleum Science and Engineering 184 (2020) 106530.

[83] R. Ahmadi, Z. Farmani, S. Osfouri, et al., Condensate blockage remediation in a gas reservoir through wettability alteration using natural $CaCO_3$ nanoparticles, Colloids and Surfaces A: Physicochemical and Engineering Aspects 579 (2019) 123702.

[84] J. Jin, Y. Wang, K. Wang, et al., The effect of fluorosurfactant-modified nano-silica on the gas-wetting alteration of sandstone in a CH4-liquid-core system, Fuel 178 (2016) 163−171.

[85] M. Aminnaji, H. Fazeli, A. Bahramian, et al., Wettability alteration of reservoir rocks from liquid wetting to gas wetting using nanofluid, Transport in Porous Media 109 (2015) 201−216.

[86] P. Esmaeilzadeh, M.T. Sadeghi, A. Bahramian, et al., Superamphiphobic surfaces prepared by coating multifunctional nanofluids, ACS Applied Materials & Interfaces 8 (2016) 32011−32020.

[87] M. Franco-Aguirre, R.D. Zabala, S.H. Lopera, et al., Interaction of anionic surfactant-nanoparticles for gas - Wettability alteration of sandstone in tight gas-condensate reservoirs, Journal of Natural Gas Science and Engineering 51 (2018) 53−64.

[88] M. Sayed, F. Liang, H. Ow, Novel surface modified nanoparticles for long-lasting mitigation of water and condensate blockage in gas reservoirs, OTC-28712-MS, in: Proceedings of the Offshore Technology Conference, Houston, Texas, April 30−May 3, 2018.

[89] A. Naghizadeh, R. Azin, S. Osfouri, et al., Wettability alteration of calcite and dolomite carbonates using silica nanoparticles coated with fluorine groups, Journal of Petroleum Science and Engineering 188 (2020) 106915.

[90] M. Sayed, H. Ow, R. Raini, et al., Wettability alteration using functionalized nanoparticles with tailored adhesion to the rock surface for condensate banking mitigation, Canadian Journal of Chemical Engineering (Special Issue)(2021).

[91] Y. Li, Y. Wang, Q. Wang, et al., Achieving the super gas-wetting alteration by functionalized nano-silica for improving fluid flowing capacity in gas condensate reservoirs, ACS Applied Materials & Interfaces 13 (2021) 10996−11006.

[92] B. Zoghbi, M.M. Fahes, H. Nasrabadi, Identifying the optimum wettability conditions for the near-wellbore region in gas-condensate reservoirs, SPE-134966-MS, in: Proceedings of the PE-Tight Gas Completions Conference, San Antonio, Texas, November 2−3, 2010.

[93] O. Ajagbe, M. Fahes, Establishing screening criteria for field application of wettability alteration in gas-condensate reservoirs, Journal of Petroleum Science and Engineering 193 (2020) 107342.

[94] I. Nowrouzi, A.K. Manshad, A.H. Mohammadi, Wettability alteration and enhanced gas condensate recovery by treatment of carbonate reservoir rock using supercritical R134A and R404A gases, Journal of Petroleum Exploration and Production Technology 10 (2020) 3751−3766.

[95] S.-A. Hoseinpour, M. Madhi, H. Norouzi, et al., Fluoro-chemical foam injection for wettability alteration of sandstone and carbonate reservoirs rocks towards gas-wet state, Journal of Natural Gas Science and Engineering 91 (2021) 103943.

[96] Y. Wang, Y. Li, Q. Wang, et al., The influence of fluorochemical-modified graphene oxide on the gas-wetting alteration of reservoir cores, Colloids and Surfaces 620 (2021) 126565.
[97] G. Lopez, M.B. Myers, Q. Xie, et al., Wettability alteration using benzoxazine resin: a remedy for water blockage in sandstone gas reservoirs, Fuel 291 (2021) 120189.
[98] E. Arjomand, Implications of wettability alteration of sandstone rocks using non-fluorinated $scCO_2$-based silylation, Ph.D. thesis, Curtin University, 2020.
[99] Z. Liu, Y. Sun, W. Guo, et al., Effects of gas injection combined with wettability alteration on boundary water inrush, Journal of Natural Gas Science and Engineering 90 (2021) 103920.
[100] V. Kumar, V.S.S. Bang, G.A. Pope, et al., Chemical stimulation of gas/condensate reservoirs, SPE-102669-MS, in: Proceedings of the SPE Annual Technical Conference and Exhibition, San Antonio, Texas, September 24–27, 2006.
[101] A. Delavarmoghaddam, S.A. Mirhaj, P.L.J. Zitha, Gas condensate productivity improvement by chemical wettability alteration, SPE-122225-MS, in: Proceedings of the Eighth European Formation Damage Conference, Scheveningen, The Netherlands, May 27–29, 2009.
[102] M. Sheydaeemehr, B. Sedaeesola, A. Vatani, Gas-condensate production improvement using wettability alteration: a giant gas condensate field case study, Journal of Natural Gas Science and Engineering 21 (2014) 201–208.
[103] Z. Sakhaei, M. Mohamadi-Baghmolaei, R. Azin, et al., Study of production enhancement through wettability alteration in a super-giant gas-condensate reservoir, Journal of Molecular Liquids 233 (2017) 64–74.
[104] M. Mohamadi-Baghmolaei, Z. Sakhaei, R. Azin, et al., Modeling of well productivity enhancement in a gas-condensate reservoir through wettability alteration: a comparison between smart optimization strategies, Journal of Natural Gas Science and Engineering 94 (2021) 104059.
[105] O. Ajagbe, M. Weiss, M. Fahes, Economic evaluation of wettability alteration towards the alleviation of condensate blocking, SPE-189529-MS, in: Proceedings of the SPE International Conference and Exhibition on Formation Damage Control, Lafayette, Louisiana, February 7–9, 2018.
[106] M. Weiss, Optimizing wettability to maximize production from retrograde condensate reservoirs - A simulation approach, M.S thesis, University of Oklahoma, 2017.
[107] O. Ajagbe, Evaluating treatment parameters toward optimizing production from gas reservoirs using wettability alteration, M.S thesis, University of Oklahoma, 2019.
[108] M. Butler, J.B. Trueblood, G.A. Pope, et al., A Field demonstration of a new chemical stimulation treatment for fluid-blocked gas wells, SPE-125077-MS, in: Proceedings of the SPE Annual Technical Conference and Exhibition, New Orleans, Louisiana, October 4–7, 2009.
[109] B.N. Al Ghamdi, B.H. Al-Malki, A. Al-Kanaan, et al., Field implementation of condensate bank removal using chemical treatment, IPTC-16545-MS, in: Proceedings of the International Petroleum Technology Conference, Beijing, China, March 26–28, 2013.
[110] A.M. Al-Yami, F.A. Gomez, K.I. AlHamed, et al., A successful field application of a new chemical treatment in a fluid blocked well in Saudi Arabia, SPE-168086-MS, in: Proceedings of the SPE Saudi Arabia Section Technical Symposium and Exhibition, Al-Khobar, Saudi Arabia, May 19–22, 2013.

[111] A. Restrepo, A. Ocampo, M. Arenas, et al., Wettability modifier field trial in a sandstone condensate system: facts and gaps of a promising technology, SPE-152310-MS, in: Proceedings of the SPE Latin America and Caribbean Petroleum Engineering Conference, SPE, Mexico City, Mexico, April 16–18, 2012.

[112] D. López, R.D. Zabala, C. Matute, et al., Well injectivity loss during chemical gas stimulation process in gas-condensate tight reservoirs, Fuel 283 (2021) 118931.

[113] O. Glaso, Generalized minimum miscibility pressure correlation, SPE Journal 25 (1985) 927–934.

[114] R.B. Alston, G.P. Kokolis, C.F. James, CO_2 minimum miscibility pressure: a correlation for impure co2 streams and live oil systems, SPE Journal 25 (1985) 268–274.

[115] T. Ahmed, H. Nasrabadi, A. Firoozabadi, Complex flow and composition path in CO_2 injection schemes from density effects, Energy & Fuels 26 (2012) 4590–4598.

[116] M. Mehana, M. Fahes, L. Huang, The density of oil/gas mixtures: insights from molecular simulations, SPE Journal 23 (2018) 1798–1808.

[117] M. Ayub, M. Ramadan, Mitigation of near wellbore gas-condensate by CO_2 huff-n-puff injection: a simulation study, Journal of Petroleum Science and Engineering 175 (2019) 998–1027.

[118] M. Al-Wadhahi, F.H. Boukadi, A. Al-Bemani, et al., Huff'n puff to revaporize liquid dropout in an Omani gas field, Journal of Petroleum Science and Engineering 55 (2007) 67–73.

[119] O.S. Alade, M. Mahmoud, A. Hassan, et al., Evaluation of kinetics and energetics of thermochemical fluids for enhanced recovery of heavy oil and liquid condensate, Energy & Fuels 33 (2019) 5538–5543.

[120] A. Al-Nakhli, A. Hassan, M. Mahmoud, et al., Removal of condensate banking from different formations using thermochemical treatment, SPE-197847-MS, in: Proceedings of the Abu Dhabi International Petroleum Exhibition and Conference, Abu Dhabi, UAE, November 11–14, 2019.

[121] A. Hassan, M. Abdalla, M. Mahmoud, et al., Condensate-banking removal and gas-production enhancement using thermochemical injection: a field-scale simulation, Processes 8 (2020) 727.

[122] A. Hassan, M. Mahmoud, A. Al-Majed, et al., Performance analysis of thermochemical fluids in removing the gas condensate from different gas formations, Journal of Natural Gas Science and Engineering 78 (2020) 103333.

[123] H.V. Fairbanks, W.I. Chen, Ultrasonic acceleration of liquid flow through porous media, CONF-701103, in: Proceedings of the Sixty-Third National Meeting of The American Institute of Chemical Engineers, Chicago, IL, November 29, 1970.

[124] A.C.T. Aarts, G. Ooms, Net flow of compressible viscous liquids induced by travelling waves in porous media, Journal of Engineering Mathematics 34 (1998) 435–450.

[125] S. Karami, S. Dehaghani, A. Hossein, et al., Condensate blockage removal using microwave and ultrasonic waves: discussion on rock mechanical and electrical properties, Journal of Petroleum Science and Engineering 193 (2020) 107309.

CHAPTER

Foam-assisted liquid lift

8

Fenfen Huang[1] and Kees C.A.M. Veeken[2]

[1]CandV Consulting and Services, Sugar Land, TX, United States

[2]Mature Gas Well Consultancy, Gieten, Drenthe, The Netherlands

Chapter outline

Flow Assurance. DOI: https://doi.org/10.1016/B978-0-12-822010-8.00001-5

8.1 Introduction

Liquid accumulation is a widely encountered challenge in natural gas production and transportation, causing significant back pressure and reduction of capacity and reserves. This is most notable in depleting gas wells where the gradual reduction of reservoir pressure (P_{res}) causes a decline of the gas production rate (Q_{gas}). As soon as the gas rate declines below the minimum stable gas rate (Q_{min}), the gas flowing to surface is no longer able to efficiently lift associated water and/or condensate. As a result, liquid will accumulate in the wellbore increasing the hydrostatic pressure and hence the bottomhole pressure (BHP). This is referred to as liquid loading, which triggers intermittent production, defers production, and eventually, causes production to stop. Note that Q_{min} is also referred to as the critical gas rate or the liquid-loading gas rate.

A similar increase of back pressure due to liquid accumulation, causing production deferment, can happen in gas flowlines and pipelines in the areas across terrain undulations and risers. When a gas production system experiences such liquid accumulation, this may also exacerbate corrosion risks or cause operational issues related to excessive liquid slugging.

Foam-assisted lift (FAL) is one of the deliquification techniques that are routinely applied to mitigate liquid loading in gas wells, in order to sustain gas well capacity and maximize gas well reserves. Other deliquification techniques are surface compression (SC), velocity string (VS), plunger lift (PL), gas lift (GL), and downhole pumping (DP) (rod, hydraulic jet, electric submersible, etc.). When applied in oil wells, these same techniques are labeled as artificial lift instead of deliquification. The difference is that, in gas wells, the purpose is to prevent excessive liquid holdup in the gas stream, while in oil wells the purpose is to reduce the BHP to boost oil production. Note that liquid loading occurs if Q_{gas} drops below Q_{min} regardless of the liquid production rate: it happens even in gas wells producing very little liquid.

Since the 1960s, surfactants have been used to create foam in various oil and gas field applications. For over 50 years, foam was used to assist solids circulation and reduce fluid loss during drilling and/or wellbore cleanout, to reduce fluid efficiency and accelerate cleanup in hydraulic fracturing, and to boost secondary oil recovery in water injection schemes [1]. In the early 2000s, foam became very popular for deliquification of gas wells in the United States, facilitated by high gas prices and an abundant choice of commercial surfactants to foam well fluids covering a wide range of compositions [2].

During FAL, the surfactant is introduced downhole to create foam to prevent liquid loading: the minimum stable gas rate (Q_{min}) or gas velocity that is required to carry light foam to surface is much less compared to heavier water or condensate. Successful FAL requires that the surfactant reaches a suitable downhole location where agitation with gas and liquid results in a stable enough foam, in which the bubble film contains the liquids. The surfactant or foaming agent is deployed either as a solid soap stick or as a liquid "foamer."

The solid surfactant can only be applied intermittent, which means batch treatment, while the liquid surfactant can be applied either intermittent (batch) or continuous. Intermittent application is especially suited for kicking off wells that do not require FAL after they have been kicked off but also applies when liquid loading is "slow" and the production deferment in between periodic batch treatments can be kept small. Periodic application of solid foam sticks benefits from automatic launching devices that enable dosing at the desired frequency. In all other cases, continuous injection of liquid foamer is preferred. In continuous liquid foamer applications, the required dosage of foamer is pumped via the annulus of packer-less wells or packer wells with injection valves, or via a capillary string inside the completion.

FAL can be a cost-effective method to sustain production from low rate gas wells as it involves modest capital investment (CAPEX) and operational expenditure (OPEX). Also, the well response to foamer treatment is easy to monitor and assess within a matter of hours to days. Besides stand-alone, FAL can be applied to improve the performance of other deliquification methods such as GL. Also, multifunctional foamer products are available to resolve other issues besides liquid loading such as corrosion and the deposition of scale or paraffin.

The success of FAL is governed by the efficacy of surfactant injection for removing liquid from the well and reducing the minimum stable gas rate, and the economics of FAL compared to other deliquification techniques. The efficacy of FAL is characterized by the reduction of the minimum stable gas rate (Q_{min}), which depends on the surfactant type and concentration, the bottomhole temperature (BHT) and pressure, and the water and condensate production and composition.

Foam performance typically decreases with increasing temperature and therefore the chemical stability and efficacy of surfactants should be confirmed in the laboratory at representative temperature conditions with representative water and condensate. Water and condensate react very differently with the surfactant: water is usually easy to foam while condensate will generally act as a natural defoamer.

For a foamer to be effective, it needs to create stable foam whose liquid film holds both the water and the condensate. Therefore, applying FAL in wells producing a large fraction of condensate compared to water [wells with low basic sediment and water (BS&W)] can be problematic, requiring the selection of surfactants that tolerate condensate. Also, irrespective of condensate, not all surfactants are tolerant to high water salinity.

The economics of FAL is governed by the installation CAPEX and operating OPEX, where the OPEX is dictated by the surfactant injection rate and price. Conventional foamers for treating wells with high BS&W cost significantly less than unconventional foamers for treating wells with low BS&W. Also, the required surfactant concentration tends to increase as the BS&W decreases. OPEX includes the cost of chemical inventory management and performance monitoring and optimization. The CAPEX consists of the cost of the chemical injection pump (or launcher for solid soap sticks) and storage tank, and, if required, the capillary string installation.

8.2 Liquid loading and deliquification

8.2.1 Liquid loading

In depleting gas reservoirs, the capacity of producing gas wells (Q_{gas}) decreases as the reservoir pressure (P_{res}) decreases. Above the so-called minimum stable gas rate or liquid loading rate (Q_{min}), the gas velocity is sufficient to transport associated free liquid (water and condensate) up to the surface by means of annular flow where most of the liquid moves up the tubing as a liquid film along the tubing wall and the remainder as droplets entrained in the gas stream. This annular multiphase flow regime causes minimum liquid holdup and therefore creates minimum hydrostatic and friction pressure losses across the wellbore between bottomhole and surface.

When Q_{gas} reduces below Q_{min}, the flow regime changes to churn flow and slug flow, which are characterized by a much higher liquid holdup causing much more pressure loss across the wellbore. This is illustrated by the relation between the BHP and the gas rate (Q_{gas}) at a certain wellhead or tubing head pressure (THP) in Fig. 8.1. BHP gradually increases with increasing Q_{gas} above Q_{min} due to friction in annular flow but also increases with decreasing Q_{gas} below Q_{min} due to increasing liquid holdup in churn, slug, and eventually bubble flow: Q_{min} coincides with the gas rate, at which BHP achieves its minimum value. The relation between BHP and Q_{gas} at a constant THP is referred to as a vertical lift performance (VLP) curve.

The well capacity declines significantly below Q_{min} due to the increasing BHP, which is referred to as liquid loading. Fig. 8.2 shows two examples of such liquid loading behavior, which tends to be intermittent in character and consists of repeating cycles of "good" and "bad" production. Liquid loading is easy to

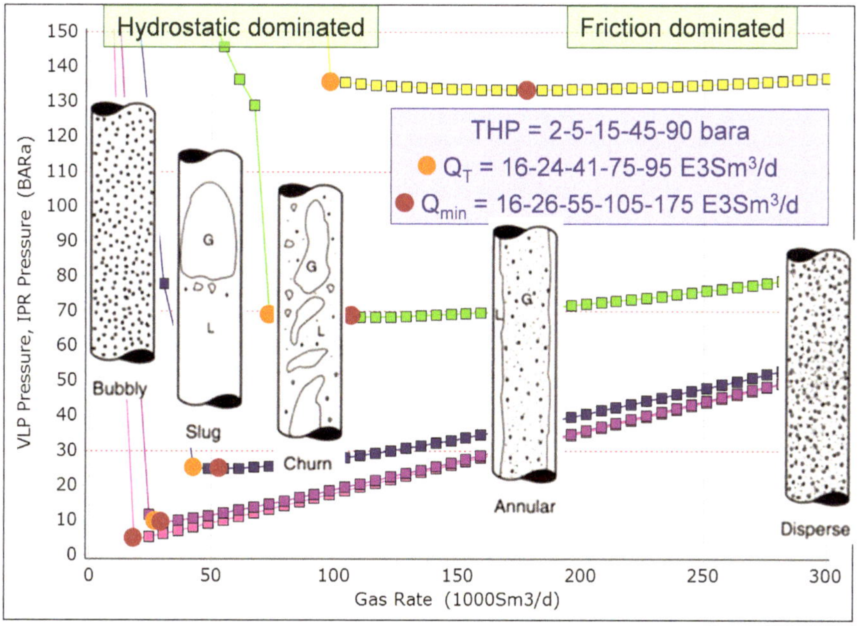

FIGURE 8.1

Vertical lift performance curves, showing relation between bottomhole pressure (BHP) and gas rate Q_{gas} at varying wellhead pressure (THP), illustrated by respective multiphase flow regimes.

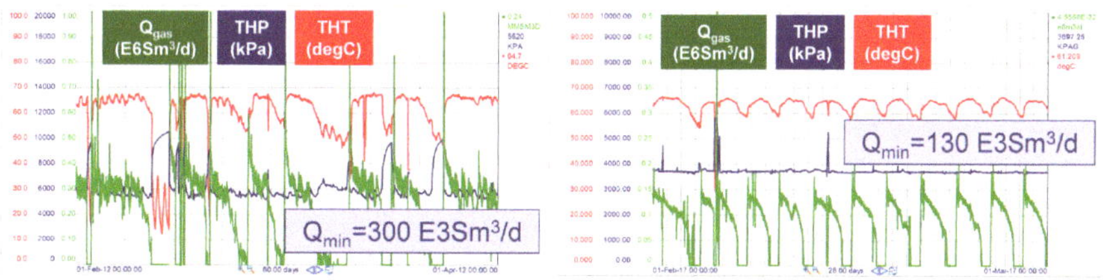

FIGURE 8.2

Field examples of intermittent production due to liquid loading.

detect as a rapid reduction of the real-time wet gas rate or wellhead temperature, or as a rapid increase of the annulus pressure in case of a packer-less completion.

The intermittent production cycle is driven by a recharge of the reservoir pressure during the "bad" production period. Following sufficient pressure recharge, the well is again able to sustain a "good" production period above or at Q_{min}. Long-term, this recharging diminishes and the uptime of such intermittent wells decreases; eventually "good" production ceases to exist entirely.

8.2.1.1 Turner criterion

Fig. 8.3 illustrates how in the most efficient annular flow regime the liquid in the wellbore is transported upward as liquid droplets carried by the gas stream and as a liquid film on the tubing wall dragged upwards by the same stream of gas. Droplet or film reversal occurs when the gravity force on the liquid droplet or film exceeds the drag force exerted on the droplet or film by the flow of gas.

Turner et al. [4] asserted that droplet reversal best matched their field data and derived expressions for the critical gas velocity (V_c) and the corresponding gas rate (Q_T) that describe the balance between gravity and drag forces for water droplets at wellhead conditions. They introduced the following expression for the critical gas velocity:

$$V_c = 1.593\left(\frac{\rho_L - \rho_G}{\rho_G^2}\sigma\right)^{1/4} \tag{8.1}$$

where V_c is the critical gas velocity (ft./s), σ is the surface tension between the liquid and gas (dynes/cm), ρ_L is the liquid density (lbm/ft.^{3}), and ρ_G is the gas density (lbm/ft.^{3}). For a well producing both water and condensate, the critical

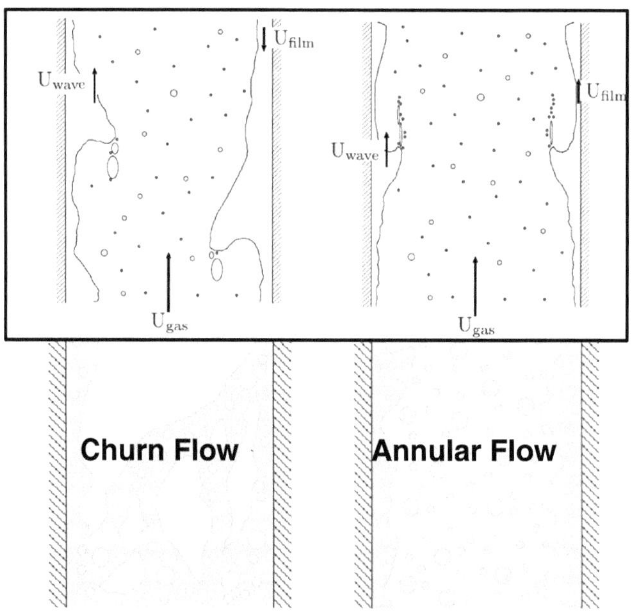

FIGURE 8.3

Schematic illustration of liquid transport in a vertical gas well by droplet and film flow.

Reproduced with permission from J.M.C. Van 't Westende, Droplets in annular-dispersed gas-liquid pipe-flows (Ph.D. Dissertation), Delft University of Technology, 2008 [3].

gas velocity based on water is significantly higher than for condensate, due to the higher density (ρ_L) and surface tension (σ) of water compared to condensate. Hence, the following expressions all assume that liquid loading is initiated and dictated by water.

After inserting default values for σ, ρ_L, and ρ_G, and a 20% upward adjustment to better match field data, V_c and Q_T were written as:

$$V_c = 5.321\frac{(67-0.0031\text{THP})^{1/4}}{(0.0031\text{THP})^{1/2}} \tag{8.2}$$

$$Q_T = \frac{0.0890\ \text{THP}\cdot\text{ID}^2}{(\text{THT}+460)Z}\frac{(67-0.0031\ \text{THP})^{1/4}}{(0.0031\ \text{THP})^{1/2}} \tag{8.3}$$

where Q_T is the critical gas rate (MMscf/d), THP is the wellhead pressure (psia), THT is the wellhead temperature (°F), ID is the tubing diameter (in.), and Z is the gas deviation factor (−).

Later work, amongst others by Van 't Westende [3] and Van Nimwegen et al. [5], revealed that liquid loading is actually caused by film reversal. The process is illustrated in Fig. 8.3. As the gas rate diminishes, the liquid film thickens and starts moving down rather than up; this causes the flow regime to change from annular to churn and slug flow, and results in liquid accumulating in the wellbore. This transition between flow regimes results in the VLP's of Fig. 8.1. Irrespective of the multiphase flow fundamentals, Eq. (8.3) remains a valid engineering expression to describe the onset of liquid loading.

Veeken and Belfroid [6] demonstrated in the laboratory that by making the tubing wall hydrophobic instead of standard hydrophilic, they could reduce the liquid-loading gas rate by about 50%. This further demonstrates that liquid loading is governed by film reversal: the hydrophobic coating prevents forming a liquid film and results in droplet transport only, which reversal is observed at about half the gas rate of Eq. (8.3).

8.2.1.2 Modified Turner criterion

Rewriting Eq. (8.3) into mainly metric units produces:

$$Q_T = \text{TC}\cdot\text{ID}^2\cdot\text{THP}^{0.5}/[(273+\text{THT})\cdot Z] \tag{8.4}$$

where Q_T is the critical gas rate (10^3Sm^3/d), ID is the completion size (in.), THP is the wellhead pressure (bara), THT is the wellhead temperature (°C), TC is the so-called Turner constant (−) with a default value of 285, and Z is the gas deviation factor with a default value of 1. Veeken et al. [7] found Eq. (8.4) to be optimistic and introduced a modified Turner expression that better matched actual liquid loading data:

$$Q_{min} = \{(1-2.95\text{E}-3\cdot Q_T) - [(1-2.95\text{E}-3\cdot Q_T)^2 + 8.68\text{E}-6\cdot {Q_T}^2)]^{0.5}\}/(-4.34\text{E}-6\cdot Q_T) \tag{8.5a}$$

for $Q_T < 330\ 10^3 Sm^3/d$, and

$$Q_{min} = 2 \cdot Q_T \tag{8.5b}$$

for $Q_T > 330\ 10^3 Sm^3/d$.

Fig. 8.1 illustrates that compared to Q_T, Q_{min} also better describes the minimum of the VLP's generated by commercial wellbore simulators. Fig. 8.4 presents Q_T and Q_{min} as function of THP for different combinations of ID and THT assuming TC of 285 and Z of 1. Note that Q_T and Q_{min} are almost the same below ~30 $10^3 Sm^3/d$ (~1 MMscf/d) and that the value of THT is of secondary importance.

Eq. (8.5a and 8.5b) is based on wellhead conditions and can therefore only approximate the minimum of a VLP that characterizes the pressure drop across the entire wellbore [8]. Therefore, deviations exist between actual VLP's and Eq. (8.5), for example, in the case of tapered tubing completions and horizontal wells, between wells where fresh water condenses near the surface and wells where significant formation water is produced downhole, and between wells producing little or lots of condensate. In all cases, the TC can be used to match Eq. (8.5a and 8.5b) against either model VLP or a field observed Q_{min}; TC tends to range between 220 and 330, straddling the default 285.

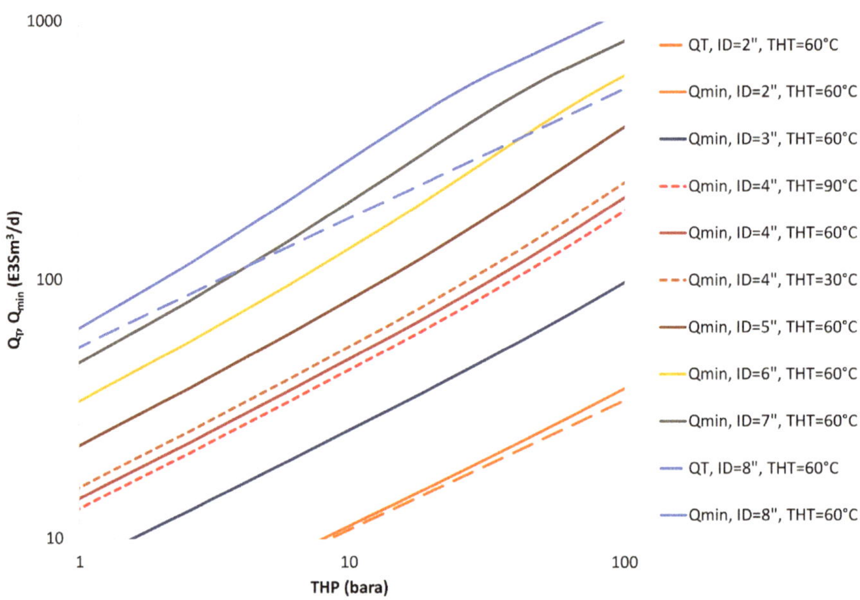

FIGURE 8.4

Liquid-loading gas rate versus tubing head pressure at different tubing diameter and wellhead temperature, according to Eq. (8.5).

In some cases, the gas production rate is insufficient during initial production only, for example, during cleanup of a well after a significant well intervention (stimulation, cleanout, recompletion), or when opening up a well that is located close to a water producing zone and suffers from cross flow during shut-in periods. Liquid loading then ceases as soon as the gas rate recovers above Q_{min} because deliquification is only required during that short initial production period.

8.2.2 Continuous deliquification

The objective of continuous gas well deliquification measures is to permanently reduce the minimum stable gas rate Q_{min} to prevent liquid loading and sustain gas capacity. Referring to Eq. (8.4), Q_{min} can be permanently reduced as follows:

- Reduce THP by SC: Surface gas compressors are generally installed in central gathering locations to serve as many wells or fields as possible, but can also be implemented on individual wells when these are located far from a gathering facility [9].
- Reduce ID by VS: This entails any activity that reduces the tubing ID, either as part of a tubing workover or by installing a smaller insert string within the existing tubing.
- Reduce TC by FAL: Foam is created by downhole injection of a surfactant into a turbulent mixture of gas and water; the foam can be produced to surface at much lower gas velocity than regular water, which translates into a significant reduction of TC. The surfactant is injected continuously as a liquid.

Other means of deliquification described by Lea et al. [2] are:

- Increase Q_{gas} by GL: Rather than reducing Q_{min}, the gas rate from the reservoir is supplemented by injecting dry lift gas from surface so Q_{gas} always exceeds the original Q_{min}. Dry lift gas injection also reduces the average liquid–gas ratio (LGR) and hence reduces the wellbore hydrostatic pressure loss as well; lift gas is generally delivered by an existing gas export compressor.
- Make Q_{min} irrelevant by DP: Rather than relying on gas to lift liquid to surface, gas and liquid are separated downhole and the liquid is lifted to surface by a pump suspended on a separate pump string. The associated dry gas will then flow to the surface without liquid, via the new annulus between the tubing and pump string. The required downhole separation of liquid and gas limits this pump application to Q_{gas} less than $0.3 \cdot Q_{min}$.
- Increase initial Q_{gas} by PL: This is an advanced type of intermittent production that utilizes the transient peak gas rate when opening up the well, to push the slug of liquid that accumulated on top of the check valve and plunger to the surface. The piston-shaped plunger fits loosely inside the production tubing and shuttles up and down between wellhead and bottomhole, which prevents gas slipping past the liquid slug when lifting the

liquid to surface. The high initial gas rate when opening up the well is delivered by decompression of the reservoir gas that gets compressed inside the wellbore and the (hydraulic or natural) fracture as the pressure builds up during the shut-in period. This decompression peak gas rate must be large enough for long enough (10–20 minute) to push the plunger plus liquid slug to the surface, which can be achieved by providing sufficient storage volume and/or high enough buildup pressure. The decompression gas is produced from the reservoir during the shut-in period at a downhole gas rate (far) below Q_{min}. Hence, the average gas rate produced during a PL cycle can be significantly less than Q_{min}.

The ultimate objective of continuous deliquification is to sustain stable gas production until it reaches the minimum feasible economic gas rate (Q_e). The Q_e is governed by the well OPEX and the gas price and will therefore depend on local conditions; practical values range from less than 1 $10^3Sm^3/d$ in a low-cost onshore environment to more than 100 $10^3Sm^3/d$ in a high-cost subsea setting. Q_e can be reached by combinations of the above deliquification measures, where the optimum combination achieves Q_e whilst spending as little CAPEX and OPEX as possible.

Table 8.1 presents an example matrix that outlines deliquification scenarios and their operating envelopes assuming a 4.5″ tubing completion, based on Veeken and Al-Kharusi [10]. Similar scenarios and envelopes apply for other fields but can differ depending on the specifics of reservoir, well, and facilities [11]. Some generic conclusions do apply:

- SC down to the lowest economic THP should be considered, a given SC reduces BHP independent of Q_{min} (need low THP to achieve low BHP irrespective of Q_{min}).
- A VS is an ideal (first) deliquification measure as long as the inflow performance is not too good ($A > 10$ in Table 8.1 where A is the Darcy inflow resistance in $bar^2/10^3Sm^3/d$, see [7] and [10]).
- FAL competes with GL, either as a first measure when inflow is good ($A < 10$) or as a secondary measure after installing a VS when inflow is moderate ($10 < A < 100$).
- In case of poor inflow ($A > 100$), the default solution consists of a VS followed by PL (VS + PL), except when LGR is excessive.
- A DP is applicable in the case of a liquid-rich low-rate well only.

The choice between continuous FAL and GL should be based on a comparison of their technical and commercial merits. GL has the technical benefit that it can achieve any Q_{min} independent of LGR, BS&W, and BHT. The reduction of Q_{min} by FAL can reach up to 50%–80% but can also be compromised by excessive liquid production (high LGR), a large fraction of condensate (low BS&W), or high BHT. FAL has the commercial benefit that the cost is contained while the GL line connecting the well to a potential source of lift gas can be long and cost prohibitive.

Table 8.1 Example operating envelope of continuous deliquification scenarios for 4.5″ tubing completion; cut-off values for A, LGR, BS&W and BHT are notional depending on local conditions.

Measure	A ($bar^2/10^3Sm^3/d$)	LGR ($m^3/E6Sm^3$)	BS&W (%)	BHT (°C)
SC	Any	Any	Any	Any
FAL	<10	<1000	>50	<150
GL	<10	Any	Any	Any
VS	>10	Any	Any	Any
VS + FAL	>10	<1000	>50	<150
VS + GL	>10	Any	Any	Any
VS + PL	>100	<1000	Any	Any
DP	$Q_{gas} < 0.3 \cdot Q_{min}$	>1000	High	<200

In summary, continuous FAL is applicable in case of good-to-moderate inflow performance, especially, if the GL is unattractive due to the long distance between well and source of lift gas and lack of compressor ullage. Continuous FAL is applied either stand-alone or in combination with a VS (VS + FAL).

8.2.3 Intermittent deliquification

Some gas wells only need lift assist during kick-off to initiate or restore stable production. The cost of such intermittent deliquification could be much lower than of continuous deliquification. Intermittent deliquification primarily applies to cleanup after stimulation, workover, or long shut-in periods where large volumes of (usually water-based) liquid need to be back-produced to increase gas production to its full potential.

The same intermittent strategy can be applied if the production decline during liquid loading is slow enough. In that case, one can continue production while the well is slowly liquid loading until the liquid holdup in the wellbore becomes excessive. Then, deliquification is applied to lift the slug of liquid and restore full capacity. In practice, the liquid loading production decline period ranges between 1 and 100 hours, while a liquid slug can be lifted within several hours. Hence, intermittent deliquification is feasible if the production decline is slow enough. Such periodic deliquification causes deferment but may be preferred due to the associated cost saving compared to continuous deliquification.

Intermittent deliquification options are as follows:

- SC can be applied intermittent to lift liquid slugs from liquid loading wells in case the available compressor capacity is insufficient to lift all relevant wells continuously at the same time. Shields et al. [12] describe this procedure for an offshore gas field where multiple wells benefit in turn from a (too) small booster compressor.

- FAL can be applied intermittent both to assist kick-off and to periodically lift liquid slugs. Intermittent FAL consists of the batch application of either solid or liquid surfactant. A solid foam stick can be dropped down the wellbore while the well is still producing. When pumping a batch of liquid surfactant down the wellbore, the well needs to be shut in to allow the liquid to reach the bottom, unless a VS is present and the surfactant can be pumped down the backside.
- Coiled tubing nitrogen lift constitutes GL that is applied for short periods to kick-off wells. The lifting period could be as short as 1 hour, long enough to evacuate a slug of liquid from the wellbore, up to several days or weeks, long enough to clean up the liquid in the near-wellbore area to restore near-wellbore gas saturation and productivity. Such a coiled tubing-based nitrogen lifting operation is too costly to apply on a regular basis to periodically lift liquid slugs. When a VS is present, the nitrogen can also be pumped down the backside, eliminating the cost of coiled tubing.
- A hydraulic jet pump deployed on coiled tubing presents a DP solution applied for short periods to kick-off wells, from which a large volume of liquid needs to be lifted to initiate and/or restore gas production. Such a coiled tubing-based jet pumping operation is too costly to apply on regular basis to periodically lift liquid slugs.
- PL is essentially an intermittent deliquification technique designed to periodically lift liquid slugs. It is listed and has been described as continuous deliquification because of its high intermitting frequency up to once per hour. Also, PL is not suitable for kick-off operations as the size of the liquid slug and the gas rate will be unknown and variable, and therefore, the PL cycle cannot be adequately controlled.

In summary, intermittent FAL can be used both (1) to kick-off wells that suffer from liquid loading after stimulation, workover, or a long shut-in period as long as they still produce sufficient gas to create downhole foam and (2) to periodically lift liquid slugs from liquid loading wells provided that the required frequency is manageable, for example, less than once per week. The operating envelope of intermittent FAL is less restrictive than of continuous FAL as sketched in Table 8.1. The intermittent batch treatments are low cost and flexible. Limitations do apply but can be managed by trial-and-error rather than by specifying their limitations up front. Therefore, intermittent FAL can be tried and tested at higher LGR, lower BS&W, and higher BHT than established and customary for continuous FAL. However, it is still advised to carry out laboratory testing as described in Section 8.6.

8.3 Foam-assisted lift

8.3.1 Foam-assisted lift performance

By capturing free water and condensate within a lattice of foam bubbles, these liquids are easier to transport up to the surface, which translates into a reduction of Q_{min}.

Hence, FAL can restore and/or sustain stable gas production while reservoir pressure depletion continues, thus increasing the ultimate recovery of reservoir gas and associated condensate. A larger reduction of Q_{min} results in a larger volume of incremental produced gas and hence increases the potential benefit of FAL. Therefore, the efficacy of FAL is best described by the potential reduction of Q_{min}.

In principle, any reduction of Q_{min} could be economic as long as the associated additional reserves are more than sufficient to cover the cost. However, when considering long-term continuous deliquification, FAL is compared with other deliquification techniques and the reduction of Q_{min} needs to be at least a factor 2 to be competitive. Successful field applications of FAL indicate a maximum reduction of Q_{min} between 3 and 5. Unfortunately, no methods yet exist to determine the maximum possible reduction of Q_{min} up front.

The standard laboratory tests presented in Section 8.6 are useful for screening and ranking foaming surfactants but the results do not reveal the actual field efficacy. The gas velocity in these small-scale laboratory tests is $10x-100x$ lower than in real gas wells. Hence, the tests simulate bubble flow rather than the critical transition between annular flow and churn or slug flow. Additionally, field data highlights whether FAL successfully restores stable production at a gas rate less than the original Q_{min} without surfactant but seldom reveals the Q_{min} that could ultimately be achieved with FAL and how this Q_{min} varies as a function of the surfactant concentration.

Large scale vertical flow loop testing at representative gas velocities has been performed amongst others by Van Nimwegen et al. [13] and Ajani et al. [14] to better understand the changes in flow regime and VLP associated with FAL: they measured the VLP and the liquid holdup across a large vertical section of transparent tubing while flowing a mixture of air, water, and foam, and visualized the multiphase flow behavior with cameras. Fig. 8.5 shows VLP's at different

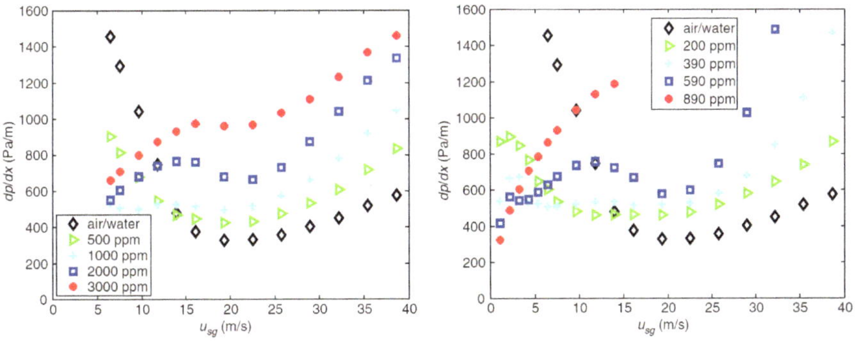

FIGURE 8.5

VLP's for two types of surfactant at different surfactant concentrations (including zero) that illustrate reduction of minimum stable velocity.

Reprodiced with permission from A.T. van Nimwegen, The Effect of Surfactants on Gas-Liquid Pipe Flows (PhD dissertation), Delft University of Technology, 2014 [13].

surfactant concentrations that demonstrate a significant reduction of the velocity, at which the VLP reaches its minimum value, equivalent to a large reduction of $Q_{\min}$. The visual observations, illustrated in Fig. 8.6, confirms that the foam film is much more stable than the original water film and, as a result, the stable annular flow regime is sustained at much lower gas velocities.

Besides a reduction of the hydrostatic pressure component at low gas velocity, Fig. 8.5 also indicates a significant increase of the frictional pressure component at high gas velocity. This is supported by VLP field data that suggests that the frictional pressure drops about doubles in the presence of FAL. Referring to Fig. 8.5, this would correspond to the green triangles or blue plusses. None of the field data resembles anything like the dark blue squares or red circles.

Veeken et al. [15] carried out a campaign of instrumented field trials specifically designed to quantify the reduction of $Q_{\min}$ across a wide range of well conditions using multiple foamers. Fig. 8.7 demonstrates the change in well response

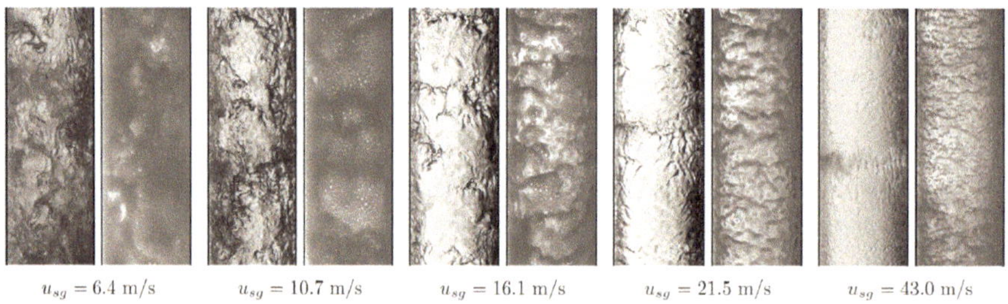

FIGURE 8.6

Summary of flow visualization with (left) and without (right) surfactants in 50 mm pipe.

Reproduced with permission from A.T. van Nimwegen, The Effect of Surfactants on Gas-Liquid Pipe Flows (PhD dissertation), Delft University of Technology, 2014 [13].

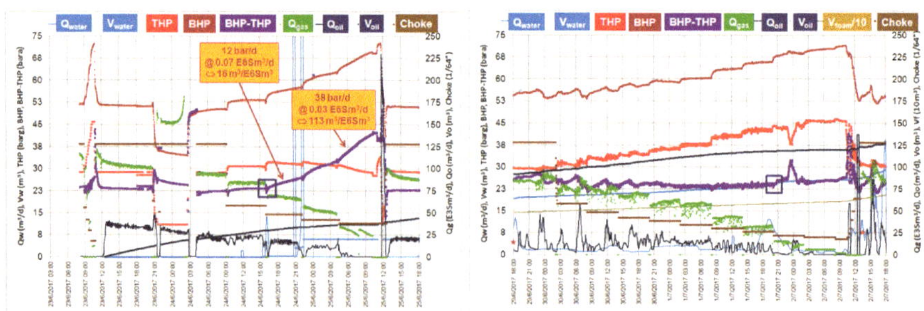

FIGURE 8.7

Dedicated well test to measure impact of FAL on $Q_{\min}$: Left side without foamer, Right side with foamer.

caused by surfactant injection. The blue squares highlight where the wellbore pressure drop in purple starts to increase due to liquid loading, which happens at Q_{gas} of 75 $10^3Sm^3/d$ without surfactant and at 25 $10^3Sm^3/d$ with surfactant, presenting a reduction of Q_{min} by a factor 3.

Such dedicated field trial data is rare but similar information could be extracted by monitoring wells with FAL long enough until they decline far enough to show liquid loading again. Unfortunately, such long-term field data is seldom collected or reported.

McWilliams and Gonzales [16] guesstimated the potential reduction of Q_{min} by modifying foam parameters in the Turner expression for the critical gas velocity in Eq. (8.1). A realistic twofold reduction of the surface tension (σ) and an optimistic 10-fold reduction of the liquid density (ρ_L) together result in a modest twofold reduction of V_g and Q_{min}. A fivefold reduction of Q_{min} could be achieved by assuming a further 30-fold reduction of the so-called Weber number that describes the maximum droplet size. As highlighted before, the droplet reversal theory that underlies Eq. (8.1) does not describe reality, and therefore, manipulating Eq. (8.1) does not necessarily produce credible results.

Soni and Kelkar [17] matched the VLP data of wells with FAL by assuming that the surfactant reduces the gas–water surface tension 70-fold from ~70 down to ~1 mN/m. In Eq. (8.1) such a 70-fold reduction of surface tension results in a threefold reduction of Q_{min}. This is a theoretical construct as laboratory testing indicates a maximum reduction of surface tension by a factor of 2–3 to ~30 mN/m.

Van't Westende et al. [18] summarizes the most recent flow loop test results and multiphase flow modeling attempts based on film flow and film flow reversal. They conclude that more work is required to arrive at a realistic description of FAL outflow behavior with parameters that can be determined by means of standard laboratory performance testing.

8.3.2 Foam-assisted lift operating envelope

FAL presents a powerful deliquification solution but conditions exist for which FAL is not ideal or suitable. Table 8.1 presents notional boundary values for BS&W (>50%), BHT (<150°C), and LGR (<1000 $m^3/E6Sm^3$). BS&W is important because conventional surfactants are designed to foam the water phase where the hydrocarbon condensate phase will act as a natural defoaming agent. Thus, it is more difficult to foam well liquid when the condensate cut increases hence BS&W decreases. At low BS&W, FAL requires a higher dosage of a more expensive (e.g., fluorinated) surfactant and may only achieve a limited reduction of Q_{min} (i.e., twofold instead of fivefold) at a much higher cost. Generally, at BS&W below 50%, FAL becomes more challenging.

Similarly, most surfactants do not survive and perform at a BHT above 150°C. A BHT above 150°C requires dedicated, more expensive, and less effective surfactants. As LGR increases, the surfactant consumption and cost increases. Hence, a well with high LGR can respond favorably to FAL but will require more surfactant.

Besides efficacy and cost, surface processing can pose restrictions as well, for example, related to the compatibility between the surfactant and other process chemicals or to overboard disposal of the produced water in offshore conditions. These will be elaborated in Section 8.6.

8.3.3 Foam-assisted gas lift

In practice, FAL is used in combination with other deliquification techniques such as SC and VS to deliver the ultimate goal of stable production at the lowest possible economic gas rate. That ultimate goal is usually achieved by a sequence of measures, as outlined by Veeken and Al Kharusi [10]. Preferably, FAL would be scheduled following SC and, if applicable, VS installation to minimize the duration and rate of foamer injection. Ideally, FAL should serve as the final deliquification measure either stand-alone or in combination with other deliquification techniques to improve their efficacy.

A well-known combination is foam-assisted gas lift (FAGL) where surfactant is added to the lift gas stream to reduce Q_{min} and hence the volume of lift gas required to sustain stable flow. FAGL decreases the lift gas and power requirement and therefore, cost. FAGL has been applied successfully in gas wells [19,20] and also in oil wells [20–22]. In oil wells, FAGL reduces slugging and the wellbore hydrostatic and thus increases oil production. Most benefit is achieved in oil wells that produce at high BS&W and experience severe slugging. Note that adding a liquid surfactant to dry lift gas causes evaporation of the solvent phase and hence introduces the risk of solids deposition (gunking) across the injection valve. The surfactant must be designed and tested to prevent this [22].

The FAL part of FAGL mainly serves to reduce the lift gas requirement, in case of limited compressor ullage and/or excessive wellbore friction.

8.3.4 Surface foam-assisted lift

Foaming can also be deployed at the surface to reduce slugging and back pressure in flowlines and pipelines. This applies to both gas and oil production systems [23–26]. The reduction in back pressure increases production while the reduction of slug size and frequency mitigates processing upsets in the receiving separation system. Adding surfactant reduces the gas rate at which the flow regime changes from favorable to unfavorable, this time from stratified-wavy to stratified-slugging.

8.4 Foam-assisted lift application

This Chapter does not compare FAL against other deliquification measures. It assumes that FAL is actively considered and covers the following four steps towards application and operation:

1. Diagnose and forecast well performance (a) to confirm it as a valid candidate for continuous or intermittent FAL, (b) to decide what date FAL is required, and (c) to specify its key operating parameters.
2. Carry out laboratory testing to select the surfactant.
3. Carry out field testing to confirm FAL efficacy.
4. Install and operate FAL to mature maximum FAL benefit at minimum cost.

Steps (2) and (3) can be skipped in case representative offset well experience has already confirmed the surfactant type and FAL efficacy.

8.5 Well performance

8.5.1 Collect well data

Well and production data must be collected to diagnose and forecast well performance, and to support laboratory testing. A typical data set consists of the following:

- Well completion diagram: tubing and casing size, depth, and deviation.
- Well model: inflow and outflow performance described by static and flowing pressure, temperature and gas rate at the reservoir, bottomhole and surface conditions.
- Production data: historic and current gas, condensate and water rate [or gas rate, condensate–gas ratio (CGR), and water–gas ratio (WGR)], and wellhead pressure and temperature.
- Gas and condensate gravity and composition (and condensate sample for lab testing).
- Water gravity and composition (and water sample for lab testing).
- Production logging and well test data.

Production data preferably includes real-time gas rate, wellhead pressure, and wellhead temperature data as that data are most valuable for diagnosing well performance (see Fig. 8.2). Less frequent data such as monthly average data can be used as well but is more difficult to interpret.

8.5.2 Diagnose and forecast liquid loading

Based on the well completion and/or well model, the liquid-loading gas rate $Q_{\min}$ can be estimated using Eq. (8.5), illustrated in Fig. 8.4, or by picking the minimum of the simulated wellbore VLP at the current THP, sketched in Fig. 8.1.

If the production data indicates that the well is still producing stable above $Q_{\min}$, the start date of liquid loading can be predicted by forecasting when the declining gas rate will reach $Q_{\min}$. If the production data indicate that the well is already intermitting, the historic intermittent production pattern should confirm whether the intermitting is consistent with liquid loading starting at about Q_{min}. The orange well in Fig. 8.8 starts intermitting at ~130 $10^3Sm^3/d$, which is 27% below but still compatible with the Q_{min} from Eq. (8.5a and 8.5b) of 178 $10^3Sm^3/d$. Eq. (8.5a and 8.5b) is a wellhead equivalent value and in this case does not account for the bottom section of tubing with a smaller ID of 3.74″ instead of 4.67″. The blue well starts intermitting at ~100 $10^3Sm^3/d$, which is 80% higher and incompatible with the Q_{min} from Eq. (8.5a and 8.5b) of 56 $10^3Sm^3/d$: well surveillance indeed revealed that the intermittent behavior was due to halite scaling rather than liquid loading.

In the absence of gas rate data, other parameters can be used to diagnose liquid loading as well, such as:

- In many cases, the THT presents an excellent proxy of the gas rate. This is illustrated in Figs. 8.2 and 8.8 where THT varies in sync with Q_{gas}.
- In case, significant back pressure exists across surface choke and/or flowline, a reduction of THP mirrors a reduction of Q_{gas}. Note that this data are not available in most liquid loading wells that tend to be operated with full open choke without significant back pressure.

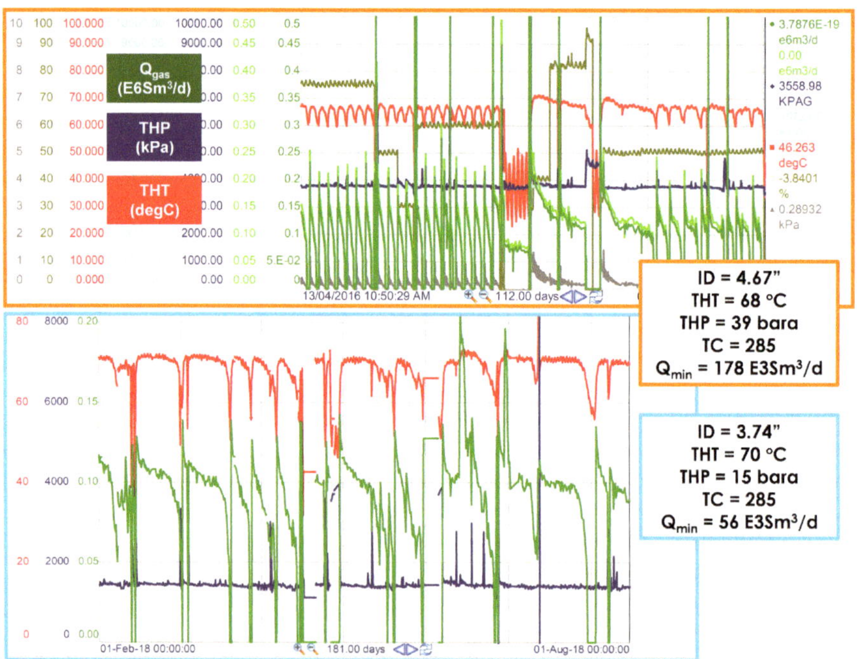

FIGURE 8.8

Real-time production data for two wells: Orange well is intermittent due to liquid loading while blue well is intermittent due to halite scaling.

- In case of a packerless completion, an increasing differential between increasing annulus pressure and constant wellhead pressure indicates an increasing wellbore hydrostatic due to liquid loading, see orange well in Fig. 8.9.
- When shutting in a well after liquid loading, the surface pressure buildup signature will exhibit a change that corresponds to the drainage of the liquid column, see blue wells in Figs. 8.9 and 8.10.
- When water production is metered, liquid loading will express itself as a significant drop in WGR: WGR may not be zero due to water condensation at lower separator temperature. Note that frequent or continuous water metering is rare.

8.5.3 Predict foam-assisted lift operating parameters

Intermitting liquid loading can be treated using either continuous or intermittent FAL. That choice depends, amongst others, on the time it takes for the gas rate to decline when the well loads up. The liquid loading halftime T_{LL} in days, during

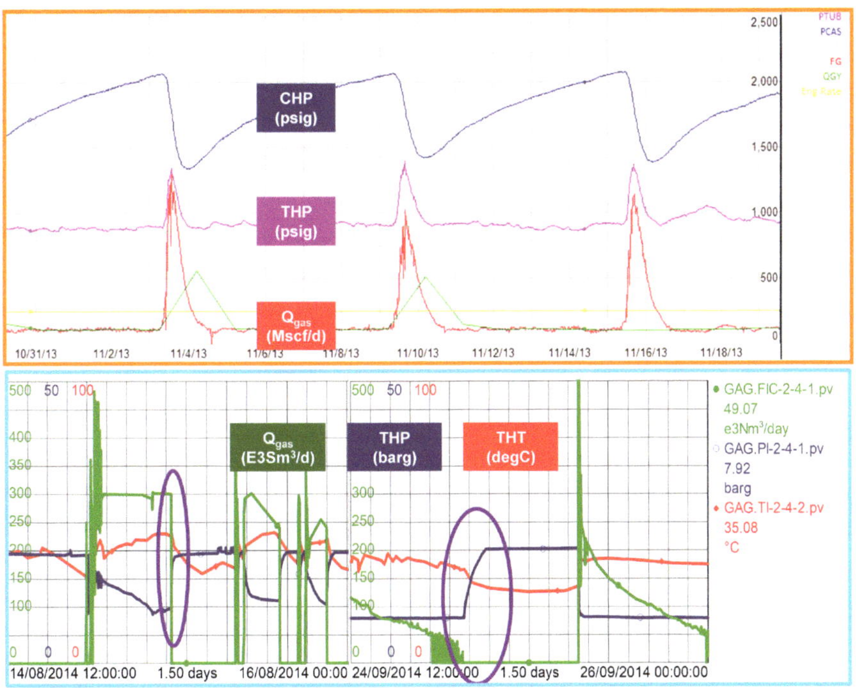

FIGURE 8.9

Orange well shows liquid loading and unloading through increasing and decreasing difference between CHP and THP; blue well shows liquid loading through a change in THP buildup signature.

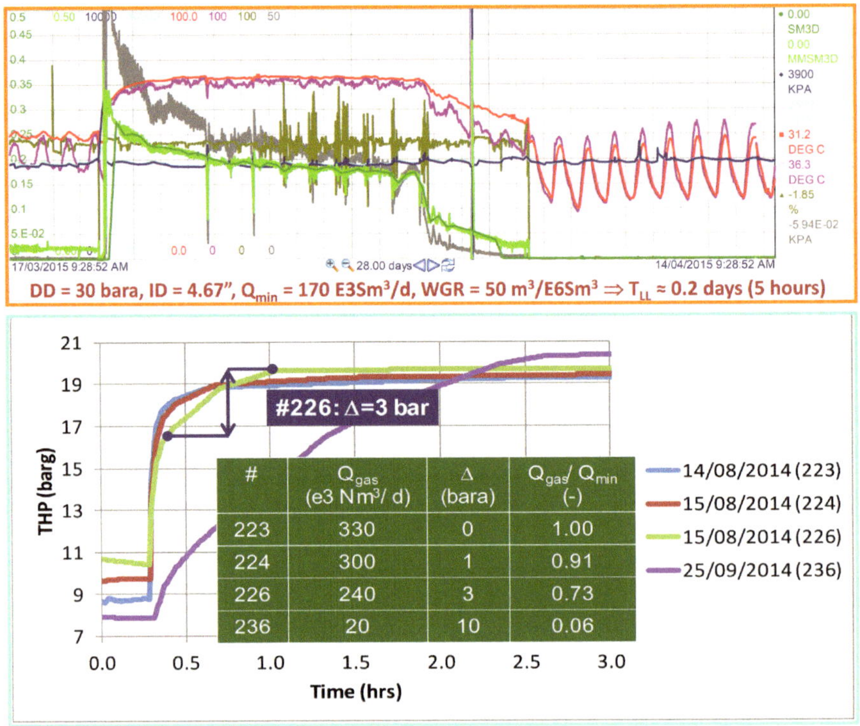

FIGURE 8.10

Orange well production decline indicates TLL of 0.2 days compared to calculated TLL of 0.2 days; blue well shows HH of 30 m (3 bara) for surface pressure buildup #226 compared to maximum HH in #236 of 90 m (9 bara).

which 50% of production capacity is lost, can either be distilled from actual production data or inferred using the following expression based on accumulating the equivalent liquid column and hydrostatic head inside the wellbore:

$$T_{LL} \approx 2.53 \cdot DD \cdot ID^2/(Q_{min} \cdot WGR) \quad (8.6)$$

Where the drawdown (DD) is the difference between reservoir pressure (P_{res}) and flowing BHP in bara. For example, with DD of 50 bara, ID of 4 in., Q_{min} of 50 $10^3Sm^3/d$, and WGR of 10 $m^3/E6Sm^3$, T_{LL} becomes 4 days. The larger the WGR is, the faster the well loads up and the faster the gas rate declines. The orange well in Fig. 8.10 shows an example comparing the observed and calculated T_{LL}.

The amount of required surfactant can be calculated based on the well performance. In case of continuous FAL, the surfactant should be supplied at a constant rate Q_{foam} in L/d, as follows:

$$Q_{foam} \approx 1E-6 \cdot Q_{gas} \cdot LGR \cdot C_{foam} \quad (8.7)$$

Where C_{foam} is the target surfactant concentration on total liquid (water and condensate) in ppm. For example, with Q_{gas} of 100 $10^3Sm^3/d$, LGR of 100 $m^3/E6Sm^3$, and C_{foam} of 1000 ppm, Q_{foam} becomes 10 L/d.

In case of intermittent lifting, the required surfactant volume V_{foam} in L must match the liquid column that needs to be lifted:

$$V_{foam} \approx 0.507E-6 \cdot HH \cdot ID^2 \cdot C_{foam} \tag{8.8}$$

where HH is the height of the liquid column in m. For example, with HH of 100 m, ID of 4 in., and C_{foam} of 10,000 ppm, V_{foam} becomes 8.1 L.

The height of the liquid column HH in Eq. (8.8) can be derived using one of the following methods:

Method A: Take the DD in bara used in Eq. (8.6) and convert it into the equivalent liquid column height in m by multiplying by 10.

Method B: Take the difference between the closed-in wellhead pressure (CITHP) immediately after shut in and the CITHP that corresponds to a gas-filled wellbore: the difference corresponds to the hydrostatic pressure of the liquid column [27]. The CITHP of the gas-filled wellbore is either measured after the pressure buildup is completed or calculated based on the latest reservoir pressure.

Method C: Measure the liquid holdup depth in the shut-in well using acoustics or wireline. This number is impacted by the time of data collection as the liquid level can drop overtime as the liquid drains away into the reservoir.

In producing wells, HH based on method A presents the maximum possible column height and always exceeds HH based on method B that characterizes the actual liquid column height immediately after shut in, as illustrated by the blue well in Fig. 8.10. In turn, HH based on method B will always exceed HH based on method C as part of the liquid column will drain away before executing the actual surveillance activity: in non-tight gas wells, a liquid loaded column drains away within a few hours to days. Usually, the DD is known and can be used to derive a conservative value of HH and V_{foam}. An exception is wells that await kick-off, where the liquid column could be a lot higher than HH based on method A when additional liquid was introduced either by well intervention or by formation water production.

In case of a one-time kick-off operation using FAL, the gas rate after is expected to exceed Q_{min} and hence to be stable. This expectation of a stable gas rate could be based on stable historic production data immediately prior to the event that triggered the kick-off (stimulation, workover, shut in) or inferred from a representative well model, in case of a new well with yet uncertain inflow performance.

FAL for kick-off only works if the well produces a sufficient gas rate to create and lift foam. FAL will not work in case of a completely dead well that does not exhibit any gas production. The minimum required gas rate to enable kick-off can be approximated by the so-called Taylor gas bubble rise velocity [28]. This Taylor velocity describes the rise of elongated gas bubbles through a liquid-filled wellbore: such bubbles can provide some vertical lift of the liquid column on top, which can be significantly improved by turning this liquid into foam. The corresponding Q_{Taylor} in $10^3\text{Sm}^3/\text{d}$ can be expressed as follows:

$$Q_{\text{Taylor}} \approx 0.0076 \cdot \text{ID}^{2.5} \cdot \text{BHP} \tag{8.9}$$

For example, for ID of 4″ and BHP of 50 bara, Q_{Taylor} is 12 $10^3\text{Sm}^3/\text{d}$ while for ID of 2″, and BHP of 100 bar, Q_{Taylor} is 4 10^3Sm3/d.

As a rule of thumb, Q_{Taylor} is 10x smaller than Q_{min}, which means that FAL for kick-off stands a chance as long as Q_{gas} prior to kick-off and cleanup exceeds 10% of Q_{min}. In wells where the initial gas rate is (far) less than Q_{Taylor}, other techniques such as GL or DP are more appropriate as they do not rely on an initial influx of reservoir gas.

8.6 Laboratory testing

Once a gas well is diagnosed as a candidate for FAL, the right foamer(s) should be selected based on a comprehensive laboratory testing program. Key performance indicators that need to be assessed are:

- Foamer performance.
- Physical properties and stability.
- Compatibility with other chemicals, injection equipment, and surface processing.

Various types of surfactants exist that can be used as a foaming agent, see Table 8.2. These are grouped according to the charge they carry: amphoteric (charge depends on pH of solution), cationic (positive charge), anionic (negative charge), nonionic (no charge), or a combination thereof. The resulting foaming performance varies significantly due to their specific nature and properties. The application domain of these surfactant types is as follows:

- Anionic surfactants usually have a low tolerance to high total dissolved solids (TDS) brine and hydrocarbons and are not suitable for wells producing high-salinity water or at low BS&W.
- Cationic surfactants could potentially cause compatibility issues with additives such as scale inhibitors (typically negatively charged). The cost of cationic surfactants is high, which makes them less suitable as foaming agent.
- Nonionic surfactant solutions in water can become cloudy when the temperature is above a certain threshold (cloud point), which renders them unsuitable for downhole application.
- Amphoteric surfactants have good tolerance to high TDS brine, liquid hydrocarbon, and high temperature, which makes them the most versatile and suitable surfactant for FAL. Combinations of amphoteric and anionic surfactants are utilized to boost foaming performance due to their synergetic effect [29].

Table 8.2 Examples of various types of surfactants used as foaming agents.

Surfactant type	Charge	Example	Chemical structure
Anionic	Negative	Sodium C14-olefin sulfonates	
Cationic	Positive	Cocoalkyltrimethyl ammonium chloride	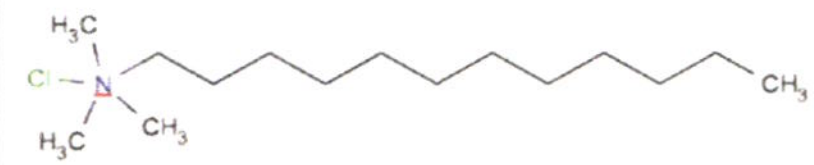
Nonionic	Neutral	Decyl alcohol ethoxylate	
Amphoteric	pH dependent	Lauramidopropyl hydroxysultaine	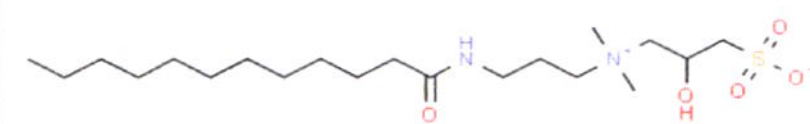

Besides active surfactant, foamer products can also contain organic solvents, corrosion inhibitors, and scale inhibitors. Organic solvents such as alcohols, glycols, or glycol ethers are often added to either improve foaming performance or to depress the freezing point of liquid foamers to ensure flowability and pumpability in colder climates.

8.6.1 Foamer performance

There are several laboratory tests that evaluate and compare the foaming performance of specific surfactants. The most used tests are modified ASTM D892 methods to evaluate foam formation and foam stability.

8.6.1.1 Sparging test (modified ASTM D892)

The Modified ASTM D892 sparging test was designed to evaluate the foaming characteristics of crude oil but is also used to evaluate the foam performance for gas well deliquification. This test consists of injecting a fixed gas rate (e.g., air at 1 L/min) through a frit into a certain amount of representative well liquid (e.g., 200 mL) with the dissolved surfactant, heated to represent wellbore temperatures (usually limited to 90°C) in a measuring cylinder (e.g., 1000 mL), see Fig. 8.11. The time required to build up a certain foam height is recorded. Then the gas supply is removed, and the foam half-life or total foam disappearance time is recorded as well. Table 8.3 presents the criteria that are used for buildup and halftime to screen candidate surfactants and to establish the required surfactant concentration.

This sparging test characterizes the foam formation and foam buildup under dynamic conditions, but only evaluates foam stability after agitation is removed, which differs from the constant dynamic condition foam experiences in producing wells. That was the reason to introduce the dynamic foaming test described below.

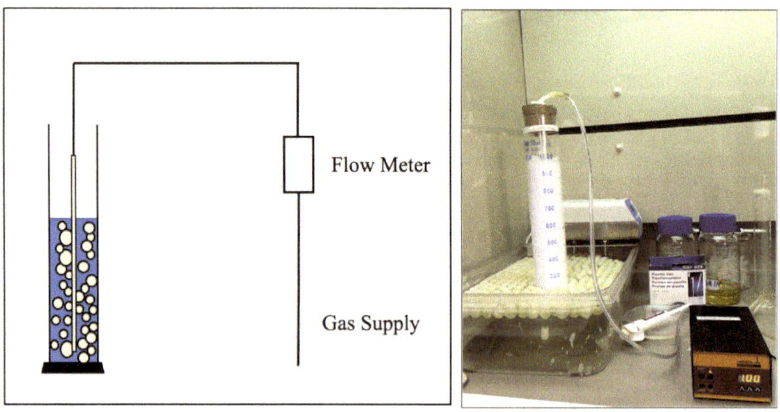

FIGURE 8.11

Schematic and picture of sparging test.

Table 8.3 Foam performance screening criteria based on sparging test results.

Foam buildup time (s)	Ranking	Foam halflife time (s)	Ranking
<80	Good	> 180	Good
80< X <120	Moderate	60< X <180	Moderate
> 120	Poor	<60	Poor

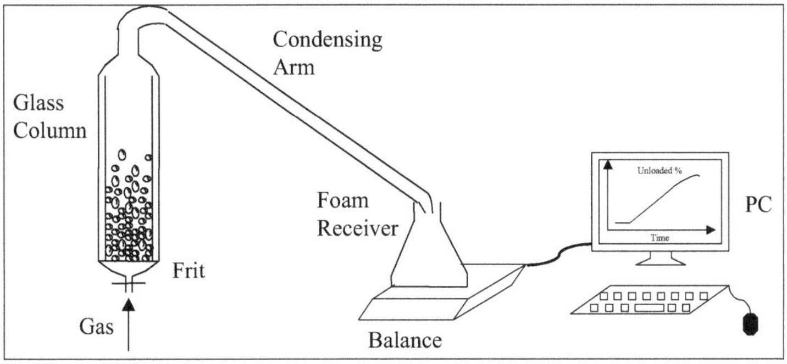

FIGURE 8.12

Schematic representation of standard dynamic foaming test apparatus for measuring liquid unloading efficiency.

Reproduced with permission from P.F. Huang, D. Nguyen, Optimized foamers for natural gas well deliquification: A statistical design approach, Fuel 97 (2012) 523–530 [29].

8.6.1.2 Dynamic foaming test (modified ASTM D892)

The dynamic foaming test is used to evaluate the overall unloading efficiency of a foaming agent. The unloading efficiency is governed by foam formation and foam stability in the presence of continuous agitation; it combines the impact of foam buildup and halflife under constant dynamic conditions that better simulate the downhole situation.

Fig. 8.12 shows a standard dynamic foaming test apparatus in orange, which is a modified version of a setup from the US Bureau of Mines [29]. This test involves supplying a certain gas rate (typically 1–2 L/min) through a frit into a column containing a fixed amount of liquid pretreated with a certain dosage of surfactant. As the foam is created and lifted to the top of the column, it enters a condensing arm, and to a foam receiver sitting on a balance. The balance is connected to a data recording computer that captures the weight of the collected liquid in real-time.

The liquid unloading efficiency of the foamers in percentage is calculated using Eq. (8.10):

$$\text{Unloading efficiency} = \text{unloaded weight}/\text{initial weight} \tag{8.10}$$

where the unloaded weight is the weight of the fluid collected in the foam receiver and the initial weight is the weight of the total fluid loaded in the test column at the beginning of the foaming test. The liquid unloading efficiency presents a straightforward measure of foaming efficacy and an unloading efficiency of more than 50% is typically considered acceptable.

The fluid makeup can be adjusted to reflect the condensate-water mixture that the well produces. For example, 20% condensate and 80% brine corresponds to 80% BS&W. If possible, fresh field fluids should be used for testing to best represent the characteristics of the produced fluids. Synthetic formation water can be used as an alternative. The liquid sample can be heated to represent wellbore temperatures but in the standard setup, this is limited to 90°C.

The advanced setup used by Omrani et al. [30] caters for temperatures up to 160°C, pressures up to 18 bara, and sparging rates up to 1 Sm^3/min (1000 L/min). They investigated the influence of sparging rate, BS&W, temperature, and pressure for a range of surfactants and surfactant concentrations. Their results illustrate several key points:

- The sparging gas velocity needs to exceed 0.05 m/s to achieve the best unloading efficiency.
- The unloading efficiency increases with increasing surfactant concentration with most increase up to 2000 ppm and little increase beyond.
- The maximum unloading efficiency using water foamers decreases with decreasing BS&W or increasing fraction of condensate and reduces from 100% to 50% as BS&W decreases from 100% to 50%.
- The unloading efficiency decreases significantly with increasing BHT and reduces more than tenfold comparing 25°C and 120°C.

The ambient pressure did not have a significant impact on the unloading efficiency but was essential for testing at temperatures above 90°C, to reduce evaporation. Hence, an advanced setup is required to test at representative BHT above 90°C that cannot be achieved using the standard setup at atmospheric conditions.

The sparging gas velocity in both standard and advanced setup is 10–100 times lower than the critical gas velocity in natural gas wells, which means the tests represent a bubble flow regime rather than the more relevant annular, churn, and slug flow regime. Also, the measuring cylinders are at least 1000 times shorter than actual field completions. Nonetheless, experience has learned that these tests do provide relevant information on foamer performance: good foam performance in these lab tests generally translates into adequate foam performance in the field. These tests are very useful for screening and selecting foamers for field use, but do not provide the relation between Q_{min} reduction and surfactant concentration.

8.6.2 Secondary performance

The primary indicator for a successful FAL surfactant is the foamer performance in lab testing. However, there are several secondary performance indicators that need to be taken into account as well to ensure successful field application. These include:

- Physical properties and stability.
- Compatibility with existing chemical injection program.
- Compatibility with injection system.
- Compatibility with separation process.

8.6.2.1 Physical properties and stability

The foamer viscosity increases as temperature decreases and in regions with low ambient temperatures, the liquid foamer viscosity in storage should still be low enough to ensure pumpability. At the same time, the foamer must remain stable both (short term) at elevated downhole temperature and (long term) at ambient storage temperature. Specifically, the foamer should not decompose and separate, and must remain clear without solids precipitating. Viscosity and high-temperature stability testing are a routine part of surfactant screening.

Many foamers tend to be (slightly) corrosive when in contact with (carbon) steel. Hence, either the injection skid and well materials need to be corrosion resistant or the surfactant needs to be specified and designed as fully noncorrosive. Depending on the equipment material, corrosion testing may be required.

The foamers need to be safe to handle, inject, produce, and dispose and hence need to comply with local HS&E requirements for chemical storage, usage, and disposal. This specifically applies to the disposal of produced water into surface water, for example, overboard into the sea. Usually, strict regulations apply regarding the potential toxicity of residual surfactant for aquatic life.

8.6.2.2 Compatibility with existing chemical injection program

If there is any existing chemical injection program already in place, it is advisable to check the compatibility of the foamer with those other chemicals to rule out potential incompatibility issues either downhole or in the surface processing system. Other chemicals in use could be scale and corrosion inhibitors, demulsifiers, and clarifiers. Multifunctional combinations of a foamer with scale and/or corrosion inhibitor could present the most cost-effective solution for such situations [31].

8.6.2.3 Compatibility with injection system

This specifically applies to FAGL when foamer is injected at the surface into the stream of dry lift gas. In that case, the solvent will (partly) evaporate, which introduces the risk that gunk is deposited in either GL conduit or GL mandrel and injection valve. Hence, the foamer solvent package must take evaporation into account and relevant testing is advised to confirm suitable performance [21,22,32].

Table 8.4 Antifoam performance screening criteria based on re-sparging test results (assumes 200 mL initial fluid).

Foam height	Ranking
No foam	Good
200–250 mL	Moderate
250–350 mL	Doubtful
>350 mL	Poor

8.6.2.4 Compatibility with separation process

A gas well equipped with FAL will produce foam and water-condensate emulsion to the surface. Also, the produced liquid stream could still contain a residual surfactant. Depending on the surface flowline, pipeline configuration, and operating conditions, some of the foam, emulsion, and surfactant could ultimately reach the production separation and processing facilities.

When foam enters the facilities, it may be carried over into the gas processing stream where it could create serious upsets, for example, by depositing solids inside the glycol reboilers. Likewise, stable emulsion and residual surfactant can upset water and condensate processing. Specifically, the condensate content of the produced water (oil-in-water or OIW) could increase significantly. In an offshore environment where produced water is disposed overboard, FAL may increase the OIW above the allowed maximum (e.g., 10 ppm), potentially jeopardizing the license to operate.

Long flowlines and pipelines can neutralize the contaminated liquid before it reaches the facilities. Also, if FAL is applied on a minority of wells, dilution may further mitigate the contamination risks. However, conditions will exist where the problem needs to be tackled proactively. The first mitigation measure is the injection of antifoamer or defoamer to kill any residual foam and neutralize any residual foamer. A second mitigation measure entails demulsifier and/or water clarifier to ensure the quality of disposal water and export condensate.

The standard sparging test setup in Fig. 8.11 can also be used to establish the performance and required concentration of the antifoamer: antifoamer is added to the foam column and the solution with foamer and antifoamer is then re-sparged. The performance of the antifoamer is judged by the height of the re-sparged foam column, as illustrated in Table 8.4.

Foam performance testing also indicates whether foaming results in stable condensate–water emulsion and whether a demulsifier can break this emulsion. The laboratory screening can help to determine the required dosage of demulsifier.

8.7 Foam-assisted lift field testing

Once surfactants successfully pass the laboratory testing screening criteria, they can be considered for field application. There is no need for dedicated field testing if relevant offset experience already indicates adequate foamer performance and absence of processing issues. If such information is not available, one could decide to install FAL in selected wells to gain that field experience "on the fly." In most cases, one would prefer to carry out some type of field testing first, to confirm field foam performance and to ensure that FAL does not trigger serious operational issues. The results of such intermittent (batch) or continuous field trials can also be used to optimize the choice of foamer and antifoamer chemicals.

8.7.1 Batch foam-assisted lift trial

In a batch trial [33], foamer is deployed intermittent in a well that experiences liquid loading or kick-off issues. The batch surfactant either consists of foam sticks that are dropped into the well or of liquid surfactant flushed down by a displacement fluid such as water. A batch trial generally runs as follows:

1. Shut-in well and drop sticks or pump liquid (surfactant and displacement). The volume of surfactant is calculated as per Eq. (8.8) with a default target surfactant concentration of 10,000 ppm or 1%.
2. Keep well shut in for 8–24 hours so the surfactant reaches bottom and mixes with the remaining wellbore liquid, and to allow (partial) drainage of the liquid column and recharge of the near-wellbore reservoir pressure.
3. Open up well and monitor well performance to evaluate foamer response.

The foamer performance is judged by comparing the well performance after opening up with surfactant to opening up without surfactant, preferably based on real-time gas production data or, alternatively, real-time wellhead temperature. Foam performance is deemed favorable if the gas rate increases significantly over a significant time period. Most of the surfactant is produced to the surface during the initial kick-off period that lasts 2–6 hours. Unless kick-off results in a stable gas rate that exceeds $Q_{\min}$, the period of production improvement will be limited as the well will start loading again as soon as the surfactant and foam are gone.

No response doesn't necessarily mean that the foamer is noneffective. It may require increasing the foamer dosage to make the well responds to the treatment. Also, these batch jobs may need to be repeated several times (3–5) before achieving kick-off. During the prior liquid loading period, a mass of water may have accumulated around the wellbore that reduces gas productivity and it may take several unloading cycles to restore the gas potential. In the worst case, liquid loading has diminished inflow performance to such an extent that despite good foamer performance, sufficient liquid cannot be unloaded, and gas inflow performance cannot be restored. This may happen if the gas rate remains less than Q_{Taylor}, see Eq. (8.9).

Sometimes, the success rate of batch treatments can be improved by alternating production (rocking) and/or restricting production (choking). The objective of these procedures is to achieve a better mixing of surfactant and wellbore liquid and to avoid spending the stored well energy too quickly.

Constant and close monitoring of the well and surface system parameters are integral part of the field test to ensure that surface processing proceeds without interruption. It is a good practice to have antifoamer on site as contingency while conducting the field trial and optimizing the foamer program. The need for antifoamer becomes less as the length of flowline and pipeline becomes longer as this provides more time for foam and emulsion to break down. In case of offshore production with overboard water disposal, the risk of water quality going off-spec is significant; a test separator could be utilized to divert foam treated fluids to reduce the foam tendency and water quality issue of the total system. A higher dosage of emulsion breaker and water clarifier may be needed to ensure uninterrupted topside operation.

8.7.2 Continuous foam-assisted lift trial

Liquid surfactant is pumped into the well continuously either directly downhole via an internal capillary string or external control line, or indirectly from surface dripping down inside a connected but closed-in annulus. The injection depth can be chosen in case of an internal capillary string but is otherwise fixed. The foamer injection rate is selected to achieve a specific surfactant concentration, typically in the range of 1000–10,000 ppm, as per Eq. (8.7). Well performance is best monitored based on wet gas rate and, alternatively, wellhead temperature.

There are two types of continuous FAL trial:

1. Inject foamer into liquid loaded/loading gas well: success is defined as restoring stable gas production at a gas rate below the original Q_{min}. By varying the foamer rate the minimum required surfactant concentration can be established. The chance of success can be improved by cycling the well: surfactant is continuously injected downhole while the liquid in the well moves up and down as the well is opened and closed, and the reservoir pressure is able to recharge during the shut-in periods.
2. Inject foamer into a stable gas well while step-wise reducing the choke size to trigger liquid loading: by comparing the well performance at reducing choke size and increasing wellhead pressure without and with foamer injection, the maximum possible reduction of Q_{min} can be established, together with the required surfactant concentration.

While testing, the surface facilities need to be monitored for upsets; antifoamer, demulsifier, and water clarifier may be required to mitigate adverse consequences of FAL on gas or liquid processing and quality. Surveillance includes regular sample collection and analysis at critical locations.

Most field trials are type (1) as they address wells that are intermitting or producing very little due to liquid loading. These trials are designed to confirm that FAL can be applied successfully but do not establish the maximum future reduction of Q_{min}. Type (2) is the most elaborate and specifically designed to quantify the achievable reduction of Q_{min} [15].

Besides reducing Q_{min} or the gas rate, at which the VLP reaches its minimum BHP value, the creation of foam in the wellbore will also change the hydrostatic and frictional pressure components. The wellbore pressure loss was measured during a series of tests and analyzed by Veeken and De Vries [34]. Data analysis did not reveal any change in hydrostatic but did suggest a doubling of the frictional pressure component. Generally, such doubling has minor impact on well capacity and does not affect the decision to apply FAL. The large-scale flow loop tests show an even more significant increase of friction [13].

8.8 Foam-assisted lift application

Once a candidate well is confirmed based on a diagnosis of its well performance, and a candidate foamer has been identified based on comprehensive laboratory testing, one has to decide on the specifics of the FAL installation and operation. Key decisions are:

- Continuous or intermittent FAL.
- Solid or liquid foamer.
- Liquid foamer agent via backside or capillary.
- Capillary depth, size, and material.
- Surface injection system.
- Antifoamer, demulsifier, and/or water clarifier.

8.8.1 Continuous or intermittent foam-assisted lift

Continuous FAL is considered the best technical solution by far. Veeken and De Vries [34] compared the average success rate of batch or intermittent FAL with continuous FAL and found 50% for intermittent compared to 90% for continuous FAL. The reason is that continuous FAL deposits the surfactant, exactly where and when required, whether the well is flowing or closed, while pumping a batch of liquid foamer from surface into a closed-in well offers little control, and hence does not necessarily result in a fair distribution of surfactant across the liquid to be lifted. For example, the liquid column that caused the liquid loading event could have drained away into the reservoir before the liquid foamer reaches bottom. Therefore, batch FAL can be unsuccessful where continuous FAL shows good performance.

Note that intermittent FAL in liquid loading wells generally requires more foamer volume. The batch foamer volume in Eq. (8.8) is generally based on a target surfactant concentration of 10,000 ppm while the surfactant concentration in continuous FAL applications tends to be of the order of 1000 ppm or lower [34]. This needs to be taken into account when comparing the chemical cost and impact on surface processing and water disposal of intermittent versus continuous FAL.

8.8.2 Solid or liquid foamer

When intermittent or batch FAL is proven to work, it may actually be the preferred solution for reasons of cost efficiency and ease of implementation. This certainly applies when FAL is required only occasionally for kick-off but may also apply to wells that exhibit intermittent liquid loading behavior. When using a liquid foamer, the intermittent liquid loading cycle must be long enough. Eq. (8.6) can be used to assess the required FAL frequency: intermittent FAL is deemed an option if that frequency is less than once every 2–4 days because every batch job requires the well to be shut in for about 12–24 hours. A higher batch frequency, such as once every 1–2 days, causes excessive deferment.

In case of solid foam sticks, there is no need for shut-in periods as the foam stick should drop down despite gas flowing up. This removes the concern about the length of the liquid loading cycle as sticks can be dropped at any frequency using an automated foam-stick launcher. The volume of surfactant delivered by a typical foam stick (1.25″ diameter, 15″ length) is about 0.3 L, which makes this method especially suited for low rate wells with low LGR. Eq. (8.7) can be used to calculate the minimum required amount of surfactant or number of foam sticks. For example, assuming Q_{gas} of 10 $10^3Sm^3/d$, LGR of 100 $m^3/E6Sm^3$, and C_{foam} of 1000 ppm corresponds to a practical 1 L/d or three sticks per day while the number of sticks increases to an impossible 375 per day for Q_{gas} of 50 $10^3Sm^3/d$, LGR of 500 $m^3/E6Sm^3$, and 5000 ppm.

The solid foam sticks can stand-up high at completion accessories such as landing profiles, side-pocket mandrels, etc. Hence, foam sticks are best suited in case of slick completions without upsets.

8.8.3 Annulus or capillary

Liquid foamer can be delivered downhole in a controlled manner by pumping it through a dedicated capillary string or pumping it down a dead annulus, as illustrated in Fig. 8.13. The capillary string is either external control line strapped onto the original tubing completion or installed as and when needed inside the original tubing or retrofit VS (Fig. 8.13A–D). The dead annulus is either the original (external) A-annulus or is the new (internal) annulus created between VS and original tubing (Fig. 8.13E and F).

8.8.3.1 External annulus (A-annulus)

This option is seldom used. First, external or A-annuli are not always accessible deep enough due to the presence of a packer or cement. Second, downhole injection valves between tubing and annulus that qualify as a pressure barrier are seldom installed and difficult to maintain. Third, the A-annulus casing is usually carbon steel and therefore susceptible to foamer corrosion, creating corrosion products that will block the injection valve.

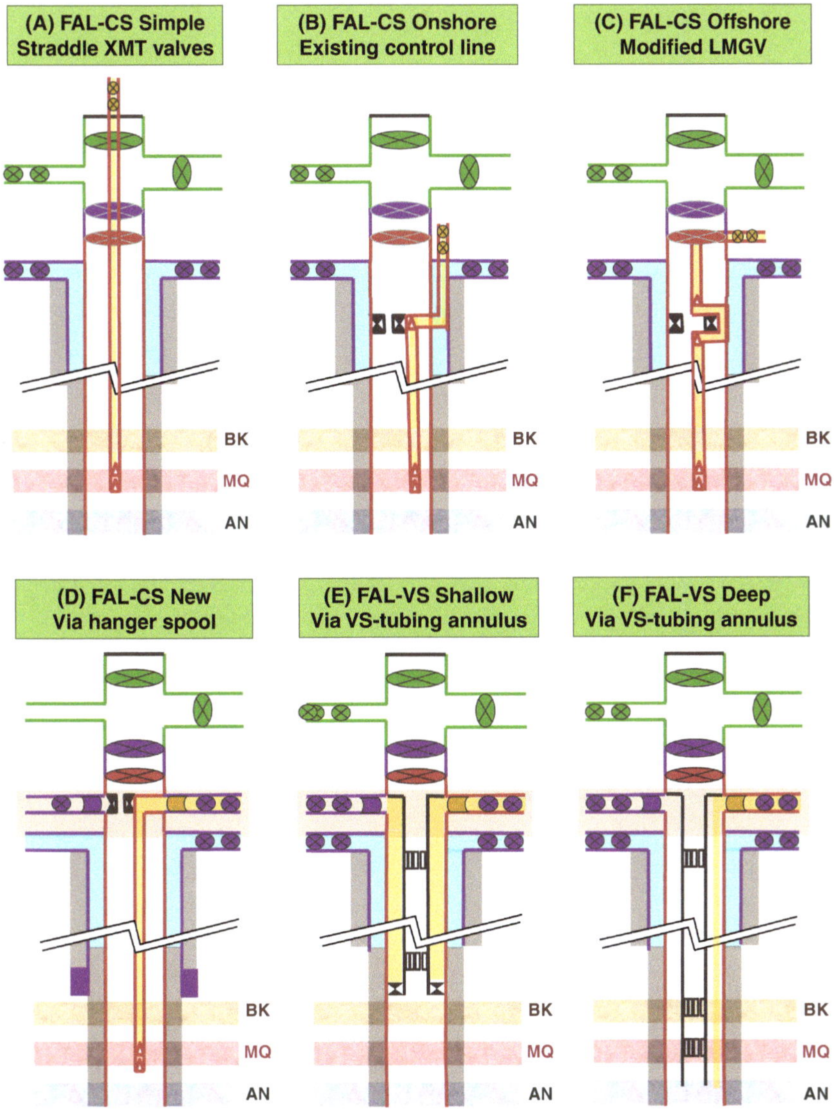

FIGURE 8.13

Downhole delivery of liquid foamer via capillary string or dead annulus (example reservoir consisting of three separate intervals): (A) via capillaty straddling XMT valves, (B) via capillary tied to SSSV control line, (C) via capillaty tied to modified LMGV, (D) via capillary tied to new hanger spool, (E) via VS-tubing closed annulus tied to new hanger spool, (F) via VS-tubing open annulus tied to new hanger spool.

8.8.3.2 Internal annulus (new annulus)

When available, this option can save the cost of installing a capillary string. It requires an insert or VS installed from surface to allow access to this internal (new) annulus (Fig. 8.13E and F). The dead annulus is shut in, gas filled, and connected to the flowing part of the completion. The liquid foamer that is injected at the wellhead drips down the walls of the original tubing and insert string before reaching the string end or exit where it enters the combined flow of gas and free water and condensate. The dripping down takes time: assuming a fall rate of 100 m/h, the well response to changes in foamer injection rate can take 1–2 days. Part of the foamer solvent will evaporate and some of the foamer may plate out on the tubing wall. Nonetheless, the process has proven to be effective and is preferred when such an insert string and dead annulus is present from the surface down to a suitable foamer injection depth.

8.8.3.3 External control line

Sometimes, control line or capillary tubing is strapped onto the original tubing completion to provide a future conduit for a production chemical such as a scale or corrosion inhibitor. The same capillary can also be used to inject foamer or a combination product, for example, corrosion or scale inhibitor. This control line option could be preferred in case the injection valve depth is suitable.

8.8.3.4 Internal capillary

In absence of a suitable dead annulus or external control line, a capillary string can be installed inside the wellbore to deliver the foamer at a suitable depth (Fig. 8.13A–D). There are different solutions for different local requirements as sketched in Fig 13A–D and described below:

(A) Hanging the capillary string from the top of the Christmas (Xmas) tree (Fig. 8.13A [35,36] and Fig. 8.14). This presents the easiest and most used solution. One drawback is that the capillary string blocks all safety-critical Xmas tree valves. Hence, leak-testing these valves requires the string to be removed and reinstalled. The Xmas tree valves must be selected to cut standard 0.25 and 0.375″ capillary to ensure they function in case of an emergency.

(B) Hanging the capillary string from a hanger that is set across the existing subsurface safety valve (SSSV) profile (Fig. 8.13B [37–39] and Fig. 8.15). This solution was developed to allow foamer injection whilst retaining the functionality of the SSSV. It involves locking out the existing tubing-retrievable SSSV, if present, and installing a wireline-retrievable SSSV with a back pressure valve and capillary below. The back-pressure valve is set such that sufficiently high pressure is created to open the SSSV before allowing any foamer injection down the capillary to bottom. Hence, the foamer also serves as hydraulic control

FIGURE 8.14

Capillary string deployed from the top of Xmas tree.

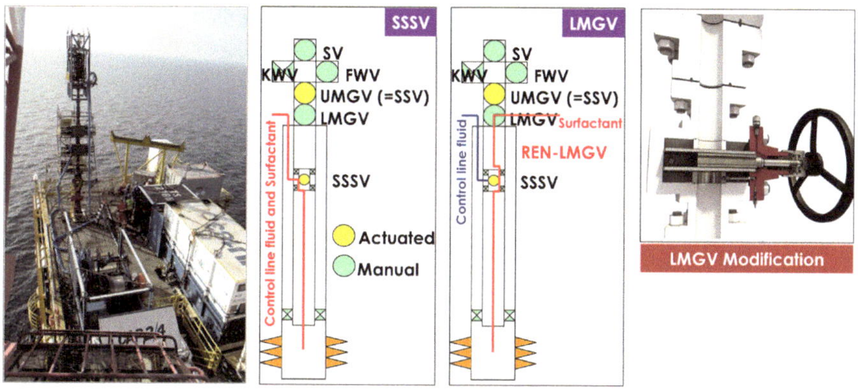

FIGURE 8.15

Capillary string deployed from SSSV profile and/or modified LMGV.

line fluid. The main drawback is the meantime to failure of the back pressure valve that varies between 0.5 and 5 years.

(C) Hanging the capillary string from the lower master gate valve (LMGV) (Fig. 13C [37,40], and Fig. 8.15). This solution was originally developed as a contingency to operate for SSSV's with blocked or leaking control lines. The same setup can be used to inject the foamer separate from the hydraulic control line fluid thus retaining the SSSV functionality whilst eliminating the need for a back pressure valve. This solution requires modifying the LMGV and installing a capillary string hanger.

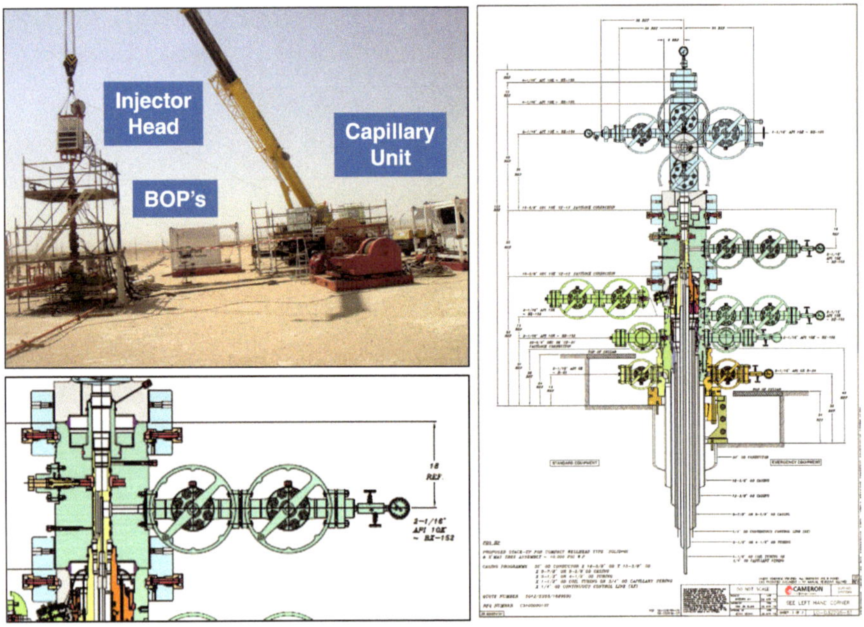

FIGURE 8.16

Capillary string deployed from additional hanger spool.

(D) Hanging the capillary string from an additional hanger spool (Figs. 8.13D and 8.16). An additional hanger spool needs to be inserted below the Xmas tree. This involves modification of the hookup as the Xmas tree will need to be raised.

Solutions (A) and (D) are most straightforward and preferred, but do not offer the SSSV functionality that is mandatory in offshore and some onshore wells. Solutions (B) and (C) can provide a working SSSV. Note that more solutions exist such as the Y-configuration presented by Bolding et al. [41].

With exception of (A) all capillary solutions involve a hanger that is exposed to solids production. These hangers may experience erosion in case of significant production of sand or proppant. High strength and corrosion-resistant material such as Inconel provides the most protection against erosion.

8.8.4 Capillary specifications

8.8.4.1 Depth of capillary

In almost all continuous FAL applications, the surfactant is introduced into the flow of gas and liquid at one specific depth. When injecting via a dead annulus, this depth is controlled by the original or retrofit VS completion details. For example, in the case of a VS, the depth of that string will dictate where foaming first occurs. Similarly, in case of an external control line, the injection depth is also fixed. However, when installing an internal capillary, the injection depth can be deliberately chosen for maximum impact.

When the producing interval is short (less than 30 m or 100 ft.), the injection location is the best chosen short distance (10 m or 30 ft.) above the top of the inflow zone. This location prevents any erosion by produced solids that could impinge sideways on the string. It also avoids trapping the string in produced solids (sand, proppant) or scale (halite, calcite). The location above the producing interval ensures maximum agitation and hence best creation of foam. The wellbore section that is not being treated by foamer is too short to significantly impact outflow performance.

The situation changes when the producing interval is long (more than 100 m or 300 ft.). In that case, liquid loading causes significant hydrostatic pressure across the untreated inflow section and could impair or kill production from deeper zones. In such cases, velocity strings are being installed as deep as practical to straddle the entire producing interval [42]. Similarly, capillary strings should be installed deeper as well but not all the way down because agitation becomes zero near the bottom of the producing interval. Hence, one should pick a location where production is still significant enough to ensure that the foamer will cause foaming and liquid lift. In practice, one would set the capillary above the deepest zone of significant inflow, provided that the prior mentioned risks of solid erosion and solid fill can be managed.

In late field life, liquid loading is often accompanied by another flow assurance issue that consists of downhole deposition of halite (NaCl salt) caused by evaporative scaling. This halite scale will block production and needs to be prevented or removed to realize full production potential. Similar to deliquification, halite mitigation usually consists of batch pumping fresh water from the surface or continuously pumping fresh water either via a dead annulus of a capillary string. Considering the last option, the injection depth should be below the depth where halite is deposited.

Hence, when simultaneously mitigating liquid loading and halite scaling, halite deposition may dictate a deeper injection point, also in case of a shorter interval. Such combined treatment consists of pumping foamer diluted in water. The required water rate is 10–100 times larger than the foamer rate. This may dictate the use of a larger capillary string (e.g., 3/8″ instead of 1/4″, or 1/2″ instead of 3/8″) and a larger injection pump delivering a higher rate at higher pressure.

Distributed surfactant injection systems have been proposed to promote foaming at multiple inflow locations. This introduces unnecessary complexity as all foamer will eventually traverse upward and treat the entire wellbore located above the injection point.

8.8.4.2 Capillary size and material

The capillary string needs to be large enough to deliver the required amount of surfactant to the chosen downhole location. Eq. (8.7) can be used to

Table 8.5 Maximum vertical depth as function of capillary string specifications.

OD/ID \\ YS (kpsi)	90	105	120
0.25″/0.18″	4380 m	4760 m	5330 m
0.375″/0.245″	5105 m	5724 m	7087 m

estimate that foamer injection rate. Experience tells that most injection rates are within the range of 1–100 L/d and can be accommodated by a common 0.25″ OD control line or capillary string. Higher injection rates generally require 0.375″ OD string, especially when FAL is combined with water injection to prevent and dissolve halite. The combination of a low enough injection rate and a large enough capillary string size will eliminate the surface injection pressure when the foamer hydrostatic pressure exceeds the foamer friction pressure plus the downhole (depleted) reservoir pressure. In many cases, the foamer will siphon into the capillary string without requiring any pumping pressure.

The capillary string material is chosen for yield strength and corrosion resistance. Yield strength is required to deploy the string to its target depth. Table 8.5 illustrates how the maximum depth of deployment is impacted by yield strength; depths over 5 km can only be reached using high-strength material and/or a larger string size. Corrosion resistance is essential because most foamers are slightly corrosive and will contain dissolved oxygen as well.

The foamer becomes less or not corrosive after exiting the capillary string into the wellbore due to dilution in gas and liquid. The outside of the capillary string is exposed to wellbore fluid and needs to resist the ambient chloride level and temperature. Stainless steel has been shown to easily crack and fail. Inconel alloys are favorite as they offer superior corrosion resistance but also most expensive. Super duplex material is generally adequate but has been shown to fail under certain downhole conditions as well [36].

The capillary string hanger and bottomhole assembly material must be corrosion resistant as well. If there is a likelihood of solid production, a stronger hanger material may be selected to maximize erosion resistance. The capillary bottomhole assembly is standard equipped with two check valves (or one check valve and one back-pressure valve) to prevent backflow of gas up the capillary during installation and operation. These valves will start leaking after 1–10 years and can trigger well intervention for repair.

8.8.5 Surface injection system

The surface injection system depends on whether FAL uses solid or liquid foamer and whether the injection is continuous or intermittent.

8.8.5.1 Solid injection

Foam sticks can be inserted into the well manually or using an automated launching system (Fig. 8.17). One needs to take the safety aspects of soap stick deployment into consideration. Especially the manual means of deployment is prone to error and accident. The automated launching system relies on proper operation of sealing control valves.

The number of sticks that can be deployed is limited by the number of sticks in the launcher and the supply schedule. The practical number of sticks is limited to 10–30 per day, which corresponds to 3–10 L/d surfactant.

8.8.5.2 Continuous liquid injection

The surface liquid injection system must deliver a constant flow of foamer at sufficient rate and pressure. The injection system could be designed for a single (e.g., onshore) well or could be laid out to simultaneously serve a collection of (e.g., offshore) wells. The system includes a storage tank and an injection pump. In most cases, the injection pump is designed for individual wells as distributing the flow from a single pump across multiple wells introduces control complications whilst not saving much cost. The storage tank should be shared if practical as this simplifies and reduces supply operations. Fig. 8.18 shows examples of single-well injection skids catering for foamer, antifoamer, and/or water injection.

FIGURE 8.17

Automated foam stick launcher.

https://www.aquasolwelding.com/solutube (left), https://www.prosealinc.com/model-9e-18e-solar-powered (right).

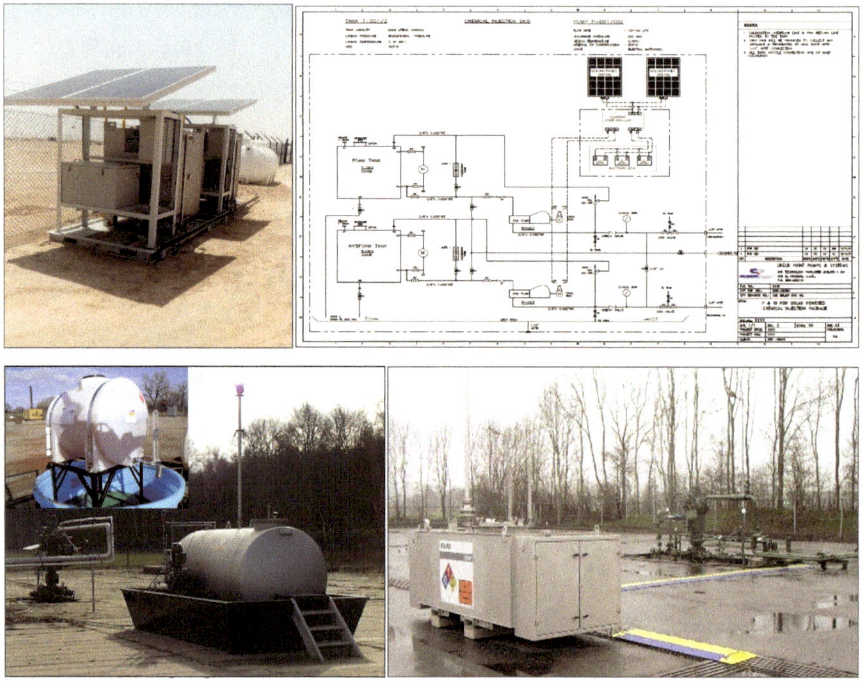

FIGURE 8.18

Single well injection skids.

Note that water would be injected to flush the string prior to a long shut-in period to prevent foamer drying out and plugging the string.

The continuous FAL injection pump is typically low rate (1–100 L/d) and medium pressure (20–200 bara), similar to injection pumps used for other process chemicals such as scale and corrosion inhibitors. The pumps are pneumatic or electric piston-type requiring only small amounts of instrument air or (solar) electric power. The material of pump and pipework should be 316 L stainless steel for corrosion resistance. The pump should feature means of controlling and measuring foamer consumption and must include nonreturn or check valves at the outlet to prevent back-flow of high-pressure gas from the well.

The pump pressure preferably exceeds the CITHP so injection can be resumed in case the string has been invaded with reservoir gas after a long shut-in period. Lower injection pressure should still work as long as the check valves are functioning: in that case, the gas content of the string can be lubricated to foamer by repeated bleed down and fill up. Depending on pump type, an anti-siphon valve may be required to prevent uncontrolled dumping of foamer down the capillary string when the capillary goes on vacuum.

The size of the storage tank depends on the number of wells served and the supply frequency. Onshore, regular supply may not be a big issue but offshore, logistics may dictate less frequent supply visits and keeping a larger stock. The storage tank should not be oversized to prevent storing foamer for more than 1 year at hostile conditions (temperature, sunlight, dust, etc.). Generally, these storage tanks are not provided with nitrogen blanketing, which means that the foamer will contain dissolved oxygen.

The foamer fluid must be kept clean and free of solids and bacteria. The storage tank should be regularly inspected to make sure no mineral solids or bacterial slime are accumulating and must be cleaned if necessary. The injection skid must be equipped with fine filters to avoid solids or residue entering the capillary string where they can cause plugging or damage of the downhole check valves. The injection skids and pumps are generally made of stainless steel; carbon steel must be avoided for the risk of corrosion. The atmospheric storage tank can also be made of glass fiber material.

8.8.5.3 Intermittent liquid injection

In case of intermittent FAL, the pumping job typically consists of 10–100 L of foamer. When there is no dead annulus, control line, or capillary string to pump through, the foamer is pumped into the closed-in wellbore from top chased by 50–200 L of water to help flush it down. The pumping time should be limited to reduce the associated cost of pumping services and to reduce the time that the well is shut-in waiting for the foamer plus water to reach bottom. Hence, the 100 L/d pump capacity quoted for continuous FAL is not sufficient and 1–10 m3/d capacity is more appropriate. The required pump pressure is equal to the CITHP and generally ranges between 50 and 200 bara. Such intermittent pumping services will need to be shared between customers. If demand is high, a dedicated resource could be contracted specifically for this type of FAL treatment (Fig. 8.19).

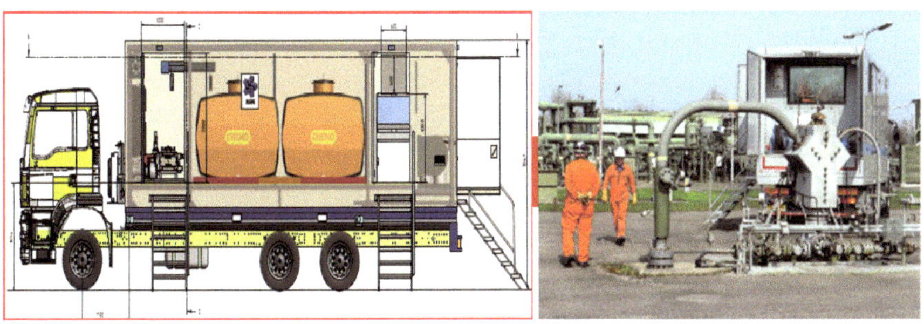

FIGURE 8.19

Dedicated batch FAL pump truck.

Rauf et al. [43] described squeeze treatments where large volumes of foamer, water, and compressed gas are injected into fractured reservoir to remove deep halite scale damage and provide a long-lasting release of surfactant for liquid lift. The large liquid volume of 5–50 m^3 is bullheaded down the well using a regular stimulation pump truck with about 1 m^3/min pump capacity. The same type of truck is used to carry out fresh water batch injection treatments for removing downhole halite scale [44].

8.8.5.4 Antifoamer, demulsifier, and clarifier

As highlighted before, FAL results in the production of residual surfactant and foam, first into the flowline and eventually into the surface processing system. When foam is carried over into the gas stream, it can cause serious upsets of the glycol dehydration system. When surfactant is present in the liquid stream, it can result in emulsion and excessive OIW that could compromise oil export and water disposal, the latter specifically in an offshore setting with an OIW limit of 10 ppm or less.

These surface processing issues are exacerbated by higher surfactant concentration and shorter distance between well and separator. Long flowlines and pipelines between wells and facilities will help kill the surfactant and foam. Diluting the production of FAL-treated wells with nontreated wells also helps reduce the impact of the surfactant and foam. Experience to data indicates that no issues are expected for the combination of long flowlines and surfactant concentrations below 1000 ppm. Most processing upsets occur where the distance between wells and separators is very short.

The surface system performance needs to be closely monitored especially in the early days of FAL application. Samples should be regularly collected at relevant locations and tested for the presence of foam, emulsion, and OIW. If necessary, antifoamer, demulsifier, and clarifier can be injected at suitable surface locations to remediate foam and emulsion issues. Also, the foamer type can be changed to improve the situation. This specifically applies to excessive OIW that may be difficult to rectify any other way. The use of these additional process chemicals increases the operating cost of FAL.

8.9 Foam-assisted lift operation

With the exception of an isolated FAL kick-off job, any other continuous or intermittent FAL operation must be sustained for several years to deliver the expected benefits. Such a long-term operation involves regular performance and system monitoring and optimization. The necessary long-term commitment is articulated below, illustrated by field examples.

The time that FAL should last can be estimated by decline curve analysis where the stable gas well rate is extrapolated into the future. The FAL duration is

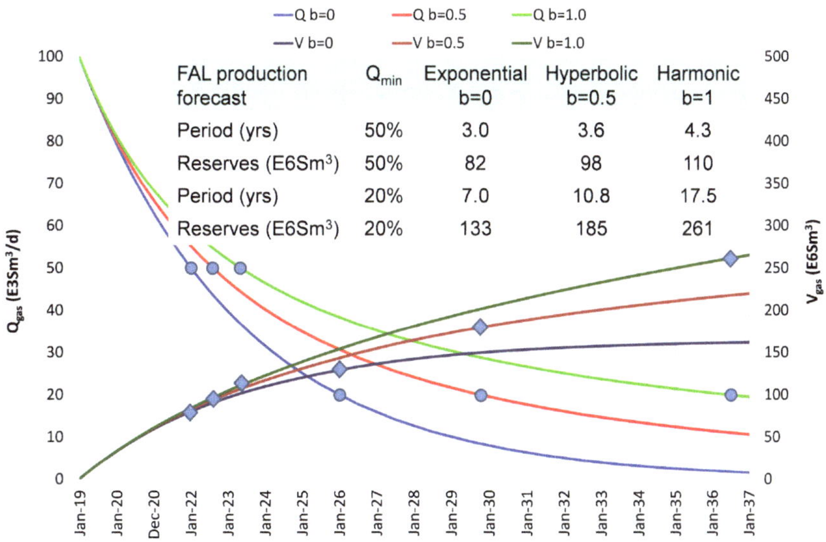

FAL production forecast	Q_{min}	Exponential b=0	Hyperbolic b=0.5	Harmonic b=1
Period (yrs)	50%	3.0	3.6	4.3
Reserves (E6Sm³)	50%	82	98	110
Period (yrs)	20%	7.0	10.8	17.5
Reserves (E6Sm³)	20%	133	185	261

FIGURE 8.20

Comparing exponential (b = 0), hyperbolic (b = 0.5), and harmonic (b = 1) production decline.

the period between the date that Q_{gas} reaches Q_{min} without FAL and the date that Q_{gas} reaches the reduced Q_{min} with FAL. The more FAL reduces Q_{min}, the longer FAL will last, and the more benefits it will generate.

Decline curve analysis assumes a certain type of decline such as exponential, hyperbolic, or harmonic. Exponential decline is fastest and hence dictates the shortest FAL period and lowest reserves benefit, while harmonic decline is slowest and results in the longest FAL period and largest reserves benefit. From experience, the future production of mature and depleted gas wells is best described by an exponential decline. Fig. 8.20 illustrates the difference between the different types of decline and demonstrates the impact on both FAL period and FAL benefit. Exponential decline results in the most conservative and robust production forecast and project economics. Note that a fivefold reduction of Q_{min} generates about two times more reserves than a twofold reduction and takes three times as long to realize.

8.9.1 Optimizing foam-assisted lift

The successful response to FAL falls in one of the following three categories:

1. By means of intermittent FAL, the well is able to kick-off and clean up to a stable gas rate above Q_{min} without FAL. Fig. 8.21 shows an example of such

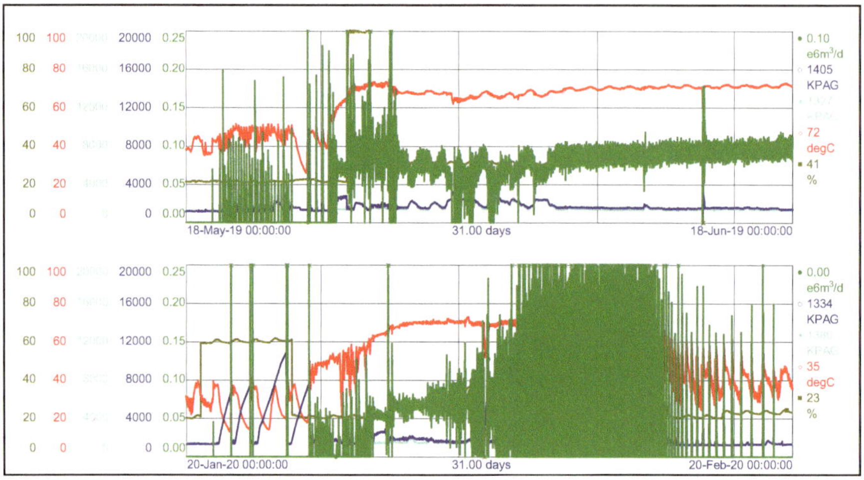

FIGURE 8.21

Intermittent kick-off of water producing gas well: first kick-off in 2019 restores stable production but second kick-off in 2020 does not.

kick-off and cleanup. In this case, the intermittent FAL treatment was carried out using a temporary capillary string, which allowed continuous foamer injection over a period of several days, during which the well was cycled on and off. The problem in this well is increasing formation water production that gradually increases both WGR and A. The kick-off in 2019 is able to restore the stable production but the kick-off in 2020 does not achieve the stable production as well performance further deteriorated (WGR and A increased) in the meantime.

2. By means of intermittent FAL, the well is able to continue production at a reasonable gas rate below Q_{min}. Fig. 8.22 shows an example of such intermittent FAL, followed by continuous FAL. After pumping a batch of liquid foamer, the gas rate reduces to half within 1 week, after which the next weekly batch treatment helps to remove the accumulated liquid to restore liquid-free gas production. The BHP data reflects this gradual liquid loading and rapid liquid removal process during intermittent FAL and shows how the subsequent continuous FAL prevents liquid loading and thus achieves a permanent low BHP.
3. By means of continuous FAL, the well is able to produce at a stable rate below Q_{min}. Figs. 8.22–8.25 present examples where FAL restores stable production.

Fig. 8.23 is a gas well producing at very high WGR that cannot produce at full potential without continuous FAL. As soon as the foamer injection stops, the gas rate declines within a matter of hours. In this well, 8 years of continuous FAL

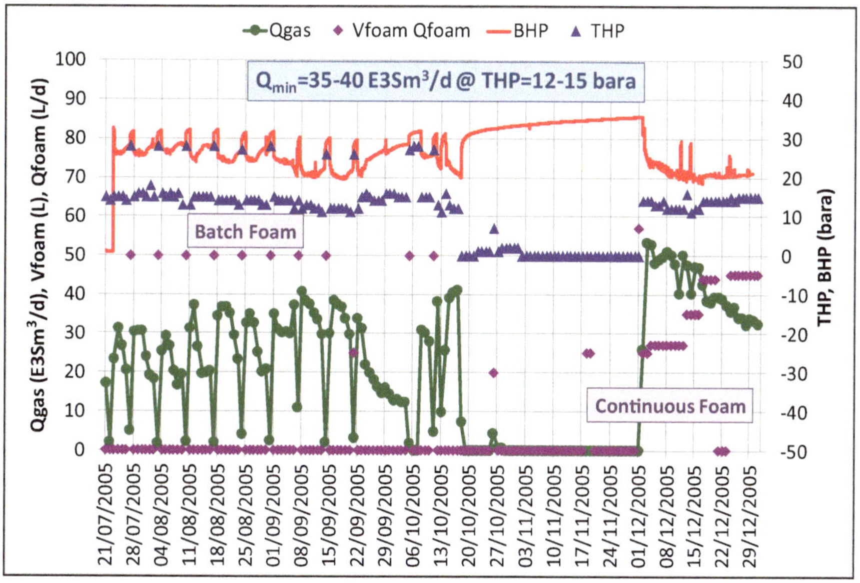

FIGURE 8.22

Gas well response during intermittent FAL followed by continuous FAL.

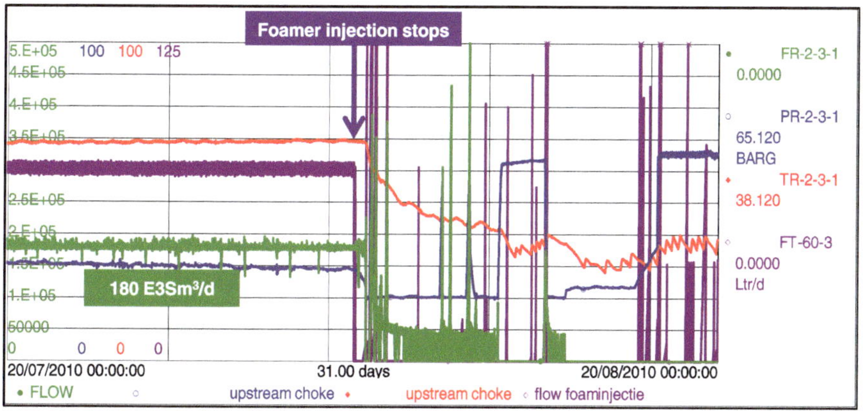

FIGURE 8.23

Gas well response during and after continuous FAL.

enabled 200 E6Sm^3 (7.1 Bscf) incremental production. Fig. 8.24 shows the production of an offshore gas well without and with FAL. Continuous FAL increases average gas production 9 times and liquid production 43 times [45]. In the example of Fig. 8.25, continuous FAL increases gas production 1.5 times and condensate and water production 4.0 times.

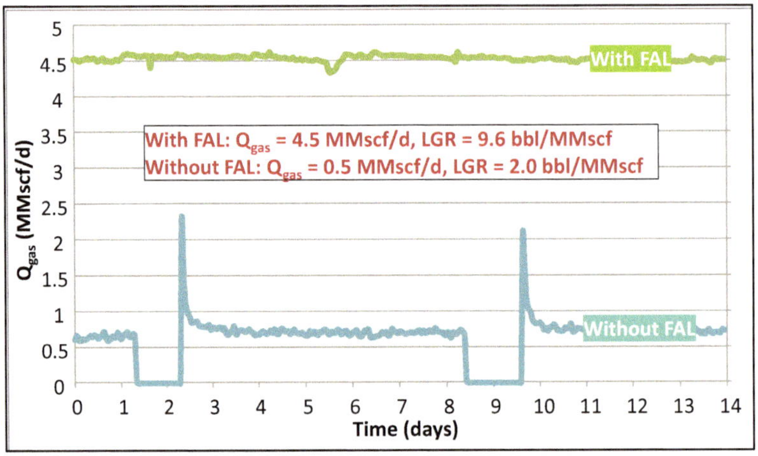

FIGURE 8.24

Gas well response without and with continuous FAL.

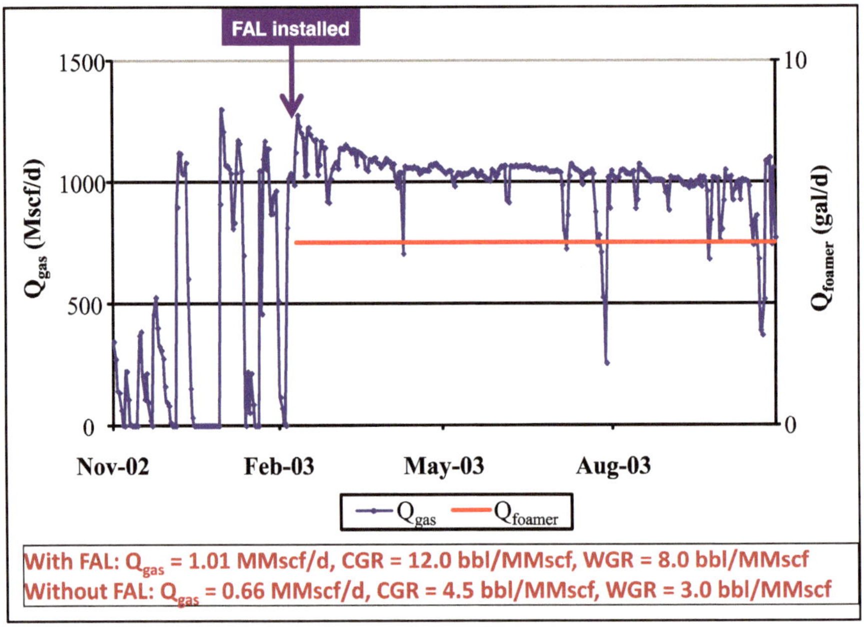

FIGURE 8.25

Gas well response without and with continuous FAL.

Figs. 8.23–8.25 illustrate that liquid loading triggers a significant loss of well capacity but does not entirely eliminate production. The residual production without FAL corresponds to either slug flow where some liquid still reaches surface but the flow regime increases liquid holdup and BHP which decreases Q_{gas} [46], or metastable flow where all liquid drains away downhole but a column of bubbly liquid remains, increasing BHP and reducing Q_{gas} [47]. Based on liquid production, Fig. 8.24 without FAL corresponds to metastable flow (43 times less liquid) while Fig. 8.25 prior to FAL resembles slug flow (4 times less liquid).

Veeken and De Vries [34] analyzed the continuous FAL performance in a collection of wells and concluded that continuous FAL was successful in 90% of wells and able to reduce Q_{min} 2–3 times. A similar conclusion was reached by Veeken et al. [15] after carrying out extensive testing of continuous FAL in five wells. In each well, Q_{min} was successfully reduced at least two times using one of the available foamers. The foamers used in the above studies were selected based on extensive lab testing, which further supports the use of such prescreening even though the testing situation does not represent reality in terms of flow regime.

Ideally, the foamer injection rate is tuned such that it sustains Q_{min} just below the producing Q_{gas}. Injecting not enough surfactant does not reduce Q_{min} enough leading to liquid loading while injecting too much surfactant wastes costly chemical and may cause or exacerbate surface processing issues. With progressive depletion, Q_{gas} slowly declines, which means that the surfactant concentration should be slowly increased so Q_{min} can be reduced in sync.

Veeken and De Vries [34] recorded successful surfactant concentrations between 500 and 15,000 ppm, depending on well characteristics. The required concentration increases with the required reduction of Q_{min} but also depends on BS&W (lower BS&W requires higher concentration), condensate type (different condensates can cause different degrees of anti-foaming), and formation water composition (more salinity and hardness require higher concentration). The calculated surfactant concentration also reflects uncertainty in the assumed CGR and WGR when well test data is not available or outdated.

In practice, the optimum foamer injection rate could vary between 200 and 20,000 ppm, which needs to be established through field optimization, starting for example at 2000 ppm for continuous FAL and 10,000 ppm for intermittent FAL. Note that a surfactant concentration as low as 100 ppm can already make a difference. In the example of the high WGR gas well in Fig. 8.23, the surfactant concentration was about 600 ppm, while a successful FAGL operation by Martins et al. [22] applied FAGL in a high water cut oil well with a foamer concentration of about 400 ppm.

8.9.2 Sustaining foam-assisted lift

The value of FAL can only be realized if the success of FAL is sustained over the entire FAL period. Field experience indicates that this requires consistent attention for all operating aspects. The most critical aspects are discussed below.

8.9.2.1 Injection rate

The foamer supply must be uninterrupted while the well is being produced and foamer injection must stop when the well is shut in for a longer period. This requires automation and operator attention. The drivers are to minimize the consumption of foamer chemical, to prevent capillary string blockage, and repair to minimize well OPEX. Optimizing the foamer injection rate is well worth the effort; this requires both proactive variation of the foamer rate and diligent evaluation of the associated well response.

8.9.2.2 Capillary string

The capillary string surface pressure needs to be monitored. When injection is stopped and the surface pressure is observed to increase, reservoir gas is entering the capillary string through leaking check valves or hanger seals. In that case, the string may need to be pulled and repaired. The back-pressure valve installed to operate the SSSV (Fig. 8.13B) will periodically fail and shut in the well; the string will need to be pulled and repaired to restore production.

In case of a long shut-in period, the capillary string should be flushed and displaced with water or glycol to prevent plugging of the string due to foamer solvent evaporation and gunking. Solids production (sand, proppant) of deposition (mineral scale such as halite) can erode or trap the capillary string necessitating intervention. It is good practice to carry out a capillary string pull test every year to make sure that the string is not locked in place. Downhole scale deposition can occur either independent of FAL or could be associated with the specific foamer being used.

8.9.2.3 Compatibility

Some foamers have been observed to cause downhole scaling when exiting the capillary string, caused by a reaction with reservoir fluids, most notably formation of water components. This scale will eventually trap the string and/or block production. Such a stuck string and wellbore blockage can be difficult to recover from. Acid injection can help dissolve the scale. The downhole FAL deposits in Fig. 8.26 consist of white Powellite or $CaMoO_4$ where the Mo was added to the foamer for corrosion inhibition and reacted with formation water components at the ruling BHT of 135°C.

Downhole gunking has also been observed when the foamer solvent evaporates and foamer residue deposits on wellbore and capillary surfaces. Water injection can help dissolve this solid foamer.

When first introducing foamer, foam in the flowline can mobilize and carry stagnant liquids, solids, emulsions, and corrosion products down the processing system. This could cause some issues in separating and treating this initial batch of gas, liquids, and solids but these issues should disappear after the surface system reaches a new equilibrium.

FIGURE 8.26

Downhole foamer deposits above injection point.

FIGURE 8.27

Foam and emulsion in flowline sample: foam on top disappears overtime but water-condensate emulsion requires demulsifier treatment to remove.

The presence of surfactant and foamer in the surface system can cause immediate problems in gas and liquid separation and processing, such as foam carry over, high water-in-condensate, and excessive OIW. These need to be managed by changing the foamer product or by introducing antifoamer and/or demulsifier at suitable locations. Fig. 8.27 shows the changes in time and with demulsifier of a flowline sample taken during continuous FAL [14].

The FAL surfactant can impact the performance of other surface-active process chemicals. Most relevant, the surface system integrity can be compromised if the foamer surfactant significantly reduces the corrosion inhibitor performance. Laboratory testing is essential to quantify this risk, which needs to be managed either by changing the foamer and corrosion inhibitor combination or by increasing the corrosion inhibitor dosage.

8.10 Case studies of successful foamer applications

8.10.1 Optimize horizontal wells with batch foamer treatment

Aided by advanced horizontal drilling and completion technologies, fueled by high oil price, there were drilling and production campaigns in oil-rich shale play in United States in 2010s. In the early production stage, a high throughput horizontal well could produce a few million cubic feet of natural gas and more than 1000 barrels of liquid hydrocarbon on a daily basis. However, horizontal wells were more susceptible to liquid loading compared to vertical gas wells, causing sharp production decline within the first few years if not treated. There are a few reasons for early liquid loading in horizontal wells and the main one is the complex and undulating wellbore trajectory of a horizontal well, which could cause earlier onset of liquid loading in horizontal wells [48]. In addition, horizontal wells in liquid hydrocarbon-rich shale play produce a large amount of fluids, which has a greater tendency to accumulate alone the long lateral wellbore and causes production decline earlier. How to effectively lift the large amount of loading liquids becomes critical in maintaining profit production and operation for these wells.

The horizontal lateral, well deviation, and presence of downhole packer limited the choice of artificial methods such as rod pump. The high oil cut of these wells makes traditional foamer not effective as they are designed for water-rich fluids. Chemical foamer can be a viable solution to deliquefy horizontal wells with a large amount of liquid hydrocarbon production only if the cost of the program is not prohibitive. The traditional foamer to foam oil is typically based on fluorinated surfactant, which is cost-prohibitive to treat a large amount of fluids. Other oil foamer such as silicone-based product would downgrade the crude value as it could upset refinery process. It takes innovative chemical foamer solution to be able to foam fluids with high oil cut while still meeting the financial goals of the investment. The case study below demonstrated the successful foamer batch treatment to enhance the production of horizontal wells.

8.10.1.1 Background and challenges

An operator was producing from a large number of horizontal wells in a liquid-rich shale play in United States. These wells were producing 2–3.5 million cubic feet per day of natural gas, and 600–1200 barrels per day of condensate prior to liquid loading. Liquid loading dramatically decreased the production, causing gas rate drop to 100–500 mcf/d and oil rate to 5–200 bbl/d. Due to the presence of hydrogen sulfide gas (H_2S) and packer, lack of GL mandrels, there were limited artificial lift options for the operator. The operator was resorting to frequent swabbing to keep these wells producing and cost of swab job could vary from $13,000 to $26,000. Furthermore, swabbing liquid from sour well exposed staff to hazardous H_2S, which was a huge safety concern for the operator. Limited benefit of swabbing was realized, and production diminished within a short period of time before another swabbing job was required. Some wells ceased production due to lack of effective artificial lift program [49].

The chemical supplier used a thorough chemical foamer selection process and the dynamic foaming test method to design and select foamer for each candidate well. Table 8.6 illustrates the performance of two foamers on fluids samples collected from five different wells in a liquid-rich shale play in United States. These wells made a good amount of condensate, ranging from 60% to 100% before ceased production. The produced water had medium salinity ranging from 25,000 to 60,000 ppm TDS.

Each oil sample had its own characteristics, which impacted foaming performance of the foamer. Foamers 1 and 3 were innovative foamer products based on conventional surfactants with synergistic formulation blends and are refinery friendly. Also, lab simulating testing revealed low foam tendency on the topside. However, as a best practice, defoamer was made available as start-up contingency during field trial.

8.10.1.2 Results

Three wells were chosen for field test, prior to that all are ceased production due to liquid loading. Well A repeatedly loaded up and had required multiple swab to kick off following previous load up. It had to be swabbed frequently. The swab cost more than 10,000 per treatment. Less than 1 month of production was realized after each swab.

Before each treatment, static fluid level was obtained within 3 days prior to field trial to estimate the fluid volume, which was used to calculate the needed

Table 8.6 Foamer recommendation based on dynamic foaming test.

Well	Oil Cut (%)	Recommended foamer	Recommended foamer dosage (ppm)
Well 1	85	Foamer 1	29,000
Well 2	60	Foamer 1	29,000
Well 3	100	Foamer 3	48,000
Well 4	100	Foamer 1	58,000
Well 5	100	Foamer 3	48,000

amount of foamers. The 50 gallons to 2 bbls of foamers were batched down the tubing followed by 2 bbls of potassium chloride solution flush, then the wells were shut in to allow to soak and build up pressure. In 24 hours post-treatment, the well was reopened, and the flow back fluids were monitored. What also were monitored were oil rates, gas rates, and topside separation.

Immediate positive response from the three foamer treated wells were observed as shown in Fig. 8.28. Well A responded well to the first foamer batch treatment with a large amount of oil was unloaded from the well before the well production was interrupted by central compressor downtime. As a result, well A had to be shut in causing wells to load up again. The second foamer batch treatment was carried out after compression was back up running and the well was unloaded successfully again. The second batch of foamer helped realized nearly 30 days production before its effectiveness was worn off. The third foamer batch treatment was conducted and the well responded positively again.

For well B, the previous artificial lift remedies used included swab and soap stick. When the soap stick was in use, slugging production pattern and liquid loading were noted within two weeks even after optimization (date not shown).

As shown in Fig. 8.29, after foamer was batched down to well B, sustained production more than 1 month was achieved for this well. The second batch was pumped down and the well responded positively.

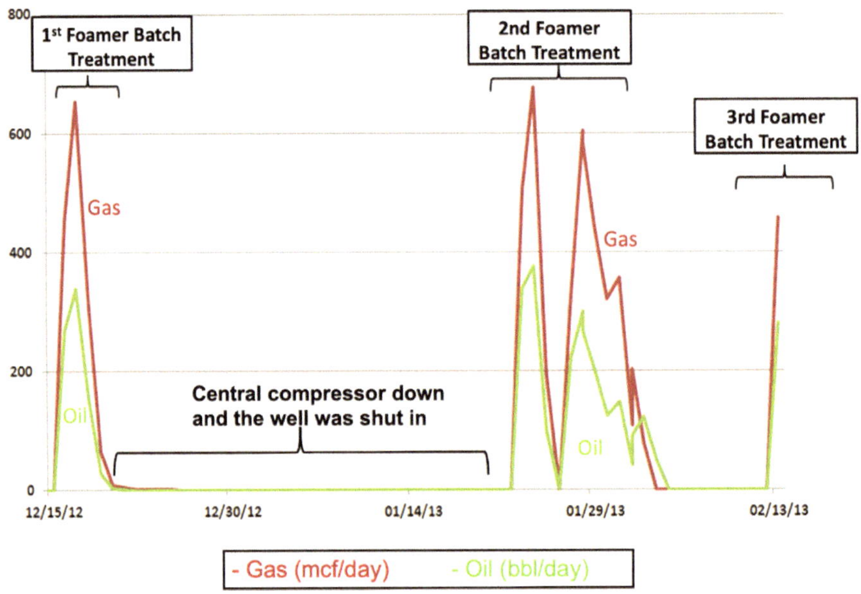

FIGURE 8.28

Results of three foamer batch treatments for well A.

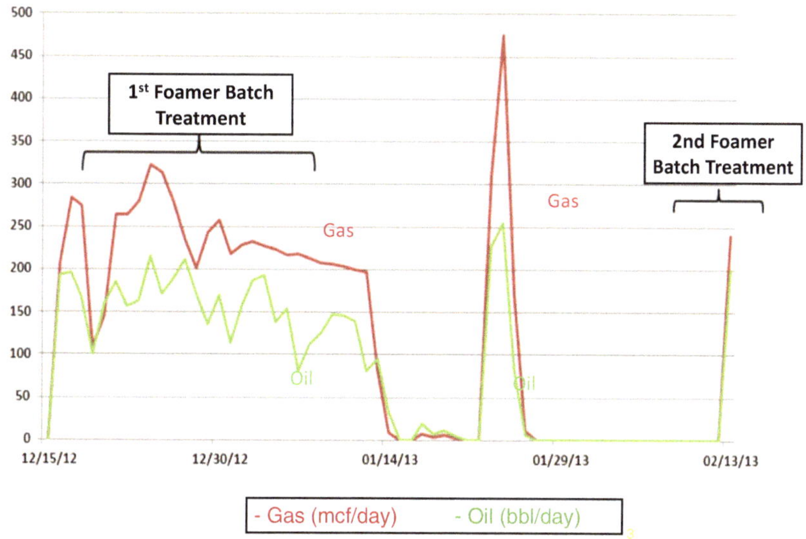

FIGURE 8.29

Results of three foamer batch treatments for well B.

well C, ceased production due to liquid loading. Attempts to cycle the well could not bring the well back to production prior to foamer treatment. The batch foamer treatment was able to sustain production for more than 1 month (Fig. 8.30).

Overall, the batch foamer treatment realized 200–400 bopd oil and 200–800 Mscf/d gas. There was significant cost saving to the operator using foamer as an alternative to swab. What noteworthy is that there was no upset topside caused by foamer application hence no need to apply defoamer on the surface process. Some foamer-treated wells had more than 1 month sustained production.

8.10.2 Continuous foam in conjunction with dry-gas-lift

This case study demonstrates the benefits of combining foam with GL to improve the overall artificial lift efficiency and enhance production [19].

8.10.2.1 Background

A producer in the North America was extracting gas in an aging field, using gas-lift to unload liquid from the tubing to enhance production. As the gas production continues to decline, it was becoming more and more challenging to balance between reinjecting gas for GL operation and saleable gas for revenue. As a result, some wells though on GL, were not optimized to produce at their most potential due to the limited gas supply for GL. The operator attempted to use conventional foamers and injected them into the lift gas, which was hot and dry. The

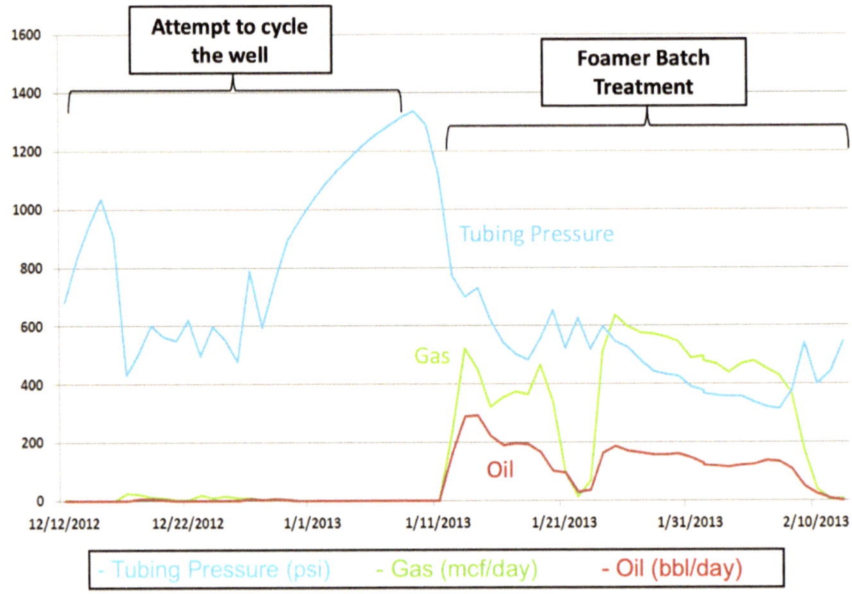

FIGURE 8.30

Production rate and tubing pressure of well C, before and after batch foamer treatment.

conventional foamer flashed off a large amount of solvent, leaving solid deposit to cause operational issues. The operator then worked with a chemical vendor who specifically designed a foamer, which not only had a good foaming performance under the simulated field conditions but also passed compatibility and stability test to work in conjunction with dry gas. A schematic illustration of GL mandrel in the production tubing is shown in Fig. 8.31. The foamer was injected to the GL mandrel continuously to maximize the benefit of foamer program.

8.10.2.2 Results

When GL foamer was injected into the dry gas-lift mandrel, the enhanced production was observed (Fig. 8.32). While experimenting with conventional foamer not designed for hot dry gas-lift mandrel injection, sharper production decline was noticed, indicating the inefficiency of the foamer, which also caused operational challenges.

As the treated well was optimized to maximize the production using the dry-gas compatible foamer, stabilized water production was noted too. With use of the dry-gas compatible foamer, the lift performance was enhanced for the foamer-treated gas-lift wells, the operator could divert gas to other loaded wells in the field, improving the overall operation, and profit of their production in that field.

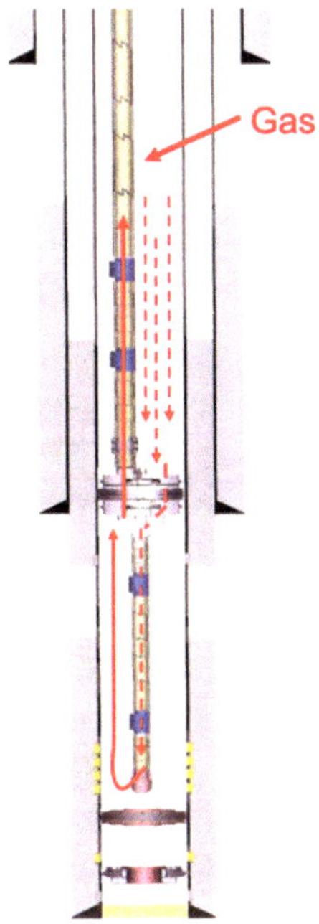

FIGURE 8.31

Schematic illustration of gas lift mandrel.

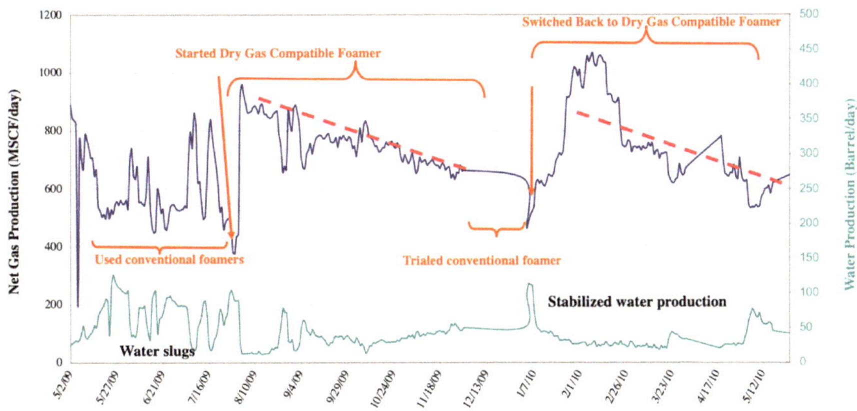

FIGURE 8.32

Enhanced gas production and stabilized water production achieved by combining compatible foamer with dry gas lift.

8.10.3 Innovative foamer as sustainable deliquification solution

Coal seam gas (CSG) production was a vital energy source, especially in parts of the world like Australia. As part of the CSG well lifecycle, dewatering can be required at start-up or in maturation, requiring a form of artificial lift. Chemical foamers were a well-proven technology for this purpose. Coal bed methane produces a large amount of water, and reuse is an important environmental consideration, especially for areas with water restrictions such as Australia, where government authorities require the produced water to be treated to a level to enable beneficial reuse in various applications such as agriculture, mining, irrigation, etc.

8.10.3.1 Background

A prominent CSG producer in the Surat Basin of Queensland Australia was experiencing reduced gas flow in several wells due to liquid loading after 5 years of production. The produced water was treated with reverse osmosis (RO) membrane to enable reuse. RO membranes are sensitive to chemicals and typical foamers employed to accelerate the water unloading generally fouls and destroys the membrane. This limited the choice of compatible foamer to support the sustainable operation. Conventional foamers failed the membrane compatibility test (Fig. 8.33), so an innovative foamer product was required [51].

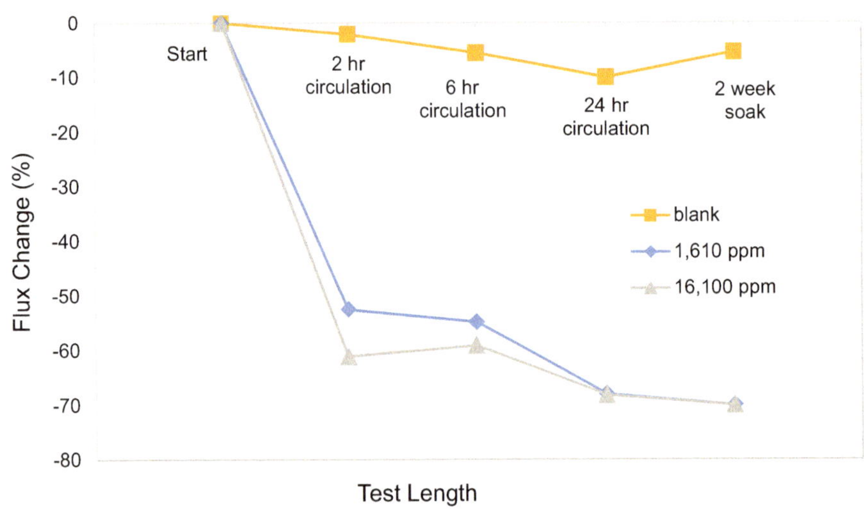

FIGURE 8.33

Best-in-class conventional foamer A negatively impacted the flux performance of RO membrane at its typical dosage of 1610 and 16,100 ppm.

Reproduced from F. Huang, D. Nguyen, Foamer for downhole injection, Patent US8950494, 2015 [50].

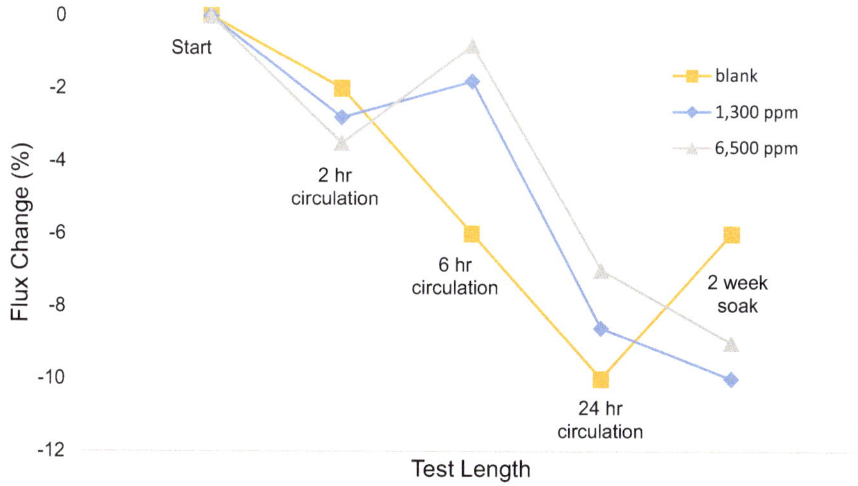

FIGURE 8.34

Flux performance of RO membrane compatible foamer B at its typical dosages of 1300 and 6500 ppm.

Reproduced from F. Huang, D. Nguyen, Foamer for downhole injection, Patent US8950494, 2015 [50].

The chemical and service provider studied the RO membrane properties and developed an innovative foamer that was not negatively impacting the flux performance of the RO membrane as illustrated in Fig. 8.34. The foamer was also effectively foaming brine with fine solids presence.

8.10.3.2 Results

The RO membrane-compatible foamer was the chosen foamer for field trial based on lab results. The candidate well was suffering a production decline prior to foamer injection. Foamer B was injected into the well at an initial rate of 24 L/d for 12 hours. As this product is proven to be RO membrane-compatible, the operator had no concerns about the chemical going through the water treatment plant.

The production pattern changed upon injection of foamer (Fig. 8.35), clearly identifying the uplift from the foamer treatment compared to the 3 months historical production data. As the treatment period progressed, the production rate reached a steady state and the production increase sustained.

Throughout the application, foamer treat rate optimization took place resulting in the final treat rate being 12 L/d. There was no upset on the water treatment facility. Foamer B was proven to be compatible with RO membranes and a sustainable deliquification solution with excellent unloading performance. The well response was immediately realized, halting the production decline, and returning rates to its optimal levels.

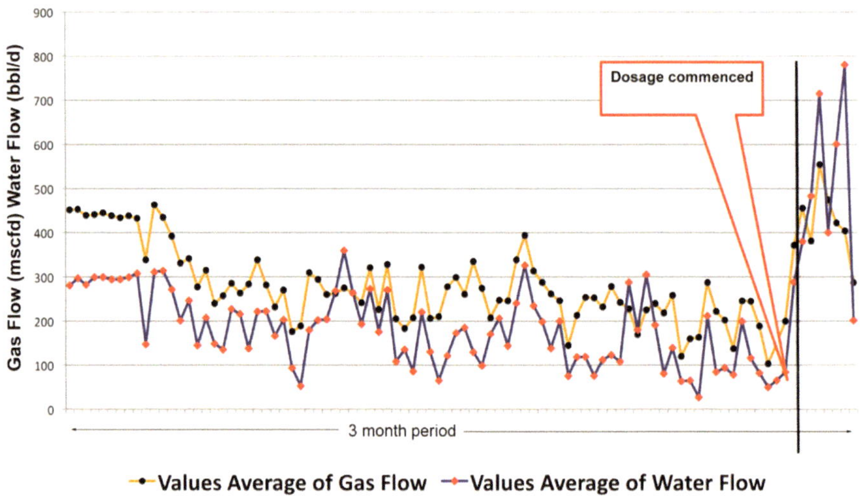

FIGURE 8.35

Average gas and water rates before and after foamer injection for well 1.

8.10.4 Use foam to remove liquid in subsea flow lines and enhance production

As the reservoir pressure of subsea gas wells depletes, gas rate continues to drop while the liquid rate may increase. As a consequence, liquid accumulation can occur in the undulating and long subsea flow lines, which in-turn will block the flow of the gas and adversely hinder the production. In addition, the liquid accumulation in the subsea flow lines causes subsea riser flow instability, leading to undesirable operational disadvantages such as cyclic slugging on the surface, upset in the surface separation, off-specification produced water quality, etc. The liquid accumulation also poses a severe corrosion threat to production assets if the liquid contains water with a high TDS level. Hence, efficient and effective removal of liquid from the subsea flow lines is critical, not only in extending the tail-end production of the asset but also in minimizing operational challenges in offshore production.

Choking, pigging, and GL are the traditional methods to deliquefy subsea multiphase flow lines and mitigate flow instability. Choking and pigging are carried out at the expense of reduced or temporary loss of production. GL often requires the wells to produce enough to sustain continuous flow GL, which can be challenging for an asset that already suffers from production drop caused by liquid loading.

This case study demonstrated a novel foamer could be a cost-effective alternative, to remove the liquid accumulation in the subsea flow lines, reduce the back pressure, and enhance the production of the asset [26].

8.10.4.1 Background

An operator producing natural gas from 16 subsea wells located in 500–1500 m of water in the Krishna–Godavari basin offshore eastern India was plagued with production problems related to accumulation of produced liquids in the production system. Natural depletion of the reservoir was causing water cuts to increase and resulted in liquid accumulation in the production streams of well A and well B (Fig. 8.36). The operator was transporting produced gas and liquids from well A and well B through 10 in. (7 degree slope) flow lines into a subsea manifold and then through 16-in. subsea pipeline (2 degree slope) to a subsea pipeline end manifold (PLEM). Furthermore, the increased back pressure in the flowline of these two wells, in turn, significantly reduced the natural gas production rates and threatened the economic viability of the offshore asset. According to the operator, a stratified flow regime existed in these two near horizontal legs of the production system, with a top gaseous phase flowing at a faster rate than a bottom liquid phase. Well A was found to be the weakest well having a continuous depleting production pattern due to loading of produced water at the flow line. The production was approximated around 0.32 MMscf/d with an increasing pressure at the wellhead, which is mostly due to the increase in pressure at the manifold and deep water pipeline end manifold (DWPLEM). This will create a negative pressure, which in turn will make the unloading job easier for well A. The operator needed to find a way to resolve pressure anomalies underlying the liquid loading and restore the production of well A and well B.

Based upon information provided by the operator, the chemical and service provider concluded the stratified flow between the wells and PLEM was a significant contributing factor to

Decreased gas production rate. If liquid loading could be remediated in the subsea flow lines, then pressure anomalies would be resolved and well A and well B could be optimized.

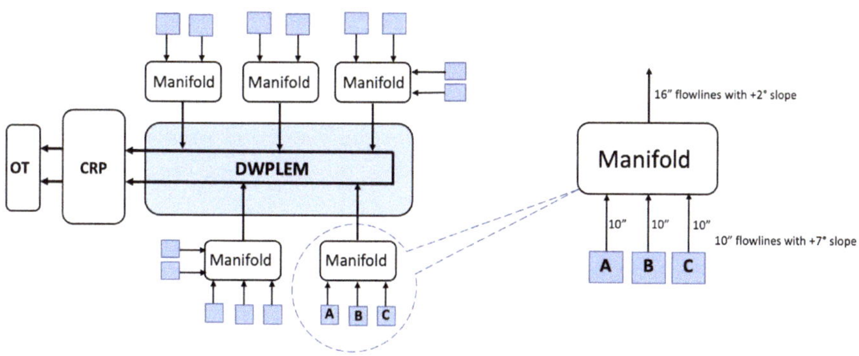

FIGURE 8.36

Schematic illustration of deepwater gas field. *OT*, Onshore terminal; *CRP*, central riser platform; *DWPLEM*, deep water pipeline end manifold.

Sampled produced fluids was glycol rich due to an existing hydrate-inhibiting program utilizing glycol. No downhole injection provision was available. A spare umbilical cord could be utilized to inject foamer into the flowline. The chemical vendor utilized a rigorous process to develop and qualify a novel chemical foamer suitable for subsea application via umbilical injection. A novel chemical foamer, effective in foaming glycol-rich fluid, was recommended for field trial. This foamer also passed stringent tests and rated as umbilical cord compatible. The recommended injection rate of the foamer was 3500 ppm for well A and 5000 ppm for well B. well A is considered the weaker well compared to well B in terms of BHP. The primary idea was to use more concentration at well B, the stronger well, which in turn would keep unloading the flow line between subsea manifold and DWPLEM, which is a common trunk for both wells. This will create a negative pressure, which in turn will make the unloading job easier for well A. Defoamer would also be injected at MRU in the upstream of the Flash drum and in the upstream of TEG dehydration column to mitigate any foaming tendency in the top side processes.

8.10.4.2 Results

The field trial was conducted with foamer being injected through unused umbilical lines connected upstream of the chokes of well A and well B (Fig. 8.37). It was envisioned the amount of shear and agitation generated while foamer and fluid passed through the choke would be sufficient to create foam and unload the liquids from the flow lines. At the same time, a defoamer was being injected from central riser platform (CRP) at a rate of 300 L/d. After 67 hours initiation of foamer injection, the operator began observing drops in the flowline pressure near the well heads, followed by a pressure drop at the subsea manifold (Fig. 8.38).

As expected, the back-pressure reduction and removal of liquid from the flowlines allowed higher production rates from both wells. After 6 hours into the field trial, both wells responded well to the foamer treatment, with increasing gas rates (Fig. 8.39; Table 8.7). After 18 hours into the field trial, the back pressure of well A was reduced by 3 bar, which was 4.5% reduction. The 149 MMscf/d gas production rate increase was realized, which was 41% increase. And 3.3% back

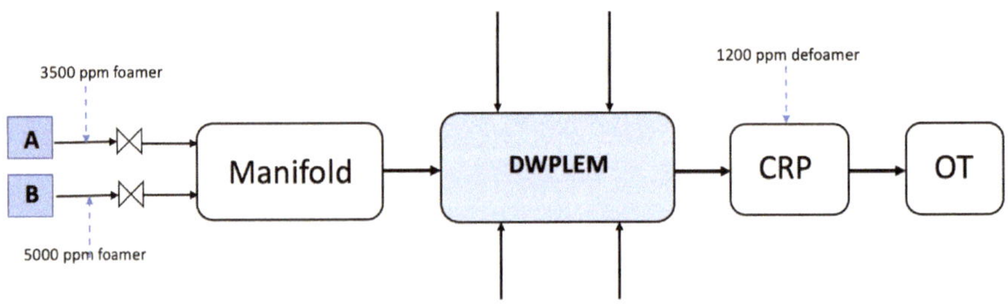

FIGURE 8.37

Schematic illustration of field trial injection points of foamer and defoamer.

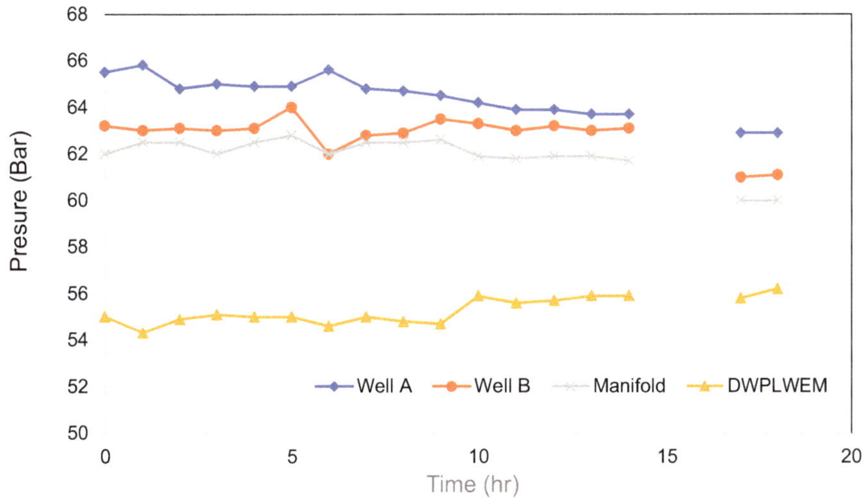

FIGURE 8.38

First 18 h of pressure data following foamer injection.

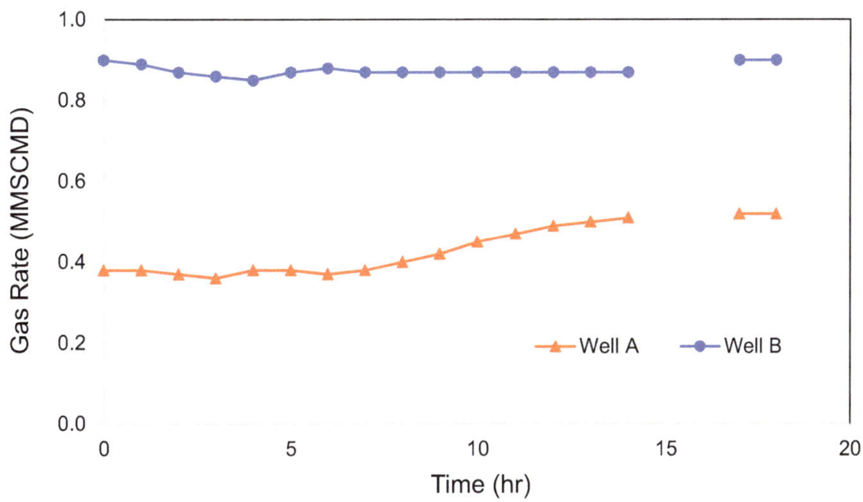

FIGURE 8.39

First 18 h of gas production rates following foamer injection.

pressure reduction was achieved for well B. Gas rate of well B increased slightly by 6 MMscf/d (0.6%). This was likely due to the fact that well B was already producing at its optimum stage. Back pressure at subsea manifold was also reduced.

Table 8.7 Pressure and gas rate change at 18 h following foamer injection.

	Pressure difference (Bar)	Pressure change (%)	Gas rate difference (MSCFD)	Gas rate increase (%)
Well A	−3	−4.5	149	41
Well B	−2.1	−3.3	6	0.6
Subsea manifold	−1.59	−2.6	—	—
DWPLEM	1.34	+2.4	—	—

There was a slight increase of back pressure at DWPLEM, which was likely caused by fluids-containing residual foam at DWPLEM.

After 14 hours of substantial unloading and significant drop in pressure at wellhead, reduced pressure drop in between manifold and DWPLEM and consequent increase in the volumetric flow rates from both wells, the foamer injection rate was optimized and reduced to 1700 and 3000 ppm respectively along with defoamer to 150 L/d. Well A and well B both have maintained increased production rates following field trial. The long-term foamer continuous injection was implemented after the successful field trial.

This first-of-its-kind application proved the novel chemical foamer could be an effective remedial method to relief flow line liquid accumulation, hence optimize the production of the asset.

8.10.5 Foam-assisted lift to optimize mature oil wells

Contrast to the abundant amount of literature on utilizing FAL to reduce critical lift rate and enhance the production of mature gas wells, there was a scarce amount of case studies on leveraging FAL to optimize mature oil well production. The earliest reported case study that could be found in literature was from Bernadiner [52], in which foamed GL was achieved by injecting foamer into the gas flow and through the GL valves of mature oil wells with depleted reservoirs. The article showed combing FAL and GL could increase oil rate by 31%,13%, and 122% for three tested wells.

The second reported successful case study on using foamed GL for oil wells that could be found in literature was from Farag et al. in 2016 [53]. Their article reported a successful field trial of utilizing FAL in a GL oil well with a high water cut (>70%) in the Netherlands operated by the Nederlandse Aardolie Maatschappij in 2015. Though already on GL, the well was slugging and had unstable production before the foamer treatment. Foamer was injected to the well via temporarily installed capillary string into the well at the top of the inflow zone. The well production was stabilized by foamer treatment and oil rate increased. At an optimum concentration of 800 ppm of foamer, a gain of more than 100% oil production was realized without the need of using antifoam on the topside. Encouraged by these positive results, a permanent capillary string was installed in the same well in 2017. Continuous foamer injection and longer test helped to reduce the optimum foamer concentration to around 400 ppm with stabilized oil production.

In 2017, Ahmad et al. [54] conducted a field trial in an oil field in Malaysia operated by Shell, during which they injected foamer directly into the GL mandrel of a mature oil well. At a dosage of 7500 ppm foamer, an increase of 22% of gross production was observed. However, likely due to the high dose concentration of foamer, an antifoam was injected to the topside separating facility to avoid upsets, simultaneously.

Martins et al. [22] investigated the impact of different foamer injection methods on the performance of foamed GL in mature wells. They performed two trials in the same well between November 2016 and February 2018. The first trial took place in November 2016 and followed a similar process used by Farag [53]. A tubing inserted capillary string was utilized to inject the foamer. Initially, the capillary string was set at the target depth at the top of the inflow zone. After optimization, a gross production gain of 158% was reported with 350 ppm foamer. There was no need to use antifoam on the surface. While maintaining the optimized foamer dosage of 350 ppm, they tested the impact of different injection depths of capillary string. It was demonstrated that the well performance remained virtually unchanged until the capillary string injection depth was set at the GL injection valve. If the injection depth was set above the GL injection value, performance of foamed GL was slowly reduced.

The second trial took place in the same well in February 2018, in which they injected a different foamer directly into the GL line. More than triple production gain was observed during the field trial. After optimization, 700 ppm foamer concentration was found optimal in this test versus the 350–400 ppm required for the capillary string injection method. The higher dosage requirement is in line with the lab tests results predicting a 25%–50% dose rate requirement increases due to the different chemistry.

Martin's work proved again that foamed GL could significantly enhance the production of mature oil wells. Similar results could be realized by injecting foamer either via capillary string or via the GL line. The latter injection method allows significant saving on installation and maintenance cost and should be considered as a foamer delivery method.

8.11 Remaining challenges

In general, FAL has been proven as a cost-effective remedial deliquification solution and is widely applied in mature gas well and high water cut liquid hydrocarbon wells due to its low cost, quick response, and product availability. Amidst the growing popularity and significant progress made over the last two decades, there are some remaining challenges, as listed below, to be overcome.

- Product chemistry: Identifying economical foamer products for wells with high oil or condensate cut remains a challenge for the industry. It requires a high dose of conventional foamer or an expensive unconventional surfactant to foam the liquid and remove it from wellbore, resulting in high operating cost that may not be economically viable.
- Application process: The discussed foamer selection process has a few critical steps to increase the chance of success in the field. Though the laboratory testing helps to reveal the top-performing foamer products and recommended dosage, the prediction is more qualitative than quantitive, which renders field testing, performance monitoring and continuous optimization essential for successful application.
- Operation: After a successful foamer treatment, a well would respond positively to the treatment with realized production gains and increased liquid rates. However, the efficacy of the foamer treatment would decrease or cease if foamer concentration in the treated liquid drops below the optimum dosage rate. Thus, to maintain the benefits of foamer application, either new batch treatment or continuous injection are required. There was effort to conduct foamer squeeze to achieve deep penetration of foam and longer-lasting treatment result, however, as the foamer squeeze involves pumping foamer with other additives such as scale inhibitor, chelating agents, mutual solvent, in addition with natural gas in a few steps, makes it difficult to really understand the benefit of the foamer chemistry alone [55]. High operation cost involved in foamer squeeze application makes it less attractive compared to foamer batch treatment.
- Residual monitoring: Due to the nature of the foamer application, foamer chemistry residual would be carried downstream of the process such as topside separation, transport pipelines, and making the impact of foamer global. This would require the monitoring of the whole production system and optimization if needed.

Nomenclature

BHP	bottom hole pressure
BS&W	basic sediment & water
CAPEX	capital investment
CGR	condensate–gas ratio
CITHP	closed-in wellhead pressure
CSG	coal seam gas
CRP	central riser platform
DP	downhole pump
DWPLEM	deep water pipeline end manifold

$10^3Sm^3/d$	1000 standard cubic meters per day
FAL	foam-assisted lift
FAGL	Foam-assisted gas lift
GL	gas lift
ID	internal diameter
L/d	liters per day
LGR	liquid-gas ratio
LMGV	lower master gate valve
Mscf/d	thousand standard cubic feet per day
MMscf/d	million standard cubic feet per day
OIW	oil in water
OPEX	operational expenditure
OT	onshore terminal
PL	plunger lift
PLEM	pipeline end manifold
ppm	parts per million
RO	reverse osmosis
SC	surface compression
SSSV	subsurface safety valve
TC	turner constant
TDS	total dissolved solids
THP	tubing head pressure
THT	wellhead temperature
VLP	vertical lift performance
VS	velocity String
WGR	water−gas ratio
Xmas	Christmas

References

[1] J.L. Eakin, Foaming Agents for Removing Problem Liquids from Gas Wells, US Department of the Interior, Bureau of Mines, Washington, DC, 1965.

[2] J.F. Lea, H.V. Nickens, M.R. Wells, Gas Well Deliquification, second ed., Gulf Professional Publishing, Houston, 2008.

[3] J.M.C. Van 't Westende, Droplets in Annular-Dispersed Gas-Liquid Pipe-Flows (Ph.D. dissertation), Delft University of Technology, 2008.

[4] R.G. Turner, M.G. Hubbard, A.E. Dukler, Analysis and prediction of minimum flow rate for the continuous removal of liquids from gas wells, Journal of Petroleum Technology 21 (1969) 1475−1482.

[5] A.T. Van Nimwegen, L.M. Portela, R. Henkes, The effect of surfactants on vertical air/water flow for prevention of liquid loading, SPE Journal 21 (2016) 488−500.

[6] C.A.M. Veeken, S.P.C. Belfroid, New perspective on gas-well liquid loading and unloading, SPE Production & Operations 26 (2011) 343−356.

[7] C.A.M. Veeken, B. Hu, W. Schiferli, Gas-well liquid-loading-field-data analysis and multiphase-flow modelling, SPE Production & Operations 25 (2010) 275−284.

[8] R.P. Sutton, S.A. Cox, J.F. Lea, et al., Guidelines for the proper application of critical velocity calculations, SPE Production & Operations 25 (2010) 182−194.

[9] L.K. Harms, Wellhead compression on tight gas wells in the long run: A follow-up case history on seven years of success in Lobo, SPE-138488-MS, in: Tight Gas Completions Conference, San Antonio, TX, 2–3 November, 2010.

[10] C. Veeken, D. Al Kharusi, Selecting artificial lift or deliquification measures for deep gas wells in the Sultanate of Oman, SPE-198080-MS, in: SPE Kuwait Oil & Gas Show and Conference, Mishref, Kuwait, 13–16 October, 2019.

[11] L.K. Harms, Better results using integrated production models of gas wells, SPE-93648-MS, in: SPE Production and Operation Symposium, Oklahoma City, Oklahoma, 17–19 April, 2005.

[12] C.A. Shields, M. Dollard, S. Sridhar, et al., Use of sonar metering to optimize production in liquid loading prone gas wells, SPE-166652-MS, in: SPE Offshore Europe Oil and Gas Conference and Exhibition, Aberdeen, UK, 3–6 September, 2013, Journal of Petroleum Technology 66 (2014) 139–142.

[13] A.T. van Nimwegen, The Effect of Surfactants on Gas-Liquid Pipe Flows (Ph.D. thesis), Delft University of Technology, 2014.

[14] A. Ajani, M. Kelkar, Pressure drop prediction in vertical wells under foam flow conditions, SPE-181237-MS, in: SPE North America Artificial Lift Conference and Exhibition, The Woodlands, Texas, 25–27 October, 2016.

[15] K. Veeken, K. Hinai, A.A. Shanfari, et al., Evaluating performance of foam-assisted lift in Sultanate of Oman by dedicated field testing, SPE-188223-MS, in: Abu Dhabi International Petroleum Exhibition & Conference, Abu Dhabi, UAE, 13–16 November, 2017.

[16] J.P. McWilliams, D. Gonzales, Downhole capillary surfactant injection system pilot on low pressure gas wells in the San Juan Basin, SPE-94293-MS, in: SPE Production Operations Symposium, Oklahoma City, Oklahoma, 16–19 April, 2005.

[17] S. Soni, M. Kelkar, Pressure-drop predictions in tubing in the presence of foamers, SPE-120042-MS, in: SPE Production and Operations Symposium, Oklahoma City, Oklahoma, 4–8 April, 2009.

[18] J.M.C. Van't Westende, R. Henkes, A. Ajani, et al., The use of surfactants for gas well deliquification: a comparison of research projects and developed models, BHR-2017-161, in: 18th International Conference on Multiphase Production Technology, Cannes, France, 7–9 June, 2017.

[19] F. Huang, D. Nguyen, Use of foamer in conjunction with dry-gas-lift system for gas well deliquification, 5th European Conference on Gas Well Deliquification, 28–30 September, 2010, Groningen, The Netherlands.

[20] I. Tayyab, M. Farooq, M. Uddin, et al., Combination of foam assisted lift & gas lift (FAGL) to de-liquefy gas wells, SPE-178259-MS, in: SPE/IADC Middle East Drilling Technology Conference and Exhibition, Abu Dhabi, UAE, 26–28 January, 2016.

[21] A. Farag, T. Robertson, M. Kerem, et al., Foam assist in a gas-lifted oil well, SPE-184217-MS, in: SPE Middle East Artificial Lift Conference and Exhibition, Manama, Bahrain, 30 November – 1 December, 2016.

[22] A. Martins, M. Marino, M. Kerem, et al., Foam assisted gas lift: The impact of different surfactant delivery methods on oil well performance, SPE-195462-MS, in: 81st EAGE Conference and Exhibition, London, UK, 3–6 June, 2019.

[23] J. Outlaw, R. Guliuzo, S. Lehrer, et al., Three successful black oil foamer applications in deepwater GoM, OTC-29394-MS, in: Offshore Technology Conference, Houston, Texas, 6–9 May, 2019.

[24] O. Ugoala, A. McMahon, G. Forsyth, et al., Extending the life of subsea wells with long tiebacks by foamer injection, in: DEVEX Conference, Aberdeen, UK, 18–19 May, 2016.

[25] S. Ahmed, D.A. Bt Awang Bakar, M.M. El Ashmawy, et al., Application of foam to manage liquid holdup in deepwater subsea networks, SPE-172798-MS, in: SPE Middle East Oil & Gas Show and Conference, Manama, Bahrain, 8–11 March, 2015.
[26] F. Huang, Subsea application of a foamer to enhance production. in: SPE Production Chemistry and Chemical Systems Workshop, New Orleans, LA, 25 March, 2015.
[27] K. Veeken, A. Mittal, S. Bertsché, Early detection of gas well formation damage, water breakthrough and liquid loading by use of surface pressure build-up data, SPE-174221-MS, in: SPE European Formation Damage Conference and Exhibition, Budapest, Hungary, 3–5 June, 2015.
[28] F. Viana, R. Pardoi, R. Ya'Nez, et al., Universal correlation for the rise velocity of long gas bubbles in round pipes, Journal of Fluid Mechanics 494 (2003) 379–398.
[29] P.F. Huang, D. Nguyen, Optimized foamers for natural gas well deliquification: A statistical design approach, Fuel 97 (2012) 523–530.
[30] S. Omrani, R.K. Shukla, F. Vercauteren, et al., Improving the foamer selection prcedure for gas well deliquification application, SPE-181592-MS, in: SPE Annual Technical Conference and Exhibition, Dubai, UAE, 26–28 September, 2016.
[31] J.J. Wylde, N. Turner, M. Austill, et al., Development, testing and field application of a novel combination foamer-iron sulfide scale inhibitor-corrosion inhibitor in East Texas, SPE-184584-MS, in: SPE International Conference on Oilfield Chemistry, The Woodland, Texas, 3–5 April, 2017.
[32] O. Morinello, A. Allegrucci, M. Buccolini, et al., Field application of foamer for gas lift enhancement: Lesson learnt and way forward, OMC-2019-1186, in: 14th Offshore Mediterranean Conference and Exhibition, Ravenna, Italy, 27–29 March, 2019.
[33] W. Schinagl, S.R. Green, A.C. Hodds, et al., Most successful batch application of surfactant in North Sea gas wells, SPE-108380-MS, in: SPE Offshore Europe Oil and Gas Conference and Exhibition, Aberdeen, U.K, 4–7 September, 2007.
[34] K. Veeken, G. de Vries, Evaluating 10 years of foam-assisted lift, in: 9th European Gas Well Deliquification Conference & Exhibition, Groningen, The Netherlands, 22–24 September, 2014.
[35] J. Adwadzi, J. Babbitt, S. Holland, et al., Downhole capillary soap injection improves production, SPE-52153-MS, in: SPE Mid-Continent Operations Symposium, Oklahoma City, OK, 28–31 March, 1999.
[36] R.B. Sullivan III, D.L. Hinson, D.E. Hendrix, Benefits and challenges associated with cra injection tubing in corrosive gas wells, SPE-91692-MS, in: SPE Annual Technical Conference and Exhibition, Houston, TX, 26–29 September, 2004.
[37] D. Klompsma, B. Lugtmeier, R. Lacy, et al., The development of a complete system to allow continuous downhole chemical injection without affecting the subsurface and surface safety systems, SPE-132659-MS, in: SPE Asia Pacific Oil and Gas Conference and Exhibition, Brisbane, Australia, 18–20 October, 2010.
[38] P. Imbò, D. Magnani, M. Di Iorio, et al., Innovative capillary deliquification safety system resolves liquid loading problems with a cost-effective solution that maximizes production while maintaining well-safety requirements, SPE-135537-MS, in: SPE Annual Technical Conference and Exhibition, Florence, Italy, 19–22 September, 2010.
[39] R. Rahman, D. Poirault, S. Warsito, et al., A full field downhole foam assisted lift application for enhance gas recovery, SPE-170634-MS, in: SPE Annual Technical Conference and Exhibition, Amsterdam, The Netherlands, 27–29 October, 2014.
[40] B. Bremner, L.M. Murphy, R.J. Affinito, et al., The first through-tubing capillary string installation in the UK sector of the North Sea, SPE-130672-MS, in: SPE/

ICoTA Coiled Tubing and Well Intervention Conference and Exhibition, The Woodlands, Texas, 23–24 March, 2010.
[41] J.L. Bolding, L.E. Hartman, S.J. Szymczak, Resurrecting a low-pressure gas well offshore: through-tubing foamer injection via a capillary tubing system and a specialized WRSCSSV, SPE-110086-MS, in: SPE Annual Technical Conference and Exhibition, Anaheim, California, 11–14 November, 2007.
[42] C.E. Ogugbue, C.R. Hopkins, L.R. Greenly, et al., Lowering tubing in jonah field offset production decline, SPE-154394-MS, in: SPE Americas Unconventional Resources Conference, Pittsburgh, Pennsylvania, 5–7 June, 2012.
[43] O. Rauf, S. Gerdes, N.R. Lummer, Foam and scale inhibitor squeeze with natural gas: A lifeline for water flooded and damaged gas well, SPE-165098-MS, in: SPE European Formation Damage Conference & Exhibition, Noordwijk, The Netherlands, 5–7 June, 2013.
[44] C. Veeken, A. Alawi, K. Ardia, et al., Innovative use of capillary string for halite cleanout in Sultanate of Oman, SPE-197842-MS, in: Abu Dhabi International Petroleum Exhibition & Conference, Abu Dhabi, UAE, 11–14 November, 2019.
[45] Vogelij, N.A., Veeken, C.A.M., Islamov, et al., Four years of continuous foamer application in Southern North Sea Offshore gas wells, SPE-180031-MS, in: SPE Bergen One Day Seminar, Grieghallen, Norway, 20 April, 2016.
[46] M. Marino, W. Schiferli, C.A.M. Veeken, et al., Shifting Turner's paradigm: Gas wells can produce in slug flow, SPE-187291-MS, in: SPE Annual Technical Conference and Exhibition, San Antonio, TX, 9–11 October, 2017.
[47] N. Dousi, C.A.M. Veeken, P.K. Currie, Modelling the gas well liquid loading process, SPE-95282-MS, in: SPE Offshore Europe Oil and Gas Exhibition and Conference, Aberdeen, UK, 6–9 September, 2005.
[48] D. Jackson, C. Virues, D. Sask, Investigation of liquid loading in tight gas horizontal wells with a transient multiphase flow simulator, SPE-149477-MS, in: Canadian Unconventional Resources Conference, Alberta, Canada, 15–17 November, 2011.
[49] F. Huang, L. Thomas, Foamer application to optimize the production of horizontal gas wells, in: 11th Gas Well Deliquification Workshop, Denver, Colorado, 17–20 February, 2013.
[50] F. Huang, D. Nguyen, Foamer for downhole injection, Patent US8950494, 2015.
[51] F. Huang, D. Nguyen, M. Willis, The development of a sustainable deliquification solution, in: 7th European Gas Well Deliquification Conference, Groningen, the Netherlands, 24–26 September, 2012.
[52] M.G., Bernadiner, Foamed gas lift, SPE-21639-MS, in: Production Operations Symposium, Oklahoma City, OK, 7–9 April, 1991.
[53] A. Farag, T. Robertson, T. Kerem, et al., Foam assisted in a gas-lifted oil well, SPE-184217-MS, in: SPE Middle East Artificial Lift Conference and Exhibition, Manama, Bahrain, 20 November - 1 December, 2016.
[54] S.A. Ahmad, S. McGregor, D. Ekol, et al., Increasing production via foam assisted gas lift in a mature oil well, in: SPE Symposium: Production Enhancement and Cost Optimization, Kuala Lumpur, Malaysia, 7–8 November, 2017.
[55] O. Rauf, N.R. Lummer, S. Gerdes, Foam and scale inhibitor squeeze with natural gas: a lifeline for water flooded & damaged gas wells, in: SPE European Formation Damage Conference & Exhibition, Noordwijk, The Netherlands, 5–7 June, 2013.

CHAPTER

Corrosion inhibition

9

Jeremy Moloney[1], Dharmendr Kumar[2], Venkata Muralidhar K[2] and Thunyaluk Pojtanabuntoeng[3]

[1]*ChampionX, Sugar Land, TX, United States*

[2]*TCS Research, Tata Consultancy Services Ltd., Pune, Maharashtra, India*

[3]*Curtin University, Perth, WA, Australia*

Chapter outline

Flow Assurance. DOI: https://doi.org/10.1016/B978-0-12-822010-8.00006-4

9.1 Corrosion inhibitors

A chemical substance that minimizes or prevents corrosion, upon its addition in small quantities to an environment, is called a corrosion inhibitor. Corrosion inhibitors are effective and specific to certain environments, that is, a proven corrosion inhibitor for a given metallic material in one environment may not provide suitable inhibition for the same metallurgy in another environment. The corrosion inhibition by an inhibitor molecule and the extent of corrosion inhibition in a given environment depends on various factors, such as:

1. Nature and composition of the surface of metallic material
2. Composition of the bulk of the metallic material
3. Composition of the environment that surrounds the metallic material
   a. Physical: distribution of various phases such as particulate matter, gaseous phases, liquid phases, intermixing and dissolution of various phases, etc.
   b. Chemical: pH of the solution, aqueous or nonaqueous, activity of aggressive species, activity of oxidants, etc.

4. Flow conditions
5. Temperature

Basically, corrosion is an electrochemical process that has the following essential components [1]:

1. Cathodic reaction (reduction, i.e., gain of electrons)
2. Anodic reaction (oxidation, i.e., loss of electrons)
3. Electrically conducting path between cathodic and anodic reaction sites (for electron transfer)
4. Path for ionic conductivity, that is, the electrolyte itself

It is imperative that a corrosion inhibitor inhibits the activity of one or more of the key constituents of the corrosion process. Typically, irrespective of the electrolyte, the inhibitor molecules are active species that reduce the metal corrosion. Based on their nature, corrosion inhibitors can be classified into environmental conditioners or scavengers, and interface inhibitors [2,3]. The latter include cathodic, anodic, and mixed inhibitors.

9.1.1 Environmental conditioners/scavengers

As explained earlier, the presence of aggressive species affects corrosion drastically. The concentration of aggressive species and their activity impacts the corrosion rate (CR). A chemistry which modifies the environment by alleviating the effect of aggressive species on the metal corrosion, either by scavenging or by entrapping the aggressive species or by any other means, can be called an environmental conditioner/scavenger.

9.1.2 Interface inhibitors

9.1.2.1 Cathodic inhibitors

Chemistries that inhibit the corrosion by affecting the cathodic reaction either by retarding the reaction rate (poisoning) or by blocking the reaction sites (precipitation or film formation) are called cathodic inhibitors. They may block the reaction sites by adsorption as well, that is, physisorption or chemisorption, apart from film formation.

9.1.2.2 Anodic inhibitors

Inhibitors that mitigate the corrosion by retarding the anodic reaction are called anodic inhibitors. Anodic inhibitors may form insoluble or partly soluble substances on the metal surface by combination with ions from metal surface, that is, passivation, or they may block the anodic reaction sites through precipitation or film formation or even adsorption, that is, physisorption or chemisorption.

9.1.2.3 Mixed inhibitors

Chemistries that do not exhibit predominantly either of the above behaviors are called mixed inhibitors. Mixed inhibitors reduce the metal corrosion predominantly by adsorption and film formation. They are classified based on their interaction modes such as physisorption, chemisorption, and film formation.

Physisorption or physical adsorption is the interaction and attachment of molecule onto the metal surface driven by electrostatic forces. Inhibitors that act through physisorption are least effective in the category of mixed inhibitors as the interaction is physical in nature, weak and dependent on the surface charge of the metal as well. Depending on the extent of physisorption, physisorbed molecules tend to desorb with an increase in temperature, suggesting that the protection is weak and temporary. Physisorption depends on several aspects such as molecular weight of the inhibitor, charge on the molecule, surface charge, electrolyte, diffusivity, etc.

Chemisorption or chemical adsorption is the interaction of the molecule with the metal surface and is chemical in nature, implying a bond formation. Inhibitors that protect the metal through chemisorption tend to offer better protection than those with physisorption and, to a certain extent, an increase in temperature does not drive the molecule to desorb. Generally, with an increase in temperature, physisorption tends to weaken while chemisorption tends to strengthen.

Film formation can also result in efficient inhibition. The interaction between the inhibitor molecules and metal surface is chemical to a large extent and hence, stronger. The protection increases with increasing film thickness, as the film separates the metal surface from the electrolyte. However, the film only offers protection as long as it is adherent and impervious to the electrolyte. If the film is porous, it can, in some instances, result in an acceleration of metal corrosion rather than inhibiting it, since highly active species can permeate and reach the metal surface to oxidize the metal, often locally, which is undesirable. Alternatively, if the film is impervious but nonadherent, the film can grow to a critical thickness after which it becomes detached from the surface leaving it exposed to the electrolyte, which can also result in accelerated corrosion or thinning of the metal component.

9.2 Mechanism of corrosion inhibition

9.2.1 Environmental conditioners/scavengers

In neutral or basic conditions in the presence of dissolved oxygen, oxygen reduction is a common cathodic reaction, which is inhibited by oxygen scavenging. The presence of hydrogen sulfide in the electrolyte aggravates corrosion, and scavenging hydrogen sulfide reduces the extent of corrosion. Sulfite, carbohydrazide, hydrazines, among other chemistries, are used as oxygen scavengers [2], while hydrogen sulfide scavengers can include, but not limited to, triazines, ferrous gluconate, chelating agents like ammoniated-EDTA (Ethylene Diamine Tetra Acetic acid), ammoniated-HEDTA (Hydroxyethyl EDTA), etc. [4,5].

9.2.2 Interface inhibitors

The interface inhibitors reduce corrosion by modifying the interface between the metal surface and the electrolyte. Usually, any electrochemically active material or metal when exposed to an electrolyte, exhibits a surface charge, either positive or negative. Depending on the surface charge, an electrical double layer gets established, irrespective of the electrolyte (unless it is nonpolar), and the constituents of the solution including the solution dipole compensate/reciprocate for the surface charge of the metal. For example, if aqueous solutions are electrolytes, the solution dipole is the water dipole (Fig. 9.1). If the metal surface is positively charged in the system with water as electrolyte, then the oxygen (negative part) of the water molecule comes closer to the metal surface, along with hydroxide ions (Fig. 9.2). Conversely, if the metal surface is negatively charged, the hydronium ion and the hydrogen atoms (positive part) of the water molecule come closer to the metal surface [1].

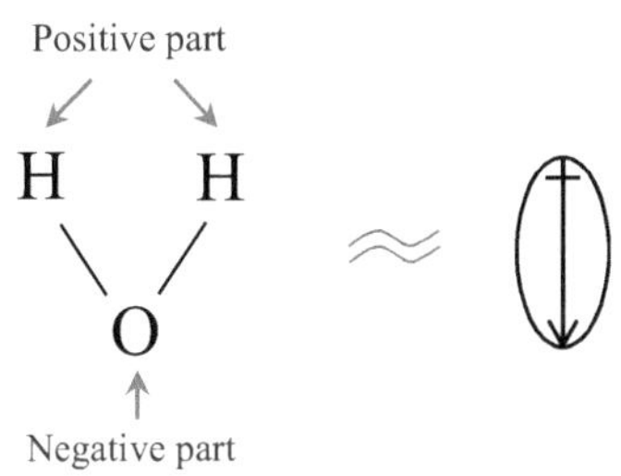

FIGURE 9.1

Illustration of water dipole.

Based on [1].

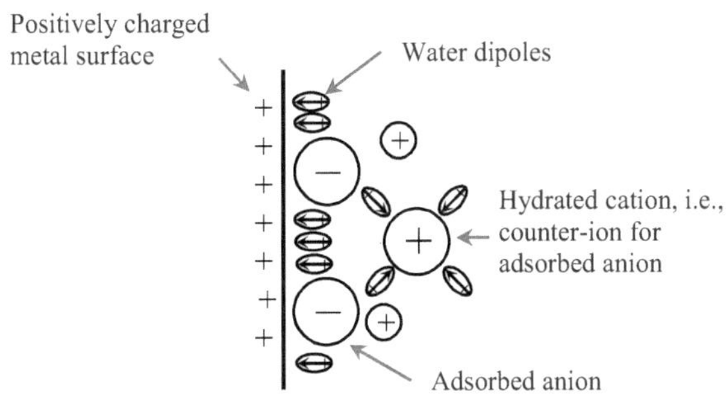

FIGURE 9.2

Positively charged metal surface and double layer (Bockris-Devanathan-Muller Model) for blank solution.

Based on [1].

The electrical double layer is the cause of potential difference across the interface which can be exploited or altered using species that affect the length and charge in the electrical double layer. The process of corrosion involves the dissolution of metal into the electrolyte in which the metal ions (cations) pass through the double layer. Similarly, the reactive oxidant species from the electrolyte pass through the electrical double layer to the metal surface to facilitate the cathodic reaction.

The inhibitor molecules often have positively charged and negatively charged parts within them, which tend to manifest depending on the electrolyte. The charges on the inhibitor molecule alter the electrical double layer and thereby the corrosion process itself. Such effects can be understood by correlating the elements of the representative electric circuit for the system such as solution resistance, charge transfer resistance, inductance, double layer capacitance, etc., through electrochemical impedance spectroscopic studies on the system.

9.2.2.1 Cathodic inhibitors

Cathodic inhibitors inhibit the corrosion by impeding the cathodic or reduction reaction. Hydrogen evolution is the predominant cathodic reaction in anaerobic acidic solutions and cathodic chemistries tend to impede it by poisoning. Oxygen reduction is the main cathodic reaction in aerobic neutral or alkaline solutions, which is often retarded by precipitation [2,3]. Typically, the action of cathodic inhibitors is as seen in polarization curves, given in, Fig. 9.3. For a system containing a cathodic inhibitor, the slope of the cathodic branch of the polarization curve decreases, often with a concomitant shift of corrosion potential to lower values. Ultimately, the value of corrosion current reduces in the presence of cathodic inhibitor, implying a decrease in corrosion rate.

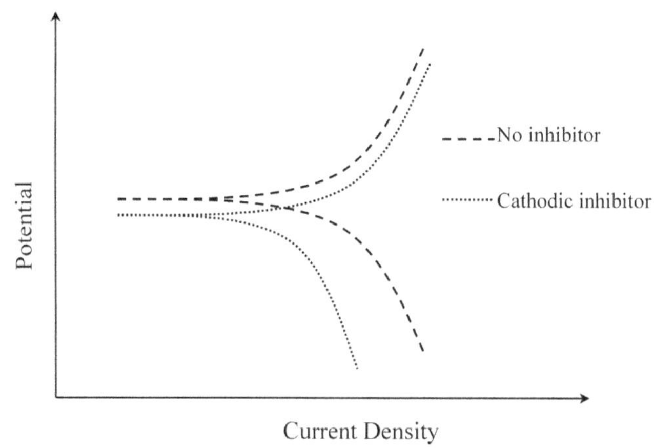

FIGURE 9.3

Illustration of polarization curve in the presence of a cathodic inhibitor.

9.2.2.2 Anodic inhibitors

Anodic chemistries tend to inhibit the metal corrosion by forming adherent films or hinder the anodic reaction by elevating the activation barriers. Basically, they inhibit the oxidation reaction [2,3]. Hence, the polarization curves with anodic inhibitors may have passive regions or the slope of anodic branch increases, often with an associated shift of corrosion potential to higher values (Fig. 9.4). Ultimately, the value of corrosion current reduces in the presence of anodic inhibitor, implying a decrease in corrosion rate.

9.2.2.3 Mixed inhibitors

These inhibitors act by physisorption or chemisorption or film formation (Fig. 9.5). At first, the molecule interacts with the metal surface and then adsorbs on it.

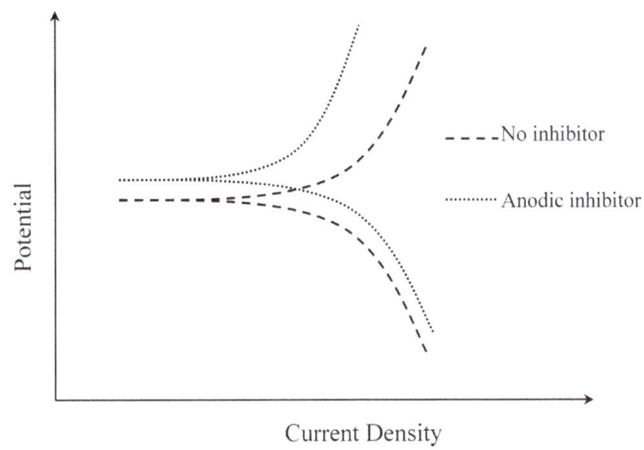

FIGURE 9.4

Illustration of polarization curve in the presence of an anodic inhibitor.

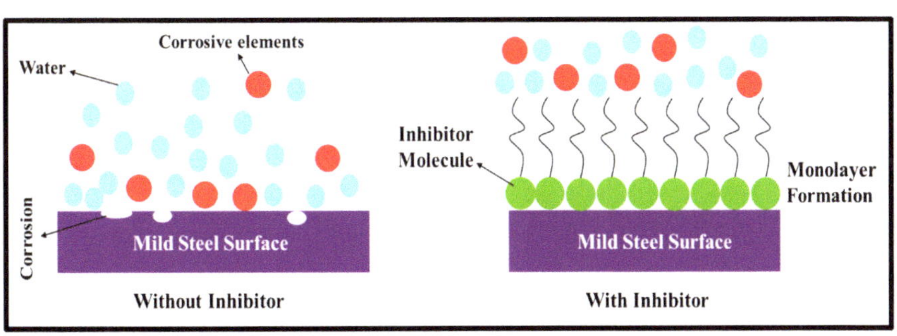

FIGURE 9.5

Corrosion inhibition mechanism of inhibitor molecules.

Hence, the metal surface charge is an important aspect for these inhibitors, which is determined by the difference between Potential for Zero Charge (PZC) and corrosion potential (E_{corr}). PZC is the potential at which the metal surface charge is zero and is dependent on the system including the electrolyte.

The metal surface charge is positive, if $E_{corr} > \text{PZC}$, while it is negative, if $E_{corr} < \text{PZC}$, and zero if $E_{corr} = \text{PZC}$. Antropov rational corrosion potential is the term that easily signifies the difference between E_{corr} and PZC [2,6]. Antropov rational corrosion potential is E_{corr}-PZC. In other words, in a given system, if Antropov rational corrosion potential is positive, the metal surface is positively charged, while if it is negative, the metal surface is negatively charged, but if it is zero, the metal surface has no charge.

Depending on the surface charge of the metal, the inhibitor molecules interact with the surface with their reciprocating portions. That means, if the surface charge is negative, the positive portion/part of the inhibitor molecule interacts with the metal surface, that is, the dipole gets oriented according to the metal surface charge. Such reciprocating portions, in other words, anchoring groups contain functional groups that have higher charge density than other portions of the molecule, and help the molecule to adsorb. If the inhibitor is organic in nature, the anchoring group contains a heteroatom, which accumulates the charge to reciprocate with the charged metal surface. Typically, the heteroatoms in such chemistries include phosphorous (P), sulfur (S), nitrogen (N), and oxygen (O). The inhibitor molecules adsorb sometimes not by direct interaction with metal surface, but through an intermediate layer of charged species, for example, cationic surfactants, halide ions, etc. The adsorption of the inhibitor molecule is governed by several factors, a few of which are as follows [6–9]:

1. Chemistry of the surface
2. Charge on the surface
3. Charge density on the anchoring group
4. Charge density on the heteroatom
5. Electronegativity of the heteroatom
6. Polarizability of heteroatom (ionic radii, atomic numbers, electronegativity difference between heteroatom and surface atom, effective nuclear charge, vacant orbitals, lone pairs, etc.)
7. Polarizability of the anchoring group and the molecule
9. Contribution of electron cloud by each heteroatom to overall reactivity/nucleophilicity.
9. Charge distribution within the molecule
10. Bulkiness of the molecule
11. Solubility of the molecule in the electrolyte
12. Electrolyte properties

For the adsorption onto the metal surface, the inhibitor molecule interacts weakly with the metal surface and displaces the solution dipoles (e.g., water dipoles) from the metal surface. Since this occurs through electrostatic interactions, it is weak, reversible, and physical in nature, and hence is called physisorption.

With effective interactions and favorable conditions, if chemical bonds form between inhibitor molecule and the metal, chemisorption occurs, that is irreversible and stronger than physisorption. As chemical bonds are formed, all the necessary factors for bond formation affect the inhibition efficiency. The inhibition efficiency of organic molecules with similar structures or chain, but different heteroatoms tend to increase with increasing polarizability (e.g., O < N < S < P) and decreasing electronegativity (e.g., O > N > S > P). Essentially, the inhibition efficiency improves with the covalency of the bond formed between the inhibitor and the metal surface.

Hence, it can be deduced that the open-circuit potential (OCP) value of the system with physisorbed inhibitor tends to fluctuate more due to adsorption and desorption cycles, while that with chemisorbed inhibitor tends to fluctuate less due to a chemical bond formation, which is stronger than electrostatic forces.

In addition, molecules with longer lipophilic chains, which assist in protecting the metal surface from direct contact with electrolyte, tend to form micelles at critical concentrations. Depending upon the concentration, electrolyte, etc., such molecules can protect the metal in various forms, such as bilayer structures, cylindrical structures, micelles, hemispherical micelles, etc., as well as in the monomer form itself.

With the above in mind, one can understand that corrosion inhibition involves complex interfacial science and is highly system dependent. Additionally, there are cases where molecules known to be inhibitors, act as complexing agents or even flotation collectors and selective flocculants. In such cases, the molecules accelerate the corrosion. For instance, chitosan and thiobarbituric acid are inhibitors for steel corrosion, whereas with copper, animal-derived chitosan and thiobarbituric acid act as a selective flocculant and a complexing agent, respectively. Urea, a known corrosion inhibitor, is also known to accelerate aluminum corrosion in nitric acid, if present in quantities >2% [10].

9.3 Measurement of corrosion inhibition

It is important to understand or assess the effectiveness of the molecule employed as a corrosion inhibitor, in mitigating the metal corrosion, which is considered in more detail in Section 7, Corrosion Rate Measurement Techniques, but quickly referred to here. The effectiveness of the corrosion inhibitor is expressed as corrosion inhibition efficiency and can be found with certain degrees of accuracy across various preliminary screening or assessment methods, some of which are given in Table 9.1.where CR_b is the baseline CR of the metal under blank conditions that is, without a corrosion inhibitor, CR_{inh} is the CR of the metal with corrosion inhibitor, R_{p0} is the linear polarization (LP) resistance without corrosion

Table 9.1 Corrosion inhibition efficiency determination using various preliminary methods.

	Evaluation method	Expression for inhibition efficiency (%)	Remarks
1	Gravimetric tests	$\frac{CR_b - CR_{inh}}{CR_b} \times 100$	• Slow • Destructive • Mostly accurate • Sufficiently sensitive to know the effect of concentration • Limited insights into inhibition mechanism • Includes the incubation time and averages out the corrosion rate/inhibition over the time considered • Best for ranking the inhibitors/concentrations
2	Linear polarization	$\frac{R_p - R_{p0}}{R_p} \times 100$	• Very fast • Nondestructive • Not so accurate • Mildly sensitive to know the effect of concentration • No insights into inhibition mechanism • Depends on the OCP of test area, allowed incubation time, exposed area, etc. • Not very good for ranking the inhibitors/concentrations
3	Potentiodynamic polarization	$\frac{i_{corr_0} - i_{corr}}{i_{corr_0}} \times 100$	• Very fast • Destructive • Not as accurate as gravimetric tests as it is an accelerated test • Sensitive to know the effect of concentration • Meaningful insights into inhibition mechanism • Depends on the scan rate, exposed area, etc. • Good for ranking the inhibitors/concentrations
4	Electrochemical impedance spectroscopy	$\frac{R_{ct} - R_{ct_0}}{R_{ct}} x100$	• Fast • Nondestructive • Not accurate, but gives an approximate to actual behavior

(*Continued*)

Table 9.1 Corrosion inhibition efficiency determination using various preliminary methods. *Continued*

Evaluation method	Expression for inhibition efficiency (%)	Remarks
		• Sufficiently sensitive to know the effect of concentration • Meaningful insights into inhibition mechanism • Depends on OCP of test area, allowed incubation time, scanned frequencies, etc. • Not very good for ranking the inhibitors/concentrations

inhibitor, R_p is the LP resistance with corrosion inhibitor, i_0 is the corrosion current without corrosion inhibitor, i is the corrosion current with corrosion inhibitor, R_{ct0} is the charge transfer resistance without corrosion inhibitor, R_{ct} is the charge transfer resistance with corrosion inhibitor.

9.4 Oilfield corrosion inhibitor chemistry examples

Knowledge of the chemistries, structures, and inhibition mechanisms of known corrosion inhibitors, help us understand and improve/design the corrosion inhibitors. The oil and gas industry is one of the largest consumers of corrosion inhibitors needed to attend to a variety of service conditions and requirements. High pressures and temperatures coupled with highly acidic environments make the corrosion more complex and severe [11]. Depending on the nature of the electrolyte and prevailing conditions, various organic molecules have been identified as corrosion inhibitors, such as, but not limited to, amines, amides, quaternary ammonium compounds, imidazolines, aromatic amines, phosphate esters, thiols, etc. Table 9.2 shows some classes of inhibitor molecules along with their molecular structures and some contributing factors to corrosion inhibition.

9.5 Molecular modeling of corrosion inhibitors

Generally, experimental methods are used to evaluate the performance of corrosion inhibitors for metals and alloys [56–59]. However, such experimental approaches are not only expensive and time-consuming, but also can be deficient in determining the underlying inhibition mechanisms at the atomistic and molecular scale,

Table 9.2 Some inhibitor classes with their molecular structures and contributing factors to corrosion inhibition.

Inhibitor Class	General structure	Important features	Contributors to inhibition
Amines [12–21]	$R_1{-}NH_2$ $R_1{-}NH{-}R_2$ $R_1{-}NR_3{-}R_2$ $R_{1,2,3}$–Aliphatic or aromatic groups	• Predominantly bases • Can have individual or combinations of primary, secondary and tertiary amines • Can have aliphatic and/or aromatic groups • Can be heterocyclic compounds	• Interaction of electron cloud on nitrogen atom with metal surface • Formation of layered structures with the help of anions between metal surface and amines (synergistic effects) • Sharing π electrons of aromatic rings with metal surface, that is, vacant orbitals • The electron effects such as inductive, mesomeric play vital role in the case of secondary and tertiary amines
Amides [8–10,22–27]	$R_1-C(=O)-NH_2$ $R_1-C(=O)-NH-R_2$ $R_{1,2}$–Aliphatic or aromatic groups	• Can have aliphatic and/or aromatic groups • Can be heterocyclic compounds	• Interaction of electron clouds of predominantly nitrogen, and oxygen with the metal surface • Act as chelates • Formation of micelles, layered structures on top of anions which adsorb on metal surface • Resonance between lone pair of nitrogen and π electrons reduces availability of electrons for sharing with metal
Quaternary Ammonium Compounds [6,7,28–32]	$R_1R_2R_3R_4N^+\ X^-$ $R_{1,2,3,4}$–Aliphatic or aromatic groups	• Predominantly cationic surfactants • Can have aliphatic and/or aromatic groups • Can be heterocyclic compounds	• Adsorption of cationic part of inhibitor over already adsorbed halide or anion (strong synergistic effects between cationic part and anions) • Not only adsorption through halide or anion, but the cationic part may also adsorb directly on metal, along with adsorption through anion, that is, two adsorption types are possible simultaneously

			• Formation of micelles, layered barrier structures on top of anions which adsorb on metal surface • Hydrophilic part approaches metal for adsorption, while hydrophobic part extends into solution and repels the polar groups, thereby shielding the metal • Presence of aromatic rings allows for interaction of π electrons of the rings with vacant orbitals of metal and predominates the adsorption through cationic part, hence acting as anodic inhibitor
Imidazolines [33–36]	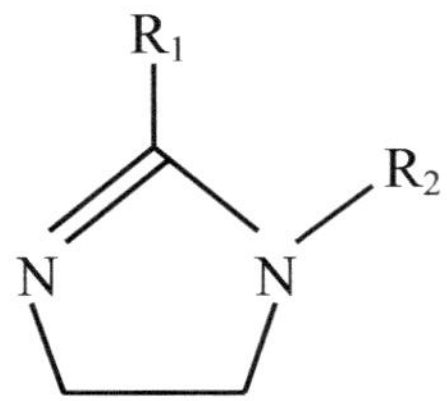 $R_{1,2}$–Aliphatic or aromatic groups	• Heterocyclic compounds • Can have aliphatic and/or aromatic groups/chains	• Interaction of electron cloud of both the nitrogen atoms with metal surface • N-C-N is of p- π conjugated system that helps in sharing π electrons of the ring with vacant orbitals of metal • Adsorption of inhibitor over already adsorbed halide or anion • Geometric blocking effect in the presence of corrosion products • Plugging of the defect sites of corrosion product layers to make corrosion product layer thick and strong • Formation of shielding hydrophobic layers through interacting aliphatic chains, which can sometimes be as continuous/impervious as film

(Continued)

Table 9.2 Some inhibitor classes with their molecular structures and contributing factors to corrosion inhibition. *Continued*

Inhibitor Class	General structure	Important features	Contributors to inhibition
Aromatic Amines [37–43]	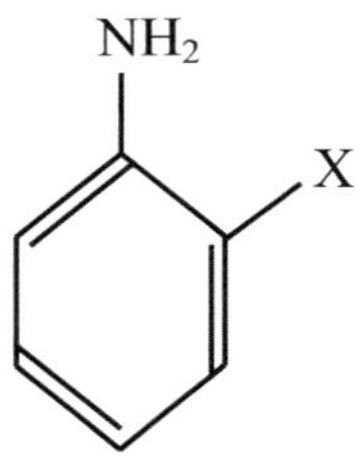 X–Aliphatic or aromatic groups	• Basic in nature • Can have various substitutions on ortho-, meta-, para-positions • Substitutions greatly change the nature and basicity of molecule	• Adsorption is possible in neutral or protonated state through nitrogen atom and through delocalized electrons of ring • Adsorption of inhibitor over already adsorbed halide or anion • Can act as a corrosion accelerator at certain concentrations • Inhibition efficiencies fluctuate with increasing concentrations of anions and protons • In monomer molecules with no substitutions, the interaction of delocalized π electrons of aromatic ring with metal causes more adsorption than that through nitrogen atom • Monomers substituted with electron releasing groups favor the interaction with metal through nitrogen atom as well • Steric effects and hetero atoms in substitutions greatly affect basicity which in turn affects inhibition • Diazotization and inhibition depending on electrolyte and substitution at o-, m-, p- positions, for example, in HNO_3 • Polymers tend to physisorb more efficiently due to abundance of π electrons, which help easy and efficient adsorption • Monomers cause surface passivation due to electropolymerized film • Polymers prevent formation of passive films • Site blocking effects; formation of metal-inhibitor complex and adsorption of this complex as a layer • Mostly mixed inhibitors

Phosphate Esters [44–48]	$O=P(OR_1)(OR_2)(OR_3)$ $R_{1,2,3}$–Aliphatic or aromatic groups	• Acidic in nature • Can have aliphatic and/or aromatic groups/chains	• Generally mixed inhibitors • Chemisorbed as unionized molecules or anionic surfactants • Formation of micelles and/or insoluble phosphate scales • Notable for oxygen corrosion inhibition • Geometric blocking in cathodic conditions by film formation and lipophilic chains protruding out from above the metal surface • Deactivation coverage in anodic conditions by forming scales of insoluble complexes with metal or oxides/oxyhydroxides of metal, along with protrusion of lipophilic chains from above the metal surface • Synergistic effects with imidazolines • Phosphorus delivery for inhibition depends on the structure of entire ester molecule
Fatty Acids [49,50]	$R_1-C(=O)-OH$ R_1–Aliphatic or aromatic groups	• Weak acids • Have long-chain aliphatic and/or aromatic groups	• Vertical adsorption is the only route for chains with single π bond, which adsorb via acid group • Both horizontal (via acid group and π electrons) and vertical (via acid group) adsorption are possible for chains with more than one π bond; bent chains are often the case to avoid π electron repulsion in vertical adsorption • Molecules that form complexes/salts with fatty acids, such as amines, often plug between the vertically adsorbed fatty acid chains and create interchain cross linking • Formation of surface metallic soaps • With salts of fatty acids, inhibition is in two steps, that is, adsorption of fatty acid onto metal surface and surface aggregation/layer growth with the help of the salt forming group

(*Continued*)

Table 9.2 Some inhibitor classes with their molecular structures and contributing factors to corrosion inhibition. *Continued*

Inhibitor Class	General structure	Important features	Contributors to inhibition
Thiols [51–55]	R_1–S–H R_1–Aliphatic or aromatic groups	• Acidic in nature • Exhibits tautomerism • Can have aliphatic and/or aromatic groups/chains • Contains various subclasses such as, mercaptobenz-(imidazoles/triazoles/thiazoles/oxazoles), thioureas, etc.	• Adsorption of cationic form of inhibitor (protonated species/form) along with adsorbed halide ions (strong synergistic effects between cationic form and anions), that is, coadsorption; or direct adsorption of the cationic form/protonated species on the metal surface; and blocking the sites • Presence of aromatic rings/π bonds governs the orientation • Thione form tend to inhibit corrosion better than thiol form

which is imperative for design and development of novel, more efficient, and also, possibly, more environmentally friendly corrosion inhibitor products. With the advent and increased availability of high-performance computing, molecular modeling techniques such as density functional theory (DFT) and molecular dynamics (MD) have played an important role in this regard [60–62]. In this section, a few of these techniques and approaches are discussed as well as how they have been and can be used for screening and design of inhibitor chemistries.

9.5.1 Computation of quantum chemical descriptors

Quantum chemical descriptors (QCD) are important parameters to elucidate the physicochemical properties of inhibitor molecules. These parameters of the inhibitors are influenced by their spatial as well as electronic structures. The frontier molecular orbitals such as the highest occupied molecular orbital (HOMO) and the lowest unoccupied molecular orbital (LUMO) plots depict the local electron density distribution in the molecule [63]. To investigate the effect of molecular structure and electronic properties of inhibitors on their inhibition efficiencies, the QCD like the HOMO, LUMO, HOMO-LUMO gaps ($E_{gap} = E_{LUMO} - E_{HOMO}$), dipole moment ($D$), ionization potential (I), electron affinity (A), global hardness (η), absolute electronegativity (χ), global softness (σ), fraction of electron transferred (ΔN), global electrophilicity (ω), and electron back-donation ($\Delta E_{back\text{-}donation}$) are generally computed.

The HOMO and LUMO of the inhibitor molecules are important parameters for defining the chemical reactivity of the molecule [64]. It is reported in the literature that the E_{HOMO} of the molecule is associated with its electron-donating capacity, while the E_{LUMO} is associated with its electron-accepting capacity [65]. Higher HOMO energy and lower LUMO energy generally lead to higher corrosion inhibition efficiencies [66]. The other important parameter that measures the chemical reactivity of the molecule toward its adsorption on the metal surface is the E_{gap}. Generally, the lower the value of E_{gap}, the higher will be the chemical reactivity and consequently better inhibition efficiency [67].

The ionization potential (I), electron affinity (A), global hardness η, absolute electronegativity (χ), global softness σ, fraction of electrons transferred ΔN, global electrophilicity (ω), and electron back-donation ($\Delta E_{back\text{-}donation}$) can be calculated from the following relations Eqs. (9.1)–(9.8):

$$I = E(N-1) - E(N) \tag{9.1}$$

$$A = E(N) - E(N+1) \tag{9.2}$$

where $E(N-1)$, $E(N)$, and $E(N+1)$ are the energies of the molecule after losing one electron (cation), in its basal state (neutral) and after gain of one electron (anion), respectively.

$$\chi = \left(\frac{I+A}{2}\right) \tag{9.3}$$

$$\eta = \left(\frac{I - A}{2}\right) \tag{9.4}$$

$$\sigma = 1/\eta \tag{9.5}$$

$$\omega = \frac{\mu^2}{2\eta} \tag{9.6}$$

$$\Delta N = \frac{\varnothing_m - \chi_{\text{inh}}}{2(\eta_{\text{m}} + \eta_{\text{inh}})} \tag{9.7}$$

$$\Delta E_{\text{back-donation}} = -\frac{\eta}{4} \tag{9.8}$$

where Φ_m and χ_{inh} indicate the work function of the metal surface and the absolute electronegativity of the inhibitor molecule, respectively; η_m and η_{inh} indicate the global hardness of the metal surface ($\eta_m = 0$) and the inhibitor molecule, respectively; μ is the chemical potential and given by the negative of electronegativity ($\mu = -\chi$) [68].

Ionization potential (I), like the HOMO, is an important descriptor of the reactivity of the atoms and molecules [69]. It measures the amount of energy required to remove one electron from a molecule [70]. The lower the value of I the easier it is to remove an electron from a molecule and, thus higher is the electron-donating capability. On the other hand, a higher value of I indicates higher stability and chemical inertness. Thus molecules with a lower value of I generally correlate with a better corrosion inhibition efficiency.

Electron affinity (A) is defined as the change in energy of a molecule when an electron is added to the molecule to form a negative ion. The higher value of A of the molecule creates a partial negative charge on the molecule, which tends to result in a stronger interaction of a molecule with the positively charged metal surface [70]. Thus the enhanced overall adsorption of the molecule results generally results in increased corrosion inhibition efficiency.

The dipole moment (μ) can be used to identify the direction of corrosion inhibition. The dipole moment measures the polarity in a bond and it is related to the electron distribution in a molecule [71]. It is reported in the literature that the adsorption of polar compounds possessing high dipole moments on the metal surface should lead to better inhibition efficiency [72]. Comparison of dipole moments with experimental inhibition efficiency indicates that the percentage inhibition efficiencies of the molecule increase with an increasing value of the dipole moment, which could be related to the dipole-dipole interactions of the molecules and the metal surface [73]. Thus molecules with higher dipole moment values would be expected to work as better corrosion inhibitors. However, in many studies, dipole moment is shown to follow an inverse trend with corrosion inhibition efficiency so caution needs to be taken when selecting molecules based on dipole moments alone [54].

According to Sanderson's electronegativity equalization principle, a molecule with high electronegativity attains equalization spontaneously hence, low reactivity is expected which, in turn, tends to result in low inhibition efficiency [74]. It has been reported that a lower electronegativity correlates with better inhibition efficiency [70].

The global hardness (η) and global softness (σ) are important features of the molecule to measure the stability and chemical reactivity. It is known from the literature that global hardness indicates the resistance toward the deformation or polarization of the electron cloud of the atoms, ions, or molecules [75]. Conversely, the molecule with high value of global softness tends to react with and adsorb on the metal surface more easily. The molecule with the lower value of global hardness (and the higher value of global softness) is expected to show higher corrosion inhibition efficiency [76].

The fraction of electrons transferred (ΔN) from the inhibitor molecule to the metal surface was calculated using Eq. (9.7) as reported by Kokalj [77]. If ΔN <0, electron transfer takes place from the metal to inhibitor molecule and vice versa if $\Delta N >0$ [78]. The ΔN value shows the electron-donating capacity of the molecule. It has been reported that the inhibition efficiency of the molecule increases with increasing electron-donating capacity [76]. However, it has also been noted in the literature that there is not a regular trend in the inhibition efficiency with increasing values of ΔN [79].

The global electrophilicity index, ω, shows the ability of the inhibitor molecules to accept electrons for the partially filled orbitals of the metal. It is a measure of the stabilization in energy after a system accepts an additional amount of electron charge ΔN from the environment [80]. A good electrophile is characterized by high value of ω while a good nucleophile is characterized by a low value of ω. Thus the molecule with the lower value of ω is expected to show higher inhibition efficiency [70].

The electron back donation ($\Delta E_{\text{back-donation}}$) indicates that since $\eta > 0$, $\Delta E_{\text{back-donation}} < 0$, the charge transfer to a molecule, followed by a back-donation from the molecule is energetically favorable [81]. It is directly proportional to global hardness and implies that a less negative value of $\Delta E_{\text{back-donation}}$ should result in better corrosion inhibition efficiency. Thus the inhibitor molecule with less negative $\Delta E_{\text{back-donation}}$, is expected to have the highest inhibition efficiency.

Trends of corrosion inhibition efficiencies have been correlated with QCDs of various molecules. Kumar et al. [82] reported QCDs such as E_{HOMO}, E_{LUMO}, and E_{gap} for Henna extracts and correctly predicted the inhibition efficiency trend. Wazzan and Al-Mhyawi [76] studied the 4-(4-nitrophenyl)-2-oxo-1,2,5,6-tetrahydrobenzo[h]quinoline-3-carbonitrile (NPQC) and 4-(furan-3-yl)-2-oxo-1,2,5,6-tetrahydrobenzo[h]quinoline-3-carbonitrile (FQC) using quantum chemical approach and predicted that NPQC is a better corrosion inhibitor compared with FQC. Zhang et al. [83] evaluated the corrosion inhibition performance of four 1-R-2-undecyl-imidazoline compounds [R = CH_2COOH (A), CH_2CH_2OH (B), $CH_2CH_2NH_2$ (C), and H (D)] for carbon steel using QCDs such as E_{HOMO},

E_{LUMO}, and global hardness, etc., and predicted that corrosion inhibition efficiency varies in the following order: A > B > C > D. Obot and Gasem [84] theoretically evaluated the corrosion inhibition performance of some pyrazine derivatives and predicted that the corrosion inhibition efficiency trend of 2-methylpyrazine (MP), 2-aminopyrazine (AP) and 2-amino-5-bromopyrazine (ABP) using quantum chemical parameters such as E_{HOMO}, the energy gap ($E_{LUMO} - E_{HOMO}$), dipole moment, hardness, softness, and electrophilicity especially in the neutral molecules. Similarly, in several other studies correlations have been reported.

Winkler et al. [85] observed that the correlation between frontier orbital parameters and inhibitor efficiency (IE) disappears when applied to a larger data set of small organic corrosion inhibitors. Also, as articulated by Kokalj et al. [86], there are two major drawbacks in relying solely on this approach of computing the QCDs to predict inhibition efficiency trends—(A) it provides little physical insight into the actual corrosion inhibition mechanism and (B) the correlation between the computed electronic parameters and the experimentally established inhibition efficiencies may be lacking in some cases. Therefore, solely relying on quantum chemical parameters for design of corrosion inhibitor may lead to misleading results.

9.5.2 Metal inhibitor interactions using density functional theory

In order to overcome the shortcomings associated with assessing the inhibitor performance by a simple calculation of the molecule's quantum chemical parameters and to understand the interaction mechanisms between the inhibitor molecule and the metal surface, adsorption studies can be carried out using first principles DFT. There are several computational chemistry software packages such as Quantum Espresso, VASP, Gaussian, Siesta, etc., available for carrying out DFT computations. Numerous papers have been published using DFT as a theoretical tool to compute the interaction energies of inhibitor molecules with the metal surface. The general procedure to carry out such adsorption studies is shown in Figs. 9.6 and 9.7. Where E_{int} is the interaction energy, $E_{complex}$ is the total energy of inhibitor-metal complex, E_{slab} is total energy of the slab, and E_{inh} is the total energy of the inhibitor.

The more negative the interaction energy, the stronger is the adsorption of the molecule on the metal surface and therefore the better the corrosion inhibition [82].

Generally, several initial orientations (such as horizontal, vertical, and inclined) of inhibitors are considered on the metal surface to identify the most energetically favored configurations. The QCDs for henna extracts showed that the corrosion inhibition efficiency varies in the following order lawsone > gallic acid > D-glucose [82]. The computed interaction energy also follows the same trend as observed in QCDs and experimental inhibition efficiency. The optimized structures of lawsone, gallic acid, and D-glucose on the metal iron surface show that there is chemical interaction between henna extracts and the iron [82].

Optimize the bulk metal or alloy ground-state structure under consideration

↓

Create various crystallographic surfaces of the metal and identify the most stable surface (having the lowest total energy)

↓

Optimize the inhibitor molecule using same level of theory

↓

Place the molecule on the top of the most stable surface and insert sufficient vacuum (typically > 15 Å) along the z-direction to avoid self-interaction

↓

Finally, relax the inhibitor-metal complex

↓

Compute interaction energy using the following expression:
$E_{int} = E_{complex} - E_{slab} - E_{inh}$

FIGURE 9.6

General procedure to carry out adsorption studies.

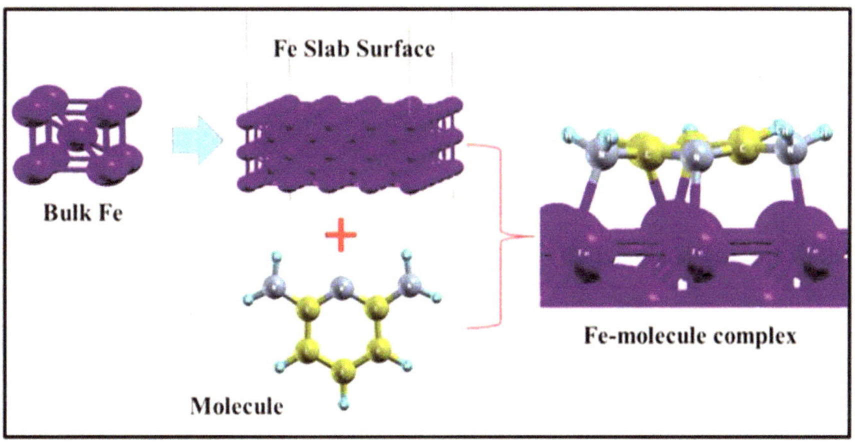

FIGURE 9.7

Adsorption computation methodologies.

Guo et al. [87] carried out extensive DFT simulations using density functional tight-binding (DFTB) method to understand the corrosion inhibition mechanism of three chalcone derivatives: 3-(4-(dimethylamino)phenyl)-1-(4-hydroxyphenyl)

prop-2-*en*-1-one (Inh1), 3-(3-hydroxyphenyl)-1-(4-hydroxyphenyl)prop-2-*en*-1-one (Inh2), and 1-(4-hydroxyphenyl)-3-phenylprop-2-*en*-1-one (Inh3) which were already shown to be good corrosion inhibitors under the specific conditions with inhibition efficiencies 91.3%, 87.0%, and 85.9%, respectively. The computed HOMO-LUMO gaps for Inh1, Inh2, and Inh3 were 2.56, 2.57, and 2.46, respectively, suggesting that the inhibition efficiency varies in the following order: Inh3 > Inh1 > Inh2 which is not in line with the experimental inhibition efficiencies [88]. The adsorption of all the inhibitors was carried out on Fe (110) surface and the magnitude of computed interaction energy was found to be maximum for Inh2 and minimum for Inh1 [87]. Thus, both the computed HOMO-LUMO gaps and interaction energies do not follow the expected trend. These results signify the importance of environmental conditions on the inhibition efficiency.

To address the problem of environmental conditions, implicit solvation models for plane-wave DFT codes were developed by Fattebert and Gygi [89]. The implicit solvent theory (IST) describes a solvent system as a continuum model. There are several ISTs and some of them are already implemented in Quantum ESPRESSO. Nishihara and Otani implemented one such IST, the reference interaction site method (RISM) [90]. The RISM is a classical solution theory and solution is treated using a classical force field. The RISM has been successful in describing the solvation structures around various isolated or biological molecules using a classical force field [88] or Gaussian-based ab-initio calculations [91].

Furthermore, an effective corrosion inhibitor molecule does not only adsorb on the material surface but can also form a dense, compact, and defect-free protective layer on it. Several studies have been reported in the literature where the effect of surface coverage has been studied. Kumar et al. [92] studied the surface coverage effect of amino acids on copper surface and reported that the intermolecular interactions play a significant role in formation of a protective layer. Jiang and Adoms [93] investigated the adsorption mechanism of benzotriazole on Cu (111) surface and reported that the physisorption of the molecules and the formation of an intermolecular H-bonding network results in a thin and protective film.

9.5.3 Molecular dynamics study

Although DFT has proven to be a powerful tool in qualitatively predicting the corrosion inhibition trends of novel molecules, it is currently quite expensive. It can only be used for small systems containing 100–200 atoms as extremely complex and time-consuming computations accompany it. To overcome these drawbacks, MD simulations have been extensively used for corrosion inhibition studies. Using MD, one can not only simulate larger systems but also for longer times such as microseconds to milliseconds apart from explicit inclusion of environmental conditions. A MD calculation typically involves several steps, which are shown in Fig. 9.8.

These steps are carried out until the forces on each atom fall below the threshold criterion. Interaction energies in the gas phase are computed using the same

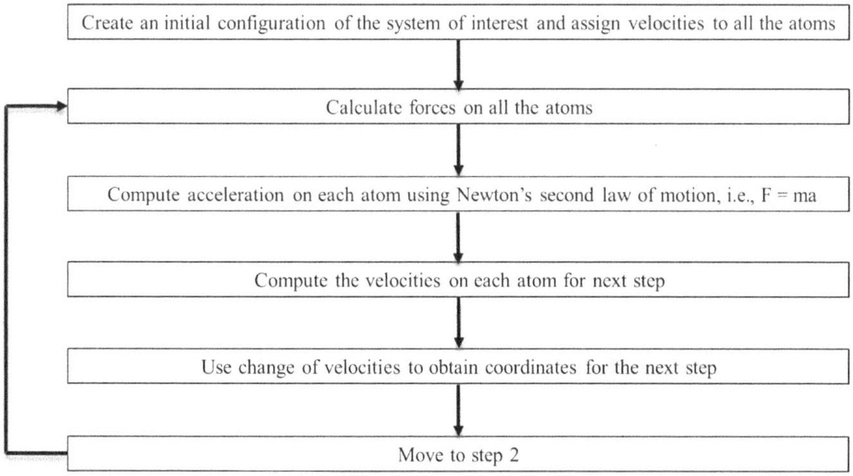

FIGURE 9.8

Steps involved in a molecular dynamics' calculation.

correlation reported in the DFT section, while those in the solvent phase are computed using the following correlation [Eq. (9.9)]:

$$E_{int-sol} = E_{tot} - [E_{mol} + N_{*}E_{sol} + E_{surf} + E_{mol-sol} + E_{sol-surf} + E_{sol-sol}] \tag{9.9}$$

where $E_{int\text{-}sol}$ is the interaction energy in the solvent phase, E_{tot} is the total energy of the complex, E_{mol} is the total energy of the molecule, E_{sol} is the total energy of one solvent molecule, E_{surf} is the total energy of the surface, $E_{mol\text{-}sol}$ is the interaction energy between the solvent molecule and the inhibitor molecule, $E_{sol\text{-}surf}$ is the interaction energy of the solvent with the surface, $E_{sol\text{-}sol}$ is the interaction energy of the solvent molecule with another solvent molecule, and N is the number of solvent molecules.

Guo et al. [94] investigated the corrosion inhibition performance of 1,3-thiazole (TA), 2-amine-1,3-thiazole (2-ATA), and 4-amine-1,3-thiazole (4-ATA) using QCD and MD. Using computed QCD such as E_{HOMO}, E_{LUMO}, ΔE, η, χ, S and ΔN, the authors predicted that the inhibition efficiencies vary in the following order: 4-ATA > 2-ATA > TA. Additionally, in order to better understand the corrosion inhibition mechanism of all the three inhibitors with Fe (110) surface and to compute the interaction energies, MD was used. It was observed that the interaction between inhibitor and Fe (110) surface occurs mostly through heteroatoms and p-electrons of the thiazole moiety and the magnitudes of interaction energy vary in the following order: 4-ATA > 2-ATA > TA, in line with the experimental inhibition efficiencies [94].

Similarly, Obot and Gasem [84] evaluated the corrosion inhibition performance of MP, AP, and ABP using quantum chemical calculation and MD simulations. The computed E_{HOMO}, the energy gap (E_{LUMO} - E_{HOMO}), dipole moment, hardness, softness, and electrophilicity showed that the corrosion inhibition efficiency varies in the following order ABP > AP > MP. Furthermore, to understand the corrosion inhibition mechanism, MD simulations were conducted. The computed binding energies vary in the following order ABP > AP > MP, which was also consistent with the experimental inhibition efficiencies [84].

Although MD is useful in simulating larger systems, it is difficult to gain information about interaction mechanisms, as conventional MD does not consider bond breaking and bond formation. To overcome these shortcomings, reactive force field (ReaxFF) has been developed. ReaxFF has capabilities of classical MD as well as DFT, that is, it can be used to simulate larger systems with accuracy close to DFT [95]. Kumar et al. [96] evaluated ReaxFF's developed by Monti et al. [97] for imidazole derivatives and copper systems against DFT and successfully captured experimental and DFT adsorption energy trends using ReaxFF.

9.5.4 Monte Carlo simulations

Monte Carlo (MC) simulations use stochastic methods to generate new configurations of a system of interest. The steps involved in MC simulations are given in Fig. 9.9.

The steps are followed until the system reaches equilibrium. The interaction energy is computed using the expression given in the DFT section.

Erdoğan et al. [98] studied corrosion inhibition performances of novel quinolone derivatives, namely, 2-amino-7-hydroxy-4-phenyl-1,4-dihydroquinoline-3-carbonitrile (Q1), 2-amino-7-hydroxy-4-(p-tolyl)-1,4-dihydroquinoline-3-carbonitrile (Q2), 2-amino-7-hydroxy-4-(4-methoxyphenyl)-1,4-dihydroquinoline-3-carbonitrile) (Q3), and 2-amino-4-(4-(dimethylamino)phenyl)-7-hydroxy-1,4-dihydroquinoline-3-carbonitrile (Q4), against the corrosion of iron. The computed QCDs such as E_{HOMO}, E_{LUMO}, HOMO-LUMO energy gap (ΔE), chemical hardness (η), softness (σ), electronegativity (χ), chemical potential (μ), electrophilicity (ω), nucleophilicity

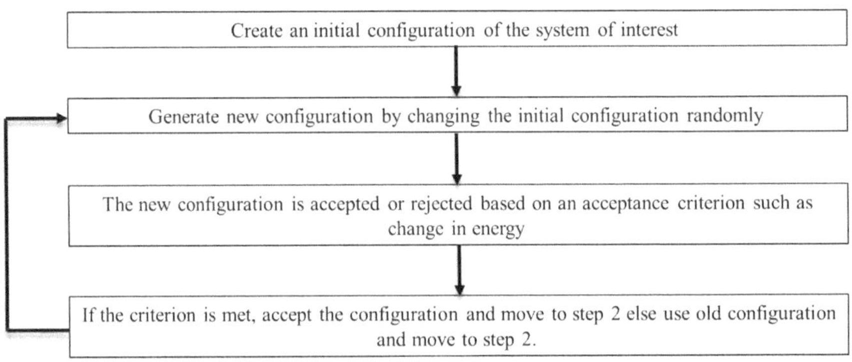

FIGURE 9.9

Steps involved in Monte Carlo simulations.

(ε), electrons transferred from inhibitors to metal surface (ΔN), etc., were computed and found that corrosion inhibition efficiency varies in the following order: Q4 > Q3 > Q2 > Q1 [99]. Adsorption study of Q1, Q2, Q3, and Q4 on Fe (110) surface was carried out using MC Simulations. The magnitude of computed interaction energy was found to be a maximum for Q4 and minimum for Q1 and varies in following order: Q4 > Q3 > Q2 > Q1, which is in line with the QCD and experimental inhibition efficiencies [99]. Similarly, Khaled et al. [100] studied the adsorption of alanine on iron surface using MC simulations. It was observed that alanine adsorbs on Fe surface through the nitrogen/oxygen atoms with the lone pair electrons in its molecule [100].

Guo et al. [101] investigated the corrosion inhibition performance of 4-chloroacetophenone-O-1'-(1', 3', 4'-triazolyl)-metheneoxime (CATM), 4-fluoro-acetophenone-O-1'-(1', 3', 4' triazolyl)-metheneoxime (FATM), and 3,4-dichloro-acetophenone-O-1'-(1', 3', 4'-triazolyl)-metheneoxime (DATM) using DFT approach. The computed QCD such as electrophilicity index, HOMO-LUMO energy gap, and global softness, etc., display the trend FATM < CATM < DATM, whereas electronegativity and the fraction of electrons transferred display the opposite trend, that is, FATM > CATM > DATM. Therefore, it is difficult to predict the trend of novel molecules based only on QCDs. To further investigate the inhibition efficiency trend, adsorption energy of all three molecules on Fe (110) surface was computed using MC simulations. The adsorption energy magnitude was found to be a maximum for DATM and a minimum for FATM which is in agreement with the experimental inhibition efficiencies, that is, DATM > CATM > FATM. Obot et al. [99] also carried out QCD and MC simulations of triazine derivatives. Both the QCDs and adsorption energy (both in vacuum and gas phase) accurately predicted the experimental inhibition efficiency trends.

9.5.5 Synergistic effect study using molecular modeling

Corrosion inhibitor active chemistries are generally used in conjunction with other chemistries, some of which are referred to in the oilfield chemical industry as intensifiers or synergists. When two corrosion inhibitor chemistries are used together or in conjunction with an intensifier, synergistic effects can be seen, resulting in an increase in corrosion inhibition performance of the mixed chemistries as compared to the individual ones. A few papers are reported in the literature related to molecular modeling of synergistic effects of corrosion inhibitors. Han et al. [102] established the concept of molecular orbital matching principle to identify the synergism between molecules. According to the molecular orbital matching principle, molecules having HOMO energy closer to metal Fermi energy donate electrons to the metal surface and molecule having LUMO energy closer to metal Fermi accepts electrons from the metal surface, thus leading to synergistic interactions. Synergism also exists if the interaction energy of the mixed inhibitor is more than the sum of the interaction energies of the individual molecules. The intermolecular interactions also play an important role in the synergism.

Qiang et al. [103] studied the synergistic effect of tartaric acid (TA) with 2,6-diaminopyridine (DAP) on the corrosion inhibition of mild steel in 0.5M HCl. DAP and TA are low cost and environmentally friendly molecules. DAP was shown to exhibit a maximum inhibition efficiency of 74.5% at a concentration of 10.0 mM in 0.5M HCl, while TA shows comparatively lower efficiencies of 68.8% at a concentration of 5 mM in 0.5M HCl. On the other hand, when DAP and TA were used in a mixture, the inhibition efficiency increased to 92.1%. Although DAP and TA are not good corrosion inhibitors individually, corrosion inhibition efficiency increases significantly when used together due to the synergism. We have used DFT computations to understand the synergistic effects between DAP and TA and to build a molecular modeling framework to predict synergism between new molecules. From the molecular orbital plot (Fig. 9.10), it is evident that the HOMO energy of DAP is close to the Fermi energy of Fe while the LUMO energy of TA is close to the Fermi energy of Fe. Hence, as per the molecular orbital matching principle, when DAP and TA co-adsorb, synergism is gained. Subsequently, interaction energy computations were carried out for the individual molecules as well as the mixture (1:1 ratio). It can be seen from Table 9.3 that the magnitude of interaction energy for the mixed system is more than the sum of individual interaction energies (magnitude). It is also evident from Table 9.3 that the computed interaction energy follows the same trend as observed in experimental inhibition efficiency. As indicated by Fig. 9.11 (A) and (B), both DAP and TA chemically bond with the iron surface. Additionally, it can be seen from Fig. 9.11 (C) that the intermolecular interactions

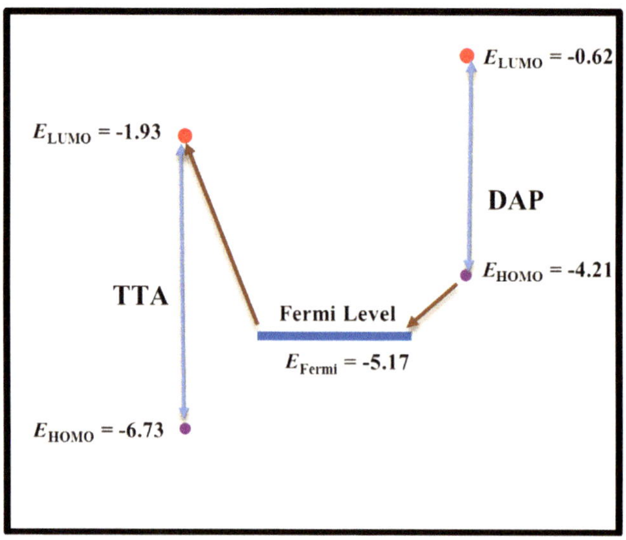

FIGURE 9.10

Frontier molecular orbital energy positions of DAP + TTA molecules (Fermi energy of Fe (001) = −5.17).

Table 9.3 Computed interaction energies of DAP, TA and DAP-TA (1:1 mix).

	Interaction energy (kcal/mol)		
Molecule	**Flat**	**Vertical**	**Inhibition efficiency (%)**
DAP	−55.45	−39.80	74.5
TA	−40.92	−17.97	68.8
DAP-TA (mix)	−105.53		92.1

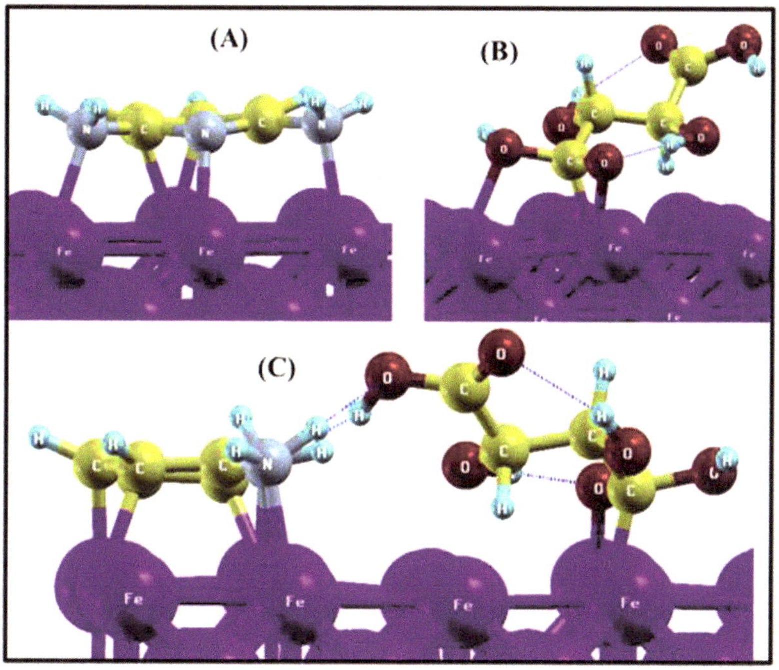

FIGURE 9.11

Optimized adsorption geometries of DAP, TTA, and DAP + TTA on Fe (001) surface.

are mainly driven by hydrogen bonding (indicated by the dotted lines) which significantly contributes to the synergism.

9.5.6 Screening of corrosion inhibitors using machine learning

Atomic scale modeling of materials has played a crucial role in better understanding the origins of a materials' functionality as well as design of novel materials for various applications. Quantum mechanics based methods such as DFT, in combination with molecular simulation techniques such as MC and/or MD have played a central role in describing the atomic structures, physical and chemical properties of materials. Recently, machine learning (ML) and artificial intelligence (AI) have emerged as powerful tools for automating various processes as well as for predictive modeling, utilizing historical, and multivariate data. Inspired by the success of data-driven modeling and optimization of various business and industrial processes, the chemical and material science community has started to embrace ML to address various problems such as the discovery of novel materials, prediction of material properties like band-gaps, bulk and shear moduli, molecular structure, structure-property relations, and tuning and defining of force-fields.

The success of the quantitative structure–activity relationship (QSAR) in predicting corrosion inhibition efficiencies has stimulated the interest of researchers. The major advantage of QSAR approach lies in the fact that it does not require any experimental properties or process parameters; rather, it utilizes only chemical structure parameters to predict corrosion inhibition efficiencies. However, it requires corrosion inhibition efficiencies of various molecules in order to build the predictive QSAR model. Once the model is built, it can be utilized to predict the inhibition efficiencies of unknown molecules. QSAR-based methods utilize molecular descriptors to predict inhibition efficiencies. To date, there are more than 5000 molecular descriptors being reported and computed using different software [104]. The first step in predictive model building is the identification of the most important descriptors. The general procedure involved in QSAR predictive modeling is summarized in Fig. 9.12.

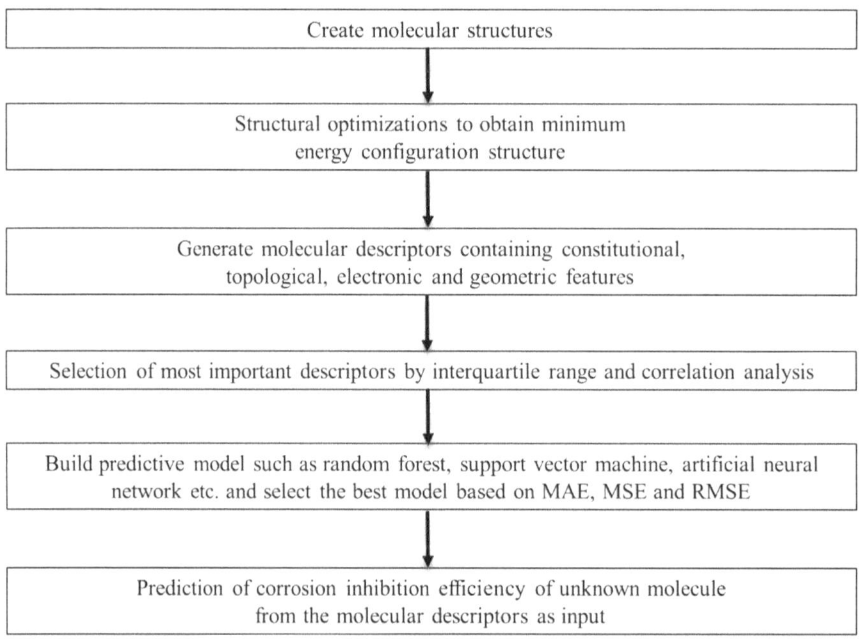

FIGURE 9.12

General procedure involved in quantitative structure–activity relationship predictive modeling.

Several studies have been reported in literature for different classes of molecules. Zhao et al. [105] built a nonlinear quantitative structure–activity relationship model using support vector machine and successfully predicted the inhibition efficiencies of five newly designed amino acids. Liu et al. [106] carried out QSAR modeling for benzimidazole derivatives and built an optimal support vector machine (SVM) model having 11 molecular descriptors. Using the SVM model, the authors predicted the inhibition efficiencies of six newly designed benzimidazole molecules out of which two of them were having inhibition efficiencies more than 90%. Olasunkanmi et al. [107] built a QSAR model to predict inhibition efficiency and found that inhibitive performances of the molecules correlated well with their molecular weights, energy gap, fraction of electron transferred (ΔN), and binding energy. Similarly, Murulana et al. [108] also built a QSAR model and found that inhibition efficiency increases with the increase in ω, E_{LUMO}, and η and decreases with a decrease in both dipole moment (μ) and partition coefficient (logP). While most authors have developed QSAR models, which are applicable only for a particular class of molecule; recently, Agarwal et al. [109] have developed a robust computational predictive model that is able to predict inhibition efficiency of the molecules across multiples classes with reasonable accuracy.

9.6 Corrosion inhibitor performance evaluation

Carbon steel is the most used metal in oil and gas operations [110]. However, it is prone to corrosion under most oil and gas production, transportation, and storage conditions and requires appropriate corrosion protection. For internal corrosion, the use of suitable corrosion inhibitors can suppress corrosion to a manageable level, avoiding the need to use more expensive corrosion-resistant alloys (CRAs) or dehydration equipment. Most corrosion inhibitors used for oilfield applications are film-forming organic inhibitors [110]. They can be injected concurrently with the production stream at a relatively low dosage rate (continuous treatment) or intermittently at high concentration (batch treatment). The ultimate goal is to reduce CR to an acceptable level without compatibility issues or impacting downstream processes.

Before a chemical product can be approved for field use, a corrosion inhibitor qualification program is conducted which involves a range of performance and compatibility assessments. Corrosion inhibitor qualification is relevant not only for new oil and gas fields but also existing ones where conditions have changed over time. Changes in environmental and operational parameters leading to different corrosivity, lifetime extension requirements, environmental concerns, discovered damages, and change of a chemical supplier can all necessitate a requalification of corrosion inhibitors. Fig. 9.13 presents a typical process of a corrosion inhibitor selection and qualification after the need for an evaluation program is identified.

This section reviews test methodologies for performance evaluation and compatibility tests. Various monitoring techniques to assess corrosion inhibitor

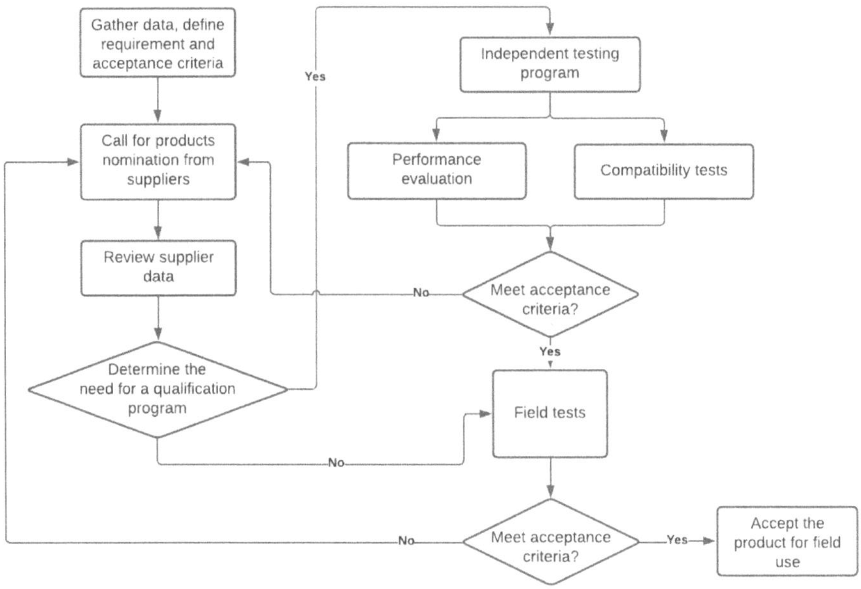

FIGURE 9.13

Corrosion inhibitor selection and qualification process.

performance are also presented, along with more advanced surface characterization techniques used in mechanistic studies. It should be noted that, because a laboratory study cannot entirely replicate complex field conditions, inhibitor performance in the field cannot be precisely estimated. As such, the laboratory study systematically compares the behavior of various inhibitor candidates at various concentrations to assess how they respond to critical parameters, such as temperature, solution chemistry, and flow conditions, etc. Before selecting the best product for final field deployment after laboratory testing and assessment, a further dose rate optimization based on monitoring or inspection in a focused field trial may be required.

Various test methodologies have been developed and utilized to replicate a wide range of corrosion phenomena anticipated in services. These tests aim to examine the behavior of a corrosion inhibitor under simulated conditions and to ultimately demonstrate that the required designed life of the equipment can be met, which includes both general and localized corrosion. A commonly used target that an effective inhibitor is expected to reduce the CR of carbon steel to is 0.1 mm/year [111]. However, the acceptance criteria may vary from asset to asset. In this section, a variety of test methods to evaluate the corrosion inhibitor performance and so-called secondary properties are presented. The methodologies presented here are not an exhaustive list as there are variants to these as well as other options, but they constitute the more common tests currently used in research laboratories and the industry. It is also appreciated that bacteria and their byproducts can also induce corrosion but, since it is believed to be more effective to treat the cause of a problem rather than the symptoms and that these microbiological issues are most effectively attended to by biocide rather than by corrosion inhibitors, these corrosion issues are not considered here.

Inhibitor testing can be as simple as a glass cell immersion evaluation or as complex as high-pressure and high-temperature flow loop test. Commonly, when there are many nominated chemicals to be screened, simplified tests are initially preferred to reduce cost and time. After narrowing down to a few promising products, additional performance evaluation tests can be incorporated which are designed to mimic as closely as possible the field operating conditions. Corrosion inhibitor concentration profiles of leading candidate products along with underdose and overdose assessments can also assist in the decision-making process in deciding to opt for one product over another. It is, however, not always practical to replicate every nuance of the field conditions. Attempts should be made to identify and mimic the critical corrosion-influencing factors. These are often temperature, pressure (partial pressure of corrosive gases), solution chemistry, the presence of sand/deposits, and hydrodynamics.

Computation fluid dynamic modeling can be performed before the performance testing to identify flow regimes, hydrodynamic factors, and potential stagnating areas, where solids and debris may accumulate with associated underdeposit corrosion (UDC) risk.

To obtain meaningful results from any corrosion tests, four main aspects should be considered when designing a corrosion experiment. These are (1) metal samples, (2) solution chemistry, (3) test parameters, and (4) assessment techniques. Details of each facet are described below.

9.6.1 Metal samples

The selection of metal samples for use in any corrosion tests is critical. Steel coupons machined from an actual pipe sample to be protected by the inhibitor are the most representative. Not only can the chemical composition and microstructure of the steel affect the uninhibited CR, but they can also influence inhibitor adsorption and the resulting efficacy. Lopez et al. reported that benzimidazole promoted corrosion on quenched and tempered steel while showing an inhibitive effect on annealed samples in pH 6, 5% NaCl saturated with CO_2 [112]. The authors attributed the findings to the differing level of cementite in the microstructure. On the contrary, Paolinelli reported that imidazoline-based commercial corrosion inhibitor performed better on quenched and tempered steel than on annealed samples [113]. To date, the effects of steel microstructure on the inhibitor behavior are yet to be fully understood and are often not considered in laboratory tests. This is one reason why field testing is important.

In addition, it has been shown that the presence of welds can affect the inhibition efficiency. Therefore the metal samples used in the performance evaluation tests should incorporate welds. Further details on test methodologies to study preferential weld corrosion (PWC) are given later (see Section 10.6.3.10).

9.6.2 Solution chemistry

The test solution chemistry should replicate that of the field as much as practical. The corrosivity of the aqueous phase largely depends on its composition, acid gas (CO_2 and/or H_2S) partial pressure, organic acid concentration, and temperature. A scale prediction software may be used to determine the scaling tendency of test brine composition. The test solution containing bicarbonate and divalent ions such as calcium may, in some instances, form scale and precipitates, as do the presence of other scaling ions at sufficient concentrations. Separating the scaling species of the brine into two portions (e.g., one with bicarbonate and the other with calcium) and mix them after sparging with carbon dioxide is one approach that can minimize unwanted precipitation. In systems with high scaling brines, a scale inhibitor (SI) may also need to be deployed so such tests including a SI, which would not only inhibit the scaling in the corrosion test but also better simulate the produced fluids, can be considered. Further, the presence of other chemicals used to treat noncorrosion problems can impact the corrosivity. For example, thermodynamic hydrate inhibitors (glycols) can reduce the CR of carbon steel in sweet (CO_2-containing environment) [114–116]. However, under sour (H_2S-containing environments), localized corrosion and pitting remained persistent in the presence of glycols [117].

Typically, the oil and gas production environment is anoxic and so dissolved oxygen must be removed from the test solution by presparging with the test gas which, in most cases, is CO_2, H_2S, or a mixture of both. The deoxygenation step can also be carried out with inert gas, such as nitrogen or argon, before the test gas is introduced. The use of inert gases for deaeration is sometimes preferred in instances where even a small amount of corrosion leading to increased levels of dissolved iron in the brine in the "pretest" stage may result in test artifacts being generated. For example, if carbon dioxide sparged fluids are transferred to a closed inventory test cell containing a test steel coupon in preparation for a high-temperature test (e.g., 200°C), the corrosion that takes place during the heating stage to test temperature may generate sufficient dissolved iron to result in iron carbonate formation causing a reduction in the observed CR. The presence of oxygen contamination not only enhances corrosion by promoting an additional cathodic reaction (oxygen reduction reaction) but oftentimes negatively impacts the inhibition efficiency of typical oilfield products. Continuous monitoring of dissolved oxygen is recommended especially in high flow conditions, such as in a jet impingement cell (JIC) test [118].

Furthermore, several oilfield production chemicals may be used concurrently. Complex chemistries of other chemical packages at varying dose rates can impact the performance of the inhibitors. Xiong et al. reported that corrosion inhibitor performance was affected by the presence of a biocide but unaffected by a specific

SI [119]. Corrosion inhibitors can have an adverse effect on the performance of hydrate inhibitors and vice versa [120,121]. Unsaturated fatty acids and their derivatives (imidazolines) are subject to vulcanization in the presence of elemental sulfur which could arise from H_2S mixed with oxygen [122]. To assess the performance of a corrosion inhibitor chemistry under the most representative field environment, other production chemicals such as biocides, SIs, oxygen scavengers, paraffin inhibitors, asphaltene inhibitors, thermodynamic hydrate inhibitors, for example, methanol and glycol, low dosage hydrate inhibitors including antiagglomerant and kinetic hydrate inhibitors, etc., should also be incorporated in the test methodology. Likewise, the impact of the corrosion inhibitor on the performance of these other process chemicals in their respective tests should also be considered.

9.6.3 Test parameters

Direct variables such as temperature and pressure can be closely replicated in the laboratory. Test cells such as those shown in references [123,124] can be used at atmospheric pressure while pressure vessels as used in reference [125] simulate more elevated pressure environments. However, the flow dynamics and regimes in the fields cannot always be directly simulated in the laboratory.

While flow loops tend to mimic the hydrodynamics of pipelines reasonably well other types of laboratory apparatus (e.g., rotating cage autoclave), which may need to be used due to availability, resource or space restrictions, are less aligned so hydrodynamic parameters such as shear stress, mass transfer coefficient and Froude's number (in case of slug flow), etc., are used to correlate the flow conditions. Although it has been shown that shear stress had little effect on the adhesion of inhibitor molecules under typical operating conditions [126], hydrodynamics impacts other processes, for example, the mass transfer of cathodic reactants and various species to and from the metal surface, which affects corrosivity, erosion rates, and precipitation of solids including corrosion products.

Hydrodynamic parameters are not constant over the length or circumference of pipelines as they are affected by topography, pipe geometry (bends, elbows, flanges, joints), etc. Flowrates, temperature, and pressure also change over the length of a pipeline. These variations can affect both performance and solubility of chemicals. Depending upon the conditions, scales such as iron carbonate or calcium carbonate and various iron sulfides may exist on the metal surface which in turn can impact the adsorption and effectiveness of corrosion inhibitor chemistries [127]. Furthermore, where unique corrosion scenarios, such as UDC, weld corrosion, top-of-the-line corrosion (TLC), etc., are suspected, it is important to mimic the environment and materials to understand the behavior of the nominated corrosion inhibitors. This section introduces more specialized experimental setups designed to simulate various complex corrosion scenarios.

Another important consideration when designing a corrosion test is the exposure duration. Types of corrosion, measurement methods, and experiment setup determine the appropriate exposure duration. For instance, accelerated electrochemical tests, such as potentiodynamic polarization (PP), typically require only a few hours of exposure. Long exposure time (weeks to months) is often required to evaluate localized corrosion behavior, while shorter test duration (days to weeks) may be acceptable for general CR determination. It has been proposed that a minimum of 2 weeks of exposure is required to reach a stable pit growth stage [128]. Additionally, pit propagation is not a linear process. It may take several days for a pit to initiate and its rate can change with time—both increasing and decreasing. Therefore tests should ideally be performed at various exposure times to assess the trend of pit growth. Maintaining the solution chemistry, especially in longer term tests, is imperative. Selecting the appropriate metal surface area to solution volume ratio in closed-inventory tests can minimize the buildup of dissolved metallic ions produced from the corrosion process, which may lead to an unwanted corrosion product scale formation. By doing so some consideration may need to be given to possible unintended consequences. For example, reducing the size of the working electrode (WE) may assist in reducing the dissolved iron in the test fluids but, if too small (such as opting to use a single microelectrode), it may also result in a reduction in the cathodic to active anodic pitting surface area which may in turn hinder pit propagation resulting in artificially lowering the extent of pitting taking place.

Selection of test methodologies depends on the risks associated with the application, time, resources, equipment availability, and the complexity of test conditions. Some of the common test methodologies used for corrosion inhibitor performance evaluation are summarized in Fig. 9.14 and the details are given herein. It should be noted that Fig. 9.14 is not all inclusive since deviations may be needed for specific cases. For example, dissolved oxygen may be needed to be introduced in the test gas to evaluate corrosion inhibitor performance under simulated field conditions where oxygen ingress is occurring.

9.6.3.1 Wheel box test (or rotating wheel test)

During the preliminary screening of corrosion inhibitor candidates where several inhibitor chemistries are considered, their performances are often ranked in the so-called wheel box test. The apparatus consists of a temperature-controlled oven with rotating racks housing small test cells containing a metal sample immersed in the test solution. The method has also been used to determine the persistency of corrosion inhibitors [129]. However, the wheel box test has limitations due to the uncontrollable nature of the flow [130]. Corrosion rates are estimated by measuring the mass loss of the coupon over a defined time period. The wheel box has been used as a coarse screening approach to remove poorly performing options before other, more robust and more resource-intensive, test methodologies are employed to identify the best performing inhibitor for use in a given environment.

9.6.3.2 Bubble tests/kettle tests

Another commonly employed test to screen corrosion inhibitors is the bubble test, sometimes referred to as a kettle test, where stationary metal electrodes are

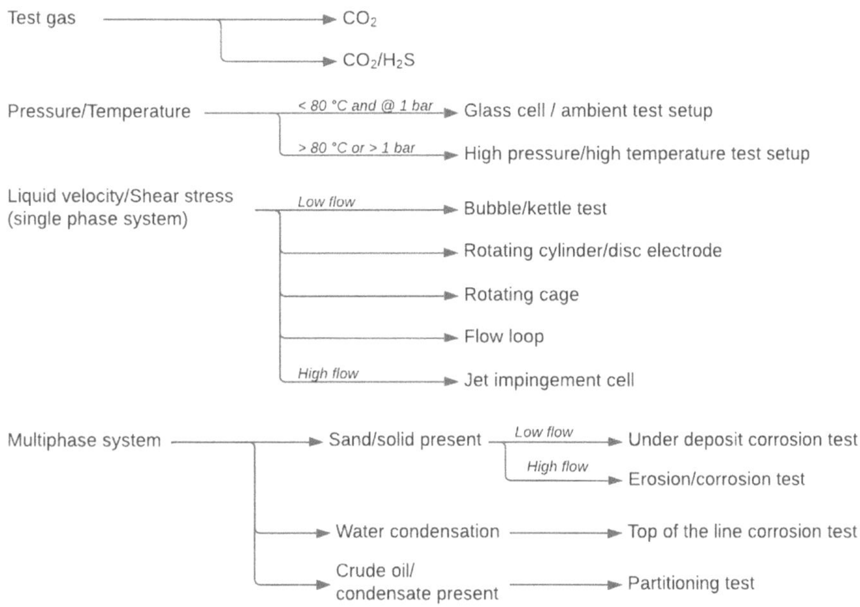

FIGURE 9.14

Test methodology selection.

immersed in the test brine solution, often within a glass cell on a stirred heater plate, although other configurations are also possible. If required, a hydrocarbon phase (oil, condensate, kerosene, etc.) is added on top of the test brine at a comparable oil to water ratio observed in the field. The presence of the hydrocarbon phase enables the effects of partitioning on the inhibition efficiency to be determined. The required inhibitor is generally added at desired dose to the hydrocarbon phase and the solution is gently stirred with a magnetic bar to minimize temperature gradients in the cell. The bubbling of the test gas through a glass frit toward the bottom of the aqueous phase can also provide agitation to the fluids but, if simulating a more stagnant environment is desired, the fluids can also be blanketed with the flowing test gas. Corrosion rates are monitored using electrochemical techniques, usually LP test, for about 24 hours or until a steady-state corrosion rate is obtained.

9.6.3.3 Rotating cylinder electrode

The controlled flow and well-defined hydrodynamics, and the uniform current-potential distribution of the rotating cylinder electrode (RCE) make it one of the more commonly used apparatus for corrosion and corrosion inhibition studies under flowing conditions than the wheel box or bubble cell tests. Fig. 9.15 shows RCE setups for use under atmospheric pressure and at elevated temperature and pressure. Walsh et al. provided a thorough review of the RCE and its applications

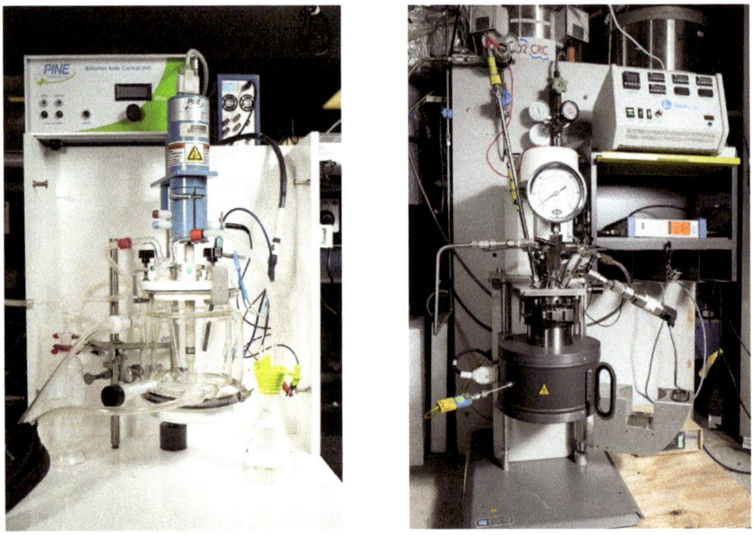

(A) (B)

FIGURE 9.15

Photographs of rotating cylinder electrode setups for (A) atmospheric pressure and (B) at high pressure/high temperature.

Photographs from Curtin Corrosion Centre.

in corrosion engineering [131]. In a typical RCE setup, a carbon steel electrode is mounted on a rotating shaft, which provides an electrical connection for electrochemical monitoring. The shaft connects to a modulated speed rotator which controls the rotational speed of the electrode. As the electrode rotates, the wall shear stress exerted on the electrode surface for a single-phase flow can be calculated using Eq. (9.10) [132]. It is assumed that the hydrodynamic condition in a straight pipe can be mimicked by maintaining the similar hydrodynamic parameters such as wall shear stress or mass transfer coefficient in the RCE setup [132].

$$\tau_{\mathrm{RCE}} = 0.0791 Re^{-0.3} \rho r^2 \omega^2 \tag{9.10}$$

where τ_{RCE} is the shear stress at the electrode surface (pa), Re is Reynolds number (dimensionless), ρ is density (kg/m^3), r is electrode radius (m), ω is angular velocity of the electrode (rad/s).

The RCE setup has been extensively used to investigate corrosion inhibitor performance [133–136]. Gulbrandsen et al. used an RCE setup to demonstrate the effects of oxygen contamination on the efficiency of imidazoline, quaternary ammonium, and amine-based corrosion inhibitors [135]. Tan et al. conducted electrochemical noise (EN) measurement using a dual RCE setup to monitor the desorption of an imidazoline corrosion inhibitor as the electrode was transferred from an inhibited to an uninhibited solution [136].

9.6.3.4 Rotating disk electrode

Rotating disk electrode (RDE) is employed to investigate the effects of hydrodynamics on the kinetics of the corrosion process. An RDE setup is more commonly used in fundamental research investigations rather than in industry application. An RDE setup is similar to that of an RCE (see Section 7.3.3, RCE). The primary difference is the electrode geometry. The test specimen electrode is a attached to the end of the rotating shaft, facing downward. As the shaft rotates, an upward flow of the electrolyte and chemical species occurs. Contrary to an RCE, an RDE maintains a laminar flow regime over large rotational speeds and is used to mimic conditions under low shear stress [130,133]. Electrochemical measurements are performed on an RDE allowing real-time corrosion monitoring and corrosion mechanism study.

9.6.3.5 Rotating cage

The RC apparatus was developed to simulate the hydrodynamic condition of a straight horizontal pipe. Papavinasam et al. compared corrosion (both uniform and localized) on carbon steel in several common corrosion test setups and concluded that the RC produced corrosion attack that closely resembled those found in pipelines [130]. They also found that the RC generated the worst form of corrosion damage, namely pitting corrosion, which is considered by many to provide a conservative approach for corrosion inhibitors qualification [130].

The setup consists of a rotating shaft connected to a motor and a controller, a RC, a test vessel, and steel coupons. Multiple coupons are mounted in between 2 nonmetallic and inert disks, forming a cage. Flat and relatively large coupons are used where weldment specimens may also be incorporated. Fig. 9.16 shows a schematic of a RC assembly. Corrosion activity and inhibitor performance are assessed using nonelectrochemical-based techniques such as gravimetry and microscopy. Although the large surface area enables statistical analysis of pitting corrosion, care must be taken that the test brine volume to metal surface area ratio is such that a large build-up of dissolved iron within the closed-inventory test is minimized to lessen test artifacts such as iron carbonate formation and its associated effects (e.g., reduced CR), which may not be seen in the field.

As the cage rotates, different areas of the test coupons experience varying hydrodynamic effects. The leading edge tends to suffer higher wall shear stress than the bulk surface and the trailing edge as determined electrochemically using the Lévêque Equation (correlation between current density and shear stress) [138,139]. The shear stress and consequently mass transport of reactive species are not uniform across the surface since they depend on cage geometry, rotational speed, coupon thickness, the presence and location of baffles, solution properties, etc. [140], while the environment on the inside face of the coupon can be quite different to that of the outside coupon face. The poorly defined hydrodynamics can complicate the interpretation of the results. The average wall shear stress is commonly determined on the outside face of the coupon according to Eq. (9.11),

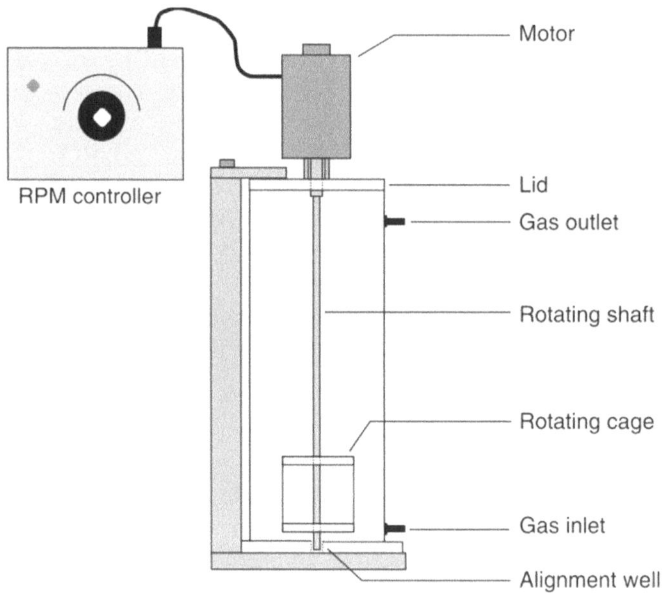

FIGURE 9.16

A schematic of a rotating cage setup.

Reproduced with permission from S. Papavinasam. Rotating cage and jet impingement techniques, in: L. Yang (Ed.), Techniques for Corrosion Monitoring, first ed., Woodhead publishing, 2008, pp. 322–343.

which is similar to that of the RCE [130,140]. The exponent of the angular velocity of the RC is 2.3 while that of the RCE is 2. As a result, it is proposed that the RC apparatus can attain, theoretically, a higher shear stress range than the RCE.

$$\tau_{RC} = 0.0791 Re^{-0.3} \rho r^2 \omega^{2.3} \tag{9.11}$$

where τ_{RCE} is the shear stress at the electrode surface (pa), *Re* is Reynolds number (dimensionless), ρ is density (kg/m^3), r is electrode radius (m), ω is angular velocity of the electrode (rad/s).

However, Senior et al. argued that the average shear stress produced by RC was actually comparable to that produced by RCE [141]. A recent computation fluid dynamic (CFD) study of the RC proposed Eq. (9.12) for calculating the shear stress of the RC [141,142]. The decrease in the exponent of the angular velocity and the coefficient resulted in much lower average shear stress than that calculated by Eq. (9.11).

$$\tau_{RC} = 0.0261 Re^{-0.3} \rho r^2 \omega^2 \tag{9.12}$$

where τ_{RCE} is the shear stress at the electrode surface (pa), *Re* is Reynolds number (dimensionless), ρ is density (kg/m^3), r is electrode radius (m), ω is angular velocity of the electrode (rad/s).

Other authors have cautioned the use of the RC due to the experimental artifacts it may produce. A vortex is formed as the cage rotates and at high rotational speed, the length of vortex may be sufficient to expose the coupons to the gas phase or, if using, the oil phase, invalidating the test. To minimize this error, a large solution volume (e.g., 5 L [142]) is suggested. The CR is also highly sensitive to the cage design [143], making interlaboratory comparison challenging.

9.6.3.6 Jet impingement cell

For a study of corrosion inhibitors under extreme flow conditions, exceeding the capability of the RCE, RDE, and RC, a JIC can be used. The history of the JIC development has been summarized [137]. A stream of high-velocity water flows through a small diameter nozzle impinging the corrosion test specimen placed perpendicular to and few millimeters away from the nozzle. The jet stream can be a single phase or multiphase with solid particles to create mechanical damage. Fig. 9.17 schematically depicts the jet direction and the variation of flow characteristics with respect to the flat surface radius.

As the liquid jet impacts the surface and spread outwardly, three distinct hydrodynamic zones are developed, that is, stagnation (A), turbulence (B), and wall jet zone (C), as shown in Fig. 9.17A. Region B is commonly used for corrosion studies under high turbulent conditions [137,144]. Wall shear stress in this region obtained with JIC can be as high as 1000 Pa [118]. The shear stress at different zones is determined using Eq. (9.13). It should be noted that the transition from one zone to another is not distinctive but gradual. Therefore the ring design, as shown in Fig. 9.17B, is recommended to ensure the electrode is in the desired

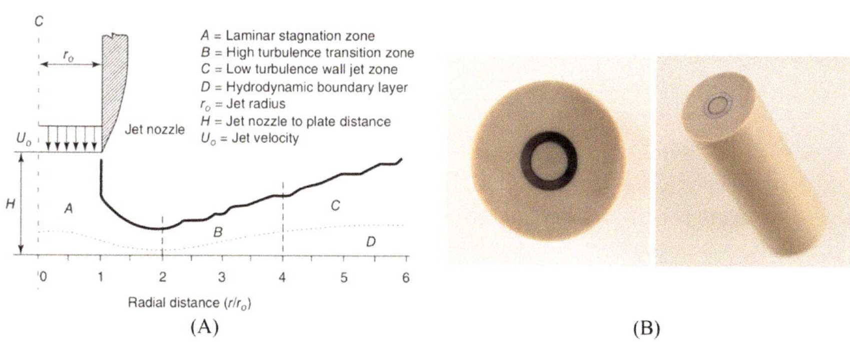

FIGURE 9.17

Schematic of (A) hydrodynamic characteristics of jet impingement on a flat plate. (B) Top and side views of a jet impingement cell corrosion test specimen.

Reproduced with permission from S. Papavinasam. Rotating cage and jet impingement techniques, in: L. Yang (Ed.), Techniques for Corrosion Monitoring, first ed., Woodhead publishing, 2008, pp. 322–343 (A). Photographs from Curtin Corrosion Centre (B).

area of shear stress. A disk shape has also been used, but, in this case, the hydrodynamics parameters are averaged over multiple hydrodynamic zones.

$$\tau_{JIC,turb} = 0.179\rho U^2 Re^{-0.182}\left(\frac{r}{r_0}\right)^{-2} \quad (9.13)$$

where $\tau_{JIC,turb}$ is the shear stress at the electrode surface (Pa), Re is Reynolds number (dimensionless), ρ is density (kg/m^3), r is electrode radius (m), U is jet velocity (m/s), $\frac{r}{r_o}$ is the ratio between the nozzle diameter and the radial distance of the ring electrode.

John et al. demonstrated that the width of the concentric ring electrode influenced the mass transfer coefficient of hydrogen ion (as in the case of CO_2 corrosion) and the limiting current density of cathodic reaction [144]. The average shear stress and mass transfer of corrosive species over the wider electrode (3 mm) are less than that over the narrower one (1 mm) at the same jet velocity.

Corrosion inhibitor performance in a JIC test is monitored electrochemically. Hu et al. used JIC to simulate erosion-corrosion failures found in carbon steel pipework carrying crude oil and process fluids at 95% water cut (i.e., the portion of water out of the total volume of fluids produced) with 500 mg/L sand loading [145]. The authors also used the JIC to qualify a corrosion inhibitor to be used in such conditions. Hu et al. demonstrated the corrosion inhibitor and synthetic hydrocarbon (ISOPAR-M) synergistic effects in reducing the erosion-corrosion damage on API-5L-X65 steel [146]. The CFD model clarified that the presence of hydrocarbon reduced the viscosity of the resulting emulsion, which in turn decreased the impact angle of sand particles and, consequently, lessened the mechanical damage.

The interface between the metallic and nonmetallic holders may induce crevice corrosion. It is important to note that if crevice is present, the test becomes invalid. Besides, the small electrode surface area is not suitable for localized corrosion assessment. It is also crucial to avoid oxygen contamination in a JIC test, as discussed in Section 10.6.2.

9.6.3.7 Flow loops

In certain scenarios, the delivery of the corrosion inhibitor to the steel surface can determine the success of chemical treatment, particularly in multiphase flow regimes. For that purpose, large-scale flow loops demonstrate advantages over benchtop apparatus because the flow regime generated in the flow loop is more representative of field conditions. However, the flow loops are commonly confined to operate at moderate temperature and pressure ranges. Due to such limitations, extreme conditions simulating production tubings, glycol regeneration, and reclamation systems where high temperature and high pressure are anticipated are perhaps best investigated in the smaller scale setups previously described. Additionally, in comparison with other laboratory testing techniques, flow loops are expensive to operate and the cleaning process to avoid inhibitor contamination is time, labor, and resource-intensive. Several studies compared corrosion results from flow loops and other benchtop apparatus, as summarized by Papavinasam [130]. Correlations among techniques appear inconsistent. For example, results from flow loops agreed with RCE or RC in some studies, while disagreed in others.

A number of flow loops are described in the literature and the following provides examples of two different types:

1. A single-phase flow loop: This type of flow loop is generally less complex and less expensive to operate than flow loops catered for multiphase flow. An example of this type of flow loop is described in [147] where it was used to evaluate corrosion inhibition of carbon steel and weld materials in a CO_2 environment by quaternary ammonium and imidazoline chemistries. This flow loop was made of UNS S31600 with two test sections housing corrosion coupons. The test solution was circulated with a centrifugal pump and kept deaerated with pressurized CO_2 and N_2 in the overhead space in the solution reservoir. Corrosion specimens were installed at the bottom (6 o'clock position) of the flow loop.
2. A wet gas flow loop: This type of flow loop is used to simulate two-phase (gas-aqueous) flow, for instance, in a TLC study. One example of the wet gas flow loop consists of a 4′ diameter UNS S31600/03 pipe with the TLC test section, a 1000 L tank filled with the test solution, a gas blower to circulate the test gas saturated with water vapor, and a liquid pump [148]. In this case, the TLC test section could house 1¼′ diameter carbon steel samples or flat carbon steel slabs. The flow loop could operate at high pressure up to 8 bar, at gas temperature up to 90°C and a gas velocity up to 20 m/s. External cooling water through a copper coil generated a temperature gradient causing water to condense at the sample surface.

Though the flow loop is highly utilized for TLC mechanistic studies, corrosion inhibitor studies using the flow loop are less common due to its operational complexity and cleaning difficulty. However, when the transport of a corrosion inhibitor was investigated by foam [149] or by liquid entrainment [150], experiments were found to be best conducted in the flow loop.

9.6.3.8 Under-deposit corrosion

UDC occurs when solids settle onto the metal surface. These solids can be inert (not electrically conductive), such as sand, and certain minerals, or conductive, such as iron sulfide scales, elemental sulfur, and magnetite. The presence of solids can diminish the effectiveness of inhibitors through parasitic absorption and blocking effects [151]. Reportedly, corrosion inhibitors can preferentially adsorb at the deposits' surface, making it less available to the metal surface. In addition, the surface area of some deposits can be very large, in some cases $>10\ m^2/g$, making this process of inhibitor scavenging highly efficient. In the case where the deposit is a semiconductor, corrosion can be accelerated due to a galvanic type of mechanism. Hence, the inhibitor performance can be reduced and localized corrosion promoted [152].

Several test methods have been developed to study UDC for various objectives and these have been thoroughly reviewed by Brown and Moloney [150]. A simple UDC test involves in situ corrosion monitoring with electrochemical techniques on a steel specimen uniformly covered with a selected deposit material [153]. In such a test, the inhibitor performance is obtained. Additional surface characterization techniques are required to better understand possible underlying UDC mechanisms. A more complex experimental setup, such as those used in [154], consisted of coupling bare steel with deposit covered steel, to evaluate UDC at the interface between covered/uncovered steel. Galvanic current measurement using a zero-resistance ammeter (ZRA) was used to elucidate if localized corrosion propagated by the galvanic coupling mechanism.

A coupled multielectrode array system (CMAS) or wire beam electrode (WBE) has been used to study the localized nature of UDC. Their working principle is essentially the same. They are made of multiple pins or wires mounted in an inert resin or polymeric materials such as PEEK, as shown in Fig. 9.18. It is important to ensure the absence of crevice at the pins/resin interface. The pins are often machined from carbon steel rod to 1 mm diameter or this can be customer-made. The surface exposed to the test environment is ground to obtain a flat surface. CMAS (or WBE) can be tested either as a single electrode, with all the electrodes electrically coupled together, for an average CR determination using LP or electrochemical impedance spectroscopy (EIS) technique, or as individual pins in galvanic corrosion measurements with a ZRA. Corrosion current of individual pins can be measured against an external electrode [111] or the remaining pins [154,155]. Relative susceptibility to corrosion of these pins can then be presented as a map, as illustrated in Fig. 9.19 [154]. It was shown that corrosion did not occur uniformly under the deposit. In this case, two carbon steel pins (electrode no. 45 and 66) exhibited lower electrode potential and higher anodic current density. Subsequent surface analysis confirmed greater material loss at these two electrodes.

Hinds and Turnbull [111] developed 24-pin CMAS to study the behavior of two commercial corrosion inhibitors on carbon steel under a 1 cm layer of 100–300 μm acid-washed SiO_2. The authors pointed out that though the average CR under inhibition was low (0.07 mm/year), a pit penetration rate as high as 0.4 mm/year was experienced, highlighting the importance of the localized nature of the UDC and the appropriate selection of test methodologies.

Since parasitic absorption of inhibitors by deposits is one of the explanations used to describe UDC, the residual concentration of inhibitor is of interest. UV-Spectrophotometry has been used to determine the residual corrosion inhibitor remaining in the solution [153]. Suarez et al. used UV-Spectrophotometry to demonstrate preferential adsorption of surfactant inhibitors, cetylpyridinium chloride monohydrate (CPC), 1-Dodecylpyridinium chloride hydrate (DPC), and 2-mercaptopyrimidine (MPY) by different inert deposits, namely, Al_2O_3, $CaCO_3$, and SiO_2.

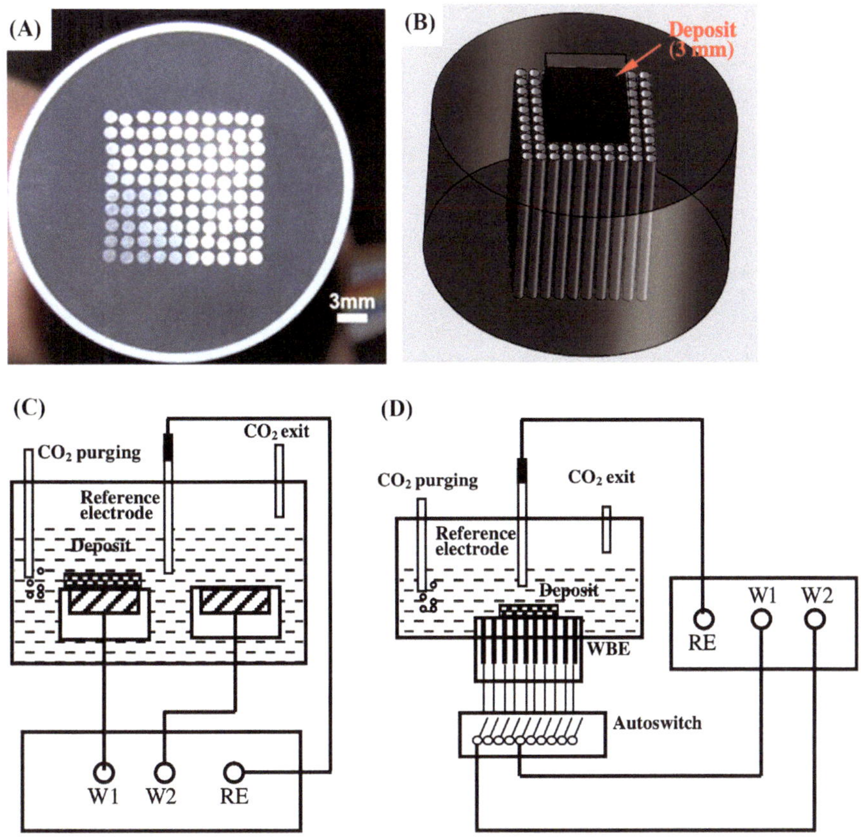

FIGURE 9.18

(A) A photograph of wire beam electrode (WBE) with a 10 × 10 electrode array. (B) The schematic diagram of WBE covered by mixed deposit. (C) Schematic diagram of galvanic corrosion measurement between the deposit-covered electrode and bare electrode. (D) Schematic diagram of potential and galvanic current measurement.

Reproduced with permission from G.A. Zhang, N. Yu, L.Y. Yang, et al., Galvanic corrosion behavior of deposit-covered and uncovered carbon steel, Corrosion Science 86 (2014) 202–212.

9.6.3.9 Top-of-the-line corrosion

TLC is observed at the top section of wet gas pipelines, usually between 10 and 2 o'clock position, of pipelines operating in a stratified flow regime where water condensation occurs [156]. The TLC rate is generally proportional to the water condensation rate (WCR) [156,157]. The condensed water is saturated with corrosive species, such as acid gases (CO_2 and H_2S) and volatile organic acids, but without the alkalinity from produced water, it can be very corrosive. Transport of

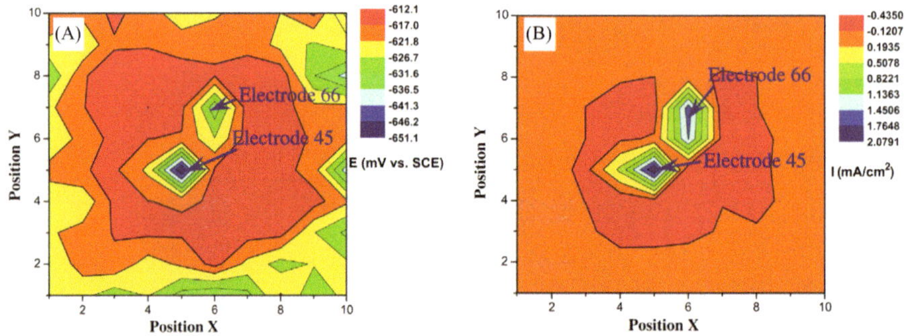

FIGURE 9.19

(A) Corrosion potential and (B) galvanic current distribution of X65 wire beam electrode in CO_2-saturated formation water at 60°C after 72 hour under 3 mm thick of mix deposit (sand: clay: ferrous carbonate:ferrous sulfide:elemental sulfur at the weight ratio of 3.75:0.75:3:1:0.3.).

Reproduced with permission from G.A. Zhang, N. Yu, L.Y. Yang, et al., Galvanic corrosion behavior of deposit-covered and uncovered carbon steel, Corrosion Science 86 (2014) 202–212.

a conventional corrosion inhibitor to the top section of the pipe is limited and so TLC can be a difficult problem to control. While operators tend to manage TLC by engineering design (such as the use of CRA where high WCR is expected), research into special means of delivering corrosion inhibitors to the top part of pipelines have been explored, namely, volatile corrosion inhibitor (VCI) [124,158], foam [159], batch treatment [160], etc.

Several test methodologies have been developed to investigate TLC ranging from small glass cells to large-scale flow loops. The detail of the TLC flow loop was described earlier (see Section 7.3.7, Flow loops). There are many variations of small-scale TLC test setups where the gas flow rate cannot be controlled. Some are operated at atmospheric pressure, while others are designed for elevated pressure conditions [123,124,161,162]. These setups can facilitate water condensation on a steel specimen using an external cooling liquid [124,161], or thermoelectric cooling element [162]. The WCR can be determined mathematically [157] or measured experimentally [163], by measuring the accumulated condensed liquid in a collection cup placed below the steel specimen [123,161]. To prevent measurable re-evaporation and for analysis of its chemistry, the condensed water is transferred to another container at room temperature.

Amine-based chemicals such as morpholine and diethylamine have been studied for their efficacy in mitigating TLC in a glass cell setup [161]. The same experimental design and procedure were also utilized to evaluate thiols [124]. The performance of both amines was determined by the weight loss technique. To analyze the change in pH of condensed water due to the addition of a VCI, a separate experiment was performed with a stainless-steel sample instead of a carbon steel to avoid the influence of ferrous ions from a corroding specimen on the liquid pH.

High-pressure/high-temperature autoclaves with a U-shape carbon steel tube as a TLC sample have also been used to study VCI [163]. In this setup, the gas phase was stagnant. Qin et al. [164] developed a high-pressure condensation autoclave that enabled the specimen rotation speed of up to 1.72 m/s. Four rectangle samples were fixed vertically onto a rotating shaft. However, this form of specimen arrangement does not realistically mimic the condensation process, which may affect the test results. It was shown that the condensed water droplet formation and movement is important in defining TLC behavior, as it affects the replenishment of corrosive species and corrosion inhibitor [148,157].

Corrosion rate and IE under TLC environment are usually measured by gravimetry (or weight loss) technique. Recent research led to the development of a two-electrode system that enabled electrochemical measurements under TLC [165]. This consisted of a 1 mm pin inserted into the center of a 25 mm cylindrical rod. Both were machined from UNS G10300 carbon steel and were electrically insulated by an epoxy resin. The electrodes were mounted to expose their surfaces as a ring disk. The pin, or small 1 mm diameter disk, served as the WE while the surrounding area of the large ring was used as a counter/reference electrode (RE). The TLC rate measured in-situ with LP and EIS techniques compared well with the gravimetry result.

9.6.3.10 Dissimilar metals

When dissimilar metals are electrically connected and exposed to a corrosive environment, galvanic corrosion can occur. The less noble metal corrodes at an increased rate while corrosion on the other metal is reduced. Several scenarios involving dissimilar metals where the inhibitor performance should be evaluated are, for instance, the transition between cladded and uncladded section [128] and between welds and parent metal pipes PWC [166,167]. Typically, the weld material is designed to be more cathodic than the parent metal to avoid a large cathode to anode ratio, which would accelerate the dissolution at the weld. However, preferential adsorption of a corrosion inhibitor on the parent metal over the weld metal has been reported, leading to greater corrosion risk at welds [166].

Fig. 9.20 show a segmented PWC sample used in corrosion and inhibitor studies. The segments were taken from distinctive zones of a welded pipe, that is, parent metal, weld metal, and heat-affected zones. The specialized samples enable (1) the CR measurements of individual segments and (2) galvanic current measurements between segments [168].

It was also proposed that, in addition to the galvanic effects from different chemical compositions and microstructures of parent metal and weld material, the weld bead can disrupt the flow causing the local shear stress at the weld to increase by several orders of magnitude [166]. Therefore high shear stress apparatus such as a submerged impinging jet, as used by Barker et al. [166] has been employed. In this apparatus, two corrosion samples, machined from weld and parent metal, respectively, were impacted with a high-velocity water jet. However, the hydrodynamic parameters could not be obtained from this apparatus. Based on the measured galvanic currents between the two materials, the study concluded that the presence of sand and high-velocity water influenced PWC.

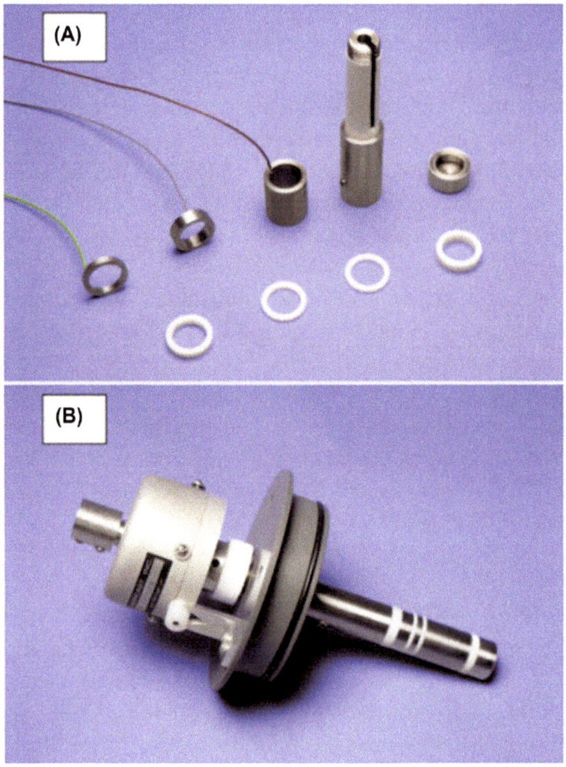

FIGURE 9.20

Photographs of segmented corrosion samples for preferential weld corrosion study. (A) component parts and (B) a complete assembly.

Reproduced with permission from K. Alawadhi, M.J. Robinson, Preferential weld corrosion of X65 pipeline steel in flowing brines containing carbon dioxide, Corrosion Engineering Science and Technology, 46 (2011) 318–329.

9.7 Corrosion rate measurement techniques

9.7.1 Gravimetry (weight loss measurements)

In the case of using gravimetry to determine the CR, preweighed metal specimens with a known surface area are exposed to a corrosive environment for a period of time. After exposure, corrosion products or precipitates formed on the metal surfaces are carefully removed, which in some cases can be assisted by using an inhibited acid solution [169]. The difference between the weight of the metal specimens before and after exposure is determined and the corrosion rate and IE are calculated based on Eqs. (9.14) and (9.15), respectively.

$$\mathrm{CR} = \frac{K \times W}{A \times t \times D} \tag{9.14}$$

where CR is corrosion rate in mm/year, W is the mass loss in grams, A is the exposed surface area in cm^2, t is the exposure time in hours, D is density in g/cm^3, and K is a constant 8.76×10^4 for carbon steel. Note, care must be taken to use the correct units. The constant K can be adjusted if the CR is required for other metals or alloys or needed in other units [61].

$$\% \text{ Inhibition efficiency} = \frac{CR_b - CR_{inh}}{CR_b} \times 100 \tag{9.15}$$

where CR_b is the baseline corrosion rate under the blank condition without a corrosion inhibitor and CR_{inh} is the corrosion rate after adding a corrosion inhibitor.

Weight loss corrosion rate and %IE calculated from Eqs. (9.14) and (9.15) assume that corrosion proceeds uniformly. Therefore %IE should be used cautiously since localized corrosion may also occur and perhaps even be promoted in the presence of ineffective corrosion inhibitors. It is not uncommon to obtain satisfactory inhibition efficiency, while localized corrosion is manifested [170]. For example, in one instance, it was found that an amine-type corrosion inhibitor promoted the formation of $CaCO_3$ scale in a wet gas transportation line but since the protective scale was nonuniform, localized attack leading to through-wall penetration occurred where the scale was absent [170].

Additionally, while it is popular to present the IE as a percentage protection, it should be noted that it is the inhibited corrosion rate that is most important. A seemingly high level of 90% protection may not be sufficient under highly corrosive conditions. For example, if the uninhibited corrosion rate is 10 mm/year, 90% protection would result in a corrosion rate of 1 mm/year and this rate would be too high under most circumstances. Alternatively, if the uninhibited corrosion rate was 1 mm/year then 90% protection (0.1 mm/year) may be more acceptable. However, an acceptable inhibited corrosion rate is often specific to an asset and its desired lifetime, so it is important to understand the objectives of the operator in each circumstance.

Several experimental factors can impact the results; namely, the design of the specimen, its dimension, metal surface area to test solution ratio, and exposure time. The appropriate exposure time depends on the aggressiveness of the environment. Short exposure times may be used if a high corrosion rate is expected, and vice versa. The required exposure time to obtain a meaningful weight loss has been previously suggested and given in Eq. (9.16) [171].

$$\text{Exposure time (h)} = \frac{2000}{\text{Expected corrosion rate (mpy)}} \tag{9.16}$$

The test duration and the ratio between the specimen surface area and solution volume should be such that solution corrosivity and corrosion mechanisms, as may be determined by pH, conductivity, the concentration of dissolved ions, and corrosive gases, remain constant throughout the exposure. As a guideline, the test solution volume to the surface area ratio should be at least $0.2\ mL/mm^2$ for exposure up to 30 days [172].

Weight loss measurement is useful when corrosion assessment via electrochemical means is not suitable, particularly when the test condition involves a discontinuous aqueous phase. A common example is the multiphase flow environment, for example, TLC or oil wetting surfaces. With relatively long exposure time, the tendency for pitting/localized corrosion may also be acquired. The weight loss technique gives an integrated degradation rate over a long period of time so information regarding the initial corrosion process might be missed. However, this can be complemented by nondestructive electrochemical techniques.

9.7.2 Electrochemical techniques

Since corrosion is an electrochemical process, electrochemical techniques can be used to monitor the corrosion process and study the corrosion inhibition mechanism. In brief, electrochemical techniques utilize the established relationships between potential and current to derive the rate and mechanism of corrosion. Depending on the applied potential, electrochemical techniques can be categorized into the following three categories, based on the size of the potential windows, and associated common techniques used in inhibitor studies:

1. No applied potential: EN, ZRA, corrosion potential (E_{corr})
2. Small applied potential: LP, EIS
3. Large applied potential: PP, cyclic potentiodynamic polarization (CPP)

9.7.2.1 Basic instrumentation

An extensive review of instrumentation for electrochemical tests can be found in previous publications [173,174]. In brief, electrochemical measurements for corrosion studies are usually performed in a three-electrode setup, which comprises the following:

1. The working electrode (WE): This is made from the material under investigation. Depending on the test setup used, the shape and dimension of the WE may vary. The design and preparation of the WE can impact the test results. An appropriate exposed surface area should be used. A large surface area is suitable when low currents are expected. Small surface areas are often beneficial in low conductivity electrolytes to minimize experimental errors caused by the ohmic drop. Before experimentation, the WE is usually ground or polished to remove mill scale and other surface contamination, and to obtain a reproducible surface profile. The surface roughness of a carbon steel electrode was shown to influence the corrosion test results [175–177]. Rough surfaces (high R_a) exhibited greater initial metal loss. It is critical to have a consistent electrode preparation procedure for reproducibility. The WE surface should be free of polishing particles or greasy residue from handling. It can be rinsed, cleaned, and degreased in an ultrasonic bath. Solvents such as acetone, or alcohols such as ethanol and isopropyl alcohol are commonly used to remove moisture and greasy residues from WE surfaces.

2. The counter electrode (CE): The purpose of a CE is to complete the electrochemical circuit by enabling current flow through the cell. Common CEs used in corrosion studies are platinum and graphite. Other materials such as stainless steel and titanium alloys have been used [174]. It is desirable that the CE is inert in the test environment. However, when electrochemical tests are conducted in the field or in an industry laboratory, a simpler setup of two or three identical carbon steel electrodes are used. To avoid contamination by the reaction products generated at the CE, a different compartment separated by a porous membrane may be used. The surface area of the CE is usually larger than that of WE to reduce the current density at CE to avoid hindering the current flow at the WE which can lead to artifacts and misleading corrosion results.
3. The reference electrode (RE): The WE potential is measured and controlled against that of the RE. Common REs used in corrosion studies are the calomel electrode $Hg|Hg_2Cl_2$ in saturated KCl (SCE, +0.24 V vs standard hydrogen electrode, SHE), Ag|AgCl in 3M KCl (+0.21 V vs SHE) and $Cu|sat.CuSO_4$ (+0.32 V vs SHE). Contamination of filling and test solutions can affect the potential reading. For low conductivity solutions, a Luggin capillary filled with a conductive medium can be used to reduce the distance and impedance between the WE surface and RE. Consistent care and maintenance of REs are important to ensure quality readings. When exact electrochemical potentials versus a reference scale are not needed but simply a corrosion rate measurement is being sought or in environments not conducive for the intricacies of standard REs, pseudo RE employing a range of materials including carbon steel, stainless steel, Hastelloy, etc., have been used.
4. Potentiostat: A computerized potential and current control instrument regulates and records the potential and current of the WE in an electrochemical experiment. The technical requirements of a potentiostat for corrosion studies have been reviewed previously [174].

9.7.2.2 Corrosion potential measurement

The mixed potential theory defines the corrosion potential as a potential at which the cathodic current equals the anodic current. The current at this potential is the corrosion current. Thus changes in cathodic or anodic reaction, for example, as a result of the inhibitor addition, can influence the corrosion potential and current. Therefore as mentioned in Section 2.2, Interfacial Inhibitors, changes in the corrosion potential can indicate a difference in the corrosion mechanism, such as that caused by corrosion inhibition. A shift in E_{corr} in the positive direction suggests the net suppression of the anodic reaction, for instance, due to the formation of a protective film [178]. This trend manifests when an inhibitor is added to protect carbon steel from CO_2 corrosion [179]. A negative shift generally indicates that the cathodic reaction is preferentially inhibited. Little to no changes in E_{corr} tends to indicate a mixed inhibition of both the anodic and cathodic reactions and/or that the inhibitor has a geometric blocking effect. In such cases, the inhibition efficiency has been linked to the surface coverage [112]. However, the variation in E_{corr} is mainly qualitative and further evaluations using complementary techniques should be carried out to elucidate the exact inhibition mechanisms taking place. E_{corr} alone does not provide sufficient about the IE—only information of the inhibitor mechanism.

9.7.2.3 Linear polarization

In this technique, a narrow potential scan, typically ± 10 mV with respect to the corrosion potential, is applied to the electrode, and the responding current is recorded. Fig. 9.21A is a typical current density (i) versus potential (E) curve obtained from a corroding carbon steel electrode at one moment in time. If this scan is repeated over time, a corrosion rate vs time can be created as shown in Fig. 9.21B.

The polarization resistance (R_p) is obtained from the slope of E-i plots [Eq. (9.17)]. The corrosion current density (i_{corr}) is then derived from R_p using the Stern-Geary relationship [Eq. (9.18)]. Faraday's law is used to convert the corrosion current density to a corrosion rate as explained in ASTM G59 [180]. Further details and theories behind this technique can be found elsewhere [174,181].

$$R_p = \left(\frac{dE}{di}\right)_{E_{corr}} \tag{9.17}$$

$$i_{corr} = \frac{B}{R_p} \tag{9.18}$$

where R_p is polarization resistance, in ohm-cm^2, i_{corr} is corrosion current density in A/cm^2 and B is the Stern-Geary coefficient, in V. B is calculated according to Eq. (9.19). The anodic (b_a) and cathodic (b_c) Tafel slopes can be obtained from

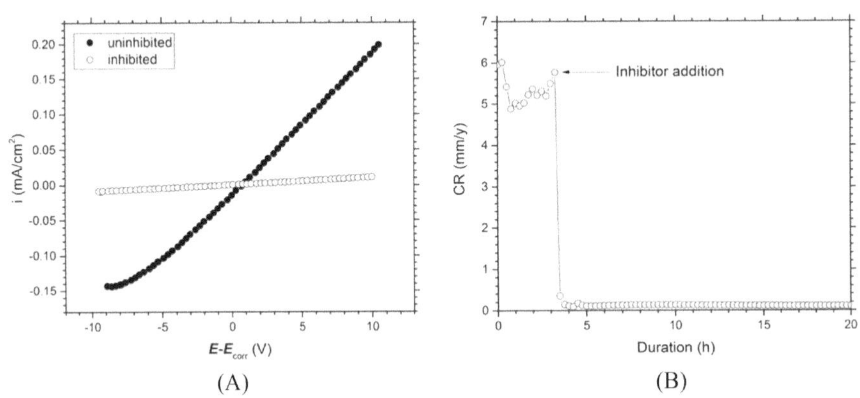

FIGURE 9.21

(A) i-E graph showing the change in slope ($1/R_p$) after an inhibitor was added. (B) corrosion rates calculated from R_p versus time (carbon steel UNS G10300, 77 °C, standard brine solution (3 wt.% NaCl + 0.01 wt.% $NaHCO_3$), 1.5 bar pCO_2, 100 ppm of a commercial film forming corrosion inhibitor, B value = 14 mV).

the potentiodynamic curves performed under the same experimental conditions by measuring the gradient of the curves (known as Tafel slopes).

$$B = \frac{b_a b_c}{2.303(b_a + b_c)} \tag{9.19}$$

In many cases, B values are selected based on previous experience or literature reports because a true linear Tafel region does not exist in the inhibited solution to obtain b_a and b_c experimentally. It should be noted that this does not mean they are correct. The Tafel behavior requires the system to be under charge transfer control. In the presence of corrosion inhibitor, transport of corrosive substances may become the limiting step making Tafel parameters irrelevant [180]. In CO_2 corrosion environments, B values ranging from 14 to 26 mV have been used.

Since the applied voltage is small, this technique is considered nondestructive, which does not inadvertently affect or change the electrode surface. As such, it allows consecutive measurements to be performed on the same electrode to monitor the changes in corrosion activity at the electrode surface. Fig. 9.21B shows the time variation of corrosion rate of carbon steel UNS G10300 which was initially exposed to an uninhibited, deaerated standard brine solution for 2 hours, followed by an addition of a corrosion inhibitor. The sudden drop in CR after a corrosion inhibitor was added indicated how readily the added inhibitor adsorbed and provided corrosion mitigation.

R_p may be experimentally affected by scan rates [182], the conductivity of test solutions [182], and corrosion product build-up [180]. In nonaqueous solutions, such as those containing organic solvents (glycols or alcohols), R_s may be high and not negligible, particularly in relation to the baseline CR and so it should be subtracted from the experimentally obtained R_p to avoid underestimating the CR. Several methods can determine R_s, that is, EIS, positive feedback, and current interruption techniques. Additionally, like the gravimetry technique, the results from the LP technique cannot distinguish between uniform and localized corrosion.

9.7.2.4 Electrochemical impedance spectroscopy

In this technique, an alternating potential with a small excitation voltage ($< \pm 10$ mV) is applied to the steel specimen over a range of frequencies, typically from 100 kHz to 1 mHz. The sum of the current response, as a result of polarization and ohmic resistance, capacitance, and inductive impedances of a given electrochemical system, is recorded.

Fig. 9.22A shows typical EIS plots obtained on a corroding carbon steel in a MEG-water mixture (80 wt.% MEG + 20% simulated condensed water). A single semicircle is often observed in the Nyquist plot (a plot between Z_{Re} and $|-Z_{im}|$ on a linear scale). The relationship between the total impedance, phase angle, and frequency are shown in the Bode plots using a semilogarithmic scale (Fig. 9.22B and C). When a corrosion inhibitor is added, the charge transfer resistance increases and the CR decreases, which is represented by a larger magnitude of the semicircle in the Nyquist plot (Fig. 9.22A) and an increase in impedance in the Bode impedance plot

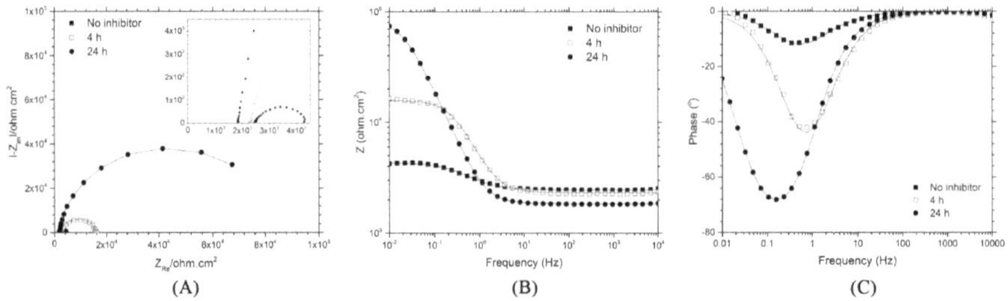

FIGURE 9.22

Electrochemical impedance spectroscopy plots of carbon steel X65 in lean MEG solution (3% NaCl and 0.01% $NaHCO_3$) with and without a commercial film-forming corrosion inhibitor. (A) Nyquist, (B) and (C) Bode plots.

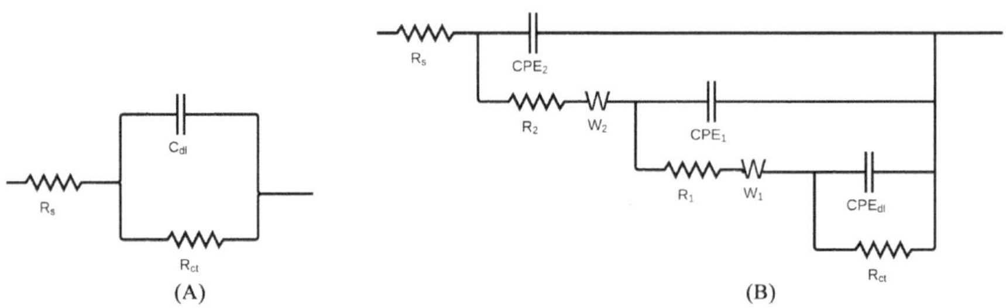

FIGURE 9.23

(A) Randle circuit and (B) complex circuit representing multilayers inhibitor film.

(Fig. 9.22B). The increase in magnitude for the inhibited solution from 4 to 24 hours is due to the growth and protectiveness afforded by the inhibitor film.

In addition to determining corrosion rates and inhibition efficiency, the EIS technique has been used to elucidate the inhibition mechanism electrochemically as electrical components in electrical equivalent circuit (EEC) can represent certain electrochemical components/processes. An appropriate EEC is used to deconvolute the measured total impedance using electrical components such as a resistor, capacitor, and inductor. These components can have physical meanings and represent certain processes occurring in the corroding system. Common circuits used in corrosion inhibitor studies are shown in Fig. 9.23. Fig. 9.23A represents a freely corroding steel (uninhibited) while Fig. 9.23B typically explains the existence of a porous bilayer inhibitor film with diffusion effects within the pores.

The EIS technique is particularly useful for solutions with low conductivity. The ohmic resistance of the test solution can overwhelm the charge transfer resistance, and, consequently, the corrosion rate can be severely underestimated. A constant phase element is commonly used to account for nonideal capacitance behavior. The increase in the number of time constants is often attributed to multiple adsorbed layers. A system with a single time constant (parallel arrangement between a resistor and a capacitor) indicates the existence of a double-layer capacitance. However, when multiple time constants appear, the presence of inhibitor layer(s) has been proposed [179]. With an appropriate EEC, the fitted parameters have been used to estimate the percentage inhibitor adsorption and thickness of the protective layer [134,179]. Inhibition mechanisms may then be proposed based on the fitted parameters. The capacitance values derived from EIS measurement have been used to obtain the potential of zero charge (PZC) to help determine whether anions or cations are more likely to adsorb on the metal surface [124,161].

9.7.2.5 Electrochemical noise

The techniques described previously are more suited for uniform or general corrosion evaluation. For localized corrosion monitoring, the EN method has been used. Several possible configurations have been summarized [182]. However, the most typical EN test setup consists of two nominally identical WEs connected via a ZRA to measure current fluctuations (electrochemical current noise, ECN). The potential fluctuations (electrochemical potential noise, EPN) of the electrode pair are measured against a standard laboratory RE.

The addition of a corrosion inhibitor has been observed to lead to an increase in the noise resistance (R_n), which is the ratio between the standard deviations of ECN and EPN [183]. Tan et al. found that R_n increased with time after an imidazoline-based corrosion inhibitor was added to a corroding carbon steel in a 3% NaCl solution saturated with CO_2 at room temperature. R_n readily declined as the steel sample was removed from an inhibited to an uninhibited solution, suggesting inhibitor desorption [136]. R_n was found to be comparable to R_p and corrosion current [136]. However, Scully and Tan et al. [182,184] argued that the correlation between R_n and corrosion current lacks electrochemical foundation and is not universally applicable to all corroding systems. The abrupt spike in ECN followed by a decay in ECN that also corresponds to the same EPN trend has been linked to localized corrosion events [185], while the permanent increase in ECN suggested a desorption of corrosion inhibitor [186]. The analysis and postprocessing of EN data are complicated. Therefore it has been suggested that EN should be used as qualitative means in corrosion monitoring [180]. Readers are referred to [182,187] for in-depth discussion of the EN fundamentals. While not EN per se, a localized corrosion monitoring technique has been proposed for the in-situ assessment of pitting corrosion and its inhibition [188]. In this work, potential transients, taken to be pitting events, were detected and converted to a current equivalent using periodic LP resistance measurements. In turn, the pit transient current equivalents were converted to pit metal volume and mass loss using Faraday's laws and estimations for pit depths made using assumed pit diameter to depth ratios taken from previous assessments. Under the conditions assessed, it was shown that the metal volume loss from the pitting events using this technique matched up well to the metal volume loss calculated from interferometric appraisal of the metal surface.

9.8 Additional performance evaluations

9.8.1 Partitioning

Under low flow or laminar flow conditions in which there are separate hydrocarbon and water phases, it is important that the corrosion inhibitor does not remain in the hydrocarbon phase but adequately transfers to the water phase where the corrosion on the metal surface is taking place. The way a chemical distributes itself among the different liquid phases is known as partitioning and it is important to understand this behavior when selecting a corrosion inhibitor. The ratio of the chemical in equal volumes of the water phase and the oil phase is known as the partitioning coefficient. It should be noted that the amount or percentage of a corrosion inhibitor partitioning to the water phase does not necessarily directly equate to performance. For example, a corrosion inhibitor with a lower partitioning to the water phase resulting in a lower amount of product in the brine may result in better corrosion inhibition performance than a different product comprising different active chemistries with a higher water partitioning.

There are several methods for determining the partitioning behavior but one of the most useful is to assess the efficacy of a corrosion inhibitor product in an actual corrosion experiment. For this, a single test cell containing both phases along with a probe or coupon at the bottom of the vessel in the water phase can be used in which the chemical is dosed into the oil phase with little or no stirring. Alternatively, a separate partitioning stage can be utilized in which the aqueous and hydrocarbon phases dosed with a corrosion inhibitor are vigorously mixed together. The two phases are then allowed to separate after which the aqueous phase is later used for a corrosion test. Ideally, the ratio between aqueous and oil phases should be similar to that in the field.

The percentage partitioning of the same corrosion inhibitor is known to vary with temperature, salinity, and oil types among other factors [128]. Therefore the field conditions, including using the field oil, should be mimicked as closely as possible. The presence of other oilfield chemicals or solvents can also affect the partitioning of corrosion inhibitor components so these should also be included in the laboratory evaluations if the field is operating or planning to operate with these chemistries. In addition to using a partitioning stage before corrosion testing, partitioning studies are often carried out to measure and assess the concentration of inhibitor in the water phase—known as the corrosion inhibitor residual (see Section 11.2, Corrosion Inhibitor Residual Measurements). This is helpful for determining an approximate injection rate needed in the field under various operating environments to gain the minimum effective concentration (MEC) in the water phase to mitigate corrosion to an acceptable level based on the laboratory evaluations. In turn, the actual injection rate may need optimizing based on the actual residuals measured in the field or information gained from various forms of monitoring, inspections, etc.

9.8.2 Persistency

Inhibitor persistency is the ability to maintain uniform adsorption on the metal surface to protect it against corrosion. Desorption or detachment is influenced by many parameters, such as the type of adsorption of the inhibitor film, the depletion of corrosion inhibitor chemistry in the bulk water phase disrupting the equilibrium between the molecules at the metal surface and those in the brine, the chemical composition of the multiphase flow, the wall shear stress/mass-transport coefficient and the erosive nature of the flow [186]. Good film persistency is required whether the inhibitor is delivered by continuous injection at a relatively low dose or by batch treatment at high concentration. Inhibitor film persistency is vital in multiphase flow regimes, where the presence of nonaqueous phases (gas, hydrocarbons, and solids) as well as during times of no chemical availability (such as pump outages or logistical issues) may result in inhibitor desorption and loss of protection.

Various test environments represent certain aspects of field conditions. In one investigation, a JIC was used to evaluate three commercial corrosion inhibitors; namely, an oleic imidazoline, an alkyl phosphonate ester, and methoxylated amide in combination with a quaternary amine [186]. In this work, the inhibitor film persistency at the metal surface was examined surface by (A) liquid jet mimicking a single-phase condition, (B) by liquid jet + ultrasonic cavitation simulating a slug flow, and (C) by gradual dilution to reduce the bulk inhibitor concentration. ZRA measurements were conducted to continuously monitor the fluctuation of potential and current (with respect to a Pt CE) responses to the flow interruptions. The rise in current and the anodic shift in corrosion potential indicated an increase in corrosion activity following the detachment of the inhibitor [186].

The film persistency test intended to represent a batch treatment is typically carried out in 2 steps, that is, the filming of electrodes followed by an electrode wash step [189]. The first step involves exposing the electrode to an inhibited solution at the expected field concentration or dilution for the expected contact time. The wash step is performed either by transferring the electrode to another test cell containing an uninhibited brine [189,190] or gradually replenishing the test cell with an uninhibited brine [186]. The desorption of an inhibitor film generally manifests as a rise in the corrosion rate. Fig. 9.24 shows a time variation of corrosion rates determined by LP of carbon steel UNS G10220 in a standard brine solution (3 wt.%NaCl with 0.01 wt.% $NaHCO_3$). The sharp drop in the corrosion rate after about 4 hours was due to the exposure of the steel specimen to the highly inhibited solution during the filming step. The spikes after about 20 and 26 hours into the test were when the electrode was transferred to an uninhibited brine. The corrosion rate consistently rose after the second fluid exchange and, after about a day of exposure to the uninhibited solution, had approximately

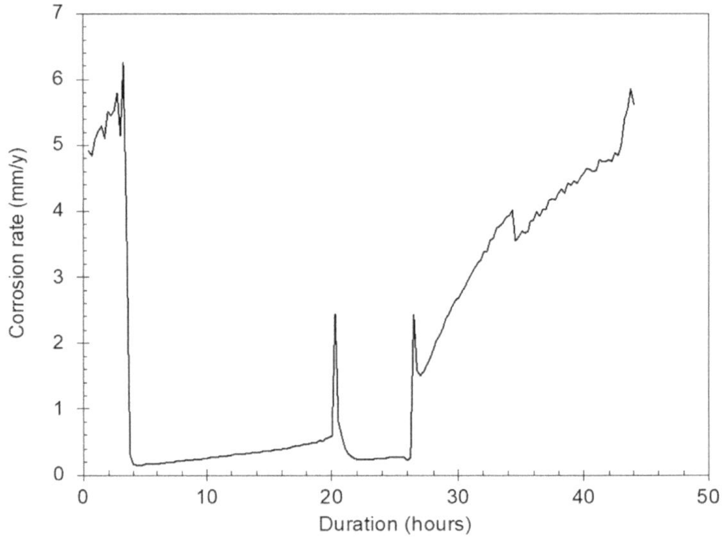

FIGURE 9.24

Corrosion rate with time trend of carbon steel in standard brine solution at 70°C showing a drop in the corrosion rate after an inhibitor treatment and a gradual rise in the corrosion rate due to film desorption after a second fluid change at about 26 hours.

reached the blank corrosion rate at the start of the experiment indicating that the inhibitor had essentially completely desorbed from the metal surface [189].

9.8.3 Pitting tendency analysis

Pitting corrosion on carbon steel is mostly associated with the presence of nonuniform surface films. In the oil and gas industry, the surface films can be corrosion scales, such as iron carbonate and iron sulfides in sweet and sour corrosion, respectively, inhibitor films or mixtures of these. It is believed that pitting corrosion is stochastic [191]. Therefore corrosion coupons with a large surface area are preferable from the statistical viewpoint. Additionally, pit propagation is known to be not linear with time. Extrapolating the pit penetration rate from a short-term exposure test can overestimate or underestimate the steady-state pit growth. Instead, multiple tests at various exposure intervals are recommended [128].

As previously mentioned, many assessment techniques (weight loss and most electrochemical techniques) do not reveal the extent of pitting corrosion. Hence, corrosion coupons should be closely inspected and analyzed for any evidence and severity of pitting corrosion after all corrosion tests. The extent of pitting corrosion can be quantified by pit morphology (depth vs width or diameter), pit density, and pitting ratio. The pitting ratio is the ratio between the pit penetration rate and uniform CR. A pitting ratio larger than five is considered true pitting corrosion [192]. ASTM G46 describes general guidelines and standard practice to apply a statistical approach to pitting analysis [192], although operators may have their own requirements for their specific assets.

In many cases, corrosion products/scales from the specimen surfaces need to be removed to expose any underlying damage and corrosion pits. ASTM G1 describes procedures and cleaning solutions for various alloys. Clarke's solution (inhibited acid solution, 20 g Sb_2O_3 + 50 g $SnCl_2$ + 1 L concentrated HCl) is commonly used on carbon steel specimens. A manual evaluation of the pitting can be conducted using a metallurgical optical microscope while scanning electron microscopy (SEM) can provide much higher magnification assessment.

Scanners and profilometers, including laser options, visible light microscopy, and vertical scanning interferometry, allow a comprehensive microscopic assessment of the metal surface to determine the extent of localized attack [193], as shown in Fig. 9.25. In this example, a 3D image of a corroded surface with an

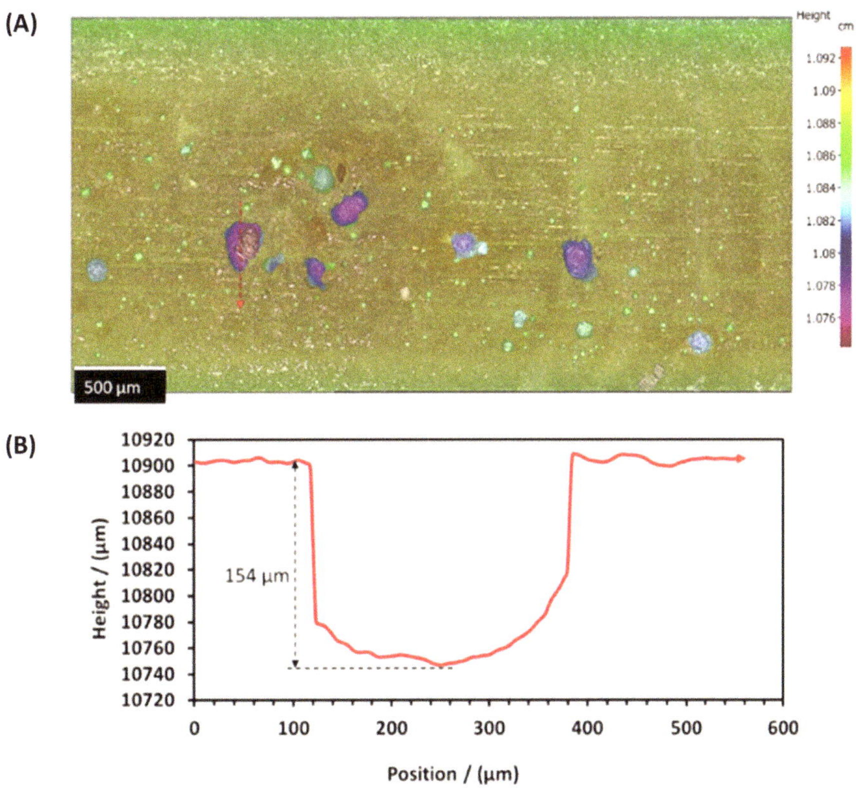

FIGURE 9.25

(A) An optical microscopic image showing pitting corrosion on carbon steel (B) A corresponding surface profile showing the depth of a pit.

Reproduced with permission from W. Li, B.F.M. Pots, X. Zhong, et al., Inhibition of CO_2 corrosion of mild steel – Study of mechanical effects of highly turbulent disturbed flow, Corrosion Science 126 (2017) 208–226.

approximated area of 6 mm^2 (2 × 3 mm^2) was constructed showing several corrosion pits according to the color scale. Profile analysis was performed on selected pits to quantify the pit depth. Menendez has reviewed the differences in operating principles of various profilometers [194]. Recent developments in analysis software and automation technology enable surface analysis to be performed over large surface areas and even on cylindrical shape samples. Furthermore, the available software has improved such that pit counts and distribution of pit depth over large surface areas can be automated [188,195].

9.9 Surface characterization

Advancements in surface analytical tools enable researchers to investigate the underlying corrosion and inhibition mechanisms. This section introduces analytical techniques commonly used in corrosion and inhibitor studies and the type of information gained from these techniques. In most studies, more than one technique is utilized to obtain complementary information. Where possible, the outcomes and practical applications of these measurements are discussed.

9.9.1 Atomic force microscopy

Atomic force microscopy (AFM) is a scanning surface technique with nanoscale resolution. The AFM probe (in many cases a silicon tip on a free-moving cantilever controlled by a piezoelectric element) scans across a small area of a sample. Information such as surface topography, physical and mechanical properties can be obtained. As such, among many applications, AFM is increasingly utilized in the field of corrosion science. With its nanoscale resolution, inhibitor molecules can be mapped on various surfaces and environments, for instance, on inert surfaces (such as mica, gold, etc.), and inhibited and uninhibited carbon steel. It should be noted that performing AFM technique on an actively corroding steel is challenging due to the increased surface roughness as the steel corrodes.

Its unique ability to investigate the solid/liquid interfaces at the nanoscale has improved the understanding of the interaction of inhibitor molecules and steel surfaces. Xiong et al. used AFM to demonstrate that the lateral force required to remove the adsorbed imidazolium chloride ions from an X65 substrate in deionized water saturated with CO_2 can be as high as 100 MPa [126]. They noted that the shear stress commonly observed in oil and gas pipelines does not exceed 1 kPa and hence the authors postulated that the typical fluid flow in these pipelines could not remove the inhibitor film from steel surfaces.

AFM can be combined with other techniques to elucidate the inhibition mechanism. Pandarinathan et al. conducted electrochemical measurements (OCP, LP, and EIS) while mapping the inhibited surface of UNS G10300 with the AFM under a CO_2 environment in a specially designed flow-through cell. The authors

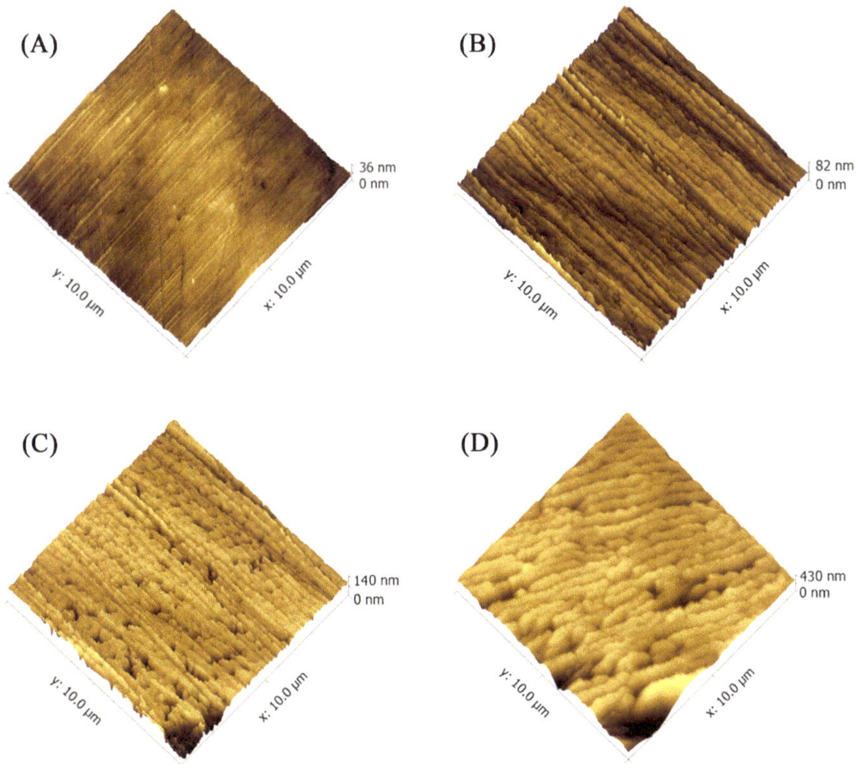

FIGURE 9.26

In situ atomic force microscopy topography of carbon steel in brine with DPC at 500 ppm. Sequence shows (A) untreated bare steel before immersion and after immersion for (B) 5 minute, (C) 30 minute, and (D) 4 hour.

Reproduced with permission from V. Pandarinathan, K. Lepkova, S. Bailey, et al., Adsorption of corrosion inhibitor 1 dodecylpyridinium chloride on carbon steel studied by in situ AFM and electrochemical methods, Industrial and Engineering Chemistry Research 53 (2014) 5858–5865.

demonstrated the changes in the inhibitor structure from spherical to cylindrical aggregates as its concentration was increased above the critical micelle concentration (CMC) [196]. Fig. 9.26 shows the adsorption process of 500 ppm DPC onto carbon steel surface over a period of 4 hours [196].

9.9.2 Fourier transform infrared spectroscopy

The Fourier transform infrared (FTIR) spectroscopy technique is used to characterize functional groups and chemical bonds of liquid and solid substances. Infrared (IR) spectra can be used to fingerprint chemicals. Its applications range from quality control to inhibition mechanistic studies. For instance, Moloney

[121] employed FTIR to determine if the inhibitors developed for sour services suffered thermal degradation.

Ex-situ FTIR measurement can elucidate the mechanism of adsorption (physisorption vs chemisorption) [197]. It is common to accept that chemisorption is likely when the presence of adsorbed inhibitor is evident ex situ [198,199], for example, detection of $-CH_3$ or $-CH_2$ on the metal surface in the case of fatty acid amine and imidazoline/ester inhibitor [197]. Lepkova et al. utilized FTIR to detect the presence of an inhibitor at the surface and to reveal the orientation angle of an adsorbed DPC molecule on a carbon steel surface in a noncorroding environment [198].

9.9.3 X-ray photoelectron spectroscopy

X-ray photoelectron spectroscopy (XPS) is a surface analysis technique highly sensitive to the first few atomic layers with a sampling depth of approximately 3 nm [200]. The XPS technique reveals what elements exist on the surface and the nature of chemical bonding based on the position of the binding energy. Monoatomic depth profiling can be performed whereby the surface is bombarded with an inert gas such as argon to gradually remove chemical species from the top surface. Performing a high-resolution XPS scan sequentially after a set period of surface sputtering generates a depth profile of the studied elements. For instance, Lopez et al. utilized XPS and SEM to study the inhibitor adsorption mechanism [201]. The lack of substrate signal (iron peaks) implies that the inhibitor forms a thick ($>2-3$ nm sampling depth) and uniform film. On the contrary, when the substrate signals are evident, nonuniform adsorption or a thin layer of inhibitors have been proposed [202].

9.9.4 Others

There are several other advanced techniques, tools, and approaches employed in inhibitor studies. For instance, the electrochemical quartz microbalance (EQCM) has been used to study the growth of an inhibitor adsorption layer on a metal surface [203]. The small change in mass per unit area of the electrode, either as mass loss from corrosion or mass gain by inhibitor adsorption, is a function of resonant frequencies measured on the electrode.

9.10 Compatibility tests

An effective corrosion inhibitor should mitigate corrosion risks to an acceptable level but, in doing so, it should not introduce other problems. A chemical product usually comes into contact with many materials during transportation, storage, and injection into the system. It is essential to evaluate and confirm the corrosion inhibitor's compatibility with the various metals, plastics, and elastomers and if deemed incompatible, then to select the most appropriate material for containers, seals, pumps, etc., or, alternatively, another inhibitor product may be required for safe use. The application of the corrosion inhibitor should not cause problems on downstream activities, such as fluid separation, refinery processes, etc. The risk of product incompatibility should be contemplated when selecting and recommending a product for field deployment.

9.10.1 Compatibility with metallic materials

In their neat form, corrosion inhibitors may be corrosive to various metallic materials handling these chemicals such as transportation containers, storage tanks, umbilicals, capillaries, injection quills, etc. Such components can be susceptible to both localized and general corrosion from the corrosion inhibitor product. For instance, neat amine chloride and quaternary ammonium compound inhibitors are corrosive to UNS S31600 [122]. For metallic material compatibility, the susceptibility to various forms of localized corrosion (e.g., pitting, crevice, and stress corrosion cracking, etc.) should be evaluated. However, the primary purpose of these tests is to rank inhibitor candidates since it cannot be guaranteed that pitting will or will not occur in service.

Commonly, immersion tests in which a coupon of the metallurgy of interest is exposed to the neat corrosion inhibitor product at a specific temperature for a set duration are used to assess the neat chemical compatibility. These tests are usually long, perhaps several weeks to months, so gaining data can be time-consuming and in some instances challenging, especially when a decision on the specific material of construction of a component is needed quickly. However, some chemical products are organic in nature which may not be an effective electrolyte and are not conducive to electrochemical tests. In such cases, immersion tests are necessary.

CPP is used to determine the susceptibility to pitting corrosion. It is critical to ensure that the specimen is free of crevice or the CPP result is invalid. An example of a commercial test setup specifically designed to avoid crevices is FlexCell. Fig. 9.27 presents the typical CPP curves of systems not susceptible and susceptible to pitting corrosion, respectively. The arrows indicate the potential scan direction, that is, forward and reverse potential sweep. When a material is not prone to pitting corrosion, the reverse scan yields lower current density compared to the same potential on the forward scan, that is, negative hysteresis (Fig. 9.27A). On the contrary, for a system prone to the passive film breakdown and, consequently localized corrosion, a sharp rise in the current density is evident with greater current density in the reverse scan than that of the forward scan, that is, positive hysteresis (Fig. 9.27B). The key parameters derived from this technique are (A) OCP, (B) breakdown potential or pitting potential (E_b), and (C) repassivation potential (E_{rep}). Large E_b with respect to OCP generally implies a superior resistance to pitting corrosion. Moloney et al. conducted CPP measurements and

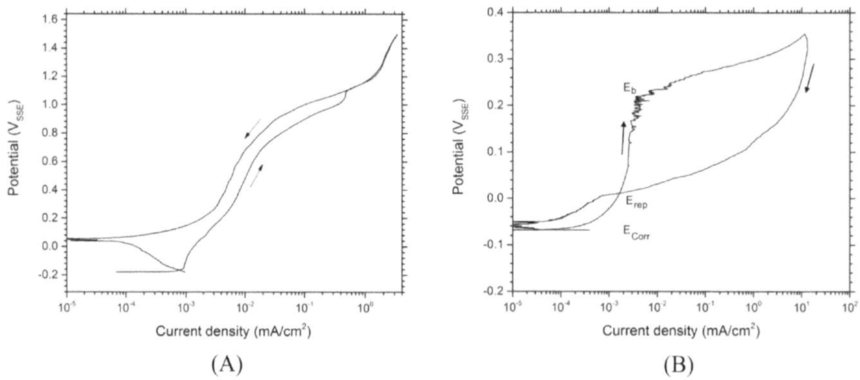

FIGURE 9.27

Typical cyclic potentiodynamic polarization curves of (A) a system not prone to pitting corrosion (negative hysteresis). (B) A system prone to pitting corrosion (positive hysteresis). Arrows indicate the potential scan direction.

three-month immersion tests of UNS S31603 in 13 production chemicals, including corrosion inhibitors [204]. The authors proposed that a chemical has a high likelihood of compatibility with UNS S31603 when E_{pit} is greater than 200 mV vs OCP and the current density determined at 150 mV above OCP is less than 5 μA/cm^2. The system that met these criteria did not develop pitting corrosion in a three-month immersion test.

The presence of crevices is almost inevitable in engineering design. Crevices are narrow gaps between two surfaces, such as flanges and gaskets, etc. Recently, standard crevice corrosion test methods were reviewed by Iannuzzi et al. [205]. Crevice formers are used to create crevices in a controlled and reproducible manner. They are made of inert polymers, such as PTFE, PVDF, and PCTFE, or ceramic, or PTFE-tape covered ceramic. They can be a flat surface [206] or serrated [207]. The tightness between the crevice former and the metal specimen is controlled by the applied torque, which has been shown to influence the test results [208]. Test reproducibility seemed to increase when a disk spring assembly, such as that shown in Fig. 9.28A, are used to maintain a constant force [206]. A serrated crevice former is shown in Fig. 9.28B [209]. Creviced specimens are then immersed in the test environment, in this case a corrosion inhibitor, for a certain period of time. After exposure, visual observation and microscopy analysis are performed to quantify the crevice attack. Alternatively, electrochemical techniques can be used to study the crevice corrosion susceptibility. For instance, OCP of the crevice assembly can be monitored. A sudden change in OCP can indicate the initiation of crevice corrosion. Readers can further review the application of electrochemical techniques in assessing crevice corrosion summarized by Iannuzzi [205].

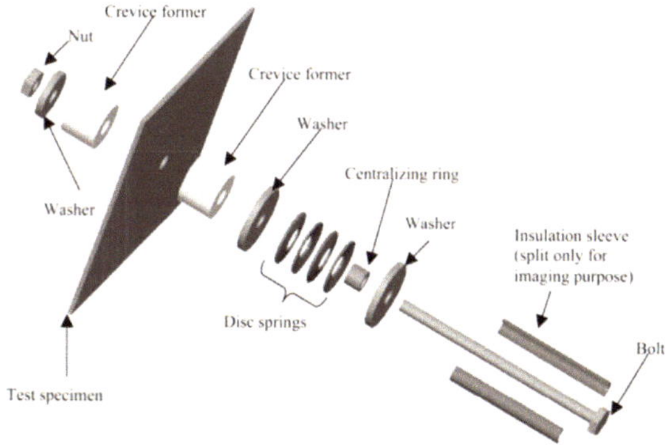

FIGURE 9.28

Photographs and schematic of crevice corrosion tests. (A) disk-spring multiple crevice assembly, (B) Serrated TFE-fluorocarbon crevice formers.

Reproduced with permissions from B. Espelid, Crevice formers for specimens of plate material, in: U. Kivisäkk, B. Espelid, D. Féron (Eds.), Methodology of Crevice Corrosion Testing for Stainless Steels in Natural and Treated Seawaters: (EFC 60), Maney Publishing, Leeds, 2010 and R.G. Kelly, J.S. Lee, Localized corrosion: crevice corrosion, in: K. Wandelt (Ed.), Encyclopedia of Interfacial Chemistry, Elsevier, Amsterdam, 2018, pp. 291–301.

9.10.2 Compatibility with nonmetallic materials

A wide variety of nonmetallic materials are used in the oil and gas industry, including elastomers, plastics (thermosetting and thermoplastic), and composite materials. They are used in various applications (valves, seals, gaskets, linings, etc.) and are exposed to environments that could accelerate their degradation. These materials may be incompatible with produced hydrocarbons and production chemicals (including corrosion inhibitors) that they come into contact with causing swelling, shrinkage, brittleness, softening, and crazing [210,211]. Ultimately, this may result in leakages and spills of hazardous and toxic chemicals, including the oil and gas itself. The standard information on chemical resistance/compatibility charts is generally too broad to be useful. It was also pinpointed that the chemical composition of the same polymer type can vary between vendors and/or batches [210]. As such, before the corrosion inhibitor chemical products are approved for field use, compatibility tests with nonmetallic materials are necessary.

Immersion tests are typically performed where test specimens are immersed in the production chemical of interest for a defined period of time, and preferably at the anticipated exposure temperature. Afterward, the relative changes in appearance, physical and mechanical properties are recorded. Some of the key properties commonly quantified in a chemical compatibility test are changes in yield strength, % elongation, hardness, volume, dimension, and mass loss/gain. The change in dimension can indicate shrinkage/swelling of the materials after exposure.

Hardness measurement on polymers and elastomers is often performed with a durometer (an indentation-based method) following ASTM D2240 Rubber Property—Durometer Hardness [212]. Other available test methods involve an impresser (ASTM D2583) and pencils (ASTM D3363). ASTM D2240 notes that the hardness value measured from a durometer cannot be correlated with the fundamental properties of a material. This notion is also applied to other hardness tests [210]. The results are empirical, and the method is primarily for control purposes by quantifying the relative changes in the hardness resulting from chemical exposure.

Mechanical properties of nonmetallic materials are influenced by absorbed fluids and temperature [211]. Produced hydrocarbons and certain components in production chemicals can affect the tensile properties (elongation, yield, and ultimate tensile strength) of nonmetallic materials. ASTM D638 [213] and D412 [214] outline standard test methods for tensile properties measurements for plastics and rubber, respectively.

9.10.3 Foaming and emulsion tendency

As a requirement of their function, corrosion inhibitors are commonly surfactant in nature and formulated products often have several surface-active components in them, which can cause foaming and emulsions to form leading to oil/water separation issues downstream. Thus the tendency for a corrosion inhibitor to cause foaming and/or create an emulsion needs to be evaluated. There are a number of tests, methodologies, and variations that can be employed to assess the emulsion and foaming tendency. Typically, these require the mixing of the hydrocarbon and aqueous phases with the corrosion inhibitor, and other production chemicals intended to be used for a defined duration under simulated temperature and shear rate conditions. During the test, the behavior of the fluids is observed and it is compared with a reference test (or blank) condition in which no corrosion inhibitor is present. The synthetic hydrocarbon may be used when field hydrocarbon cannot be obtained. However, it is preferable to use field hydrocarbons in these tests, since they may contain naturally occurring surfactants that can affect interfacial properties between the phases and, consequently, emulsion and foaming tendency. The test results may be influenced by the ratio of aqueous and hydrocarbons phases. Thus the water cut replicating the field should be used, taking into account that this may change over the period that the corrosion inhibitor is expected to be deployed. It is also advisable to assess a range of dose rates, not just that at which the corrosion inhibitor has been qualified for from the various performance tests, to account for both overdose and underdose situations. Furthermore, it is recommended that any other process chemicals at the dose rates that they are being applied at in the system (or planned to be injected) are also included along with the corrosion inhibitor since the combination of the chemicals can result in more severe effects than by either alone.

There are various methodologies to run emulsion tendency evaluations including agitation using handshaking, mechanical mixers, or via pressure

(A) (B)

FIGURE 9.29

Photographs of (A) an emulsion test and (B) a foaming test.

release. Fig. 9.29A shows graduate cylinders containing emulsion mixtures at various time intervals. After mixing, the emulsion mixture is often poured into a graduate cylinder. The changes in the volume of hydrocarbon, aqueous, and emulsion phases with time as emulsion breaks are recorded along with visual observations of the various phases.

It was reported that using different mixing apparatus can lead to contradicting results [215]. The droplet size distribution of the emulsion phase may be used as a controlled parameter. Seltzer bottles produced an emulsion with a small droplet size distribution, which required a longer time to separate. On the contrary, the emulsion produced by handshaking contained large droplet sizes leading to reduced separation times [215]. The concentrations of oil in water and water in oil may be analyzed after the emulsion test is complete. The hydrocarbons carried over into the water phase can cause environmental concerns while excessive water in the oil phase can contaminate the downstream processes.

As with the emulsion tendency evaluations, foam can be generated in various ways including bubbling gas into aqueous and hydrocarbons phases along with the corrosion inhibitor and any other process chemicals for a period of time or via pressure drop approaches. The changes in the volume of hydrocarbon, aqueous, and foam phases with time as the foam collapses are recorded. Fig. 9.29B shows graduate cylinders containing foam mixtures at various time intervals.

The results obtained from both tests are purely comparative among different inhibitors. Pacheco et al. proposed a residence time of 10 minutes or less as an acceptance criterion [128], although this is typically dependent on the specific system in which the corrosion inhibitor is to be deployed as well as operator preference.

9.10.4 Physical properties, product stability, and additional considerations

Corrosion inhibitors are only effective if they can be applied to the operating system at the correct dose rates at the right locations. If the chemicals cannot be injected correctly then even the best performing product is essentially useless. Thus physical properties, such as freeze point, flash point and viscosity, and stability of the corrosion inhibitor under the ambient transportation, storage, and delivery systems are of critical importance. For example, a product with a pour point of −10°C would freeze in −20°C to −40°C conditions and would not be able to be injected. These low temperatures are common in winter climates experienced in some parts of the world such as those in various Canadian, European, and Russian regions. Even when above the freezing point, high viscosity products may also exceed pump limitations resulting in system injection difficulties. In high-temperature locations, corrosion inhibitors with a low flash point may experience solvent losses. In addition to health, safety, and chemical handling concerns, the loss of solvent may lead to product composition changes. Furthermore, exposure to extreme temperatures and temperature fluctuations can impact corrosion inhibitor product stability in which separation, gelling, particle generation, solid drop-out, etc., can occur, which must be assessed accordingly during product development and recommendation.

In some instances, corrosion inhibitor products need to be injected into locations great distances away from the chemical storage tank. For example, the chemical product may need to be transferred via many kilometers of subsea umbilicals or thousands of meters of downhole capillaries. To ensure the suitability of use for such systems, a corrosion inhibitor candidate needs to be stable under extreme conditions to avoid product degradation and potentially costly blockages of the narrow bore tubing delivery lines. In other instances, the only means of applying the corrosion inhibitor is via a gas stream, such as that for gas lift applications, so appropriate product properties need to be considered, accordingly. Before product recommendation, it is critical to conduct assessments such as dynamic cold centrifuge stability tests, high-pressure viscosity measurements, hydrate resistance tests, gunking evaluations, elevated pressure, and thermal stability flow loop tests and product cleanliness evaluations as well as the pertinent material compatibility previously described.

When used in downhole applications or closed-loop glycol regeneration/reclamation system, the operating temperatures may exceed 200°C. Under these conditions, the inhibitor's thermal stability is critical to ensure that the inhibitor remains effective and demonstrates that chemical decomposition, phase separation, and solid precipitation do not occur. Greg and Ramachandran used Carbon^{13} NMR to verify the stability of individual components of two commercial corrosion inhibitors following exposure to a temperature of 115°C for 24 hours [216]. It should be noted that, in some instances, degraded corrosion inhibitors may still provide corrosion inhibition. Such thermally aged products or solutions prepared with simulated aqueous composition and anticipated dosage should be reevaluated in the various corrosion performance tests to confirm that the inhibitor's effectiveness has not been compromised.

Further laboratory testing may include assessments under any other condition (s) to take into account the particular location in which the corrosion inhibitor may be deployed or the specific operating environment that it will need to perform in to ensure the suitability of the chemistry and verify no additional problems are caused due to its use. Some of these may include, but are not limited to, performance, stability, or compatibility evaluations in glycol regeneration systems, amine units, condensate stabilization processes, water disposal or reinjection locations, reservoir core, refinery, and downstream operations, etc. Additional aspects that also need contemplating when selecting and recommending corrosion inhibitors for use include local environmental and regulatory considerations where specific chemistries may not be permitted for use or that certain ecotoxicity, biodegradation, and bioaccumulation specifications may need to be met in certain countries or regions.

9.11 Field performance evaluation

As previously mentioned, regardless of how sophisticated the laboratory setup is, it cannot fully replicate all aspects of the actual field operating conditions. Therefore, the ultimate corrosion inhibitor qualification is to observe how it performs under field conditions. Before a full-scale implementation of the corrosion inhibitor, it is recommended to first conduct smaller sized assessments such as using side-stream units, field fluid compatibility assessments including emulsion, and foaming tendency with the produced water and hydrocarbon, and a field trial using select wells or pipelines. In doing so, any issues can be seen and addressed (perhaps through formulation optimization or alternative product options) before full commitment is made. Online corrosion monitoring is often used for this and for optimizing inhibitor dosage rate. As such, corrosion monitoring and complimenting information from water analysis is vital. The inhibitor product performance is determined to be effective if corrosion is suppressed to an acceptable rate and if there are no adverse effects upon its addition in the system.

9.11.1 Corrosion monitoring

Field corrosion monitoring utilizes techniques such as weight loss coupons, electrical resistance (ER), and electrochemical. A nonintrusive approach such as ultrasonic testing (UT) probes has also been introduced [208,217]. The most commonly used monitoring method in the field is weight loss because of its low cost. The method does not require electrical connections and a data logger, making it simple to operate. In this method, corrosion coupons are introduced to the environment and retrieved after a set duration. Corrosion coupons are fabricated from the same material grade as the system components. The visual appearance and weight are recorded before exposure [218].

For the most part, the benefits and drawbacks of this technique, when conducted in the field, are similar to laboratory testing. Periodic retrievals of the coupons in the field conditions are labor intensive and are not without risk to the personnel who are breaking into high-temperature and pressure systems. Choosing appropriate exposure duration is critical to obtaining representative assessment. Initial CR tend to be high before gradually declining. As such, a short exposure time can overestimate the CR. On the contrary, too long an exposure in corrosive conditions can cause significant metal loss and detachment of coupons by mechanical action [219]. Depending on the corrosivity, a two- to twelve-month exposure period is recommended for CO_2 environments [219]. Standard practice is outlined in NACE SP0775–2018 "Preparation, Installation, Analysis, and Interpretation of Corrosion Coupons in Oilfield Operations".

ER probes are the second most used corrosion monitoring technique in the field. Responses to corrosion associated with environmental changes are monitored in real time. ER probes rely on the relationship between the dimension and electrical resistance of the sensing element, which is made from the same alloy as that of the system component. As metal loss occurs either through corrosion and/or erosion, the geometry and electrical resistance of the probe element changes following the relationship are shown in Eq. (9.20).

$$R = r\frac{L}{A} \tag{9.20}$$

where R is electrical resistance (ohm), r is resistivity (ohm.cm), L is element length (cm), and A is element cross-sectional area (cm^2).

The metal loss is estimated from the change in electrical resistance of the sensing element in relation to that of the reference element. The sensing element of an ER probe is made from a metal that responds to the environment the same way the system component does. Various shapes are available, such as wires, strips, loops, etc., and the selection is usually based on the application, namely, anticipated corrosivity and the probe sensitivity [220]. Thin element probes generally provide faster response times but have a shorter life than the thicker ones. Interpreting results from an ER probe is relatively simple. However, the occurrence of localized corrosion and the formation of electrically conductive scale on the sensing element, such as sulfide scales in H_2S-containing environment, can lead to misinterpretation of results. Therefore the probe should also be visually inspected after retrieval.

Field corrosion monitoring can also be done with the previously mentioned electrochemical techniques. The most commonly used electrochemical techniques in the field corrosion monitoring are LP and galvanic monitoring [110]. EIS and EN techniques are less utilized in the oil and gas fields because interpreting these results is complex [221]. The working principles of these techniques are described previously in Section 2.4 (Measurement techniques). For LPR probes, as a conductive electrolyte is required to facilitate the flow of current between the WE and CE, these probes may not be suitable or give erroneous data when exposed to crude oil, condensate, or a thin or a discontinuous layer of water. In addition, in sour system where conductive iron sulfide scales are expected, electrochemical probes may not be suitable as adjacent electrodes may be short-circuited.

The readings from any monitoring probes are only representative of the corrosivity at that particular location. Therefore the success of a monitoring program depends largely on the selection of the monitoring location. To aid the decision, corrosion modeling in conjunction with fluid flow modeling is often used to determine locations associated with the highest corrosion risks, where the monitoring is recommended. Often, these are areas associated with high shear stress and low points where water and solid accumulate. However, this is sometimes not an easy task as the suitable monitoring location may not be accessible or that the highest CR location changes with time. In some instances, quantified CR are a true indication of the actual rates. Nonetheless, readings obtained from these probes should be treated simply as an indication of changes in the environment corrosivity. Differences in metallurgy, appropriate selection of B-value (if CR are determined electrochemically), and relatively shorter exposure time can affect the measured CR.

9.11.2 Corrosion inhibitor residual measurements

Regular measurement of the concentration of the inhibitor in the aqueous phase, known as residual monitoring, is a standard practice as part of an asset integrity program. The concentration of the active compounds above a certain threshold may indicate that an inhibitor dosage was adequate and that the metal surface preceding the measurement point was exposed to the corrosion inhibitor. A minimum corrosion inhibitor residual threshold is often set based on the dose rate required to pass all the various laboratory assessments. In some cases, a maximum chemical residual threshold may also be incorporated which should not be exceeded to minimize potential problems associated with overdosing. It should be noted that the detection of the corrosion inhibitor product simply indicates its presence at whatever value is measured but does not indicate or verify performance or that the system is adequately inhibited from corrosion. Regular residual monitoring in combination with corrosion monitoring assists in injection dose rate optimization.

A common field method for corrosion inhibitor residual measurement is based on the colorimetric techniques, assessing color development as a result of a reaction between a complexing agent and a specific chemistry present in the inhibitor formulation. This method is often referred to as a dye transfer technique. One example of this approach is when methyl orange is used to react with quaternary ammonium compounds, which are active chemistries in some commercial corrosion inhibitors. A solvent, in this case chloroform, is used to extract the resulting complex between methyl orange and the target compound from the brine solution. The spectrophotometry technique is employed to measure the absorbance, which will vary as a function of the inhibitor concentration. Benefits of this technique include its field applicability, simplicity in use, its low cost and that it is quick to conduct. However, there are several limitations associated with the methyl orange technique that have been reported, including interferences from other nitrogen-containing compounds and certain chemicals in the oilfield matrix. Some active oilfield corrosion inhibitor chemistries are not detectable by methyl orange. Additionally, the need to use chloroform, which is toxic, is a concern from a health and safety perspective [222] while its use may be limited in certain locations. An operation shutdown due to an error in the residual corrosion inhibitor measurement has been reported [223]. The inhibitor residual determined by the methyl orange technique was underestimated. Consequently, the inhibitor was overdosed, causing the formation of severe emulsion. The issue was further investigated by comparing the in-field technique with liquid chromatography-mass spectroscopy (LC-Ms) technique. The results show that, in this case, the methyl orange technique underestimated the concentration of inhibitor residual by 36%. The results suggested that the methyl orange technique should be complemented with laboratory validation, field corrosion monitoring, equipment inspection, and fluid sampling.

Chromatography and mass spectrometry techniques provide molecular specificity and high-resolution measurements. Water samples from oilfield brines or those extracted from solids are analyzed and compared against "the fingerprint" generated from the neat chemicals. Key markers are generally those present as active compounds in the inhibitor formulation [223]. Electrospray-mass spectrometry (ES-Ms) was introduced as another technique for qualitative analysis of oilfield production chemicals, including corrosion inhibitors [224]. Electrospray is an ionization technique that utilizes electrical energy to ionize molecules for mass spectroscopy analysis. The coupled LC/ES-Ms is highly sensitive and robust method compared with the LC/UV technique in detecting imidazolines and quaternary ammonium chemistries from production fluids [224]. The coupled LC/ES-Ms has been employed to quantify the residual concentration of quaternary ammonium and imidazolines-based inhibitors after the transient adsorption (within three hours) to plastic, glass, and metals [225]. Seven corrosion inhibitor active species were targeted. Selective loss of corrosion inhibitor chemistries was noted, confirming the high sensitivity associated with the LC/ES-Ms technique. Moreover, the coupled LC/ES-Ms can reveal the structure of imidazolines present in the corrosion inhibitor [226]. Previous work by the same authors suggests that the LC/ES-Ms technique is a promising tool to elucidate the surface adsorption mechanism of corrosion inhibitor mixtures. Drawbacks associated with such techniques are that they are laboratory based and so can result in a long turnaround time from gaining a field water sample and having it shipped and analyzed, they involve expensive analytical equipment and they need expert users to conduct the analysis.

Recently, surface-enhanced Raman scattering (SERS) spectroscopy has been used in low-level detection of oilfield corrosion inhibitors. The technique overcomes the inherently weak Raman signal when a conventional Raman spectroscopy is conducted. A colloidal of metallic nanoparticles or rough metal surfaces are used to enhance the Raman scattering effect, hence amplifying its signal. The SERS technique has been developed into an onsite corrosion inhibitor residual measurement in which a proprietary metal nanoparticle solution is mixed with inhibited brine solutions and in turn the inhibitor concentration is determined using a portable Raman spectroscopy [227].

9.11.3 Water chemistry analysis

Water samples are routinely collected and analyzed for their chemical compositions and physical properties. Trends are evaluated and any deviation from the norm may require attention. Regarding corrosion and corrosion inhibition, the determination of the concentration of iron in the water phase, known as iron counts, can be used to estimate the CR. Additionally, corroding carbon steel is usually the source of manganese found in produced water although there are exceptions to this. The concentration of manganese in steel is typically 0.5%–1.5% that of the iron concentration [110]. An increased level of manganese can, therefore indicate active corrosion in the system. The concentration of iron in the water samples is influenced by corrosion activity upstream of the sampling point and the inherent iron ions level in the reservoir or by iron scales precipitation. Additionally, it is unknown where exactly in the system the iron is dissolving from. Ideally this would be perfectly uniform to be able to determine the CR and wall metal loss of the system but it is unlikely that this would be the case. Hotspots of higher levels of general corrosion or regions of localized activity in the system would not be able to be identified from dissolved iron or manganese analysis alone which highlights the importance of assessing trends over time. In wet gas pipelines where full or partial pH neutralization is a mitigation strategy, pH is raised by bases such as MDEA or NaOH to promote iron carbonate precipitation. In a H_2S environment, iron sulfide precipitation is kinetically fast, hence iron concentration is not a reliable indicator for corrosion activity. The precipitation of carbonate or sulfide scales reduces the iron ions concentration in the bulk liquid and therefore can lead to estimates of CR that are too low. Manganese-to-iron ratio greater than 0.02 (1:50) typically indicates iron precipitation [228].

Errors in water samples analysis are possible for several reasons. For instance, corrosion of the sampling point, incompatible sample containers, inappropriate sampling points, and flow disturbance during sample collection. Further detail of the standard practice for iron counts measurement is explained in NACE SP0192 [228]. It is emphasized that iron count analysis and trends should not be used alone but in conjunction with other monitoring techniques in an integrity management program to help gain a fuller picture of the system's health from a corrosion perspective and efficacy of the corrosion mitigation strategy in place.

9.12 Case studies

This section gives examples from the literature of the field application of corrosion inhibitors under an array of sweet (carbon dioxide containing) and sour (hydrogen sulfide containing) oilfield operating systems.

9.12.1 Field application of corrosion inhibitors in sweet (carbon dioxide containing) systems

9.12.1.1 The presence of solids

The presence of flowing solid particles can aggravate internal corrosion in the form of erosion-corrosion. A case study by Hu et al. [145] reported a failure investigation of a welded carbon steel pipework retrieved from an offshore facility, which produced 95% water cut and the maximum sand production was 500 mg/L. The pipeline was already treated with a commercial corrosion inhibitor. Static corrosion tests were conducted, but the results were not comparable with the metal loss experienced in the field. The discrepancy was attributed to the strong influences of the flow and sand in the system. Subsequently, complex but representative jet impingement tests were conducted. The results showed that the existing chemical treatment was not effective as the inhibition efficiency was 11.5% and 7.8% for the parent metal and the weld material, respectively. The ineffectiveness of the chemical treatment at that time is likely to explain the observed corrosion failures. A new chemical product was nominated and evaluated. The JIC results showed an increased inhibition efficiency with no evidence of localized corrosion. This study demonstrated that using the wrong inhibitor in the field may not provide much benefit, the importance of mimicking the field conditions as closely as possible during the laboratory assessment and that carbon steel in conjunction with a suitable corrosion inhibitor can be a practical approach to manage erosion-corrosion while providing an overall cost-benefit over the use of corrosion-resistant alloys.

9.12.1.2 Presence of additional production chemicals

Not only should the incompatibility between a corrosion inhibitor and other production chemicals be contemplated, but the synergy between them should also be considered and if present be highlighted since overall performance can be improved leading to operational cost savings [229]. Crude oil arriving at the

Safaniya Arab heavy wet crude handling facilities contained a large percentage of water-oil emulsion, which needed to be separated. After separation, the water phase was reinjected to enhance production. Oil contamination in the water phase can cause technical issues, such as a rise in injection pressure, contamination of reinjection wells, increasing shut-in time, and loss of crude oil, etc. Additionally, water contamination in crude oil can lead to corrosion problems at downstream facilities. Kima [229] reported the effects of optimizing the corrosion inhibitor and demulsifier chemical products used in the Safaniya Arab heavy wet crude handling facilities. Although there were no details given or discussion of the chemistries of the incumbent and the new products, the new treatment strategy included the reduction in the demulsifier consumption along with increasing the corrosion inhibitor dosage. Emulsion tendency tests were conducted on the new chemical package and dose rates at the operating temperature of the facilities (12°F or 49°C). Oil in water content in both the produced water and wastewater, CR using LPR and ER probes as well as corrosion coupons, residual corrosion inhibitor concentrations, and iron counts were monitored after the new chemical treatment regime was commenced. The in-situ corrosion monitoring results demonstrated that the CR was mitigated by >99% (a reduction from 290 to 1.1 mpy). With the improved corrosion inhibition efficiency, the oxygen scavenger, which was previously injected in conjunction with the corrosion inhibitor, was no longer required. As a result, a cost saving of $2MM per year was achieved from the reduced consumption rate of demulsifier, the omission of the oxygen scavenger product, and reduced costs associated with oil-water separation issues, workovers, well stimulations, etc.

9.12.1.3 Corrosion inhibitor chemical squeeze application

In an offshore oilfield in Abu Dhabi, several wells experienced water breakthrough with water cuts as high as 60%, enhancing the corrosion risk to the assets. CO_2 corrosion was expected since the field produced 1.5%–3% mole CO_2 [230]. The integrity of the production tubing was critical because of safety concerns as well as the likelihood of production interruptions. However, since there were many wells, replacing the production tubing with a corrosion-resistant alloy would significantly increase the CAPEX costs. Hence, the use of lower grade steels such as carbon steel or 13%Cr steel in conjunction with a corrosion inhibitor was considered a cost-effective approach. While the chemistry of the selected corrosion inhibitor was not disclosed, the treatment strategy was described. As these wells were not equipped with capillary lines for chemical injection, a technique called "corrosion inhibition chemical by a squeeze to formation" was employed whereby the inhibitor product was squeezed into the reservoir formation. Subsequently, the chemistry desorbed from the reservoir rock into the produced fluids. As the production fluids, which now contained the corrosion inhibitor, traveled up the production tubing, a corrosion film was formed on the tubing surface. In other words, the inhibitor was essentially delivered via a pseudo batch treatment. The effectiveness of this mitigation program was monitored by installing and retrieving corrosion coupons every two months. The results showed that the treatment was effective for about four months, during which the CR was reduced to approximately 5 mpy. At the time of the publication in 2020, it was reported that approximately 30 well strings were treated on an annual basis since 1994. The author assumed that the treatment program could prevent corrosion failures of up to three well strings. Based on avoiding production downtime of 6 months, the annual net cost saving was estimated to be $38.4M.

9.12.1.4 Corrosion inhibition under high-velocity environments

High flow velocity environments can be challenging for chemical inhibitors to control corrosion to acceptable levels. In the past, the ability of corrosion inhibitors to provide protection under such operating conditions had been an issue with a long history in the gas production facilities in the Nederlandse Aardolie Maatschappij gas field in the North East of The Netherlands [231]. Soon after the Groningen gas field started production, it was recognized that corrosion inhibition was required to protect the various facilities. For many years there was a rule of thumb that the maximum production velocity for inhibited flowlines was 20 m/s but in the ninety's laboratory testing using an oil-soluble corrosion inhibitor (Product A) showed that the critical velocity could be increased to about 30 m/s using a dose rate of 300 ppm (on total liquids using a 90% water cut at 70°C) compared with about 20 m/s using a dose rate of 150 ppm (on total liquids using a 50% water cut at 45°C). The concern at the time of the work leading to the publication was that the need to maintain gas production at a certain level resulted in much higher velocities. Rather than spend more time and expense conducting laboratory testing it was preferred and decided to carry out two field trials since these would be under real field situations without compromises and about the same cost. From the fieldwork at the two locations, it was shown that using a modified oil-soluble inhibitor (Product B, which contained the same esters as Product A but with an additional amine) at an appropriate dose rate, CO_2 corrosion in high-velocity wet gas/condensate systems can be effectively mitigated by corrosion inhibition up to flow velocities of 50 m/s [231]. It was also iterated that it should not be assumed that the corrosion inhibition observed at the high velocities is generally acceptable. System-specific attributes (such as the presence of sand, etc.) can impact chemical performance.

9.12.1.5 Tubing displacement for downhole corrosion inhibitor application

The corrosion of downhole tubulars can be of considerable concern in high water cut production wells with appreciable a carbon dioxide content [232]. Typically, as in many cases in the oil and gas industry, the two main corrosion control options for tubing in such environments include either a suitable CRA or carbon steel with an applicable corrosion inhibitor program. Oftentimes, the latter approach is found to be more cost-effective if corrosion can be mitigated to acceptable levels. These were also options available to the Kuwait Oil Company when there was an increased occurrence of corrosion-induced failures of downhole tubulars, many of which were producing fluids having high water cuts in the range of 30%–40% along with 2%–3% CO_2, in the Sabriya oil field located in North Kuwait [232]. After an initial laboratory testing program of candidate corrosion inhibitors, a leading product was identified and, after verifying compatibility with other production chemicals being deployed in the system, a field trial was conducted on a selected production well. The corrosion inhibitor treatment procedure used was the tubing displacement technique where a batch of chemical was displaced through the tubing to the base of the well, which was shut-in for 24 hours before the production was restarted. The volume of the chemical pill was calculated based on the chemical supplier's recommendations in order to give a 3 mil (75 um) thick active corrosion inhibitor chemistry film on the inside of the tubing for the desired treatment lifetime. A downhole corrosion monitoring system (DCMS), which operated on the ER principle, downhole coupons, and fluid sampling and assessment were used to assess the efficacy of the corrosion inhibitor application. The coupon data gained showed a significant reduction in the corrosion, especially localized attack, as a result of the chemical inhibitor application and the complimentary dissolved Fe and Mn analysis were found to be a convenient approach to determine both the effectiveness and longevity of the downhole treatment. However, while these monitoring techniques were found beneficial, problems were experienced with the DCMS-ER technique and it was recommended that further development was required with such an approach before it could be considered a reliable option for downhole CR measurement under harsh conditions. It was also stressed that corrosion monitoring in surface flowlines is not useful for downhole conditions. Overall, it was concluded that the tubing displacement technique using an applicable corrosion inhibitor is a viable option for downhole corrosion control of crude oil wells [232].

9.12.1.6 Top-of-the-line corrosion inhibition

TLC is regularly mitigated through batch corrosion inhibitor application. Oftentimes, this preferred approach is not due to the limitations of the commonly used continuous corrosion inhibitor chemistries (applied to treat for bottom-of the-line corrosion issues) to control the actual corrosion taking place but more as a result of the limitations in their transport to the top of the pipeline. However, in some instances, the recommended batch inhibitor treatment frequency cannot be adhered to for one reason or another, which was the situation for one particular field case [158]. In this example, the efficiency of the incumbent batch inhibitor treatment was shown to be approximately 80% in the first month which then significantly decreased to 50% if the application was not performed for a month while after two months there was essentially no corrosion protection at all. While such data verified that the established monthly batch corrosion inhibitor treatment frequency used by various operators is appropriate, due to the increasing number of platforms and platforms this goal was challenging to achieve in this particular field where the average application was 3–4 times a year [158]. The injection of a continuous product to not only attend to the corrosion occurring at the bottom-of-the-line but also that at the top-of-the-line, perhaps through the use of a volatile type of corrosion inhibitor, would be attractive to overcome the limitations of the field operations. Following satisfactory laboratory testing in independent laboratories, two products were selected for field trial with the aim to gain a comparable efficacy to that of monthly batch treatments. A cooled ER type probe incorporating a highly responsive system (resulting in 100 × better sensitivity than that a standard ER probe) was installed at an appropriate location. While over the 9 months or so field evaluation it was not possible for the operating environment to be kept constant because of the changing conditions of the wells feeding the pipe monitoring location, efficiencies of the inhibitor products were made compared with stabilized one-week baseline CR. The field conditions ranged from 50 to 90°C, 44%–70% CO_2 with line pressures 25–30 barg, gas production 3–12 MMSCF/day, and water production 30–60 bbl/day without any condensate being produced. The field trial confirmed the possibility of TLC mitigation by continuous corrosion inhibitors and that an efficiency of 70% could be achieved [158]. While it was also highlighted that quite high dose rates were needed when using a continuous volatile inhibitor with a minimum of 500 ppm, which was at least a factor of five times greater than the existing bottom-of-the-line injection rates, and an upgrade of the existing injection facilities and/or product reformulation would be required, it is also worth considering the potential operational benefits of the reduced chemical volumes, reduced chemical handling and consumption as well as the reduced deferred or lost production of not having to perform a monthly batch inhibitor application.

9.12.2 Field application of corrosion inhibitors in sour (hydrogen sulfide containing) systems

9.12.2.1 *Corrosion inhibition facilitates successful operations under high hydrogen sulfide conditions*

The combination of high levels of hydrogen sulfide, high temperatures, and high-water production creates the potential for severe sour corrosion [233]. Nevertheless, gas and condensate can be safely and successfully produced through carbon steel infrastructure under such environments by the utilization of an appropriate corrosion inhibitor. Furthermore, using appropriate corrosion monitoring and assessment can help in identifying better corrosion inhibitor options and injection rate optimization to find the most economical solution for a particular system, an example of which is illustrated in the corrosion inhibitor testing program at the Kaybob South field, west-central Alberta, Canada [233]. At the time of publication in 2001, the Kaybob South field, which had been operating for approximately 30 years, was producing about 1200 10^3 m^3/day of sour gas, 600 m^3/day condensate, and 1400 m^3/day of produced water with acid gas contents ranging from 7% to 19% H_2S and 0.5%–5% CO_2. Given the severity of the operating conditions, the corrosion could only be managed through the deployment of specially designed corrosion inhibitors which also had to be stable at temperatures up to 90°C. For several years a specific corrosion inhibitor had been successfully used and injection rates optimized as far as possible using coupon monitoring but given the chemical spend of $1.2 MM ($Cdn) per year, which accounted for about 25% of the total operating costs, a corrosion inhibitor testing program was initiated with the objective to improve inhibitor economic effectiveness by 15%. Such efficiencies can be gained in several ways including the use of a more expensive chemical that is effective at lower injection rates as well as less expensive options. The field-testing apparatus for assessing three chemical products and their MEC included a skid-mounted corrosion monitoring pot complete with the appropriate piping and hammerlock fittings to tie it into wellheads or flow lines at meter-proving valve locations. Additionally, the unit was fitted with three protruding-electrode EN probes that were installed via high-pressure access fittings. EN monitoring, with a two-week time requirement per round of testing, was preferred over the more traditional weight loss coupons where the time requirement would have been at least eight weeks. Furthermore, the EN methodology also allowed real time and continuous data to be gained giving a higher degree of confidence than the one data point gained at each dose injection rate if weight loss coupons were used. From the field corrosion inhibitor testing program, an alternative product was identified that, although having a comparable cost to the incumbent, had a MEC that was 17% lower, which in turn saved the field about $75,000 per year on the well selected for the inhibitor testing. The deployment of the alternative corrosion inhibitor was then expanded throughout the Kaybob South field in which injection rates were reduced where possible by 15% resulting in additional cost savings of $134,000 per year [233]. This study demonstrated the benefits of conducting in-field corrosion inhibitor assessments which not only included significantly reducing chemical costs but also increasing confidence inhibitor performance and better understanding of how inhibitors work in the actual system. Furthermore, it also successfully exhibited the EN assessment technique to be a viable option for field corrosion inhibitor testing and assessment with several advantages over traditional testing methods [233].

9.12.2.2 Presence of elemental sulfur

One of the most aggressive and rapid forms of corrosion is that associated with elemental sulfur but, as a 20-year success story from the LaBarge operation demonstrates [234], with a robust corrosion inhibitor and integrity management program in place, high acid gas, highly sour well streams producing high quantities of elemental sulfur can be safely and reliably produced using carbon steel infrastructure. Production from the LaBarge field, located in the Bridger Teton National Forest, which is an area of outstanding natural significance, presents unique challenges regarding corrosivity, environmental protection, and asset preservation. The production of 720 MMSCFD high acid gas content sour gas (~5% H_2S and ~66% CO_2) with very little condensate and with nearly 2 kg/MSCF elemental sulfur constitutes an appreciable corrosion challenge. While the use of alloy 625 clad carbon steel flowlines was considered in the initial conceptual studies, the long lead times for this material would have adversely impacted the start data of gas production and so resulted in the decision to use carbon steel wet gas lines. After a laboratory program assessing 16 corrosion inhibitor products with mixed results, one product was identified that successfully mitigated the blank general CR of ~0.8 mm/year and localized CR as high as 5 mm/year in the presence of adherent sulfur to 0.025 mm/year general CR with no visible pitting after 30-day exposures. The MEC was found to be 1000 and 1500 ppm in nonsulfur and sulfur-containing LaBarge environments. The initial field treatment rate was set at 3000 ppm of the continuous corrosion inhibitor, or three times the rate deemed necessary from the laboratory tests, while all flowlines were pigged every 6 weeks with poly cup pigs and batch corrosion inhibitor. In addition, periodic sulfur remediation treatments involving a batch treatment of diaryl disulfide with corrosion inhibitor addition into the wellbore helped to attend to sulfur deposition. Intense corrosion monitoring and inspection work, which included ultrasonic and radiographic inspection, in conjunction with visual inspection of removed spool pieces allowed the performance of the inhibition program to be realized and in turn, as confidence grew in its success, the treatment was optimized in most cases to the laboratory determined MECs. Overall, "the performance of this chemical inhibition system has been outstanding, both in its level of corrosion protection and overall reliability" [234]. The success was attributed to the attention to detail to the design and day-to-day production operations as well as the corrosion inhibition technology and, based on the experience at LaBarge, demonstrated that carbon steel could be inhibited economically at temperatures up to 135°C (275°F) and total acid gas contents of up to 70%.

9.12.2.3 Effective pipeline cleaning complements corrosion inhibitor application

Kapusta et al. [235] shared the experiences and learnings from 15 years of operation of the Fairway Field, an offshore sour gas field in Mobile Bay, Alabama. The production rate of 200 MMSCFD gas with 5000 BPD water had initial operating conditions comprising wellhead temperatures of 230°F–250°F (110°C–120°C) although at times could be as high as 300°F (145°C), a wellhead pressure of 650 bar and a gas composition of 1.6% H_2S and 3.5% CO_2. The internal corrosion management, particularly the corrosion inhibitor product selection, design, and operation of the system, was crucial given that not only are such conditions very corrosive to carbon steel but also because of the environmental sensitivity of the offshore location. A laboratory testing program assessing 14 corrosion inhibitors, including both continuous and batch application products, from four chemical company suppliers was initially conducted under simulated conditions where blank sour CR of up to 3 mm/year were inhibited to less than 0.03 mm/year after a week of exposure with 250 ppmv corrosion inhibitor. The field corrosion inhibition program was designed to mitigate against the three main risks envisaged of general corrosion, TLC, and UDC. From the laboratory evaluation, the top-of-the-line tests showed very little corrosion (< 0.1 mm/year), which was attributed to the formation of a protective iron sulfide scale from the rapid saturation of the condensed water with corrosion products. From experience obtained from the operation of other sour gas systems, the worst corrosion was expected under debris or deposits accumulating at the bottom of the pipelines. Since previous studies had shown difficulties of corrosion inhibitor products penetrating deposits, it was deemed that the most effective way to control this problem was to keep the pipelines clean and free of debris. With this in mind, the field implementation of the inhibition program involved three key aspects including a continuous corrosion inhibitor injection, batch corrosion inhibitor treatment, and pipeline cleaning [235].

The continuous injection of a water soluble or highly dispersible corrosion inhibitor product at a preliminary dose rate of 750 ppmv was initiated for the protective film formation stage over the first three months of operation after which an optimized rate of 250 ppmv was used for the maintenance dose in the gas pipeline and 500–750 ppmv for the produced water pipeline, given the much higher chloride concentrations. The batch corrosion inhibitor treatment using an oil-soluble chemical product diluted in diesel applied between two tight pigs with an estimated contact time of between 15 and 30 seconds was initially performed weekly for the first few months of operation and then optimized to a monthly and quarterly schedule for the gas pipeline and produced water pipeline, respectively [235]. Periodic cleaning with scraping pigs to remove debris was first carried out on a weekly basis to determine a baseline of the solids produced. After three months, the cleaning schedule was then optimized to a monthly schedule and then later to a quarterly frequency due to the small amount of solids being gained in the pig trap. The corrosion monitoring, which is an integral part of a corrosion management program, involved visual verification of the inhibitor product supply and injection system, field signature method units at the inlet and outlet of the gas and produced water pipelines, a corrosion spool installed upstream of the corrosion inhibitor injection to monitor, assess and track inhibited CR via UT inspection, retrievable coupons at the inlet and outlet of the gas and produced water pipelines, dissolved manganese concentration monitoring, corrosion inhibitor residual monitoring and periodic intelligent pig inspection of the gas and produced water pipelines. Similar to the LaBarge success story, the key learning from the operations at the offshore Fairway Field is that a well designed and implemented corrosion management system using carbon steel in conjunction with corrosion inhibitor can effectively mitigate the corrosion risks for the safe and efficient production and transportation of sour wet gas. The combination of consistent chemical inhibitor application and adequate cleaning resulted in a sour wet gas pipeline being in an excellent, practically "as new" condition without any notable wall loss after 15 years of service [235]. Another learning experience was gained from a failure and leak in the water disposal line and that corrosion problems occur when the inhibition program is not consistently maintained and when the corrosivity of the pipeline changes from the introduction of aerated water or acids. To remedy this incident, after the failed section was replaced, the brush pig cleaning and batch treatment frequency were increased to three weekly schedule in combination with corrosion rate and chemical residual monitoring [235].

9.12.3 Learnings from the literature case studies

The case studies presented demonstrate that carbon steel downhole tubulars, pipelines, and assets can be safely and efficiently operated under a range of oilfield operating environments in which multiple different corrosion mechanisms are occurring or potentially can exist through the prudent deployment of corrosion inhibitor chemistries. Many other case studies are also available in the literature. Effective corrosion inhibitor programs usually commence with a thorough laboratory testing, qualification, and selection of the most appropriate chemical product based on a thorough understanding of the production conditions and the corrosion processes taking place or expected to occur for a given operating system. It is sensible to also have an appreciation of other activities taking place such as the deployment of production chemicals for other purposes which could also impact a corrosion inhibitor's performance, problems associated with oxygen ingression which may best be remedied by identifying and attending to the source of entry, corrosion issues caused by bacteria and/or its byproducts which, in most instances, may be more effectively overcome with proper biocide treatment and monitoring, etc. It is seen that corrosion inhibitors can be successfully applied to a system using various approaches downhole or topside including continuous, batch, squeeze, tubing displacement, etc., or a combination of these and that the optimum approach for one system may be different from another. The successful application of a chemical product assumes the optimal equipment and infrastructure is in place to be able to deliver and apply the corrosion inhibitor at the appropriate levels including having robust delivery and storage tanks, reliable and well-maintained injection pumps with sufficient redundancy, the right types, and placement of injection quills, specially designed downhole equipment [236], pig launchers, traps, and receivers, etc. Besides the main production fluids of oil and/or gas and water, many systems can also contain debris including sand, iron sulfide and elemental sulfur and that a key learning from some of the corrosion inhibitor success stories is that keeping a system clean through periodic scraping pigging activities is important for an efficient chemical treatment program to not only minimize threats such as UDC or erosion corrosion but also reduce parasitic losses of the chemical itself.

Monitoring, data management, assessment, and review are key aspects of any chemical program and integrity management plan but rather than rely on one approach in isolation multiple facets should be considered to give a more complete picture of the corrosion control approaches. Corrosion inhibitor residual monitoring, fluid analysis to assess dissolved iron and manganese trends and ratios, various corrosion monitoring techniques and inspection approaches all have their pros and cons but conducted in conjunction with each other helps to give an overall depiction of the health of the system and effectiveness of the operations and activities being performed. Residual monitoring is used to confirm that the corrosion inhibitor is being injected and passing through the system at the recommended rates as well as highlight inadequate concentrations, but it does not provide any information about the performance of the product or the state of the pipeline condition. Inadequate corrosion inhibitor concentrations include both excessive as well as insufficient chemical quantities [236]. If a deficiency in corrosion inhibitor is detected, then the system may not be effectively protected while too much product is not only an obvious waste of money but can lead to downstream problems including emulsions and, in some instances, can lead to a reduction in corrosion inhibition. There has been controversy regarding monitoring of wet gas carbon steel pipelines [236]. Using conventional monitoring approaches such as weight-loss coupons and ER probes has been questioned due to the difficulty in predicting where the corrosion is taking place as well as accessibility to locations where the highest corrosion rate may be occurring, which may also change over time. In such cases, assuming reliable data, such monitoring could be used for monitoring corrosion trends with time, understanding that this is location specific but complement with the other available monitoring (including chemical and dissolved ion residuals) and inspection assessments. This also highlights potential areas for innovation and development to advance more efficient and effectual monitoring techniques and/or perhaps a more cost-effective, less time-, and laborintensive system-wide inspection approaches.

9.13 Summary

This chapter summarizes the fundamentals of corrosion inhibitor chemistries and broad inhibition mechanisms along with a review of common assessment techniques encompassing performance evaluation and other essential properties. Additionally, various approaches to molecular modeling are presented. While much progress has been made in corrosion inhibitor chemistry research and applications, some areas need further attention, as highlighted below.

Molecular modeling techniques such as DFT, classical MD, and MC can play an important role in the design and screening of novel corrosion inhibitor chemistries. Through the calculation of QCDs such as HOMO, LUMO, and the HOMO-LUMO gap, one can obtain insights into the chemical reactivity of an inhibitor molecule. Using DFT, one can compute the interactions between the molecule and the metal surface in vacuum and, also implicitly, in a solvent environment. MD and MC can be used for large-scale simulations of molecule-metal surface interactions.

While molecular modeling techniques have made significant contributions toward corrosion inhibitor research, there is considerable scope for further improvement. It is believed that future research in this field will be increasingly focused on the development of computational approaches for design of inhibitors for high-temperature applications (e.g., using Reax force-fields); modeling interactions of inhibitor molecules with defect surfaces (vacancies, impurities, etc.), pitting and localized corrosion inhibition, surfaces covered with oxide, carbonate/carbide (sweet corrosion), sulfide layers (sour corrosion), surfaces with alloying additions; considering explicitly the solvent environments, ionic species, etc. Since corrosion inhibition via organic inhibitors is primarily about the formation of a stable, adherent film on the metal surface (instead of adsorption of a single molecule), it is expected that more research focused on detailed coverage studies and molecular modeling of self-assembled monolayers of corrosion inhibitor molecules on metal surfaces will be conducted. Employing molecular modeling for designing and developing combinations of inhibitor chemistries with enhanced synergism would seem an important area for future research, both from an academic as well as an industrial perspective while further advancements in AI- and ML-based approaches are also expected to have an impact on expediting corrosion inhibitor research.

Reviewing the current research and field techniques commonly used in assessing and qualifying corrosion inhibitors' performance, it is seen that significant progress has been achieved owing to technological advancements in many areas. For instance, extreme conditions that closely mimic harsh and complex exposure environments can be simulated while advanced surface characterization at a molecular level can be incorporated to provide insight into inhibition mechanisms. Nonetheless, there are still research areas where further investigation can improve the understanding of corrosion inhibition mechanisms and consequently inhibition efficiency. For example, advanced characterization techniques have been shown to provide valuable information but many of the mechanistic studies that adopted those techniques mostly investigated model inhibitor molecules (as a single compound) on clean, finely polished surfaces. On the contrary, commercial corrosion inhibitors are oftentimes proprietary blends of multiple active chemistries that provide corrosion mitigation properties, commonly along with additional components to enhance the inhibition performance, augment partitioning to the water phase, improve product stability and physical properties, etc. With this in mind, the effects of supposed nonactive ingredients on corrosion inhibition need to be elucidated.

A corrosion inhibitor investigation often uses a sample machined from commercial steels produced in relatively small dimensions in many cases. However, the inhibitor needs to perform on a field component which may be machined or formed from a different process. Metallurgical factors such as microstructure, chemical composition, and contaminations may not be well represented in the laboratory. This discrepancy can lead to unexpected inhibitor performance as inhibitors are sensitive to small changes in the steel's composition and physical properties. Improving mechanistic understanding of how an inhibitor might respond to the aforementioned metallurgical factors might aid in the inhibitor selection and qualification processes. In addition to progressing such laboratory to field correlations, other areas that would benefit advancements include in situ or online pitting or localized corrosion, options for more extensive or holistic field corrosion monitoring as well as chemical residual detection and tracing.

From a corrosion inhibitor development perspective (both in terms of continuous as well as batch treatments), higher performance options for more challenging environments such as higher temperature operations, UDC, TLC, systems containing high quantities of elemental sulfur and polysulfides, as well as more persistent film-forming products would be advantageous to end-users. Sustainability, reducing the carbon footprint, environmental, social, and governance factors, initiatives, and responsibilities are at the forefront of most organizations' business so innovations in more environmentally acceptable chemistries and more versatile products that perform at lower dose rates across multiple operating environments to help reduce chemical handling and logistics would likely appeal to most operators and the oil and gas industry as a whole.

Acknowledgments

The authors thank Dr. Katerina Lepkova, Professor Brian Kinsella, Dr. William Durnie, Dr. Bill Hedges, Mr. Vinay Jain, Dr. Beena Rai, Dr Gautam Shroff, and Mr. K. Ananth Krishnan for the fruitful discussions had along with their valuable feedback and also appreciate the support of ChampionX, Tata Consultancy Services and Curtin University on this chapter.

Nomenclature

AFM	atomic force microscopy
BPD	barrels per day
CAPEX	capital expenditures
CE	counter electrode
CFD	computational fluid dynamics
CMAS	coupled multielectrode arrays
CMC	critical micelle concentration
CPP	cyclic potentiodynamic polarization
CR	corrosion rate
CRA	corrosion resistant alloy
DCMS	downhole corrosion monitoring system
DFT	density functional theory
DFTB	density functional tight binding
DPC	1-dodecylpyridinium chloride hydrate
ECN	electrochemical current noise
EEC	electrical equivalent circuit
EIS	electrochemical impedance spectroscopy
EN	electrochemical noise
EPN	electrochemical potential noise

EQCM	electrochemical quartz microbalance
ER	electrical resistance
ES-Ms	electrospray-mass spectrometry
FSM	field signature method
FTIR	Fourier-transform infrared spectroscopy
HOMO	highest occupied molecular orbital
IE	inhibition efficiency
IR	infrared spectroscopy
JIC	jet impingement cell
LC-Ms	liquid chromatography-mass spectrometry
LP	linear polarization
LUMO	lowest unoccupied molecular orbital
M	thousands
m/s	meters per second
MAE	mean absolute error
MC	Monte Carlo
MD	molecular dynamics
MDEA	methyldiethanolamine
MEC	minimum effective concentration
MIC	microbiologically influence corrosion
MM	millions
MMSCFD	million standard cubic feet per day
MSCF	thousand standard cubic feet
mpy	mils per year
MSE	mean squared error
NMR	nuclear magnetic resonance
OCP	open-circuit potential
PCTFE	polychlorotrifluoroethylene
PEEK	polyether ether ketone
ppmv	parts per million by volume
PTFE	polytetrafluoroethylene
PVDF	polyvinylidene fluoride
PWC	preferential weld corrosion
PZC	potential of zero charge
QCD	quantum chemical descriptors
QSAR	quantitative structure-activity relation
RC	rotating cage
RCE	rotating cylinder electrode
RDE	rotating disk electrode
RE	reference electrode
ReaxFF	reactive force field
RISM	reference interaction site method
RMSE	root mean square error
SCE	saturate calomel electrode
SEM	scanning electron microscopy
SERS	surface enhanced Raman scattering
SHE	standard hydrogen electrode
TLC	top-of-the-line corrosion

UDC	under-deposit corrosion
UNS	unified numbering system
UT	ultrasonic thickness measurement
VCI	volatile corrosion inhibitor
WBE	wire beam electrode
WCR	water condensation rate
WE	working electrode
XPS	X-ray photoelectron spectroscopy
ZRA	zero resistance ammeter

References

[1] E. McCafferty, Introduction to Corrosion Science, Springer, New York, 2010.

[2] R. Revie, Uhlig's Corrosion Handbook, third ed., John Wiley & Sons, New York, 2011.

[3] S. Papavinasam, Corrosion Control in The Oil and Gas Industry, Elsevier Science, Burlington, 2013.

[4] D.R. Wilson, Hydrogen sulphide scavengers: recent experience in a major North Sea field, SPE-36943-MS, in: European Petroleum Conference, Milan, Italy, 22–24 October, 1996.

[5] M. Amosa, I. Mohammed, S. Yaro, Sulphide scavengers in oil and gas industry - a review, NAFTA 61 (2010) 85–98.

[6] R. Fuchs-Godec, The adsorption, CMC determination and corrosion inhibition of some N-alkyl quaternary ammonium salts on carbon steel surface in 2 M H2SO4, Colloids and Surfaces A: Physicochemical and Engineering Aspects 280 (2006) 130–139.

[7] L. Niu, H. Zhang, F. Wei, et al., Corrosion inhibition of iron in acidic solutions by alkyl quaternary ammonium halides: correlation between inhibition efficiency and molecular structure, Applied Surface Science 252 (2005) 1634–1642.

[8] F. Kandemirli, S. Sagdinc, Theoretical study of corrosion inhibition of amides and thiosemicarbazones, Corrosion Science 49 (2007) 2118–2130.

[9] A. Yıldırım, S. Öztürk, M. Çetin, Novel amide-based cationic surfactants as efficient corrosion inhibitors for carbon steel in HCl and H_2SO_4 media, Journal of Surfactants and Detergents 16 (2012) 13–23.

[10] D. Singh, M. Singh, R. Chaudhary, et al., Inhibition and polarization studies of some substituted urea compounds for corrosion of aluminium in nitric acid, Electrochimica Acta 26 (1981) 1051–1056.

[11] M. Finšgar, J. Jackson, Application of corrosion inhibitors for steels in acidic media for the oil and gas industry: a review, Corrosion Science 86 (2014) 17–41.

[12] J. Roque, T. Pandiyan, J. Cruz, et al., DFT and electrochemical studies of tris(benzimidazole-2-ylmethyl)amine as an efficient corrosion inhibitor for carbon steel surface, Corrosion Science 50 (2008) 614–624.

[13] S. Granese, B. Rosales, Inhibition of the corrosion of iron and steel in acid pickling media by nitrogen containing compounds, Key Engineering Materials 20–28 (1991) 2733–2746.

[14] K. Babić-Samardžija, K. Khaled, N. Hackerman, N-heterocyclic amines and derivatives as corrosion inhibitors for iron in perchloric acid, Anti-Corrosion Methods and Materials 52 (2005) 11–21.

[15] S. Cardoso, J. Gomes, L. Borges, et al., Predictive QSPR analysis of corrosion inhibitors for super 13% Cr steel in hydrochloric acid, Brazilian Journal of Chemical Engineering 24 (2007) 547–559.

[16] A. Alsabagh, M. Migahed, H. Awad, Reactivity of polyester aliphatic amine surfactants as corrosion inhibitors for carbon steel in formation water (deep well water), Corrosion Science 48 (2006) 813–828.

[17] F. Jjunju, S. Maher, D. Damon, et al., Screening and quantification of aliphatic primary alkyl corrosion inhibitor amines in water samples by paper spray mass spectrometry, Analytical Chemistry 88 (2016) 1391–1400.

[18] E. Stupnišek-Lisac, A. Brnada, A. Mance, Secondary amines as copper corrosion inhibitors in acid media, Corrosion Science 42 (2000) 243–257.

[19] G. Sığırcık, D. Yildirim, T. Tüken, Synthesis and inhibitory effect of N,N'-bis(1-phenylethanol)ethylenediamine against steel corrosion in HCl Media, Corrosion Science 120 (2017) 184–193.

[20] G. Gao, C. Liang, H. Wang, Synthesis of tertiary amines and their inhibitive performance on carbon steel corrosion, Corrosion Science 49 (2007) 1833–1846.

[21] K. Khaled, K. Babić-Samardžija, N. Hackerman, Theoretical study of the structural effects of polymethylene amines on corrosion inhibition of iron in acid solutions, Electrochimica Acta 50 (2005) 2515–2520.

[22] E. Ebenso, U. Ekpe, B. Ita, et al., Effect of molecular structure on the efficiency of amides and thiosemicarbazones used for corrosion inhibition of mild steel in hydrochloric acid, Materials Chemistry and Physics 60 (1999) 79–90.

[23] E. Reyes-Dorantes, J. Zuñiga-Díaz, A. Quinto-Hernandez, et al., Fatty amides from crude rice bran oil as green corrosion inhibitors, Journal of Chemistry 2017 (2017) 1–14.

[24] R. Abdel Hameed, M. Alfakeer, M. Abdallah, Inhibiting properties of some heterocyclic amide derivatives as potential nontoxic corrosion inhibitors for carbon steel in 1.0 M sulfuric acid, Surface Engineering and Applied Electrochemistry 54 (2018) 599–606.

[25] J. Fang, J. Li, Quantum chemistry study on the relationship between molecular structure and corrosion inhibition efficiency of amides, Journal of Molecular Structure: THEOCHEM 593 (2002) 179–185.

[26] A. Yıldırım, M. Çetin, Synthesis and evaluation of new long alkyl side chain acetamide, isoxazolidine and isoxazoline derivatives as corrosion inhibitors, Corrosion Science 50 (2008) 155–165.

[27] S. Öztürk, A. Yıldırım, M. Çetin, et al., Synthesis of quaternary, long-chain n-alkyl amides and their corrosion inhibition in acidic media, Journal of Surfactants and Detergents 17 (2014) 471–481.

[28] H. El Dahan, T. Mohamed, S. Abo El-Enin, Efficient quaternary ammonium salt as corrosion inhibitor for steel pickling in sulphuric acid media, Anti-Corrosion Methods and Materials 46 (1999) 358–363.

[29] R. Zvauya, J. Dawson, Inhibition studies in sweet corrosion systems by a quaternary ammonium compound, Journal of Applied Electrochemistry 24 (1994) 943–947.

[30] A. Badawi, M. Hegazy, A. El-Sawy, et al., Novel quaternary ammonium hydroxide cationic surfactants as corrosion inhibitors for carbon steel and as biocides for sulfate reducing bacteria (SRB), Materials Chemistry and Physics 124 (2010) 458–465.

[31] A. El Maghraby, T. Soror, Quaternary ammonium salt as effective corrosion inhibitor for carbon steel dissolution in sulphuric acid media, Advances in Applied Science Research 1 (2010) 143–155.

[32] H. Ma, S. Chen, B. Yin, et al., Impedance spectroscopic study of corrosion inhibition of copper by surfactants in the acidic solutions, Corrosion Science 45 (2003) 867–882.

[33] P. Okafor, X. Liu, Y. Zheng, Corrosion inhibition of mild steel by ethylamino imidazoline derivative in CO2-saturated solution, Corrosion Science 51 (2009) 761–768.

[34] G. Zhang, C. Chen, M. Lu, et al., Evaluation of inhibition efficiency of an imidazoline derivative in CO2-containing aqueous solution, Materials Chemistry and Physics 105 (2007) 331–340.

[35] J. Cruz, R. Martínez, J. Genesca, et al., Experimental and theoretical study of 1-(2-ethylamino)-2-methylimidazoline as an inhibitor of carbon steel corrosion in acid media, Journal of Electroanalytical Chemistry 566 (2004) 111–121.

[36] D. Wang, S. Li, Y. Ying, et al., Theoretical and experimental studies of structure and inhibition efficiency of imidazoline derivatives, Corrosion Science 41 (1999) 1911–1919.

[37] S. Hosseini, M. Salari, E. Jamalizadeh, et al., Electrochemical and quantum chemical studies of aromatic amines on the steel corrosion in acid solution, Corrosion 68 (2012) 600–609.

[38] M. Quraishi, D. Jamal, Dianils: new and effective corrosion inhibitors for oil-well steel (n-80) and mild steel in boiling hydrochloric acid, Corrosion 56 (2000) 156–160.

[39] K. Khaled, N. Hackerman, Investigation of the inhibitive effect of ortho-substituted anilines on corrosion of iron in 1 M HCl solutions, Electrochimica Acta 48 (2003) 2715–2723.

[40] C. Jeyaprabha, S. Sathiyanarayanan, G. Venkatachari, Polyaniline as corrosion inhibitor for iron in acid solutions, Journal of Applied Polymer Science 101 (2006) 2144–2153.

[41] M. Desai, V. Shah, Aromatic amines as corrosion inhibitors for 70/30 brass in nitric acid, Corrosion Science 12 (1972) 725–730.

[42] M. Mahida, H. Chaudhari, Aromatic amines as corrosion inhibitors for zinc in hydrochloric acid, Journal of Chemical and Pharmaceutical Research 4 (2012) 5195–5201.

[43] R. Vashi, S. Desai, P. Desai, Nitro aniline as corrosion inhibitor for Zinc in nitric acid, Journal of Environmental Research and Development 3 (2008) 97–104.

[44] X. Gao, S. Liu, H. Lu, et al., Corrosion inhibition of iron in acidic solutions by monoalkyl phosphate esters with different chain lengths, Industrial & Engineering Chemistry Research 54 (2015) 1941–1952.

[45] O. Yépez, N. Obeyesekere, J. Wylde, Development of novel phosphate based inhibitors effective for oxygen corrosion. Part I: rotating cage autoclave study, NACE-6163, in: Proceedings of the Corrosion Conference and Expo, Dallas, Texas, 15–19 March, 2015.

[46] O. Yépez, N. Obeyesekere, J. Wylde, Development of novel phosphate based inhibitors effective for oxygen corrosion. Part II: electrochemical characterization, NACE-6162, in: Proceedings of the Corrosion Conference and Expo, Dallas, Texas, 15–19 March, 2015.

[47] W. McGregor, Novel synergistic water soluble corrosion inhibitors, SPE-87570-MS, in: Proceedings of the SPE International Symposium on Oilfield Corrosion, Aberdeen, UK, 28 May, 2004.

[48] G. Bota, P. Jin, F. Farelas, et al., Phosphate esters inhibitors in naphthenic acid corrosion - An experimental evaluation, NACE-13491, in: Proceedings of the Corrosion Conference and Expo, Nashville, Tennessee, 24–28 March, 2019.
[49] T. Szauer, A. Brandt, On the role of fatty acid in adsorption and corrosion inhibition of iron by amine—fatty acid salts in acidic solution, Electrochimica Acta 26 (1981) 1257–1260.
[50] V. Abbasov, H. Abd El-Lateef, L. Aliyeva, et al., Efficient complex surfactants from the type of fatty acids as corrosion inhibitors for mild steel C1018 in CO_2-environments, Journal of The Korean Chemical Society 57 (2013) 25–34.
[51] M. Makhlouf, S. El-Shatory, A. El-Said, The synergistic effect of halide ions and some selected thiols as a combined corrosion inhibitor for pickling of mild steel in sulphuric acid solution, Materials Chemistry and Physics 43 (1996) 76–82.
[52] E. Azzam, M. Hegazy, N. Kandil, et al., The performance of hydrophobic and hydrophilic moieties in synthesized thiol cationic surfactants on corrosion inhibition of carbon steel in HCl, Egyptian Journal of Petroleum 24 (2015) 493–503.
[53] D. Hmamou, M. Aouad, R. Salghi, et al., Inhibiting effects 4,5-Diphenyl-1H-Imidazole-2-Thiol for C38 steel in 1 M HCl: electrochemical study, Journal of Chemical and Pharmaceutical Research 4 (2012) 3489–3497.
[54] M. Mahdavian, S. Ashhari, Corrosion inhibition performance of 2-mercaptobenzimidazole and 2-mercaptobenzoxazole compounds for protection of mild steel in hydrochloric acid solution, Electrochimica Acta 55 (2010) 1720–1724.
[55] M. Gholami, I. Danaee, M. Maddahy, et al., Correlated ab initio and electroanalytical study on inhibition behavior of 2-mercaptobenzothiazole and its thiole–thione tautomerism effect for the corrosion of steel (API 5 L X52) in sulphuric acid solution, Industrial & Engineering Chemistry Research, 52, 2013, pp. 14875–14889.
[56] S. Deng, X. Li, X. Xie, Hydroxymethyl urea and 1,3-bis(hydroxymethyl) urea as corrosion inhibitors for steel in HCl solution, Corrosion Science 80 (2014) 276–289.
[57] T. Laabaissi, F. Benhiba, Z. Rouifi, et al., New quinoxaline derivative as a green corrosion inhibitor for mild steel in mild acidic medium: electrochemical and theoretical studies, International Journal of Corrosion and Scale Inhibition 8 (2019) 241–256.
[58] B. Rani, B. Basu, Green inhibitors for corrosion protection of metals and alloys: an overview, International Journal of Corrosion 2012 (2012) 1–15.
[59] N. Gupta, C. Verma, R. Salghi, et al., New phosphonate based corrosion inhibitors for mild steel in hydrochloric acid useful for industrial pickling processes: experimental and theoretical approach, New Journal of Chemistry 41 (2017) 13114–13129.
[60] I. Obot, D. Macdonald, Z. Gasem, Density functional theory (DFT) as a powerful tool for designing new organic corrosion inhibitors. Part 1: an overview, Corrosion Science 99 (2015) 1–30.
[61] M. Murmu, S. Saha, N. Murmu, et al., Effect of stereochemical conformation into the corrosion inhibitive behaviour of double azomethine based Schiff bases on mild steel surface in 1 mol L − 1 HCl medium: an experimental, density functional theory and molecular dynamics simulation study, Corrosion Science 146 (2019) 134–151.
[62] E. Oguzie, Y. Li, S. Wang, et al., Understanding corrosion inhibition mechanisms - experimental and theoretical approach, RSC Advances 1 (2011) 866–873.
[63] M. Özcan, İ. Dehri, M. Erbil, Organic sulphur-containing compounds as corrosion inhibitors for mild steel in acidic media: correlation between inhibition efficiency and chemical structure, Applied Surface Science 236 (2004) 155–164.

[64] H. Wang, X. Wang, H. Wang, et al., DFT study of new bipyrazole derivatives and their potential activity as corrosion inhibitors, Journal of Molecular Modeling 13 (2006) 147–153.

[65] S. Song, S. Choi, J. Kim, et al., Effect of Molecular Weight of Polyethylene Glycol on Copper Electrodeposition in the Presence of Bis-3-sulfopropyl-disulfide, International Journal of Electrochemical Science 11 (2016) 10067–10079.

[66] K. Khaled, Modeling corrosion inhibition of iron in acid medium by genetic function approximation method: a QSAR model, Corrosion Science 53 (2011) 3457–3465.

[67] Y. Karzazi, M. El Alaoui, A. Belghiti, et al., Theoretical investigation on the corrosion inhibition of mild steel by piperidine derivatives in hydrochloric acid solution, Journal of Chemical and Pharmaceutical Research 6 (2014) 689–696.

[68] R. Parr, R. Donnelly, M. Levy, et al., Electronegativity: the density functional viewpoint, The, Journal of Chemical Physics 68 (1978) 3801–3807.

[69] R. Pearson, Absolute electronegativity and hardness: application to inorganic chemistry, Inorganic Chemistry 27 (1988) 734–740.

[70] P. Udhayakala, A.M. Samuel, T.V. Rajendiran, et al., DFT study on the adsorption mechanism of some phenyltetrazole substituted compounds as effective corrosion inhibitors for mild steel, Der Pharma Chemica 5 (2013) 111–124.

[71] G. Gece, The use of quantum chemical methods in corrosion inhibitor studies, Corrosion Science 50 (2008) 2981–2992.

[72] H. Ashassi-Sorkhabi, B. Shaabani, D. Seifzadeh, Corrosion inhibition of mild steel by some Schiff base compounds in hydrochloric acid, Applied Surface Science 239 (2005) 154–164.

[73] B. Mert, M. Erman Mert, G. Kardaş, et al., Experimental and theoretical investigation of 3-amino-1,2,4-triazole-5-thiol as a corrosion inhibitor for carbon steel in HCl medium, Corrosion Science 53 (2011) 4265–4272.

[74] P. Geerlings, F. De, Proft, Chemical Reactivity as Described by Quantum Chemical Methods, International Journal of Molecular Sciences 3 (2002) 276–309.

[75] S. Pilli, T. Banerjee, K. Mohanty, HOMO–LUMO energy interactions between endocrine disrupting chemicals and ionic liquids using the density functional theory: evaluation and comparison, Journal of Molecular Liquids 207 (2015) 112–124.

[76] N. Wazzan, S. Al-mhyawi, Application of newly quiniline-3-carbonitriles as corrosion inhibitors on mild steel in 1.0 M HCl: electrochemical measurements, HF and DFT/B3LYP calculations, International Journal of Electrochemical Science 12 (2017) 9812–9828.

[77] A. Kokalj, On the HSAB based estimate of charge transfer between adsorbates and metal surfaces, Chemical Physics 393 (2012) 1–12.

[78] B. Xu, Y. Ji, X. Zhang, et al., Experimental and theoretical studies on the corrosion inhibition performance of 4-amino-N, N-di-(2-pyridylmethyl)-aniline on mild steel in hydrochloric acid, RSC Advances 5 (2015) 56049–56059.

[79] I. Ahamad, R. Prasad, M. Quraishi, Thermodynamic, electrochemical and quantum chemical investigation of some Schiff bases as corrosion inhibitors for mild steel in hydrochloric acid solutions, Corrosion Science 52 (2010) 933–942.

[80] Y. Karzazi, M. Mcharfi, A. Dafali, et al., A theoretical investigation on the corrosion inhibition of mild steel by piperidine derivatives in hydrochloric acid solution, Journal of Chemical and Pharmaceutical Research 6 (2014) 689–696.

[81] B. Gómez, N. Likhanova, M. Domínguez-Aguilar, et al., Quantum chemical study of the inhibitive properties of 2-pyridyl-azoles, The Journal of Physical Chemistry B 110 (2006) 8928–8934.

[82] D. Kumar, V. Jain, B. Rai, Unravelling the mechanisms of corrosion inhibition of iron by henna extract: A density functional theory study, Corrosion Science 142 (2018) 102–109.

[83] J. Zhang, G. Qiao, S. Hu, et al., Theoretical evaluation of corrosion inhibition performance of imidazoline compounds with different hydrophilic groups, Corrosion Science 53 (2011) 147–152.

[84] I. Obot, Z. Gasem, Theoretical evaluation of corrosion inhibition performance of some pyrazine derivatives, Corrosion Science 83 (2014) 359–366.

[85] D. Winkler, M. Breedon, A. Hughes, et al., Towards chromate-free corrosion inhibitors: structure–property models for organic alternatives, Green Chemistry 16 (2014) 3349–3357.

[86] A. Kokalj, S. Peljhan, M. Finšgar, et al., What determines the inhibition effectiveness of ata, btah, and btaoh corrosion inhibitors on copper? Journal of the American Chemical Society 132 (2010) 16657–16668.

[87] L. Guo, C. Qi, X. Zheng, et al., Toward understanding the adsorption mechanism of large size organic corrosion inhibitors on an Fe(110) surface using the DFTB method, RSC Advances 7 (2017) 29042–29050.

[88] F. Hirata, P. Rossky, An extended rism equation for molecular polar fluids, Chemical Physics Letters 83 (1981) 329–334.

[89] J. Fattebert, F. Gygi, First-principles molecular dynamics simulations in a continuum solvent, International Journal of Quantum Chemistry 93 (2003) 139–147.

[90] S. Nishihara, M. Otani, Hybrid solvation models for bulk, interface, and membrane: reference interaction site methods coupled with density functional theory, Physical Review B 96 (2017) 115429.

[91] H. Sato, A. Kovalenko, F. Hirata, Self-consistent field, ab initio molecular orbital and three-dimensional reference interaction site model study for solvation effect on carbon monoxide in aqueous solution, The Journal of Chemical Physics 112 (2000) 9463–9468.

[92] D. Kumar, N. Jain, V. Jain, et al., Amino acids as copper corrosion inhibitors: a density functional theory approach, Applied Surface Science 514 (2020) 145905.

[93] Y. Jiang, J. Adams, First principle calculations of benzotriazole adsorption onto clean Cu(1 1 1), Surface Science 529 (2003) 428–442.

[94] L. Guo, X. Ren, Y. Zhou, et al., Theoretical evaluation of the corrosion inhibition performance of 1,3-thiazole and its amino derivatives, Arabian Journal of Chemistry 10 (2017) 121–130.

[95] A. van Duin, S. Dasgupta, F. Lorant, et al., A reactive force field for hydrocarbons, The Journal of Physical Chemistry A 105 (2001) 9396–9409.

[96] D. Kumar, V. Jain, B. Rai, Imidazole derivatives as corrosion inhibitors for copper: a DFT and reactive force field study, Corrosion Science 171 (2020) 108724.

[97] S. Monti, C. Li, V. Carravetta, Reactive dynamics simulation of monolayer and multilayer adsorption of glycine on Cu(110), The Journal of Physical Chemistry C 117 (2013) 5221–5228.

[98] Ş. Erdoğan, Z. Safi, S. Kaya, et al., A computational study on corrosion inhibition performances of novel quinoline derivatives against the corrosion of iron, Journal of Molecular Structure 1134 (2017) 751–761.

[99] I. Obot, S. Kaya, C. Kaya, et al., Theoretical evaluation of triazine derivatives as steel corrosion inhibitors: DFT and Monte Carlo simulation approaches, Research on Chemical Intermediates 42 (2015) 4963–4983.

[100] K. Khaled, N. Abdelshafi, A. El-Maghraby, et al., Alanine as corrosion inhibitor for iron in acid medium: a molecular level study, International Journal of Electrochemical Science 7 (2012) 12706–12719.

[101] L. Guo, S. Zhu, S. Zhang, et al., Theoretical studies of three triazole derivatives as corrosion inhibitors for mild steel in acidic medium, Corrosion Science 87 (2014) 366–375.

[102] P. Han, Y. He, C. Chen, et al., Study on synergistic mechanism of inhibitor mixture based on electron transfer behavior, Scientific Reports 6 (2016) 33252.

[103] Y. Qiang, L. Guo, S. Zhang, et al., Synergistic effect of tartaric acid with 2,6-diaminopyridine on the corrosion inhibition of mild steel in 0.5 M HCl, Scientific Reports 6 (2016) 33305.

[104] V. Kamath, A. Pai, Application of molecular descriptors in modern computational drug design - an overview, Research Journal of Pharmacy and Technology 10 (2017) 3237–3241.

[105] H. Zhao, X. Zhang, L. Ji, et al., Quantitative structure-activity relationship model for amino acids as corrosion inhibitors based on the support vector machine and molecular design, Corrosion Science 83 (2014) 261–271.

[106] Y. Liu, Y. Guo, W. Wu, et al., A machine learning-based QSAR model for benzimidazole derivatives as corrosion inhibitors by incorporating comprehensive feature selection, Interdisciplinary Sciences: Computational Life Sciences 11 (2019) 738–747.

[107] L. Olasunkanmi, B. Moloto, I. Obot, et al., Anticorrosion studies of some hydantoin derivatives for mild steel in 0.5 M HCl solution: experimental, quantum chemical, Monte Carlo simulations and QSAR studies, Journal of Molecular Liquids 252 (2018) 62–74.

[108] L. Murulana, M. Kabanda, E. Ebenso, Investigation of the adsorption characteristics of some selected sulphonamide derivatives as corrosion inhibitors at mild steel/hydrochloric acid interface: experimental, quantum chemical and QSAR studies, Journal of Molecular Liquids 215 (2016) 763–779.

[109] A. Agarwal, P. Rathore, V. Jain, et al., In-silico model for predicting the corrosion inhibition efficiency of steel inhibitors, NACE-13329, in: Proceedings of the Corrosion Conference and Expo, Nashville, Tennessee, 24–28 March, 2019.

[110] R. Heidersbach, Metallurgy and Corrosion Control in Oil and Gas, first ed., John Wiley & Sons, Inc, New Jersey, 2011.

[111] G. Hinds, A. Turnbull, Novel multi-electrode test method for evaluating inhibition of underdeposit corrosion - part 1: sweet conditions, Corrosion 66 (2010) 1–11.

[112] D.A. Lopez, S.N. Simison, S.R. de Sanchez, The influence of steel microstructure on CO_2 corrosion. EIS studies on the inhibition efficiency of benzimidazole, Electrochimica Acta 48 (2003) 845–854.

[113] L.D. Paolinelli, T. Pérez, S.N. Simison, The effect of pre-corrosion and steel microstructure on inhibitor performance in CO_2 corrosion, Corrosion Science 50 (2008) 2456–2464.

[114] T. Pojtanabuntoeng, M. Salasi, R. Gubner, The influence of mono ethylene glycol (MEG) on CO2 corrosion of carbon steel at elevated temperature (80°C to 120°C), NACE-4176, in: Proceedings of the Corrosion Conference and Expo, San Antonio, Texas, 9–13 March, 2014.

[115] E. Gulbrandsen, J.H. Morard, Why does glycol inhibit co_2 corrosion?, NACE-221, in: Proceedings of the Corrosion Conference and Expo, San Diego, CA, 22–27 March, 1998.
[116] T. Pojtanabuntoeng, B. Kinsella, H. Ehsani, et al., Assessment of corrosion control by pH neutralisation in the presence of glycol at low temperature, Corrosion Science 126 (2017) 94–103.
[117] J. Kvarekvål, An electrochemical study of corrosion inhibition of carbon steel in sour glycol solutions, NACE-2447, in: Proceedings of the Corrosion Conference and Expo, Orlando, FL, 17–21 March, 2013.
[118] E. Gulbrandsen, A. Grana, Testing of carbon dioxide corrosion inhibitor performance at high flow velocities in jet impingement geometry. Effects of mass transfer and flow forces, Corrosion 63 (2007) 1009–1021.
[119] Y. Xiong, C. Li, S. Desai, et al., The importance of chemical compatibility in corrosion inhibitor performance, IPTC-18435-MS, in: Proceedings of the International Petroleum Technology Conference, Doha, Qatar, 6–9 December, 2015.
[120] J. Moloney, W.Y. Mok, C.G. Gamble, Corrosion and hydrate control in wet sour gas transmission systems, SPE 115074, in: Proceedings of the SPE Asia Pacific Oil & Gas Conference and Exhibition, Perth, Australia, 20–22 October, 2008.
[121] J. Moloney, Corrosion inhibitor development for sour gas pipeline systems using either monoethylene glycol or kinetic hydrate inhibitor for hydrate control, Paper No. 14065, in: Proceedings of the 15th Middle East Corrosion Conference and Exhibition, Manama, Bahrain, 2–5 February, 2014.
[122] S.D. Kapusta, Corrosion inhibitor testing and selection for E&P: a user's perspective, NACE-16, in: Proceedings of the Corrosion Conference and Expo, San Antonio, Texas, 25–30 December, 1999.
[123] M.M. Islam, T. Pojtanabuntoeng, R. Gubner, Condensation corrosion of carbon steel at low to moderate surface temperature and iron carbonate precipitation kinetics, Corrosion Science 111 (2016) 139–150.
[124] Z. Belarbi, T.N. Vu, F. Farelas, et al., Thiols as volatile corrosion inhibitors for top-of-the-line corrosion, Corrosion 9312 (2017) 892–899.
[125] H. Zhang, X. Pang, K. Gao, Localized CO_2 corrosion of carbon steel with different microstructures in brine solutions with an imidazoline-based inhibitor, Applied Surface Science 442 (2018) 446–460.
[126] Y. Xiong, B. Brown, B. Kinsella, et al., Atomic force microscopy study of the adsorption of surfactant corrosion inhibitor films, Corrosion 70 (2014) 21–22.
[127] M. Javidi, R. Chamanfar, S. Bekhrad, Investigation on the efficiency of corrosion inhibitor in CO_2 corrosion of carbon steel in the presence of iron carbonate scale, Journal of Natural Gas Science and Engineering 61 (2019) 197–205.
[128] J.L. Pacheco, F.C. Ibrahim, R.J. Franco, Testing requirements of corrosion inhibitor qualification for pipeline applications, NACE-10325, in: Proceedings of the Corrosion Conference and Expo, San Antonio, Texas, 14–18 March, 2010.
[129] NACE, Wheel Test Method Used for Evaluation of Film-Persistent Corrosion Inhibitors For Oilfield Applications, NACE International, Houston, 2017.
[130] S. Papavinasam, R.W. Revie, M. Attard, et al., Comparison of laboratory methodologies to evaluate corrosion inhibitors for oil and gas pipelines, Corrosion 59 (2003) 897–912.
[131] F.C. Walsh, G. Kear, A.H. Nahlé, et al., The rotating cylinder electrode for studies of corrosion engineering and protection of metals - an illustrated review, Corrosion Science 123 (2017) 1–20.

[132] ASTM, G170-06 Standard Guide for Evaluating and Qualifying Oilfield and Refinery Corrosion Inhibitors in the Laboratory, ASTM International, West Conshohocken, 2012.
[133] P. Altoe, G. Pimenta, C.F. Moulin, et al., Evaluation of oilfield corrosion inhibitors in CO_2 containing media: a kinetic study, Electrochimica Acta 41 (1996) 1165–1172.
[134] Y.J. Tan, S. Bailey, B. Kinsella, An investigation of the formation and destruction of corrosion inhibitor films using electrochemical impedance spectroscopy (EIS), Corrosion Science 38 (1996) 1545–1561.
[135] E. Gulbrandsen, J. Kvarekvål, H. Miland, Effect of oxygen contamination on inhibition studies in carbon dioxide corrosion, Corrosion 61 (2005) 1086–1097.
[136] Y.J. Tan, S. Bailey, B. Kinsella, The monitoring of the formation and destruction of corrosion inhibitor films using electrochemical noise analysis, Corrosion Science 38 (1996) 1681–1695.
[137] S. Papavinasam, Rotating cage and jet impingement techniques, in: L. Yang (Ed.), Techniques for Corrosion Monitoring, first ed., Woodhead publishing, 2008, pp. 322–343.
[138] L. Chaal, B. Albinet, C. Deslouis, et al., Wall shear stress mapping in the rotating cage geometry and evaluation of drag reduction efficiency using an electrochemical method, Corrosion Science 51 (2009) 1809–1816.
[139] G. Schmitt, P. Plagemann, C. Deslouis, et al., Quantifying local wall shear stresses in the rotated cage, NACE-04727, in: Proceedings of the Corrosion Conference and Expo, New Orleans, Louisiana, 28 March–1 April, 2004.
[140] ASTM, G202-09 Standard Test Method for Using Atmospheric Pressure Rotating Cage, ASTM International, West Conshohocken, 2009.
[141] N. Senior, P. Boisvert, A. Runstedtler, Simulating liquid pipeline flow using the rotating cage method, NACE-6107, in: Proceedings of the Corrosion Conference and Expo, Dallas, Texas, 15–19 March, 2015.
[142] A. Runstedtler, J. Huang, P. Boisvert, et al., Parametric wall shear stress characterization of the rotating cage test method, Corrosion 75 (2019) 580–587.
[143] S. Papavinasam, A. Doiron, R.W. Revie, Effect of rotating cage geometry on flow pattern and corrosion rate, NACE-03333, in: Proceedings of the Corrosion Conference and Expo, San Diego, California, 16–20 March, 2003.
[144] D. John, B. Kinsella, S. Bailey, et al., Flow dependence of carbon dioxide corrosion rates and the interference of trace dissolved oxygen, Paper No 07315, in: Proceedings of the Corrosion Conference and Expo, Nashville, Tennessee, 11–15 March, 2007.
[145] X. Hu, R. Barker, A. Neville, et al., Case study on erosion - corrosion degradation of pipework located on an offshore oil and gas facility, Wear 271 (2011) 1295–1301.
[146] X. Hu, K. Alzawai, A. Gnanavelu, et al., Assessing the effect of corrosion inhibitor on erosion – corrosion of API-5L-X65 in multi-phase jet impingement conditions, Wear 271 (2011) 1432–1437.
[147] E. Gulbrandsen, A. Dugstad, Corrosion loop studies of preferential weld corrosion and its inhibition in CO2 environments, NACE-05276, in: Proceedings of the Corrosion Conference and Expo, Houston, Texas, 3–7 April, 2005.
[148] M. Singer, Study of the localized nature of top of the line corrosion in sweet environment, Corrosion 73 (2017) 1030–1055.

[149] I. Jevremovic, M. Singer, M. Achour, Evaluation of a novel top-of-the-line corrosion (TLC) mitigation method in a large-scale flow loop, Corrosion 71 (2015) 389–397.

[150] N. Jauseau, F. Farelas, M. Singer, et al., Investigation of the role of droplet transport in mitigating top of the line corrosion, Corrosion 74 (2018) 873–885.

[151] B. Brown, J. Moloney, Under-deposit corrosion, in: A.M. El-Sherik (Ed.), Trends Oil Gas Corrosion Research and Technologies: Production and Transmission, Woodhead publishing, Cambridge, 2017, pp. 363–383.

[152] B. Brown, A. Saleh, J. Moloney, Comparison of mono to diphosphate ester ratio in inhibitor formulations for mitigation of under-deposit corrosion, Corrosion 71 (2015) 1500–1510.

[153] E.M. Suarez, L.L. Machuca, B. Kinsella, et al., CO_2 corrosion inhibitors performance at deposit-covered carbon steel and their adsorption on different deposits, Corrosion 75 (2019) 1118–1128.

[154] G.A. Zhang, N. Yu, L.Y. Yang, et al., Galvanic corrosion behavior of deposit-covered and uncovered carbon steel, Corrosion Science 86 (2014) 202–212.

[155] Y. Tan, Y. Fwu, K. Bhardwaj, Electrochemical evaluation of under-deposit corrosion and its inhibition using the wire beam electrode method, Corrosion Science 53 (2011) 1254–1261.

[156] M. Singer, Top of the line corrosion, in: A.M. El-Sherik (Ed.), Trends Oil Gas Corrosion Research and Technologies: Production and Transmission, Woodhead publishing, Cambridge, 2017, pp. 385–408.

[157] Z. Zhang, D. Hinkson, M. Singer, et al., A mechanistic model of top-of-the-line corrosion, Corrosion 63 (2007) 1051–1062.

[158] S. Punpruk, M. Thammachart, Y. Gunaltun, Field testing of volatile corrosion inhibitors and evaluation of batch treatment efficiency by cooled probe, NACE-10096, in: Proceedings of the Corrosion Conference and Expo, San Antonio, Texas, 14–18 March 2020.

[159] I. Jevremovic, M. Singer, M. Achour, et al., A novel method to mitigate the top-of-the-line corrosion in wet gas pipelines by corrosion inhibitor within a foam matrix, Corrosion 69 (2013) 186–192.

[160] Y. Gunaltun, L. Payne, A new technique for the control of top of the line corrosion: TLCC-PIG, NACE-03344, in: Proceedings of the Corrosion Conference and Expo, San Diego, California, 16–20 March, 2003.

[161] Z. Belarbi, F. Farelas, M. Singer, et al., Role of amines in the mitigation of CO_2 top of the line corrosion, Corrosion 72 (2016) 1300–1310.

[162] T. Pojtanabuntoeng, M. Singer, S. Nesic, Water/hydrocarbon co-condensation and the influence on top of the line corrosion, NaCE-11330, in: Proceedings of the Corrosion Conference and Expo, Houston, Texas, 13–17 March, 2011.

[163] M.C. Oehler, S.I. Bailey, R. Gubner, et al., Testing of generic volatile inhibitor compounds in different top-of-the-line corrosion laboratory test methods, NACE-01483, in: Proceedings of the Corrosion Conference and Expo, Salt Lake City, Utah, 11–15 March, 2012.

[164] H. Qin, L. Xu, W. Chang, et al., Top of the line corrosion under low temperature and high condensation rate conditions, NACE-11328, in: Proceedings of the Corrosion Conference and Expo, Houston, TX, 13–17 March, 2011.

[165] M.M. Islam, R. Gubner, T. Pojtanabuntoeng, Development of an electrochemical method to study top-of-the-line corrosion, NACE-9106, in: Proceedings of the Corrosion Conference and Expo, New Orleans, LO, 26–30 March 2017.

[166] R. Barker, X. Hu, A. Neville, The influence of high shear and sand impingement on preferential weld corrosion of carbon steel pipework in CO_2-saturated environments, Tribiology International 68 (2013) 17–25.
[167] M. Sephton, P.C. Pistorius, Localized corrosion of carbon steel weldments, Corrosion 56 (2000) 1272–1279.
[168] K. Alawadhi, M.J. Robinson, Preferential weld corrosion of X65 pipeline steel in flowing brines containing carbon dioxide, Corrosion Engineering Science and Technology 46 (2011) 318–329.
[169] ASTM, G1 Standard Practice for Preparing, Cleaning, and Evaluating Corrosion Test, ASTM International, West Conshohocken, 2017.
[170] H. Mansoori, R. Mirzaee, F. Esmaeilzadeh, et al., Pitting corrosion failure analysis of a wet gas pipeline, Engineering Failure Analysis 82 (2017) 16–25.
[171] M.G. Fontana, Corrosion Engineering, third ed., McGraw Hill, India, 2005.
[172] ASTM, G31-12 Standard Guide for Laboratory Immersion Corrosion Testing of Metals, ASTM International, West Conshohocken, 2012.
[173] C. Lefrou, R.P. Nogueira, F. Huet, et al., Electrochemistry, in: L.L. Shrier (Ed.), Shrier's Corrosion Handbook, Elsevier, Amsterdam, 2010, pp. 13–51.
[174] F. Mansfeld, Electrochemical methods of corrosion testing, in: S.D. Cramer, B.S. Covino Jr (Eds.), ASM Handbook Volume 13A - Corrosion: Fundamentals, Testing, and Protection, ASM International, West Conshohocken, 2003, pp. 446–462.
[175] S.K. Kim, I.J. Park, D.Y. Lee, Influence of surface roughness on the electrochemical behaviour of carbon steel, Journal of Applied Electrochemistry 43 (2013) 507–514.
[176] B. Evgeny, T. Hughes, D. Eskin, Effect of surface roughness on corrosion behaviour of low carbon steel in inhibited 4 M hydrochloric acid under laminar and turbulent flow conditions, Corrosion Science 103 (2016) 196–205.
[177] Y. Li, Y.F. Cheng, Effect of surface finishing on early-stage corrosion of a carbon steel studied by electrochemical and atomic force microscope characterizations, Applied Surface Science 366 (2016) 95–103.
[178] V. Jovancicevic, S. Ramachandran, P. Prince, Inhibition of carbon dioxide corrosion of mild steel by imidazolines and their precursors, Corrosion 55 (1999) 449–455.
[179] M.P. Desimone, G. Grundmeier, G. Gordillo, et al., Amphiphilic amido-amine as an effective corrosion inhibitor for mild steel exposed to CO_2 saturated solution: polarization, EIS and PM-IRRAS studies, Electrochimica Acta 56 (2011) 2990–2998.
[180] R.H. Hausler, Corrosion inhibitors, in: R. Baboian (Ed.), Corrosion Tests and Standards: Application and Interpretation, second Ed., ASTM International, West Conshohocken, 2005, pp. 480–499.
[181] ASTM, G59–97 Standard Test Method for Conducting Potentiodynamic Polarization Resistance Measurements, ASTM International, West Conshohocken, 2014.
[182] J.R. Scully, Electrochemical tests, in: R. Baboian (Ed.), Corrosion Tests and Standards: Application and Interpretation, second ed., ASTM International, West Conshohocken, 2005, pp. 107–130.
[183] J. Cui, D. Yu, Z. Long, et al., Application of electrochemical noise (EN) technology to evaluate the passivation performances of adsorption and film-forming type corrosion inhibitors, Journal of Electroanalytical Chemistry 855 (2019) 113584.
[184] Y.J. Tan, S. Bailey, B. Kinsella, Factors affective the determination of electrochemical noise resistance, Corrosion 55 (1999) 469–475.

[185] Y. Hou, C. Aldrich, K. Lepkova, et al., Detection of under-deposit corrosion in a CO_2 environment by using electrochemical noise and recurrence quantification analysis, Electrochimica Acta 274 (2018) 160–169.

[186] B. Heeg, D. Klenerman, Persistency of corrosion inhibitor films on c-steel under multiphase flow conditions. Part II: optical SHG and electrochemical studies, Corrosion Science 40 (1998) 1313–1329.

[187] R.A. Cottis, Interpretation of electrochemical noise data, Corrosion 57 (2000) 265–285.

[188] J.J. Moloney, W.Y. Mok, C.M. Menendez, In situ assessment of pitting corrosion and its inhibition using LCM, NACE-10156, in: Proceedings of the Corrosion Conference and Expo, San Antonio, Texas, 14–18 March, 2010.

[189] R. De Marco, W. Durnie, A. Jefferson, et al., Persistence of carbon dioxide corrosion inhibitors, Corrosion 58 (2002) 354–363.

[190] Y. Chen, G.R. Ruschau, J.A. Beavers, et al., Corrosion and inhibitor film persistency under flow conditions, NACE-02503, in: Proceedings of the Corrosion Conference and Expo, Denver, Colorado, 7–11 April, 2002.

[191] C. Compère, P. Jaffré, D. Festy, Aging of type 316 L stainless steel in seawater: relationship between open-circuit potential, exposure time, and pitting potential, Corrosion 7 (1996) 496–501.

[192] ASTM, G46–94 Standard Guide for Examination and Evaluation of Pitting Corrosion, ASTM International, West Conshohocken, 2018.

[193] W. Li, B.F.M. Pots, X. Zhong, et al., Inhibition of CO_2 corrosion of mild steel – Study of mechanical effects of highly turbulent disturbed flow, Corrosion Science 126 (2017) 208–226.

[194] C.M. Menendez, Application of optical profilometric techniques to corrosion monitoring, NACE-2506, in: Proceedings of the Corrosion Conference and Expo, Orlando, Florida, 17–21 March, 2013.

[195] B. Holme, O. Lunder, Characterisation of pitting corrosion by white light interferometry, Corrosion Science 49 (2007) 391–401.

[196] V. Pandarinathan, K. Lepkova, S. Bailey, et al., Adsorption of corrosion inhibitor 1 dodecylpyridinium chloride on carbon steel studied by in situ AFM and electrochemical methods, Industrial and Engineering Chemistry Research 53 (2014) 5858–5865.

[197] W. Durnie, B. Kinsella, R. de Marco, et al., A study of the adsorption properties of commercial carbon dioxide corrosion inhibitor formulations, Journal of Applied Electrochemistry 31 (2001) 1221–1226.

[198] K. Lepkova, W. Van Bronswijk, V. Pandarinathan, et al., Synchrotron infrared microspectroscopy study of the orientation of an organic surfactant on a microscopically rough steel surface, Vibrational Spectroscopy 68 (2013) 204–211.

[199] J. Wang, B. Hou, J. Xiang, et al., The performance and mechanism of bifunctional biocide sodium pyrithione against sulfate reducing bacteria in X80 carbon steel corrosion, Corrosion Science 150 (2019) 296–308.

[200] K. Asami, K. Hashimoto, X-ray photoelectron spectroscopy for corrosion studies, Langmuir 3 (1987) 897–904.

[201] D.A. Lopez, W.H. Schreiner, S.R. de Sanchez, et al., The influence of inhibitors molecular structure and steel microstructure on corrosion layers in CO_2 corrosion. An XPS and SEM characterization, Applied Surface Science 236 (2004) 77–97.

[202] A. Swift, A.J. Paul, J.C. Vickerman, Investigation of the surface activity of corrosion inhibitors by XPS and time-of-flight SIMS, Surface and Interface Analysis 20 (1993) 27–35.

[203] M. Finsgar, EQCM and XPS analysis of 1,2,4-triazole and 3-amino-1,2,4-triazole as copper corrosion inhibitors in chloride solution, Corrosion Science 77 (2013) 350–359.

[204] J. Moloney, Y. De-abreu, J. Helander, Accelerated stainless steel 316 L material compatibility assessment of chemical products using potentiodynamic polarization, NACE-5547, in: Proceedings of the Corrosion Conference and Expo, Dallas, Texas, 15–19 March, 2015.

[205] M. Iannuzzi, E. Hornus, M. Salasi, Crevice corrosion, in: R. Baboian (Ed.), Corrosion Tests and Standards: Application and Interpretation,, second Ed., ASTM International, West Conshohocken, 2005, pp. 221–232.

[206] B. Espelid, Crevice formers for specimens of plate material, in: U. Kivisäkk, B. Espelid, D. Féron (Eds.), Methodology of Crevice Corrosion Testing for Stainless Steels in Natural and Treated Seawaters: (EFC 60), Maney Publishing, Leeds, 2010.

[207] ASTM, G48-11 Standard Test Methods for Pitting and Crevice Corrosion Resistance of Stainless and Related Alloys by Use of Ferric Chloride Solution, ASTM International, West Conshohocken, 2015.

[208] B. Cai, Y. Liu, X. Tian, et al., An experimental study of crevice corrosion behaviour of 316 L stainless steel in artificial seawater, Corrosion Science 52 (2010) 3235–3242.

[209] R.G. Kelly, J.S. Lee, Localized corrosion: crevice corrosion, in: K. Wandelt (Ed.), Encyclopedia of Interfacial Chemistry, Elsevier, Amsterdam, 2018, pp. 291–301.

[210] M.S. Schilling, Experience with nonmetallic materials in petroleum refinery applications, NACE-05574, in: Proceedings of the NACE Corrosion Conference and Expo, Houston, Texas, 3–7 March, 2005.

[211] K.A. Esaklul, J. Mason, Nonmetallics applications in oil and gas production (pipes, liners, rehabilitations), in: A.M. El-Sherik (Ed.), Trends Oil Gas Corrosion Research and Technologies: Production and Transmission, Woodhead publishing, Cambridge, 2017, pp. 627–660.

[212] ASTM, D2240-15 Rubber Property - Durometer Hardness, ASTM International, West Conshohocken, 2015.

[213] ASTM, D638-14 Standard Test Method for Tensile Properties of Plastics, ASTM International, West Conshohocken, 2014.

[214] ASTM, D412-16 Standard Test Methods for Vulcanized Rubber and Thermoplastic Elastomers, ASTM International, West Conshohocken, 2016.

[215] J. Jackson, R. Harrington, D. Manko, Emulsion tendency studies - understanding method, inhibitors and water cut, NACE-01223, in: Proceedings of the Corrosion Conference and Expo, Salt Lake, Utah, 11–15 March, 2012.

[216] M.R. Gregg, S. Ramachandran, Review of corrosion inhibitor developments and testing for offshore oil and gas production systems, NACE-04422, in: Proceedings of the Corrosion Conference and Expo, New Orleans, Louisiana, 28 Mar-2 April, 2004.

[217] P. Schempp, K. Preub, M. Troger, About the correlation between crude oil corrosiveness and results from corrosion monitoring in an oil refinery, Corrosion 72 (2016) 843–855.

[218] S. Papavinasam, Electrochemical polarization techniques for corrosion monitoring, in: L. Yang (Ed.), Techniques for Corrosion Monitoring, first ed, Woodhead Publishing, Cambridge, 2008, pp. 49–85.
[219] B. Hedges, K. Sprague, T. Bieri, et al., A review of monitoring and inspection techniques for CO2 & H2S corrosion in oil & gas production facilities: location, location, location!, NACE-06120, in: Proceedings of the Corrosion Conference and Expo, San Diego, California, 12–16 March, 2006.
[220] C.S. Brossia, Electrical resistance techniques, in: L. Yang (Ed.), Techniques for Corrosion Monitoring, first ed., Woodhead publishing, Cambridge, 2008, pp. 277–292.
[221] B. Hedges, S. Papavinasam, et al., Monitoring and Inspection Techniques for Corrosion in Oil and Gas Production, NACE-5503, in: Proceedings of the Corrosion Conference and Expo, Dallas, Texas, 15–19 March, 2015.
[222] A. Son, Developments in the laboratory evaluation of corrosion inhibitor: a review, NACE-07618, in: Proceedings of the Corrosion Conference and Expo, Nashville, Tennessee, 11–15 March, 2007.
[223] J. Anthony, R. Harrington, C. Durnell, et al., Lessons learned from residual corrosion inhibitor measurements by LC-MS in a mature North Sea gas/condensate/MEG production system, SPE-190902-MS, in: Proceedings of the SPE International Oilfield Corrosion Conference and Exhibition, Aberdeen, UK, 18–19 June, 2018.
[224] M.A. Gough, G.J. Langley, Analysis of oilfield chemicals by electrospray-mass spectrometry, Rapid Communication in Mass Spectrometry 13 (1999) 227–236.
[225] T. Jackson, J. Vargas, Selective loss of bulk corrosion inhibitor species when exposed to glass, plastic, and steel, NACE-14999, in: Proceedings of the Corrosion Conference and Expo, Virtual, 14–18 June, 2020.
[226] P. McCormack, P. Jones, S.J. Rowland, Liquid chromatography/electrospray ionisation mass spectrometric investigations of imidazoline corrosion inhibitors in crude oils, Rapid Communication in Mass Spectrometry 16 (2002) 705–712.
[227] S. Murugesan, Z. Liu, P. Dhulipala, et al., Nanotechnology-based field method for detection and monitoring of corrosion inhibitors, NACE-14825, in: Proceedings of the Corrosion Conference and Expo, Virtual, 14–18 June 2020.
[228] NACE, SP0192 standard practice monitoring corrosion in oil and gas production with iron counts, NACE International, Houston, 2012.
[229] E.A. Kima, Oil field chemicals synergy upstream Arab heavy crude handling facilities, SPE-154099-MS, in: Proceedings of the SPE International Conference and Exhibition on Oilfield Corrosion, Aberdeen, UK, 28–29 May, 2012.
[230] N. Odeh, K. Al Azeezi, K. Aldhaheri, et al., Assessment of a 30 years corrosion management in oil producers under carbon dioxide environment in a giant field offshore Abu Dhabi case study, SPE-202739-MS, in: Proceedings of the Abu Dhabi International Petroleum Exhibition & Conference, Abu Dhabi, UAE, 9–12 November, 2020.
[231] B.F.M. Pots, E.L.J.A. Hendriksen, H.A.M. de Reus, et al., Field study of corrosion inhibition at very high flow velocity, NACE-03321, in: Proceedings of the Corrosion Conference and Expo, San Diego, California, 16–20 March, 2003.
[232] A. Al-Shamari, S. Al-Sulaiman, A. Jarragh, et al., Efficiency of downhole corrosion inhibitor treatment in sweet crude producer wells, NACE-11269, in: Proceedings of the Corrosion Conference and Expo, Houston, TX, 13–17 March, 2011.

[233] E.E. Barr, A.H. Greenfield, L. Pierrard, Application of electrochemical noise monitoring to inhibitor evaluation and optimization in the field: results from the Kaybob South sour gas field, NACE-01288, in: Proceedings of the Corrosion Conference and Expo, Houston, Texas, 11–16 March, 2001.

[234] T.G. Martin, S.N. Smith, J. Bondos, Materials and corrosion history with LaBarge Madison production: a 20 year story of success, NACE-08634, in: Proceedings of the Corrosion Conference and Expo, New Orleans, Louisiana, 16–20 March, 2008.

[235] S. Kapusta, D. Raghu, J. Richard, Managing corrosion in sour gas systems: testing, design, implementation and field experience, NACE-08641, in: Proceedings of the Corrosion Conference and Expo, New Orleans, Louisiana, 16–20 March, 2008.

[236] M.J.J.S. Thomas, Corrosion inhibitor selection - Feedback from the field, NACE-00056, in: Proceedings of the Corrosion Conference and Expo, Orlando, Florida, 26–31 March, 2000.

CHAPTER

Microbial control

10

Xiangyang Zhu
Saudi Aramco, Dhahran, Saudi Arabia

Chapter outline

10.1 Introduction

Crude oil is a complex mixture of hydrocarbons and other organic compounds, and its compositional and physical properties vary widely from different reservoirs around

Flow Assurance. DOI: https://doi.org/10.1016/B978-0-12-822010-8.00004-0

the world. Hydrocarbon-metabolizing microorganisms are widely distributed in each stage of oil processing from exploration, transportation, processing and storage. In addition to the indigenous microorganisms in the reservoir, microorganisms can be introduced into oil reservoirs and related downstream operation processes via drilling and fracturing operation, stimulation, water flooding of the reservoir, crude processing, distribution, and storage, and hydrotesting [1–4]. The presence of water in crude oil from the production wells, and the drop-out and accumulation of water and solids in the flowlines, transmission pipelines, and storage tanks render the microbial control a very complex and challenging issue in the petroleum industry. Proper management of microbial issues has direct impact on the asset integrity and reliability of pipeline operations and crude processing and storage facilities with significant economic and safety implications. In order to reduce the damaging effect of microbial growth, the oil industry typically uses a variety of organic and inorganic chemicals (biocides) to treat the affected systems and to control and mitigate microbiologically influenced corrosion (MIC), reservoirs souring caused by biological generation of hydrogen sulfide, and pipe plugging and formation damage due to biofouling.

These detrimental effects of microorganisms in oil industry start with water and inorganic and organic solids drop-out and accumulation, and biofilm formation on the surface of metal structures of oil industry system or equipment. Exopolysaccharides (EPS) secreted by some microbes help the initial attachment of microorganisms to metal surface, staring the process of microbial colonization and the formation of biofilms. This process depends on the surface characteristics, including metal surface free energy, roughness, hydrophobicity, and metallurgical features. The thick gelatinous biofilm layer, together with other inorganic insoluble chemicals (e.g., corrosion products, scale precipitates), and metabolic products of microbial activity (e.g., organic and inorganic acids, enzymes, hydrogen sulfide), provides protection from physical disturbances and antimicrobial compounds, and superior access to energy sources as compared to that for bacteria in the bulk fluid phase [5,6]. Heterogeneous biofilms contain individual cells as well as microcolonies of bacterial cells separated by interstitial channels that allow diffusion and transport of energy sources and waste products [7]. The composition of microbial community in the biofilm and its metabolic activity largely depend on the extracellular conditions such as temperature and oxygen gradient, electron donors and acceptors, accumulation and removal of metabolites, pH, and salinity, etc. [8].

The microbial community in the biofilms often displays the cooperative metabolic activities within the biofilms. Through synergistic interactions of different groups of microorganisms, biofilms influence the metal surface in many different ways. For instance, aerobes are active in the interface of top layer of biofilm and the surrounding liquid/air, consume oxygen, provide a microenvironment for the anaerobes deep in the biofilms and on the metal surface. The resulting differential aeration cells contribute to the metal corrosion. The aggressive metabolites (e.g., sulfide and acids) produced by microbial communities in the biofilms also directly or indirectly enhances the rate of metal corrosion. Corrosion was estimated (in 2001) to cost the

United States oil and gas industry about $13.4 billion/year, and MIC was estimated to account for as much as $2 billion/year [9]. In addition to the corrosion failures of equipment and facilities caused by microbial activities, the examples of other costly microbial problems in oil and gas operations include injectivity loss of the water injection wells, oil production loss due to formation rock plugging, reservoir souring due to biogenic H_2S production, increased crude processing cost and decreased product value, decreased pipeline efficiency, reduced heat transfer efficiency of heat exchangers, and biodeterioration of refined products [5,10–12]. This chapter will cover the major microorganisms encountered in oil and gas industry and their detrimental impacts on the oil industry systems and facilities, and the applications of a variety of organic and inorganic chemicals (biocides, preservatives,), biological means (bacteriophage), and physical processes (ultraviolet (UV) disinfection, filtration, and sonication) for microbial control and mitigation in the affected systems. In particular, the chapter focused on the biocide classification, selection, performance evaluation, best treatment practices, and residual and treatment effectiveness monitoring in the field applications.

10.2 Major microorganisms in oil and gas industry

Many microorganisms occurring in natural environment of oil and gas industry can cause corrosion, biofouling, and reservoir souring. Some common groups of these microorganisms include sulfate-reducing bacteria (SRB), sulfate-reducing archaea (SRA), sulfur-oxidizing bacteria (SOB), methanogens, acid-producing bacteria (APB), iron-oxidizing bacteria (IOB), iron-reducing bacteria (IRB), manganese-oxidizing bacteria (MOB), nitrate-reducing bacteria (NRB), and slime-forming bacteria [5,13,14]. Each of these physiological groups of microorganisms may contain hundreds or thousands of individual species.

MIC is an electrochemical and chemical process influenced by the presence or activities of microorganisms [15,16]. The activities of individual microbial species or a synergistic group of microbes alter the rates and types of electrochemical and chemical reactions on metal surface, resulting in various types of corrosions such as pitting corrosion, crevice corrosion, under-deposit corrosion, and enhanced galvanic and erosion corrosion [15,17]. Microbial corrosion has significant impact on the oil and gas and other industries, as it degrades the integrity, safety, and reliability of pipeline operations and crude processing facilities, resulting in high operation and maintenance costs, and production interruption [18,19]. However, the mere presence of the given classes of organisms associated with MIC does not indicate that MIC is occurring. Neither does show that the presence of a given type of microorganisms establishes a cause-and-effect relationship between the bacteria and metal corrosion [20]. An effective MIC management program should include routine monitoring of microbial activities, periodic assessment of microbial risks, and an effective mitigation program, such as scraping and biocide treatments.

MIC is rarely associated with one single mechanism or one single species of microorganism [21,22]. In fact, in a natural environment, it is always microbial communities containing many different types of microbes that cause the MIC. For instance, at the low spots of the pipelines with accumulated water and solid deposits, SRB, SRA, methanogens, APB, and IOB frequently coexist in a symbiotic relationship. IOB use molecular oxygen as electron acceptor and create an anaerobic environment, promoting the growth of APB, SRB, SRA, and methanogens. APB are capable of producing organic acids (e.g., lactate, propionate, acetate, butyrate), which are utilized by various microorganisms including SRB, SRA, and methanogens. Some SRB species produce hydrogen, CO_2, and acetate by fermentation, and methanogens and some other SRB species consume these compounds for growth. In addition, many SRB species and methanogens are capable of consuming cathodic hydrogen (cathodic depolarization), which depolarizes the galvanic cell, greatly increasing the corrosion rate.

MIC diagnosis has traditionally been achieved by looking into three evidences, that is, (1) unique corrosion pit morphologies such as localized, circular, and ring-like pit; (2) specific corrosion products and deposits such as mackinawite, greigite, and pyrite from SRB-related corrosion; and (3) the presence of corrosion-causing microorganisms or their by-products. Traditionally, the presence of corrosion-causing microorganisms is determined using culture-dependent growth methods such as most probable number (MPN) method. The culture-dependent methods underestimate the size and misrepresent the composition of microbial communities [23,24]. Some microorganisms (e.g., methanogens) are extremely difficult to grow in laboratory conditions. Finally, many samples from a corrosion failure investigation are often already compromised, and cannot be analyzed by conventional cultivation-based methods. Therefore, conventional microbiology methods often fail to provide a critical piece of evidence required for a firm diagnosis of MIC, i.e., the presence of corrosion-causing microorganisms in the failed metal samples.

Many culture-independent molecular microbiology techniques have been developed to circumvent problems associated with cultivation-based methods. In particular, real-time quantitative polymerase chain reaction (qPCR) methods have been widely used to diagnose MIC by detecting and quantifying various types of corrosion-causing microorganisms by targeting the small sub-unit ribosomal RNA (16S rRNA) genes, or functional genes that are uniquely present in the specific groups of bacteria [14,25–27]. Molecular microbial analyses have provided a solid piece of evidence to firmly diagnose the MIC in oil and gas industry systems.

10.2.1 Sulfate-reducing bacteria and archaea

SRB are a group of morphologically and phylogenetically diverse anaerobes that carry out dissimilatory reduction of sulfate to sulfide. Although SRB are often considered to be strictly anaerobic, some SRB species have significant tolerance to oxygen and can survive and quickly recover from brief oxygen exposure

[28–30]. At low dissolved oxygen concentrations, some SRB are able to respire with Fe^{3+} or even oxygen with hydrogen acting as an electron donor [31]. SRB are present in a variety of environments, including soils, domestic, industrial, and mining wastewaters, oil- and gas-bearing formations, storage tanks and oil and gas pipelines [32,33]. The general concern about SRB is that they can produce hydrogen sulfide (H_2S), which directly attacks metal surfaces leading to anaerobic corrosion and pit formation [34]. However, it is clear that sulfide is not only produced by SRB, H_2S can also be produced by SRA, methanogens and fermentative microorganisms in oilfield systems [35].

Dissimilatory sulfate reduction is found in primary four bacterial lineages and restricted to a single archaeal genus, *Archaeoglobus*. Under the bacterial lineages, SRB belong to the Phylum Proteobacteria, Firmicutes, Thermodesulfobacteria, and Nitrospirae. The largest group is Phylum Proteobacteria, of which almost all SRB fall into the Class Deltaproteobacteria, with three different Orders, Desulfovibrionales, Desulfobacterales, and Syntrophobacterales. The genera Desulfovibrio, Desulfomicrobium, Desulfobacterium, and Desulfobulbus are the major representatives of this lineage. The second largest group of SRB is under the Phylum Firmicutes, represented by genera Desulfosporosinus and Desulfotomaculum. SRB under the Phylum Thermodesulfobacteria and Nitrospirae are thermophiles, including the genera Thermodesulfobacterium and Thermodesulfovibrio. SRB are mainly Gram-negative, mesophilic or thermophilic bacteria [36]. The exception is the genus *Desulfotomaculum* under the Phylum Firmicutes, which is Gram-positive spore forming SRB, and many of them are also thermophiles [37,38].

SRB are capable of utilizing a broad range of organic compounds of low molecular weight as electron donors, including volatile fatty acids, alcohols (ethanol, propanol, butanol, etc.), and hydrocarbons [28,39]. Acetate, lactate, pyruvate, butyrate, ethanol, and hydrogen are among the most commonly used substrates by SRB [39]. SRB use sulfate and other sulfur compounds (sulfite, thiosulfite, trithionate, tetrathionate, and elemental sulfur) as final electron acceptors in respiration processes [40,41], and are capable of producing large amount of biogenic H_2S [42].

In addition to SRB, SRA under the Domain Archaea are also capable of reducing sulfate. SRA are heterotrophic hyperthermophiles, restricted to only one Archaea genus, *Archaeoglobus*, which includes five species, *Archaeoglobus fulgidus*, *Archaeoglobus lithotrophicus*, *Archaeoglobus infectus*, *Archaeoglobus profundus*, and *Archaeoglobus veneficus*. Archaeoglobus species grow anaerobically at extremely high temperatures between 60°C and 95°C, with optimal growth at 83°C [43]. They can be found in hydrothermal vents, oil deposits, and hot springs [44]. Archaeoglobus members are capable of oxidizing many different organic carbon sources, including fatty acids, amino acids, aldehydes, and organic acids, using sulfate as the final electron acceptor.

Together, the term sulfate-reducing prokaryotes (SRP) is used to describe SRB and SRA. SRP are known to be the major contributors to MIC in oil and gas production systems [21,45–47], reservoir souring and formation damage, injectivity loss of water injection wells [48], and other problems of great economic impact in oil and gas industries by producing H_2S and formation of iron sulfide [33,49]. Suspended iron sulfide particulates in sour oil or gas processing system can plug wellbores or filtration equipment and stabilize emulsions leading to poor separation and separated fluid quality. Iron sulfide can also bind with organic substrates (e.g., asphaltenes, waxes or crude oil) and other inorganics solids (e.g., sand, scale particles or corrosion debris) to produce black, sticky solids, plugging flowlines and production facilities [50]. Diverse species of SRP have been isolated from crude oil reservoir and formation waters with wide range of salinity (0%–17% salt content) and temperature (4°C–80°C) [45]. The mesophilic Desulfovibrio and thermophilic, spore-forming Desulfotomaculum are the most abundant SRB associated with biofilms formation and microbial corrosion in oil processing facilities, storage tanks, and pipelines of oil and gas industries [37,51,52]. Desulfovibrio species incompletely oxidize organic compounds to acetate and use energy from the oxidation of hydrogen, lactate, and pyruvic acid for growth.

Thermophilic SRB include Thermodesulfobacterium, Thermodesulfovibrio, and Desulfotomaculum genera. Thermodesulfobacterium is the most thermophilic of all known SRB. The representative species are *Thermodesulfobacterium commune*, *Thermodesulfovibrio yellowstonii*, *Desulfotomaculum geothermicum*, respectively. *T. commune*, isolated from algal-bacterial mats associated with volcanic activity in Yellowstone National Park, is Gram-negative, strictly anaerobic, straight rod-shaped SRB, and proliferates between 45°C and 85°C with an optimum growth temperature of 70°C. *T. commune* uses pyruvate, lactate or H_2 as electron donors and sulfate or thiosulfate as electron acceptors for growth and sulfide formation [53]. *T. yellowstonii*, initially isolated from thermal vent water in Yellowstone Lake, is Gram-negative, curved rod-shaped SRB, grows between 40°C and 70°C with optimal growth at 65°C. *T. yellowstonii* utilizes lactate, pyruvate, hydrogen and acetate, or formate and acetate as electron donors, and sulfate, thiosulfate and sulfite as electron acceptors [54]. *D. geothermicum*, isolated from geothermal ground water, is strictly anaerobic, thermophilic, spore-forming SRB, with optimum temperature, pH and NaCl concentration for growth at 54°C, pH 7.3–7.5, and 25–35 g NaCl/L. The strain grows with sulfate as electron acceptor and lactate, ethanol, fructose and saturated fatty acids (up to C_{18}) as electron donors and carbon sources. Thermophilic SRB are related to the processes of in situ oil transformation, capable of incompletely biodegrading hydrocarbons to compounds such as acetate, lactate, ethanol, butanol, and carboxylic acids in the presence of sulfates, resulting in the presence of biogenic H_2S in crude oil reservoirs and in formation waters [55,56]. Biogenic sulfide production causes health and safety complications, environmental hazards, and huge economic losses due to corrosion of equipment and reservoir souring.

Many researchers have reported that the corrosion rates caused by SRB under laboratory conditions are much lower than the rates under field conditions

[30,57,58], and the rate usually cannot be maintained at high level for long period of time. In anaerobic soil, the maximum corrosion rate of steel and iron by SRB was reported to be 0.7–7.4 mm/year [57,59]. The existence and activity of SRB causes the average corrosion rate of steel exposed to anaerobic soil to be more than 20 times higher than that of the control case [57]. SRB influence the corrosion of iron mainly by two primary processes in sulfate-containing anoxic environments [60,61]. One is that SRB corrode iron/steel structures indirectly through corrosive H_2S, produced by SRB as a dissimilatory product from sulfate reduction (chemical microbially influenced corrosion, or CMIC). The other is that some SRB species can corrode iron/steel via direct electron uptake from the metal itself (electrical microbially influenced corrosion, or EMIC) [60]. It is estimated that the activities of SRB are the cause of corrosion in 77% of oil producing wells in the US [62].

The production of H_2S can also cause other serious problems to the oil and gas industry. H_2S formed deep in the formation can lead to reservoir souring and reduced product quality. Biomass and formation of insoluble FeS scale by reaction with metal (mainly iron) ions can lead to plugging in injection wells and in the formation rocks, causing injectivity loss and lower well production rate. In the pipelines, together with slime, insoluble FeS can form scale or deposits and reduce pipeline oil flow [56,63].

10.2.2 Methanogens

Methanogenic archaea or methanogens are a diverse group of organisms that can live in a wide range of environments. They have been found in a range of extreme environment on Earth—buried under kilometers of ice in Greenland and living in hot, dry desert soil. They produce methane by degradation of organic compounds under anaerobic conditions. Their growth and activity are influenced by physical and chemical factors such as temperature, salt content, and pH. Methanogens are very sensitive to very low concentrations of oxygen even in the range of several parts per million [64]. From phylogenetic taxonomy, all methanogens belong to the Phylum Euryarchaeota under the Domain Archaea, including four classes, Methanomicrobia, Methanobacteria, Methanococci, and Methanopyri [65]. The largest group is Class Methanomicrobia, of which the majority of methanogens fall into the Orders Methanosarcinales and Methanomicrobiales. Genera Methanosarcina, Methanococcoides, Methanolobus, Methanococcoides, and Methanothrix are the major members in the Order Methanosarcinales. Order Methanomicrobiales mainly consists of genera Methanoculleus, Methanogenium, Methanomicrobium, Methanolinea, Methanoregula, Methanospirillum, Methanocorpusculum. The second largest group of methanogens is under the Class Methanobacteria, represented by genera Methanobrevibacter, Methanobacterium, Methanothermobacter, and Methanosphaera. Class Methanococci includes methanogens in the genera Methanocaldococcus, Methanococcus, and Methanothermococcus. The small Class Methanopyri is mainly represented by the genus *Methanopyrus*.

Methanogens are an important group of microorganisms in crude oil reservoir [66,67]. They have been enumerated from mild saline to hyper saline oil well waters in the mesophilic and thermophilic conditions. Most methanogens are mesophilic organisms, although some are extremophiles. For instance, *Methanoculleus submarinus*, isolated from methane hydrate-bearing sediments with an in situ temperature approximately 15°C–16°C, is irregular coccus-shaped mesophilic methanogen, with optimum growth at temperature 45°C, salinity 6–23 g/L NaCl, and pH 4.8–7.7 [68]. Thermophilic species of genera Methanothermococcus and Methanothermobacter have been isolated from offshore and onshore reservoirs around the world [69]. *Methanopyrus kandleri* is a hyperthermophilic rod-shaped methanogen, isolated from a hydrothermally heated deep-sea sediment. *M. kandleri* grew at pH 5.5–7.0, temperature 84°C–110°C, and NaCl 0.2%–4%, with optimum at pH 6.5, temperature 98°C, and NaCl 2%, respectively [70]. *Methanococcus vulcanicus* is a coccus-shaped methanogen, isolated from a deep-sea hydrothermal chimney sample collected on the East Pacific Rise at a depth of 2600 m. This methanogen showed the optimum growth at pH 6.5, temperature 80°C (in a range of 49°C–89°C), and NaCl concentration 25 g/L [71]. At the other end of the temperature spectrum, *Methanococcoides burtonii* and *Methanogenium frigidum* are psychrophilic methanogenic archaea, which live at extremely low temperatures. *M. burtonii* and *M. frigidum* were first isolated from the anoxic hypolimnion of Ace Lake, Antarctica, where the temperature remains permanently at 1°C–2°C. Their optimum and maximum temperatures for growth were 23°C and 29°C, and 15°C and 18°C, respectively [72,73].

The process of biogenic methane formation is known as methanogenesis, where a small number of simple substrates are used to form methane, including $H_2 + CO_2$, acetate, formate, methylated compounds (methanol, methylamines, dimethylsulphur), and primary and secondary alcohols [74]. There are three distinct pathways for methanogenesis [75,76]. Acetoclastic methanogenesis, responsible for two-thirds of biogenic methane production and occurring only in two genera Methanosarcina and Methanosaeta in Order Methanosarcinales, splits acetate to form CO_2 and CH_4 ($CH_3COOH \rightarrow CO_2 + CH_4$). Methylotrophic methanogenesis, found in Order Methanomicrobiales, can generate CH_4 from a variety of chemical compounds containing a methyl group such as methanol, methylsulfides, or methylamines ($CH_3OH + H_2 \rightarrow CH_4 + H_2O$). The third pathway for methane formation, hydrogenotrophic methanogenesis, occurs in most methanogens. In hydrogenotrophic methanogenesis, hydrogen serves as a source of electrons and energy while the CO_2 (or CO or formate) serves as the source of carbon and electron acceptor ($4H_2 + CO_2 \rightarrow CH_4 + 2H_2O$). Hydrogenotrophic methanogens are found in Order Methanopyrales, Methanococcales (e.g., genus Methanococcus), Methanobacteriales (e.g., genera Methanothermobacter, Methanobrevibacter), and Methanomicrobiales (e.g., genus Methanoculleus). While hydrogenotrophic methanogens can cause mild steel corrosion by direct elemental iron oxidation, acetoclastic methanogens can syntrophically thrive with SRB [77].

Methanogenesis is often limited by the supply of fermentation products (hydrogen or acetate). SRB also uses hydrogen or acetate as an electron source.

However, relative to SRB, methanogens are less competitive for these substrates [78,79]. Thus, methanogenesis in the ecosystem with high concentration of sulfate will be inhibited, such as in the case of the reservoir with seawater injection for pressure maintenance [80]. Generally, in most environments, there is little to no overlap between the zone of methanogenesis and the zone of sulfate reduction [81]. However, in the absence of sulfate, SRB and methanogens can coexist with increased methane production. This is explained by the fact that some SRB can ferment organic acids such as lactate and formate to produce H_2 and acetate, which can be used by acetoclastic and hydrogenotrophic methanogens [79,82,83].

Methanogens are reported to be associated with metal corrosion [5,84–86]. Methanogens can efficiently consume hydrogen from the surface of metals, enhance the cathodic reduction of proton, and thereby accelerate anodic metal dissolution (cathodic depolarization) [20,87].

10.2.3 Acid-producing bacteria

APB are present in a variety of environments, including oil and gas formations, processing facilities, transportation pipelines, soils, and domestic, industrial, and mining wastewaters [88,89]. They produce inorganic or organic acids as byproducts of their metabolism. Inorganic acids include nitric acid (HNO_3), nitrous acid (HNO_2), sulfurous acid (H_2SO_3), sulfuric acid (H_2SO_4), and carbonic acid (H_2CO_3). The SOB, for example, *Acidithiobacillus thiooxidans*, *Acidithiobacillus ferrooxidans*, *Thiomonas intermedia*, and *Thiomonas perometabolis*, produce H_2SO_4 and H_2SO_3 by oxidation of sulfur, H_2S, or other condensed sulfur mixture [18,90,91].

Other major acid producers include Bacillus, Clostridium, Streptococcus, Pseudomonas, Micrococcus, Lactobacillus, and Neisseria [4,92]. Bacillus sp. are versatile chemoheterotrophs capable of degrading many organic materials, and produce a number of volatile compounds and organic acids such as malic, lactic, acetic, citric, succinic, propionic, and butyric acids [93,94]. The majority of bacilli are mesophiles, with temperature optima between 30°C and 45°C, but some are thermophiles with optima as high as 65°C. Clostridium sp. are also capable of degrading large biological molecules and produce end products such as butyric acid, acetic acid, butanol and acetone, and large amounts of gas (CO_2 and H_2).

In natural environments, different types of microorganisms frequently coexist in symbiotic relationships. For instance, the organic acids, produced by APB from the fermentative metabolism can be utilized by various microorganisms, including SRB and methanogens [56,95,96]. In addition, these biogenic acids increase the concentration of protons, which can then be reduced at the cathode, generating hydrogen. SRB and methanogens consume hydrogen as an electron source from the surface of metals, enhance the cathodic proton reduction (cathodic depolarization), and thereby accelerate anodic metal dissolution or corrosion [18,20,56].

Inorganic or organic acids produced by APB cause metal corrosion (general corrosion, pitting attack, and stress corrosion cracking, etc.) by either direct reaction with metal or disrupting the protective surface oxides films and calcium scales [15,20,33,97]. Biofilm on metal surface can trap the acidic metabolites and create a microenvironment with drastically low pH, compared to the pH in the surrounding bulk aqueous phase, disrupting the protective film on metal surface and resulting in an elevated corrosion [8,98]. The large amount of organic acid produced also promotes metal depolarization starting the local corrosion process [99].

10.2.4 Iron- and manganese-oxidizing bacteria

IOB are chemo-lithotrophs deriving energy from the oxidation of ferrous ion (Fe^{2+}) with the formation of ferric oxide (Fe^{3+}), typically found in fresh water from rivers, lakes, marine environment, and oil production. They are generally filamentous, and frequently surrounded by a sheath usually encrusted with iron, manganese, or both [100]. It is difficult to isolate and culture IOB in the laboratory. Gallionella, Sphaerotilus, Leptothrix, Thiobacillus, Crenothrix, and Clonothrix are among the most common genera of IOB. The detection is usually dependent on diagnostic liquid cultures, which is very difficult even for experienced microbiologists. Microscopic identification of IOB is based on their distinctive morphologies, such as the extracellular twisted ribbon-like stalks formed by *Gallionella ferruginea*, but it is also quite difficult for an experienced analyst. Several direct and indirect tests for the presence of IOB are described in NACE Standard [15]. However, these techniques are not capable of quantifying metal-oxidizing bacteria. A new technique called qPCR is now available for quick detection and quantification by targeting 16S rRNA gene of Leptothrix, Sphaerotilus, and Gallionella [25,26]. The presence of IOB within tubercles associated with localized corrosion is considered a positive indication of MIC.

In an aerobic environment, IOB oxidize soluble ferrous ions (electron donors) from the bulk medium or the substratum to insoluble iron oxides and hydroxides such as Fe_2O_3 and $Fe(OH)_3$ [101]. They deposit iron oxides/hydroxides on the surfaces and produce orange-red tubercles at a rate hundreds of times higher than the abiotic process, which creates conditions that are favorable for localized corrosion [102]. The dense deposits formed by IOB deprive oxygen from the area immediately under the deposits, which becomes a relatively small anode compared to the large surrounding oxygenated cathode [103]. In addition to large cathode:anode surface area, the formation of differential aeration cell is also favorable for pitting corrosion and crevice corrosion [104].

IOB have been widely reported to be involved in MIC of stainless steel exposed to well water, chlorinated drinking water, or relatively low chloride waters [102,104–107]. IOB might be the initial colonizers on stainless steel surface. The deposits of iron hydroxides can overcome the passivation of stainless steel and initiate biocorrosion process [8,107], and also create a low-oxygen microenvironment that promotes colonization by other corrosive microorganisms, such as SRB [106,108]. Chloride (Cl^-) ions from the electrolyte migrate to the anode to neutralize buildup of charge, forming extremely aggressive ferric chlorides (very low local pH value, ~1.0), which causes pitting corrosion on surfaces of both carbon steel and austenitic stainless steel in particular [109]. Pitting corrosion of IOB involves the conventional features of differential aeration cell, a large cathode-to-anode surface area, and the development of acidity and metallic chlorides [15]. Accumulation or inorganic fouling (iron oxides and hydroxides) caused by IOB activity can also lead to problems to industrial equipment such as blockages in oil pipelines [100,110]. Author investigated a pinhole leak of a newly constructed 10″ stainless steel (TP 316 L) blowdown pipe within two months after draining the hydrotesting water and inert gas lay-up. Visual inspection on the pipe section revealed clear evidence of stagnant water at the bottom of the pipe (6 o' clock position). Microbial analysis detected high number of general bacteria and corrosive IOB present in the internal pipe surface swab, condensate water, and the demineralized water used for hydrotesting.

MOB can be isolated from different habitats. Identification and quantification of MOB is generally based on the agar spread plate procedure [111]. Some bacteria species from genera Pseudomonas, Flavobacterium, Sphingomonas, Leptothrix, Noviherbaspirillum has capability to catalyze the oxidation of soluble manganese (Mn^{2+}) to insoluble manganese oxide (Mn_2O_3, MnOOH, Mn_3O_4, and MnO_2), and form brown-black manganese oxide deposits on metal surface [111,112]. The rate of manganese oxidation by MOB is up to 5 orders higher than that of abiotic process [113,114]. Manganese oxides can elevate corrosion potential, which may initiate localized corrosion of passive metals [101,115]. In addition, manganese oxides can also serve as a cathode to support corrosion at an oxygen depleted anode within the deposits, resulting in under-deposit corrosion, pitting, and crevice corrosion [15,116].

10.3 Biocide classification

Uncontrolled microbial growth in oilfield systems can lead to costly problems such as fouling, plugging, corrosion and souring. It also directly impacts the safe operation, economic performance, and environmental compliance in all areas of production, transportation, refining and distribution. In oil and gas industry, the issues caused by microorganisms (bacteria, fungi, yeasts, algae) are typically mitigated through chemical (biocides and biostats), mechanical (pigging or scraping), physical (UV, filtration, sonication), and electrochemical (cathodic protection) methods. Biocides are used in all stages of oilfield development, from the initial drilling and workover, production of oil and gas, to storage and transport of crude oil and its processed products.

According to the EU Biocidal Products Regulation 528/2012 (EU BPR), biocides are chemicals intended to destroy, deter, render harmless, or exert a controlling effect on any harmful organisms. The chemical structure of the biocides determines its mode of actions on microorganisms, but the biocidal effect or biostatic effect is also highly dependent on its applied concentration and the contact time with microorganisms. There are many biocidal substances in the market, but their use is heavily regulated by various regulatory bodies. In oil and gas industry, only a few biocidal actives are commonly used in the commercial biocidal products. Sometime, several biocidal actives are blended into a single biocide product to increase its overall effectiveness by taking advantage of synergistic effects of different actives. Moreover, other chemicals such as surfactants may also be added to biocidal products for better penetration to the biofilm, increasing the efficacy of biocides in killing microorganisms.

Biocides can be broadly classified into two groups, nonoxidizing or organic biocides, and oxidizing or inorganic biocides, based on the mode of actions. They are applied to two distinct systems in oil and gas industry [117]. Nonoxidizing or organic biocides are generally used in the anaerobic systems such as crude production systems, crude pipelines, and injection water pipelines, which account for the majority of the systems in oil and gas industry. However, there are systems that contain or are exposed to oxygen such as the facilities in the injection water treatment plants, wash water in crude processing facilities, hydrotesting water, make-up water in drilling operations, and open-air storage tanks. In these systems, oxidizing or inorganic biocides are often used, or sometimes preferred.

10.3.1 Nonoxidizing biocides

The most commonly used nonoxidizing or organic biocides in oil and gas industry are formaldehyde, glutaraldehyde, tetrakis hydroxymethyl phosphonium sulfate (THPS), quaternary ammonium compounds (Quats), 2,2-dibromo-3-nitrilopropionamide (DBNPA), and isothiazolones. Theses biocides have broad-spectrum antimicrobial properties by targeting different molecules on the microbes. They interfere with cell metabolism, structure, and DNA translation. Cross-linking of exposed amine groups on proteins, oxidation of cell walls, disruption of metabolic proteins and the cytoplasmic membrane are among the most common modes of actions [117,118]. Kill efficiency of biocidal products is affected by many factors such as water chemistry, salinity, temperature, hydrogen sulfide, and oxygen concentration. The incompatibility with other oilfield chemicals may also impact the kill efficiency of biocides [119]. Compared to oxidizing biocides, nonoxidizing biocides are more specific, longer acting and less corrosive.

10.3.1.1 Formaldehyde and glutaraldehyde

Formaldehyde and glutaraldehyde are aldehyde compounds with a similar mode of action in microbial control. The kill mechanism is by reacting with primary and/or secondary amines in proteins or DNA, causing cross-linking of proteins in the cell membrane and disrupting cellular processes and cell metabolism [110]. Formaldehyde and glutaraldehyde are thermally unstable, losing biocidal efficacy at temperature above 50°C [119,120], though highly depending on pH and water salinity [121]. Formaldehyde is more toxic and considered to be "presumed human carcinogen," and is banned in many countries including the United State and EU countries for certain applications on the basis of environmental concerns and health and safety risks. Glutaraldehyde is considered to have better environmental profile [120,122], degrading readily in a broad pH range. The half-life typically is 4–12 hours depending on system conditions. Aerobic biodegradation of glutaraldehyde involves oxidation of both aldehyde groups to carboxylic acid and glutaric acid. Glutaric acid is then further mineralized to CO_2 as the final end product. Under anaerobic conditions, glutaraldehyde is transformed to hydroxypentanal and further metabolized to the di-alcohol and pentanediol [123].

Glutaraldehyde is one of the most commonly used biocide actives in oil and gas industry for microbial control. It is a broad-spectrum biocide and fast-acting biocide (3 to 4 hours, perhaps 4 to 6 hours with difficult slimes), non-ionic, non-foaming, effective over a wide pH range (typically pH 6.5–9.0). It is a good bactericide, effective against SRBs and biofilms, especially with difficult and persistent organisms due to its good penetrating ability. However, it has limited effectiveness against algae and fungi [124].

Glutaraldehyde structure in aqueous solution is not limited to the monomeric form. Instead, it can be present in at least 13 different forms depending on pH, concentration, temperature, etc. (Fig. 10.1). Glutaraldehyde rapidly auto-polymerizes, forming water-soluble dimers and trimers, and precipitates out at high temperatures ($\sim 140°C$) and/or alkaline pH. Glutaraldehyde biocide product commonly uses water or alcohols (mainly methanol and isopropanol) as carrier. The alcohols in the formulation delays polymerization during storage and increases penetration of the aldehyde into the cell [118]. The polymerization can also be minimized with the storage of glutaraldehyde biocide at acidic pH and at 4°C [125,126]. Polymerization blocks the active sites (aldehyde groups) of the glutaraldehyde molecules that are responsible for its biocidal activity [110,127]. In hydraulic fracturing application, biocidal glutaraldehyde monomer has limited time to control microbial activity in hot and/or alkaline shales, and may return along with its aqueous transformation products to the surface via flowback and produced water in cooler, more acidic, and saline shales [128]. Addition of quaternary ammonium salt and surfactants in the formulation often improves the biocidal efficacy of glutaraldehyde. Ethylenediamine-*N,N′*-disuccinic acid (EDDS) is a biodegradable chelating agent, and has been found to enhance the efficacy of glutaraldehyde in the treatment of SRB [129].

Glutaraldehyde does not react with strong acids and alkalis, is not sensitive to the presence of sulfides (H_2S, HS^-, S^{2-}), and tolerates high salinity environments. It is compatible with most other oilfield chemicals including chlorine, but its biocide efficacy can be interfered by amines, ammonia, and bisulfite-based oxygen scavengers [121,131]. It reacts violently with ammonia and amine-containing substances, which causes exothermic polymerization reaction of an aldehyde, and thus its deactivation [118]. In fracturing applications, glutaraldehyde increases the viscosity of the well treatment fluid at elevated temperatures, which could hinder flowback [120]. In oil and gas industry, this incompatibility is mitigated through the practice of chemical treatment, for example, applying

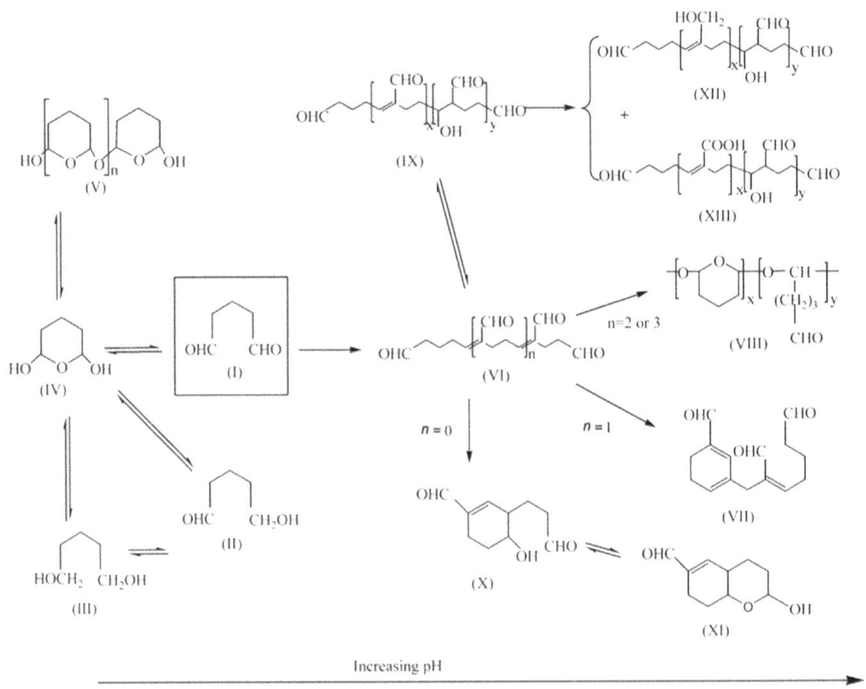

FIGURE 10.1

Possible forms of glutaraldehyde in aqueous solution.

Reproduced with permission from I. Migneault, C. Dartiguenave, M.J. Bertrand, et al. Glutaraldehyde: behavior in aqueous solution, reaction with proteins, and application to enzyme crosslinking, Biotechniques 37 (2004) 790–802 [130].

biocides and other chemicals at different locations or different time in the systems. Glutaraldehyde is less effective in penetrating biofilm and killing microorganisms in biofilms, comparing to planktonic microbes [132].

10.3.1.2 Tetrakis hydroxymethyl phosphonium sulfate

THPS is a quaternary phosphonium compound with good environmental profile, and widely used as biocide in oil and gas industry for microbial control [133]. It is degradable and non-bioaccumulative in the environment. It is a broad-spectrum biocide, effective in pH range of 3–11, with a mode of action by reacting with specific amino acids on the cellular membrane of microorganisms. In addition, THPS can block the sulfidogenic reduction pathway, which makes it highly effective in the control of SRB. In addition to biocidal effect, THPS also displays the property of iron chelator, and therefore, is used as iron sulfide scale remover in oil industry [117,119]. Compared to glutaraldehyde, THPS is more thermally stable, tolerating up to 160°C. However, similar to glutaraldehyde, it is incompatible with ammonium bisulfite oxygen scavengers and has limited penetration power to biofilms [119,134]. Campbell and Cooper [135] reported that in the presence of ammonium ions, THPS reacts with sulfide scales and soluble iron, significantly limiting its biocide efficacy.

THPS can easily degrade through the processes of hydrolysis, oxidation, photolysis, and biodegradation [136]. The breakdown products are trishydroxymethyl phosphine oxide (THPO), bishydroxymethyl phosphinic acid (BMPA), and possibly a formaldehyde adduct of trihydroxy compound (Eq. 10.1). THPO and BMPA have low aquatic toxicity and are not considered to present an environmental hazard [137,138]. Both of them will further degrade to CO_2 and inorganic matters. The degradation process is strongly affected by many factors such as temperature, pH, O_2 level, light, microbes and the presence of mild steel [136,139,140]. THPS is considered as "readily biodegradable" with 70%−80% converted to CO_2 under aerobic (21 days) and anaerobic conditions (365 days), respectively [123]. The hydrolysis half-life of THPS at 25°C is 7 days, and the photolysis half-life is less than 2 hours at pH 7 [141]. This explains, in the field application, that THPS biocides sometimes showed lower biocidal efficacy, when compared to the results obtained under the laboratory conditions, because it can be rapidly degraded to less toxic THPO and BMPA [142].

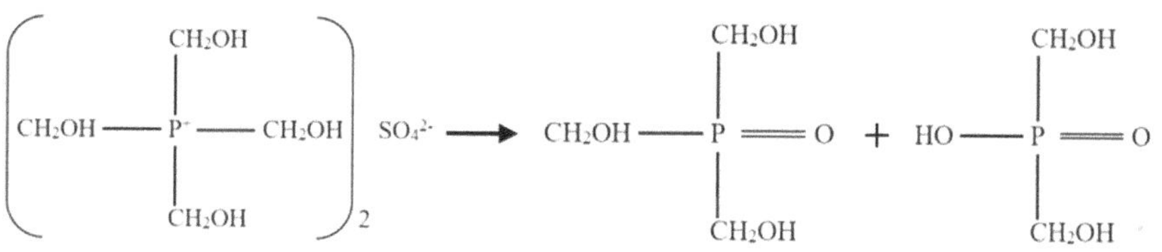

THPS THPO BMPA

(10.1)

THPS is also used to control iron sulfide precipitation for water injection and to remove deposits in oil production systems. THPS based solutions showed the performance of dissolution of iron sulfides comparable to or even better than the uninhibited hydrochloric acid, particularly in combination with ammonium chloride or organic phosphonate scale inhibitor [143,144]. Wang et al. [145] examined THPS as the scale dissolver for the scale deposits formed on downhole tubular of high temperature sour gas wells. The maximum dissolution was observed in THPS concentration of 25%−50%, with significantly enhanced dissolving power when ammonium chloride and chelating agents were added. When THPS is used at high concentration as iron chelator to remove iron sulfide scale, it can cause severe corrosion, especially in conditions of low buffering capacity of the water [145,146].

10.3.1.3 Quaternary ammonium compounds

Quats, such as alkyl dimethyl benzyl ammonium chloride and dialkyl dimethyl ammonium chloride compounds, have positively charged nitrogen ions with alkyl or aryl groups. They are common components of biocides and corrosion inhibitors because of their surfactant properties, and widely used for microbiological control in many industrial water systems [120]. Quats generally work best in an alkaline environment. As corrosion inhibitor, quats creates a thin protective layer on the metal surface, thereby reducing the interaction of oxidizing agents with steel. As biocides, quats act on the cell membrane, destabilizing the membrane and leading to the rapid decomposition of the cell [118]. Quats are not compatible with oxidizing agents, especially the chlorates, peroxides, chromates or permanganates.

Quats are effective algaecide and reasonably good bactericide, and can be applied over a wide range of pH (optimum pH 6.5–8.5). However, they tend to have only a bacteriostatic effect over Pseudomonas sp. and SRB [124]. Its excellent surface activity (as a surfactant agent) makes it highly effective to penetrate biofilms, removing biofouling and providing persistent microbial control in oilfield applications. In biocide products, quats are often used in combination with other actives such as glutaraldehyde and THPS to provide enhanced microbial control by helping biocide penetration into biofilms. On the other hand, their surfactant properties can cause problems with foaming, especially above pH 8.0, or incompatibility with anionic additives such as scale and corrosion inhibitors [147]. Quats have poor compatibility with polyanionics polymers, and are deactivated by high hardness (typically over 500 mg/L), chlorides, oil, dirt, silt and debris [124]. Quats have limited use in the water system with high total suspended solids (TSS) and total dissolved solids (TDS). Quats are usually not used during the oil exploration because they may adversely affect the permeability of the reservoir rocks. In crude oil processing, surfactant characteristic of quats may cause emulsion and foaming and interfere oil and water separation [119,147].

10.3.1.4 2,2-Dibromo-3-nitrilopropionamide

DBNPA is a bromine-based, fast-acting, nonoxidizing biocide at acidic, neutral, or alkaline pH, but interestingly, it works very similarly as many halogen-based oxidizing biocides. It displays a quick kill efficacy, by rapidly breaking down into several by-products, including ammonia, bromide ions, dibromoacetonitrile, and dibromoacetic acid when in an aqueous environment. This property allows the treated effluent to be safely discharged even in sensitive water bodies. DBNPA is not considered an oxidizing biocide, despite small release of bromine component which can oxidize bacterial substrates and enhance its biocidal activity [141]. It is very effective at low concentrations in controlling the growth of aerobic and anaerobic bacteria, fungi, and algae. The vast majority of microorganisms are killed within 5–10 minutes of contact time with DBNPA. However, the kill efficacy is significantly reduced by the presence of ammonia and any sulfur molecules (e.g., H_2S) in the system.

DBNPA is widely used for disinfection and cleaning of sulfate removing nanofiltration (NF) membrane and reverse osmosis (RO) membrane [148]. For batch treatment of NF/RO membrane, it is generally recommended to use 10–30 ppm of active ingredients for 30 minutes to 3 hours every 5 days, depending on the severity of the biofouling of the system. In case there is reducing agent (e.g., sodium bisulfite) in the feed water, higher dosage of DBNPA is required to account for deactivation by the reducing agent. DBNPA is also used as a biocide in hydraulic fracturing wells and in cooling water.

DBNPA is decomposed through hydrolysis, photolysis (UV lights), and biological pathways with the degradation rate dependent on pH and temperature [149]. The half-life of DBNPA decreases rapidly with increasing pH and temperature. The approximate half-life of DBNPA in aqueous environments is 24 hours at pH 7, 2 hours at pH 8, 15 minutes at pH 9 [123,141,149]. Short half-life may be beneficial in the application in cooling water system when required to meet strict discharge regulations. Due to its short half-life, DBNPA is not appropriate for storage tank, large pipeline network with long residence time or the systems with presence of sulfides or bisulfite-based oxygen scavengers [117,119].

10.3.1.5 Isothiazolones

Isothiazolone compounds are broad-spectrum, nonoxidizing biocides used in a variety of industrial water treatment applications for control of growth, metabolism and biofilm formation by bacteria, fungi, algae. The compounds disrupt the cell's metabolic pathways involving dehydrogenase enzymes, inhibiting critical physiological functions including growth, respiration, and ATP synthesis [150]. The death of the cells results from the destruction of protein thiols and the production of free radicals.

Compared to DBNPA, isothiazolone has a much less rapid kill rate and is more effective in waters with high organic content. They have been used for microbial control and preventing biofouling in industrial water treatment [151]. Isothiazolone compounds can be used individually or in combination with other biocides [118]. The most frequently used product is a mixture of 5-chloro-2-methyl-4-isothiazolin-3-one (CMIT) and 2-methyl-4-isothiazolin-3-one (MIT) in a ratio of 3:1 at a final concentration of 1.5% total active ingredients, stabilized with mono- or divalent cations at an acidic pH [152]. They are effective against both general aerobic and spore-forming bacteria, over a pH range of 6.5–9.0. Isothiazolines are very effective algaecides and fungicides, but only at acid to slightly alkaline pH levels [124].

For NF/RO membrane and cooling water system, the batch treatment is typically in the range of 50–120 ppm for 1.5% active isothiazoline, with a contact time of 4–6 hours [124]. Isothiazolone can also be used at low dose rates (10–20 ppm) on a continuous basis. For the best biocidal efficacy, the pH of the water should be in the range of 6–9. Isothiazolone compounds are long acting, and can be used for long term preservation (as long as 6 months) of the water system for controlling biological growth at the dosage of 500–1000 ppm. The biocidal efficacy of isothiazolones may be enhanced by various additives, including surfactants and quaternary amine [120], but can be deactivated in the environment containing reducing agents such as hydrogen sulfide.

Isothiazolones degrade in the aquatic environment rapidly with half-life of less than 24 hours. Metabolism involves the cleavage of the isothiazolone ring and subsequent oxidation to alkyl compounds and CO_2. Additionally, the isothiazolone ring is susceptible to nucleophilic attack by materials such as sulfide, bisulfite, mercaptans, organic thiols (cysteine), and amines, resulting in cleavage of the ring and the production of simple organic acids [153]. CMIT/MIT does not have any bioaccumulation potential. The degradation by-products are 4–5 orders of magnitude less toxic, and therefore, isothiazolone compounds are considered to have a good environmental profile.

10.3.2 Oxidizing biocides

Chlorine dioxide (ClO_2), hypochlorite, bromine chloride, and peracetic acid are among the common oxidizing or inorganic biocides used in oil and gas industry. They kill microbes by oxidizing organic molecules on the cell wall and cell membrane, resulting in the loss of normal enzyme activity necessary for respiration and cell metabolism [117]. This oxidation is generally nondiscriminatory and very quick, on the order of several minutes, which renders the quick kill property of oxidizing biocides. However, the kill rate and efficacy of oxidizing biocides depend on many system conditions such as pH and organic load within the system. The water pH affects the available molecules for biocidal action whether it is bromine, chlorine, or hypochlorous acid. The nondiscriminatory nature of oxidation limits the use of this group of biocides in oil and gas industry as organic molecules (e.g., crude oil) will consume the oxidizing biocides quickly, diminishing the biocide residuals for extended control of microbial activities [141]. As a result, oxidizing biocides are only used in the pure water system and only for short-term protection in oil and gas industry. Finally, oxidizing biocides are generally very corrosive to the metal structures, further limiting the use in oil and gas industry.

10.3.2.1 Chlorine dioxide

ClO_2 is a yellowish-green gas at room temperature that dissolves in water. It is approximately 10 times more soluble than chlorine and 50 times more soluble than ozone [154]. However, solubility decreases dramatically with an increase in temperature. It is a highly reactive oxidant. It can explode if the concentration in air exceeds the explosive threshold of 10% v/v. Because of safety concerns, it must be generated at the application site using a ClO_2 generator with the methods bypassing the gas phase. For the same reason, it is almost always handled as a dissolved gas in water in a concentration range of 0.5–10 g/L.

ClO_2 offers some unique advantages over chlorine gas because of its selectivity, effectiveness over a wide pH range (pH 4–10), and speed of kill [120]. Compared to chlorine, ClO_2 is superior when operating above pH 7 [155], in the presence of ammonia, amines, silica and phosphates, and for the control of biofilms in water distribution systems. It is less corrosive than chlorine [156]. It is widely used as a bleaching agent in the paper industry and for treatment of drinking water. In oil industry, due to its safety concerns, use of ClO_2 is generally limited to hydraulic fracturing applications and the front end of a water treatment plant, for example, before source water clarification or sand filters [119]. However, it is worth noting that ClO_2, as with most oxidizing biocides, is consumed almost instantaneously, leaving no residuals for extended microbial control.

10.3.2.2 Hypochlorites

Hypochlorites are the most widely used chlorine disinfectants and are available in liquid (e.g., sodium hypochlorite) or solid (e.g., calcium hypochlorite) forms. They have a broad-spectrum antimicrobial activity, do not leave toxic residues, are unaffected by water hardness, inexpensive and fast acting, and remove biofilms from surfaces [127]. However, hypochlorites are quite unstable with both the dry and the liquid forms losing their chlorine strength over short periods, often within a few days. The rate of decay of dry hypochlorite salt or hypochlorite solution increases when exposed to the atmosphere, sunlight, higher temperatures, acidic chemicals, or in contact with copper or nickel which catalyze its decomposition [157].

It is widely used as a bleaching agent and as an antimicrobial agent in water treatment. Hypochlorite is a stronger oxidizer than ClO_2, and therefore is consumed even faster after applied for water treatment. In large water treatment plant, hypochlorite is usually generated on-site by electrolysis of brine solution (sodium chloride) [156]. This eliminates transportation, storage and handling problems associated with chlorine gas. Hypochlorite is the least expensive oxidizing biocides used in oilfield applications [119].

10.3.2.3 Bromine chloride

Bromine chloride is a very reactive oxidizing agent, and exists in a liquid form at room temperature. It has been used as a biocide, specifically as an algaecide, fungicide, and disinfectant, in industrial recirculating cooling water systems. It has almost no loss in biocidal activity in the presence of ammonia. However, it is very corrosive, has high vapor pressures and fuming tendencies, and presents handling and delivery challenges that limit its widespread commercial use. Activated sodium bromide is an effective way to delivering bromine chemistry to a water system, but the method requires two separate feed systems and proper metering and mixing equipment to achieve effective results [158,159]. In addition, activated sodium bromide, generated from the reaction of bromide with an oxidant such as bleach, is unstable and must be used immediately [159].

To overcome the issues of stability and delivery of bromine chemistry, a formulation of stabilized bromine chloride has been developed [158,160]. The concentrated formulation provides biocidal compositions that maintain these high levels of active bromine for at least a two-month period, and that do not exhibit a visible or offensive vapor or odor during this period. This water-soluble bromine-based biocide is non-acidic and noncorrosive, and can be introduced to a water system using a single feed.

10.3.2.4 Peracetic acid

Peracetic acid formulation is equilibrium mixtures containing peracetic acid, hydrogen peroxide, acetic acid, water and stabilizer [119]. The biocidal activity of peracetic acid is due to the oxidation of sulfhydryl groups, disulfide bonds, and double bonds in proteins, lipids and other cellular constituents which disrupts the chemiosmotic and transport functions of the cell membrane [161].

Peracetic acid is characterized by a very rapid action against all microorganisms at low concentration (less than 5 minutes at 100 ppm) [127]. A special advantage of peracetic acid is that its degradation by-products (i.e., acetic acid, water, oxygen) are nontoxic and can easily dissolve in water. The biocidal activity of peracetic acid is not affected by pH or water hardness, and retained even in the presence of organic matter, though higher concentration is required [161].

Peracetic acid is stronger oxidizing biocide than ClO_2, and it oxidizes a broader range of targets than ClO_2. In oil and gas industry, peracetic acid has been successfully utilized in hydraulic fracturing applications to disinfect the fracking fluids in the storage tanks and ponds before or between fracking [162].

10.3.3 Preservatives

A few of the organic or nonoxidizing biocides can act as preservatives. Contrary to other commonly used biocides such as formaldehyde, glutaraldehyde, DBNPA, THPS, and Quats, which display the kill efficacy from 30 minutes to 6 hours, the preservatives typically act very slowly but provide long-term microbial control [117]. The efficacy of preservatives is measured in days or even weeks [163]. Most of the preservatives used in oil industry are formaldehyde releasers, meaning that the antimicrobial efficacy of preservatives come from the released formaldehyde through a slow hydrolysis process. Therefore, the efficacy of preservatives depends on the rates of hydrolysis to form formaldehyde, which can be significantly impacted by pH and other water chemistries [117,119]. Examples of common preservatives used in oilfields include 4,4-dimethyloxazolidine (DMO), 3,3′-methylenebis-(5-methyloxazolidine), tris (hydroxymethyl)nitromethane (THNM), *cis*-1-(3-chloroallyl)-3,5,7-triaza-1-azoniaadamantane chloride (CTAC), and dazomet. Campbell and Johnson [164] compared a single treatment of glutaraldehyde with a single treatment of a preservative chemistry, dazomet, and found that glutaraldehyde had a much more rapid reduction of viable SRBs, but dazomet was able to maintain undetectable viable SRB after 28 days.

The preservatives are often used in combination with other organic or inorganic biocide to provide quick kill of microorganisms and long-term preservation of the system [165]. For example, in reservoir souring mitigation and drilling operations, a combination of usual organic biocide and preservative biocide will provide a quick kill of microbes in the top-side facilities (e.g., injection water pipeline, drill mud storage tank) as well as lasting biocidal or biostatic effect on microbial activities for injection water and drilling mud within the reservoir [164]. Yin et al. [165] compared the performance of glutaraldehyde and preservative biocides CTAC, DMO and THNM in souring mitigation using thermophilic (*Thermotoga petrophila*) and mesophilic (*Desulfovibrio longus*) SRB under elevated (75°C) and normal (35°C) temperatures. It was found that glutaraldehyde was very effective against both mesophilic and thermophilic SRB; however, its efficacy persisted for shorter periods at 75°C compared to 35°C. In contrast, preservative CTAC, DMO and THNM showed improved performance at 75°C versus 35°C, and their efficacy persisted longer than glutaraldehyde against both thermophilic and mesophilic SRB.

10.4 Biocide selection and performance evaluation

10.4.1 Considerations for biocide selection and performance evaluation

Biocides are widely used in oil and gas industry to mitigate microbial problems in various facilities and pipeline systems. This starts with the selection of the appropriate mitigation chemistry whether it be a biocide or preservative. When selecting the biocides, it is important to understand the conditions and microbial issues in the target systems as well as the characteristics of the biocides. The following is a list of the basic but sometimes competing requirements for biocide selection [118,166]

- Broad-spectrum antimicrobial properties,
- Fast-acting and active at low concentration
- Long duration of action,
- Effective against both planktonic and sessile bacteria
- Compatible with water chemistry and system conditions,
- Compatible with other oilfield chemicals,
- Noncorrosive, nontoxic or low toxicity, or biodegradable
- Nonvolatile/flammable
- No resistance buildup of microorganisms,
- Long shelf/storage life in adverse conditions, and
- Convenience of transportation and application.

Above all these requirements, the biocide treatment program must be inexpensive. After biocide chemistry is selected based on the system conditions and microbial problems, biocide kill efficacy has to be evaluated both in laboratory and field conditions in order to optimize the treatment program.

In order to evaluate biocide efficacy and determine the optimal treatment program, a few rules have to be observed [117]. First, the selected chemistries must be tested in a system that mimics the real-world system. This ensures that the biocide chemistry is compatible with the system and waters within the system. For instance, a flow system should be used to test biocide efficacy for the applications in pipelines, process trains, and water injection system. A bioreactor testing with simulated residence time is required for the applications in storage tank and hydrotest waters. More importantly, irrespective of the method used, the test must be against the biofilms on metal coupons used in the testing [119]. The coupons have to be exposed to the conditions mimicking the actual system conditions, such as bacteria populations isolated from the actual system, presence or absence of oxygen, geochemistry of the water, and temperature. In addition, the coupons have to be exposed to the simulated conditions for enough time (usually 1–2 months for oilfield microorganisms) allowing the establishment of mature biofilms on the coupon surface for biocide evaluation. PS Biofouling Rig and CDC biofilm reactors have been used to grow biofilm on coupons for biocide kill efficacy tests [167,168].

Secondly, it has to be realized that there is no one biocide that will work in all oil and gas systems. Different biocide chemistries have pros and cons for antimicrobial treatments in a particular system. Advantages of different chemistries should be leveraged for improved microbial control and cost-effectiveness. For example, oxidizing biocides can be used in the upstream system of hydrofracturing operation due to the presence of oxygen and short residence time of water; however, nonoxidizing biocides and/or preservatives are more appropriate for downstream treatment to reservoir for prolonged and full microbial control due to the presence of significant amount of organic matters [117].

Thirdly, the biocide efficacy test should be designed to target the problem of interest directly. For reservoir souring problem, the test can be designed to include SRP, APB, and sulfide production. For MIC issue, in addition to the counts of various corrosion-related microbes such as SRP, APB and methanogens, pitting corrosion is a good indicator for biocide efficacy evaluation.

Fourthly, the notion that bacteria are more vulnerable to chemical treatments when they are growing or recovering from injury should be explored. For instance, a biocide treatment longer than weekly will have no significant impact on microbial activity or viable bacterial counts in the system [117], considering that the doubling times of most bacteria is on the order of hours.

Last but not least, the biocide performance determined in the laboratory evaluation needs to be validated and optimized in the actual field system of interest. It has to be emphasized that the biocide dosage determined in the laboratory kill test is the minimum product dosage required for microbial control. It is important to ensure this dosage/residual is still present at the end of the system of interest to provide the protection of whole system. Depending on the biocide chemistry, process and system demand, it often means that higher dosage is required at the injection point of the system. For that, residual monitoring becomes critical to confirm the amount of biocide remaining at the end of the system after a treatment is performed [119].

In a standard batch treatment program, treatment frequency, duration, and dosage are the basic components of the biocide treatment program. All these have to be evaluated in the properly designed laboratory experiments, and then validated in the actual field conditions with appropriate key performance indicators (KPI). For treatment frequency, the field operators tend to request less frequent treatment such as monthly, quarterly, or even yearly, due to some operational limitations, logistics being the most significant limitation. This is especially true for some systems with no build-in injection skid, and the operator has to bring in the biocides to the site for injection, which presents a huge logistic challenge to the operation. It is microbiologist's responsibility to educate and convince the field operators that such infrequent treatment will never work to mitigate any microbial problems in the systems. In practice, weekly treatment is the most common practice in the oil and gas industry, but the optimal treatment frequency depends on the performance of biocides selected for the system and system conditions which impact the bacteria populations and the rate of bacteria regrowth after the treatment. As a result, the biocide treatment frequency can vary in the range of twice a week to once every two weeks [119]. The simplest approach to optimize the treatment frequency is through the sessile bacteria kill test to determine the viable sessile bacteria counts after each treatment, and ensure the counts are under the pre-prescribed threshold (a KPI). Other KPIs for biocide performance can be corrosion rate, biogenic H_2S production, etc., depending on the microbial problems that the biocide treatment program is intended to address.

Treatment duration (contact time) and dosage are often interdependent. Higher dosage of biocide treatment generally requires shorter treatment duration to achieve the same kill efficiency. However, the correlation between biocide dosage and the optimal contact time is system and chemistry specific [119]. For instance, on-the-fly treatments in the upstream system of hydrofracturing operations require a biocide which can kill bacteria with very short contact time and provide a short-term protection before the other treating chemicals are added and the fluids are pumped downhole [120]. Nevertheless, the optimal duration and dosage can only be determined via laboratory evaluation, and then validated in the field conditions. Considering the time needed for biocide to penetrate and kill the bacteria in the biofilms or solid deposits, as a rule of thumb, longer treatment duration is preferred, if the logistics permit. A duration of 1–4 hours of biocide treatment is a common practice in oil and gas industry.

Bacteria colonize metal surfaces and secrete EPS, forming a slime layer or biofilm on the metal surface. For biocide dosage, it should be determined based on the performance against sessile bacteria as opposed to planktonic bacteria. The detrimental effect of microbial activities, such as corrosion and biofouling, is more often caused by the microbes in the biofilms. The environment conditions at biofilm/surface interfaces are often radically different from the bulk medium in terms of pH, dissolved oxygen, and other organic and inorganic species. It has been reported that 500–5000 times of biocides are required to kill sessile bacteria than those needed to kill planktonic cells of the same species [169]. Much of this increased resistance of biofilm bacteria to antimicrobial agents may be due to the structure of the biofilm. Poor penetration into, and diffusion limitation of the biocide chemicals within, the biofilm results in significantly less concentration of active biocide ingredients at the base of the biofilm [117,170]. Another mechanism might be the physiological properties associated with low growth rates in biofilm populations. Slowly growing and nongrowing bacteria were found to be more resistant to the antimicrobial agents [171,172]. Similar to treatment duration, various KPIs, depending on the microbial problems, should be evaluated to determine the optimal biocide treatment dosage. The selected biocide dosage should be able to control sessile microbial populations and minimize the risks caused by microbial activities, for example, corrosion, biogenic H_2S production, and biofouling [119].

10.4.2 Methods for biocide evaluation

Biocide efficacy is usually evaluated using kill tests, by comparing the counts of viable microorganisms before and after exposure to biocide treatment. It is always preferred that the kill tests are conducted against the sessile bacteria, however, planktonic bacteria kill test may be used as a prescreening step if there is a large number of biocides to be evaluated. The treatment, at minimal, consists of several biocide concentrations and contact times. In some cases, treatment temperature and frequency are the part of the test matrix. The microorganisms used in the kill tests should be isolated or enriched from the system of interest under the conditions mimicking the system's water chemistry and temperature. Depending on the microbial problems in the system of interest, the enrichment is carried out to enrich different groups of microbes for the kill tests. For practical reason of bacteria cultivation in the laboratory, general heterotrophic bacteria (GHB), SRB, and APB are among the most common bacteria consortium enriched and utilized in biocide kill tests.

There is a legitimate argument in the community of petroleum microbiologists that not all GHB, SRB, or APB will respond similarly to chemical treatments, and therefore, it is plausibly better if the biocide efficacy test is conducted against a specific species or even a specific strain of microorganism which is dominant in the system and causes the microbial problem of the interest in the system. However, such approach has not been implemented and probably will not be implemented in the foreseeable future in oil and gas industry. There are reasons for that. Modern molecular microbiology tools such as high-throughput DNA sequencing technologies, can easily determine the identities of major microorganisms in the system at the level of species or strains; however, establishment of absolute relationship between a specific species or strain of microorganism and its abundance with the microbial problems in the system is almost impossible. Targeting a specific species or strain, or even a specific group of microorganisms alone in the biocide efficacy study will most likely miss other important species, strain, or group of microorganisms, whose roles in the microbial problems are not fully understood or clearly defined in the system of interest. The most devastating microbial problems take place in the presence of microbial consortia in which many physiological types of bacteria, including SRB, APB, metal oxidizing or reducing bacteria, interact in a complex way within the structure of biofilms. As such, pure microbial strain is rarely used in any biocide efficacy tests.

There are several methods of choice for enumerating viable microorganisms before and after exposure to biocide treatment. They are discussed in the following sections.

10.4.2.1 Most probable number test

Microorganisms in oil and gas industry, including GHB, SRB, APB, NRB, and NR-SOB, were most frequently assessed with cultivation-dependent method such as the MPN technique. MPN test is semi-quantitative serial dilution method for enumerating viable microorganisms using specific culture growth media. It entails the presence of live microorganisms and requires as long as 28-day incubation period (for SRB). The technique represents the population density of only viable microorganisms that are able to grow under the physiochemical conditions that may be significantly different from the system of interest. As matter of fact, any cultivation-dependent methods, including MPN tests, can never mirror in situ conditions and the particular needs of different physiological groups of microorganisms (e.g., temperature, nutrients, syntrophic interactions). For instance, cultivation often favored fast-growing, moderate SRB, for example, Desulfovibrio spp., that were not necessarily the numerically most abundant or in situ active SRB [173]. Generally, MPN tests can only detect a small fraction (0.1%–10%) of microbial populations from environmental samples [25,174,175]. Nevertheless, MPN tests are still widely used in the field laboratories in oil and gas industries for routine monitoring of bacteria counts and evaluation of biocide treatment efficacy.

Before using MPN method for biocide kill tests, it is mandatory to first isolate or enrich bacterial consortia from the system of interest. Enrichment can start with water sample or solid deposits sample from the system, the latter being the preferred choice as the bacteria community in the solids are more relevant to the microbial problems in the system. Another consideration for the enrichment is to ensure the enrichment growth media are prepared using system water, or the synthetic water based on the geochemistry of the system water. If commercial growth media are used, the salinity of the growth media should roughly match the salinity of the system water. The incubation temperature of the enrichment culture should be similar to the system temperature to minimize the unwanted selection or redistribution of the microbial communities. However, it is important to note that the bacterial growth media that are intended to selectively support the growth of a particular type of bacteria are not completely selective, and the vast majority of microbial species cannot currently be grown in the laboratory [176,177]. As a result, the enrichment culture from a specific system is in no way the accurate representation of the microbial population in the original system; nevertheless, it is always strongly recommended for any biocide efficacy tests.

Many bacteria growth media are commercially available or can be made in the laboratory according to various recipes described in the literature [178]. It has to be noted that some microorganisms are very difficult to grow in laboratory, and the indicators for active growth are not easy to identify. Therefore, these groups of microbes, for instance, methanogens, SRA, and iron-related bacteria, are generally not used in biocide efficacy tests, even though these microorganisms are known to cause many problems in oil and gas industry. In the practice of biocide efficacy tests using MPN method, GHB, APB, SRB are the most commonly enriched consortia from the target system, and the cultures are then used for the planktonic and sessile kill tests. These groups of bacteria are easy to cultivate and have distinct indicators for positive growth in MPN vials, which are turbidity, color change (or pH change), and black iron sulfide precipitates, respectively.

10.4.2.2 Rapid sulfate-reducing bacteria detection pouch

SRB are a major concern in the oil and gas industry with significant economic and safety implication. They are known to be the major contributors to MIC in oil and gas production systems and associated with reservoir souring. Early detection and monitoring of SRB activities are critical for the field operations and the level of contamination by SRB species in the system is often used as one of the KPIs in system integrity management program.

SRB are widely used as the major target organisms in biocide performance evaluation and monitoring. The surviving SRB after biocide exposure is determined using MPN tests as a key parameter to evaluate the biocide kill efficacy. However, among many inherent drawbacks of cultivation-based methods [33,174,179], MPN tests for SRB take four weeks to deliver the results [143]. A thin-film culture device (Rapid SRB Detection Pouch) became commercially available recently for the detection and enumeration of SRB. The growth-based device is shelf-stable and self-contained, and can provide results of SRB counts within 1–3 days, a significant improvement over the conventional MPN method [180].

The device was compared to the conventional MPN method using two commercial Modified Postgate B media with matching salinity to the samples for enumeration of SRB in various samples collected from a Middle East oil company. The samples used for comparison include Wasia water, injection water, produced water from crude oil processing facility, formation water from crude oil production wells, and scraping sludge from crude pipeline. The device was inoculated with 3 mL of water sample, and incubated at system temperature. The color development in the device can be observed as early as a few hours incubation. A color comparator chart is used to provide the full range of quantitation from 10^7 to 1 SRB/mL based on the culture color of the device and the incubation hours with observed color change. The thin-film device showed a strong correlation to MPN method in enumerating SRB in a wide variety of oil industry samples. The thin-film device and MPN serial dilution method provided the same results or the results within one order of magnitude (Tables 10.1 and 10.2).

Table 10.1 Comparison of thin-film culture device and MPN method using commercial SRB media 1.

				Thin-film device		MPN test		
Origin	**Description**	**TDS (g/L)**	**Incubation temp (°C)**	**Incubation (h)**	**Results (#mL)**	**Media salinity (%)**	**Incubation (days)**	**MPN/ml**
Oilfield #1	Wasia water 1	57	30	72	<1	5.7	28	<1
	Wasia water 2	57	30	72	<1	5.7	28	<1
	Injection water 1	57	30	38	10^3	5.7	28	10^3
	Injection water 2	57	30	40	10^2	5.7	28	10^2
Crude pipeline	Pigging sample 1	—	35	24	10^4	2	28	1.5×10^4
	Pigging sample 2	—	35	24	10^5	2	28	4.3×10^5

Table 10.2 Comparison of thin-film culture device and MPN method using commercial SRB media 2.

				Thin-film device		MPN test		
Origin	**Description**	**TDS (g/L)**	**Incubation temp (°C)**	**Incubation time (h)**	**Results (#mL)**	**Media salinity (%)**	**Incubation (days)**	**MPN/ml**
Oilfield #2	Aquifer water	19	35	48	10^2	1	28	10^3
	Produced water after HPPT	19	35	48	10^2	1	28	10^2
	Produced water after WOSEP	19	35	72	<1	1	28	<1
Oilfield #3	Formation water at wellhead	165	35	72	<1	1	28	<1
	Formation water before HPPT	11	35	72	<1	1	28	<1
	Produced water after WOSEP	11	35	72	<1	1	28	10

The Rapid SRB Detection Pouch was also applied to evaluate the biocide kill efficacy on planktonic SRB. The SRB culture was enriched from a sea water injection system in Middle East region. The enrichment culture was exposed to THPS- and glutaraldehyde-based biocide products with contact time of 1, 2, and 4 hours, and at the concentration of 50, 100, and 150 ppm of biocides. The SRB culture was then inoculated into the pouch and incubated at 35°C for 48 hours. The growth and enumeration of surviving SRB in the pouch were illustrated in Figs. 10.2 and 10.3. Compared to the SRB control (10^5/mL), 2-log SRB reduction was observed when the culture was treated with THPS-based biocide for 1 hour. at the concentration of 50–150 ppm, or for 2 hours at the concentration of

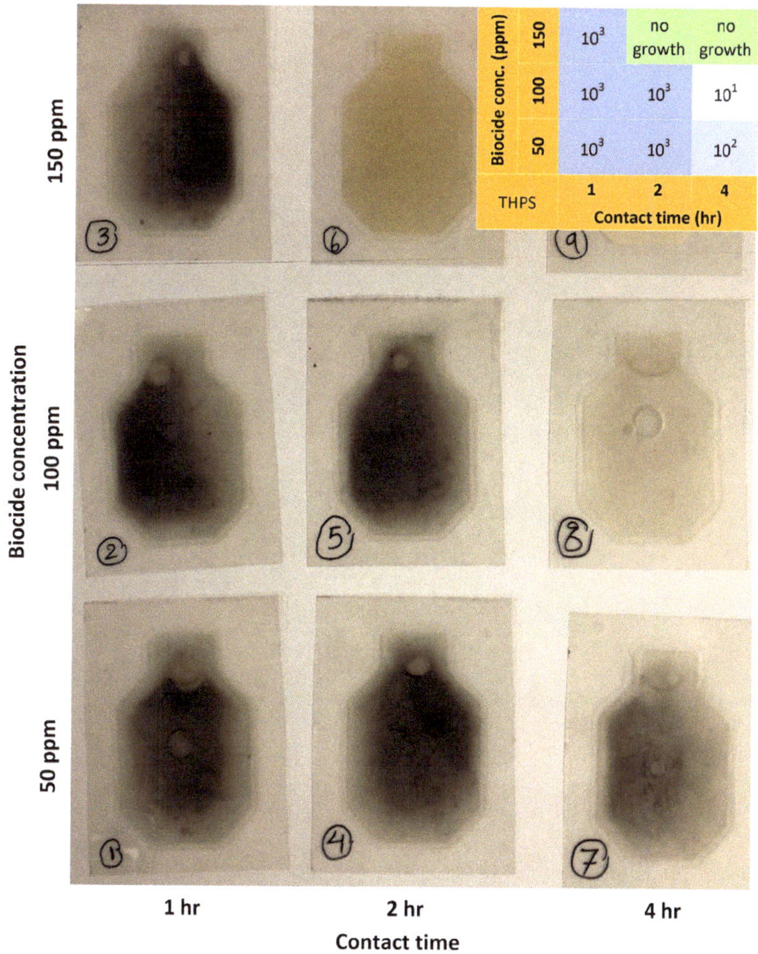

FIGURE 10.2

SRB enumeration after exposure to a THPS-based biocide product.

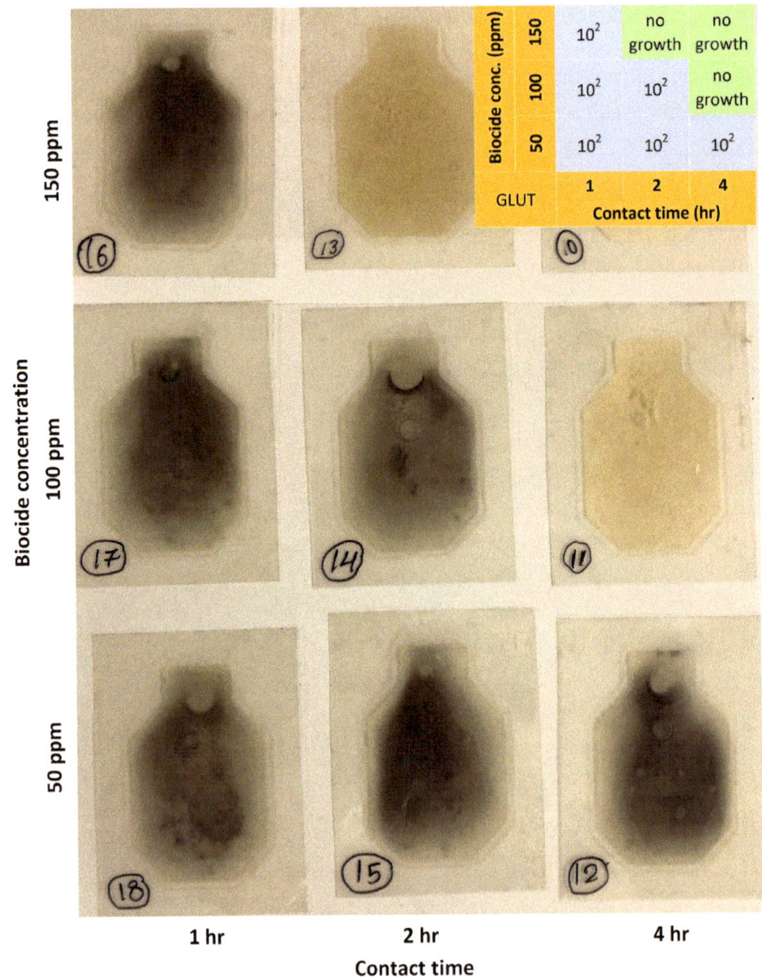

FIGURE 10.3

SRB enumeration after exposure to a glutaraldehyde-based biocide product.

50–100 ppm. If treated with THPS-based biocide for 4 hours, 3-log reduction was observed at 50 ppm, and 4-log reduction at 100 ppm. SRB growth was negative at 150 ppm with contact time 2 hours or more (Fig. 10.2). When treated with glutaraldehyde-based biocide, 3-log SRB reduction was achieved at 50 ppm with contact time 1–4 hours, 100 ppm 1–2 hours, or 150 ppm 1 hour (Fig. 10.3). At 4-hour contact time, SRB growth was negative when the biocide concentration was above 100 ppm. Similarly, at 150 ppm of glutaraldehyde-based biocide, the SRB did not grow when the contact time was 2 hours. or more.

There are two major advantages of the SRB Detection Pouch over the conventional MPN methods. One is that there is no growth media preparation required and therefore the salinity match to the sample is not a concern. The other is the quick turn-around-time (a few hours to up to three days of incubation) to obtain the enumeration results of SRB, which render it applicable for on-site routine monitoring of SRB counts or activities by field laboratory technicians. This device can also be used for evaluation of biocide kill efficacy against planktonic or sessile SRB and monitoring biocide effectiveness in the field operations.

10.4.2.3 ATP bioluminescence test

Adenosine-5′-triphosphate (ATP) is the primary energy carrier for all living organisms, and therefore, the presence of ATP is an excellent indicator of total activities from metabolically active organisms [181,182]. ATP bioluminescence test is a rapid testing technique for quantification of living cells based on the measurement of ATP found in the sample. ATP test has been widely used in food and pharmaceutical industry to monitor microbial contamination.

ATP bioluminescence is based on a chemical reaction catalyzed by luciferase enzyme (Eq. 10.2). In the reaction, luciferase catalyzes the ATP-mediated oxidation of luciferin, and the emitted light is then measured as relative light units (RLU) [183,184]. The amount of light (bioluminescence) produced is directly proportional to the amount of ATP present in the sample, which is proportional to the metabolically active organisms.

$$\text{ATP} + \text{O}_2 + \text{luciferin} \xrightarrow{\text{Mg}^{2+}\text{\& luciferase}} \text{AMP} + \text{PPi} + \text{oxyluciferin} + \text{light} \tag{10.2}$$

In the oil industry, ATP tests (first generation) have been used to detect microbial contamination in the refined fuel products and fuel associated water [185,186]. It also has been used for on-site monitoring of biocide effectiveness [187,188]. But its application in the oil and gas industry has been hindered by many interference substances present in typical oil industry samples, for example, dissolved and suspended solids, hydrocarbon contamination, and residual biocides.

The second generation ATP tests were introduced early in the 21st century and provided several advancements including enhanced resistance to interferences and high sensitivity, while maintaining the same fast response time [189]. A new line of second generation ATP test kits has been developed to overcome these interferences in different types of industry samples, for instance, low-solids water-based samples (drinking, cooling and process waters), low-solids organic-based samples (fuel, metalworking fluids, oily brine, and oilfield waters), solids samples (biofilms, corrosion products), and wastewater and bioprocessing samples [190,191]. Cellular ATP (cATP), dissolved ATP (dATP), or total ATP (tATP) are released from the water or solid samples after the interfering substance is removed, and measured with a sensitive luminometer. The result is expressed as RLU or converted to picogram of ATP, or equivalent microbial count, as a measurement of total microbial activities or counts in the sample.

The second generation ATP bioluminescence test has been widely adopted by service providers of oil and gas industry as a fast, efficient screening tool for microbial monitoring and assessment in various operations, fuels, and water systems [192,193], as well as for biocide kill efficacy tests [191,194,195].

The second generation ATP tests have been used for performance evaluation of biocide products provided from eight chemical vendors. The biocides were intended for the application in the Water Supply Pipeline of an oilfield development project in the Middle East Region. The operation temperature of the Water Supply Pipeline is 16°C–70°C. The source water is from a Wasia formation with very high salinity (TDS 100–165 g/L). The GHB consortium was enriched from the Wasia water at 40°C and 60°C, and the active cultures were exposed to various biocide concentrations for various contact times. To illustrate the application of ATP tests in biocide performance evaluation, a small data set was provided in the Figs. 10.4 and 10.5. The results clearly showed the reduction of GHB activity, as expressed in the normalized ATP% or ATP concentration (pg/mL), after exposures to 50 and 150 ppm of biocide (glutaraldehyde-based) for 1.5 hours.

The major drawback of ATP tests in the application of biocide kill efficacy test is its non-specific nature of ATP, and therefore, the ATP measurement is for the overall microbial activities rather than the specific microorganism of the concern. When SRB or APB consortia enriched from the system of interest are used in kill efficacy tests, the surviving bacteria activity after biocide treatment determined using ATP tests includes surviving SRB or APB activity as well as activity of other bacteria which survived in biocide treatment.

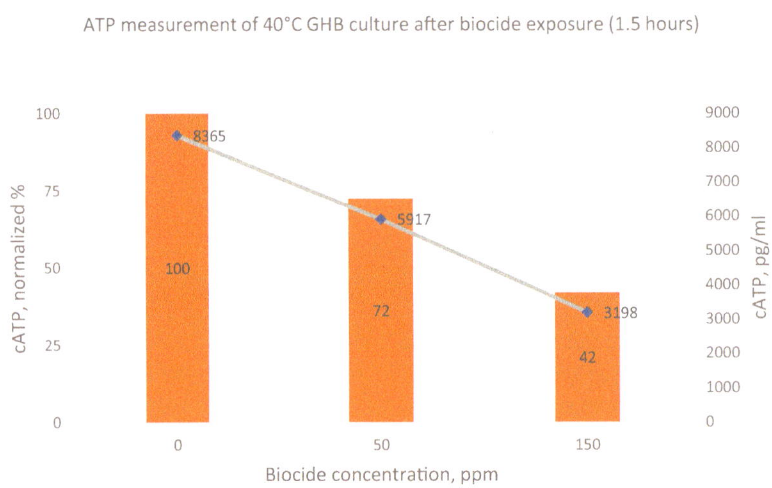

FIGURE 10.4

cATP measurement of 40°C GHB culture after biocide exposure.

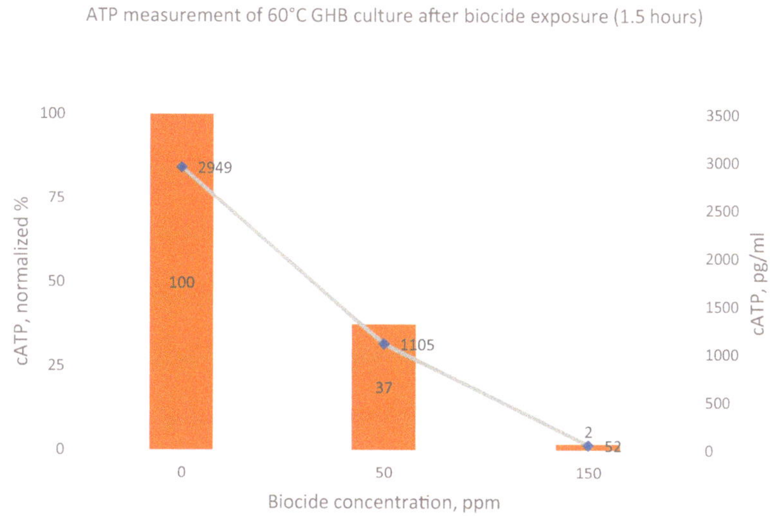

FIGURE 10.5

cATP measurement of 60°C GHB culture after biocide exposure.

10.4.2.4 Quantitative polymerase chain reaction and reverse transcriptase quantitative polymerase chain reaction

To circumvent problems associated with cultivation-based methods, many culture-independent molecular microbiology methods have been developed in the past decades for quick detection and quantification of various types of microorganisms of concerns. The most widely used molecular method in oil and gas industry are real-time qPCR method [25–27,194]. qPCR can be designed to detect and quantify the total bacteria and archaea in the environmental samples by targeting 16S rRNA genes, or a subgroup of bacteria and archaea populations, such as SRB, APB, denitrifiers, and methanogens by targeting the function genes *dsrAB*, *ackA*, *nirS*, and *mcrA*, respectively [26,27,89,196,197]. Targeting specific functional genes of microorganisms will give an indication of the presence of specific groups of bacteria, and correlate the metabolic functions (e.g., sulfate reduction, denitrification, and methanogenesis) with specific system conditions. qPCR offers a dynamic detection range of six orders of magnitude or more [26] and has the capability of high-throughput analysis, obtaining results within hours instead of days or weeks in traditional growth methods.

In biocide efficacy tests, it is required to compare the counts of viable bacteria in the culture before and after biocide exposure. This makes the conventional qPCR technique not applicable for biocide performance evaluation as the qPCR analysis quantifies the total DNA isolated from both live and dead bacteria. In this regard, a variation of conventional qPCR, reverse transcriptase qPCR (RT-qPCR), is a viable option for detecting and quantifying the microbial activities before and after the biocide exposure. Unlike the conventional qPCR which detects DNA present in actively metabolizing cells as well as in inactive and dead cells, RT-qPCR detects messenger RNA (mRNA), which is only produced by actively metabolizing cells in order for a cell to manufacture a specific protein [198]. By measuring microbial activity in actively metabolizing cells before and after exposure to biocide, the biocide kill efficacy can be determined. Furthermore, unlike the MPN method which generally only enumerates GHB, SRB and APB in biocide efficacy tests, RT-qPCR method can evaluate the kill efficacy against more specific groups of microorganisms important to the microbial problems in the system, SRA and methanogens being good examples related to microbial corrosion.

RT-qPCR is not without downside in the applications in oil and gas industry, including biocide efficacy tests. mRNA is very unstable [199,200], and its half-life varies for different environments, microorganisms, and by function and nutritional status of the cell [200,201]. Therefore, the samples must be preserved within minutes after biocide treatment or collection from the field. While this may be possible when working in a laboratory with small size of test matrix, it is not feasible or practical for large test matrix, which is often the case in biocide evaluation study, and for the performance evaluation in the field conditions. Target gene expression and mRNA extraction protocols introduce significant bias in mRNA detection and quantification [202]; therefore, the results from RT-qPCR assays targeting different loci or involving different mRNA extraction protocols are sometimes not comparable [203]. Finally, RT-qPCR is a lab-based technique and requires well-trained laboratory staff to perform the tests and interpret the data. These drawbacks have hindered the adoption of RT-qPCR technology in oil and gas industry, as a routine microbial monitoring technique or in application of biocide efficacy tests.

10.4.2.5 Flow cytometry

Flow cytometry is a laser-based technology, with high-throughput capability for qualitatively and quantitatively examining bacteria populations, and their viability and functionality in various industrial bioprocesses [204]. With fluorescent staining of the cells, flow cytometry can be used to determine total bacteria counts and the counts of live and dead bacteria in a liquid sample based on the specific light scattering and fluorescent characteristics of each cell [205]. With oligonucleotide probes targeting 16S rRNA gene and labeled with fluorescent dyes, flow cytometry can also be used to detect and quantify the specific groups of microorganisms within complex microbial populations. Applications of flow cytometry in oil and gas industry include detecting and monitoring microbial activities in various water systems (seawater, produced water, utility water, cooling water) and storage tanks (crude oil, refined fuel products), and biocide performance screening and evaluation. Keasler and team [119,206] have studied the biocide kill efficacy at the same concentration but with different contact times using live/dead cell viability staining and flow cytometry. Compared to the cluster of viable cells in the untreated sample, approximately 7% cells were immediately killed after 15 minutes of biocide treatment. After 1-hour treatment, this number increased to 50%, and by 2 hours nearly 90%. An additional 2 hours of contact time (total 4 hours) only increased the kill by an additional 2%, suggesting that the biocide achieved its maximal performance by the 2-hour time point. In their study, the kill of bacteria is defined as that the cell is depolarized and/or membrane damaged.

There are great interest and potential in flow cytometry technology in bacteria counting and sorting, and in adapting the technology for online monitoring of microorganisms of interest. The latter is especially true after the compact and benchtop flow cytometer became commercially available in last several years. However, the broad adoption of this technology in oil and gas industry remains to be seen. Flow cytometer is highly sensitive and sophisticated lab-based instrument, which requires highly trained specialist to maintain and operate the instrument, and interpret the data. Optimization of operation and measurement parameters is quite challenging due to difficulties associated with the differentiation of background signals and the real signals from small size bacteria. In biocide efficacy tests, background signals can come from many sources, for instance, particles and contaminants in the growth media prepared from the system water, precipitates due to chemical reactions and metabolic activities of the bacteria, and auto-fluorescence from biocide products and contaminants from the system water. Even though the flow cytometry may not become the method of choice for biocide performance evaluation in the near future in oil and gas industry, it shows great potential to at least complement the currently more widely used methods (e.g., MPN test, ATP measurement) for biocide selection.

10.5 Biocide treatment practices

After the biocide performance is evaluated in the laboratory and the biocide finally selected, the next is to determine how to apply the biocide to the target system. In oil and gas industry, the biocide is applied in one of the three ways, batch treatment, continuous treatment, and pulse treatment.

Biocide batch treatment is most common in oil and gas industry for various reasons. It is more cost-effective than continuous treatment in large oilfield systems over a long period of time. If applied properly, it can achieve the same goal as the continuous treatment that is to bring the microbial activities or counts under control in the target systems. In batch treatment, the biocide (250–150 ppm) is injected into the target system such as water or crude pipelines at a predetermined frequency (weekly being the most common) and for a predetermined period of time (1–4 hours being the most common). In some oil companies, two biocides with different chemistries are used, alternating on weekly basis, for the purpose to avoid or delay the development of bacterial resistance or selection against the biocides. If practical, the batch treatment of biocide should be coordinated with the scraping/pigging operation. Mechanical removal or disruption of biofilms and solid deposits through pigging operation will enhance the biocide's ability to penetrate into and kill the microorganisms in the biofilms or deposits. Biocides are often added immediately following the pig for a period of 2–8 hours [119].

The continuous treatment is generally applied at low dosage of biocides (2–50 ppm) for cost saving while providing biostatic control of microbial activities. In some cases, a combination of batch treatment and continuous treatment can be applied. High dosage of one biocide is applied on weekly/bi-weekly/monthly basis to kill microbes and remove biofilms in the target system, followed by continuous application of a low dosage of another biocide, with aim to prevent attachment of new microbes to the surface or minimize the regrowth of biofilm between the treatments [207]. It should be noted that, at low dosage of biocides, especially nonoxidizing biocides, oilfield microorganisms are more likely to adapt and develop the resistance over time, or undergo a natural selection process resulting in more resistant microbial populations in the systems [208,209].

Pulse treatment of biocide can be taken as shock treatment on a more frequent application. High concentration of biocide (250–1500 ppm) is injected on the order of minutes and the biocide pulse is evenly spaced throughout 24 hours [210]. The idea of this strategy is to add enough biocide to exceed the threshold concentration needed to kill the microorganisms. Then the treatment can be cut off during the time it takes for the microbial population to begin to recover. Pulse treatment saves biocide cost compared to the continuous treatment, but can cause operation complication. Some compromises may be needed to avoid interference or incompatibility with other oilfield chemicals.

Once a biocide has been selected and the dosage, duration, and frequency have been determined, the final step is to determine where in the system the biocide should be applied. Proper selection of biocide treatment location will have significant impact on the product performance and effectiveness of microbial control in the system of interest. To start, a comprehensive microbial survey is required to understand the microbial problems and identify the high-risk locations in the facilities and pipeline system. In general, biocide treatment location should be at the upstream of the problem location. In an old oilfield facility or pipeline, it is not uncommon that the biocide treatment is not part of the original design of the facility and pipeline. Consequently, either a significant design change is required to install the injection skid at the identified location, or the operator has to bring in the biocides to the injection site for every injection.

Since microorganisms need water and nutrients to thrive, some operation systems and facilities are particularly prone to microbial colonization and growth. In crude oil pipelines, if the production rate cannot maintain the critical flow velocity in the pipeline, water and solids will drop out of the crude and accumulate at the low points along the pipelines, providing the excellent conditions for bacteria to grow. This is also true for the export-quality, stabilized crude, which meets the tariff of sediment and water content below 0.5%. We have found in many cases that the crude oil at the production head contains very low number of microorganisms, but after long-distance traveling in flow line and trunk line and arriving at the processing facilities, it shows high microbial counts, including corrosion-related microbes, in the crude phase as well as in the water phase. In crude oil pipeline network (flow line or trunk line), if biocide treatment is determined to be necessary for mitigation of microbial issues, the best biocide injection location is at the beginning of the concerned trunk line in coordination with scraping/pigging operation. The biocide has to be water soluble and sufficient retention time must be available for the product to move to the water phase. On the other hand, biocide squeeze treatment in the reservoir formation for microbial control is generally not recommended because biocide stability is problematic under reservoir conditions (e.g., high temperature, high pressure and high salinity), biocide performance is hard to evaluate, and the treatment effect is short-lived due to mixing with formation water and migration in the reservoir liquid flow. It is also not clear whether the biocide will traverse through an entire reservoir or it is consumed in the near wellbore area.

In the crude processing facility, the microbial problems are often most severe in two water handing systems. One is the downstream water handing system and pipe after water and oil separator (WOSEP), and the other is the downstream water handing system of the desalter after the crude is washed with wash water. These two water streams will either be re-circulated back to other processing unit or combined to a disposal line for water disposal or re-injection. Commingling of formation water and produced water in the crude processing facilities creates an environment which promotes bacterial colonization and growth in all parts of the facility and pipelines. In addition, attention needs to be paid to the dead legs and bypass lines in the crude processing facilities to avoid or minimize the failures caused by microbial activities. The ideal biocide injection locations seem to be downstream of WOSEP and desalter, or upstream of the disposal pipeline. Biocide injection at the locations immediately downstream of WOSEP and desalter provides microbial control for all downstream water pipelines, and if the water is re-circulated back to other processing facilities, it also provides the protection to a certain extent. Severe internal corrosion cases of the water disposal lines have been reported in many crude oil processing facilities, and one of the major contributors is the microbial activities. Biocide injection upstream of the disposal pipeline helps ensure the disposal pipeline integrity. If this water is re-injected to the oil reservoir for pressure maintenance, uncontrolled microbial activities in this pipeline will have serious negative consequences over time, such as formation damage and reservoir souring.

Water injection or water flooding is a common practice in Middle East region to maintain reservoir pressure and oil production. In general. underground water or seawater is processed and treated in the central treatment plant and transported through pipeline network to injection wells. Depending on the size of pipeline network, the water may travel for several days before reaching the injection site.

Long residency time and abundant nutrients (especially high sulfate content in seawater) promote bacterial activities, compromising pipeline integrity and deteriorating injection water quality, which can result in injectivity decline, wellbore corrosion, formation damage, and reservoir souring. To protect the transport network integrity and ensure high injection water quality, the nonoxidizing biocide is usually injected at the beginning of transport pipeline after the water is treated and de-aerated at the treatment plant.

Depending on the biocide chemistry and system demand, it is quite possible that a single injection at the beginning of a large pipeline network may not provide adequate biocide residual at the end of pipeline for effective microbial control. Therefore, it is critical to monitor the residual concentration throughout the system, especially in a large network and at the injection well. If needed, multiple injections along the pipeline may be considered.

10.6 Biocide residual monitoring

After biocide is injected into the system of interest, it is mandatory to monitor biocide residual and the microbial counts or activities throughout the system. It is required to determine if the biocide treatment has reached the intended locations and achieved the intended microbial control. This monitoring should be performed routinely, at least every time after biocide slug passes the monitoring location. One of the monitoring locations, for a pipeline, is at the end of line to ensure sufficient biocide residual is still available for effective microbial control at this last location. The monitoring data have to be reviewed periodically and serve as the basis for any change or optimization of biocide treatment program, including biocide chemistry, treatment frequency, dosage, duration, and injection locations.

Commercially available test kits are widely available to determine the concentration of the biocide actives such as THPS and glutaraldehyde. THPS test kit is based on iodometric titration method with detection range of 0–100 ppm, which may be extended by water sample dilution. The glutaraldehyde test kit is based on its reaction with 3-methylbenzothiazol-2-one hydrazone, followed by colorimetric comparison with color comparator, or spectrophotometric measurement at 610 nm. The kit generally has detection range 1–200 ppm. The sample containing more than 200 ppm of glutaraldehyde should be diluted. The resolution of the commercial glutaraldehyde test kit is sufficient for routine monitoring, especially if the spectrophotometry is used for the final color measurement. If more accurate measurement of glutaraldehyde is desired, the high-performance liquid chromatograph (HPLC) method with UV detector (365 nm) is the method of choice for glutaraldehyde detection and quantification. The method is based on the reaction between glutaraldehyde and 2,4-dinitrophenylhydrazine to form the derivatized product glutaraldehyde bis-2,4-dinitrophenylhydrazone, which is separated on a

C_{18} reversed phase column and then detected using UV-Vis detector. However, HPLC is a laboratory-based method, and not an appropriate tool for routine monitoring of glutaraldehyde in oil and gas industry system.

However, in a large water pipeline network, it is extremely challenging to collect water samples for biocide residual measurement at downstream locations of the pipeline network after the batch treatment of the biocide product at an upstream location. This is due to the difficulties to estimate the biocide travel time in the large pipeline network because of the complexity of pipeline network (diameters, branches, etc.) and daily operation changes and fluctuations (e.g., flow rate). As a result, biocide residual concentration cannot be easily monitored in downstream locations of a complex pipeline network, especially in the remote injection wells, using manual sampling and analysis with various test kits.

An online sensor which detects THPS and/or glutaraldehyde will provide a real time monitoring of residual concentration of these biocide products in downstream locations of a large pipeline network. Recently the detection concepts which are applicable for online monitoring of THPS and glutaraldehyde were validated and the prototype of the online sensor is under the development [211,212]. The novel THPS sensing method is based on absorption measurement after reaction of an oxidizing reagent and THPS. This simple and straightforward method can be developed into an online sensor for THPS-based biocide products. The method has been validated for pure THPS and THPS-containing biocide products, in fresh water and Arabian Gulf Seawater. For online glutaraldehyde sensing, the cross-linking property of glutaraldehyde is explored to quench the fluorescence signals of custom-made nanoparticles. The glutaraldehyde sensing was validated in fresh water and Arabian Gulf seawater containing pure glutaraldehyde and glutaraldehyde-based biocide products. Both sensing methods have also been validated in a laboratory mock-up system with flow cell and appropriate optical and electronic systems, with dynamic range 10–1000 ppm of THPS or glutaraldehyde. This dynamic range perfectly covers the normal biocide treatment concentration and residual concentration encountered in the (sea)water treatment facilities and distribution networks in oil and gas industry.

10.7 Microbial monitoring for treatment effectiveness

In addition to biocide residual monitoring, microbial counts or activities also have to be monitored. It is recommended that the samples are collected and analyzed after every biocide treatment from various locations of the water systems. The microbial data after biocide treatment are essential to detect and confirm any deficiency of biocide treatment program. It has to be noted that the analysis should be targeting sessile bacteria, not the planktonic bacteria, the count of which is not reliable immediately after the biocide treatment. It is not uncommon that the bacteria count in the system water increases after the treatment, due to disruption of the biofilms by biocides. Therefore, a metal coupon sample from the biofouling rig installed at the water system is a much better sample for microbial monitoring after biocide treatment. Furthermore, microbial trend (counts or activities) in the biocide treated system is much more important than a single data point, and the same monitoring technique(s) have to be used for data comparison.

There are many conventional and molecular microbiology methods for bacteria enumeration, but for biocide treatment effectiveness, only the methods which can enumerate live bacteria are recommended. Such methods and their pros and cons have been elaborated in this chapter (Section 10.4.2). The most widely used method in the field for routine bacteria enumeration after biocide treatment is the conventional serial dilution MPN method, even though it is only capable of detecting less than 10% of microbial population in the sample [174,175]. Two most common targets of the MPN tests for routine monitoring of treatment effectiveness in the field is GHB and SRB, which take one week and four weeks of incubation, respectively, before the enumeration results become available. The long delay of the SRB results makes it impossible for the field operators to take quick countermeasures to control SRB activities in the affected system, if required.

As quick turn-around-time is preferred in routine monitoring in the field, the second generation ATP bioluminescence test is a good choice of method for monitoring the live bacteria after biocide treatment (Figs. 10.4 and 10.5). There is commercially available ATP test kit designed for sessile bacteria, and the test only takes 10–15 minutes to obtain the results. It should be noted that the ATP test is for the overall microbial activities (i.e., surviving GHB activities, or converted to equivalent GHB counts) rather than the specific group of microorganisms (e.g., SRB). A newly available Rapid SRB Detection Pouch can be used for the quick enumeration of SRB. We have compared this pouch with conventional MPN test in a wide range of samples, including injection seawater, produced water, and aquifer/formation water. In most samples, the results from SRB Detection Pouch and MPN test match exactly, with a few samples showing the different results within one order of magnitude (Tables 10.1 and 10.2). The Rapid SRB Detection Pouch can also be used for monitoring biocide effectiveness in the field operations (Figs. 10.2 and 10.3). With SRB Detection Pouch, the SRB counts are determined within 1–3 days after sample inoculation and incubation, which is a significant improvement over the conventional MPN test, which requires 28 days of incubation to obtain SRB results.

Similar to ATP test, which determines the surviving GHB activities or counts, the standard spread or pour plating method is also capable of enumerating the survived GHB after biocide treatment with 24 hours of incubation, quicker than one week of incubation with conventional MPN test. This method is relatively easy to implement in a field laboratory with minimal training of the technicians. The downside is that the standard plating method is difficult to implement for SRB enumeration in a less equipped field laboratory, as it requires anaerobic conditions to prepare the growth media plate and incubate the plates.

Culture-independent methods are generally not the methods of choice for monitoring surviving bacteria after biocide treatment, if the methods cannot differentiate the live and dead bacteria. RT-qPCR, which measures microbial activity (mRNA) in actively metabolizing cells after exposure to biocide, is one of a few valid methods for this purpose. But the implementation of RT-qPCR in a field laboratory is not easy, and actually not recommended for several reasons. Except for the equipment cost and extensive training required for the technicians, the instability and short half-life of mRNA are the major roadblock for the implementation of this method in an oilfield laboratory as a routine monitoring technique after biocide treatment. Another culture-independent method, which can detect and quantify the live bacteria after biocide treatment, is flow cytometry. However, it requires highly trained specialist to operate the flow cytometer, optimize the instrument parameters, and interpret the data. It is not practical to implement flow cytometry in a small oilfield laboratory as a routine monitoring technique.

Finally, in addition to biocide residual concentration and bacteria counts in the treated system, other system KPIs may be monitored to evaluate the effectiveness of the biocide treatment program. The two most common KPIs that are measured in a biocide program are the corrosion rate and the amount of H_2S that is produced in the system. Corrosion rate can be determined by weight loss analysis on the coupons used for sessile bacteria monitoring after biocide treatment. Lead Acetate Test Strip is a simple means to detect a few parts per million of dissolved H_2S in water.

10.8 Alternative methods for microbial control

To combat the huge economic losses related to microbial activities, biocide products with broad spectrum of kill efficacy against microorganisms have been widely used in oil and gas industry. However, there are other alternative methods that have been applied, or have potential to be applied in oil and gas industry for microbial control. For instance, molybdate, nitrate, and nitrite have been used for the control of metabolic activity of SRB and subsequent inhibition of biogenic H_2S production in oil reservoir [213,214]. This section briefly discusses the nitrate treatment, bacteriophage application, and several physical processes for microbial control.

10.8.1 Nitrate

The injection of seawater to maintain reservoir pressure during oil production is a common practice in the oil industry. The huge amount of sulfate introduced via injection seawater, together with rich organics and their degradation products as electron donors present in hydrocarbon, facilitates the active growth of SRP (i.e., SRB and SRA) in the reservoir, pipelines, as well as in pipings and top-side installations [215,216]. The activity of SRP causes severe economic problems due to the reactivity and toxicity of the produced H_2S, one of the major problems being reservoir souring. Besides the use of broad-spectrum biocides, nitrate and nitrite treatment has been used in water injection system to control SRP activities, and thus the reservoir souring and corrosion [217,218].

The mechanisms of nitrate treatment for SRB activity and souring control can be summarized as following [42,217,219–221]:

- Competitive exclusion of SRP by promoting the growth of NRB;
- Increased redox potential due to the presence of intermediates of nitrate reduction (nitrous oxide and nitric oxide). The biogenic H_2S production is suppressed when a redox potential is above -100 mV;
- Change of energy metabolism of some SRB, using nitrate as an energy source instead of sulfate;
- NR-SOB use the nitrate or nitrite to re-oxidize H_2S, resulting in the H_2S removal;
- Nitrite toxicity to dissimilatory sulfite reductase, that catalyzes final enzymatic step of sulfate reduction by SRB.

As a result of nitrate or nitrite treatment, sulfate reduction is inhibited or switched to nitrate reduction, decreasing the biogenic sulfide production in a sour system. In addition, NR-SOB further reduce the amount of sulfide by using sulfide as an energy source. While there are numerous reports that have shown the success of nitrate treatments in remediating biogenic H_2S production, several potential negative impacts of this treatment have to be carefully recognized. With nitrate injection in the water flooding system, NRB are purposefully promoted and can grow to very high concentration in short time. The consequence of large number of NRB in the system is the biofouling of near wellbore area and the plugging of the reservoir formation rock. It has been reported that the continuous nitrate injection leads to increased injection pressure at the injection well due to NRB-based biofilm build-up, and the pressure is only reduced upon batch biocide treatment [119,219]. Another concern is the increased corrosion rate after nitrate or nitrite treatment. Nitrite is an oxidizing agent, and the high nitrite concentration in the treated system, either due to incomplete reduction of nitrate or inappropriate dosage of nitrate or nitrite, can lead to accelerated corrosion rate [216,222,223]. The increased corrosion risk can also be explained by the formation of polysulfide, inorganic sulfur and thiosulfate through simultaneous oxidation of sulfide by NR-SOB [222]. In addition, the activity of NRB at high concentration was reported to increase the microbial pitting corrosion [30,224,225]. Nemati et al. [222] also reported the elevated metal corrosion rate by denitrifying bacteria in the presence of nitrate 6 to 10 mM of nitrate.

The success of nitrate and nitrite treatment depends on treatment dosage and frequency, the concentration of nitrite in the treated system, and the concentration or activity of SRP and NRB in the system [29]. As such, the nitrate and nitrite concentration, sulfate and sulfide concentration, and the concentration or activity of SRP and NRB have to be closely monitored because the system conditions have profound impact on metabolism of the microorganisms. Molecular microbiology methods have been used in oil industry to study reservoir souring and efficacy of nitrate treatment [198,226]. For instance, RT-qPCR can be used determine the SRB activities by measuring mRNA for the *dsrAB* gene responsible for sulfate reduction in SRB. In a nitrate treatment trial for souring control in two high-temperature oil reservoirs in the North Sea, SRA activity determined by RT-qPCR analysis showed a reverse relationship with nitrate concentration, and was a good indicator to predict the H_2S production from the well [227].

The treatment dosage and frequency have to be optimized for a specific water system, and adjusted according to the monitoring data to maintain proper nitrite and NRB concentration, so that the SRP activity is suppressed while not causing fouling and corrosion [29]. With these potential risks and extensive monitoring requirements, the method of nitrate or nitrite treatment for SRP control should be used with caution in oil and gas industry.

10.8.2 Bacteriophage

Bacteriophages are viruses that infect a specific bacterial cell, replicate with the infected cell, and finally lyse (or kill) the host cell. Unlike the conventional oxidizing and nonoxidizing biocides, which provide broad-spectrum kill of almost all microorganisms via various modes of actions, bacteriophages only target the specific microorganisms of interest and leave any other microbes virtually untouched. If used as a means for microbial control, the kill specificity of bacteriophages presents advantages as well as disadvantages. Using modern DNA sequencing technologies, the most abundant and most damaging microbes in a specific system can be easily determined, to the genus or species levels. With this information, specific bacteriophages can be developed to target this specific microorganism and reduce the risk. As such the bacteriophages are not toxic to any other life forms. In addition, the phage quantity will quickly drop to very low levels after all target cells are lysed. In this regard, bacteriophages can be considered as "green" biocide.

Oil and gas industry consists of operations under a wide range of conditions, from downhole reservoir (e.g., high temperature, high pressure, and high salinity) to top side processing facilities, crude and water pipelines, storage tanks, to name a few. Microbial populations are very diverse and unique in every oilfield and under different environmental conditions. This presents a huge obstacle for bacteriophage applications in oil and gas industry as a means for microbial control. All damaging microorganisms have to be identified for each and every oilfield or system, and then the specific bacteriophages have to be developed for each and every specific target microorganism group (genus or species). The changes of bacterial communities in response to the changes of the operational conditions may render the phage products ineffective for the designated system. Therefore, the commercial application of this technology ultimately relies on whether it is possible to develop the phage products with expanded host ranges [117,119].

10.8.3 Physical processes

With increasing regulatory pressure on toxic biocides, chemical free, physical processes have been introduced into oil and gas industry as the alternative water treatment methods. The physical processes reduce or eliminate the need to add biocidal chemicals, and are more environmentally friendly treatment solutions. Examples physical processes in oil and gas industry application for microbial control include UV disinfection, filtration, and sonication.

UV light (100–400 nm) is invisible to the human eyes. The UV light between 200 and 300 nm has a unique germicidal characteristic [228]. In commercial UV system, high intensity UV-C band (wavelength of 254 nm) is commonly used. At this wavelength, UV light can penetrate the cell wall of microorganisms and alter genetic material (DNA) so that their reproduction is prevented. While UV treatment has been advocated as an "alternative to biocides," perhaps due to the efficiency of UV at 99.9999% reduction in viable bacterial numbers [229], the actual disinfection performance of a UV system depends on many factors. UV intensity (total UV lamp power) and contact time, which depend on the water flow rate, determine the UV treatment dose (fluence rate, mJ/cm^2). Microbial load as well as species in the water system impact the required UV dose. Different microorganisms have different UV dose response rates. Species such as SRB are very sensitive to UV-C radiation with standard industry doses of 40 mJ/cm^2 providing a >4-log (99.99%) reduction in a single pass (0.5-second exposure) [230]. However, spore-forming bacteria may require 10-times higher dose of the UV light to be reduced 90% in counts compared to their non-spore forming equivalent strain [231]. Furthermore, UV treatment is only effective to the planktonic bacteria which have been directly exposed to the UV light. The microorganisms within the biofilm or solid deposits will not be inactivated by UV radiation due to its poor penetrating power [46,229]. Finally, water quality has significant impact on the performance of UV treatment by affecting UV transmittance (UVT%). Generally, low water quality in terms of turbidity, color and TSS results in a lower UVT and worst UV performance, because the color and particulates in the water will absorb and scatter the UV light and decrease the overall available UV radiation for disinfection [232,233]. Water with high TSS or brackish water with iron sulfide particulates are not the appropriate system using UV treatment as a sole microbial control measure. In high TSS wastewater, ultrasonic pretreatment was reported to improve the performance of UV radiation [232]. Another shortcoming of UV treatment is the UV lamp fouling. Fouling of the lamp will dramatically affect the intensity of UV lamps, which in turn will affect the overall performance of UV radiation treatment.

The advantage of UV technology for water treatment is that it does not introduce any toxic chemicals, nor produce harmful by-products in the treated water.

UV technology has been widely used in food and beverage, recreation waters, drinking water, waste water, aquaculture, and life sciences. In oil and gas industry, UV disinfection has been used in well injection (seawater or produced water), hydraulic fracturing, pipeline hydrotesting, membrane protection (RO, ultrafiltration (UF), and nano membranes), and cooling water and heat exchangers [230]. In produced water or seawater injection, both onshore or offshore, uncontrolled microbial activities especially SRB activities can lead to numerous negative impacts on the integrity of injection wells and reservoir formation, such as equipment damage due to MIC, biofouling and clogging of the injection wells, and reservoir plugging and souring. UV system can be used to disinfect the injection seawater, ensure high injection water quality and minimize negative impact of microbial activities. However, depending on the size and design of the water system, pre-filtration and biocide treatment (oxidizing or nonoxidizing) may also be required to protect the pipelines and structures upstream and/or downstream of the UV system. In shale gas operation, large volume of clean, sterile water is injected into the shale formation to release the gas trapped within the rocks. The returned flowback water or produced water is re-used on the additional fracturing stage after on-site treatment and disinfection. UV disinfection is considered the best available technique for effective chemical-free treatment of produced water for fracturing operations. An additional advantage of UV technology in oil and gas industry is that it alleviates some incompatibility concerns of different oilfield chemicals. The commonly used biocide glutaraldehyde is known to be incompatible with a common oxygen scavenger, sodium bisulfate. This incompatibility can be addressed by switching to the physical process of UV treatment as the primary disinfection technique.

While UV treatment damages cell's DNA/RNA and prevents the cell from reproducing, the cell is not immediately dead. The research has showed that UV radiation can induce a viable but nonculturable (VBNC) state in some bacteria species [234]. Several studies have demonstrated that bacteria in a VBNC state may resuscitate under suitable environments, and become metabolically active and culturable [234,235]. This poses a potential risk for the water system relying solely on UV disinfection, and strongly suggests a combined disinfection strategy for effective microbial control. Al-Gabr et al. [236] demonstrated that the combined use of UV radiation followed by chlorination had synergistic effects, and was more effective than using either UV or chlorination alone in activating Aspergillus spp. fungi in drinking water. Croese et al. [237] compared the performance of UV and THPS biocide treatment on municipal wastewater for reservoir injection, and the results showed that the treatment of wastewater with biocides seem to have better performance in terms of the percentage reduction of viable microorganisms in the system than UV treatment.

The second example of physical processes for microbial control is ultrasonic treatment. The mechanism of microbial inactivation or killing by ultrasonic treatment is mainly due to thinning of cell membranes, localized heating and production of free radicals [238,239]. Ultrasound can produce an acoustic pressure that induces cavitation bubbles in the liquid. When these micro-bubbles collapse within an extremely small interval of time (microseconds), the large amounts of energy are released at millions of such locations simultaneously [240,241]. The generated high pressures (in the order of hundreds of atmospheres) and temperatures (in the range of thousands of degrees) in these locations can destroy cell wall and cell membrane of the microorganisms, and thus inactivate or kill bacteria [242,243]. In addition, chemical species such as hydrogen peroxide and hydroxyl radicals are formed during ultrasonic process, which have biocidal effects on microorganisms [231].

Ultrasound treatment is considered as an environmentally benign technique for the mitigation of MIC by reducing the corrosion-related bacteria in the system. Ultrasonic energy is reported to have good efficacy against planktonic bacteria [239] and soft biofilm [231,244]. Ultrasound treatment has been used in control microbiological growth in cooling water systems and fracturing water used in onshore unconventional applications [119,244]. Bacterial kill tests in the simulated pipeline brine containing 20 vol.% kerosene, APB, and *Enterobacter aerogenes* showed a bacteria reduction by about three orders of magnitude and corrosion rate reduced to the level approaching that for the sterile brine [239]. However, effectiveness of ultrasonic energy to kill microorganisms within corrosion products or solids deposits is very questionable, so as to effectiveness of this method for mitigation of MIC. Finally, Ultrasonic treatment may destroy the underlying material and aggravate crevice corrosion [239].

Synergistic effect of ultrasonic pretreatment combined with UV radiation in microbial control has been reported by many researchers. Ultrasound pretreatment improved the UV disinfection efficiency through the mechanical shear effect and the presence of cavitation bubbles that can break large particles into smaller particles, thereby exposing bacteria to UV for more effective disinfection [232,245]. Results of these studies have shown that a simultaneous or sequential combination of ultrasound with UV can improve the bacterial inactivation efficiency, especially when wastewater was characterized with low UV transmittance.

Another example of physical processes is filtration, including microfiltration (MF), UF, NF, and RO [246]. The filtration membrane retains the particles depending on its pore size. MF membrane has pore size 100 nm–10 μm, and retains large colloids and bacteria; UF membrane has pore size 2–100 nm, and can also retain macro molecules and protein. RO membrane has the smallest pore size, 0.1–1 nm, can retain very small particle such as monovalent salts, and virtually only pure water can pass this membrane. NF membrane with pore size 1–2 nm is used in oil industry for sulfate removal, but it also has a microbiological benefit (Fig. 10.6).

Seawater injection is a very common practice in oilfield for pressure maintenance of the reservoir. However, massive seawater injection can create a scale problem because seawater contains up to 3000 ppm of sulfate. The formation of barium and strontium sulfate scale results in water injectivity loss, plugging of reservoir rocks and production tubing. NF membranes have been widely used to

FIGURE 10.6

Membrane classification for water purification in terms of pore size and retained species [246].

remove nearly all sulfate (~99%) from seawater for oilfield water injection operations, both offshore and onshore [247]. While the major intention of the sulfate removal system is for prevention of sulfate scale formation, one side benefit of sulfate removal from injection seawater is souring control or mitigation in reservoirs. 99% of sulfate removal by NF membranes significantly reduces the available sulfate for sulfate reduction activity of SRB, which, in turn, significantly reduces the amount of generated hydrogen sulfide in the reservoir. Consequently, the risk of reservoir souring is reduced. In addition, NF membrane is charged, has pore size 1–2 nm, and can reject large colloids, bacteria, macro molecules, proteins, and multivalent salts via size exclusion and electrostatic interactions [246], resulting in high quality injection water free of silica and bacterial materials [119]. Keasler et al. [248] reported a 90% reduction in microbes downstream of the sulfate removal membrane compared to upstream. It is noted that the main business driver to build a sulfate removal unit (SRU) by NF membrane in oil industry is always for the prevention of sulfate scale formation, not for any microbiology benefits. This is because it involves high capital investment and maintenance cost, especially membrane replacement cost.

All physical processes for microbial control have a common weakness, that is, they do not produce any residual effect on the effluent water. Therefore, the water downstream of the physical processes is not protected from the microbial activities. While the UV, ultrasound, and NF membrane can inactivate or remove the majority of microbes (up to 99.99% as in UV treatment), these processes do not guarantee that the surviving bacteria will not grow up in downstream water system. The bacteria which already colonized the downstream system will not be impacted by these physical treatments upstream. Lack of residual effect of these technologies currently prevent them from widespread use in oil and gas industry. Depending on the application of the treated water or the size of the downstream water system, other microbial control measure such as biocide chemical may still be required to supplement the physical treatment processes.

10.9 Final remarks

Microorganisms are widely distributed in every stage of oilfield operation, from the initial drilling and workover, production of oil and gas, to storage and transport of crude oil and its processed products. Uncontrolled microbial growth in oil and gas operations can lead to costly problems, for instance, corrosion failures of equipment and facilities, injectivity loss of the water injection wells, formation damage and loss of oil production, reservoirs souring, decreased pipeline efficiency, and biodeterioration of refined products.

The chapter starts with the description of the major groups of microorganisms encountered in oil and gas industries, the synergistic relationships of these microorganisms under the various environmental and operational conditions, and their detrimental impacts on the oil and gas industry systems and facilities. The chemical (biocides, preservatives, nitrate) treatment is the focus of this chapter as a major mitigation strategy for the problems caused by microbial activities. The commonly used biocides in oil and gas industry are reviewed for their mode of actions, advantages and disadvantages, and application limitations in different oil industry systems. The chapter also describes the biocide selection criteria, performance evaluation methods, typical treatment program and practice in the industry, and biocide residual and microbial monitoring for treatment effectiveness in the field applications.

The major challenge for the biocide usage is that the biocidal actives are heavily regulated, and as a result, only a few biocidal actives are commonly used in the commercial biocidal products in oil and gas industry. The development of new products in the future will be more likely focused on the biocide formulations. Blending different biocidal actives into a single product, or blending with other additives such as surfactants is one of the directions for product development, by taking advantages of synergistic effects of different chemistries and therefore, enhancing the biocidal performance for microbial control. In addition, innovative delivery mechanism is another interesting area of research with potential to provide enhanced microbial control in some unique systems, for example, protection in the near wellbore areas against biofouling and biogenic H_2S production by treating injection water with long-lasting biocide formulation and releasing mechanism. Several alternative methods including biological and physical processes are briefly discussed in the chapter for potential applications in oil and gas industry for microbial control and mitigation. However, their wide adoption in oil and gas industry remains to be seen, especially for large and complex water systems.

Nomenclature

16S rRNA 16S Ribosomal RNA
APB acid-producing bacteria
ATP adenosine-5′-Triphosphate
BMPA bishydroxymethyl phosphinic acid
cATP cellular ATP
CMIC chemical microbially influenced corrosion
CMIT 5-chloro-2-methyl-4-isothiazolin-3-one
CTAC *cis*-1-(3-chloroallyl)-3,5,7-triaza-1-azoniaadamantane chloride
dATP dissolved ATP
DBNPA 2,2-dibromo-3-nitrilopropionamide
DMO 4,4-dimethyloxazolidine
EMIC electrical microbially influenced corrosion
EPS exopolysaccharides
GHB general heterotrophic bacteria
HPLC high-performance liquid chromatograph

IOB	iron-oxidizing bacteria
IRB	iron-reducing bacteria
KPI	key performance indicators
MF	microfiltration
MIC	microbiologically influenced corrosion
MIT	2-methyl-4-isothiazolin-3-one
MOB	manganese-oxidizing bacteria
MPN	most probable number
mRNA	messenger RNA
NF	nanofiltration
NR-SOB	nitrate-reducing, sulfide-oxidizing bacteria
NRB	nitrate-reducing bacteria
pg/mL	picograms per milliliter
qPCR	quantitative polymerase chain reaction
Quats	quaternary ammonium compounds
RLU	relative light units
RO	reverse osmosis
RT-qPCR	reverse transcriptase qPCR
SOB	sulfur-oxidizing bacteria
SRA	sulfate-reducing archaea
SRB	sulfate-reducing bacteria
SRP	sulfate-reducing prokaryotes
SRU	sulfate removal unit
tATP	total ATP
TDS	total dissolved solids
THNM	tris(hydroxymethyl)nitromethane
THPO	trishydroxymethyl phosphine oxide
THPS	tetrakis hydroxymethyl phosphonium sulfate
TSS	total suspended solids
UF	ultrafiltration
UV	ultraviolet
UVT	ultraviolet transmittance
WOSEP	water and oil separator

References

[1] J.D. van Hamme, A. Singh, O.P. Ward, Recent advances in petroleum microbiology, Microbiology and Molecular Biology Reviews 67 (2003) 503–549.

[2] M. Magot, Indigenous microbial communities in oil fields, in: B. Ollivier, M. Magot (Eds.), Petroleum Microbiology, ASM Press, Washington, DC, 2005, pp. 21–34.

[3] M.E. Sindi, A.M. Ezzat, A.A. Ba Hamdan, et al., Microbial assessment and control in drilling operations, SPE/IADC-189345-MS, in: SPE/IADC Middle East Drilling Technology Conference and Exhibition, Abu Dhabi, UAE, 29–31 January, 2018.

[4] X.Y. Zhu, Microbial communities in drilling mud additive xanthan gum and their implications in the oil industry, in: Proceedings of 16th Middle East Corrosion Conference, Bahrain, 8–11 February, 2016.

[5] S.E. Agarry, K.K. Salam, A.O. Arinkoola, et al., Microbiologically influenced corrosion of mild steel in crude oil environment, European Journal of Engineering and Technology 3 (2015) 40–58.

[6] R.B. Eckert, Emphasis on biofilms can improve mitigation of microbiologically influenced corrosion in oil and gas industry, Corrosion Engineering, Science and Technology 50 (2015) 163–168.

[7] S.I. Arampatzi, G. Giannoglou, E. Diza, Biofilm formation: a complicated microbiological process, Aristotle University Medical Journal 38 (2011) 22–24.

[8] J. Telegdi, A. Shaban, G. Vastag, Biocorrosion—steel, in: K. Wandelt (Ed.), Encyclopedia of Interfacial Chemistry: Surface Science and Electrochemistry, Elsevier, Amsterdam, 2018, pp. 28–42.

[9] G.H. Koch, M. Brongers, N.G. Thompson, et al., Corrosion Costs and Preventive Strategies in the United States, FHWA-RD-01–156, Federal Highway Administration, Washington, DC, 2001.

[10] F.J. Passman, Microbial contamination and its control in fuels and fuel systems since 1980—a review, International Biodeterioration & Biodegradation 81 (2013) 88–104.

[11] L. Dorobantu, A. Yeung, J.M. Foght, et al., Stabilization of oil-water emulsions by hydrophobic bacteria, Applied and Environmental Microbiology 70 (2004) 6333–6336.

[12] D.J. Ligthelm, R.B. de Boer, J.F. Brint, et al., Reservoir souring: an analytical model for H_2S generation and transportation in an oil reservoir owing to bacterial activity, SPE-23141-MS, in: Society of Petroleum Engineers, Offshore Europe, Aberdeen, UK, 3–6 September, 1991.

[13] J. Crolet, Microbial corrosion in the oil industry: a corrosionist's view, in: B. Ollivier, M. Magot (Eds.), Petroleum Microbiology, ASM Press, Washington, DC, 2005, pp. 143–170.

[14] J. Larsen, K.B. Sorensen, B. Hojris, et al. Significance of troublesome sulfate-reducing prokaryotes (SRP) in oil field systems, NACE-09389, in: Corrosion Conference & Expo, Houston, Texas, 22–26 March, 2009.

[15] NACE, Detection, testing, and evaluation of microbiologically influenced corrosion (MIC) on external surfaces of buried pipelines, NACE TM0106–2006, NACE International, Houston, TX, 2006.

[16] J. Telegdi, A. Shaban, L. Trif, Review on the microbiologically influenced corrosion and the function of biofilms, International Journal of Corrosion and Scale Inhibition 9 (2020) 1–33.

[17] D.H. Pope, Microbiologically influenced corrosion of internal aspects of natural gas industry pipelines and associated equipment: mechanisms, diagnosis, and mitigation, in: J.G. Stoecker (Ed.), A Practical Manual on Microbiologically Influenced Corrosion, vol. 2, NACE International, Houston, 2001, pp. 13–25 (in Chapter 6).

[18] S.W. Borenstein, Microbiologically Influenced Corrosion Handbook, Woodhead Publishing, Cambridge, 1994.

[19] B. Little, J. Lee, Microbiologically Influenced Corrosion, John Wile & Sons, Hoboken, 2007.

[20] J. Horn, D. Jones, Microbiologically influenced corrosion: perspectives and approaches, in: C.J. Hurst, R.L. Crawford (Eds.), Manual of Environmental Microbiology, ASM Press, Washington, DC, 2002, pp. 1072–1083.

[21] W.A. Hamilton, Microbially influenced corrosion as a model system for the study of metal microbe interactions: a unifying electron transfer hypothesis, Biofouling 19 (2003) 65–76.

[22] H.A. Videla, L.K. Herrera, G. Edyvean, An updated overview of SRB induced corrosion and protection of carbon steel, NACE-05488, in: Corrosion Conference & Expo, Houston, Texas, 3–7 April, 2005.
[23] M.S. Osburne, T.H. Griossman, P.R. August, et al., Tapping into microbial diversity for natural products drug discovery, ASM News 66 (2000) 411–417.
[24] V. Torsvik, J. Goksoyr, F.L. Daae, High diversity in DNA of soil bacteria, Applied and Environmental Microbiology 56 (1990) 782–787.
[25] X.Y. Zhu, J.J., Kilbane II, A. Ayala, et al. Application of quantitative, real-time PCR in monitoring microbiologically-influenced corrosion (MIC) in gas pipelines, NACE-05493, in: Corrosion Conference & Expo, Houston, Texas, 3–7 April, 2005.
[26] X.Y. Zhu, H. Modi, A. Ayala, et al., Rapid detection and quantification of microbes related to microbiologically influenced corrosion using quantitative polymerase chain reaction, Corrosion 62 (2006) 950–955.
[27] X.Y. Zhu, M.A. Moniee, Molecular microbiology techniques, in: A.M. El-Sherik (Ed.), Trends in Oil and Gas Corrosion Research and Technologies: Production and Transmission, Woodhead Publishing, Duxford, 2017, pp. 513–536.
[28] A. Dolla, M. Fournier, Z. Dermoun, Oxygen defense in sulfate-reducing bacteria, Journal of Biotechnology 126 (2006) 87–100.
[29] K.U. Kjeldsen, C. Joulian, K. Ingvorsen, Oxygen tolerance of sulfate-reducing bacteria in activated sludge, Environmental Science & Technology 38 (2004) 38–43.
[30] X.Y. Zhu, H. Modi, J.J. Kilbane II, Efficacy and risks of nitrate application for the mitigation of SRB-induced corrosion, NACE-06524, in: Corrosion Conference & Expo, San Diego, CA, 12–16 March, 2006.
[31] E.E. Roden, D.R. Lovley, Dissimilatory Fe (III) reduction by marine microorganism *Desulfuromonas acetooxidans*, Applied and Environmental Microbiology 59 (1993) 734–742.
[32] R. Cord-Ruwisch, W. Kleinitz, F. Widdel, Sulfate-reducing bacteria and their activities in oil production, Journal of Petroleum Technology 39 (1987) 97–106.
[33] X.Y. Zhu, J. Lubeck, J.J. Kilbane II, Characterization of microbial communities in gas industry pipelines, Applied and Environmental Microbiology 69 (2003) 5354–5363.
[34] W.A. Hamilton, Sulphate-reducing bacteria and anaerobic corrosion, Annual Review of Microbiology 39 (1985) 195–217.
[35] P. Hugenholtz, Exploring prokaryotic diversity in the genomic area, Genome Biology 3 (2002) 1–8.
[36] H.F. Castro, N.H. Williams, A. Ogram, Minireview: phylogeny of sulfate-reducing bacteria, FEMS Microbiology Ecology 31 (2000) 1–9.
[37] C.G. Struchtemeyer, J.P. Davis, M.S. Elshahed, Influence of the drilling mud formulation process on the bacterial communities in thermogenic natural gas wells of the Barnett Shale, Applied and Environmental Microbiology 77 (2011) 4744–4753.
[38] C. Tardy-Jacquenod, M. Magot, B.K. Patel, et al., *Desulfotomaculum halophilum* sp. nov., a halophilic sulfate-reducing bacterium isolated from oil production facilities, International Journal of Systematic and Evolutionary Microbiology 48 (1998) 333–338.
[39] C. Pallud, P. van Cappellen, Kinetics of microbial sulfate reduction in estuarine sediments, Geochimica et Cosmochimica Acta 70 (2006) 1148–1162.
[40] G. Gibson, Physiology and ecology of the sulphate-reducing bacteria, Journal of Applied Bacteriology 69 (1990) 769–797.

[41] D.R. Lovley, E. Philips, Novel processes for anaerobic sulfate production from elemental sulfur by sulfate-reducing bacteria, Applied and Environmental Microbiology 60 (1994) 2394–2399.
[42] C. Hubert, G. Voordouw, Oil field souring control by nitrate-reducing Sulfurospirillum spp. that outcompete sulfate-reducing bacteria for organic electron donors, Applied and Environmental Microbiology 73 (2007) 2644–2652.
[43] H.P. Klenk, R.A. Clayton, J.F. Tomb, et al., The complete genome sequence of the hyperthermophilic, sulphate-reducing Archaeon *Archaeoglobus fulgidus*, Nature 390 (1997) 364–370.
[44] K.O. Stetter, G. Lauerer, M.N. Thomm, Isolation of extremely thermophilic sulfate reducers: evidence for a novel branch of Archaebacteria, Science (New York, N.Y.) 236 (1987) 822–824.
[45] J. dos Santos, D. Lopes, J. Da, et al., Diversity of sulfate-reducing prokaryotes in petroleum production water and oil samples, International Biodeterioration & Biodegradation 151 (2020) 104966.
[46] B.J. Little, R.I. Ray, R.K. Pope, Relationship between corrosion and the biological sulfur cycle: a review, Corrosion 56 (2000) 433–443.
[47] V. O'Dea, Understanding biogenic sulfide corrosion, Materials Performance 46 (2007) 36–39.
[48] J. Jan-Roblero, J.M. Romero, M. Amaya, et al., Phylogenetic characterization of a corrosive consortium isolated from a sour gas pipeline, Applied Microbiology & Technology 64 (2004) 862–867.
[49] M. Ayazi, S. Mirfenderski, A.A. Moghadam, et al., Study of the corrosion factors in offshore oil production units, Petroleum & Coal 48 (2006) 6–10.
[50] C.R. Jones, D. Labarre, M. Mantysaari, Pushing performance boundaries: step change efficacy of THPS based iron sulfide control technologies, in: Proceedings of 17th Middle East Corrosion Conference, Bahrain, September 30–October 3, 2018.
[51] E. Ilhan-Sungur, N. Cansever, A. Cotuk, Microbial corrosion of galvanized steel by a freshwater strain of sulphate reducing bacteria (Desulfovibrio sp.), Corrosion Science 49 (2006) 1097–1109.
[52] J.T. Rosnes, T. Torsvik, T. Lien, Spore-forming thermophilic sulfate-reducing bacteria isolated from North Sea oil field waters, Applied and Environmental Microbiology 57 (1991) 2302–2307.
[53] J.G. Zeikus, M.A. Dawson, T.E. Thompson, et al., Microbial ecology of volcanic sulphidogenesis: isolation and characterization of *Thermodesulfobacteriurn commune* gen. nov. and sp. nov, Journal of' General Microbiology 129 (1983) 1159–1169.
[54] E.A. Henry, R. Devereux, J.S. Maki, et al., Characterization of a new thermophilic sulfate-reducing bacterium *Thermodesulfovibrio yellowstonii*, gen. nov. and sp. nov.: its phylogenetic relationship to *Thermodesulfobacterium commune* and their origins deep within the bacterial domain, Archives of Microbiology 161 (1994) 62–69.
[55] R. Rabus, T.A. Hansen, F. Widdel, Dissimilatory sulfate- and sulfur-reducing prokaryotes, in: M. Dworkin, S. Falkow, E. Rosenberg, et al. (Eds.), The Prokaryotes, Springer, New York, 2006, pp. 659–768.
[56] D. Wolicka, A. Borkowski, D. Dobrzyński, Interactions between microorganisms, crude oil and formation waters, Geomicrobiology Journal 27 (2010) 43–52.
[57] S.Y. Li, Y.G. Kim, K.S. Jeon, et al., Microbiologically influenced corrosion of carbon steel exposed to anaerobic soil, Corrosion 57 (2001) 815–828.

[58] H. Tributsch, J.A. Rojas-Chapana, C.C. Bärtels, et al., Role of transient iron sulfide films in microbial corrosion of steel, Corrosion 54 (1998) 216–227.

[59] T. Hemmingsen, H. Vangdal, T. Våland, Formation of ferrous sulfide film from sulfite on steel under anaerobic conditions, Corrosion 48 (1992) 475–481.

[60] D. Enning, J. Garrelfs, Corrosion of iron by sulfate-reducing bacteria: new views of an old problem, Applied and Environmental Microbiology 80 (2014) 1226–1236.

[61] S. Kebbouche-Gana, M.L. Gana, Biocorrosion of carbon steel by a nitrate-utilizing consortium of sulfate-reducing bacteria obtained from an Algerian oil field, Annals of Microbiology 62 (2012) 203–210.

[62] S.S. Al-Jaroudi, A. Ul-Hamid, M.M. Al-Gahtani, Failure of crude oil pipeline due to MIC, Corrosion Engineering Science and Technology 46 (2011) 568–579.

[63] N. Youssef, M.S. Elshahed, M.J. McInerney, Microbial processes in oil fields: culprits, problems, and opportunities, Advances in Applied Microbiology 66 (2009) 141–251.

[64] D.A. Elias, L.R. Krumholz, R.S. Tanner, et al., Estimation of methanogen biomass by quantitation of Coenzyme M, Applied and Environmental Microbiology 65 (1999) 5541–5545.

[65] J.R. Cole, Q. Wang, J.A. Fish, et al., Ribosomal database project: data and tools for high throughput rRNA analysis, Nucleic Acids Research 42 (2014) D633–D642.

[66] N.D. Gray, A. Sherry, S.R. Larter, et al., Biogenic methane production in formation waters from a large gas field in the North Sea, Extremophiles: Life Under Extreme Conditions 13 (2009) 511–519.

[67] T.N. Nazina, A.A. Grigor'yan, N.M. Shestakova, et al., Microbiological investigations of high-temperature horizons of the Kongdian petroleum reservoir in connection with field trial of a biotechnology for enhancement of oil recovery, Microbiology (Reading, England) 76 (2007) 287–296.

[68] J. Mikucki, Y. Liu, M. Delwiche, et al., Isolation of a methanogen from deep marine sediments that contain methane hydrates, and description of *Methanoculleus submarinus* sp. nov, Applied and Environmental Microbiology 69 (2003) 3311–3316.

[69] B. Ollivier, D. Alazard, The oil reservoir ecosystem, in: K.N. Timmis (Ed.), Handbook of Hydrocarbon and Lipid Microbiology, Springer-Verlag Berlin, Heidelberg, 2010, pp. 2259–2269.

[70] M. Kurr, R. Huber, H. Konig, et al., *Methanopyrus kandleri*, gen. and sp. nov. represents a novel group of hyperthermophilic methanogens, growing at 110°C, Archives of Microbiology 156 (1991) 239–247.

[71] C. Jeanthon, S. L'Haridon, A.L. Reysenbach, et al., *Methanococcus vulcanicus* sp. nov., a novel hyperthermophilic methanogen isolated from East Pacific Rise, and identification of methanococcus sp. DSM 4213T as *Methanococcus fervens* sp. nov, International Journal of Systematic Bacteriology 49 (1999) 583–589.

[72] P.D. Franzmann, Y. Liu, D.L. Balkwill, et al., *Methanogenium frigidum* sp. nov., a psychrophilic, H_2-using methanogen from Ace Lake, Antarctica, International Journal of Systematic Bacteriology 47 (1997) 1068–1072.

[73] P.D. Franzmann, N. Springer, W. Ludwig, et al., A methanogenic archaeon from Ace Lake, Antarctica: *Methanococcoides burtonii* sp. nov, Systematic and Applied Microbiology 15 (1992) 573–581.

[74] R.K. Thauer, A.K. Kaster, H. Seedorf, et al., Methanogenic Archaea: ecologically relevant differences in energy conservation, Nature Reviews. Microbiology 6 (2008) 579–591.

[75] A.M. Alessi, B. Tao, W. Zhang, et al., A rapid, sensitive, low-cost assay for detecting hydrogenotrophic methanogens in anaerobic digesters using loop-mediated isothermal amplification, Microorganisms 8 (2020) 740–754.
[76] T. Fenchel, G. King, T. Blackburn, Bacterial metabolism, in: T. Fenchel, T. Blackburn, G. King (Eds.), Bacterial Biogeochemistry, third ed., Academic Press, Cambridge, 2012, pp. 1–34.
[77] T. Zhang, H. Fang, B. Ko, Methanogen population in a marine biofilm corrosive to mild steel, Applied Microbiology and Biotechnology 63 (2003) 101–106.
[78] S. Kalyuzhnyi, V. Fedorovich, P. Lens, et al., Mathematical modelling as a tool to study population dynamics between sulfate reducing and methanogenic bacteria, Biodegradation 9 (1998) 572–9729.
[79] L. Raskin, B.E. Rittmann, D.A. Stahl, Competition and coexistence of sulfate-reducing and methanogenic populations in anaerobic biofilms, Applied and Environmental Microbiology 62 (1996) 3847–3857.
[80] J. Chen, M.J. Wade, J. Dolfing, et al., Increasing sulfate levels show a differential impact on synthetic communities comprising different methanogens and a sulfate reducer, Journal of the Royal Society Interface 16 (2019) 20190129.
[81] K.M. Kuivila, J.W. Murray, A.H. Devol, et al., Methane production, sulfate reduction and competition for substrates in the sediments of Lake Washington, Geochimica et Cosmochimica Acta 53 (1989) 409–416.
[82] J.P. Guyot, A. Brauman, Methane production from formate by syntrophic association of *Methanobacterium bryantii* and *Desulfovibrio vulgaris* JJ, Applied and Environmental Microbiology 52 (1986) 1436–1437.
[83] M.J. McInerney, M.P. Bryant, Anaerobic degradation of lactate by syntrophic associations of Methanosarcina barkeri and desulfovibrio species and effect of H2 on acetate degradation, Applied and Environmental Microbiology 41 (1981) 346–354.
[84] C.C. Okoro, O. Amund, Souring and corrosion potentials of onshore and offshore oil-producing facilities in the Nigerian oil-rich Niger Delta, Petroleum Science and Technology 33 (2015) 1563–1570.
[85] T. Uchiyama, K. Ito, K. Mori, et al., Iron-corroding methanogen isolated from a crude-oil storage tank, Applied and Environmental Microbiology 76 (2010) 1783–1788.
[86] E. Zhou, J. Wang, M. Moradi, et al., Methanogenic archaea and sulfate reducing bacteria induce severe corrosion of steel pipelines after hydrostatic testing, Journal of Materials Science & Technology 48 (2020) 72–83.
[87] J. Larsen, K. Rasmussen, H. Pedersen, et al., Consortia of MIC bacteria and archaea causing pitting corrosion in top side oil production facilities, NACE-10252, in: Corrosion Conference & Expo, San Antonio, TX, 14–18 March, 2010.
[88] X.Y. Zhu, Rapid quantification of butyric acid-producing bacteria using real-time PCR, Patent US7276358, 2007.
[89] X.Y. Zhu, Rapid quantification of acetic acid-producing bacteria using real-time PCR, Patent US7384770, 2008.
[90] B. Huber, B. Herzog, J.E. Drewes, et al., Characterization of sulfur oxidizing bacteria related to biogenic sulfuric acid corrosion in sludge digesters, BMC Microbiology 16 (2016) 153–163.
[91] S. Okabe, M. Odagiri, T. Ito, et al., Succession of sulfur-oxidizing bacteria in the microbial community on corroding concrete in sewer systems, Applied and Environmental Microbiology 73 (2007) 971–980.

[92] P. Parthipan, P. Elumalai, O.P. Karthikeyan, et al., A review on biodegradation of hydrocarbon and their influence on corrosion of carbon steel with special reference to petroleum industry, Journal of Environment and Biotechnology Research 6 (2017) 12–33.

[93] S. Shukla, T.B. Choi, H.K. Park, et al., Determination of non-volatile and volatile organic acids in Korean traditional fermented soybean paste (Doenjang), Food and Chemical Toxicology 48 (2010) 2005–2010.

[94] Z. Yan, X. Zheng, J. Chen, et al., Effect of different bacillus strains on the profile of organic acids in a liquid culture of Daqu, Journal of the Institute of Brewing 119 (2013) 78–83.

[95] S.A. Dar, R. Kleerebezem, A.J. Stams, et al., Competition and coexistence of sulfate-reducing bacteria, acetogens and methanogens in a lab-scale anaerobic bioreactor as affected by changing substrate to sulfate ratio, Applied Microbiology and Biotechnology 78 (2008) 1045–1055.

[96] M. Holmer, E. Kristensen, Coexistence of sulfate reduction and methane production in an organic-rich sediment, Marine Ecology Progress Series 107 (1994) 177–184.

[97] G. Zellner, E. Stackebrandt, D. Nagel, et al., *Anaerofilum pentosovorans* gen. nov., sp. nov., and *Anaerofilum agile* sp. nov., two new, strictly anaerobic, mesophilic, acidogenic bacteria from anaerobic bioreactors, International Journal of Systematic and Evolutionary Microbiology 46 (1996) 871–875.

[98] D.H. Pope, Discussion of methods for the detection of microorganisms involved in microbiologically influenced corrosion, biologically induced corrosion, in: S.C. Dexter (Ed.), Biologically Induced Corrosion, NACE-8, NACE International, Houston, 1985, pp. 275–283.

[99] C.I. Pereira, M. Crespo, M. Romão, Evidence for proteolytic activity and biogenic amines production in *Lactobacillus curvatus* and *L. homohiochii*, International Journal of Food Microbiology 68 (2001) 211–216.

[100] H.A. Videla, Biocorrosion and biofouling of metals and alloys of industrial usage: present state of the art at the beginning of the new millennium, Revista de Metalurgia 39 (2003) 256–264.

[101] C.S. Chan, S.C. Fakra, D. Emerson, et al., Lithographic iron-oxidizing bacteria produce organic stalks to control mineral growth: implications for biosignature formation, ISME Journal 5 (2011) 717–727.

[102] D. Starosvetsky, J. Starosvetsky, B. Pokroy, et al., Electrochemical behavior of stainless steels in media containing iron-oxidizing bacteria (IOB) by corrosion process modeling, Corrosion Science 50 (2008) 540–547.

[103] D. Starosvetsky, R. Armon, J. Yahalom, et al., Pitting corrosion of carbon steel caused by iron bacteria, International Biodeterioration & Biodegradation 47 (2001) 79–87.

[104] R.I. Ray, J.S. Lee, B.J. Little, Iron-oxidizing bacteria: a review of corrosion mechanisms in fresh water and marine environments, NACE-10218, in: Corrosion Conference & Expo, San Antonio, TX, 14–18 March, 2010.

[105] I.G. Chamritski, G.R. Burns, B.J. Webster, et al., Effect of iron-oxidizing bacteria on pitting of stainless steel, Corrosion 60 (2004) 658–669.

[106] H. Liu, C. Fu, T. Gu, et al., Corrosion behavior of carbon steel in the presence of sulfate reducing bacteria and iron oxidizing bacteria cultured in oilfield produced water, Corrosion Science 100 (2015) 484–495.

[107] M. Moradi, J. Duan, H. Ashassi-Sorkhabi, et al., De-alloying of 316 stainless steel in the presence of a mixture of metal-oxidizing bacteria, Corrosion Science 53 (2011) 4282–4290.

[108] I.G. Chamritski, B.R. Burns, B.J. Webster, et al., The influence of iron oxidizing bacteria on corrosion of stainless steel, NACE-01254, in: Corrosion Conference & Expo, Houston, TX, 11–16 March, 2001.

[109] R.W. Revie, Uhlig's Corrosion Handbook, third ed., John Wiley & Sons Inc, New York, 2011.

[110] S. Gorman, E.M. Scott, A review—antimicrobial activity, uses and mechanism of action of glutaraldehyde, Journal of Applied Bacteriology 48 (1980) 161–190.

[111] A. Piazza, L.C. Casalini, V.A. Pacini, et al., Environmental bacteria involved in manganese(II) oxidation and removal from groundwater, Frontiers in Microbiology 10 (2019) 119.

[112] K.H. Nealson, B.M. Tebo, R.A. Rosson, Occurrence and mechanisms of microbial oxidation of manganese, Advances in Applied Microbiology 33 (1988) 279–318.

[113] Y. Bai, R. Liu, J. Liang, et al., Integrated metagenomic and physiochemical analyses to evaluate the potential role of microbes in the sand filter of a drinking water treatment system, PLoS One 8 (2013) 1–9.

[114] I.A. Katsoyiannis, A.I. Zouboulis, Biological treatment of Mn(II) and Fe(II) containing groundwater: kinetic considerations and product characterization, Water Research 38 (2004) 1922–1932.

[115] W.H. Dickinson, F. Caccavo, B. Olesen, et al., Ennoblement of stainless steel by the manganese-depositing bacterium *Leptothrix discophora*, Applied and Environmental Microbiology 63 (1997) 2502–2506.

[116] W.H. Dickinson, Z. Lewandowski, Manganese biofouling and the corrosion behavior of stainless steel, Biofouling 10 (1996) 79–93.

[117] C. Campbell, Advances in testing and monitoring of biocides in oil and gas, in: A. M. El-Sherik (Ed.), Trends in Oil and Gas Corrosion Research and Technologies: Production and Transmission, Woodhead Publishing, Duxford, 2017, pp. 489–511.

[118] A. Turkiewicz, J. Brzeszcz, P. Kapusta, The application of biocides in the oil and gas industry, Nafta-Gaz 2 (2013) 103–111.

[119] V. Keasler, R. De Paula, G. Nilsen, et al., Biocides overview and applications in petroleum microbiology, in: A.M. El-Sherik (Ed.), Trends in Oil and Gas Corrosion Research and Technologies: Production and Transmission, 2017, Woodhead Publishing, Duxford, 2017, pp. 539–562.

[120] J.K. Fink, Bacterial control, Petroleum Engineer's Guide to Oil Field Chemicals and Fluids, Gulf Professional Publishing, Waltham, 2012.

[121] H.R. McGinley, M.V. Enzien, G. Jenneman, et al., Studies on the chemical stability of glutaraldehyde in produced water, SPE-141449-MS, in: SPE International Symposium on Oilfield Chemistry, Woodlands, TX, 11–13 April, 2011.

[122] H.W. Sun, R.J. Feigal, H.H. Messer, Cytotoxicity of glutaraldehyde and formaldehyde in relation to time of exposure and concentration, Pediatric Dentistry 12 (1990) 303–307.

[123] T.M. Williams, The environmental fate of oil and gas biocides: a review, NACE-3876, in: Corrosion Conference & Expo, San Antonia, TX, 9–13 March, 2014.

[124] C. Frayne, The selection and application of nonoxidizing biocides for cooling water systems, The Analyst 8 (2001) 1–14.

[125] G.J. Jones, Letter: polymerization of glutaraldehyde at fixative pH, Journal of Histochemistry & Cytochemistry 22 (1974) 911–913.
[126] C.W. Goff, M.O. Oster, Letter: formation of 235-nanometer absorbing substance during glutaraldehyde fixation, Journal of Histochemistry & Cytochemistry 22 (1974) 913–915.
[127] W.A. Rutala, D.J. Weber, Disinfection, sterilization, and control of hospital waste, in: J.E. Bennett, R. Dolin, M.J. Blaser (Eds.), Mandell, Douglas, and Bennett's Principles and Practice of Infectious Diseases, eighth ed., Elsevier, Amsterdam, 2015, pp. 3294–3309.
[128] G.A. Kahrilas, J. Blotevogel, E.R. Corrin, et al., Downhole transformation of the hydraulic fracturing fluid biocide glutaraldehyde: implications for flowback and produced water quality, Environmental Science & Technology 50 (2016) 11414–11423.
[129] J. Wen, K. Zhao, T. Gu, et al., A green biocide enhancer for the treatment of sulfate-reducing bacteria (SRB) biofilms on carbon steel surfaces using glutaraldehyde, International Biodeterioration and Biodegradation 63 (2009) 1102–1106.
[130] I. Migneault, C. Dartiguenave, M.J. Bertrand, et al., Glutaraldehyde: behavior in aqueous solution, reaction with proteins, and application to enzyme crosslinking, Biotechniques 37 (2004) 790–802.
[131] S.L. Jordan, M.R. Russo, R.L. Blessing, et al., Inactivation of glutaraldehyde by reaction with sodium bisulfite, Journal of Toxicology and Environmental Health 47 (1993) 299–309.
[132] P.S. Stewart, L. Grab, J.A. Diemer, Analysis of biocide transport limitation in an artificial biofilm system, Journal of Applied Microbiology 85 (1998) 495–500.
[133] G.R. Lloyd, P.W. Neail, Biocides for minimum environmental impact, in: Proceedings of 2nd Australian International Oil, Gas & Petrochemical Conference, Melbourne, Australia, 23–26 November, 1993.
[134] J. Moore, V. Keasler, B. Bennett, Compatibility of tetrakis(hydroxymethyl) phosphonium sulfate (THPS) and ammonium bisulfite (ABS), NACE-10407, in: Corrosion Conference & Expo, San Antonio, TX, 14–18 March, 2010.
[135] C. Campbell, A. Cooper, Managing microbial influenced problems in the petroleum industry by understanding biocide functionality: type, application, and limitation, in: Proceedings of AWT Convention, San Diego, CA, 7–10 September, 2016.
[136] K. Zhao, J. Wen, T. Gu, et al., Mechanistic modeling of anaerobic THPS degradation in seawater under various conditions, NACE-08512, in: Corrosion Conference & Expo, Houston, TX, 16–20 March, 2008.
[137] B.L. Downward, R.E. Talbot, T.K. Haack, TetrakisHydroxymethylPhosphonium sulfate (THPS): a new industry biocide with low environmental toxicity, NACE-97401, in: Corrosion Conference & Expo, New Orleans, LA, 9–14 March, 1997.
[138] C. Jones, B. Downward, S. Edmunds, et al., THPS: a review of the first 25 years, lessons learned, value created and visions for the future, NACE-1505, in: Corrosion Conference & Expo, Salt Lake City, UT, 11–15 March, 2012.
[139] N.N. Fathima, M.C. Bose, J.R. Rao, et al., Stabilization of type i collagen against collagenases (Type I) and thermal degradation using iron complex, Journal of Inorganic Biochemistry 100 (2006) 1774–1780.
[140] K. Zhao, J. Wen, T. Gu, et al., Mechanistic modeling of anaerobic THPS degradation under alkaline condition in the presence of mild steel, Materials Performance 8 (2009) 62–66.

[141] G.A. Kahrilas, J. Blotevogel, P.S. Stewart, et al., Biocides in hydraulic fracturing fluids: a critical review of their usage, mobility, degradation, and toxicity, Environmental Science & Technology 49 (2015) 16–32.

[142] J. Wilmon, THPS degradation in the long-term preservation of sub-sea flow-lines and risers, NACE-10402, in: Corrosion Conference & Expo, San Antonio, TX, 14–18 March, 2010.

[143] R.E. Talbot, P.D. Gilbert, M.A. Veale, et al., Tetrakis hydroxymethyl phosphonium sulfate (THPS) for dissolving iron sulfides downhole and topsides - a study of the chemistry influencing dissolution, NACE-02030, in: Corrosion Conference & Expo, Denver, CO, 7–11 April, 2002.

[144] J.J. Wylde, I.G. Winning, Challenges and solutions associated with the development of an iron scale dissolver chemistry, NACE-04730. in: Corrosion Conference & Expo, New Orleans, LA, 28 March–1 April, 2004.

[145] Q. Wang, S. Shen, H. Badairy, et al., Laboratory assessment of tetrakis(hydroxymethyl) phosphonium sulfate as dissolver for scales formed in sour gas wells, International Journal of Corrosion and Scale Inhibition 4 (2015) 235–254.

[146] V. Keasler, B. Bennett, B. Bromage, et al., Bacterial characterization and biocide qualification for full wellstream crude oil pipelines, NACE-10250, in: Corrosion Conference & Expo, San Antonio, TX, 14–18 March, 2010.

[147] J.F. Kramer, A new high performance quaternary phosphonium biocide for biofouling control in industrial water systems, NACE-06093, in: Corrosion Conference & Expo, San Diego, CA, 12–16 March, 2006.

[148] U. Bertheas, K. Majamaa, A. Arzu, et al., Use of DBNPA to control biofouling in RO membranes, Desalination and Water Treatment 3 (2009) 1–3.

[149] J.H. Exner, G.A. Burk, D. Kyriacou, Rates and products of decomposition of 2,2-dibromo-3-nitrilopropionamide, Journal of Agricultural and Food Chemistry 21 (1973) 838–842.

[150] T.M. Williams, The mechanism of action of isothiazolone biocide, NACE-06090, in: Corrosion Conference & Expo, San Diego, CA, 12–16 March, 2006.

[151] T.M. Williams, Methylisothiazolone: a new biocide product for closed loop systems, NACE-07066, in: Corrosion Conference & Expo, Nashville, TN, 11–15 March, 2007.

[152] T.M. Williams, Efficacy of chloromethyl-methylisothiazolone (CMIT/MIT) biocide vs Legionella and protozoa, NACE-07433, in: Corrosion Conference & Expo, Nashville, TN, 11–15 March, 2007.

[153] T.M. Williams, A. Jacobson, Environmental fate of isothiazolone biocides, NACE-99303, in: Corrosion Conference & Expo, San Antonio, TX, 25–30 April, 1999.

[154] H. Vogt, J. Balej, J.E. Bennett, et al., Chlorine oxides and chlorine oxygen acids, Ullmann's Encyclopedia of Industrial Chemistry, Wiley-VCH, Weinheim, 2010.

[155] EPA, Alternative Disinfectants and Oxidants Guidance Manual, US Environmental Protection Agency: Office of Water, Washington, DC, 1999.

[156] S.S. Block, Disinfection, Sterilization, and Preservation, fifth ed, Williams & Wilkins, Philadelphia, PA, 2001.

[157] N.F. Gray, Free and combined chlorine, in: S.L. Percival, M.V. Yates, D.W. Williams, et al. (Eds.), Microbiology of Waterborne Diseases, second ed., Elsevier, Amsterdam, 2014, pp. 571–590.

[158] R.M. Moore Jr., C.J. Nalepa, Concentrated aqueous bromine solutions and their preparation, Patent US6068861, 2000.

[159] C.J. Nalepa, 25 years of bromine chemistry in industrial water systems: a review, NACE-04087, in: Corrosion Conference & Expo, New Orleans, LA, 28 March–1 April, 2004.

[160] C.J. Nalepa, F.D. Azomia, Development of a high activity liquid biocide for industrial water treatment, NACE-06096, in: Corrosion Conference & Expo, San Diego, CA, 12–16 March, 2006.

[161] J.F. Kramer, Peracetic acid: a new biocide for industrial water applications, NACE-97404, in: Corrosion Conference & Expo, New Orleans, LA, 9–14 March, 1997.

[162] R. De Paula, V. Keasler, J. Li, et. al., Development of peracetic acid (PAA) as an environmentally safe biocide for water treatment during hydraulic fracturing applications, SPE-164088-MS, in: SPE Conference for Oilfield Chemistry, Woodlands, TX, 8–10 April, 2013.

[163] R. De Paula, V. Keasler, T. Tidwell, Evaluation of preservative chemistries to control microbial activity during well completion, NACE-6069, in: Corrosion Conference & Expo, Dallas, TX, 15–19 March, 2015.

[164] C. Campbell, A. Johnson, Dual biocide application for optimization of microbial control in hydrofracture fluids: leveraging biocide strengths, in: Proceedings of AWT Convention, Nashville, TN, 9–12 September, 2015.

[165] B. Yin, T. Williams, T. Koehler, et al., Targeted microbial control for hydrocarbon reservoir: identify new biocide offerings for souring control using thermophile testing capabilities, International Biodeterioration & Biodegradation 126 (2016) (2016) 204–207.

[166] Y. Zhou, Bactericide for drilling fluid, Drilling Fluid & Completion Fluid 7 (1990) 10–12.

[167] V. Keasler, B. Bennett, H. McGinley, Analysis of bacterial kill vs corrosion from use of common oilfield biocides, IPC2010-31593, in: 8th International Pipeline Conference (IPC), Calgary, Canada, 27 September–1 October, 2010.

[168] T. Tidwell, R. De Paula, G. Nilsen, et al., Visualization and quantification of biofilm removal for the mitigation of MIC, NACE-6081, in: Corrosion Conference & Expo, Dallas, TX, 15–19 March, 2015.

[169] J.W. Costerton, B. Ellis, K. Lam, et al., Mechanisms of electrical enhancement of efficacy of antibiotics in killing biofilm bacteria, Antimicrobial Agents and Chemotherapy 38 (1994) 2803–2809.

[170] B.D. Hoyle, J. Alcantara, J.W. Costerton, *Pseudomonas aeruginosa* biofilm as a diffusion barrier to piperacillin, Antimicrobial Agents and Chemotherapy 36 (1992) 2054–2056.

[171] M. Brown, P.J. Collier, P. Gilbert, Influence of growth rate on susceptibility to antimicrobial agents: modification of the cell envelope and batch and continuous culture studies, Antimicrobial Agents and Chemotherapy 34 (1990) 1623–1628.

[172] R. Eng, F.T. Padberg, S.M. Smith, et al., Bactericidal effects of antibiotics on slowly growing and nongrowing bacteria, Antimicrobial Agents and Chemotherapy 35 (1991) 1824–1828.

[173] C. Whitby, T.L. Skovhus (Eds.), Introduction, in: Applied Microbiology and Molecular Biology in Oilfield Systems: Proceedings from the International Symposium on Applied Microbiology and Molecular Biology in Oil Systems (ISMOS-2), Springer, London, 2009, pp. 3–5.

[174] R.I. Amann, W. Ludwig, K.H. Schleifer, Phylogenetic identification and in situ detection of individual microbial cells without cultivation, Microbiological Reviews 59 (1995) 143–169.

[175] F. Vester, K. Ingvorsen, Improved most-probable-number method to detect sulfate-reducing bacteria with natural media and a radiotracer, Applied and Environmental Microbiology 64 (1998) 1700–1707.

[176] X.Y. Zhu, T. Zhong, Y. Pandya, et al., 16S rRNA-based analysis of microbiota from the cecum of broiler chickens, Applied and Environmental Microbiology 68 (2002) 124–137.

[177] X.Y. Zhu, J.J. Kilbane II, Molecular tools in microbial corrosion, in: R. Vazquez-Duhalt, R. Quintero-Ramirez (Eds.), Petroleum Biotechnology: Development and Perspectives, Elsevier Science, Amsterdam, 2004, pp. 219–232.

[178] NACE, Field Monitoring of Bacterial Growth in Oil and Gas Systems, NACE TM0194-2014, NACE International, Houston, 2014.

[179] H.A. Awadhi, N. Dashti, M. Khanafer, et al., Bias problems in culture-independent analysis of environmental bacterial communities: a representative study on hydrocarbonoclastic bacteria, SpringerPlus 2 (2013) 1–11.

[180] E. Brutinel, B. Wilson, K. Gangelhoff, Evaluation of a self-contained, anaerobic environment-generating culture device for detection of sulfate-reducing bacteria, NACE-13174, in: Corrosion Conference & Expo, Nashville, TN, 24–28 March, 2019.

[181] E. Amodio, C. Dino, Use of ATP bioluminescence for assessing the cleanliness of hospital surfaces: a review of the published literature (1990–2012), Journal of Infection and Public Health 7 (2014) 92–98.

[182] R. Chollet, S. Ribault, Use of ATP bioluminescence for rapid detection and enumeration of contaminants: the milliflex rapid microbiology detection and enumeration system, in: D. Lapota (Ed.), Bioluminescence - Recent Advances in Oceanic Measurements and Laboratory Applications, InTechOpen, London, 2012, pp. 99–117.

[183] B.L. Strehler, W.D. McElroy, Assay of adenosine triphosphate, in: S.P. Colwick, N. O. Kaplan (Eds.), Methods in Enzymology, Academic Press, New York, 1957, pp. 871–873.

[184] N. Thorne, J. Inglese, D.S. Auld, Illuminating insights into firefly luciferase and other bioluminescent reporters used in chemical biology, Chemistry & Biology 17 (2010) 646–657.

[185] ASTM, Standard test method for adenosine triphosphate (ATP) content of microorganisms in fuel, fuel/water mixtures, and fuel associated water, ASTM D7463-16e1, ASTM International, West Conshohocken, 2016.

[186] ASTM, Standard test method for measurement of cellular adenosine triphosphate in fuel and fuel-associated water with sample concentration by filtration, ASTM D7687-17, ASTM International, West Conshohocken, 2017.

[187] S.A. Borchardt, R.L. Wetegrove, J.D. Martens, New approaches to biocide effectiveness monitoring using on-site biocide active analysis, ATP analysis, and on-line dosage/monitoring control, NACE-97466, in: Corrosion Conference & Expo, New Orleans, LA, 9–14 March, 1997.

[188] J. Chalut, G. Small, J. Payton, Control and monitoring the effectiveness of different biocides with the use of free ATP, NACE-96276, in: Corrosion Conference & Expo, Denver, CO, 24–29 March, 1996.

[189] P.A. Whalen, D.R. Tracey, J. Duguay, Adenosine triphosphate (ATP) measurement technology, in: T.L. Skovhus, B. Højris (Eds.), Microbiological Sensors for the Drinking Water Industry, IWA Publishing, London, 2018, pp. 137–154.

[190] J.C. Cairns, P.J. Whalen, P.A. Whalen, Reagent system and process for adenosine triphosphate monitoring, Patent US7556933, 2009.

[191] E. Corrin, S. Leong, P. Whalen, Comparison of biocides using new microbial detection tool, in: Proceedings of 2nd International Symposium on Applied Microbiology and Molecular Biology in Oil Systems (ISMOS-2), Aarhus, Denmark, 17–19 June, 2009.

[192] S. Amaya Hernandez, X.Y. Zhu, G.A. Humaid, et al., ATP bioluminescence assay for onsite microbial assessment in drilling rig operations, SPE-189391-MS, in: SPE/IADC Middle East Drilling Technology Conference and Exhibition, Abu Dhabi, UAE, 29–31 January, 2018.

[193] S.J. Salgar-Chaparro, K. Lepkova, T. Pojtanabuntoeng, et al., Microbiologically influenced corrosion as a function of environmental conditions: a laboratory study using oilfield multispecies biofilms, Corrosion Science 169 (2020) 108595.

[194] V. Keasler, B. Bennett, C. Keller, et al., Expanding the microbial monitoring toolkit: evaluation of traditional and molecular monitoring methods, International Biodeterioration & Biodegradation 81 (2013) 51–56.

[195] M. Paschoalino, J. Raymond, E. Sianawati, et al., Evaluation of a glutaraldehyde/THNM combination for microbial control in four conventional oilfields, SPE-193594-MS, in: SPE International Conference on Oilfield Chemistry, Galveston, TX, 8–9 April, 2019.

[196] G. Braker, A. Fesefeldt, K.P. Witzel, Development of PCR primer systems for amplification of nitrite reductase genes (nirK and nirS) to detect denitrifying bacteria in environmental samples, Applied and Environmental Microbiology 64 (1998) 3769–3775.

[197] P.E. Luton, J.M. Wayne, R.J. Sharp, et al., The *mcrA* gene as an alternative to 16S rRNA in the phylogenetic analysis of methanogen populations in landfill, Microbiology 148 (2002) 3521–3530.

[198] A. Price, L.A. Alvarez, C. Whitby, How many microorganisms are present? Quantitative reverse transcription PCR (qRT-PCR), in: T.L. Skovhus (Ed.), Proceedings from the International Symposium on Applied Microbiology and Molecular Biology in Oil Systems (ISMOS-2), Springer, London, 2009, pp. 77–84.

[199] J.A. Bernstein, A.B. Khodursky, P.H. Lin, et al., Global analysis of mRNA decay and abundance in *Escherichia coli* at single-gene resolution using two-color fluorescent DNA microarrays, Proceedings of the National Academy of Sciences of the United States of America 99 (2002) 9697–9702.

[200] M.P. Deutscher, Degradation of RNA in bacteria: comparison of mRNA and stable RNA, Nucleic Acids Research 34 (2006) 659–666.

[201] D.W. Selinger, R.M. Saxena, K.J. Cheung, et al., Global RNA half-life analysis in Escherichia coli reveals positional patterns of transcript degradation, Genome Research 13 (2003) 216–223.

[202] T.D. Schmittgen, Real-time quantitative PCR, Methods 25 (2001) 383–385.

[203] A. Chimento, S.O. Cacciola, M. Garbelotto, Detection of mRNA by reverse-transcription PCR as an indicator of viability in *Phytophthora ramorum*, Forest Pathology 42 (2012) 14–21.

[204] M. Diaz, M. Herrero, L.A. Garcia, et al., Application of flow cytometry to industrial microbial bioprocesses, Biochemical Engineering Journal 48 (2010) 385–407.

[205] V. Ambriz-Avina, J.A. Contreras-Garduno, M. Pedraza-Reyes, Applications of flow cytometry to characterize bacterial physiological responses, BioMed Research International (2014) 461941.

[206] T. Tidwell, R. De Paula, M.Y. Smadi, et al., Flow cytometry as a tool for oilfield biocide efficacy testing and monitoring, International Biodeterioration and Biodegradation 98 (2015) 26–34.

[207] R. De Paula, V. Keasler, B. Bennett, et al., Optimization of a microbial control program in an aging Gulf of Mexico asset to minimize the risk of corrosion, NACE-1195, in: Corrosion Conference & Expo, Salt Lake City, UT, 11–15 March, 2012.

[208] S. Oh, M. Tandukar, S.G. Pavlostathis, et al., Microbial community adaptation to quaternary ammonium biocides as revealed by metagenomics, Environmental Microbiology 15 (2013) 2850–2864.

[209] A.D. Russell, Antibiotic and biocide resistance in bacteria: introduction, Journal of Applied Microbiology Symposium 92 (2002) 1S–3S.

[210] D.M. Grant, T.R. Bott, Biocide dosing strategies for biofilm control, in: Proceedings of Engineering Conferences International, Santa Fe, NM, May 18–22, 2003.

[211] X.Y. Zhu, M.A. Al-Moniee, M.A. Al-Saleh, Methods for detecting and quantifying tetrakis(hydroxymethyl)phosphonium sulfate (THPS) in biocide products, Patent Application US17/084108, 2020.

[212] X.Y. Zhu, M.A. Al-Moniee, M.A. Al-Saleh, Methods for detecting and quantifying glutaraldehyde-based products in water, Patent Application US17/107829, 2020.

[213] M.H. Isa, G.K. Anderson, Molybdate inhibition of sulphate reduction in two-phase anaerobic digestion, Process Biochemistry 40 (2005) 2079–2089.

[214] B.V. Kjellerup, R.H. Veeh, P. Sumithraratne, et al., Monitoring of microbial souring in chemically treated, produced-water biofilm systems using molecular techniques, Journal of Industrial Microbiology & Biotechnology 32 (2005) 163–170.

[215] A. Gittel, K.B. Sørensen, T.L. Skovhus, et al., Prokaryotic community structure and sulfate reducer activity in water from high-temperature oil reservoirs with and without nitrate treatment, Applied and Environmental Microbiology 75 (2009) 7086–7096.

[216] E. Sunde, T. Torsvik, Microbial control of hydrogen sulfide production in oil reservoirs, in: B. Ollivier, M. Magot (Eds.), Petroleum Microbiology, ASM Press, Washington, DC, 2005, pp. 201–213.

[217] C. Hubert, M. Nemati, G. Jenneman, et al., Corrosion risk associated with microbial souring control using nitrate or nitrite, Applied Microbiology & Biotechnology 68 (2005) 272–282.

[218] S. Myhr, B. Lillebø, E. Sunde, et al., Inhibition of microbial H_2S production in an oil reservoir model column by nitrate injection, Applied Microbiology & Biotechnology 58 (2002) 400–408.

[219] I. Davidova, M.S. Hicks, P.M. Fedorak, et al., The influence of nitrate on microbial processes oil industry production waters, Journal of Industrial Microbiology & Biotechnology 27 (2001) 80–86.

[220] E.A. Greene, C. Hubert, M. Nemati, et al., Nitrite reductase activity of sulphate-reducing bacteria prevents their inhibition by nitrate-reducing, sulphide-oxidizing bacteria, Environmental Microbiology 5 (2003) 607–617.

[221] C. Hubert, M. Nemati, G. Jenneman, et al., Containment of biogenic sulfide production in continuous up-flow packed-bed bioreactors with nitrate or nitrite, Biotechnology Progress 19 (2003) 338–345.

[222] M. Nemati, G.E. Jenneman, G. Voordouw, Impact of nitrate-mediated microbial control of souring in oil reservoirs on the extent of corrosion, Biotechnology Progress 17 (2001) 852–859.

[223] C.L. Rempel, R.W. Evitts, M. Nemati, Dynamics of corrosion rates associated with nitrite or nitrate mediated control of souring under biological conditions simulating an oil reservoir, Journal of Industrial Microbiology & Biotechnology 33 (2006) 878–886.

[224] W. Fu, Y. Li, D. Xu, et al., Comparing two different types of anaerobic copper biocorrosion by sulfate- and nitrate-reducing bacteria, Materials Performance 53 (2014) 66–70.

[225] T. Iino, K. Ito, S. Wakai, et al., Iron corrosion induced by nonhydrogenotrophic nitrate-reducing Prolixibacter sp. strain MIC1-1, Applied and Environmental Microbiology 81 (2015) 1839–1846.

[226] A. Gittel, Monitoring and preventing reservoir souring using molecular microbiological methods (MMM), in: T.L. Skovhus (Ed.), Proceedings from the International Symposium on Applied Microbiology and Molecular Biology in Oil Systems (ISMOS-2), Springer, London, 2009, pp. 103–107.

[227] J. Larsen, T.L. Skovhus, The effect of nitrate injection in oil reservoirs - Experience with nitrate injection in the Halfdan oilfield, in: T.L. Skovhus, C. Whitby (Eds.), Applied Microbiology and Molecular Biology in Oilfield Systems, Springer, London, 2009, pp. 109–115.

[228] C. Meulemans, The basic principles of UV–disinfection of water, Ozone: Science & Engineering 9 (2008) 299–313.

[229] M.W. Mittelman, Bacterial growth and biofouling control in purified water systems in biofouling and biodeterioration in industrial water systems, in: H.C. Flemming, G. G. Geesey (Eds.), Proceedings of the International Workshop on Industrial Biofouling and Biocorrosion, Springer-Verlag, Berlin, Heidelberg, 1990, pp. 133–154.

[230] ATG Evoqua, Oil, Gas & Petrochemical UV Systems, Lancashire, UK, 2019.

[231] R. Javaherdashti, Microbiologically Influenced Corrosion, Springer, London, 2008.

[232] X. Jin, Z. Li, L. Xie, et al., Synergistic effect of ultrasonic pre-treatment combined with UV irradiation for secondary effluent disinfection, Ultrasonics Sonochemistry 20 (2013) 1384–1389.

[233] M.G. Klanian, F.J. Espinosa-Faller, A. Núñez, et al., Effect of turbidity on the ultraviolet disinfection performance in recirculating aquaculture system with low water exchange, Aquaculture Research 43 (2012) 595–606.

[234] S. Zhang, C. Ye, H. Lin, et al., UV disinfection induces a VBNC state in *Escherichia coli* and *Pseudomonas aeruginosa*, Environmental Science and Technology 49 (2015) 1721–1728.

[235] A. Ohno, N. Kato, K. Yamada, et al., Factors influencing survival of Legionella pneumophila sero type 1 in hot spring water and tap water, Applied and Environmental Microbiology 69 (2003) 2540–2547.

[236] H.M. Al-Gabr, T. Zheng, X. Yu, Inactivation of *Aspergillus flavus* in drinking water after treatment with UV irradiation followed by chlorination, Science of the Total Environment 463 (2013) 525–529.

[237] E. Croese, H. de Vries, T. Robertson, et al., Oil & gas water injection treatment system case study: field trial comparing biocide/UV disinfection and new monitoring techniques for mitigating MIC and reservoir souring, NACE-2017-9114, in: Corrosion Conference & Expo, New Orleans, LA, 26–30 March, 2017.

[238] P. Butz, B. Tauscher, Emerging technologies: chemical aspects, Food Research International 35 (2002) 279–284.

[239] B.G. Pound, Y. Gorfu, P. Schattner, et al., Ultrasonic mitigation of microbiologically influenced corrosion, Corrosion 61 (2005) 452–463.

[240] P.R. Gogate, A.B. Pandit, Sonophotocatalytic reactors for wastewater treatment: a critical review, AIChE Journal 50 (2004) 1051–1079.

[241] P.R. Gogate, R.K. Tayal, A.B. Pandit, Cavitation: a technology on the horizon, Current Science 91 (2006) 35–46.

[242] R. Bhola, S.M. Bhola, B. Mishra, et al., Microbiologically influenced corrosion and its mitigation: a review, Material Science Research India 7 (2010) 407–412.

[243] P. Piyasena, E. Mohareb, R.C. McKellar, Inactivation of microbes using ultrasound: a review, International Journal of Food Microbiology 87 (2003) 207–216.

[244] S. Broekman, O. Pohlmann, E.S. Beardwood, et al., Ultrasonic treatment for microbiological control of water systems, Ultrasonics Sonochemistry 17 (2010) 1041–1048.

[245] V. Naddeoa, M. Landia, V. Belgiorno, Wastewater disinfection by combination of ultrasound and ultraviolet irradiation, Journal of Hazardous Materials 168 (2009) 925–929.

[246] Z. Yang, Y. Zhou, Z. Feng, et al., A review on reverse osmosis and nanofiltration membranes for water purification, Polymers 11 (2019) 1252–1273.

[247] M. Bader, Nanofiltration for oil-fields water injection operations: analysis of concentration polarization, Desalination 201 (2006) 106–113.

[248] V. Keasler, Real-time field monitoring to optimize microbe control, Journal of Petroleum Technology 64 (2012) 30–33.

Index

Note: Page numbers followed by "*f*" and "*t*" refer to figures and tables, respectively.

A

B

D

Q

R

Z